S0-BRB-259

HANDBOOK OF ENVIRONMENTAL RISK ASSESSMENT AND MANAGEMENT

Handbook of Environmental Risk Assessment and Management

EDITED BY
PETER CALOW
DSc, PhD, CBiol, FIBiol
Department of Animal and Plant Sciences
University of Sheffield

b
Blackwell
Science

Editorial Offices:
Osney Mead, Oxford OX2 0EL
25 John Street, London WC1N 2BS
23 Ainslie Place, Edinburgh EH3 6AJ
350 Main Street, Malden
MA 02148 5018, USA
54 University Street, Carlton
Victoria 3053, Australia
10, rue Casimir Delavigne
75006 Paris, France

Other Editorial Offices:
Blackwell Wissenschafts-Verlag GmbH
Kurfürstendamm 57
10707 Berlin, Germany

Blackwell Science KK
MG Kodenmacho Building
7–10 Kodenmacho Nihombashi
Chuo-ku, Tokyo 104, Japan

Iowa State University Press
A Blackwell Science Company
2121 S. State Avenue
Ames, Iowa 50014-8300, USA

First published 1998
Reprinted 2001

Set by Setrite Typesetters, Hong Kong
Printed and bound in Great Britain
by MPG Books Ltd, Bodmin, Cornwall

A catalogue record for this title is available from the British Library

ISBN 0-86542-732-1

Library of Congress
Cataloging-in-Publication Data

Handbook of environmental risk assessment and management/
edited by
Peter Calow.
p. cm.
Includes bibliographical references and index.
ISBN 0-86542-732-1
1. Environmental risk assessment—Handbooks, manuals, etc.
2. Environmental management—Handbooks, manuals, etc.
I. Calow, Peter..
GE145.H36 1998
333.7′14—dc21 97-12171
CIP

For further information on Blackwell Science, visit our website:
www.blackwell-science.com

The Blackwell Science logo is a trade mark of Blackwell Science Ltd, registered at the United Kingdom Trade Marks Registry

DISTRIBUTORS

Marston Book Services Ltd
PO Box 269
Abingdon
Oxon OX14 4YN
(*Orders:* Tel: 01235 465500
Fax: 01235 465555)

USA
Blackwell Science, Inc.
Commerce Place
350 Main Street
Malden, MA 02148 5018
(*Orders:* Tel: 800 759 6102
781 388 8250
Fax: 781 388 8255)

Canada
Login Brothers Book Company
324 Saulteaux Crescent
Winnipeg, Manitoba R3J 3T2
(*Orders:* Tel: 204 837-2987
Fax: 204 837-3116)

Australia
Blackwell Science Pty Ltd
54 University Street
Carlton, Victoria 3053
(*Orders:* Tel: 3 9347-0300
Fax: 3 9347-5001)

Contents

List of Contributors

MARTIN ALEXANDER PhD, *Liberty Hyde Bailey Professor, Department of Soil, Crop and Atmospheric Sciences, Cornell University, Ithaca, New York, NY 14853, USA*

JANE BEATTIE PhD, *Senior Lecturer in Psychology, Department of Experimental Psychology, University of Sussex, Brighton BNI 9QN, UK*

THOMAS A. BURKE PhD, MPH, *Johns Hopkins University, School of Hygiene and Public Health, Baltimore, MD 21205, USA*

PETER CALOW DSc, PhD, CBiol, FIBiol *Department of Animal and Plant Sciences, University of Sheffield, Sheffield S10 2UQ, UK*

FRANK J. CONSOLI *President, Consoli Consulting Company, 619 North Heilbron Drive, Media, PA 19063, USA*

VINCENT T. COVELLO PhD, *Director, Center for Risk Communication, 39 Claremont Avenue, Suite 71, New York 10027, USA*

CHRISTINA E. COWAN PhD, *Senior Scientist, Procter and Gamble, Ivorydale Technical Center, 5299 Spring Grove Avenue, Cincinnati, OH 45217, USA*

TONY J. DOBBS *Environment Business Manager, WRc plc, Henley Road, Medmenham, Marlow, Bucks SL7 2HD, UK*

MICHAEL L. DOURSON PhD, DABT, *Director, Toxicology Excellence for Risk Assessment, 4303 Hamilton Avenue, Cincinnati, OH 45223, USA*

JAMES A. FAVA PhD, *Vice President, Roy F Weston Inc, 1 Weston Way, West Chester, PA 19380, USA*

TOM C. J. FEIJTEL PhD, *Section Head, Environmental Safety, Procter and Gamble, European Technical Center, 100 Temselaan, B1853 Strombeek-Bever, Belgium*

SUSAN P. FELTER PhD, *Toxicologist, Toxicology Excellence for Risk Assessment, 4303 Hamilton Avenue, Cincinnati, OH 45223, USA*

S. ELIZABETH GEORGE PhD, *Research Biologist, US Environmental Protection Agency, Office of Research and Development, National Health and Environmental Effects Research Laboratory, Mail drop 68, Research Triangle Park, NC, 27711, USA*

JOHN H. GOULD BSc, MRSC, CChem, *Inspector of Health and Safety, Major Hazards Assessment Unit, Health and Safety Executive, Chemicals and Hazardous Installation Division, St Anne's House, Stanley Precinct, Bootle, Merseyside L20 3RA, UK*

JOHN S. GRAY PhD, DSc, *Department of Biology, University of Oslo, PB 1064, 0316 Blindern, Norway*

STUART HEDGECOTT BSc, *Environmental Toxicology Manager, WRc plc, Henley Road, Medmenham, Marlow, Bucks SL7 2HD, UK*

PETER HINDLE MA, *Director, Worldwide Technical Policy, Procter and Gamble, European Technical Center, 100 Temselaan, B1853 Strombeek-Bever, Belgium*

NORMAN J. KING BSc, PhD, *Formerly Head of Toxic Substances Division, Department of the Environment, 'The Gyles', Bledlow Road, Saunderton, Princes Risborough, Bucks HP27 9NG, UK*

OLAV KJØRVEN MA, *Environmental Assessment Specialist, The World Bank, 1818 H Street HW, Washington DC 20433, USA*

JACQUELINE PATTERSON MEn, *Director, International Toxicity Estimates for Risk Program, Toxicology Excellence for Risk Assessment, 4303 Hamilton Avenue, Cincinnati, OH 45223, USA*

DAVID PEARCE *White Friars Farm, Duddenhoe End, Saffron Walden, Essex CB11 4UU, UK*

JUDITH PETTS PhD, *Senior Lecturer, Centre for Hazard and Risk Management, Loughborough University, Ashby Road, Loughborough LE11 3TU, UK*

NICK F. PIDGEON PhD, *Senior Lecturer in Psychology, School of Psychology, University of Wales at Bangor, Bangor, Gwynedd LL57 2DG, UK*

CHARLES A. PITTINGER PhD, *Section Head, Risk Sciences, Procter and Gamble, Ivorydale Technical Center, 5299 Spring Grove Avenue, Cincinnati, OH 45217, USA*

MICHAEL QUINT *DAMES & MOORE, Booth House, 15–17 Church Street, Twickenham TW1 3NJ, UK*

KEVIN H. REINERT PhD, *Research Section Manager of Ecotoxicology and Ecological Risk Assessment, Rohm and Haas Company, 727 Norristown Rd, Spring House, PA 19477-0904, USA*

JOSEPH V. RODRICKS PhD, *Environ International Corp, 4350 North Fairfax Drive, Arlington, VA 22203, USA*

SVEN-OLOF RYDING PhD, *Director, Svenska Miljöstyrningsrådet AB, PO Box 70396, S-107 24 Stockholm, Sweden*

RAMON J. SEIDLER PhD, *Research Microbiologist, US Environmental Protection Agency, Office of Research and Development, National Health and Environmental Effects Research Laboratory, Western Ecology Division, 200 SW, 35th Street, Corvallis, OR 97333, USA*

DANIEL SIMBERLOFF PhD, *Nancy Gore Professor of Environmental Science, Department of Ecology and Evolutionary Biology, University of Tennessee, Knoxville, TN 37996, USA*

JERRY C. SMRCHEK PhD, *Biologist/ Ecotoxicologist, US Environmental Protection Agency, Risk Assessment Division (7403), Office of Pollution Prevention and Toxics, 401 M Street, Washington, DC 20460, USA*

CHRIS V. STARMER BA, MA, PhD, *Senior Lecturer in Economics, School of Economic and Social Studies, University of East Anglia, Norwich NR4 7TJ, UK*

GLENN W. SUTER II PhD, *Senior Research Staff Member, Environmental Sciences Division, Oak Ridge National Laboratory, PO Box 2008, Oak Ridge, TN 37831-6038, USA*

LIDIA S. WATRUD PhD, *Research Ecologist, US Environmental Protection Agency, Office of Research and Development, National Health and Environmental Effects Research Laboratory, Western Ecology Division, 200 SW, 35th Street, Corvallis, OR 97333, USA*

RANDALL S. WENTSEL PhD, *Chief Scientist, Environmental Technology Team, US Army Edgewood Research, Development and Engineering Centre, SCBRD-RTL, Aberdeen Proving Ground, MD 21010-5423, USA*

MAURICE G. ZEEMAN PhD, *Senior Scientist US Environmental Protection Agency, Office of Pollution Prevention and Toxics, 401 M Street, SW, Washington, DC 20460, USA*

Preface

Life is a risky business. Cliché? Yes; but also almost certainly true! Everything we do—from deciding whether to get up in the morning, what to eat, whether or not to cross the road, to choosing a partner and selecting a career—involves some kind of risk assessment, a balancing of risks with costs and benefits, and risk management decisions. Yet despite this, or maybe even because of it, risk is often hard to capture quantitatively and to make explicit in the management of our affairs. Moreover, a spate of recent, high-profile incidents has shown that while on the one hand there is an increasing call for the involvement of risk assessment procedures in the development of policies and legislation that impinge on human health and environmental protection, there is, on the other hand, a deep suspicion of outputs of risk assessment from scientists, industry and regulators that are used in this way.

This Handbook aims to present the basic principles of risk assessment in a user-friendly way, and show how they can be, and are being, brought to bear in a wide variety of management contexts. It addresses a broad range of questions. What is environmental risk assessment? How is it used in the development of environmental and human health protection policy and legislation? How should the benefits that derive from controlling risks be balanced against the costs of sacrifices made to achieve them? How is the output of risk assessment used to manage products and processes in businesses, and the quality of environmental compartments such as inland waters and the seas? These questions roughly describe, in turn, the compass of the four parts into which the volume is divided.

This span of coverage is inevitably broad, and has involved the compilation of 20 chapters by more than 30 authors. I am grateful to them all for producing chapters that will not only provide readers with the basics of the subject areas, but also take them to the cutting edge, and some even beyond with speculation on where the disciplines are likely to go over the next few years. All this has led to stimulating interactions amongst authors and between authors and Editorial Board. There has also been one sadness, and this is that one of the authors, Jane Beattie, died during the preparation of this work. I would like to take this opportunity to express sympathy to her family and colleagues.

The compilation of the Handbook would not have been possible without administrative support—from my secretarial staff, first Julie Yeardley and then Samantha Giles, and also from the editorial staff at Blackwell Science, especially Susan Sternberg. All have given me much appreciated assistance that has removed a considerable burden from the editorial task. And the editorial activity itself has been much facilitated by the useful involvement of the Editorial Board in deciding on structure, choosing authors, vetting proposals and ultimately editing contributions.

As with other Handbooks in the Blackwell's series this, with more than 500 pages, is hardly physically 'handy' to use. What we hope, though, is that by having it to hand, practitioners from industry, the regulatory community, and academia, as well as students, will find it a convenient and useful reference work. The risk business is developing apace, both in terms of the underlying principles and the putting of these into practice. There are therefore likely to be updates and new editions that follow this first production; so I end with a sincere request that readers let me have any comments on content and style that might enhance future work.

Peter Calow
Sheffield, 1997

Chapter 1
Environmental Risk Assessment and Management: the Whats, Whys and Hows?

PETER CALOW

1.1 INTRODUCTION

Measures to protect the environment, and people in it, have involved a wide variety of approaches and underlying principles, and still do. These range from those that react to problems as they occur, to those that attempt to anticipate problems (prevent rather than cure); those based only on the suspicion of a causal link between activity or substance and effect, to those that seek a detailed scientific understanding of these linkages; those based on a recognition of the potential that actions and substances have to cause harm, to those based on an assessment of those likelihoods being realized in the real world; those based only on a recognition of risks to human health and environment, to those based upon an attempt to balance these risks against benefits that are obtained in association with the activities and substances causing them. Through recent history there has been a shift in Europe and the USA from reaction and suspicion-based approaches to more pro-action and more concern with risks and their balance with costs. Still, different national approaches, different social and interest groups, and even different instruments of control within the same areas of jurisdiction lay different emphasis on each of these principles.

This Handbook is concerned with environmental risk assessment and management. It will largely cover: the way that scientific and anticipatory methodologies are used to assess risks from human activities, and the objects and the waste that emanate from them, for the environment and for people in the environment; and how this kind of understanding is used in legislation, in the practice of management in businesses, in managing problems in the various major habitats, and in managing development programmes. Increasingly, it is also becoming important to consider the way in which people handle this information on risks and take it into account in making choices. This involves psychological and economics—social science—issues, which will also be addressed in this volume.

1.2 TROUBLE WITH TERMINOLOGY

One of the problems with the 'risk discipline' is that it relates to common experiences for which we have developed ordinary language, so we all think that we know what is involved and what the key terms mean. But experiences differ, and common usages lack precision. For example, in their French origins only partial, if any, distinction is made between *hasarder* and *risquer* (Rimington 1992) and yet concepts to which some (now probably most) English-speaking scientists have associated *hazard* and *risk* are importantly distinct (see below). In consequence there has been an undercurrent of ambiguity and hence potential confusion in defining these terms.

Similarly, *environment* is a broad term open to ambiguity. The French root—*environer*, meaning to encircle and support—has at least two implications that are potentially misleading. First, it suggests that there is a sharp distinction between living systems and the supporting matrices of soil, air and water. Yet the composition of all these matrices is importantly influenced by, and in turn influences, living things. Second, there is an implication of active support from the environment to living things, which is taken by some as far as the Mother Earth (Gaia) concept (e.g. Lovelock 1988). Yet the proposal that nature is directed to specific ends, even that ecological systems operate as superorganisms directed towards self-preservation, is fundamentally at odds with the scientific approach (Calow 1992).

At a different level, in the context of risk assessment, *environment* has sometimes been intended to imply 'for humans'. So *environmental risk assessment* is taken to mean assessment of risks to human health from exposures in the environment at large. Alternatively, *ecological risk assessment* is used for the assessment of risks to non-human communities and populations (Suter 1992). Yet this can drive artificial distinctions between human organisms and other living things and between the routes of exposure to which both are subject.

In this Handbook, therefore, the following definitions are emphasized.

Hazard intended as potential to cause harm.

Risk intended as the likelihood of that potential being realized (for a somewhat broader view, see Chapter 11).

Environmental intended to refer to the routes of exposure for both humans and wildlife. So ecological risk assessment—for ecological systems—is a subset of environmental risk assessment.

1.3 REFINEMENT OF CONCEPTS

It is probably not too broad a generalization to say that all we do and make have potential to cause harm to human health and ecological systems. Some things are potentially more harmful than others and we have developed standard methods (test systems) for assessing these potentials and ranking them. Describing this potential and carrying out the tests are referred to, respectively, as *hazard identification* and *hazard assessment*. Thus, in considering the potential ecological harm from commercial chemicals, they are screened for effect against standard species that are presumed to be ecologically relevant—in dose (concentration)–response tests that identify how much chemical is required to result in a specified harmful effect (often to kill 50% of the population). The lower the concentration having the effect, the greater the *potential* to cause harm. Some of the environmental protection legislation is driven by these kinds of results, with some account being taken sometimes of what might be described as potential exposure criteria, such as likelihood to persist in the environment and likelihood to bioaccumulate (as indicated by physicochemical properties, or the results from simple measurements such as the extent to which substances partition between octanol and water in standardized conditions). In considering the potential harm that might be caused by introduced organisms ('natural' or 'engineered'), we might be interested in their potential capacity to spread and multiply.

However, even the most hazardous substances or organisms are very unlikely to cause harm if they exist in small quantities and are kept in strict security. Therefore, the likelihood of potential harm being realized depends not only on this potential, but also on the circumstances that lead to a particular exposure scenario. Assessing the likelihood of harm, to both humans and ecological systems, requires a bringing together of an understanding of hazard with an understanding of exposure to the target systems. This is risk assessment.

Identification of the target system is also important. Thus, species differ in their sensitivities to chemicals and radiation (e.g. humans are often less sensitive than other species) and in exposure (e.g. aquatic animals are unlikely to be subject to the same risks from chemicals as soil-dwelling animals) (see Calow 1992, 1993, 1994). As far as humans are concerned, we know what the targets are and what we want to protect: health and survival. The same is not always the case for ecological systems. Sometimes we do have specific systems in mind—a particular endangered species or habitat; a particular ecosystem downstream of a factory effluent or downwind of a power station. But often, risk assessment of existing and new commercial chemicals, and genetically modified organisms that are going to be used broadly, are concerned with protecting ecological systems in general and, by definition, targets here are less clearly defined.

1.4 HOW RISK ASSESSMENT IS CARRIED OUT

Formally, the potential to cause harm is gauged by characterizing the relationship between level of application of a contaminant or activity and an effect. This is a dose (concentration)–response assessment. The effect should be measured in the target of interest or a surrogate (e.g. laboratory rodents for humans; 'base set' organisms for ecological systems) and might be in terms of a property that is identified for protection (assessment end-point: e.g. survival, reproduction potential or ecosystem productivity) or again a surrogate (measurement end-points: e.g. physiological/biochemical responses known or thought to be related to survival, reproduction potential or production). There are uncertainties associated with each and all these measurements due to observational errors, environmental and biological variability, and complex interactions between all these aspects. This uncertainty should be expressed as variability about the concentration/dose producing a particular effect.

Judgements still have to be made on the level of effects, often non-zero, that are deemed tolerable. These can be based on scientific evidence (e.g. that x% of species can be lost from a community without impairing productivity), on socio-political grounds (e.g. that it is necessary to cull cattle at a level over and above scientific requirements in order to allay public fears about infection from bovine spongiform encephalopathy (BSE)), arbitrary criteria, and a combination of all these aspects.

The likelihood of a defined adverse effect occurring depends upon likely exposure of the target to the contaminant or activity in the environment. This depends in turn on such factors as the patterns and levels of production, use and disposal of contaminants, their physicochemical properties, and the properties of the environment most likely to be exposed. These predicted environmental levels, or concentrations or doses, are also subject to uncertainties of the same kind as described above for the dose/concentration-response analysis and so, properly, should be expressed as probability distributions.

The probability of an adverse effect then depends upon the likelihood of exposure exceeding critical effect levels. This is illustrated schematically and simply in Fig. 1.1. These simplified distributions capture the basis of the formal analysis, but miss out on a couple of important features.

1 Rarely can all the uncertainty be quantified. We might know that it is there, but because of complexities and ignorance we might be unable precisely to take it into account. And this is often used to make a distinction between risk and uncertainty. Risk assessment recognizes a range of possible outcomes, the relative likelihood of which can be predicted on the basis of the full understanding of scenarios. Uncertainty assessment recognizes some outcomes as being a possibility but it cannot quantify them; it usually involves computing likelihoods on the basis of presumptions about what might happen and/or by introducing 'noise' to risk assessments in a 'controlled way', such as by simulation.

2 Time is not explicitly included. This ought to be represented as a third dimension, especially on the exposure graph because exposures certainly alter through time. But, equally, the condition of individuals, populations, and communities might also alter through time in a way that affects their susceptibly (e.g. age structure for populations, and species composition for communities).

Rarely, of course, do we have such detailed information and understanding of these probability distributions of effects and exposures and their dynamic behaviour. More often we have only qualitative and/or semiquantitative indications of exposure and effects. Various numerical scores are sometimes used to represent 'high', 'medium', and 'low' levels of exposure, effect and risk. For example, in constructing a risk profile of their businesses as part of an environmental management system, managers were asked to score for potential to cause problems and the likelihood of that being realized (Calow & Streatfeild 1994). They did this by reference to tables that define scenarios in terms of scores on a 1 (good) to 5 (bad) range. For *potential* (analogous to *hazard*), managers considered each element of a process in terms of the extent that it might cause problems for the environment and from environmental legislation under normal, abnormal and emergency operating scenarios. The default—for ignorance—was 5, because ignorance is risky. For *likely occurrence* (analogous to *exposure*), managers were invited to consider, again with respect to tables, the extent to which systems (both hard- and software) were in place to address the potential

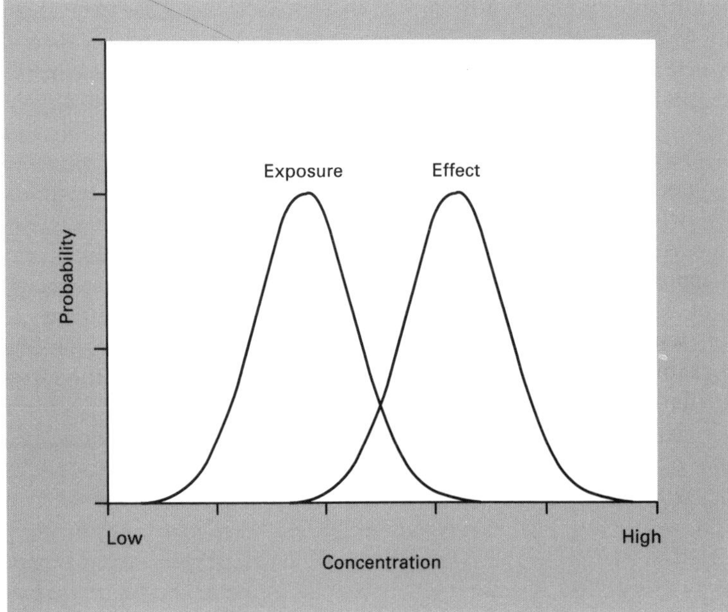

Fig. 1.1 Risk assessment involves comparing likely exposures with likely effects. Both distributions are represented hypothetically as symmetrical, but they need not be so. Precise probability of effects could, in principle, be computed from the extent of overlap of distributions. (After Calow & Forbes 1997.)

problem (the default was average—i.e. score 2.5—if the system was only designed to be in legal compliance). The 'hazard' and 'exposure' scores were multiplied together to give a risk assessment (minimum 1 to maximum 25) for comparison across the process to identify 'hot spots' for action. These are arbitrary numbers intended as a basis for management decisions.

In chemical control legislation, on the other hand, the predicted no-effect concentration (PNEC) is obtained by taking lowest observed effect measures (EC_{50} values) and dividing by arbitrary 'safety' (uncertainty) factors. Similarly, predicted environmental concentrations (PEC) are derived from simple models that predict levels in environmental compartments based upon levels of production and physicochemical properties of substances. The risk assessment then entails a comparison of PEC with PNEC (van Leeuwan & Hermens 1995). If the ratio of PEC:PNEC is less than unity it is presumed that the probability of effect is low, but because the detailed knowledge involved in Fig. 1.1 is lacking, a precise value cannot be ascribed to this.

1.5 RISK ASSESSMENT LEADING TO, OR LED BY, RISK MANAGEMENT?

The main reason for wanting to specify risks is so that they can be managed, reduced or removed. It is generally accepted that the assessment process should be separated from management decisions. The rationale is that the assessment should be as objective as possible, depending upon scientific criteria rather than political or social views and judgements. The latter are involved in making decisions about management, and to be included in the risk assessment could make it self-fulfilling in terms of political and social aspirations. It is for these reasons that we have separated those chapters with an 'assessment bias' (Parts 1 and 2) from those with a 'management bias' in (Part 4) of this Handbook.

Yet the distinction is not as clear as this. First, and most importantly from the perspective of ecological risk assessment, decisions have to be taken about what to protect prior to an assessment being carried out. Second, decisions also have to be taken about to what level protection should be exerted so that appropriate threshold levels can be defined. Both of these issues often involve not only scientific but also socio-political considerations (see above). Again, throughout this Handbook the extent of the overlap of assessment and management criteria will be apparent.

Similarly, management decisions often involve balancing the advantages to environment and human health of different options with their consequences for other social benefits. As will be argued in Part 3, these are most appropriately expressed and analysed in terms of economic considerations. But these are not without a scientific and ecological basis. For example, the economic value of protecting a particular species, habitat or ecological process can be considered, in part, in terms of the services they provide to the human economy through the provision, for example, of clean air, clean water and good quality soil. Working out the part that these services play in the economy—e.g. in the yield of crops, the provision of raw materials to industry and the support of good health—is a problem for economics, but working out how species, habitats, and ecosystems affect these services is a problem for ecology; so there is an interface between disciplines here. Moreover, understanding how ecological systems relate to services and consequences for the economy may determine which ecological targets are protected and hence which should be the focus of attention in a risk assessment. This again emphasizes how management criteria can, or indeed should, influence assessment criteria.

1.6 PRECAUTION WITH RISK ASSESSMENT

Risk assessment is often presented as being in conflict with the precautionary approach, but much of the argument has been driven by lack of clarity in what is meant by the 'precautionary principle' (PP). In fact probably there have been as many definitions of PP as articles that have been written about it (O'Riordan & Cameron 1994).

One extreme form of PP (that might be described as 'deep green') is that we should seek to avoid (or, in the most extreme case, ban) all activities and outputs that might cause harm to human health and ecological systems. Apart from the philosophical problems of proving negatives (no harm—see Chapter 17), this would be unreasonable and unrealistic. Probably all we do as humans has some negative effect on the environment.

Another interpretation of PP is that in implementing controls we should not necessarily wait until we have a complete understanding of the relationship between an adverse effect and putative causes. It is often argued that this is appropriate if the effect is serious and/or there is good *prima facie* evidence that the nominal causative agent is the actual one. But the latter judgement will depend upon scientific input and so there is not necessarily any controversy here. If the effect is serious and an agent is identified as the very likely cause, but without the full rigour of scientific evidence, then there may be good grounds for action. Chapter 7 explores how weight of evidence, based upon strong inference rather than the full force of the scientific, experimental approach, is often used in retrospective risk assessment.

A variant of this interpretation of PP is that because the world is a complex place, there will always be uncertainty associated with a scientific analysis of the relationship between causes of problems and their effects. We should recognize this by building 'safety margins' or uncertainty assessments into our risk assessments, especially in designating levels of no concern. In fact, this is a frequent feature of the scientific approach, as will be clear from Chapters 2, 3 and 4. It is still important to recognize, though, that safety (sometimes called application or uncertainty) factors, by definition, depend upon subjective and not objective judgements: for example, based on how much we feel we do not know about the situation under scrutiny or judgements about how serious a mistake could be. This is another reason why risk assessment and risk management activities cannot be kept so neatly apart (see above).

Yet another variant of PP is that the world is so complex that science will never be able to come to terms with predicting environmental risks, so why bother? It is more expeditious and safer to ignore the science and presume the worst. But this is a return to the 'deep green' version of PP and is most certainly at odds with the approach being advocated in this Handbook.

1.7 ORGANIZATION OF THIS BOOK

As already indicated, this Handbook makes the not-too-sharp distinction between assessment procedures and management in Parts 1 and 4, with Parts 2 and 3 bridging this gap: Part 2 is concerned with the way in which risk assessment is used in policy and legislation as a *prelude* to decisions about risk management; Part 3 is concerned with how risks have to be balanced against other considerations (economic and psychological) in coming to a decision about risk management.

The 'big concept' in environmental protection is sustainable development: the proposition that pathways of economic development can be found that leave the environment more or less intact, or if not that provide alternatives for future generations. By assessing likely impacts of our activities and emanations on the environment, and trying then to manage them, there is a presumption that understanding environmental risk will contribute to sustainability. This is explored further throughout the Handbook but especially in the final chapter of Part 4.

1.8 REFERENCES

Calow, P. (1992) Can ecosystems be healthy? Critical considerations of concepts. *Journal of Aquatic Ecosystem Health*, **1**, 1–15.

Calow, P. (1993) *Handbook of Ecotoxicology*, Vol. I. Blackwell Scientific Publications, Oxford.

Calow, P. (1994) *Handbook of Ecotoxicology*, Vol. II. Blackwell Scientific Publications, Oxford.

Calow, P. (1997) *Controlling Environmental Risks from Chemicals: Principles and Practice*. Wiley, Chichester.

Calow, P. & Forbes, V.E. (1997) Science and subjectivity in the practice of ecological risk assessment. *Environmental Management*, in press.

Calow, P. & Streatfeild, C. (1994) *A DIY Environmental Risk Profile*. Sheffield Regional Green Business Club, Sheffield. (ISBN 0 9524211 00)

Lovelock, J. (1988). *The Age of Gaia*. Oxford University Press, Oxford.

O'Riordan, T. & Cameron, J. (1994) *Interpreting the Precautionary Principle*. Cameron & May, London.

Rimington, J.D. (1992) Overview of risk assessment. In: *Risk Assessment*, Part I. *Proceedings of an International Conference*, pp. 20–27. Health and Safety Executive, HMSO, London.

Suter, G.W. (1992) *Ecological Risk Assessment*. Lewis Publishers, Chelsea, MI.

van Leeuwen, C.J. & Hermens, T.L.M. (eds) (1995). *Risk Assessment of Chemicals*. Kluwer Academic Publishers, Dordrecht.

Part 1
Risk Assessment

The scientific principles of risk assessment are most easily appreciated with respect to chemical substances and so the first two chapters address this respectively for human health (Chapter 2) and ecological (Chapter 3) systems. They illustrate the systematic and stepwise approach, define the jargon that is becoming associated with this, and demonstrate how it is becoming assimilated and codified in regulatory frameworks especially for the USA and Europe. Most of the legislation, however, is concerned with controlling controlled releases. And yet spectacular impacts to human health and ecological systems often arise from uncontrolled accidents. These certainly command public attention. Here the principles of risk assessment are similar, but the predicted environmental concentrations and their likelihood of occurrence depend upon likelihood of failure, and this is dealt with in Chapter 4.

The following Chapters (5 and 6) then go on to show how the principles of risk assessment can and indeed are being applied to the somewhat more difficult circumstances involving introduced organisms. The main difference between inanimate and animate is, of course, the capacity to reproduce and evolve, and this introduces complications for assessing the risks from introduced organisms, both 'engineered' and 'natural', for human health and environment that are carefully explained and explored in Chapters 5 and 6.

Assessing risks cannot always be done prospectively—before release of a substance or organisms. Indeed, in a complex industrial society we create things as both products and wastes, from which we may not even begin to recognize the potential to cause harm until evidence starts appearing in the world around us. Recognizing this information and dealing with it systematically and as objectively as possible is referred to as epidemiology in human health circles and as ecoepidemiology, or retrospective risk assessment, in ecological circles. These areas are dealt with in Chapters 7 and 8.

Chapter 2
Assessing Risks to Human Health from Chemicals in the Environment

SUSAN P. FELTER, MICHAEL L. DOURSON AND JACQUELINE PATTERSON

2.1 INTRODUCTION

Humans are exposed to a multitude of potentially hazardous chemicals in their indoor and outdoor air, food, soil and ambient and drinking waters. Risk assessment is the process whereby scientists evaluate the toxicity data for chemicals to which humans are, or may be, exposed, and attempt to identify and quantify potential risks to health. The risk assessment process is also used to estimate levels of intake via the various media that are expected to be 'safe'. These values are then used in conjunction with information on exposure in order to determine acceptable levels for concentrations of hazardous chemicals in environmental media.

The process of human health risk assessment was first described as a four-component paradigm by the National Research Council (NRC) of the National Academy of Sciences (NAS) in 1983 and was subsequently updated in 1994. This chapter follows the NAS paradigm and introduces each component with an excerpt from the NAS (1994) publication, *Science and Judgment in Risk Assessment*. The focus is primarily on the risk-assessment methods used by national or health agencies, such as the International Programme on Chemical Safety (IPCS) or the US Environmental Protection Agency (USEPA). However, scientists from other groups have made contributions to the field, especially in the area of research to improve the standard methods.

2.2 HAZARD IDENTIFICATION

> *Hazard Identification* entails identification of the contaminants that are suspected to pose health hazards, quantification of the concentrations at which they are present in the environment, a description of the specific forms of toxicity (neurotoxicity, carcinogenicity, etc.) that can be caused by the contaminants of concern, and an evaluation of the conditions under which these forms of toxicity might be expressed in exposed humans ...
>
> NAS, 1994

2.2.1 Hazard identification of non-cancer end-points

Hazard identification is generally the first step of the risk assessment process, in which it is determined if there is a potential cause for concern over human exposure to an agent. This involves an evaluation of the appropriateness, nature, quality and relevance of scientific data on the specific chemical; the characteristics and relevance of the experimental routes of exposure; and the nature and significance to human health of the effects observed. The USEPA, for example, has developed hazard identification guidelines for developmental and reproductive toxicity that carefully address these issues (USEPA 1991, 1994a). Table 2.1 gives a brief list of considerations.

Much of the process of hazard identification for non-cancer end-points depends on professional judgement as to whether or not an observed effect, or collection of effects (or syndrome), constitutes an adverse response. This is not always easy, and often requires the views of experts in the subject area, because although many effects are clearly adverse (e.g. fatty infiltration of the liver), many

Table 2.1 Considerations in characterizing hazard. (Information from USEPA 1995b.)

What are the key toxicological studies and of what quality?
Are the data from laboratory or field studies? Single or multiple species?
For cancer: Was there a single or multiple tumour site(s)? Benign or malignant? Was the maximum tolerated dose achieved?
For other-than-cancer: What end-points were observed and what is the basis for the critical effect? Other supporting studies? Conflicting?

Besides for the critical effect, are there other end-points of concern?

What are the significant data gaps?

What are the available epidemiological or clinical data?
- What types of studies were used (i.e. ecological, case–control, cohort)?
- Describe the degree to which exposures were described adequately, to which confounding factors were accounted for adequately, and to which other causal factors were excluded

Were there non-positive animal or human data?

How much is known about the biological mechanism of action and how does this aid in the interpretation of the data?

Summarize the hazard identification and discuss the confidence in the conclusions, alternative conclusions that are also supported by the data, significant data gaps and highlights of any major assumptions

others are of uncertain toxicological consequence (e.g. decrease in body weight gain).

Because toxic chemicals often elicit more than one adverse effect, the process of hazard identification for non-cancer toxicity includes an evaluation of the target organ or 'critical' effect; i.e. the first adverse effect or its known precursor that occurs as the dose rate increases. This is shown hypothetically in Fig. 2.1, where several effects are evoked from chemical exposure:

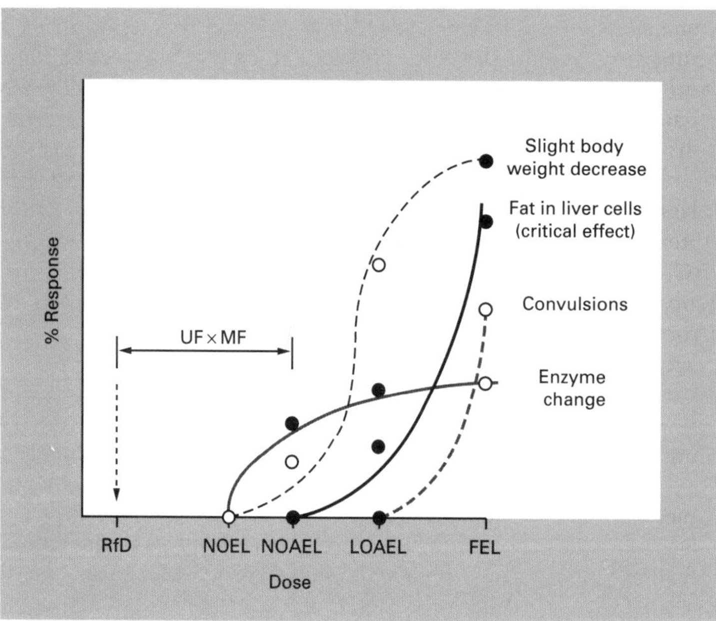

Fig. 2.1 The judgement of the critical effect and its NOAEL (no-observed-adverse-effect level), along with the appropriate uncertainty factor (UF) and modifying factor (MF), leads to the estimation of the RfD (reference dose). NOEL, no-observed-effect level; LOAEL, lowest-observed-adverse-effect level; FEL, frank effect level.

enzyme change, slight decrease in body weight, fatty infiltration of the liver and convulsions. Enzyme change and slight decrease in body weight are judged not to be adverse effects. Fatty infiltration of the liver is judged to be the critical effect.

The judgement of whether an effect is adverse or critical may change among toxicity studies of different durations, and may be influenced by toxicity in other organs or by toxicokinetics. A good example of this is increased liver weight due to a proliferation of smooth endoplasmic reticulum with chemical exposure. Such an effect may be judged as not adverse if the parent chemical is the toxic moiety and such an increase is likely to quicken its metabolism, or may be judged to be adverse if a metabolite is the toxic moiety (Farland & Dourson 1992). The distinction of adverse compared with non-adverse effects and the choice of critical effects in the hazard identification component of the paradigm is then used as a basis for the dose–response assessment.

2.2.2 Hazard identification of carcinogens

Hazard identification of carcinogens refers to the process of determining if a compound has the potential to elicit a carcinogenic response in humans. Many types of information may be used to determine the overall weight-of-evidence of carcinogenicity: epidemiological information, chronic animal bioassays, mechanistic data, mutagenicity tests, other short-term tests, structure–activity relationships, metabolic and pharmacokinetic properties, toxicological effects and physical and chemical properties.

The first organization to develop a classification scheme for carcinogenicity was the International Agency for Research on Cancer (IARC) in 1978. Based on a strength-of-the-evidence approach (evidence coming from human or laboratory animal data or short-term studies), chemicals were placed in one of three categories.

Group 1: carcinogenic to humans.
*Group 2**: probably carcinogenic to humans.
Group 3: cannot be classified as to its carcinogenicity to humans.

In 1986, the USEPA published general guidelines to be used by Agency scientists in developing and evaluating risk assessments for carcinogens (USEPA 1986). Based on the weight-of-evidence from epidemiological and laboratory animal bioassays, chemicals are placed in one of six categories. Supporting data (e.g. mutagenicity data, mechanistic data) may then be used to move a chemical up or down in the ranking. These categories are modelled after those used by IARC.

Group A: carcinogenic to humans.
Group B†: probably carcinogenic to humans.
Group C: possibly carcinogenic to humans.
Group D: not classifiable as to human carcinogenicity.
Group E: evidence of non-carcinogenicity for humans.

In April 1996, the USEPA proposed revisions to the carcinogen risk assessment guidelines. In contrast to the concise alpha-numeric classification system of 1986, the guidelines proposed advocate the development of a more comprehensive characterization of the carcinogenic hazard in the form of a narrative. Within this context, a cancer hazard characterization should include all information relevant to the weight-of-evidence for carcinogenicity, not just tumour data in humans and animals. This means that mechanistic data can play an integral role in the hazard identification step for carcinogenicity, and may also influence the choice of a dose–response model. Another change is that the hazard characterization can provide specific information about the conditions under which a chemical is likely to be carcinogenic; for example, it may be likely to be carcinogenic by the route of inhalation but not by ingestion. These proposed changes in the USEPA methods reflect a general movement in the field of cancer risk assessment to include more chemical-specific data and to move away from the use of default positions wherever possible.

* Group 2 includes subgroups 2A (for chemicals having limited evidence of carcinogenicity in humans) and 2B (for chemicals having sufficient evidence of carcinogenicity in laboratory animals, and inadequate evidence in humans).

† Group B includes subgroups B1 (for chemicals having limited evidence of carcinogenicity in humans) and B2 (for chemicals having insufficient human data but sufficient animal data).

In addition to what has been described here for the USEPA, other groups have published carcinogen classification schemes along similar lines. Moolenaar (1994) has provided a summary and comparison of several international classification schemes, including eight governmental agencies and two independent organizations. These classification schemes have anywhere from two to six distinct categories with varying degrees of emphasis on mechanistic data. In addition, Ashby *et al.* (1990) have recommended an eight-category system.

Common to many of these groups, the determination of carcinogenic hazard includes a determination of whether the incidence of tumour types observed to occur in laboratory animals is statistically significantly elevated over that observed in controls. Two forms of statistical tests are used to answer this question: trend tests, which look for an overall trend of increasing tumour incidence with increasing dose; and pairwise comparison tests, which directly compare the tumour incidence in an individual dose group with that seen in controls.

Determination of the mechanism by which a chemical causes cancer in laboratory animals also provides information about the potential for human carcinogenicity relevant to the hazard identification process. The potential for the same or a related mechanism to be operative in humans provides the basis for extrapolation from other animal species to estimate the risk of cancer to humans. For some specific tumour types, or mechanisms of carcinogenicity, there are indications that tumours observed in laboratory animals may have no relevance or limited relevance to human carcinogenicity. Tumour types that are included in this group include kidney tumours in male rats that are caused by the accumulation of a male-rat-specific protein (alpha$_{2u}$-globulin); liver tumours in male B6C3F1 mice; thyroid follicular cell tumours; and bladder tumours related to the formation of silicate-containing precipitate and crystals (e.g. as seen in saccharin-induced bladder cancer in rats).

2.3 DOSE–RESPONSE ASSESSMENT

> *Dose–Response Assessment* entails a further evaluation of the conditions under which the toxic properties of a chemical might be manifested in exposed people, with particular emphasis on the quantitative relation between the dose and the toxic response. The development of this relationship may involve the use of mathematical models. This step may include an assessment of variations in response, for example, differences in susceptibility between young and old people.
>
> NAS, 1994

2.3.1 Non-cancer end-points

The 'Safe' dose approach

Dose–response assessment follows hazard identification in the risk assessment process. Dose–response assessment involves the quantitative evaluation of toxicity data to determine the likely incidence of the associated effects in humans. The information available for dose–response assessment ranges from well-conducted and well-controlled studies on human exposures, and epidemiology studies with large numbers of subjects, well-characterized exposures, and supportive studies in several animal species, to a lack of human and animal toxicity data with only structure–activity relationships to guide the evaluation. In any case, scientists should consider all pertinent studies in this process; even a single human case study can provide useful information. However, only data of sufficient quality, as judged by experts, should be used in the dose–response assessment of a chemical. Table 2.2 lists some questions to be considered in characterizing the dose–response relationship for an agent.

Most non-cancer effects resulting from exposure to toxic agents are thought to be associated with a threshold; i.e. an exposure exists below which toxicity does not occur. The dose–response component of risk assessment involves the quantitative evaluation of toxicity data to determine a level of exposure for humans that is considered by risk assessors to be below the threshold for toxicity for sensitive subgroups.

Health agencies throughout the world support the use of a 'safe' dose concept, and define terms

Table 2.2 Considerations in characterizing the dose-response assessment. (Information from USEPA 1995b.)

Data
Which data were used to develop the dose-response curve? Would the results have been different if based on a different data set?

If animal data were used, which species were used—most sensitive, average of all species, or other? Were any studies excluded and why?

If epidemiological data were used, were they only the positive, all studies, or a combination? Were any studies excluded and why?

Was a meta-analysis performed to combine the studies?
If so, what approach was used?

Models
What model was used to develop the dose-response curve? The rationale for this choice? Is chemical-specific information available to support this approach?

For non-cancer end-points how was 'safe' dose calculated? What assumptions and/or uncertainty factors were used? For benchmark doses, what model was used and why?

For cancer end-points, what dose-response model was used and why was it selected? Would other models have provided as plausible results?

Discuss the route and level of exposure observed in the data compared with anticipated human exposure. If data are from a different route, are pharmacokinetic data available to extrapolate across routes? How far is the extrapolation from the observed data to environmental exposures and what is the impact of this extrapolation?

Toxicity values
Summarize the risk value and discuss the confidence in the value. Can a range of values be provided? What are the results of different approaches or models?

and conditions for use. This 'safe' or subthreshold dose often goes by different names, such as: Health Canada's Tolerable Daily Intake or Concentration (TDI or TDC) (Meek *et al.* 1994); IPCS's Tolerable Intake (TI) (IPCS 1994); US Agency for Toxic Substances and Disease Registry's (ATSDR's) Minimum Risk Level (MRL) (Pohl & Abadin 1995); USEPA's Reference Dose (RfD) (Barnes & Dourson 1988; Dourson 1994) or Reference Concentration (RfC) (Jarabek 1994; USEPA 1994b); or the World Health Organization's Acceptable Daily Intake (ADI) (Lu 1985, 1988). Many of the underlying assumptions, judgements of critical effect, and choices of uncertainty factors (or safety factors) are similar among health agencies in estimating these subthreshold doses.

One of the best-known methods is that used by the USEPA to derive reference doses (RfDs) and reference concentrations (RfCs), which are subthreshold exposures for non-cancer toxicity. They are defined as: '... an estimate (with uncertainty spanning perhaps an order of magnitude) of a daily exposure to the human population (including sensitive subgroups) that is likely to be without an appreciable risk of deleterious effects during a lifetime' (Barnes & Dourson 1988).

The subthreshold dose approach starts with an identification of the critical effect(s), as described in Hazard Identification (2.2.2). The critical dose is then chosen. All groups rely on the experimental dose that represents the highest level tested at which the critical effects were not demonstrated as this critical dose. This dose is often called the no-observed-adverse-effect level (NOAEL), or the no-observed-effect level (NOEL). If a NOAEL is not available, the use of a lowest-observed-adverse-effect level (LOAEL) is often used as the critical dose.

Human data are preferred in the determination of an RfD or RfC. However, in the absence of these data, animal data are closely scrutinized. Risk assessment scientists seek to identify the animal model that is most relevant to humans, based on the most defensible biological rationale. In the absence of a clearly most relevant species, the critical study and species that shows an adverse effect at the lowest administered dose are generally selected. This is based on the assumption that, in the absence of data to the contrary, humans may be as sensitive as the most sensitive experimental animal species. Uncertainty factors (UFs) are then used as divisors to this critical dose (NOAEL or LOAEL) to determine the subthreshold dose. These factors are considered as reductions in the dose rate to account for several areas of scientific uncertainty inherent in most toxicity databases. As shown in Table 2.3, these areas include interhuman variability (designated as H); extrapolation from experimental animals to humans (designated as A); extrapolation from subchronic to chronic exposure (designated as S); extrapolation from an experimental LOAEL to NOAEL (designated as L); and how to account for the lack of a complete database. In addition to these UFs, several groups also use a modifying factor that can be used to account for uncertainties not explicitly dealt with by the standard factors.

All groups occasionally use a factor less than 10 or even a factor of 1, if the existing data reduce or obviate the need to account for a particular area of uncertainty.* For example, the use of a 1-year rat study as the basis of an RfD may reduce the need for a tenfold factor for the area of subchronic-to-chronic extrapolation to threefold, because it can be demonstrated empirically that 1-year NOAELs for rat are generally closer in magnitude to chronic values than are 3-month NOAELs. Lewis *et al.* (1990) investigate this concept of variable uncertainty factors more fully through an analysis of expected values.

The choice of appropriate uncertainty and modifying factors reflects a case-by-case judgement by experts and should account for each of the applicable areas of uncertainty (described in Table 2.3) and any nuances in the available data that might change the magnitude of any factor. Several reports describe the underlying basis of uncertainty factors (Zielhuis & van der Kreek 1979; Dourson & Stara 1983) and research into this area (Calabrese 1985; Hattis *et al.* 1987; Hartley & Ohanian 1988; Lewis *et al.* 1990; Renwick 1991, 1993; Calabrese *et al.* 1992; Dourson *et al.* 1992; Calabrese & Gilbert 1993; Kroes *et al.* 1993; Abdel-Rahman & Dourson 1995).

The scientific strengths and limitations of this approach have been discussed in the literature (Munro & Krewski 1981; Lu 1983, 1985, 1988; Krewski *et al.* 1984; Crump 1984, 1986; Dourson *et al.* 1985, 1986; Barnes & Dourson 1988; Kimmel & Gaylor 1988). The scientific strengths, in brief, are that all toxicity data are reviewed in the choice of the NOAEL for the critical effects, and that uncertainties in the entire data base can be factored into the resulting value of the subthreshold dose through the use of professional judgement as to the appropriate uncertainty and modifying factors.

The limitations, in brief, are that the NOAEL is restricted by the choice of dose-spacing and the number of animals, as well as factors that influence the quality of the study. Studies with wide dose-spacing and a low number of animals per dose group can lead to a more poorly characterized subthreshold dose as compared to studies with tighter dose-spacing and more animals per dose group (see, e.g., Hattis *et al.* 1987; Leisenring & Ryan 1992). The NOAEL is also not generally influenced by the nature of the dose–response curve. Uncertainty factors, although considered necessary and perhaps to reflect accurately the potential underlying areas of uncertainty, are quite imprecise. Nor does the subthreshold-dose approach enable an estimate of risks at exposures greater than the subthreshold dose.

Scientists are developing methods that address some of these latter limitations (e.g. DeRosa *et al.* 1985; Dourson *et al.* 1985; Kimmel & Gaylor 1988; Kimmel *et al.* 1988; Hertzberg 1989; Hertzberg & Dourson 1993; Renwick & Walker 1993; Faustman *et al.* 1994; Allen *et al.* 1994a,b). Two of these methods are described briefly here.

* The usual intermediate factor used is 3 because it is the approximate logarithmic mean of 1 and 10. The choice of 3, instead of 5 for example, reflects both the expected precision of the UFs (about 1 digit, log base 10) and the view that it is not generally possible to be more precise in considering the nuances of these areas of uncertainty than about half-way.

Table 2.3 Description of typical uncertainty and modifying factors in the development of subthreshold doses for several groups*.

Uncertainty factors (UFs)†	Guidelines ‡	UF value				
		Health Canada Agency§	IPCS	RIVM	USATSDR	USEPA
Interhuman (or intraspecies)	Generally use when extrapolating from valid results from studies of prolonged exposure to average healthy humans. This factor accounts for the variation in sensitivity among humans and is composed of toxicokinetic and toxicodynamic components	1–100	10 (3.16 × 3.16)	10	10	10
Experimental animal to human	Generally use when extrapolating from valid results of long-term studies on experimental animals when results of studies of human exposure are not available or are inadequate. This factor accounts for the uncertainty in extrapolating animal data to humans and is also composed of toxicokinetic and toxicodynamic uncertainties		10 (2.5 × 4.0)	10	10	10
Subchronic to chronic	Generally use when extrapolating from less than chronic results on experimental animals or humans. This factor accounts for the uncertainty in extrapolating from less than chronic NOAELs or LOAELs to chronic NOAELs or LOAELs	1–100	1–100	10	NA¶	≤10
LOAEL to NOAEL	Generally use when extrapolating a LOAEL to a NOAEL. This factor accounts for the experimental uncertainty in developing a subthreshold dose from a LOAEL, rather than a NOAEL			10	10	≤10
Incomplete database	Generally use when extrapolating from valid results in experimental animals when the data is 'incomplete'. This factor accounts for the inability of any single study to adequately address all possible adverse outcomes			NA	NA	≤10
Modifying factor	Generally use upon a professional assessment of scientific uncertainties of the study and data base not explicitly treated above (e.g. the number of animals tested)	1–10	1–10	NA	NA	0< to ≤10

* Source: Dourson (1994), Jarabek (1994), IPCS (1994), Meek *et al.* (1994) and Rademaker & Linders (1994).

† Note: The maximum uncertainty factor used with the minimum confidence data base is generally 10 000. LOAEL, lowest-observed-adverse-effect level; NOAEL, no-observed-adverse-effect level.

‡ Professional judgement is required to determine the appropriate value to use for any given UF. The values listed in this table are nominal values that are used frequently by these agencies.

§ Abbreviations used: IPCS (International Programme on Chemical Safety); RIVM (The Netherlands National Institute of Public Health and Environmental Protection); USATSDR (US Agency for Toxic Substances and Disease Registry); USEPA (US Environmental Protection Agency).

¶ ATSDR develops minimum risk levels (MRLs) for specified durations of exposure, and generally does not extrapolate among durations. Therefore, an uncertainty factor for extrapolation between subchronic and chronic exposures is not used.

Benchmark dose

Another form of quantitative risk assessment of non-cancer end-points is the benchmark dose (BMD) method. The USEPA (1995a) has defined the BMD as: 'a statistical lower confidence limit for a dose that produces a predetermined change in response rate of an adverse effect ... compared to background.'

This method, which was first described by Crump (1984) and Dourson *et al.* (1985), was developed in an attempt to remedy some notable shortcomings of the use of a NOAEL in the subthreshold-dose approach described above. For example, the NOAEL is limited by the experimental doses chosen by the investigators in the toxicity studies. The larger the dose spacing, the less accurate the experimental NOAEL (or LOAEL) is apt to be. Also, the slope of the dose–response curve provides valuable information that is not used explicitly in this approach (although it may influence the choice of uncertainty factors). The BMD method attempts to use more of the available dose–response information by fitting a mathematical model to the data and then determining the dose associated with a specified incidence of adverse effect. In this way, the BMD is not limited to the experimental doses chosen by the investigators.

Although the BMD method offers some advantages over the NOAEL, it can be used only in cases where data are available that are suitable for modelling. It is not, therefore, a replacement for the NOAEL, but should be considered as an additional method that may offer advantages for some risk assessments.

There are a number of decisions to be made in applying the BMD method, for example: which mathematical model to use; what degree of confidence limit to use; what incidence rate to predetermine as the benchmark response (e.g. a 1%, 5% or 10% incidence of an effect). For more information, the reader is referred to a guidance document on the use of the benchmark-dose approach in risk assessment that was issued by USEPA's Risk Assessment Forum (USEPA 1995a).

Categorical regression

Another method that has been proposed for quantitative dose–response analysis for non-cancer effects is that of categorical regression. This involves statistical regression on severity categories of overall toxicity (Hertzberg & Miller 1985; Hertzberg 1991; Hertzberg & Wymer 1991). By assigning severity categories, all adverse effects may be taken into account rather than just the critical effect. Categorical regression also allows use of group data (i.e. at the dose-group level, not individual animals) as well as toxicity data from multiple studies. The results of the regression can then be used to develop a subthreshold dose much as the BMD is used.

Categorical regression can also provide information about relative risks from exposures exceeding the RfD. The NOAEL and BMD approaches are limited in that they are focused on the determination of 'safe' and 'acceptable' intake levels (i.e. they are point estimates designed to be below the population threshold for toxicity). In situations where exposures may exceed these levels, however, information is needed to help determine the urgency of a situation. Herein lies one of the advantages of categorical regression because it provides information about increasing toxicity with increasing dose rate. If human data are available, categorical regression can be used to actually estimate potential risk above the RfD or RfC. With only animal data, categorical regression can help prioritize risks based on how quickly the toxicity severity changes with dose.

2.3.2 Cancer end-points

The elicitation of a carcinogenic response traditionally has been presumed to occur without a threshold. Because of this, it has often been assumed by several regulatory agencies (e.g. USEPA 1986; Rademaker & Linders 1994) that any dose of a carcinogen is associated with some increased risk. As a result, dose–response assessment for carcinogens often focused on determining a *de minimis* risk level, frequently expressed as the risk of one-in-a-million, by using a linear model to extrapolate risks down to low-dose levels. Other groups, such as Health Canada (Meek *et al.* 1994), do not advocate extrapolation beyond the range of observable data, but rather use a margin-of-exposure approach (described more fully below). In EPA's 1996 proposed revisions to the carcinogen risk assessment guidelines, an option of using a margin-of-exposure (MOE) approach is also described.

Exposure to a carcinogenic agent often causes more than one tumour type. Similar to the process used for the evaluation of non-cancer toxicity, the risk assessor must evaluate the data to determine which end-point(s) occur(s) at the lowest dose. Unless there are data to support otherwise, it is generally assumed that humans may be as sensitive as the most sensitive animal model. After identifying the study(ies) that is(are) most appropriate for developing a quantitative risk estimate, the next step is to transform the doses to which the animals were exposed into human equivalent doses. For example, in the absence of a chemical-specific model, the USEPA (1996) recommends the use of a cross-species scaling factor of (body weight)$^{3/4}$ for oral exposures, and default methodology (USEPA, 1994a) for estimating respiratory deposition and absorption of particles and gases for inhalation exposures. Finally, the dose–response data are modelled to determine the carcinogenic potency of the chemical at low doses.

Estimating risk with mathematical models

Dose–response assessment for carcinogens is concerned with estimating the central estimate and/or the upper confidence bound for carcinogenic risk associated with environmental exposures. Alternatively, risk managers may be interested in setting standards for exposures by various media (e.g. air, drinking water) based on a carcinogenic risk level that is considered to be *de minimis* (e.g. one-in-a-million excess risk). The cancer bioassays generally used in the dose–response assessment, however, are performed in laboratory animals at very high doses relative to levels at which humans may actually be exposed. These high doses are necessary in order to produce a statistically measurable effect given the relatively small number of animals used. The nature of the curve at levels of exposure below the lowest experimental dose is not known. Many models have been developed to estimate cancer risk in this low-dose region.

A common model used to perform this extrapolation is adapted from the multistage model. This model assumes that cancer is the result of a sequence of changes in a cell or organ and that exposure to a carcinogen can increase the transition rate between these stages, resulting in malignancy (Armitage & Doll 1954, 1961; Crump *et al.* 1976). The 95% upper confidence limit of the linear component of this model (often referred to as the q_1^*) has been used by the USEPA (1986) as an upper bound estimate of cancer potency because it is numerically more stable than a central estimate and also is in keeping with the low-dose linear approach adopted for cancer-risk assessments. Although there are no data to demonstrate that the linearized multistage model is more appropriate than any other model, it has been used as the default because it provides a plausible and stable upper bound estimate that is not likely to underestimate the cancer risk. The USEPA and others recognize that at very low doses the response could be as low as zero.

In the USEPA's 1996 revised guidelines, it is proposed that the dose–response assessment be considered as a two-step process. The first step is to fit a model to data in the observed range only. If sufficient data are available, a biologically based model is the preferred approach. Also, there may be cases in which data other than tumour incidence (e.g. information on DNA adducts) can be used to extend dose–response below the observable range. The outcome of the first step is the estimation of an ED_{10} or LED_{10}. The ED_{10} (effective dose at the 10% level) is the dose associated with a 10% increase over background in the end-point being measured (e.g. tumour incidence or other). The LED_{10} is the lower 95% confidence limit on this dose.

The second step under the USEPA's 1996 revised guidelines is to use an extrapolation procedure to estimate risk in the low-dose region (i.e the range of human exposure) if it is appropriate to do so. For cases where data do not support the development of a biologically based model, and for which the dose–response relationship is thought to be linear, the proposed guidelines suggest terminating the model in the range of experimental data, and drawing a straight line to the origin. Whereas the linearized multistage model may still be used in modelling the data in the experimental range, justification for using this (or any other) model needs to be provided.

For chemicals that have a non-linear dose–response relationship, the proposed guidelines advocate the use of an MOE analysis. The MOE is the LED_{10} (or another predetermined starting point within the range of observation) divided by the exposure of concern. The risk manager then decides whether the margin of exposure is large enough to satisfy management policy criteria. The proposed guidelines suggest that a factor of 100 be used as a science policy default position to reflect allowances for intra- and interspecies variability. Chemical-specific data can then be used to adjust this factor upward or downward as appropriate.

This type of MOE analysis is similar to that used currently by Health Canada (Meek *et al.* 1994). Main differences between the MOE approaches of the agencies are twofold: Health Canada determines a TD_{05} (the dose associated with a 5% increase over background of tumour incidence), whereas the USEPA is proposing determination of an ED_{10}. Secondly, the USEPA uses the 95% lower confidence bound on the ED_{10} whereas Health Canada uses the central estimate.

Approaches with uncertainty factors

A number of organizations have developed other methods for quantitative dose–response assessment for carcinogens. Moolenaar (1994) has described the similarities and differences between approaches used by the USEPA, the UK, Denmark, the European Union (EU), The Netherlands, and Norway. He points out that the USEPA is the only organization to have described carcinogenic risk in terms of an 'upper bound' (i.e. the 95th percentile of the slope of the dose–response curve in the low-dose region).

For example, each of the aforementioned groups has developed separate methods for dealing with genotoxic versus non-genotoxic carcinogens. Norway does not perform low-dose extrapolation for any carcinogens, but rather uses the TD_{50} to determine a potency classification for a carcinogen. For non-genotoxic carcinogens, the UK, the EU and The Netherlands use a subthreshold dose approach: they set ADIs using the method described above for non-cancer toxicity. For genotoxic carcinogens thought to have no threshold, The Netherlands extrapolates linearly from the lowest experimental dose having an increased incidence of tumours. Clearly, there are many variations in the ways that dose–response assessment for carcinogens can be, and are, performed. A common theme among all of these groups is that the mechanism by which the agent is believed to cause cancer is playing a greater role in the way in which the dose–response assessment is approached.

2.4 EXPOSURE ASSESSMENT

> *Exposure assessment* involves specifying the population that might be exposed to the agent of concern, identifying the routes through which exposures can occur, and estimating the magnitude, duration, and timing of the doses that people might receive as a result of their exposure.
>
> NAS, 1994

Environmentally relevant routes of exposure for humans are inhalation, oral, and dermal. An exposure assessment may include a component for each, such as an assessor would conduct when investigating the potential impact of a point source of pollution. In such a multimedia investigation, an exposure assessment is initiated by estimating the amount and rate at which a toxic agent is released from a given source. Fate and transport models are then used to estimate the movement of the agent through environmental media to which humans may be exposed. A number of models is available for use in estimating transport and fate; many of these are described in the USEPA's exposure assessment guidelines (USEPA 1992). Table 2.4 provides guidance in characterizing the exposure assessment step.

An exposure assessment may also be focused on one particular medium and one route of exposure, for example, the oral intake of a chemical from drinking water. This type of exposure assessment may be used, for example, to determine whether there is sufficient exposure of humans to a chemical in a given medium to warrant regulation.

Exposure can be determined directly, through

Table 2.4 Considerations in characterizing exposure. (Information from USEPA 1995b.)

What are the most significant sources and pathways of environmental exposure, presently and in the future (if appropriate)?
Are there data on other sources of exposure and what is the relative contribution of different sources of exposure?
Describe the populations that were assessed, including highly exposed groups and highly susceptible groups
Describe the basis for the exposure assessment, including any monitoring, modelling, or other analyses of exposure distributions such as Monte Carlo.
Describe the range of exposures to 'average' and 'high-end' individuals, the general population, high exposure group(s), children, susceptible populations
How was the central tendency estimate developed? What factors and/or methods were used in developing this estimate?
Are there highly exposed subgroups and how are they accounted for?
Is there reason to be concerned about cumulative or multiple exposures?
What are the results of different approaches, i.e. modelling, monitoring, probability distributions?
What are the limitations of each and the range of most reasonable values?
What is the confidence in the results obtained and the limitations to the results?

personal monitoring devices, or indirectly, through environmental monitoring. If environmental monitoring is used, then the assessor must estimate the extent to which individuals may be exposed to the media for which monitoring data are available. Risk assessment scientists often use default values for these assessments (e.g. assuming an inhalation rate of 20 m^3 per day or consumption of 2 l of water daily).

A need exists to estimate the distribution of exposures that may result to individuals and populations. For example, the USEPA (USEPA 1992) recommends assessing exposure to the total population, and also for assessing the upper end of the exposure distribution; i.e. a 'high-end exposure estimate' and a 'theoretical upper bounding estimate'.

2.5 RISK CHARACTERIZATION

> *Risk characterization* involves integration of information from the first three steps to develop a qualitative or quantitative estimate of the likelihood that any of the hazards associated with the agent of concern will be realized in exposed people. This is the step in which risk-assessment results are expressed. Risk characterization should also include a full discussion of the uncertainties associated with the estimates of risk.
>
> NAS, 1994

Risk characterization is the final step of the risk assessment process, in which information from the hazard identification, dose–response and exposure steps are considered together to determine and communicate the actual likelihood of risk to exposed populations. The risk characterization discussion includes an evaluation of the overall quality of the data, the specific assumptions and uncertainties associated with each step, and the level of confidence in the resulting estimates.

Specific key qualities, or attributes, of risk characterizations have been identified (AIHC 1992; USEPA 1995b). These attributes include transparency in decision making, clarity in communication, consistency and reasonableness. Exercising transparency and clarity result in scientific conclusions being identified separately from policy judgements. In addition, default values, assumptions and uncertainties are disclosed so that the end-user can better identify what is based on data and what is assumed. Greater consistency in the terminology used, along with definitions and assumptions, will provide for better comparability across assessments. In addition, use of standard descriptors, such as those outlined in the USEPA's Exposure Assessment Guidelines (USEPA 1992), reduce confusion and lead to greater understanding. Lastly, the risk characterization should be reasonable and balanced in its presentation. The information and conclusions should be presented in such a fashion that they are clearly understood by the intended audience. The ultimate goal of risk characterization is to provide the decision makers with enough information, presented in a comprehensible fashion, that they understand what is known and unknown about the risk to human health from the situation being evaluated, thereby leading to the best possible risk-based decisions.

In order for risk assessors to meet this goal, they must understand the need for the risk assessment and its intended end-use. Risk assessors should meet with the decision makers and engage them in the process throughout. Involving the risk managers and decision makers will help the risk assessor to meet a level of detail and analysis appropriate for the situation (e.g. initial screening versus national regulation). By communicating with the end-user, risk assessors can ensure that the risk manager will comprehend the results of the analysis.

Involvement of the end-user supports taking an iterative approach to the risk assessment. For example, if the risk assessment is for a contaminated site, it would be very useful to have the exposure assessment scientists involved in developing a monitoring plan and in reviewing initial results so that the monitoring could be refined to collect the most useful data. Only if a first, conservative screening indicates that some level for concern is warranted would a more in-depth analysis be pursued. The iterative approach to risk assessment assures better use of limited resources to address problems.

2.5.1 Uncertainty and variability

The field of risk assessment is increasingly utilizing uncertainty and sensitivity analyses to better assess risks to human health. Critical to a complete risk characterization is a full discussion of the uncertainty within each analysis and that related to the overall assessment. Uncertainty discussions are important because they form the basis for the overall judgement as to the adequacy of the data and conclusions drawn from it. In addition, highlighting of uncertainties can identify areas where the collection of additional data may reduce the uncertainty and strengthen the risk assessment. An uncertainty discussion includes the quality and quantity of data available (toxicity and exposure), identification of data gaps, use of default assumptions and parameter values, and the uncertainties in the models used.

Crucial to a discussion of uncertainty is maintaining a clear distinction between uncertainty and variability within each step of the process. The USEPA distinguishes between these two concepts in its risk characterization guidance (USEPA 1995b). Variability describes inter-individual, spatial or temporal differences within an animal or human population or within monitoring data. It reflects the heterogeneity of the data. Uncertainty, on the other hand, applies to areas for which data are unknown. There are uncertainties associated with both dose–response or fate and transport models; an uncertainty analysis would evaluate the basis for the model and validation of the model.

Given the extensive use of modelling to estimate exposure in particular, the uncertainty related to the chosen parameters can have a great impact on the resulting risk estimates. Risk assessors must be careful to identify the parameter values and their sources so that others can evaluate their appropriateness and impact on the final results. Risk assessors can use probability density functions and/or likelihood distributions to characterize uncertainty quantitatively. Monte Carlo analysis is one statistical procedure used.

To summarize a risk characterization, the risk assessor should consider questions such as those listed in Table 2.5. These questions can be used to help outline a discussion of risk conclusions and comparisons. These questions build upon those in Tables 2.2–2.4 relating to hazard identification, dose–response assessment, and exposure assessment, respectively.

2.6 SUMMARY

This chapter has outlined a process to assess the

Table 2.5 Questions to assist in developing a risk characterization summary. (Information from USEPA 1995b.)

Risk conclusions
What is the overall picture of risk and the specific risk estimates and/or ranges?
For the hazard identification, dose–response and exposure assessment steps:
What are the major conclusions and strengths?
What are the major limitations and uncertainties?
What are the science policy options, and what other alternatives were considered?

Risk context
What are the qualitative characteristics of the hazard (e.g. voluntary versus involuntary, one population segment versus another)? Comment on any risk perception studies related to this type of hazard

What are the alternatives to this hazard and how do the risks compare?

How does this risk compare with other risks?

Are there significant community concerns which influence public perception of risk? Are there perceived or actual inequities in distribution of risks and benefits?

Other information
Have other risk assessments been done on this chemical and were there significantly different conclusions?

human health risks from exposure to chemicals in the environment. The methods used to assess and characterize these risks are being improved and expanded upon. Recently, the area of quantifying uncertainty has greatly expanded, along with developing better procedures to use the available data more fully. Improved methods to characterize and communicate the results of the risk assessment process are also being explored. Although significant research has been done to develop methods to characterize exposure and health effects, much more is needed to provide risk decision makers with the accurate estimates they need to make reasonable and cost-effective decisions. Risk assessors will always be faced with gaps in data and scientific understanding. How they deal with these uncertainties will determine how useful risk assessment will be to decision makers and the public.

2.7 REFERENCES

Abdel-Rahman, M.S. & Dourson, M.L. (Guest Eds) (1995) *Proceedings of EPA Uncertainty Factor Workshop. Journal of Human and Ecological Risk Assessment*, **15**(5).

AIHC (1992) *Improving Risk Characterization. Summary of Workshop held September 26 and 27, 1991.* American Industrial Health Council.

Allen, B.C., Kavlock, R.J., Kimmel, C.A. & Faustman, E.M. (1994a) Dose–response assessment for developmental toxicity. II. Comparison of generic benchmark dose estimates with no observed adverse effect levels. *Fundamental and Applied Toxicology*, **23**(4), 487–495.

Allen, B.C., Kavlock, R.J., Kimmel, C.A. & Faustman, E.M. (1994b) Dose–response assessment for developmental toxicity. III. Statistical Models. *Fundamental and Applied Toxicology*, **23**(4), 496–509.

Armitage, P. & Doll, R. (1954) The age distribution of cancer and a multistage theory of carcinogenesis. *British Journal of Cancer*, **8**, 1–12.

Armitage, P. & Doll, R. (1961) Stochastic models for carcinogenesis. In: *Proceedings of the Fourth Berkeley Symposium on Mathematical Statistics and Probability*, Vol. 4, pp. 19–38. University of California Press, Berkeley.

Ashby, J., Doerrer, N.G., Flamm, F.G. *et al.* (1990) A scheme for classifying carcinogens. *Regulatory Toxicology and Pharmacology*, **12**, 270–295.

Barnes, D.G. & Dourson M.L. (1988) Reference dose (RfD): description and use in health risk assessments. *Regulatory Toxicology and Pharmacology*, **8**, 471–486.

Calabrese, E.J. (1985) Uncertainty factors and inter-individual variation. *Regulatory Toxicology and Pharmacology*, **5**, 190–196.

Calabrese, E.J. & Gilbert, C.E. (1993) Lack of total independence of uncertainty factors (UFs): implications for the size of the total uncertainty factor. *Regulatory Toxicology and Pharmacology*, **17**, 44–51.

Calabrese, E.J., Beck, B.D. & Chappell, W.R. (1992) Does the animal to human uncertainty factor incorporate interspecies differences in surface area? *Regulatory Toxicology and Pharmacology*, **15**(2), 172–179.

Crump, K.S. (1984) A new method for determining allowable daily intakes. *Fundamental and Applied Toxicology*, **4**, 854–871.

Crump, K.S. (1986) Letter to the Editor: reply. *Fundamental and Applied Toxicology*, **6**, 183–184.

Crump, K.S., Hoel, D.G., Langley, C. & Peto R. (1976) Fundamental carcinogenic processes and their implications for low dose risk assessment. *Cancer Research*, **36**, 2973–2979.

DeRosa, C.T., Stara, J.F. & Durkin, P.R. (1985) Ranking of chemicals based upon chronic toxicity data. *Toxicology and Industrial Health*, **1**(4), 177–192.

Dourson, M.L. & Stara, J.F. (1983) Regulatory history and experimental support of uncertainty (safety) factors. *Regulatory Toxicology and Pharmacology*, **3**, 224–238.

Dourson, M.L., Hertzberg, R.C., Hartung, R. & Blackburn, K. (1985) Novel approaches for the estimation of acceptable daily intake. *Toxicology and Industrial Health* **1**(4), 23–41.

Dourson, M.L., Hertzberg, R.C. & Stara, J.F. (1986) Letter to the Editor. *Fundamental and Applied Toxicology*, **6**, 182–183.

Dourson, M.L., Knauf, L.A. & Swartout, J.C. (1992) On Reference Dose (RfD) and its underlying toxicity data base. *Toxicology and Industrial Health*, **8**(3), 171–189.

Dourson, M.L. (1994) Methods for establishing oral reference doses (RfDs). In: *Risk Assessment of Essential Elements* (eds W. Mertz, C.O. Abernathy & S.S. Olin), pp. 51–61. ILSI Press, Washington DC.

Farland, W. & Dourson, M.L. (1992) Noncancer health endpoints: approaches to quantitative risk assessment. In: *Comparative Environmental Risk Assessment* (ed. R. Cothern, pp. 87–106. Lewis Publishers, Boca Raton. FL.

Faustman, E.M. Allen, B.C., Kavlock, R.J. & Kimmel, C.A. (1994) Dose–response assessment for developmental toxicity. I. Characterization of database and determination of no observed adverse effect levels. *Fundamental and Applied Toxicology*, **23**(4), 478–486.

Hartley, W.R. & Ohanian, E.V. (1988) The use of short-term toxicity data for prediction of long-term health effects. In: *Trace Substances in Environmental Health—XXII*, University of Missouri, May 23–26 (ed. D.D. Hemphil), pp. 3–12.

Hattis, D., Erdreich, L. & Ballew, M. (1987) Human variability in susceptibility to toxic chemicals—a preliminary analysis of pharmacokinetic data from normal volunteers. *Risk Analysis*, **7**(4), 415–426.

Hertzberg, R.C. (1989) Fitting a model to categorical response data with application to species extrapolation of toxicity. *Health Physics*, **57**, 405–409.

Hertzberg, R.C. (1991) Quantitative extrapolation of toxicological findings. In: *Statistics in Toxicology* (eds D. Krewski & C. Franklin). Gordon and Breach Science Publishers, New York.

Hertzberg, R.C. & Miller, M. (1985) A statistical model for species extrapolation using categorical response data. *Toxicology and Industrial Health*, **1**, 43–57.

Hertzberg, R.C. & Wymer, L. (1991) Modeling the severity of toxic effects. In: *Proceedings' Papers from the 84th Annual Meeting and Exhibition*, June 16–21, British Columbia. Air and Waste Management Association.

Hertzberg, R.H. & Dourson, M.L. (1993) Using categorical regression instead on a NOAEL to characterize a toxicologist's judgment in noncancer risk assessment. In: *Proceedings, Second International Symposium on Uncertainty Modeling and Analysis* (ed. B.M. Ayyub), pp. 254–261. IEEE Computer Society Press, Los Alamitos, CA.

IARC (1978) *Monographs on the Evaluation of the Carcinogenic Risk of Chemicals to Humans*, Vol. 17. International Agency for Research on Cancer, Lyon.

International Programme on Chemical Safety (IPCS) (1994) *Environmental Health Criteria No. 170: Assessing Human Health Risks of Chemicals. Derivation of Guidance Values for Health-based Exposure Limits.* World Health Organization, Geneva.

Jarabek, A.M. (1994) Inhalation RfC methodology: dosimetric adjustments and dose–response estimation of noncancer toxicity in the upper respiratory tract. *Inhalation Toxicology*, **6**(Suppl.), 301–325.

Kimmel, C.A. & Gaylor, D.W. (1988) Issues in qualitative and quantitative risk analysis for developmental toxicity. *Risk Analysis*, **8**(1), 15–20.

Kimmel, C.A. Wellington, D.G., Farland, W. *et al.* (1988) Overview of a workshop on quantitative models for developmental toxicity risk assessment. *Environmental Health Perspectives*, **79**, 209–215.

Kroes, R., Munro, I. Poulsen, E. (eds) (1993) Scientific Evaluation of the Safety Factor for the Acceptable Daily Intake. *Food Additives and Contaminants*, **10**(3), 269–373.

Krewski, D., Brown, C. & Murdock, D. (1984) Determining 'safe' levels of exposure: safety factors or mathematical models? *Fundamental and Applied Toxicology*, **4**, 5383–5394.

Leisenring, W. & Ryan, L. (1992) Statistical properties of the NOAEL. *Regulatory Toxicology and Pharmacology*, **15**(2), 161–171.

Lewis, S.C., Lynch J.R. & Nikiforov, A.I. (1990) A new approach to deriving community exposure guidelines from no-observed-adverse-effect levels. *Regulatory Toxicology and Pharmacology*, **11**, 314–330.

Lu, F. 1983. Toxicological evaluations of carcinogens and noncarcinogens: pros and cons of different approaches. *Regulatory Toxicology and Pharmacology*, **3**, 121–132.

Lu, F. 1985. Safety assessments of chemicals with thresholded effects. *Regulatory Toxicology and Pharmacology*, **5**, 460–464.

Lu, F. (1988) Acceptable Daily Intake: inception,

evolution, and application. *Regulatory Toxicology and Pharmacology*, **8**, 45–60.

Meek, M.E., Newhook, R., Liteplo, R.G. & Armstrong, V.C. (1994) Approach to assessment of risk to human health for priority substances under the Canadian Environmental Protection Act. *Environmental Carcinogenesis and Ecotoxicology Reviews*, C**12**(2), 105–134.

Moolenaar, R.J. (1994) Carcinogen risk assessment: international comparison. *Regulatory Toxicology and Pharmacology*, **20**, 302–336.

Munro, I.C. & Krewski, D.R. (1981) Risk assessment and regulatory decision making. *Food and Cosmetics Toxicology*, **19**, 549–560.

NAS (National Academy of Sciences) (1983) *Risk Assessment in the Federal Government: Managing the Process*. National Academy Press, Washington, DC.

NAS (National Academy of Sciences) (1994) *Science and Judgment in Risk Assessment*. National Academy Press, Washington, DC.

Pohl, H.R. & Abadin, H.G. (1995) Utilizing uncertainty factors in minimal risk levels derivation. *Regulatory Toxicology and Pharmacology*, **22**, 180–188.

Rademaker, B.C. & Linders, J.B.H.J. (1994) *Progress Report 3: Estimated-concentrations-of-no-concern of polluting Agents in Drinking Water and Air for Humans*. National Institute of Public Health and Environmental Protection, Bilthoven, The Netherlands.

Renwick, A.G. (1991) Safety factors and establishment of acceptable daily intake. *Food Additives and Contaminants*, **8**(2), 135–150.

Renwick, A.G. (1993) Data derived safety factors for the evaluation of food additives and environmental contaminants. *Food Additives and Contaminants* **10**(3), 275–305.

USEPA (1986) *Guidelines for Carcinogen Risk Assessment*. US Environmental Protection Agency, 51 Federal Register 33992, pp. 1–17.

USEPA (1991) *Guidelines for Developmental Toxicity Risk Assessment*. Federal Register, Vol. 56, No. 234, pp. 63798–63826.

USEPA (1992) *Guidelines for Exposure Assessment*. US Environmental Protection Agency, 57 Federal Register, 22888–22938.

USEPA (1994a) *Methods for Derivation of Inhalation Reference Concentrations and Application of Inhalation Dosimetry*. US Environmental Protection Agency, Office of Health and Environmental Assessment, Washington, DC, EPA/600/8-90-066F.

USEPA (1994b) *Draft Revisions to the Guidelines for Carcinogen Risk Assessment*. US Environmental Protection Agency, Office of Research and Development, Washington, DC, External Review Draft.

USEPA (1995a) *The Use of the Benchmark Dose Approach in Health Risk Assessment*. Risk Assessment Forum, February, US Environmental Protection Agency, Washington, DC, EPA/630/R-94/007.

USEPA (1995b) *Policy and Guidance for Risk Characterization*. Memo from Carol M. Browner, Administrator to Assistant Administrators, March 21. US Environmental Protection Agency, Washington, DC.

USEPA (1996) *Proposed Guidelines for Carcinogen Risk Assessment*. US Environmental Protection Agency, Office of Research and Development, Washington, DC, EPA/600/P-92/003C.

Zielhuis, R.L. & van der Kreek, F.W. (1979) The use of a safety factor in setting health based permissible levels for occupational exposure. *International Archives of Occupational Environmental Health*, **42**, 191–201.

Chapter 3
Assessing Risks to Ecological Systems from Chemicals

JERRY C. SMRCHEK AND MAURICE G. ZEEMAN

3.1 INTRODUCTION

3.1.1 Introduction to chemical review

Organic and inorganic chemicals provide the raw materials for modern chemical industries and are important in the operation of industrialized societies. Nearly 6 trillion pounds (2.7 trillion kg) of industrial chemicals alone are produced or imported in the USA each year, and 72 000 different chemicals circulate through the USA economy (Inform 1995). Thousands of these industrial chemicals are used to produce a variety of products and are used as feedstocks or reactants for the synthesis of additional chemicals.

Industrial chemicals are used widely, for example as solvents, adhesives, dyes and plastics. They can be split conveniently into new and existing chemicals. New chemicals are those that are synthesized for a new use or application. Existing chemicals are those already listed on a national inventory; these are used in commerce, often for a number of years. Industrial chemicals are only occasionally pesticides, but are more often their synthetic precursor(s).

Pesticides are chemicals used to control organisms occurring on agricultural products and crops, and serve to protect plants, animals, and humans. Worldwide, an estimated 5 billion pounds (2.3 billion kg) of 1600 different pesticides are applied yearly; less than 0.1% of these applied pesticides ever reach their target pests (Pimentel 1995). Pesticides can be categorized by toxic mode of action against a group of organisms, i.e. into herbicides, fungicides, insecticides and rodenticides. There are also pesticides that are categorized by uses, for example as fumigants, growth regulators and repellants.

Various procedures and methods have been developed in the industrial countries to assess the hazard and risk of new and existing chemicals and pesticides on ecological systems and non-target organisms. These countries include those in Europe, the Western Hemisphere and Australasia; members of the Organization for Economic Co-operation and Development (OECD), and countries belonging to the European Union (EU). In this chapter we:

1 Describe and compare the ecotoxicological procedures, methods, approaches, and some of the underlying theoretical bases of these hazard and risk assessment schemes.

2 Briefly discuss and contrast international activities in chemical regulation and in ecological risk assessment.

3 Use and discuss the OECD Screening Information Data Set (SIDS) programme as an example of international chemical assessment and regulation.

4 Describe the parts of ecological risk assessment (e.g. hazard, exposure, risk characterization, risk management) and how this process is applied to the evaluation of industrial chemicals and pesticides.

5 Present practical guidance on how to assess hazard and risk, based on our experiences with chemicals and pesticides at the US Environmental Protection Agency (USEPA).

Disclaimer. This document has been reviewed by the Office of Pollution Prevention and Toxics (OPPT), US Environmental Protection Agency, and approved for publication. Approval neither signifies that the contents necessarily reflect the official views and policies of the Agency, nor does mention of trade names or commercial products constitute endorsement or recommendation of these products for use.

3.1.2 Ecological systems and current assessment approaches

Top-down approach

An ecological system is defined here as equivalent to an ecosystem; i.e. a discrete unit that consists of living and non-living parts interacting to form a stable system (Tansley 1935; Allaby 1994). An ecosystem represents a conceptual view of a plant and animal community (Walker 1989); assessing the risk of chemicals to ecosystems and to landscapes or drainage basins composed of ecosystems is an important and appropriate goal for studying widespread pollution problems (Cairns & Cherry 1993; Holl & Cairns 1995). Levin *et al.* (1984) and Cairns and co-workers (e.g. Buikema *et al.* 1982; Matthews *et al.* 1982) have presented a strong argument for using an ecosystem perspective to evaluate toxic effects, one reason being that measurements of effects upon individuals in the laboratory do not readily translate into or predict field effects on natural populations, communities or ecosystems. Also, indirect effects on the structural (e.g. diversity) and functional (e.g. system rate changes) integrity of ecosystems would be accounted for by this approach.

However, this 'top-down' approach remains difficult to carry out. Ecosystems consist of a web of synergistic interactions and interdependencies among their multiple components and levels. These relate to processes such as photosynthesis, respiration, energy transfer, nutrient cycling and biological or ecological regulatory processes (Matthews *et al.* 1982; Perry 1994). It is currently difficult to assess hazard and risk of chemicals on this web of functions and as a result to protect organisms present in these ecosystems (Chapter 1). It is also laborious and difficult to derive numerical criteria that describe a 'healthy' ecosystem or even operationally define a 'normal' or 'unpolluted' ecosystem (Peters 1991; Calow 1992; Forbes & Forbes 1993). Ecosystem health may even be a meaningless descriptor because health is a property some believe should be restricted only to organisms (Whittaker 1957; Suter 1993a). Furthermore, ecosystems do not behave like organisms or possess the properties of organisms. Integrated monitoring that is comprised of experimentation and observation may be the best way to assess ecosystem health (Chapman 1995b).

Landscapes are often composed of several to many ecosystems. It is very difficult to quantify and interpret changes in spatial, functional and structural parts of landscapes (Holl & Cairns 1995). Developing accurate ecotoxicological stress indicators for landscapes is also difficult.

Even when an ecosystem functional impairment or 'effect' is detected, it is often difficult to justify and implement preventive or corrective measures such as deriving concern concentrations or levels that would be used to trigger regulatory or protective actions. For example, assume the primary production of a particular ecosystem is reduced by 25%, and it is determined that the reduction is caused by the introduction of an industrial chemical or pesticide. It is difficult, using currently accepted hazard/risk-assessment methods and theory, to determine if this decrease is or is not significant, in relation to the natural temporal variability present in this ecosystem. It is also currently hard to determine how this reduction can serve as a trigger for conducting further testing (see section 3.5.1) or as a stimulus for regulatory action.

Ecosystem-level end-points must be identified and ecosystems must be protected (Suter 1993b). Microcosm/mesocosm test methods may be advantageous in addressing these problems. Experiments can be conducted to determine if these functional or process reductions are really important and if similar test end-points to those identified for an ecosystem (e.g. species richness) can also be used in microcosms.

Bottom-up approach

Another approach is to determine the hazard and risk of chemicals based on more manageable or observable biological units, namely the individual animal or plant or 'small' populations of these organisms (Warren 1971). In this 'bottom-up' approach the hazards and risks identified are extrapolated from organisms to populations, communities or even to ecosystems by using various factors (e.g. application, assessment or uncertainty factors) from relatively simple, homogeneous, 'clean' laboratory toxicity tests.

The 'bottom-up' approach to ecotoxicology originated, in part, from research undertaken by biologists in the late nineteenth and early twentieth centuries, who studied the effects of industrial discharges on the distribution and abundance of organisms (Marsh 1907; Ellis 1937; Warren 1971). With the expansion of the chemical industries in the developed countries from the 1920s onward, increased chemical discharges and emissions resulted; these discharges visibly affected aquatic and terrestrial ecosystems. The toxic effects of this pollution to individual organisms were noticeable (e.g. dead fish, 'excessive' growth of algae) and there was a societal impetus to study these problems. For example, Carpenter (1924, 1925, 1927, 1930) and Jones (1938) studied the effects on fish of heavy metal discharges in the UK. The effects of pollutants on physiological, cellular, and behavioural processes of individual organisms were also of interest at this time and remained so until at least the 1970s. (See Chapter 9 of this volume for further discussion.)

Single species toxicity or bioassay test methods were first developed to directly determine the toxic effects of chemicals to fish and daphnids (Ellis 1937; Anderson 1944, 1946; Hart *et al.* 1945; Doudoroff *et al.* 1951; American Public Health Association *et al.* 1955, 1960; Henderson & Tarzwell 1957; Henderson 1960). Standardizing these methods was (and still is) an important consideration (Zeeman 1997).

In contrast to ecosystem protection in the top-down approach, the bottom-up approach is important because Darwinian evolution and theory is viewed as the central organizing principle in biology today. Most importantly, natural selection acts only for the benefit of individuals, in terms of their reproductive success (Gould 1995). As Gould notes (p. 9), balanced ecosystems and well-designed organisms are 'side consequences'. It is the success and welfare of individuals that is most significant, not the communities and ecosystems in which they exist. A result of this thinking may be to believe that even with all the pollution occurring today, or which has occurred in the past, ecosystems are resilient to a certain extent and will 'adjust' to some other structural or functional (process) state (assessed as being 'degraded' or otherwise), in spite of efforts made to prevent and control pollution and protect ecosystems.

3.1.3 Ecotoxicology: a synthesis of approaches

At the present time, the field of ecotoxicology combines both of these approaches. Ecological variables are used to assess toxicity (Cairns 1989). Ecotoxicology can be viewed as a 'merging' of the fields of ecology and toxicology (Chapman 1995b) with the assessment of the effects of toxic substances on entire ecosystems, not just on isolated parts of ecosystems, as a goal (Cairns & Pratt 1993). The current interest in ecological risk assessment and the integration of risk-assessment concepts with aquatic toxicology methods are important outcomes of this merger.

3.2 INTERNATIONAL REGULATION OF CHEMICALS

3.2.1 International activities: general aspects

Over the past 20–30 years, international efforts to regulate new and existing chemicals, as well as pesticides (or plant-protection products and biocides) and to assess their hazard and risk have been extensive and ongoing. The USA, EU and OECD have been particularly active in this area (Zeeman 1997). The OECD was established in 1960 and is made up of 25 member countries in Europe, North and Central America, and Australasia (Table 3.1), which have 'advanced market economies' (Grandy 1995).

Some of the logic and rationale used in chemical regulation, as well as the toxicity tests performed, is similar from one country to another. An international chemical company may, therefore, complete toxicity tests for an industrial chemical or pesticide to be marketed in one country, and also send these test data to regulatory authorities in another country. It is important that all countries adhere to the 1981 OECD Council 'Decision Concerning the Mutual Acceptance of Data in the Assessment of Chemicals', or the MAD decision, so as to decrease unnecessary testing, control regulatory costs, and promote information exchange (Grandy 1995). Chemicals and pesticides are often of interest in several to many countries

Table 3.1 Countries with identified chemical control programmes (number and percentage of countries out of total in each region) and those countries that are members of the Organization for Economic Co-operation and Development (OECD) and EC/EU. (After USEPA 1995a.)

Europe (23 of 44 countries in region or 52%)		
Albania	Greece*†	Portugal*†
Austria*†	Hungary	Russia
Belgium*†	Ireland*†	Spain*†
Czech Republic	Italy*†	Sweden*†
Denmark*†	Latvia	Switzerland*
Finland*†	Luxembourg*†	United Kingdom*†
France*†	Netherlands*†	Yugoslavia‡
Germany*†	Norway*	
Australasia (12 of 58 countries or 21%)		
Australia*	Indonesia	New Zealand*
China	Japan*	Singapore
Cyprus	Philippines	South Korea
India	Papua New Guinea	Thailand
Africa (2 of 53 countries or 4%)		
Namibia		
Nigeria		
Western Hemisphere (7 of 35 countries or 17%)		
Bolivia	Mexico*	
Brazil	USA*	
Canada*	Venezuela	
Jamaica		

* Member of OECD (OECD also includes Iceland and Turkey, for a total of 25 countries).
† Member of EC/EU. There are 15 members as of 1997.
‡ The latest environmental policy document is dated June 1993. It is not known if the chemical control programme applies to the separate countries that formerly made up Yugoslavia.

and are commonly marketed in several countries or even on a world-wide basis.

Lists of toxic chemicals and pesticides or priority pollutants of current interest in various countries are not presented in this chapter. Even though such lists may be informative, they may change, as chemicals change in priority, or are placed on, or taken off such lists. Many of the lists are periodically updated or revised, thus becoming obsolete. Also, it is difficult to track efficiently the hundreds of chemicals of interest to particular countries and regulatory agencies.

Instead, industrial chemicals and pesticides are considered here more in a generic sense; for example as substances with 'high' log K_{ow} (equilibrium constant for octanol water partition coefficient; Calow 1993a) of 'moderate' aquatic toxicity, or being 'poorly' soluble. Specific values will be given to define these generic terms, when possible.

Specific lists of priority industrial chemicals or pesticides are in current references such as Miller (1993) and Hedgecott (1994), but even these will soon become obsolete, if they are not so already. Also, regulatory authorities in particular countries or international organizations such as OECD can be contacted for requesting the latest lists and information. The OECD would have lists of industrial chemicals and pesticides of interest to several countries, for example those chemicals reviewed in the Screening Information Data Set programme (see section 3.3).

3.2.2 Comparisons of international activities

World-at-a-Glance Report

The Office of Pollution Prevention and Toxics (OPPT) of the USEPA first developed the World-at-a-Glance (WAAG) Report in 1990; it was recently updated (USEPA 1995b). This compares and highlights international chemicals programmes and laws in countries throughout the world.

Interest in chemical programme development has increased, in part because of the implementation of the London Guidelines for Prior Informed Consent (PIC) by the United Nations Environmental Programme (UNEP) and the Food and Agricultural Organization (FAO) of the United Nations (UN). As work proceeds towards establishing a binding PIC convention and developing model chemical legislation, this interest is expected to increase. Of the 190 world countries, 113 (or 59%) already participate in the UNEP/FAO PIC programme by having a designated national authority.

Furthermore, 44 (23%) of the 190 world countries (166 were contacted in preparing the updated WAAG report), have chemical control provisions as part of their chemical laws or programmes

Table 3.2 Key features of chemical legislation and programmes in 44 world countries. (From USEPA 1995a.)

	Region			
Key feature	Europe	Australasia	Africa	Western Hemisphere
Confidential business information protection provisions	All except Czech Republic, Hungary, Latvia, Russia and Yugoslavia	All except India, Philippines, New Zealand, Singapore and Thailand	None*	All except Bolivia, Brazil, Jamaica and Venezuela
Enforcement provisions	All except Belgium, Czech Republic, Greece, Hungary, Ireland, Italy, Latvia, Luxembourg and The Netherlands	All except India	All†	All except Brazil and Venezuela
Import/export programmes	All except Latvia and Russia	All, and including Kuwait	None	All except Bolivia, Brazil and Venezuela
Pollutant release and transfer registry	None	None	None	Only Canada and the USA
Information collection provisions	All except Albania, Czech Republic, Finland, Hungary and Latvia	All except China, Cyprus, India, Indonesia and Philippines	All	All except Bolivia, Brazil, Jamaica and Venezuela
Emergency planning	All except Albania, Austria, Czech Republic, Greece, Hungary, Latvia, Norway, Portugal, Russia, Spain, Sweden, Switzerland and Yugoslavia	Only New Zealand	None	Only Canada and the USA
New chemical programme	All except Albania, Latvia, Norway and Yugoslavia	All except China, India, Indonesia, New Zealand, Papua New Guinea, Singapore and Thailand	None	Only Canada and the USA

(Continued)

Table 3.2 *Continued.*

Key feature	Region: Europe	Australasia	Africa	Western Hemisphere
Chemical substance inventory	All except Albania, Hungary, Latvia, Norway and Yugoslavia	All except India, Indonesia, Papua New Guinea, New Zealand, Singapore and Thailand	All	Only Canada and the USA
Pollution prevention programmes	All except Albania, Czech Republic, Finland, Germany, Hungary, Italy, Latvia, Russia, Spain and Yugoslavia	None	None	Only Canada and the USA
Testing requirements	All except Greece and Latvia	All except India, Papua New Guinea, New Zealand, Singapore and Thailand	None	Only Canada and the USA

* None, none of the legislation or programmes of the countries in the particular region has this key feature.
† All, all of the legislation or programmes of the countries in the region have this key feature.

(USEPA 1995b). These 44 countries, listed in Table 3.1, have the ability to ban, severely restrict, or otherwise control chemicals in commerce. Specific key features or characteristics of chemical legislation and programmes of these countries are highlighted in Table 3.2.

The update publication (USEPA 1995b) also contains information on applicable chemical laws and regulations, as well as contacts for each country (Appendices I and II, respectively). Some of this information is also available from the USEPA, OPPT. For those new countries (e.g. the former Soviet Union and Yugoslavia) the status of chemical laws is in flux and information in the updated publication may not be entirely accurate or complete.

3.2.3 Regulatory activities in the USA and Canada

United States of America

The impetus for USA non-pesticide chemical regulation is the Toxic Substances Control Act (TSCA), Public Law 94-469. A number of new requirements and authorities were established for identifying and controlling toxic chemical hazards to the environment. Various sections of TSCA give OPPT the authority to regulate new industrial chemicals (Section 5) by the submission, and approval or rejection of pre-manufacture notifications (PMNs) and existing industrial chemicals (Sections 4, 6 and 8) by means such as risk reduction, production triggers, labelling and regulatory rules and orders to determine if these chemicals present or will present an unreasonable risk of injury to human health or the environment (Smrchek *et al.* 1993; Zeeman & Gilford 1993; Zeeman 1997).

Existing chemicals approved for use in the USA are those listed in the Toxic Substances Control Act Chemical Substance Inventory or TSCA Inventory, whereas new chemicals are reviewed on a case-by-case basis and may later be added to the current inventory of over 72000 chemicals. Over 26000 new chemicals have been reviewed since 1979 and over 10000 of these have been placed on the TSCA Inventory (Zeeman 1995a). Additional information on TSCA regulation of chemicals is given in Nabholz (1991), Smrchek *et al.* (1995), and Zeeman (1995a, 1997). Toxicity test data are requested and used to assess and evaluate ecological hazard and risk of industrial chemicals.

The Pollution Prevention Act of 1990 (Public Law 101–508) is a proactive method of regulation of chemicals, in contrast to the more reactive regulatory approach taken by TSCA. The Act considers measures to prevent or reduce pollution at the source (site of production). Source reduction prevents, lowers or limits the amount of hazardous substances, pollutants or contaminants entering aquatic or terrestrial environments before possible adverse effects may occur (and be subject to regulation).

The authority and basis for regulation of pesticides in the USA has rested with USEPA, in the Office of Pesticide Programs (OPP) under the Federal Insecticide, Fungicide, and Rodenticide Act (FIFRA), Public Law 95-396 (as amended, 7 U.S.C. 136 *et seq.*). The USEPA, under this Act, must determine whether or not a pesticide can be registered or reregistered for a particular use (Urban & Cook 1986; USEPA 1994d). Environmental risk assessments are developed for pesticides and risk reduction/mitigation is promoted (SETAC 1994).

The USEPA has been involved in the development and use of water quality criteria since the 1970s and the origin of these criteria dates back to the early 1900s in the USA (Marsh 1907; Shelford 1917; USEPA 1977). Criteria are reviewed and published pursuant to section 304(a) of the US Federal Water Pollution Control Act (Public Law 92-500), as amended by the Clean Water Act in 1977 (Public Law 95-217). A national goal is to prohibit discharges of toxic pollutants in harmful amounts. Some industrial chemicals and many pesticides have published criteria. Criteria are specified as concentrations of water constituents which, if not exceeded, are expected to result in an aquatic ecosystem suitable for 'higher uses'. These concentrations cannot be considered 'safe' levels for the survival and reproduction of all aquatic organisms at all times in a given ecosystem. Instead, they are intended to protect and ensure a reasonable degree of safety to essential and significant life in water, direct users of water, and to life that is dependent on water for its existence, or that may consume any edible portion of such life intentionally or unintentionally (USEPA 1977). Criteria should be based on toxicity data from many plants and animals occupying various trophic levels (USEPA 1980).

There are 175 industrial chemicals and pesticides and their acute and chronic toxicity thresholds listed in a criteria summary table (dated May, 1991). Criteria, when toxicity test data are available, are listed for both freshwater and marine environments and are acute values, based on lethal concentration or effective concentration to 50% of test organisms (LC_{50} or EC_{50}), and chronic values, based on the maximum acceptable or allowable toxicant concentration (MATC). These criteria can serve as a valuable resource in assessing the hazard of specific industrial chemicals and pesticides. Detailed guidance on derivation and use of criteria is given in USEPA (1980, 1991c).

Canada

Substances are considered for systematic assessment if Federal, provincial or international programmes (e.g. OECD), or members of the public in Canada, have identified them as potentially harmful to the environment (Environment Canada 1995). A substance is considered toxic if, after scientific assessment and based on Federal programme decisions, it either conforms to or is equivalent to 'toxic' as defined in Section 11 of the Canadian Environmental Protection Act (CEPA). New substances are assessed under the New Substances Notifications Regulations of CEPA.

Existing chemical substances meeting all criteria of toxicity (as defined by CEPA), bioaccumulation (bioaccumulation factor (BAF), defined as the ratio of tissue chemical residue to chemical concentration in an external environmental phase such as food or sediment (Rand 1995); bioconcentration factor (BCF), defined as the degree to which a chemical can be concentrated in the tissues of an aquatic organism as a result of exposure to a waterborne chemical (Rand 1995); or log K_{ow}), persistence (half-life), and anthropogenicity follow

Track 1—virtual elimination from the environment of substances that result predominantly from human activity. Substances not meeting all the above criteria follow Track 2—in which lifecycle management occurs to prevent or minimize environmental releases (SETAC 1995). A priority chemical substances list is available and was published in 1989 (see list in Hedgecott 1994). (See Chapter 10 of this volume for additional discussion of Canadian chemical legislation.)

An important Canadian resource on the hazard of industrial chemicals and pesticides is the publication *Canadian Water Quality Guidelines*, first published in 1987 and updated annually (Environment Canada 1987). Guidelines for the protection of aquatic life (Chapter 3) and parameter-specific background information (Chapter 6) are particularly useful. A wealth of information is summarized on aquatic toxicity, uses and production, sources and pathways of chemicals entering the aquatic environment, environmental concentrations, and forms and fate in the aquatic environment of certain chemicals and pesticides.

3.2.4 European Community/European Union regulatory activities (with a note on the UN)

The EEC (European Economic Community), later the EC (European Community), was established in 1957 to serve as an umbrella organization to deal with common problems among member states. Furthermore, the EC and six of the seven members of the European Free Trade Association (EFTA) agreed to form the European Economic Area (EEA). As part of this agreement EFTA members will harmonize their environmental legislation with legislation of the EC (Handley & Knight 1995). The EC has now become the European Union (EU), after the Maastricht Treaty on European Union entered into force.

Chapter 9 presents a comprehensive description of EU regulatory activities. Information presented there will not be repeated here.

Existing and new chemicals

Various EU Directives and Regulations have been established that apply to both existing and new industrial chemicals. For new chemicals, a notification system prior to marketing is in place (Calow 1993b; Blok & Balk 1995; SETAC 1996). This system is similar to that present in the USA. Ecotoxicity test data are sent in with European new chemical notifications (in the USA, testing is not required for new chemical PMN submissions).

Classification/labelling of industrial chemicals

Since 1967, within both the EU and OECD, work has progressed on the classification and labelling of dangerous industrial chemicals pursuant to EC Directive 67/548, Annex I, VI. Various classification systems will be harmonized among the member countries. Also, commonly agreed to environmental effects criteria have or will be developed and adopted by the EU and the Nordic countries (Pedersen *et al.* 1994). These criteria now pertain only to the aquatic environment, as criteria for the terrestrial/soil environment remain to be agreed upon. For the aquatic criteria, chemicals are classified and given numbered 'R' designations (e.g. R50) or risk phrases. There are also numbered 'S' designations (i.e. safety phrases) that deal with handling and disposal procedures that should be used. Dozens of chemicals are being considered and classified. However, this is a difficult and long process, especially when data are lacking, as is often the case for some chemicals.

Pesticides

Procedures and approaches used in approving pesticides vary from one member state to another in the EU and from country to country outside the EU (Greig-Smith 1992; Blok & Balk 1995). This has occurred even though OECD guidelines (see section 3.2.7) are widely followed and incorporated into hazard assessment procedures. Maximum pesticide residue levels were implemented by Council Directive 76/895 (with amendments 80/428, 81/36, and 82/528). Also, the classification system of Directive 79/831 (sixth amendment) was made to apply to pesticides by implementation of Directive 78/631 (as amended by Directive 81/187). Council Directive 79/117 (amended by Directive 90/533) prohibited several dangerous pesticides, e.g. DDT (dichlorodiphenyltrichloroethane), chlordane, heptachlor (see Blok & Balk 1995).

Council Directive 91/414 is perhaps the most important pesticide Directive, because it deals with pesticide use, marketing and registration. Also, there is a list in this Directive of registered active ingredients accepted for use by EU member states.

Pesticide registration consists of a two-tiered approval system (Greig-Smith 1992). First, there will be a single overall approval for an active ingredient to be incorporated into plant protection products. Second, there will be individual approvals for each product to be used in EU member countries. There will be mutual acceptance of separately approved products between members, which will increase consistency and reduce unnecessary testing.

The European and Mediterranean Plant Protection Organization (EPPO) and member states have been jointly involved since 1989 in harmonizing, risk assessment approaches due to pesticide use (Greig-Smith 1992). Separate, yet mutually interdependent, risk assessment schemes have been developed, e.g. for terrestrial vertebrates, aquatic organisms, earthworms, honeybees, soil mesofauna, and soil microorganisms (Greig-Smith 1991).

Surface waters

Some efforts have been directed toward improving surface water quality (Blok & Balk 1995). Council Directive 76/464 is concerned with the reduction or elimination of water pollution caused by dangerous substances. Two lists are established; member states are to take steps to eliminate pollution from substances on the first list (e.g. organohalogen and organophosphorus compounds, carcinogenic substances and some metals). Pollution from substances on the second list should be reduced and these include other metals, cyanides, organic silicon compounds, petroleum hydrocarbons and inorganic phosphorus and ammonia compounds. Member states are required to establish pollution reduction programmes.

EC/EU member state activities

Specific chemical regulatory activities have also occurred in many of the individual member countries. They will not be discussed further here. The reader is referred instead to Lundahl (1979), OECD (1989), Greig-Smith (1992), Bayer & Fleischhauer (1993), UK Department of the Environment (1993), Ahlers *et al.* (1994), Klopffer (1994), and USEPA (1995b).

3.2.5 ECETOC activities

The European Centre for Ecotoxicology and Toxicology of Chemicals (ECETOC), Brussels, Belgium, is a non-profit, advisory organization made up of 60 chemical companies. Reviews are developed by industry task forces of scientists working on specific technical issues and problems. Several technical reports have been published on hazard identification, risk assessment, evaluation of aquatic toxicity data and on environmental exposure assessments (ECETOC 1993a,b, 1994). Data on 6000 industrial chemicals have been collected. From this list, priority chemicals lists will be developed.

3.2.6 United Nations: IPCS and IRPTC regulatory activities

The UN has several bodies that are involved with international regulation of chemicals. The joint International Programme on Chemical Safety (IPCS) is a part of the UN World Health Organization (WHO). The IPCS has published information on evaluating the toxicity of chemicals, with emphasis so far in the human health area (IPCS 1978, 1990). The IPCS is working cooperatively with the OECD. Screening Information Data Set (SIDS) programme, in order to use SIDS information (see section 3.3.2) to prepare assessments of the health and environmental impacts of existing chemicals.

The International Register of Potentially Toxic Chemicals (IRPTC) is part of the United Nations Environmental Programme (UNEP), and will serve as an archive for data collected and generated in the SIDS programme. Information on the IRPTC will be disseminated in a series of documents published as part of the International Programme on the Sound Management of Chemicals (IOMC). These publications will include the completed initial assessments of high-production-volume chemicals and other nonconfidential data associated with SIDS chemicals (see section 3.3.2).

3.2.7 Regulatory activities in the OECD

Historical perspective/current activities

The OECD has been involved in hazard and risk assessment activities since 1975. At that time the OECD Chemicals Groups discussed the consequences of regulatory measures taken by some member countries in the area of reducing the adverse effects of chemicals 'on the ecosystem'. Differences in testing requirements and possible economic losses due to these measures were a concern (OECD 1979). In response an OECD Chemicals Testing Programme (later called the OECD Expert Group on Ecotoxicology) was established.

The Programme concentrated on the impact of chemicals on aquatic organisms representing different ecosystem functions (e.g. processes such as primary production, secondary production, consumption, decay and mineralization) in a bottom-up approach.

For the past several years, OECD efforts in hazard/risk assessment have been led first by the Hazard Assessment Advisory Body (HAAB), and now the Risk Assessment Advisory Body (RAAB) of the OECD Chemicals Programme. The purpose of this body is to serve as the focus for promoting an awareness of, and improving procedures in, hazard and risk assessment used in member countries. To the extent possible, these procedures are harmonized and as a result member countries are assisted in protecting the environment from potentially harmful effects of chemicals (OECD 1995a). The RAAB also is interested in specific topics such as testing of sparingly soluble substances, statistical analyses of aquatic toxicity data, SARs (structure–activity relationships) and QSARs (quantified structure–activity relationships). A continuing activity is to harmonize assessment reports and provide guidance on industrial chemicals and pesticides to EU countries. These reports consist of both those submitted by industry and data review reports of the individual country.

The OECD has published a series of useful workshop reports and monographs on important topics, which include ecological effects assessment (OECD 1989), extrapolation of laboratory aquatic toxicity data to the real environment (OECD 1992a), the use of QSARs in aquatic effects assessment (OECD 1992b), effects assessment of chemicals in sediments (OECD 1993), guidance for aquatic effects assessment (OECD 1995a), and environmental hazard/risk assessment (OECD 1995b).

Test guidelines

In 1981, OECD published a series of ecotoxicity test guidelines as well as guidelines on physical–chemical properties, degradation and accumulation, and health effects (OECD 1984). The ecotoxicity guidelines were updated in 1984 and 1992, and work continues on revising other guidelines (e.g. the *Daphnia magna* chronic toxicity reproduction test).

Efforts to develop harmonized policies and practical tools, such as test guidelines for managing chemicals and protecting the environment, were later organized under the umbrella of the OECD Chemicals Programme (Grandy 1995). There is a Test Guidelines Programme as part of the RAAB that is involved in developing new test guidelines (e.g. *Lemna* test, chironomid sediment test) and revising/updating current OECD tests.

The OECD is actively involved in the development of ecological effects test guidelines suitable for the pesticide registration process. Work is also proceeding in the areas of registration data requirements, hazard and risk assessment methods, reregistration of existing pesticides, and risk reduction practices due to pesticides (Grandy 1995).

OECD member country activities

Specific chemical regulatory activities in OECD member countries are not discussed here. Refer instead to the references cited in the previous section on EU member state activities and to Chapter 9 of this volume. Many countries are members of both the EU and OECD (Table 3.1). For activities and legislation in countries that are members of OECD only (e.g. Japan, Australia), refer to Chemical Products Safety Division (1977), Fujiwara (1979) and USEPA (1995b).

There are numerous technical reports, unpublished draft documents and discussion papers written by one or more OECD member countries to answer specific questions in hazard and risk assessment or to lay out how hazard and risk assessment is conducted in one or more member countries. Important recent examples include several publications by the Nordic Council of Ministers (Lander *et al.* 1994; Pedersen *et al.* 1994). The Nordic Council was formed in 1952 to promote cooperation between the governments of Denmark, Iceland, Norway and Sweden (Finland joined in 1955).

Moreover, some of these publications are in the area of terrestrial effects assessment. This area is not as well developed as aquatic effects assessment activities, but is receiving increasing interest (this is discussed further in section 3.5.1). Specific terrestrial topics and areas of interest (currently unpublished) include general guidance on terrestrial effects assessment, selection of a set of standardized laboratory toxicity tests for the hazard assessment of chemical substances in terrestrial ecosystems, ecotoxicological test methods using terrestrial arthropods, and development of terrestrial ecotoxicology test guidelines.

3.3 A CHEMICAL REGULATION EXAMPLE: OECD'S SCREENING INFORMATION DATA SET

3.3.1 Background

The OECD has undertaken the enormous task of assessing the safety of nearly 100000 industrial chemicals that are now in commercial use worldwide. It was first decided in 1987 (OECD Council Decision-Recommendation C[87]90[Final]) and later in 1990 (Council Decision C [90]163[Final]) to compile, review and assess data and information on selected existing chemicals in a systematic manner. Since 1988, these efforts have centred on existing high-production-volume (HPV) chemicals. High-production-volume chemicals are of mutual concern to member countries and are often marketed in several countries or throughout the world. Member countries work cooperatively and divide work responsibilities by sponsoring a certain number of HPV chemicals for review. Chemicals have been identified for their potential hazards and risks, which could be reduced by appropriate management or control practices.

For each chemical, information is collected in a uniform and consistent way, and a Screening Information Data Set (SIDS) is completed. Testing is conducted if needed, and an initial assessment of potential hazard is made. These chemicals are considered in groups of approximately 50 in review 'phases'.

The SIDS process is intended to avoid duplication in information gathering or testing conducted in response to regulatory requirements of each individual country. Also, existing chemicals will be investigated more effectively, testing costs ultimately will be reduced, and procedures used by one member country for the initial assessments of chemicals will be understood better by, and applied more easily to, other countries. This can result in harmonized or commonly shared procedures and thus a reduction in duplicative testing by member countries and chemical companies.

The OECD has prepared a SIDS Manual (latest version, May 1996; available from OECD, Environment Directorate, Environmental Health and Safety Division) that presents an introduction, background and current description of the SIDS programme. Information is presented on chemical selection, sponsor country responsibilities, data evaluation, gathering and assessing information, preparation and review of SIDS documents, testing plans and on post-SIDS activities.

3.3.2 The SIDS process

Background

Each OECD member country is responsible for a portion of the industrial chemicals identified, and makes information available to other member countries by completing SIDS Dossier documents. Information from governmental and public sources is collected. Chemical companies play a significant role here, by making available unpublished information in their files. Often they are the lead organization identified for a chemical. The information is reviewed, testing is conducted as needed, and the SIDS Dossier is completed, sometimes in a collaboration between the chemical company (or several companies) and the sponsor government. The SIDS programme is also carried out in collaboration with other organizations, including the EU, IPCS and IRPTC.

High-production-volume chemicals, i.e. those defined as being manufactured by, or imported into, any one OECD member country in excess of 10 000 tonnes or in two or more countries in excess of 1000 tonnes are the focus of this process.

In 1990, 1592 HPV chemicals were identified, and of these, 648 chemicals with little or no available information were given the highest priority for review. The remaining 954 SIDS chemicals were sorted into those with either sufficient information available (i.e. having the Screening Information Data Set) to permit a detailed assessment of hazard or those of low concern. The current 1995 HPV list, available from the OECD Secretariat, contains about 2500 chemicals available for SIDS assessment or testing, or both.

Basic minimum information in SIDS, needed for review of each HPV chemical, includes chemical identity, physical–chemical properties, sources and levels of exposure, environmental fate and pathways, human health data and ecotoxicity data. These data are similar to the Minimum Premarketing set of Data (MPD) for new EU chemicals.

Reported ecotoxicity data in SIDS consists of results from one or more acute fish tests, an acute and sometimes a chronic daphnid test, one or more algal toxicity tests, and tests with terrestrial organisms (soil-dwelling organisms, plants and avian acute dietary) should this exposure be indicated. Other optional tests that can be included are data from other aquatic toxicity tests, fish chronic tests, bacterial toxicity tests, and from avian reproduction tests. The types of tests reported in SIDS are dependent on the exposure pathways of the HPV chemical. Aquatic exposure is likely for most chemicals. If 'significant' terrestrial exposure is expected, then terrestrial testing should be conducted or reported in the SIDS process.

Chemical review

Information available on each chemical is collected, reviewed, assessed for quality, and a SIDS Dossier is prepared by the sponsor country for review and comment by other member countries. The SIDS Dossier lists all the information available (data elements) on physical–chemical properties, exposure, fate and effects of the HPV chemical. Recommendations for further action to clarify certain issues, for example develop a testing plan to conduct additional ecotoxicity tests and fill missing information gaps or to obtain exposure information from other member countries, are made. Additional testing needs are identified, agreed to, and the testing is conducted according to the OECD Test Guidelines (OECD 1984) and the principles of good laboratory practice (GLP).

After all or most testing is completed and information on exposure and/or effects is collected, a SIDS Initial Assessment Report (SIAR) and a full SIDS Dossier are prepared. Five provisional guidance documents for sponsor countries to follow when preparing SIARs and recommendations are included in the SIDS Manual. The SIAR document has a discussion of the identity, exposure, toxicity, and an overall assessment of the chemical, along with conclusions and recommendations (e.g. post-SIDS work or other activities). The full SIDS Dossier is an updated, revised version of the original SIDS Dossier, containing the results of completed tests and other missing information.

The SIAR is circulated for further review and comment. At a general meeting of OECD member countries (SIDS Initial Assessment Meeting, or SIAM) the chemical is discussed and conclusions on potential risk are reached. These conclusions are based on an initial assessment of the effects and exposure data and may, for example, be placed in several general categories.

1 The chemical may be found to present a low potential for risk to humans and the environment. Low priority for further work is given to these chemicals.

2 It may be concluded that the chemical presents a potential for risk to humans or the environment or both. Thus, there is a priority for undertaking one or more, or any combination of actions. These include post-SIDS testing, additional exposure analysis, and in-depth assessment of any concerns identified. Alternatively, limited exposure of the chemical or current risk-reduction measures currently taken by member countries may make further action(s) unnecessary or of low priority, in spite of the potential for risk.

3 The chemical may be found to have a low priority for further work, but exposure to humans and the environment should nevertheless be avoided due to an identified potential for risk.

4 The chemical may be found to present a potential for risk to humans or the environment or both. Further risk management actions are necessary, depending on the exposure situation and current risk-reduction measures. These further actions could include pollution prevention and reduction activities, post-SIDS testing, meetings of interested parties, and national and international risk-reduction activities.

After the chemical has been reviewed, and conclusion(s) reached on its risk, a recommendation for additional information, data analysis and follow-up or post-SIDS work (e.g. testing) may be made and agreed upon by member countries. This follow-up work is a member-country–industry collaboration. After such information is generated (usually by industry) and reviewed, the chemical is again discussed at a Post-SIDS Assessment Meeting, similar in format to the SIAM. Conclusions and recommendations similar to those already mentioned may be made. Member countries may wish to develop common, consistent or harmonized risk-reduction practices. Follow-up activities by IRPTC and IPCS also may be recommended.

3.3.3 SIDS data collection and quality

Guidance is offered in the SIDS Manual on evaluating and documenting the quality of collected SIDS test data. The sponsor country for a particular HPV chemical is responsible for this important task. Test data summaries should be referenced adequately and reported in sufficient detail to provide information to peer reviewers on methods used, end-points evaluated, and a description of the results of testing.

Peer review and approval of test results by experts is critical to achieving the goals of the SIDS process. In practice there are significant problems (described below), especially with respect to review of older or unpublished work.

Reporting

Chemical information is entered on a reporting form (Revised HPV-Form 1) or in a computerized version of this form (HEDSET). Each of these are part of the SIDS Dossier. The HEDSET version is convenient to use and is compatible with most personal computers, as well as conforming with EU requirements for collecting information on existing chemicals. The HPV-1 form includes a wider range of data categories than the minimum required in the SIDS, which will help in formulating testing plans and obtaining an even more comprehensive assessment of hazard. For example, there is a category for reporting the results of a fish chronic test, even though it is not required in the SIDS.

When little or no information is available, for example for physical–chemical properties or for acute aquatic toxicity, SAR/(Q) SAR (also abbreviated as SAR/QSAR, (Q) SAR or QSAR) calculations or estimates are appropriate as a replacement (see section 3.5.1). Testing would be appropriate to fill any missing information in the SIDS Dossier, or to replace information that is incorrect or inadequate. If there is uncertainty over whether or not to test, the bias is always to test. The QSAR analyses can also serve as a validity check of test data. See an example of this analysis for two SIDS chemicals, 1,4-diethylbenzene and isobutene, in Zeeman *et al.* (1995).

Important considerations apply to the question of data adequacy in the SIDS documents. If other studies are available and the results are consistent or similar, or if the results suggest a 'very low concern', then questionable SIDS test data may be considered minimally adequate and retesting may not be required. Screening level toxicity tests are considered to be less 'rigorous' than definitive tests and an 'approximate test value' may be adequate for the SIDS, when using a weight-of-evidence approach to determining hazard. Analogues or homologue test data are considered if possible. When a group of related chemicals is being compared, test results from the worst case, or the most toxic, should be used in the comparisons.

Quality and validity of test data

A very important problem, mentioned briefly above, is that the quality and validity of all SIDS test data are difficult to determine within the current reporting format and SIDS guidance, or are not known. Limited space (blank lines to be filled in or Yes/No alternatives to be selected and checked off) is provided on the form or in HEDSET and there is usually only a very brief description given of the toxicity test method and whether or not GLPs were followed. Furthermore, the reference where the test is described in more detail is sometimes in an obscure or unpublished, or unavailable, source, such as a chemical company technical or project report. Thus, the validity of the tests reported is questionable or difficult to verify. It must be assumed or taken on faith that the sponsor country reviewed the quality of each test for a chemical adequately.

Determining whether or not a test is valid is extremely important (see Table 3.4), because valid toxicity tests form the foundation for assessing SIDS chemical hazard and risk. Without evaluating the complete report or reference it is not possible in many cases to determine if a test is valid. If the chemical has certain physical–chemical properties, for example, log octanol/water partition coefficients (log K_{ow} or P_{ow}) of ≥5.0 (for neutral organic chemicals) and 'low' water solubility, acute testing may be found to be misleading and such test results should be considered invalid. For some classes of chemicals with log K_{ow} values of >5 but <8, it has been found that toxic effects may not be expressed for several weeks or months. Chemicals with log K_{ow} values >8 are not expected to be toxic at saturation (Nabholz *et al.* 1993a). For some classes of chemicals, acute test values may be obtained that are above the water solubility value and thus would be considered unreliable and invalid. For other chemicals, known as persistent bioaccumulators, acute testing is difficult or impossible; instead, chronic testing should be conducted.

Thus, the SIDS process could be improved and SIDS peer review strengthened by making the complete test reports available, and in the future offering more detailed guidance to take into account that, for some chemicals, modifications in SIDS reporting requirements may be necessary.

3.3.4 Results as of January 1996

As of mid-January 1996, 277 chemicals are being reviewed in the SIDS programme. Sixty-eight of these chemicals have been assessed and 13 of these have SIARs and SIDS Dossiers published. In 1992 the United Nations Conference on Environment and Development (UNCED) concluded that international chemical risk assessments should be strengthened. This would be accomplished by evaluating, by the year 2000, several hundred priority chemicals and contaminants that were of global significance. SIDS Chemicals are to make up a significant part of these priority chemicals. Current selection and assessment criteria are to be used to complete evaluations of 200 SIDS chemicals by 1997 and 300 more by the year 2000 (for a total of 500).

3.4 THE ECOLOGICAL RISK ASSESSMENT PROCESS

3.4.1 Introduction to ecological risk assessment

Ecological risk assessment is a process that evaluates the probability or likelihood that adverse ecological effects will occur (or have occurred or are occurring) as a result of exposure to stressors from various human activities (USEPA 1986, 1992; Norton *et al.* 1992; Suter 1993b, 1995). Stressor is a term describing something chemical, physical or biological in nature, which can cause adverse effects on non-human ecological components ranging from organisms, populations and communities, to ecosystems. Thus, industrial chemicals or pesticides are examples of chemical stressors that are introduced into the environment intentionally or unintentionally from human activities and cause possible adverse (toxic) effects.

Ecological risk assessment is a means for regulatory authorities to assess the risk in general, and the toxic effects in particular, of stressors on non-human organisms. The process is of increasing interest to, and now commonly used by, ecotoxicologists, environmental scientists and ecologists around the world. (See Chapter 10 in this volume for further discussion of the application of risk assessment to North American policy and legislation.)

New publications on ecological risk assessment are now appearing frequently, and are too numerous to mention all of them here. Suter (1993b), however, is a useful general reference because it also discusses many of the toxicology concepts associated with hazard assessment (e.g. uncertainty, end-points, assessment methods). Some publications have differences (e.g. in terminology and in descriptions of the process) from that of the USEPA framework document presented here (e.g., Greig-Smith 1992; Zeeman & Gilford 1993; Calow 1994; Grandy 1995; Handley & Knight 1995; Suter 1995; van Leeuwen & Hermans 1995). For example, sometimes the term 'risk assessment' is used interchangeably with the term 'risk characterization' or 'hazard assessment' (Calow 1994; Grandy 1995; see also Chapter 1).

3.4.2 Framework for ecological risk assessment

Description of EPA process

Figure 3.1 illustrates the framework for ecological risk assessment used in this chapter; it is taken from the EPA publication entitled *Framework*

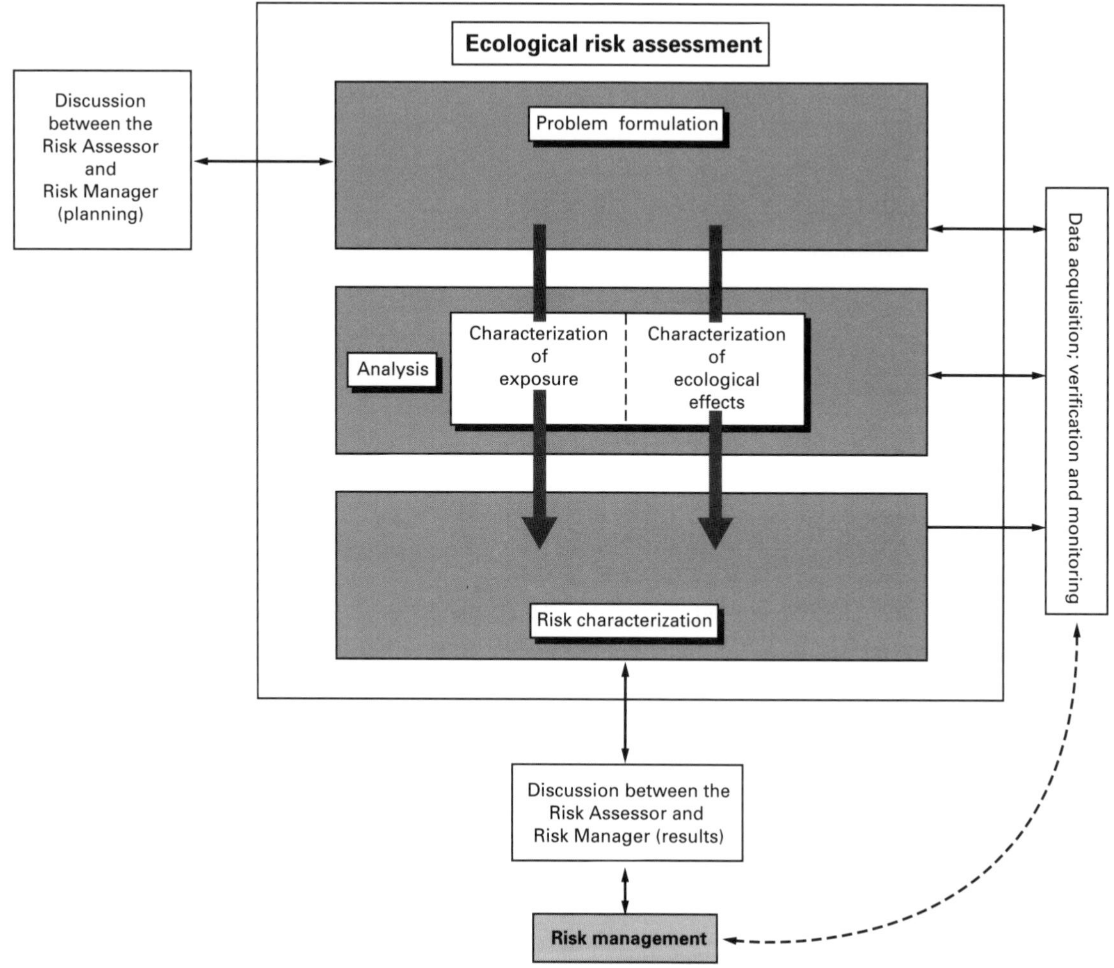

Fig. 3.1 The US Environmental Protection Agency (EPA) framework for ecological risk assessment. (From USEPA 1992.)

for Ecological Risk Assessment (USEPA 1992). This publication represents the initial step in a long-term USEPA programme to develop risk assessment guidelines for ecological effects. It is a document based on a consensus of USA experts in ecotoxicology, aquatic/terrestrial toxicology and ecological effects. The 1983 National Research Council's (NRC) (part of the National Academy of Sciences or NAS) risk assessment paradigm was viewed originally as a possible foundation for ecological risk assessment (NAS 1983). Later, it was concluded that a single paradigm (Fig. 3.1) could accommodate all the different types of ecological risk assessments but that the NRC paradigm would need to be modified to fulfil this role. The framework is discussed in USEPA (1992) and is intended mainly for EPA risk assessors, risk managers and EPA contractors. However, the framework may be useful to others involved in risk and hazard assessment. It is similar to assessment schemes followed by other countries.

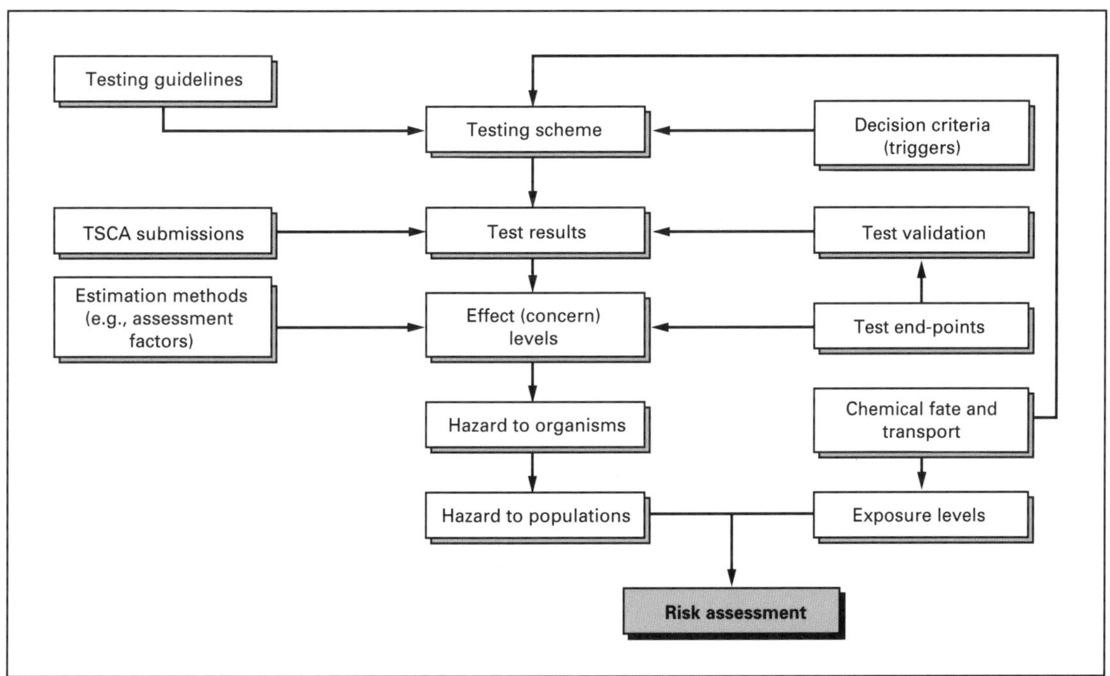

Fig. 3.2 Ecotoxicology testing and assessment flow chart for industrial chemicals, US Environmental Protection Agency (EPA). (From Smrchek *et al.* 1993.)

Substantive risk assessment guidelines and detailed guidance will be developed by USEPA over the next several years to expand upon and replace or supplement the framework document. A draft document entitled *Guidelines for Ecological Risk Assessment*, which proposes basic principles and terminology, has been circulated for public comment in preparation for later revision and release by Autumn (1997).

Hazard assessment process for USA industrial chemicals

The ecotoxicology testing and hazard assessment process used for industrial chemicals is presented in Fig. 3.2 (from Smrchek *et al.* 1993). It contains all the important components of assessing the hazard and risk of industrial chemicals and pesticides. This figure is in agreement with Fig. 3.1 even though the terminology differs somewhat. Earlier efforts by OPPT and OPP in hazard and risk assessment helped form the basis for the Agency framework document. The Effect Levels, and Hazard to Organisms boxes (Fig. 3.2) correspond to the Characterization of Ecological Effects box in Fig. 3.1, whereas the Chemical

Fate and Transport, and Exposure Levels boxes (in Fig. 3.2) correspond to the Characterization of Exposure box. The Risk Assessment and Hazard to Populations boxes in Fig. 3.2 include Risk Characterization processes.

Also, Fig. 3.2 is similar to other hazard and risk assessment flow diagrams used in several other countries. See, for example, Fig. 9 in UK Department of the Environment (1993), Fig. 1 in Ahlers *et al.* (1994), Fig. 1 in EC (1994), Fig. 1 of ECETOC (1993b), Figs 2 and 1.1 of OECD (1989 and 1995b, respectively).

3.4.3 Problem formulation

The first step in ecological risk assessment (or in any hazard assessment) is formulation of the problem, for example, is the industrial chemical or pesticide toxic to organisms, and if so, to what degree or extent? This step includes a preliminary characterization of effects and exposure (USEPA, 1992). The feasibility, scope and objectives of the ecological risk assessment are defined in a management context by examining data needs, policy and regulatory issues, time and budget available for the assessment, and the maximum level of uncertainty that can be accepted and will still allow a decision to be made (Moore & Biddinger 1995).

Problem formulation identifies the key factors, early on, that are to be considered. Measurement end-points, i.e. the measureable ecological characteristic(s) such as reproductive effects in a fish chronic test, are selected. Also, assessment end-points, i.e. an explicit expression of an environmental value(s) to be protected (Suter 1990), such as viability of fish populations, are chosen. It is important that there be close interaction between risk assessors (ecotoxicologists) and risk managers at this stage to ensure the success of the risk assessment and final decision-making.

3.4.4 Ecological effects, exposure, and risk characterization

Ecological effects (hazard assessment)

Characterization of ecological effects, often used interchangeably with hazard assessment, is one-half of the analysis phase (Fig. 3.1). First, the hazard is identified or data on the effects of the stressor (industrial chemical or pesticide) in a specific sequence of steps are categorized (see Table 3.8). Second, the evidence is weighed on whether or not an identified hazard is likely to be of practical significance. Both processes require a combination of knowledge and expert judgement. The tools, methods and underlying logic of hazard assessment or ecological effects characterization are described first (section 3.5.1) in order to aid understanding.

Exposure

Characterization of exposure is the other half of the analysis phase. The exposure profile characterizes the ecosystems or environmental compartments in which the stressor may occur and the biota that may be exposed to the stressor (section 3.5.2).

Risk characterization

Risk characterization is the final phase of risk assessment and integrates the information (profiles) generated in the ecological effects (hazard) and exposure characterizations (Rodier & Zeeman 1994). Risk is expressed as a qualitative or quantitative estimate, by using comparison methods, for example, the quotient method (see section 3.5.1 and Chapter 1).

Risk characterization includes a discussion of the scope of the assessment, identifies key issues, and discusses sources of uncertainty and variability. Also, it presents risk conclusions and information on the strengths and limitations of the risk assessment. Risk characterization should include a summary of the key issues and conclusions of each of the components of the risk assessment (Fig. 3.1). The likelihood of harm should be described and there should be some consistency in general format, while at the same time recognizing that each situation is unique in some ways.

The specific risk(s) identified should, at least in a qualitative sense, be compared with other similar risks, for example, comparisons with other chemicals or situations. The limitations of such comparisons also should be discussed. Risk characterization is also a key part of risk communication (Chapter 20). The latter is an interactive process in which information and expert opinion is exchanged among individuals, groups and institutions.

Risk assessments and all components should be transparent, clear and consistent. The conclusions drawn from the science are identified separately from policy judgements and the assumptions used in arriving at these conclusions should be described clearly.

Risk management

Risk characterization leads into risk management by providing the basis for discussions between the risk assessor and risk manager (USEPA 1992; Rodier & Zeeman 1994). These discussions are held to ensure that the results of the risk assessment are presented clearly and completely, and in an unbiased manner. Risk management is a policy-based activity that defines end-points and questions from risk assessment so as to protect human health and the environment (ecosystems) (SETAC 1994). Risk management is a decision-making process that attempts to minimize risks without harm to other societal values (NAS 1983).

As a result of these discussions, timely regulatory decision-making and mitigation measures may then occur, for example, a reduction or restriction in use, a ban or remediation. Follow-on activities to the risk assessment may be identified (e.g. monitoring, collection of additional data to reduce uncertainty, stakeholders meetings).

3.4.5 Current USEPA hazard assessment for industrial chemicals

Each box in Fig. 3.2 will be discussed here in order to aid understanding of how hazard assessment or characterization of ecological effects is carried out. A similar approach is applicable to pesticides (see Fig. 1 in SETAC 1994).

Description

The testing scheme (Fig. 3.3) forms the central core of concepts and reasoning used to assess the results of tests, which are conducted to determine the toxic effects to aquatic and terrestrial organisms of an industrial chemical or which are submitted to OPPT pursuant to TSCA. Information on the fate and transport of the chemical will directly influence how the testing scheme is used. Available test guidelines are used in the testing scheme as a means to recommend and require completion of the tests needed to give an indication of the toxicity of the chemical, and to fill data gaps. Decision criteria or triggers (Table 3.3) are stopping points within the scheme where a decision must be made either to proceed with further testing based on previous test results, or to stop testing and determine the hazard to organisms more intensively.

Any tests that are received as completed test reports, obtained from published and unpublished literature sources, or from various databases, are reviewed and must be validated by determining if certain criteria are met (Table 3.4). These validated tests are used to establish sound, reliable test measurement end-points. Estimation or extrapolation methods that have been developed, such as assessment factors, or uncertainty factors (but not safety factors!), are applied to these measurement end-points and serve to address uncertainties and to determine concern levels or concentrations. The latter are then used to estimate hazard to populations of representative organisms. Population hazard along with exposure levels (predicted environmental concentrations or PECs; expected environmental concentrations or EECs) based on fate and transport information, as well as on exposure models are used to characterize the risk of the chemical (Zeeman & Gilford 1993). Various tools and methods are used in hazard and risk assessment.

3.5 RISK ASSESSMENT TOOLS AND METHODS

3.5.1 Hazard assessment tools and methods

Tiered testing scheme

The basis of any hazard assessment method for industrial chemicals and pesticides is often some

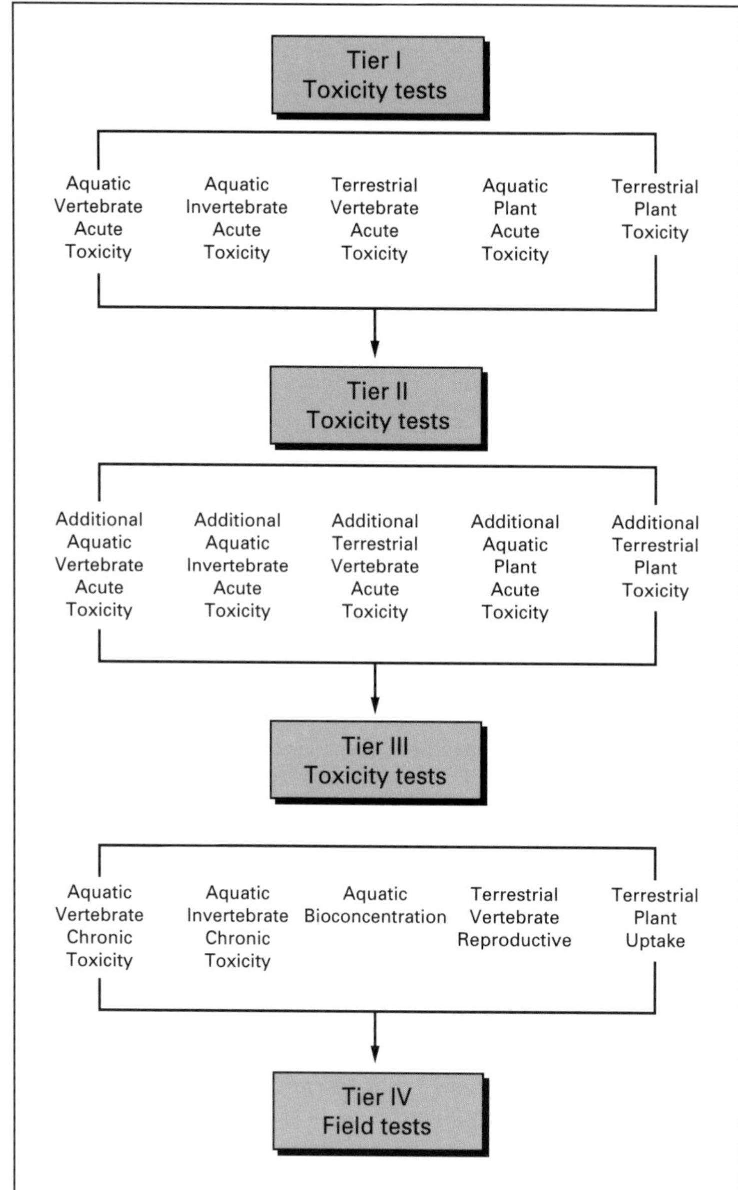

Fig. 3.3 Testing scheme for determining environmental effects of industrial chemicals. (From Smrchek *et al.* 1993.)

type of tiered or phased testing scheme. Such schemes have been recommended for a number of years (Cairns *et al.* 1978; Maki 1979; Levin *et al.* 1984; Cairns & Dickson 1995). These schemes have been developed in some countries throughout the world and the former are applicable and relevant to many other countries. Examples of countries using these schemes are Germany, The Netherlands, the United Kingdom, and the USA (Maki 1979; Urban & Cook 1986; ECETOC 1993b; Smrchek *et al.* 1993; UK Department of the Environment 1993; Zeeman & Gilford 1993; USEPA 1994e). Also, the EU (in Directive 79/831) and OECD have advocated the use of tiered or stepwise approaches (OECD 1989, 1995b; EC 1994; Blok & Balk 1995).

The scheme illustrated in Fig. 3.3 is used when USA regulatory authorities must determine toxi-

Table 3.3 Decision criteria used to enter a tiered testing scheme applied in assessing the ecological hazard or effects of an industrial chemical or pesticide, or to move from one tier to another. (After Smrchek *et al.* 1993 and Zeeman & Gilford 1993.)

Decision criteria or triggers*	Appropriate tier†
Tier I	
'High' production level	Enter Tier I
Chemical may be 'toxic'; $LC(EC)_{50} \leq 100\,mg\,l^{-1}$; test(s) valid?	Enter Tier I
Chemical has 'moderate' or 'high' acute toxicity; $LC(EC)_{50} \leq 100\,mg\,l^{-1}$	Enter Tier I and/or Tiers II and III
Toxicity information is incomplete; more test results needed	Enter Tier I; complete base set testing
QSAR (or analogue chemical) analyses indicate 'some' toxicity	Enter Tier I; complete base set testing
Chemical is persistent (half-life is >4 days)	Enter Tier I
Chemical is bioaccumulative (log K_{ow} or log P_{ow} >3.5)	Enter Tier I
Predicted or monitored environmental concentrations are 'high' or 'significant'	Enter Tier I
Tier II	
Results of acute base testing (most sensitive test species) indicate high acute toxicity; $LC(EC)_{50} \leq 1\,mg\,l^{-1}$	Proceed to Tier II and/ or conduct chronic testing (Tier III)
Chemical will partition to a specific environment (e.g. estuaries)	Proceed to Tier II; conduct additional acute testing (estuarine spp.)
Confirmation of high acute toxicity is needed (e.g. to plants)	Proceed to Tier II; conduct additional acute testing
Chronic effects are indicated in chemical or analogue analyses	Proceed to Tier II or III
Testing indicates 'low' toxicity; $LC(EC)_{50} > 100\,mg\,l^{-1}$	Stop—exit Tier I or II
Tier III	
Chemical is persistent (half-life is >4 days)	Proceed to Tier III
Chemical is bioaccumulative (log $K_{ow} \geq 4.2$)	Proceed to Tier III; conduct bioaccumulation testing
Chemical has high acute toxicity based on additional acute testing; $LC(EC)_{50} \leq 1\,mg\,l^{-1}$	Proceed to Tier III; conduct chronic testing

(*Continued on p. 44*)

Table 3.3 *Continued.*

Decision criteria or triggers*	Appropriate tier†
Chronicity indicated by fish test (ratio of 24 h LC_{50}/96 h LC_{50} > 2) or by invertebrate test (24 h EC_{50}/48 h EC_{50} > 2)	Proceed to Tier III; conduct chronic testing
Chronic toxicity low (MATC > 10 μg l^{-1})	Stop—exit Tier III and complete hazard assessment
Tier IV	
Chemical is persistent, with 'significant' field exposure	Proceed to Tier IV; conduct field testing, complete hazard assessment and then risk characterization
Chemical is bioaccumulative	Proceed to Tier IV; conduct field testing; complete hazard assessment and then risk characterization
Chemical has moderate or high chronic toxicity (MATC ≤ 10 μg l^{-1})	Conduct additional chronic testing or proceed to Tier IV; conduct field testing and then complete hazard assessment and risk characterization, or stop—exit Tier III, complete hazard assessment and then risk characterization
Predicted or monitored concentration in environment through production, use and disposal will be 'high'; chemical has high chronic toxicity or field testing indicates high toxicity; chemical is persistent and/or bioaccumulative	Stop—exit Tier IV and complete risk characterization
At any tier	
Chemical will partition to sediments (based on exposure and/or physical–chemical properties)	Enter Sediment Effects Testing Scheme; tier level is dependent on completed testing

* The terms 'low, moderate, high, toxic, some and significant' should be qualitatively and quantitatively defined in the hazard assessment process.
† Risk management, risk reduction, pollution prevention, or other regulatory activities, such as restrictions on release, a chemical ban, or renewed regulatory review when certain production levels are reached, can occur as outcomes from any tier level, usually after hazard assessment and risk characterization are completed.

Table 3.4 Test elements considered by the US Environmental Protection Agency (EPA) (Office of Pollution Prevention and Toxics) in determining the validity of an aquatic ecotoxicity test and/or compliance with a pre-approved test protocol. (After Smrchek *et al.* 1993.)

---

1 Test report
- Clear objective or purpose of study is reported
- Methods, results and conclusions presented
- Note any deviations from method or pre-approved protocol

2 Test substance
- Analytical procedures described, minimum detection limits presented, and percentage recovery information given
- Chemical identified and impurities known
- Characteristics affecting test conditions reported
- Water solubility given
- Bioaccumulation potential given
- Octanol/water partition coefficient reported
- Need for solvents and their concentrations presented

3 Test procedures and conditions (must be within specified limits or ranges during the test)
- Standard, recognized procedures used
- Experimental units/test chambers adequate (no pseudoreplication)
- Test species acceptable (identification correct, organisms available, representative of organisms to be protected)
- Species size and/or age, condition and source (or strain) reported
- Acclimation, care and handling procedures for test organisms prior to testing are adequate
- Health of test organisms in stock culture(s) and controls during test acceptable
- Diet and feeding schedule acceptable for subchronic and chronic tests
- Test temperature, dissolved oxygen levels (and loading rate), pH, lighting and other environmental variables during test are acceptable
- No position effects due to placement of test units or chambers

4 Test medium and dilution water
- Medium or media correctly made (composition and ingredients)
- Quality and variation during the test described
- Any contaminants present are reported
- Limits on particulate matter, total organic carbon, chemical oxygen demand, un-ionized ammonia, residual chlorine, pesticide levels and PCBs (polychlorinated biphenyls) not exceeded
- Limits on heavy metals not exceeded
- Within specified hardness and salinity ranges during the test

5 Test concentrations/dose levels
- Spacing and number
- Replication adequate
- Measured not nominal concentrations preferred
- Maintenance of concentrations during test

6 Controls in test
- Number adequate
- Upper limit (per cent) on mortality not exceeded
- Control response acceptable

7 Test end-points and reported data
- Specified end-point(s) adequate
- Dose–response or concentration–effect relationship evident
- Sublethal effects should be minimal, i.e. no overt symptoms or signs of disease and no abnormalities in behaviour noted

---

(*Continued on p. 46*)

Table 3.4 *Continued.*

- Data (and units) clearly reported, with no missing data
- Method of data reporting should be clear
- Report any unusual observations, occurrences, or problems during the test regarding test organisms, the test substance (e.g. precipitation, layering), or test condition (e.g. test temperature exceeded upper range for *x* hours due to power failure)
- Data quality assured and good laboratory practices (GLPs) must be followed

8 Statistical analyses
- Data evaluated
- Calculation of $LC(EC)_{50}$, LOEC, and NOEC values acceptable
- Correct test(s) or procedures used, parametric or non-parametric, to support/determine if data of treatments demonstrate any significant effects when compared to controls
- Normality of test data and variances of treatments discussed
- Minimum detectable difference between treatments and controls listed
- Replication and test sensitivity discussed
- Level of significance and power of test discussed
- Data transformations made as needed
- Precision and accuracy of test data discussed

9 Results and conclusions
- Conclusions presented and supported by data
- Effects clearly attributable to exposure to test substance
- Address any indirect effects (e.g. oxygen depletion or disease outbreak) caused by the test substance
- Scientific objectivity demonstrated, with no bias in results or conclusions
- Results should be 'compatible' with other data on same substance. Agreement with other published valid data increases likelihood that new data are accurate and reliable

cological hazards, and request testing or additional tests be conducted for new or existing chemicals. Testing occurs in a series of steps, phases, levels or tiers of increasing complexity in terms of the type of test (progressing from acute to subchronic to chronic to field tests), the ability to carry out the test, the diversity of aquatic and terrestrial species tested (base set of three acute tests to additional acute tests that are exposure specific), and costs (from less costly acute tests to expensive chronic and very expensive field tests). In this weight-of-evidence approach, testing results increase and testing gaps are filled gradually so as to minimize extra costs and unneeded testing. Alternatively, the scheme may be entered at any tier and then 'ascended', for example, when acute test data already exist.

Testing schemes for industrial chemicals and pesticides

The purpose of the testing scheme for industrial chemicals is to set forth the specific evaluation strategy used when testing for the inherent or intrinsic toxicity of industrial chemi-cals to aquatic and terrestrial organisms (USEPA 1983a,b; Smrchek *et al.* 1993; Zeeman & Gilford 1993). Data are generated for small groups of test organisms and test results are used to estimate the effects on populations. The scheme is generic with a logic and a sequencing of tests that is readily understood and compatible with methods of other national and international regulatory authorities, and with standard-setting/testing organizations. The scheme is flexible in that specific tests may be added or omitted on a chemical-by-chemical basis when equivalent data are available or when a required test is not possible. The scheme may be used to test a variety of chemicals for which estimates of environmental concentrations are lacking, imprecise, or will be calculated later in the risk assessment process.

Ecotoxicity data are developed sequentially in such a way as not to unduly impede or create unnecessary economic barriers to technological innovation while providing information adequate for environmental protection. Chemical manufacturers and testing laboratories know in advance the extent and kind of testing that will be required and why. This is especially important to small companies that produce new chemicals (under Section 5 of TSCA) and which may have limited amounts of available resources or technical expertise.

A tiered testing scheme also exists for pesticides (USEPA 1994e). It is similar to the testing scheme for industrial chemicals in most respects; however, some differences between the two will be discussed.

Use and sequencing of tests

The first determination that must be made is if testing is required or needed to fill data gaps. One or more tests are then requested by the regulatory authority. Both the industrial chemical and pesticide testing schemes are organized in four levels or tiers, each with a group of tests available (Fig. 3.3). Testing usually progresses sequentially from Tier I to II, or from I to III for existing chemicals; for new chemicals, testing proceeds from Tier I to III. Tier IV field testing is done only on a chemical-specific basis and is very rarely requested or required for industrial chemicals; until recently field testing was more often requested and completed for pesticides.

Pesticide testing proceeds in a Tier 1, 2, 3, and 4 sequence. Testing can end at any tier (as with industrial chemicals) and the pesticide (or industrial chemical) will then undergo hazard and risk assessment. In contrast to the industrial chemical scheme, aquatic tests are differentiated from terrestrial (bird) tests into separate portions of the testing scheme. Also, non-target plants (algae, monocot, and dicot vascular plants) and insects (honeybees) are separate from the other two.

Tests that are required for industrial chemicals will depend on the specific concerns identified during the initial phases of the hazard assessment process (Table 3.5). Tests that are prescribed depend on the environment potentially at risk as identified in the exposure analysis, either aquatic (commonly), or terrestrial (occasionally), or both (rarely). Thus, there are aquatic and terrestrial tests on each tier.

Both schemes can also serve as a means of placing or categorizing existing test data in the correct tier to indicate the amount and kind of data available for a chemical or pesticide, when additional tests will not be requested. Gaps in test information are thus seen easily.

Tier I: Acute Tests. Tier I contains five acute tests for industrial chemicals: a 96 h aquatic vertebrate (a fish, usually fathead minnow, or rainbow trout), a 48 h aquatic invertebrate (usually *Daphnia magna*), a 96 h aquatic lower plant (green alga, e.g. *Selenastrum capricornutum*), a wild terrestrial vertebrate (substituted by a rodent oral toxicity test if available), and a terrestrial monocot, and/or dicot vascular plant test (early seedling growth). The taxonomy of *S. capricornutum* remains unsettled; the latest in a series of scientific names for this alga is *Pseudokirchneriella subcapitata*. All five tests may be completed for existing chemicals, or the first three tests listed above (called the aquatic base set) are completed for existing and new chemicals (Table 3.5).

Tier 1 of the pesticide aquatic testing scheme contains four more acute tests in addition to those in the industrial chemical Tier I: another 96 h freshwater fish (so that both bluegill (warmwater) and rainbow trout (coldwater) fish are tested), an acute estuarine/marine fish (sheepshead minnow), a 96 h estuarine/marine mollusc (eastern oyster shell deposition test), and an acute estuarine/marine shrimp (mysid, penaeid or grass shrimp) test. There is also a fish bioaccumulation test in this tier. Tier 1 of the pesticide terrestrial testing scheme has an acute avian oral test (mallard duck or quail), and two acute dietary tests (mallard duck and quail). Tier 1 of the non-target plant scheme has both terrestrial tests (seed germination, seedling emergence and vegetative vigour), and aquatic tests (a 5-day *Selenastrum capricornutum* test, and a 7-day test with *Lemna gibba*, an aquatic macrophyte).

There are also two non-target insect tests that can be completed on any pesticide tier. These tests consist of a honeybee acute contact LD_{50} test and a foliar residue bee toxicity test (USEPA 1994e).

Tests on this tier for industrial chemicals are the minimum needed to provide adequate data to assess the possibility of a chemical to cause

Table 3.5 The OPPT environmental effects testing guidelines. Methods described in Part 797, Title 40 of Code of Federal Regulations; *Federal Register* 50: 39321–39397, 27 September 1985; 51: 490–492, 6 January 1986; 52: 19056–19082, 20 May 1987; 52: 36339–36371, 28 September 1987. (After Smrchek *et al.* 1993, 1995.) Some of these guidelines are now being harmonized between the USEPA offices and with the Organization of Economic Cooperation and Development.

Test guideline	Location in the testing scheme where each test would usually be completed
Aquatic	
Algal Toxicity Test	Tier I—base set
Daphnid Acute Test	Tier I—base set
Fish Acute Test	Tier I—base set
Gammarid Acute Test	Tiers I, II
Lemna Acute Test	Tier II
Oyster Acute Test	Tier II
Mysid Shrimp Acute Test	Tiers I, II
Penaeid Shrimp Acute Test	Tier II
Daphnid Chronic Test	Tier III
Fish Early Life Stage Test	Tier III
Fish Bioconcentration Test	Tier III
Oyster Bioconcentration Test	Tier III
Mysid Shrimp Chronic Test	Tier III
Terrestrial	
Seed Germination/Root Elongation Test	Tier I
Early Seedling Growth Test	Tiers I, II
Avian Dietary Test	Tiers I, II
Avian Acute Oral Test	Tiers I, II
Bobwhite Quail Reproduction Test	Tier III
Mallard Duck Reproduction Test	Tier III
Plant Uptake and Translocation Test	Tier III
Mixed media	
Rhizobium–Legume Chronic Test	Tier III*
Generic Freshwater Microcosm Test	Tier III*
Site-specific Aquatic Microcosm Test	Tier III*
Soil Microbial Community Test	Tiers III, IV*
Soil–Core Microcosm Test	Tier III*
Aquatic provisional†	
Fish Acute Toxicity Mitigated by Humic Acid	Tier II*
Chironomid Sediment Test‡	Tiers I, II*
Tadpole/Sediment Subchronic Test	Tiers I, II*
Terrestrial provisional†	
Earthworm Toxicity Test§	Tiers I, II*
Guidelines under development	
Full Lifecycle Chronic Fish Test	Tier III
Champia parvula Marine Red Algal Test¶	Tiers II, III

* There is uncertainty over where in the industrial chemical testing scheme the test would be most appropriate.
† These methods have not yet been published in the *Federal Register* as Part 797 guidelines. They are provisional guidelines written for specific Toxic Substances Control Act, Section 4 regulatory actions (Test Rules or Consent Orders).
‡ Appeared as Section 795.135 provisional guideline; published in the *Federal Register* 56(122): 29149–29154, 25 June 1991.
§ Appeared as Section 795.150 provisional guideline; published in the *Federal Register* 56(122): 29154–29158.
¶ Guideline based on method is Steele and Thursby (1983) and USEPA (1991c).

adverse effects on organisms; if effects are found, a quantitative estimate of this ecotoxicity is determined. The measurement end-points in Tier I are those common to acute tests, and include lethality, and immobility in animals (e.g. daphnids) and changes in plant cell numbers or biomass. The algal test in Tier I is not a true acute (quantal response) test in that the test is multigenerational and more closely resembles a chronic test. Certain other parameters may be noted, such as onset of observed effects, type of effects, site and mode of action, and initial indications of chronic effects or chronicity (e.g. time to cause effects, lack of recovery, ratio of 24 h LC_{50}/96 h LC_{50} > 2 for fish). At this point a decision must be made to test further in higher tiers or to stop.

Under some conditions or situations, Tier I tests may not be required first for industrial chemicals, but may be bypassed and testing begun at a higher tier. Testing within the testing scheme does not necessarily have to proceed in a sequential manner for every chemical. For chemicals that occur only as intermediates with a short half-life, those that have very low water solubilities and are bioaccumulated in organisms or partition readily to soils and sediments, or those with high octanol/water partition coefficients such that toxic effects would not be manifested within 96 h, Tier III testing may begin instead. Furthermore, if the available ecotoxicity test data were valid (see Table 3.4) and equivalent (e.g. *Daphnia pulex* was tested instead of *D. magna*) to that of one or more Tier I tests, additional Tier I testing would be limited or possibly not required. There is less flexibility for pesticides in substituting test species, but there is concern with possible exposure and adverse effects on non-target organisms found near where the pesticide is applied.

Single species tests

Single species acute and chronic tests form the framework of both testing schemes, and the current basis for regulatory decision making for industrial chemicals and pesticides. The advantages and disadvantages of these tests have been the subject of numerous publications (see, for example, NAS 1981; Cairns 1983, 1986b; Levin *et al.*, 1984; Kimball & Levin 1985; Forbes & Forbes 1994; Hill *et al.* 1994; van Leeuwen & Hermans 1995).

Single species tests have many advantages, among which are ease of operation, low costs, short test time, ease of replication and convenience. An important advantage, which relates to harmonization of test guidelines among different countries (actively occurring now in the OECD), is that they can be standardized. Also, they are valuable as screening tools, and can be used to test chemicals quickly and efficiently for possible toxicity.

There are also several important disadvantages with these tests (see, for example, Persoone & Janssen 1994). Single species tests assess only some gross effects of chemicals on small groups of individuals. Extrapolation is necessary to predict effects on populations, or other higher levels of biological organization. Prediction of effects to communities and ecosystems is difficult (see discussion on uncertainty factors). These tests are conducted under 'clean' conditions that are much different from those variable conditions found in natural ecosystems with multiple causal and mitigating factors. Laboratory tests may be overprotective or conservative when compared with higher-level tier tests (microcosm or mesocosm tests). This can be viewed also as an advantage, as test data are derived from homogeneous conditions (worst-case), with few, if any, mitigating factors present that would affect toxicity, increase variability, and perhaps complicate testing and the interpretation of test results. Single species tests cannot integrate species interactions, assess compensatory mechanisms of ecosystems, or assess chemical effects on community-level parameters (Belanger 1994). Other disadvantages are that degradation, indirect (or secondary) effects, and cumulative effects due to multiple chemical stress are often not considered.

Additional testing needs

At every tier the PEC derived from fate testing of the industrial chemical is compared with an effect level (after dividing by an uncertainty or other factor) calculated from toxicity tests. This effect level may be based on an acute LC_{50} or EC_{50}, a chronic NOEC (No-observed-effect concentration),

LOEC (Lowest-observed-effect concentration), MATC or GmMATC (geometric mean MATC), a NEC (no-effect concentration), or on a PNEC (predicted-no-effect concentration). The NOEC and PNEC are generally preferred by European countries. The MATC and GmMATC are preferred in North America and it has been the goal of USEPA to regulate based on the MATC where warranted (USEPA 1994e). If the PEC is above the effect level there may be concern that toxicity may occur to organisms found in the environment and there may be a risk from exposure to the chemical. Additional testing may then be needed to assess this toxicity in more detail. For example, if a chemical had moderate acute toxicity, it may also have high chronic toxicity, or if it was highly toxic to a freshwater fish, it also may be highly toxic to other freshwater fish, or also to estuarine/marine fish. Information on physical–chemical properties, production and use levels, and release patterns is used, along with effects data on various test organisms, to indicate possible concern and hazard. Testing may stop if the PEC is lower than the effect level.

Similarly, the EEC of a pesticide is compared with acute end-points from Tier 1 aquatic tests. If the EEC is less than 0.1 of the acute LC_{50} value ($EEC < 0.1\ LC_{50}$), the acute risk is minimal. If $0.5\ LC_{50} > EEC \geq 0.1\ LC_{50}$ there is a potential acute risk that must be mitigated. If the $EEC \geq 0.5\ lC_{50}$ and there are field test data, there is a high acute risk to aquatic organisms and use restrictions must be taken. If the LC or EC_{50} values are $< 1\ mg\,l^{-1}$, or the $EEC \geq 0.01$ of the LC or EC_{50}, and the pesticide is applied directly to water or is continuously present in water, chronic testing is recommended. There are also other determinations that would require chronic testing. For Tier 1 terrestrial (bird) acute tests the EEC is compared with a 'Q' or quotient value (also called an LOC or level of concern). Q is the ratio of chemical exposure value(s) to toxicity value(s) and is an LD_{50}/sq ft, an EEC/LC_{50} ratio, or an $LD_{50}\ day^{-1}$ value. If $Q < 0.2$ there is minimal acute risk to birds, but if $0.5 > Q \geq 0.2$ there is a potential acute risk to birds. Values ≥ 0.5 and with field test data indicate high acute risk to birds, such that use of the pesticide must be restricted. This quotient or risk quotient method of comparing exposure with toxicity will be discussed in greater detail later in this section. Other determinations, such as the pesticide having a $K_{ow} > 1000$ would necessitate further (chronic) testing. It has been recommended that the OPP move away from the exposure and toxicity point-estimate methods used above in Tier 1, to a method using probabilistic distributions of many toxicity values (SETAC 1994).

Decision criteria or triggers

These are specific values, general conclusions or characteristics developed for a chemical from fate and effects tier testing, information and experience obtained from comparisons with similar chemicals or classes of chemicals, and from QSAR analyses. These criteria are based on those derived from the hazard assessment scheme developed by the Aquatic Hazards of Pesticides Task Group of the American Institute of Biological Sciences (Cairns *et al.* 1978; Maki 1979; Urban & Cook 1986), and from a 1982 workshop entitled *Testing Triggers Workshop* (Life Systems 1983).

Decision criteria are used at points between any two tests in a testing sequence or at points between tiers, and give an indication of which of several further actions should be taken or are appropriate to emphasize. Tests in the testing scheme are 'triggered' by the results of other tests (Fig. 3 in Smrchek *et al.* 1993). If there is a need to test the most sensitive species found in Tier I further, to verify high toxicity, then additional testing would be necessary. As an example, if *Daphnia magna* was found to be the most sensitive species of those tested in Tier I, then logically, additional testing would need to be completed in Tier II—perhaps another crustacean, or a marine crustacean acute test.

It even may be appropriate to simply stop testing if the Tier I acute EC_{50} value was very high ($> 100\ mg\,l^{-1}$) and only acute testing was warranted. Table 3.3 lists many of the decision criteria used in the industrial chemicals testing scheme. These criteria many be a single point (>4 days), a range ($1–10\ mg\,l^{-1}$), or a more subjective identifier (e.g. 'significant'). They should be used flexibly along with a certain amount of expert judgement and other information relating to the chemical.

Use of decision criteria will reduce unnecessary

testing, promote consistency to the testing scheme, and provide flexibility in designing appropriate tests (Greig-Smith 1992). However, some problems may occur if the criteria are too broad, too general, interpreted too rigidly, or if one criterion is weighted or emphasized more than another.

Tier II: Additional acute tests (industrial chemicals). This tier has an expanded set of acute single-species toxicity tests for industrial chemicals using a wider range of test organisms (Table 3.5). Included are an aquatic vertebrate (usually fish), an aquatic invertebrate, an aquatic plant (often *Lemna* spp.), a terrestrial vertebrate (usually bobwhite quail or mallard ducks), and terrestrial monocot and/or dicot vascular plants (Smrchek *et al.* 1993). The more sensitive groups or the most sensitive group to the test chemical has either already been identified or will be identified in Tier II. Also, the site where ecotoxicological activity occurs, and the range of this activity is identified. The specific organisms tested will depend on the predicted exposure and media partitioning of the chemical. For example, if daphnids are the most sensitive test species in Tier I and the chemical will partition to estuarine or marine environments, then it would be appropriate to test estuarine/marine invertebrates such as mysid shrimp, penaeid shrimp, and oysters in Tier II. Also noted are the onset of observed effects, types of effects, mode and site of action, indicators of chronic effects (chronicity), and potential for bioaccumulation and chronic effects.

Industrial chemical aquatic tests just mentioned for Tier II are combined into Tier 1 for pesticides. Estuarine/marine tests that are in Tier 1 have already been discussed above. At the pesticide Tier 2, chronic aquatic invertebrate lifecycle (usually *Daphnia magna*) and fish early-life-stage tests are conducted, similar to that recommended in Tier III for industrial chemicals (see below). Also, accumulation (fish bioaccumulation testing) of the pesticide is studied in Tier 2 and if effects on the aquatic food chain were seen then a risk assessment would need to be completed to confirm or deny these effects. Special aquatic organism tests (e.g. cholinesterase inhibition tests, metabolism studies, secondary toxicity studies, aquatic microcosm studies) are also conducted at Tier 2 (Urban & Cook 1986; USEPA 1994e). If the *Q* (LEL (lowest effective level)) ≥ 1, or if the EEC > 0.1 of the NOEL (no-observed-effect level) of either the fish early-life-stage test, or the invertebrate lifecycle test, and there is direct application of the pesticide to water, Tier 3 tests are required.

Pesticide terrestrial tests include chronic reproduction studies with quail and mallard ducks. If the quotient of EEC/LEL is ≥ 1, then Tier 3 pesticide testing is recommended. Both non-target terrestrial and aquatic plants tests are part of this tier, and EC_{50} and NOEL values are reported. Only those terrestrial species with greater than 25% adverse growth effects in Tier 1 are tested in Tier 2; seed germination, seedling emergence and vegetative vigour tests are conducted. The acute toxicity to five aquatic plant species is also evaluated in Tier 2: *Selenastrum capricornutum*, *Lemna gibba*, *Anabaena flos-aquae* (a cyanobacterium), a freshwater diatom (usually *Navicula* spp.), and *Skeletonema costatum* (a marine diatom).

Tier III: Chronic and bioconcentration tests. Testing is conducted at this tier if 'sufficient' industrial chemical toxicity is shown in lower tiers, chronicity (indicators of chronic effects, see Table 3.3) or uptake by animals or plants is possible or occurs, and effective persistence (e.g. half-lives of the chemical in water, soil/sediments, or plants) is not disproved. An acute value of $\leq 1\,mg\,l^{-1}$ for the most sensitive species tested in Tier II (or Tier I) would trigger Tier III testing. Chronic or cumulative toxicity and bioconcentration (or bioaccumulation) potential are investigated in Tier III. The organisms tested (usually fathead minnows, rainbow trout, *Daphnia magna*, and mysid shrimp) are also dependent upon sensitivity in the acute tests, and media partitioning. Test organisms are limited to those for which validated test guidelines or protocols are available (Table 3.5). If uptake by organisms is likely to occur, bioconcentration tests with fish or oysters are triggered (Table 3.3). A fish early-(partial)-life-stage test (28 or 60 days) is conducted currently, but whole-life-cycle fish tests are under development.

Some chemicals, called 'persistent bioaccumulators', have high octanol/water partition coefficients (≥ 1000), large log K_{ow} or P_{ow} values (≥ 4.2), water solubilities $< 1\,mg\,l^{-1}$, and degradation half-lives in water of from ≥4 up to >30 days, which can trigger Tier III testing. Chronic and bioconcentration testing can occur without prior acute testing. Acute testing may not be appropriate or possible for these bioaccumulative chemicals because toxic effects will not occur visibly by the end of the 48 to 96 h acute test period.

This problem illustrates the point that the physical–chemical properties of the chemical or pesticide that is assessed must be well known prior to ecotoxicity testing, such that appropriate testing programmes tailored to the chemical can be designed, and unneeded testing is eliminated. Tiered testing schemes must be flexible so as to accommodate a variety of chemicals and pesticides.

As mentioned previously, tests similar to Tier III for industrial chemicals, are placed in Tier 2 for pesticides. Tier 3 consists of a fish full-lifecycle test in which the effects of the pesticide on first- and second-generation reproduction (e.g. number of embryos hatched, number of surviving larvae hatched), on first- and second- generation juveniles, etc., are determined. Chronic aquatic (and terrestrial avian) risk is of high concern if the EEC ≥ LEL.

There are also in Tier 3, pesticide tests for terrestrial birds and non-target plants. A field screening test is conducted to determine the effects of the pesticide on birds. If effects (e.g. on mortality, diversity, behaviour) are identified, Tier 4 testing is warranted. Non-target plant tests in Tier 3 consist of phytotoxicity field studies and are conducted if greater than 25% adverse effects on terrestrial plant growth and 50% adverse effects on aquatic plant growth are expected to occur (EEC > EC_{25} terrestrial or EC_{50} aquatic value(s)).

Microcosm (multispecies) tests

Microcosms are small ecosystems held in various types of containers (Beyers & Odum 1993). They are confined, excised, assembled or isolated portions of terrestrial or aquatic ecosystems (van Voris *et al.* 1985b). These model ecosystems or micro-ecosystems are useful for studying the way natural ecosystems function and the fate of industrial chemicals and pesticides in natural ecosystems. They are also considered to be multispecies test systems because several species on different trophic levels, which interact together, are studied.

Microcosm tests are designed with varying degrees of complexity. They may be terrestrial, freshwater, estuarine or marine in design. Cairns & Cherry (1993) and Clark & Cripe (1993) recently described and compared the many kinds of aquatic test systems. Terrestrial microcosms were described by Gillett & Witt (1977), Witt & Gillett (1977), Gillett (1989), and most recently by Morgan & Knacker (1994). Other important references describing various lentic and lotic methods and systems include, Kevern & Ball (1965), Nixon (1969), Gillett & Gile (1976), Taub (1976), Crow & Taub (1979), Giesy (1980, 1985), Hammons (1981), NAS (1981), Boyle (1985), Cairns (1985, 1986a), Kimball & Levin (1985), van Voris *et al.* (1985a,b), Gearing (1989), Harrass & Sayre (1989), Beyers & Odum (1993), and Hill *et al.* (1994).

Multispecies tests have several important advantages when compared with single species tests. By studying the fate, persistence and transport of substances in microcosms, a more realistic picture is obtained of those processes as they would occur in natural ecosystems. The ecological risk assessment completed as a result of microcosm information will be more realistic and comparable with real-world systems. Species interactions, such as predator–prey relationships and changes in population abundances over time, can also be studied as well as ecosystem functions or processes, for example, primary production and nutrient cycling; the response of these processes to stressors can be studied. In general, risk may be characterized 'better' and risk management measures can be taken that potentially have a good chance to control or limit toxic effects that in microcosms are related more closely to actual ecosystem processes.

Multispecies or microcosm tests do, however, have some significant disadvantages. They are time consuming, labour intensive, and more expensive than single acute and some (not all) chronic single species tests. There is still some reluctance by testing facilities to perform such complicated tests. The complexity and variability of these systems may also make it more difficult to interpret test results, and replicability may be low. Also, few effect-end-points can be sampled non-destructively

(Morgan & Knacker 1994). Even though these systems are more realistic than single species tests, microcosms are still isolated in that they are physically enclosed and are no longer in contact with natural ecosystems (van Voris *et al.* 1985b; Beyers & Odum 1993). There remains uncertainty over how closely a microcosm should reflect the larger system it is designed to represent, mimic or resemble. It is uncertain whether it should simulate classes of ecosystems, or only a site-specific ecosystem (Suter 1993b). Perhaps most important, it would be difficult to relate microcosm test end-points and test results with initiating regulatory action(s) and making risk decisions in relation to natural ecosystem structure and function processes. For example, what decrease in some of the functional parameters or species interactions in microcosms would indicate a concern sufficient to initiate regulatory action on a natural ecosystem(s)? How can the resilience of natural ecosystems be taken into account? Other problems are discussed by Suter (1993b).

Even with these problems, microcosms are a powerful predictive, cost-effective tool (when compared with a battery of acute and chronic tests), useful in studying the fate of industrial chemicals and pesticides in natural ecosystems. Limited microcosm testing in the laboratory at, for example, Tier III (Table 3.5) can be linked to additional field and mesocosm testing at Tier IV. Decision criteria for conducting microcosm or multispecies tests have yet to be formulated, as well as triggers for moving from these tests to Tier IV tests. However, uncertainty factors to determine critical response thresholds and to deal with variability due to extrapolating from laboratory microcosms to field or natural ecosystems should be minimal, very small (< 10), or even non-existent (see Cairns 1986a for further discussion).

Microcosms should have the ability to accurately portray and predict the fate and effects of substances released to natural ecosystems. Thus, it is important that the microcosms be properly calibrated and validated against natural ecosystems. Even though similar exposure–response relationships may be demonstrated between both, differences in species sensitivity and community response may occur.

The American Society for Testing and Materials (ASTM 1993, 1995) has an aquatic microcosm method available, the Standardized Aquatic Microcosm (SAM) (Designation: E1366-91). This method was developed by Taub and co-workers (Taub 1984a,b; Taub & Crow 1978, 1980; Taub *et al.* 1987, 1989). Glass jars (with 3 l of defined medium and sediment) containing organisms from several trophic levels (10 algae, including *Selenastrum*, *Chorella* and *Scenedesmus*, and five animals, including *Daphnia magna* and *Hyalella*) are subjected to introduced substances, usually for 63 days. The method is relatively labour intensive (24–30 microcosms are run) and more difficult to interpret when compared with other microcosm methods (Shannon *et al.* 1986; Cairns & Cherry 1993). Also, there is limited ability to test many test chemical concentrations, although this is a criticism that can be made for some other methods.

The ASTM (1993, 1995) also has a terrestrial microcosm method available, the Terrestrial Soil-core Microcosm Test (Designation: E1197-87 [Reapproved 1993]. This method is based on the work of van Voris and co-workers (van Voris *et al.* 1985a,b). An intact soil-core containing natural assemblages of biota is used to test the environmental fate, effects and transport of substances that may enter terrestrial ecosystems.

Microcosm multispecies guidelines that were proposed for Code of Federal Regulations (CFR), Title 40, part 797 (Table 3.5) are available for industrial chemicals (USEPA 1987b). The generic freshwater microcosm test prescribes methodologies to predict the fate and effects of chemicals in freshwater ecosystems using a variety of microcosms, i.e. the SAM, naturally derived mixed-flask culture microcosms, and naturally derived pond microcosms. The site-specific microcosm test consists of an indigenous water column and the intact sediment core associated with it, along with associated organisms. Lastly, there are a soil microbial community test and a soil-core microcosm test that is similar to that developed by Van Voris and co-workers (see above).

Hammons (1981) summarized the use of microcosm tests in tiered testing schemes. Because of their complexity, and interpretation difficulties, these tests were seen to be of most use in the later stages of the tiered testing scheme. Thus, few chemicals will ever be tested and will include only those with great economic or commercial potential. They would serve as confirmatory tests when potential problems were identified from single species tests. This was the situation in 1981 and it remains the situation in 1997.

Several goals must be achieved before microcosms are accepted routinely for regulatory purposes (Mount 1985). Examples must be found where a seemingly adequate group of single species tests for a chemical is actually proved inadequate and where microcosm tests would better assess the hazard of this chemical. There should be agreed-upon protocols for routine testing (this has already been partially achieved), and there should be a clear rationale for applying microcosm test results to hazard and risk decision-making.

Tier IV: Field and mesocosm tests. This tier consists of actual field tests and studies using ponds and small streams, field pen studies, and assessment studies in fields and on watersheds or other large aquatic and terrestrial ecosystems. Laboratory multispecies test systems or simulated field studies include mesocosms, outdoor microcosms, limnocorrals, artificial streams and littoral enclosures. See Graney *et al.* (1994) and Hill *et al.* (1994) for extensive discussions on all aspects of mesocosm and field tests.

Use of such tests for industrial chemicals must be justified on a case-by-case basis (Smrchek *et al.* 1993). Testing is driven by test data obtained from the first three tiers. Usually Tier IV testing is not appropriate or needed for TSCA Section 4 and 5 chemicals. These tests may easily cost hundreds of thousands of USA dollars to complete over months, or several years. These costs are often too great in relation to what useful information is gained over and above the results obtained from lower tier laboratory tests to use the information effectively to make regulatory decisions.

Tier 4 aquatic field tests for pesticides are triggered by Q (LEL) values ≥ 1, when EEC ≥ 0.5 LC_{50}, and by 'unusual circumstances' (USEPA 1994e). These tests consist of a simulated field mesocosm test or a pond aquatic field study. Large quantities of data are generated, consisting of physical–chemical, sediment and atmospheric measurements, and various measurements (e.g. mortality, biomass, growth and reproduction) on biota. Phytoplankton, periphyton, zooplankton, macroinvertebrates, macrophytes and fish are monitored. The purpose of these studies is to assess the pesticide effects on populations of aquatic organisms. Mesocosm studies are conducted and used to negate a presumption of unreasonable risk, and also to define the intensity and duration of adverse effects due to pesticides, on aquatic systems (Graney *et al.* 1994). After these field tests are concluded, a risk assessment is completed. The outcomes of this assessment would be the confirmation of a potential high risk to aquatic organisms and the need for use-restrictions to be placed on the pesticide. Alternatively, a potential high risk to aquatic organisms would not be confirmed and all testing would then be concluded.

Terrestrial quantitative field tests with birds are also found in Tier 4 and are triggered when EEC ≥ 0.5 LC_{50}. Population mortality, survival, reproduction rates and survival of young, for example, are measured. If bird population effects are seen, then a risk assessment is completed, to confirm and quantify the high risk of the pesticide to birds.

Field tests (land and water) are conducted for pesticides to either refute the laboratory test-derived assumption that risks to aquatic organisms or wildlife, or both, will occur under conditions of actual pesticide use, or to obtain some quantification of what risks may occur during actual use. They are also conducted to confirm acute or chronic effects identified in Tier 3. The end-points measured are percentage mortality occurring less than a specified percentage of the time (20%) for terrestrial field tests and a 15–20% adverse effect on local populations of biota in aquatic field tests. For those pesticides with mortality or adverse effects greater than 20%, further regulatory action may be required, such as special review or other risk-reduction measures.

Future of field and mesocosm testing at the USEPA. The purposes, need for, and goals of conducting mesocosm and field tests are important (Urban & Cook 1986; Graney *et al.* 1994; Hill *et al.* 1994). There are reasons, already mentioned, that currently limit their use for industrial chemicals.

Due to a change in pesticide testing policy, and the completion and implementation of a new integrated risk assessment paradigm (SETAC 1994), mesocosm tests for pesticides are being de-emphasized and completed less often. Instead, decisions will be made in the absence of higher tier data whenever possible, and will be based on laboratory testing, incident data, and other information that can 'be collected easily' (SETAC 1994). Also, these tests should be carried out only when more cost-effective laboratory tests do not provide the information required (Hill *et al.* 1994).

EU tri-level testing scheme for industrial chemicals

The EU has a tiered testing scheme for industrial chemicals that is similar to the USEPA schemes just described. There are three levels or tiers, a 'base level' or level 0, level 1 for chemicals with ≥ 10 t year^{-1} per manufacturer, or a total market quantity of >50 t per manufacturer, and level 2 for chemicals with ≥ 1000 t year or a total market of 5000 t (Blok & Balk 1995). Handley & Knight (1995) give detailed guidance on assessing the environmental effects of existing and new chemicals for these levels.

Level 0 (Base level). Ecotoxicological studies are specified in Council Directive 67/548, Sixth Amendment (Directive 79/831), on classification, packaging and labelling of dangerous substances, as described in the Seventh Amendment (Directive 92/32). A 'base set' consisting of a 96 h LC_{50} fish acute test, a 48 h EC_{50} *Daphnia* immobilization test, a 72 h IC_{50} algal growth inhibition test, and a bacterial inhibition test are required. Depending on the production and market quantity, level 1 or 2 testing also may be triggered.

Level 1. More thorough, complex and costly tests are required at level 1 (and level 2). Based on the results of the base-set tests, some or all of the ecotoxicity tests for this level may be required (for ≥ 10 t year^{-1}) or shall be required (for ≥ 100 t). These tests include a *Daphnia magna* 21 days chronic test, further fish toxicity tests, a fish bioaccumulation test, an earthworm test and higher plant tests (Blok & Balk 1995).

Level 2. Additional tests are required to be completed at this level. These include further toxicity tests with fish, birds and other organisms. Thus, there is great flexibility in testing at this level, some amount of expert judgement is used, and extensive interaction with the regulatory authority will be necessary. Testing may be limited by the lack of available OECD test guidelines.

EU testing scheme for pesticides

Council Directive 91/414 on the use, marketing, and registration of pesticides specifies tests that are to be conducted (Blok & Balk 1995). The sets of listed data are more extensive than the base set for industrial chemicals. Effects tests on non-target organisms are also listed and numbered sequentially; the information requested is listed for both the active ingredients and for registration of the pesticide.

Many tests are required when testing the active ingredients of pesticides. They include, for birds, the acute oral toxicity, subacute toxicity (8-day dietary study on at least one bird species), and a reproductive effects test. For aquatic organisms they include fish acute, chronic reproduction and growth tests, a *Daphnia magna* acute test and an algal growth inhibition test. Non-target organisms are also tested, and these tests include acute toxicity to honeybees and other beneficial arthropods, toxicity to earthworms and other soil organisms, toxicity to soil microorganisms and effects on other non-target plants and animals that may be exposed to the pesticide.

A technical dossier must be prepared when a pesticide is being registered. Ecotoxicity tests required for birds include an acute oral toxicity test and field studies to assess the risks to birds. Also, tests on the acute toxicity of the pesticide to fish and *Daphnia magna* are to be completed, as well as fish residue studies, and special studies on aquatic organisms in applications in or near surface waters. Non-target organism tests are also important for pesticides. Possible toxic effects are to be determined for terrestrial invertebrates, honeybees, and other beneficial arthropods, for earthworms and other soil organisms, and for soil microorganisms.

Terrestrial toxicity: general comments

This topic has become increasingly important in ecotoxicology (van Leeuwen 1995), and work has progressed beyond plant, avian and mammalian toxicity testing (already discussed). Also, interest has increased in soil ecotoxicological studies, especially in Europe (see, for example, van Straalen & van Gestel 1993; Donker *et al.* 1994). Much attention has also been directed toward studying soils contaminated by industrial chemicals and pesticides, and ecotoxicological research has concentrated on finding suitable terrestrial test species that could assess and measure the adverse effects of these contaminants. Van Gestel & van Straalen (1994) present a number of criteria used for selecting test species. The most important are the function or role that the test species plays in the terrestrial ecosystem, the taxonomic representativeness of the test species, and the route of exposure.

The OECD has a representative artificial soil recommended for tests with soil organisms. Acute toxicity test methods using earthworms (*Eisenia fetida*) have been available from the USEPA (see Table 3.5) and the OECD for several years. No standardized test methods are currently available for other soil invertebrates, although promising species candidates include some nematodes, enchytraeid worms, molluscs, isopods, millipeds, orbatid mites and springtails.

Standardized test methods also exist for a few beneficial insects such as honeybees. However, additional standardized test methods are needed, because these are insects that improve the production of agricultural products by serving as pollinators (e.g. honeybees, discussed earlier) and predators or parasites of pest species (Van Leeuwen 1995). Predatory mites, spiders, parasitic Hymenoptera and various beetles are examples of the latter.

Avian, mammalian, plant and terrestrial microcosm/field tests have been discussed earlier in this chapter.

Sediment testing scheme (preliminary) for determining sediment effects

Industrial chemicals and pesticides entering aquatic environments commonly affect plants and animals occurring in the water column. However, substances with certain physical–chemical characteristics (e.g. they have low water solubility, are persistent and bioaccumulative) partition to sediments, accumulate there, and are potentially toxic to the wide variety of benthic organisms found in the sediments. Sediment toxicity problems in North America were neglected until the late 1970s (Burton 1991). Interest at that time centred on dredging and disposal of sediments and the impact of contaminated sediments. The role of sediments in influencing the fate and effects of chemicals found in aquatic systems is also an important interest area (Dickson *et al.* 1987). Now there is widespread and increasing interest in these topics throughout the world since ultimately, sediments serve as a sink or repository for substances (Cairns *et al.* 1984; Giesy & Hoke 1991). Chemicals found there are sorbed to particulate matter or occur in solution in sediment pore water or interstitial water. These chemicals may be transferred from the sediments directly to benthic organisms (Adams *et al.* 1992).

Interest in sediment toxicity assessment has also increased in Europe recently. Workshops on this topic were held in 1991 and 1993, by the Hazard Assessment Advisory Board of OECD, and SETAC-Europe, respectively (Hill *et al.* 1993; OECD 1993). Important discussions of sediment quality and toxicity assessment can be found in Burton (1992), Reynoldson & Day (1993) and Burton & MacPherson (1995).

Three types of sediment testing. Sediment toxicity testing and assessment can be divided conveniently into three types that are relevant to assessing the effects of industrial chemicals and pesticides to sediment organisms, out of the 11 methods available currently to assess the quality of sediments (USEPA 1989b).

First, there are laboratory tests (bulk sediment toxicity tests) conducted on contaminated sediments sampled and collected in the field. These contaminated sediments may contain a variety of known and unknown substances, bioavailable or not bioavailable depending on the physical–chemical conditions in the sediments. After the sediments are collected and taken to the laboratory, the physical–chemical condition of the sediments may change such that a true picture of the actual toxicity of the field sediments is difficult, if not impossible, to obtain by testing them in the laboratory. However, this method provides a direct measure of toxicity and an integrated assessment of toxicity from all the chemicals present in the contaminated sediments.

The second method, which avoids these problems, is spiked-sediment testing. Here, natural or artificial (formulated) sediments are made up with known physical–chemical characteristics, for example, percentage organic carbon, total metals, particle size and total volatile solids (examples are given in Naylor & Rodrigues 1995; Suedel *et al.* 1996). Chemicals and pesticides are introduced in known quantities into the test sediments, mixed, and used in tests to determine the toxic effects of these substances on various test organisms. Here, the problem, as with water column tests, is extrapolating laboratory results to the field, but toxicity will be related directly to the industrial chemical or pesticide tested.

A third method is to measure the toxicity of chemicals that have partitioned in the interstitial or pore water extracted from the sediments (Adams *et al.* 1992). Acute and chronic tests are conducted. It is assumed in this method that the unbound or soluble portion of the industrial chemical or pesticide is the most bioavailable and is directly responsible for toxic effects on sediment organisms. A variation on the pore-water method is the liquid phase elutriate test (Nebeker *et al.* 1984). Here, water is mixed with contaminated sediments and separated. The water portion is used later to test its toxicity to organisms.

Preliminary tiered sediment-testing scheme (industrial chemicals). The preliminary tiered sediment-testing scheme (Fig. 3.4) is organized similarly to the environmental effects water column scheme discussed previously. There are four tiers, each with several spiked-sediment and pore-water or elutriate toxicity tests. Some of the tests are currently in the development and validation process (e.g. chironomid chronic emergence test and freshwater amphipod chronic test); others are identical to water-column tests (e.g. fish acute toxicity test on Tier I). Still others, such as the freshwater chironomid acute toxicity test, the freshwater amphipod acute test and the estuarine/marine amphipod acute toxicity test, represent the current state-of-the-art in sediment-testing methods and have been conducted successfully for several years (USEPA 1994c,d).

Water-column tests are included in the sediment scheme so that both schemes can be 'tied together' or related because water exposure is common to both. Pore-water exposure and bioavailability may be significant in sediments (see Di Toro *et al.* 1991). Chemicals may partition from sediments into the overlying water column, possibly causing toxicity to organisms found there. Bioaccumulation and bioconcentration tests are found in the sediment scheme (Lee *et al.* 1993; USEPA 1994d). Additional toxicity tests with other species, such as marine polychaete annelids, and estuarine/marine molluscs and echinoderms, may be added later. For most chemicals and pesticides that will partition to the sediments, some water-column tests will be conducted to serve as a means of exposure verification and as a validation cross-check.

Guidance is also available for assessing the effects of pesticides on sediment organisms. However, this guidance has not yet been formalized into a tiered scheme. Sediment tests used for pesticides are identical to those used for industrial chemicals.

Decision criteria. Decision criteria or testing triggers, with the same purpose as those used in the environmental effects testing scheme, have not yet been finalized in the sediment scheme. There are several ways of developing such criteria. First, the same criteria could be used in both schemes, especially if the same tests are conducted in each scheme. Another method is to convert water-column values directly to sediment values. For example, a water-column acute value of $\leq 1\,mg\,l^{-1}$ is indicative of high acute toxicity and is a trigger for requiring additional tests. This value could be converted to $\leq 1\,mg\,kg^{-1}$ (dry weight) for sediment tests and used in a similar manner. Confirmatory testing should be conducted to

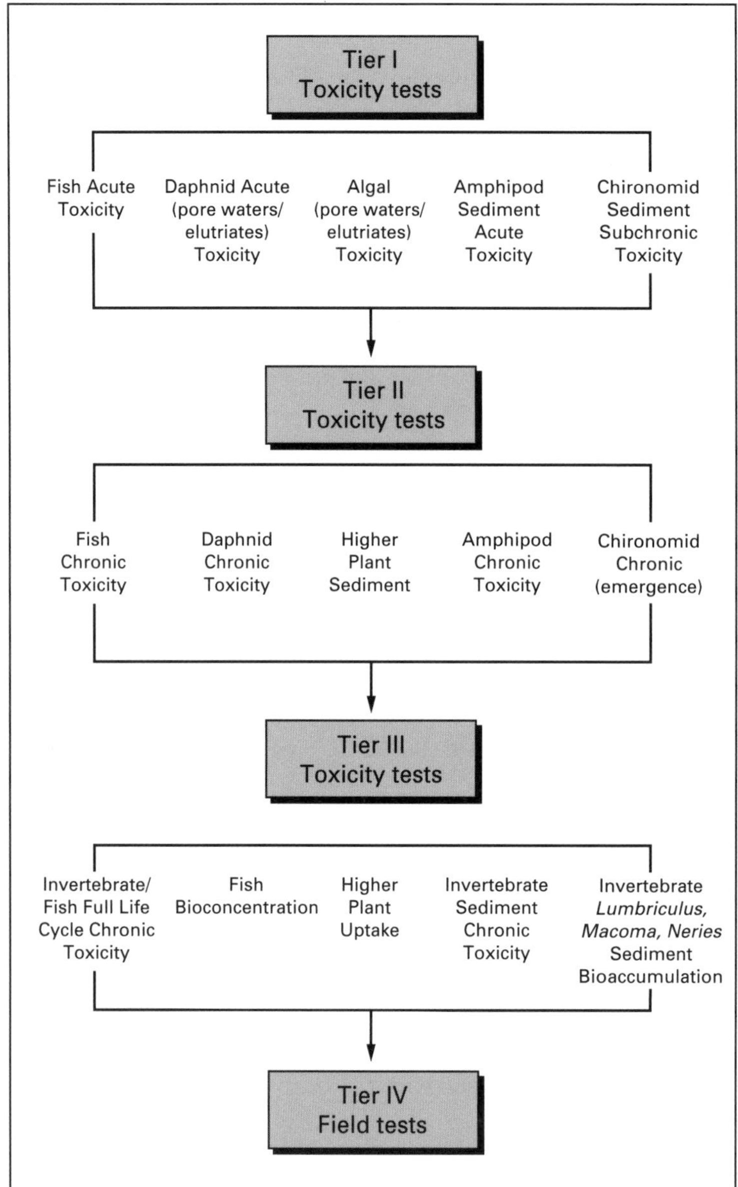

Fig. 3.4 Preliminary testing scheme for determining sediment effects of industrial chemicals.

determine if this conversion procedure is correct. A third way is to develop entirely new decision criteria, especially for completely new tests, for example, a higher plant sediment test. Many physical–chemical characteristics of the industrial chemical or pesticide tested could be used as the same triggers in both schemes.

Assessment and management of chemicals in sediments. Different alternatives can be developed and used to assess the hazard and risk of industrial chemicals and pesticides partitioning to sediments. Initially, an inventory at contaminated field sites near chemical production, use and disposal sites can be established to identify those industrial chemicals and pesticides present in sufficient quantities that may be toxic to sediment-dwelling organisms. These substances can then be tested by using tests in the scheme, and the extent and degree of their hazard could be determined. A list could be maintained of all chemicals and pesticides regulated under the applicable chemical control laws and those identified that exceed toxic threshold concentration levels at the field sites.

The introduction to commerce of new chemicals that would not partition to sediments could be encouraged through a dialogue with chemical manufacturers. New chemicals could be designed with characteristics that would prevent adsorption through biological membranes, with large cross-sectional diameters to prevent movement through cell membranes, and with functional groups embedded within the molecule to enhance rapid transformation to low-toxicity products (USEPA 1994b).

In the USA, the FIFRA pesticide law gives the EPA the authority to restrict or ban the use of pesticides that have the potential to contaminate sediments, if the risks to non-target organisms are judged to be unreasonable (USEPA 1994b). Currently, sediment toxicity problems are not addressed routinely in risk assessments for pesticide registration, reregistration and special review. There are plans to develop criteria for pesticide residues in sediments that can be used as a screening tool for the determination of 'safer' pesticides. Those placed in this category would be registered quicker and possibly replace those pesticides that are more harmful to plants and animals. For evaluating the hazard of pesticides in sediments by the USEPA it has been proposed to compare EECs in sediments to the NOEC, based on the most sensitive end-point (e.g. survival, growth), which would be determined from 10–20-day sediment toxicity tests. This proposed quotient method is a useful starting point in assessing hazard to sediment organisms but the variability in exposure concentrations and sediment-test species sensitivities must be better accounted for by conducting comparative toxicity studies and searching for new, promising sediment-test species. The quotient method of determining risk is discussed in more detail in a later section.

Test guidelines

Guidelines, test methods or test protocols, serve to support the testing schemes and are a means of developing, requesting, and receiving reliable and scientifically valid test data. Results are used to assess the hazard and risk of industrial chemicals and pesticides.

In general, the test method should be accepted by, or in the process of being accepted, by the scientific community. A detailed description of the method is either published in an independent, peer reviewed vehicle, or there is other evidence indicating an independent, unbiased scientific review of the method has been conducted. Repeatability and reproducibility of the test method should be demonstrated within and between laboratories by using ring or round-robin testing. This will provide an important indication of intra- and interlaboratory variability. The performance of the test method is demonstrated by using coded reference substances. If the test method is a substitute for an existing method, sufficient data should be provided to compare the performance of the proposed test with that of the test to be replaced or supplemented. Also, the test method must be capable of harmonization with other methods from national or international agencies or groups. Thus, standardized test methods can result from following these recommendations. Soares & Calow (1993b) have argued for the importance of standardization in the development of tests and that this should take precedence over realism in testing.

The basis for adequately completing tests in any testing scheme are test guidelines. There is a set of 13 aquatic and seven terrestrial US guidelines for industrial chemicals that are used to conduct acute, chronic and bioconcentration tests at various tiers in the testing scheme (Table 3.5). They were subjected to peer review and first published in 1985 (USEPA 1985d). They are periodically revised and updated and they have been published annually in the US Code of Federal Regulations (CFR), Title 40, parts 795 and 797 (USEPA 1987a). Some of the guidelines may be applicable to only one tier; others can be used on more than one tier. For others there may be a question of which tier(s) they would best be placed in and used. Mixed media guidelines, consisting of microcosm and microbial tests were proposed in the *Federal Register* in 1987. They are available for use but have not yet been fully integrated into the industrial chemical testing scheme (USEPA 1987b). Decision criteria need to be developed to establish when and how such tests would be triggered. Other guidelines were developed and written for specific regulatory chemical testing actions; these guidelines are provisional in nature and were also published in the *Federal Register* (Table 3.5).

There are no formal pesticide guidelines published in the CFR. Instead there currently are guidance documents, called Standard Evaluation Procedures (SEPs), which outline procedures for conducting specific acute and chronic effects tests (USEPA 1994e). There are guidance documents on conducting terrestrial field studies and ecological risk assessment (Urban & Cook 1986). Also there is a series of assessment guidelines on hazard evaluations for wildlife/aquatic organisms (subdivision E), non-target plants (subdivision J), and non-target insects (subdivision L) (USEPA 1994e).

Over the past several years the USEPA has been involved in continuing efforts to develop common or harmonized ecological effects test guidelines. These can be used to assess the toxicity of both industrial chemicals and pesticides (USEPA 1996). Duplication of testing and expenditure of unnecessary resources will thus be minimized, as identical guidelines will be used, except for those few guidelines that are specific or unique for an industrial chemical or a pesticide. Efforts have also begun, or are in progress, to harmonize several USEPA guidelines with similar OECD test guidelines.

Other EPA guidelines. The EPA has previously published a number of guidelines, some of which are mentioned here and may be useful to hazard assessors. The original USEPA publication (1975) describes acute toxicity methods with fish, macroinvertebrates, and amphibians, and serves as a basis for many other test guidelines (USEPA 1975; Zeeman 1997). The EPA has a Biological Advisory Committee that has produced a series of method manuals for measuring the acute toxicity of effluents and receiving waters to freshwater and marine organisms (USEPA 1993b), and short-term methods for estimating chronic toxicity to freshwater organisms (USEPA 1994f), and to marine and estuarine organisms (USEPA 1994g). There is also a manual on evaluating laboratories performing aquatic toxicity tests (USEPA 1991a), and good laboratory practices, which must be followed by testing laboratories, are described in the US Code of Regulations, Title 40, part 792, published annually.

Other EPA manuals, in addition to USEPA (1989b), are concerned with methods for measuring the toxicity and bioaccumulation of sediment-associated contaminants with freshwater invertebrates (USEPA 1994d), estuarine and marine amphipods (USEPA 1994c), and with general toxicity tests using marine organisms (USEPA 1978).

ASTM methods. The American Society for Testing and Materials (ASTM), Philadelphia, PA, USA, has been involved for a number of years in the development, standardization and approval of biological methods by a consensus of industry, testing facilities and government agencies. A series of annual symposia volumes on aquatic toxicology hazard assessment, environmental toxicology and risk assessment have been published since 1976. Descriptions of many test methods commonly used were first published here. Other volumes on specific topics important in ecotoxicology have been published (see Cairns & Dickson 1972; Cairns *et al.* 1976, 1978; Buikema & Cairns 1980; Boyle 1985; Cairns 1986a; Cairns & Pratt 1989).

The ASTM also publishes an annual collection (volume 11.05, section 11) of ASTM standards on water and environmental technology that includes biological and ecotoxicological test methods, practices and guides, which have undergone rigorous peer review (ASTM 1996). A complete set of acute, chronic, sediment and plant test methods, current as of 1993, is also available (ASTM 1993).

OECD methods. The OECD Guidelines for Testing of Chemicals were first published in 1981 and now number over 70 guidelines (Grandy 1995). Some of the ecotoxicity guidelines were updated in 1984 and 1992; currently there are 11 ecotoxicity guidelines (including an activated sludge test) (OECD 1984). Work continues within the Chemical Programme of OECD to develop harmonized test methods that are scientifically valid, accepted internationally, and that will avoid duplication and unnecessary testing. Grandy (1995) presents information on the structure and content of OECD guidelines and on the procedure for revising existing guidelines or proposing new ones.

Other important sources. Useful recent, comprehensive treatments on toxicity test methods and guidelines include those by Calow (1993a) and Rand (1995), and Standard Methods of the American Public Health Association, etc. (APHA *et al.* 1995). General references include Scherer (1979), USFDA (1987), Soares & Calow (1993a), Cairns & Niederlehner (1994), Menzer *et al.* (1994), Hoffman *et al.* (1995), and van Leeuwen & Hermans (1995).

Sediment-test methods and guidelines. This field is relatively new, thus there are fewer references that describe freshwater, estuarine and marine sediment methods. Specific guidance on various test methods is given in Lee *et al.* (1993), USEPA (1994c,d), and ASTM (1996). Information on conducting sediment tests is given in Dickson *et al.* (1987), Burton (1992), Burton & Scott (1992), and Hill *et al.* (1993). Sediment methods that use a diversity of organisms are reviewed in Nebeker *et al.* (1984), Giesy & Hoke (1989), Burton (1991), and Ingersoll *et al.* (1995).

Validation of tests

Any tests used to assess the hazard and risk of industrial chemicals and pesticides must be credible and of the highest quality attainable within reasonable limits imposed by available resources. This requirement is very important, because data from tests often form the basis for regulatory decisions that can have widespread economic consequences. Extensive preplanning is necessary to complete efficiently the ecotoxicity tests necessary or required for approval of the chemical substance. Reputable test data from valid, state-of-the-art methods are needed to establish that any conclusions reached, or actions taken, based on these data, are able to withstand scrutiny, possible legal challenges, or a later revision of decisions that were made earlier. By conforming to accepted and approved test guidelines there is a higher probability that the test results will be valid and variability may be lessened.

The term validation is broadly defined and has been used previously in several ways (Cairns & Dickson 1995). One is use of the term as a confirmation of predictions of responses in natural systems to potentially toxic chemicals (Cairns 1986c). These confirmations are based on test data obtained in the laboratory. The validation process has yet to be stated explicitly in these predictions and expert judgement is often needed (Cairns 1986c, 1988). Some do not even insist on field validation when it is not possible or desirable (Chapman 1995a).

Another use of the term is to determine the 'scientific soundness' of laboratory toxicity test methods and generated test data. Furthermore, the term has been used to describe the process by which the reliability and relevence of a test system is evaluated for the purpose of supporting a specific use (*ad hoc* US Interagency Coordinating Committee on the Validation of Alternative Methods, September 1995, draft). To avoid semantic confusion this term is restricted to laboratory data and the approval of new and revised test methods.

There are two major concerns when validating test methods or data (Smrchek *et al.* 1993). First, the accuracy of test data generated is important in estimating the inherent toxicity of an industrial chemical or pesticide. Inherent toxicity is defined as a toxic effect, caused by a substance, which has not been compromised, masked, or mitigated in some way by one or more factors or variables. Dilution water or test media used in a test may have contaminants that will obscure, or alter in some way (e.g. antagonistically, or synergistically) the toxic effect of the industrial chemical or pesticide tested. This may result in an over- or underestimation of the toxic effect and may have significant regulatory and economic consequences.

Second, the precision and repeatability of the test results is a reflection of the test method used. Repeating a valid test, under 'similar' test conditions, should consistently yield 'similar' test results, or test values in a similar range (e.g. ideally within a factor of two or less) time after time (Chapman 1995b).

Eight major areas are considered in determining the validity of an aquatic ecotoxicity test and the test results generated (Table 3.4). For further discussion on this topic, see Smrchek *et al.* (1993). If a valid test guideline is followed closely and a rare event does not occur during the test (e.g. electrical power disruption, a test chamber breaks), valid test results should be obtained.

Surrogate species and the most sensitive species

It is clearly impossible to test all, some, or more than a selected few of the numerous species occurring in an environment or ecosystem impacted by the production, use or disposal of an industrial chemical or pesticide. Test species therefore serve as 'surrogates' or substitutes for the large number of potentially impacted species. Debate in the past has centred on whether a few surrogate species from single species laboratory tests are capable of realistically and accurately providing predictions of effects on natural ecosystems (Tebo 1986; Cairns *et al.* 1992). Although this problem cannot be resolved here, current USEPA (and OECD) policy is to continue to use single species laboratory tests. The use of uncertainty or assessment factors (see below) attempts to minimize or lessen these difficulties in using single-species test.

As emphasis on protecting whole natural ecosystems continues to increase, it will be necessary to shift to microcosm-, multi-species, community- and ecosystem-level testing, or at least accommodate these tests in existing testing schemes. Thus, a solution to this problem may be at hand.

The number of species tested in the laboratory should define the range of response for specific subsets of the total environment (Smrchek *et al.* 1993). Based on years of assessing the toxic effects of substances, one or more of these test species will be as sensitive or more sensitive to the substance than the other test species. The USEPA testing schemes have test organisms from the major phylogenetic categories and major habitats. The species that are tested represent broad groups of organisms. A bluegill (*Lepomis macrochirus*), for example, serves as a surrogate for warm-water fish, and a mysid shrimp (*Americamysis* = *Mysidopsis bahia*) substitutes for marine or estuarine invertebrates. Sprague and Fogels (1977) consider 'a fish is a fish', and based on available data, other errors involved in extrapolating from the laboratory to the environment are greater than differences between species (Sprague 1970; Tebo 1986).

The myth of searching for the most sensitive species to all or most chemicals has been well demonstrated, and protecting the most sensitive test species will not necessarily protect all other species (Cairns 1986b,c; Cairns & Pratt 1993). As has been pointed out, it is indeed a dangerous assumption to equate the most sensitive species in a series of laboratory tests to the most sensitive species in natural ecosystems (Cairns & Niederlehner 1987). This problem is lessened by using surrogate species that attempt to duplicate the range of sensitivity or toxic response of other aquatic and terrestrial species and by using uncertainty factors to partially compensate for these differences.

Uncertainty and the use of extrapolation factors

Uncertainty has been another problem of great concern to ecotoxicologists (see, e.g., Slooff *et al.* 1986; Greig-Smith 1992; Calabrese & Baldwin 1993; Suter 1995). A residual amount of uncertainty will remain in any evaluation, however, even though it may be reduced by science (Levin *et al.* 1989). There is a tendency for regulatory authorities to expect deterministic predictions from ecotoxicologists or a simple, bottom-line single number that can be used as an indication of hazard or risk and as the basis for making management or policy decisions. Any uncertainty or 'hedging' is regarded as the result of inferior methods or due to a failure in analysis by ecotoxicologists (Reckhow 1994). Some residual uncertainty, however, will always remain because we are dealing with variable biological systems which sometimes act randomly. Thus there will always be uncertainty associated with predictions relating to the dynamics of natural ecosystems and extrapolations made from biological level of organization to another.

One important aspect of uncertainty (Forbes & Forbes 1994) is to avoid false negative results (type II error in statistical hypothesis testing), which consider the industrial chemical or pesticide to be 'safe', or with no identified adverse effects, and acceptable for production, use or disposal, when in reality it will be toxic to non-target organisms. Error can never be absent totally in hazard/risk assessment, only minimized. For protection of the diverse and variable 'environment', it is often desirable to have a slight bias toward false positives (type I error in hypothesis testing), that is restricting, controlling or even banning the production, use or disposal of a substance when it is actually not toxic, or less toxic, at a certain concentration or application rate. Thus these measures may not be necessary, but it may be toxic, however, at a higher production, use or disposal rate or level. Because of the variability of constantly changing biological systems and inherent error in laboratory toxicity testing it will be difficult to eliminate false negatives or false positives completely.

Sources of uncertainty. Ecotoxicology uncertainty may result from three main sources of variability. See USEPA (1984, 1991b). Calabrese & Baldwin (1993), and Persoone & Janssen (1994) for a more detailed discussion of sources of variability.

1 Variability due to the range of sensitivities of aquatic and terrestrial species to industrial chemicals and pesticides (intra- and intertaxa variations), also including variability in life stage and in test conditions.

2 Variability due to estimation of chronic effect levels from acute test data.

3 Variability due to extrapolations from the laboratory to the field or to natural ecosystems.

Uncertainty, with respect to determining a biological no-effect concentration, may decrease as more tests are completed, and as a chemical 'moves' through a testing scheme, i.e. acute tests at lower tiers often have more uncertainty and wider confidence limits than chronic tests at higher tiers (see Fig. 1, p. 195, in Cairns *et al.* 1978; Cairns *et al.* 1992). If, however, the testing facility is experienced in performing acute tests and completes large numbers of tests, there actually may be less uncertainty and narrower confidence limits than higher tier tests. As testing moves from chronic tests to field tests, uncertainty progressively decreases (even while variability may increase), confidence limits narrow, reliability of toxicity estimates increases, and the precision of the decision also increases (Lee & Jones 1980).

Uncertainty factors. Various factors have been developed to attempt to account for the uncertainty from the sources listed above; these have been an active area of discussion among ecotoxicologists. These factors are called 'uncertainty factors', 'assessment factors', or 'application factors' (see, e.g., OECD 1989, 1995a,b; Nabholz 1991; ECETOC 1993b; UK Department of the Environment 1993; Smrchek *et al.* 1993; EC 1994; Zeeman 1995a,b; Zeeman & Gilford 1993).

Uncertainty factors are, strictly speaking, not 'safety factors', or 'margins of safety' as used by Mount (1981), Cairns & Niederlehner (1987) or Reckhow (1994). Although the use of the term 'safety' has been used for a number of years in aquatic toxicology, it is a value-laden term that is misleading; it gives a false sense of security, and has little or no scientific meaning. It is impossible to assure that for a substance at a certain concentration, a factor can be used that gives a concentration value that is 'safe' to all organisms in the environment, or to some percentage of organisms (e.g. 95%). There probably are multiple 'safe' levels depending on the substance and on the tolerance and sensitivity of each organism. It is impossible, even with adequate resources, to determine an accurate 'safe' level for more than a few of the millions of organisms found in various ecosystems.

We prefer the term uncertainty factor (UF). The term is also used with respect to human health risk assessment in the USA, and it is defined as a factor used in operationally deriving the RfD (an estimate of daily exposure to a hazard, likely to be without risk of deleterious effects during a lifetime) from experimental data. The UF is intended to account for variation in sensitivity among humans, the uncertainty in extrapolating animal data to humans, the uncertainty in extrapolating from data obtained in a study that is less than lifetime exposure, and the uncertainty in using lowest-observed-adverse-effect-level data (LOAEL) instead of no-observed-adverse-effect-level data (NOAEL) (USHHS 1992).

Uncertainty factors are multiples of 10 usually (or sometimes of five or less), and are used to set exposure levels that are presumed to result in little or no adverse effects on populations of organisms occurring in the environment. The UF used decreases as the amount and complexity of the toxicity test information available increases (i.e. uncertainty of laboratory to field predictions decreases). If an existing chemical, for example, has only base-set toxicity test data available, a UF of 100 would be applied to the lowest measurement end-point (calculated from the most sensitive species tested) by dividing this factor into the end-point (e.g. a daphnid acute test value is 1.0 mg l^{-1}, divided by a UF of 100, results in an extrapolated final value of 0.01 mg l^{-1}). Similarly, a UF of 10 would be applied to the lowest chronic value and a UF of 1 would be applied to actual field test data (Smrchek *et al.* 1993; Zeeman & Gilford 1993; Zeeman 1995a,b). A UF of 1000 would be used if there were only very limited test data available; i.e. one or two acute test values, a single QSAR or an analogue chemical value.

The extrapolated value just mentioned is called a level of concern (LOC), concentration of concern (COC), or concern concentration (CC), but never a 'safe level'. These values are similar to the predicted no-effect concentration (PNEC) and Environmental Concern Level (ECL) used by European countries; all are levels, which if met or exceeded by predicted (PEC), expected (EEC), calculated or actual exposure concentrations, could result, or have a high probability of resulting, in adverse effects to aquatic or terrestrial species or both, from the production, use or disposal of an industrial chemical or pesticide.

Other extrapolation methods such as the Hazardous Concentration (HC_5), and the Final Chronic Value (FCV) methods are not considered here (see OECD 1995a). Both methods use the toxicity test data for all species tested to derive a maximum tolerable concentration (MTC) in contrast to the uncertainty/assessment factor approach, which uses the lowest (or assumed worst-case) toxicity value.

Assessment factors. In Europe the term assessment factor is used interchangeably with uncertainty factor (see, e.g., OECD 1989, 1995a,b; Nabholz 1991; UK Department of the Environment 1993; EC 1994). This is acceptable as long as the definition and use of the term is made clear so that any semantic confusion is minimized.

In the USA, the term assessment factor originally was developed and applied only to TSCA new industrial chemicals (USEPA 1984). Very often these chemicals have no or little test data, or the test data that are available are unreliable, or cannot be validated (Nabholz *et al.* 1993a; Smrchek *et al.* 1993; Zeeman 1995a,b; Zeeman & Gilford 1993). There is no requirement under TSCA that test data be submitted for every new chemical notification. To compensate for these lack of data, QSARs were developed and analogue chemical comparisons are also often used. Assessment factors are used to derive LOCs, and to address the sources of uncertainty listed previously. These LOCs also serve as triggers for obtaining initial ecotoxicity testing or additional testing. For example, if only one fish acute test was conducted with a new chemical and a 96 h LC_{50} of 10 mg l^{-1} was calculated, this value would indicate moderate acute aquatic toxicity, based on criteria used to rank concern. The test end-point would then be divided by an appropriate assessment factor (1000), and would result in an LOC of 0.010 mg l^{-1}. This LOC value could then be used as a trigger to require additional acute testing or chronic testing if the PECs are expected to exceed this concentration for a sufficient duration. Additional testing would not be recommended if the PEC was shown to be lower than the LOC.

Utility of uncertainty/assessment factors. Objections have been raised on the use of these factors (Zeeman 1995a). Some have argued that these factors are suitable only for use in a preliminary effects assessment, and that a more involved extrapolation method using statistical models (with acute and chronic data from several test species) is needed for completing a detailed effects assessment that would be protective of 95% of species occurring in ecosystems. The factors are viewed as too simple, with little or no scientific or theoretical foundation (Okkerman *et al.* 1991,

1993; van den Berg 1992/1993; Balk *et al.* 1993; Emans *et al.* 1993). However, when the factors used for industrial chemicals by USEPA are compared with the alternative statistical method neither was clearly better than the other (Okkerman *et al.* 1993). Moreover, Calabrese & Baldwin (1993) and Forbes & Forbes (1994) found that the statistical methods were neither more accurate nor more conservative than the simple assessment factor approach. The more complex statistical methods should be used only if they are clearly more useful and predictive, but this has not yet been shown (also see Belanger 1994). Thus, the simple uncertainty/assessment factor approach remains a useful and effective tool in addressing sources of ecotoxicological variability and uncertainty.

SAR/(Q)SAR development and use

Zeeman (1995a) describes the USA process of TSCA new chemical review and approval. Ecotoxicity test data (usually aquatic) for new chemicals are rarely submitted to USEPA. In fact, out of 15541 new chemical submissions to USEPA, from 1986 through to 1992, only 799 (average of 5.1%) had ecotoxicity data. These data gaps create problems in trying to rapidly assess the hazard and risk of new chemicals within the 90-day statutory time limit. Thus, it was necessary to develop methods, after the passage of TSCA, which could estimate the aquatic toxicity (because most releases occur in aquatic environments) and predict the ecotoxicity of new chemicals.

The SAR/(Q) SAR methods were developed to meet this need (USEPA, 1988; Auer *et al.* 1990, 1994; Nabholz 1991; Clements *et al.* 1993a,b; Nabholz *et al.* 1993a,b; Zeeman & Gilford, 1993; Zeeman *et al.* 1995). The references listed here should be consulted for more details on this topic. The SAR abbreviation has been used interchangeably with quantitative structure–activity relationships, (Q) SARs or QSARs, by some (Clements *et al.* 1993a); others have used the abbreviations separately. Here we consider SARs the same as (Q) SARs or QSARs, and will use the latter abbreviation.

Description of QSARs. It has been known for many years that there was a relationship between chemical structure, chemical properties and a specific effect. These are known as structure–activity relationships or SARs. SARs referred originally to relationships having to do only with chemical analogues.

Certain physical–chemical properties were investigated for the development of QSARs, and they included octanol/water partition coefficient, molecular weight, chemical reactivity, degree of ionizable functional group (p*Ka*), hydrophilicity, and hydrophobicity. The first two properties have been the major attributes used to correlate structure and toxic effects, and the most frequently used relationship has been the logarithm of the octanol/water partition coefficient versus the logarithm of the median acute toxicity (LC or EC_{50}) value. A regression equation relating the toxicity of a set of similar chemicals to chemical property is calculated.

Quantified structure–activity relationships must be based on valid toxicity test data, thus work has proceeded on testing various classes of industrial compounds by determining the 96h LC_{50} value of a common test organism, the fathead minnow (*Pimephales promelas*) and using this to develop QSARs. Also, the literature is searched for toxicity test data that could be used to develop or add to QSARs. To date, 140 QSARs have been developed for use in predicting the acute and chronic toxicity of many classes and subclasses of chemicals to fish, aquatic invertebrates (daphnids), and algae (Table 3.6). These QSARs are presented in an OPPT manual (USEPA 1994a) and are used to estimate the aquatic toxicity of more than 40 classes or subclasses of chemicals to at least one surrogate aquatic species (e.g. fish, aquatic invertebrates or algae, or combinations of these). A version of the QSAR manual suitable for use with a personal computer, known as ECOSAR, is also available (USEPA 1993a, 1994a). The development, updating, and use of a QSAR is a continuing process.

Hazard assessment with QSARs. Quantitative structure–activity relationships have been used to derive toxicity values to estimate the hazard of both new and existing chemicals that lack test data and to predict the toxicity of chemicals without test data or with test data that were equivocal or suspect. When QSARs are used to predict the toxicity of new chemicals, this can lead to the recommendation for testing or additional testing (Clements *et al.* 1995). After receipt of the completed toxicity test and validation of the test, the measurement end-points are compared with the QSAR predicted values. If the measurement end-points are within the acceptable range (within one order of magnitude) of the QSAR, the end-point values are added to the data used as the basis for the QSAR and it is recalculated. The QSAR will now be based on an increased number of observations. For optimal predictions, it is always preferable to have a QSAR based on many valid toxicity values, than one based on a few values. If the validated test end-points fall outside of the acceptable bounds, a new QSAR may need to be developed.

Also, large sets of existing chemicals can be screened and ranked into categories of high, medium or low acute and chronic toxicity to fish, daphnids and algae (Clements *et al.* 1993b). Over 8000 discrete (non-polymer) organic chemicals from the TSCA Inventory of over 72 000 chemicals were searched and ranked for acute and chronic toxicity. After examining the cumulative aquatic toxicity distributions of these chemicals, it was found that about 10–20% of these chemicals could have high acute toxicity to aquatic species (LC_{50} or $EC_{50} \leq 1 mg l^{-1}$) and about 15–18% could have high chronic toxicity (chronic values $\leq 0.1 mg l^{-1}$) (Zeeman 1995a).

These methods have worked well when assessing the potential toxicity and hazard of SIDS chemicals (Zeeman *et al.* 1995). Often for some of these chemicals there are gaps in the available ecotoxicity data that can be filled by QSAR generated values. When test data are available and it is not possible to validate the tests by evaluating the original studies, QSAR analyses can confirm or cross-check that toxicity test end-points calculated in these studies are reasonably correct or that there are problems with these studies. A LOC can be derived, for example, by dividing acute QSAR value(s) by UFs of 1000 or 100, and chronic values by 10.

If a chemical is a member of a particular class or group of chemicals that does not have a QSAR developed (Table 3.6) then other methods are used. Searches are undertaken to obtain toxicity data on chemical analogues or from chemical classes that are closest to the chemical and which may or may not have a QSAR developed.

In summary, the methods preferred for assessing the hazard of chemicals (in descending order) are (Zeeman 1995a) as follows.

1 Measured ecotoxicity test data.
2 QSAR predictions.
3 Nearest chemical analogue data.
4 Chemical class data.

How closely do QSAR predictions match with chemicals that have been tested for ecotoxicity? Validated test data from 462 chemicals and 920 individual acute and chronic tests were compared with their QSAR predicted values (Zeeman *et al.* 1995). Eighty-five per cent of the QSAR predictions were within an order of magnitude of the actual test data. There was an overprediction of toxicity (false positives) in 9%, and an underprediction of 6% (false negatives). Thus, the high degree of agreement determined for these predicted values when compared with actual aquatic toxicity test values indicates that SARs/(Q)SARs can be very useful in estimations of aquatic toxicity.

QSAR activities in Europe. Ecotoxicologists in Europe are increasingly using QSAR techniques. Guidance is given in OECD (1992b), EC (1994), and in Pedersen *et al.* (1994) for deriving QSAR relationships when measured toxicity data are not available.

The USEPA (OPPT) has been involved in the International Trilateral Project on High Production Volume Chemicals since 1993. Quantitative structure–activity relationship estimations have been provided for over 250 HPV chemicals and over 600 predictions for aquatic acute and chronic effects were supplied. So far, ecotoxicity assessments provided by OPPT have shown a strong correlation with those of the EU.

Results of SARs used by the USEPA were compared with the minimum pre-marketing data set (MPD) used by the EU, in a joint study (Zeeman *et al.* 1995). This comparison was undertaken in 1991–1993 for 175 chemicals submitted to and selected by the EU (Auer *et al.* 1994). Zeeman *et al.* (1995) should be consulted for a more detailed discussion of the results of this comparison. Briefly, for fish and daphnid ecotoxicity comparisons there

Table 3.6 Chemical classes and subclasses with available quantitative structure–activity relationships (QSARs) that can be used to predict ecotoxicity. (From Clements *et al.* 1995.)

Chemical class	Acute toxicity			Chronic toxicity		
	Fish	Daphnid	Algae	Fish	Daphnid	Algae
Acids	X	X	X	X	X	X
Acid chlorides	X					
Acrylates	X	X	X	X		
Acrylates, methacrylates	X					
Aldehydes	X	X	X	X		X
Amines, aliphatic	X	X	X			X
Anilines	X	X		X	X	X
Anilines, amino, *meta* or 1,3-substituted	X	X	X			
Anilines, amino, *ortho* or 1,2-substituted	X	X	X			
Anilines, amino, *para* or 1,4-substituted	X	X	X		X	
Anilines, dinitroanilines	X	X		X		
Benzotriazoles	X	X	X			
Diazoniums, aromatic	X					
Dyes, acid, Al chelate	X					
Dyes, acid, Co chelate	X	X				
Dyes, acid, Cr chelate	X	X	X	X	X	X
Dyes, acid, ≥S3	X	X	X	X	X	X
Epoxides, aziridines	X	X	X			
Epoxides, diepoxides	X	X				
Epoxides, monoepoxides	X	X				
Esters	X	X	X			X
Esters, diesters, aliphatic		X				
Esters, monoesters, aliphatic		X				
Esters, phosphate	X					
Esters, phthalate	X	X			X	
Hydrazines	X	X	X			
Hydrazines, semicarbazides, aryl, *meta/para* substituted			X			
Hydrazines, semicarbazides, aryl, *ortho* substituted			X			
Imides	X					
Ketones, diketones, aliphatic	X	X			X	X
Malononitriles	X					
Neutral organics	X	X	X	X	X	X
Nitrobenzenes, dinitrobenzenes	X	X	X	X	X	X
Peroxy acids	X	X				
Phenols	X	X	X	X	X	X
Phenols, dinitrophenols	X	X		X	X	
Polymers, polycationic	X	X	X			
Propargyl alcohols	X					
Semicarbazides, alkyl, substituted			X			
Surfactants, anionic	X	X	X	X	X	
Surfactants, cationic, ethomeen	X	X	X			
Surfactants, cationic, quaternary ammonium	X	X				
Surfactants, dialkyl, quaternary ammonium	X	X	X	X	X	X
Surfactants, non-ionic	X	X				
Thiazolinones, *iso*	X	X	X			X
Thiols (mercaptans)	X	X				
Ureas, substituted			X			
Total (140)	41	35	25	13	13	13

was 87% and 79% 'agreement' respectively, between USEPA SAR predictions and EU measured toxicity values. These high percentage agreements indicate that the SAR methods are useful for predicting the acute toxicity of chemicals to fish and daphnids.

Hazard criteria

Various criteria are used by the USEPA for determining and quantifying the hazard and ecological effects of industrial chemicals and pesticides. Table 3.7 lists some of those used to assess hazard and rank concern of industrial chemicals. The criteria are based on acute, chronic and BCF test endpoints derived from valid tests. They are based on a variety of sources that include information from the scientific literature, expert workshops, databases, and from the experience in chemical review and expert judgement of OPPT ecotoxicologists. The table serves as a means of taking LOC values and relating them to the more easily understood hazard and concern descriptors of 'high, medium and low'.

Criteria for terrestrial organisms (avian and mammalian wildlife, and plants) have not been finalized for industrial chemicals; more research in comparative toxicology is required (Smrchek *et al.* 1995). Only a few terrestrial toxicity studies have ever been submitted for USA industrial chemicals. The wildlife criteria may change when more toxicity information is received. Wildlife criteria are available for USA pesticides; they are lower than those tentatively established for industrial chemicals (USEPA 1985a,b,c; 1994e). More testing is needed on assessing effects of industrial chemicals on wildlife and plants.

An interesting, yet somewhat controversial, method of ultimately establishing hazard criteria for terrestrial soil organisms is by extrapolating or deriving NOEC, GmMATC or PNEC values from NOEC, GmMATC or PNEC values calculated from sediment tests (van Leeuwen 1995). The equilibrium partitioning method of Di Toro *et al.* (1991) is used in these extrapolations. Chemicals are assumed to be partially adsorbed to soil particles and partially in solution in the pore-waters (Lokke 1994). It is further assumed that sediment and terrestrial organisms are as sensitive to the chemical occurring in the pore-waters as are aquatic organisms exposed to the same chemical in the water column. Thus, toxicity effects data for terrestrial and aquatic organisms are normalized to the water phase common to both these groups of organisms (Lokke 1994). Perhaps uncertainty factors similar to those used for aquatic organisms could also be used to derive terrestrial LOCs. Refinement and testing of this extrapolation method with a variety of chemicals is necessary.

Toxicity as expressed by these hazard criteria alone is usually not sufficient for identifying problem industrial chemicals, however, until they are related to exposure. A chemical with high acute or chronic aquatic toxicity may not be released into ecosystems through production, use, or disposal in high enough amounts to result in adverse effects to organisms. Controls on releases of this chemical should, nevertheless, be considered. Furthermore, a chemical that is moderately toxic, based on the criteria, may also constitute a serious problem if it is released into ecosystems in large quantities or continuously, and the resulting PEC increases to a level at or above the moderately toxic value indentified previously.

NOEC (GmMATC) versus EC_x (regression). Sprague (1971) discussed the importance of sublethal or chronic effects, for example, on growth and reproduction, in studying water pollution. Aquatic toxicologists attempted, in the past, to estimate the highest concentration (threshold) that did not cause an observable change in treatment organisms when compared with the control organisms, i.e., NOEC or safe level. We have already discussed the terminology problems associated with the word 'safe'. At the present state of our knowledge We cannot determine levels or concentrations that will be completely safe to all test organisms or to all other organisms. Determining the LOEC was also of interest; MATCs were also used and represented a concentration that if not exceeded, would be 'protective' of species (Mount & Stephan 1969). A MATC is an interval bounded by the NOEC and LOEC (Mount & Stephan 1969; Suter *et al.* 1983). In the USA much research was conducted on determining MATCs; these were used in establishing water quality criteria and eventually standards for regulatory purposes (Doudoroff 1977).

Table 3.7 Hazard criteria used by the US Environmental Protection Agency (EPA) Office of Pollution Prevention and Toxics (OPPT) in assessing industrial chemicals. (After Smrchek *et al.* 1995.)

	Criteria used for ranking concern		
Hazard parameters	High	Medium	Low
Aquatic acute toxicity LC_{50}, EC_{50} ($mg\,l^{-1}$)	≤1	>1≤100	>100
Aquatic chronic toxicity MATC ($mg\,l^{-1}$)	≤0.1	>0.1≤10	>10
Log K_{ow}	≥4.2	<4.2≥3.5	<3.5
Bioconcentration factor (BCF)	≥1000	<1000 ≥100	<100
Terrestrial acute toxicity* (avian and mammalian wildlife)			
LD_{50} ($mg\,kg^{-1}$)	≤100 (≤50)†	>100 ≤2000 (>50 ≤500)†	>2000 (>500)†
LC_{50} ($mg\,kg^{-1}$ in diet or food)	≤1000 (≤500)†	>1000 ≤5000 (>500 ≤1000)†	>5000 (>1000)†
Terrestrial chronic toxicity* (avian and mammalian wildlife)			
MATC ($mg\,kg^{-1}$ in diet or food)	(≤50)†	(>50 ≤100)†	(>100)†
Terrestrial plants‡	–	–	–

* These criteria are considered to be tentative at present by the OPPT, due to limited available test information.
† Criteria in parentheses are from the Office of Pesticide Programs (OPP) (see USEPA 1985a,b,c, 1994e) and have not been finalized by the OPPT.
‡ Criteria have not yet been established for terrestrial plants by the OPPT.

Since use of a value that could be found anywhere in an interval is unwieldy, the geometric mean of the bounds (GMATC or GmMATC) is used as a point estimate of the MATC (USEPA 1980; Suter *et al.* 1983, 1987). This value may more closely estimate the actual highest NOEC than simply accepting the highest NOEC derived experimentally (Calabrese & Baldwin 1993).

In Europe the NOEC parameter is used to analyse the results of chronic toxicity tests (OECD 1989; Kooijman & Bedaux 1996). The concept of 'No-effect' and the use of NOECs has a dominant place in the risk assessment process carried out in Europe and in other European legislative frameworks (OECD 1992a; de Bruijn & van Leeuwen 1996). The NOEC is preferred over the MATC because of the difficulty in defining what concentration is 'acceptable' (OECD 1979). It was also recognized that the NOEC is imprecise, it must be used with caution, and it does not represent a 'safe' level.

Recently, the use of the NOEC has been questioned and criticized (Hoekstra & van Ewijle 1993; Pack 1993; Mitchell 1994; Noppert *et al.* 1994; Chapman 1995b; de Bruijn & van Leeuwen 1996). Many of these criticisms were based on those made in earlier critiques of chronic tests and analyses of results (Stephan & Rogers 1985; Suter *et al.* 1987). An alternative EC_x point estimation method, using regression methods is advocated in place of the NOEC (Pack 1993; Noppert *et al.* 1994; Chapman *et al.* 1996).

This issue of which test end-point to use should be resolved by careful, reasoned, unbiased debate at the international level, with eventual unanimous agreement as a goal. Several points, however, can be made to clarify and respond to some criticisms of the NOEC (and MATC) (Pack 1993).

Chapman *et al.* (1996) have stated that NOECs are inappropriate for regulatory use. However, no evidence has been found that use of the GmMATC (and NOEC) by OPPT/OPP, over the past 10 years or more has resulted in obvious problems, such as over- or underestimates of toxicity, when using this end-point from chronic fish (partial lifecycle) or daphnid tests, to derive LOCs, and to assess the hazard of industrial chemicals and pesticides. Overall 'experimental noise' associated with the test procedures may have 'masked' any problems that may have been present due to use of the GmMATC.

The important point is that the advantages and disadvantages of changing to another method of analysis must be weighed carefully against those of the current method, because with either method there will be problems in use, interpretation and acceptance by regulatory authorities. There will also be consequences on existing test guidelines and regulatory frameworks for hazard and risk assessment (de Bruijn & van Leeuwen 1996).

It is not necessary that acute and chronic effects be analysed within a common framework. They can be considered to be quite different tests that should have separate, unique methods of analyses. Also, the NOEC is not a 'safe' level; confusion in viewing it in this way should not be a reason for replacing it.

Even if the NOEC is one of the test concentrations, this is not a cause for great concern, provided a GmMATC is calculated. If two different laboratories report different NOECs, this should not be viewed as unusual and should not be a reason for replacement of the NOEC. Instead, it should be determined why the NOECs were different, and if this difference is significant. All ecotoxicity test results are variable to a certain extent and contain some experimental error. If the lowest concentration used in a chronic test produces a statistically significant effect, then the test should be repeated with test chemical concentrations adjusted downward.

Adopting an EC_x point estimation method as a NOEC replacement could result in other problems. Handling control mortality by 'minor modifications to the regression model' may be unacceptable. Control mortality is often indicative of other problems in the test, for example, health of the organisms during the test or in the stock culture. The EC_x may lie outside the range of chosen concentrations (Mitchell 1994).

It may be difficult to choose and agree on a regression model suitable to ecotoxicologists from various countries. Selecting a particular EC_x effect value, for example, an EC_1, EC_5, EC_{10}, EC_{20} or EC_{25}, as a test end-point will be difficult and model sensitivity may be most prevalent at low EC percentages. A convincing case must be made to regulators that it would be pragmatic and advantageous to change from the NOEC or GmMATC method. The percentage effect on organisms the regulatory body will be willing to 'accept' as a level of concern and stimulus for regulatory action, 1%, 5%, 10%, 20%, or 25% and why a value is chosen, must be resolved if the EC_x method is to be adopted. De Bruijn & van Leeuwen (1996) have discussed other limitations of the EC_x methods and regulatory consequences if this method was adopted.

Quotient method

For industrial chemicals and currently for pesticides the quotient method (QM) of ecological risk assessment is used to characterize ecological risk (Barnthouse *et al.* 1982; Urban & Cook 1986; Rodier & Mauriello 1993; Rodier & Zeeman 1994). This method compares the predicted, expected or measured environmental concentration with a measurement test end-point or, in many cases, an LOC:

environmental concentration/LOC = quotient

This is similar to what is done in Europe (ECETOC 1993b; UK Department of the Environment 1993; Ahlers *et al.* 1994; EC 1994; De Bruijn & Van Leeuwen 1996):

$$\frac{\text{predicted environmental concentration}}{\text{predicted no-effect concentration}} = \text{quotient}$$

If the quotient is equal to 1 or more, a risk of

equalling or exceeding the LOC or PNEC is inferred. If the quotient is less than 1 it is implied that there is less risk of reaching the concern level. This method is very simple and compatible with the exposure and ecological effects characterizations carried out by the USEPA.

There are limitations to the quotient method. It does not address taxonomic or life-stage sensitivities to an industrial chemical or pesticide, and it cannot address risks objectively at intermediate levels where the quotient is 'almost' 1 or approaches it. Pulse or short-term exposures, and incremental dosages may not be addressed with this method because it assumes exposure duration will be as long as, or longer than, toxicological test duration (Urban & Cook 1986; Rodier & Mauriello 1993). Differences between the laboratory species and field populations are not recognized and it cannot be used to estimate indirect effects, quantify uncertainties, or account adequately for ecosystem structural and functional (process) effects.

The quotient method can identify risks that exceed measurement end-points (e.g. mortality, growth and reproductive effects), but it cannot relate these risks to assessment end-points. For example, the quotient method can identify a risk to a zooplankton population from the release of a toxic substance. A daphnid would serve as a surrogate test species and chronic testing would be conducted, with an uncertainty factor of 10 applied to the chronic end-point. The quotient method cannot quantify the risk to higher trophic levels, such as fish that prey on the zooplankton. To do this, simulation models are necessary. There are numerous ecological models available, which are applicable for risk assessment; there is no single model that is suitable for all risk assessments (Rodier & Mauriello 1993). Moreover, there is no fully accepted model that meets all requirements, for example, to provide quantitative information on end-points, compatibility with available data, adequate representation of the system of concern, an ability to account for variability in data and in organismal response.

Risks to populations may be estimated by using models that can be placed in two categories (Rodier & Mauriello 1993). First, there are single-species population or demographic models that predict toxic effects of short and long exposure to toxic chemicals to specific populations. The risk analysis and management alternatives system (RAMAS) is an example of a single-species simulation model (Ferson & Akcakaya 1990).

Second, there are multispecies food-web models that are able to assess both direct chemical toxic effects on a population of interest and indirect toxic effects resulting from food-chain impacts. These food chains support the population of interest. An example of this model is ecosystem uncertainty analysis (EUA), which translates toxicity data (e.g. LC and EC_{50}) to estimates of chemical effects on biomass production in lake models. The original model used with EUA is the standard water column model (SWACOM), which simulates the seasonal production dynamics of phytoplankton, zooplankton, and fish found in a northern USA dimictic lake. The USEPA has used EUA for several industrial chemicals in which a probability is given for a certain reduction of a population occurring when exposed to a certain PEC. Work continues on evaluating and developing new ecological models, especially those with benthic organism components.

3.5.2 Exposure assessment tools and methods

Exposure is characterized or assessed by evaluating the interaction of the stressor with the ecological component (from individuals up to ecosystems) (USEPA 1992). The stressor's distribution or pattern of change is characterized, and a combination of modelling and monitoring is used. The latter includes measures of environmental release and concentrations in various media over space and time. Physical–chemical characteristics of the stressor, as affected by the ecosystem, are used in fate and transport models. In fact, the characteristics of the ecosystem can modify, by microbial biotransformation, hydrolysis, photolysis and sorption processes, the ultimate nature and distribution of the stressor. Also, the environment will affect both the bioavailability of the stressor, and the exposure of ecological components. The fate and biotransformation of chemicals as related to assessing their hazard has been the subject of much research (see, e.g., Maki *et al.* 1980; Dickson *et al.* 1982).

Exposure analyses: aquatic and terrestrial

Spatial and temporal distributions of the ecological component and the stressor are combined to evaluate exposure. Often concentrations of the industrial chemical or pesticide are measured and then combined with assumptions about co-occurrence, contact or uptake. Included here is the measurement or prediction of the movement, fate and partitioning of the industrial chemical or pesticide in the environment (Barnthouse 1992). Aquatic organisms are assumed to be exposed to the chemical by simple contact in the water column. Exposure is expressed by the concentration of the chemical in the water.

The situation is more complicated for sediment-dwelling aquatic organisms and for terrestrial organisms, especially with regard to persistence, bioavailability and other processes, which may be modified by characteristics of the sediment or soil environment. Physical and chemical conditions in the aquatic sediments, and how these conditions temporally vary, may render the industrial chemical or pesticide less bioavailable or more bioavailable. Exposure is by direct contact with the sediments or pore-waters, or both.

The industrial chemical or pesticide (or their degradation product(s)) may adsorb to soil particles in terrestrial environments and may not be bioavailable unless it is further degraded by soil microbes or ambient chemical conditions. These residues are difficult to monitor and are often considered to be 'deactivated', or not bioavailable. The degradation product(s) or metabolites may also be toxic or possibly more toxic than the parent compound. Unbound materials may volatilize into the atmosphere. Strongly adsorbed industrial chemicals and pesticides are lost mostly with sediment, and moderately adsorbed materials can be lost primarily with surface runoff water (SETAC 1994). Materials weakly or not adsorbed are lost mostly with percolating water. Of course the relative rates of precipitation and infiltration by rain into the soil will affect concentrations of the industrial chemical or pesticide. Also, properties of the soil (e.g. percentage organic matter (carbon), pH, particle size distribution, total volatile solids) will affect both adsorption and persistence. Exposure is also by direct contact with soils or water (runoff or percolating waters), or both.

Test guidelines

The USEPA Chemical Fate Testing Guidelines were originally published in 1985 (USEPA 1985d). They are divided into methods on Physical and Chemical Properties (partition coefficients, water solubility and vapour pressure), Transport Processes (in soils and sediments), and Transformation Processes (biodegradation, simulation of aerobic sewage treatment, complex formation, hydrolysis and photolysis). Like the Environmental Effects Guidelines, they have been published in the US Code of Federal Regulations, Title 40, parts 795 and 796.

Pesticide registrants normally supply the EPA with data on the physical–chemical properties, and fate and transport of the pesticide (Urban & Cook 1986). These data include, for example, colour, physical state, melting point, solubility, octanol/water partition coefficient and pH. Fate and transport test data include hydrolysis, photodegradation (in water and in soil), aerobic and anaerobic metabolism (water and soil), leaching, field dissipation (from soils, water and forests), and accumulation (in crops, fish and aquatic non-target organisms). Formal, published guidelines for these methods do not exist. Instead, there is a pesticide assessment guideline on chemistry and environmental fate (subdivision N) (USEPA 1994e). Harmonized fate guidelines for both industrial chemicals and pesticides will eventually be available.

Biomarkers

An increasing emphasis over the past several years has involved the use of biochemical, histological and physiological changes, as well as aberrations in organisms (Huggett *et al.* 1992a). These changes are called biomarkers; they are used most commonly to estimate exposure to, or the resultant effects of, chemicals at the suborganismal or organismal levels of biological organization. Delineation of basic mechanisms of action and identification of chemical residues in organisms have resulted in many potentially useful tools for detecting either exposure to, or effects of, chemicals.

Various criteria for evaluating biomarkers have been discussed by Huggett *et al.* (1992b). Some of these include the extent of biomarker response to a variety of chemicals, biomarker sensitivity compared with conventional ecotoxicity test end-points (e.g. lethality, reproduction and growth), and variability/specificity of biomarker response.

Biomarkers are useful in assessing, at the sub-organismal and organismal levels, exposure and effects of chemicals and contaminants by using additional end-points. Yet by emphasizing these levels, the use of biomarkers runs counter to the top-down approach of determining the hazard and risk of industrial chemicals and pesticides at higher levels of populations to ecosystems.

Integrating biomarkers into testing schemes and translating these effect measures (including effects on organ-specific enzymes, isoenzymes, total lipid content, hormone, free amino acid and stress protein levels) into decision criteria or testing triggers and into hazard criteria remains to be completed.

Environmental concentrations

An important objective of exposure characterization is to predict the environmental concentrations of an industrial chemical due to releases from production, use and disposal. In the USA most industrial chemical releases to the environment occur in streams, rivers, and occasionally in lakes. Releases to terrestrial ecosystems are rare, but most are to landfills and deep wells (Nabholz 1991).

Pesticides, on the other hand, are widely applied to terrestrial or aquatic ecosystems. Use information, including label rate, frequency timing, method of application, site characteristics and geographical extent of use are very important in predicting EECs of pesticides for water, soils, non-target food items and non-target organisms determined to be at risk (Urban & Cook 1986). Information on plant and soil residues is also used. Aerial transport and deposition processes, wash-off from foliage and passage through soils and sediments may release the pesticide into streams and rivers.

As a result of the new pesticide aquatic risk characterization paradigm (SETAC 1994), exposure characterizations of pesticides may change in the future. More emphasis may be placed on computer modelling (e.g. dissolved runoff, soil erosion, leaching and spray drift) for EEC estimates instead of field monitoring. Single EEC values are used only in Tier 1. They are replaced at higher tiers with temporal and spatial distributions of EECs, which relate directly to the likelihood (quantitative) of their occurrence. Aquatic levels of risk are thus assessed quantitatively and may be related to mitigation measures. Various measures taken to restrict or reduce runoff and the verification of the effectiveness of these measures are important.

Industrial chemical releases. As a first step in exposure characterization or assessment, all potential facility releases must be identified and the amount of each release must be estimated over the entire lifecycle of the chemical, from production through to disposal. These releases are expressed as kilograms of chemical released per day from one or more sites for a certain number of days per year. The type of treatment (e.g. publicly owned treatment works (POTW)) used for the chemical is determined as well as the percentage of chemical removed by this treatment. The result will be a prediction of the amount of chemical released to the environment after treatment, expressed in kilograms per day. This daily amount is mixed with the daily amount of water in the receiving stream (litres per day) to give a predicted environmental concentration (PEC). The PEC is a worst-case exposure value because no losses of the chemical are assumed to occur after discharge.

The PEC value is then used in the exposure assessment, which can be based on two methods, either the percentile stream-flow method, or the probability dilution model (PDM) (Nabholz 1991).

In the first method, PECs are estimated from releases at sites, under both low and mean stream-flow conditions. Instantaneous mixing at the discharge point, continuous release every day of the year, and no losses after discharge are assumed in this method. Stream concentrations are then calculated (Rodier & Zeeman 1994).

In the PDM instantaneous mixing and no

losses after discharge are also assumed, but more information concerning chemical release and stream flows is used (Nabholz 1991). This model predicts how many days per year the LOC, determined in the hazard assessment, is exceeded. All the daily stream flows over the whole year, the amount of released chemical per day after treatment (if any), and the estimated number of days during the year of chemical release, are used in this method.

The PECs calculated by each method are then used in the quotient method to determine risk. In the PDM model the number of days the LOC is exceeded by the PEC is an indication of risk. For example, if the LOC was derived from a 21-day daphnid chronic test, and this value is exceeded in the receiving stream 250 days per year, a significant risk would be identified for aquatic invertebrates. An exceedence of only 2 days would imply little or no risk. The LOCs from other test species could also be related to the PEC to determine the type(s) and degree of risk. Risks could be estimated at each release site by using the PDM model to determine the geographical extent of the risk from the industrial chemical. Nabholz (1991) and Rodier & Zeeman (1994) give additional guidance on how these two exposure methods are used to assess risk.

Pesticide aquatic releases. Pesticide-use information is first obtained along with information on non-target organisms. The latter includes the types of organisms, their distributions, abundances, population dynamics and natural history. Data on chemical fate and transport and on the chemistry of the pesticide are also reviewed.

Residue chemistry data from the field are used and include detailed information on pesticide residues found in plants and in soils. Normally, aquatic exposure to residues is estimated. Other chemical fate and transport data on rates of dissipation in the field and residues in fish are used.

All the previous information is used to determine EECs for water, soil and non-target food items; also a profile of non-target organisms at risk is prepared. Aquatic EECs are estimated for pesticides by using a four-level EEC tier system (Urban & Cook 1986). Risk criteria are compared with the EEC at each level. These criteria are listed in discussions of the pesticide tiered testing scheme in section 3.5.1.

At Level 1 the worst possible risk situation is considered; i.e. direct application to water. Mass balance equations for surface runoff are used to determine a 'worst-case' EEC. These equations consider maximum application rate, size of the drainage basin, percentage runoff, and surface area and depth of the water body. The worst case EEC is compared with the lowest acute aquatic LC_{50}, lowest chronic effect level, or lowest chronic no-effect level. This is similar to the risk quotient method used for industrial chemicals. If at Level 1 the EEC < risk criteria, there probably is no risk; if EEC > risk criteria, then the review proceeds to Level 2 for more detailed review, because there may be a risk.

Water and sediment surface runoff and spray drift are considered at Level 2 as the major routes of pesticide loss to the environment. The same mass balance equations that were used in Level 1 are used in Level 2. However, the variables in the equation are modified to better describe a field use site (Urban & Cook 1986). The Level 2 calculated EEC is compared with the aquatic risk criteria. If a risk is identified, the review proceeds to Level 3.

The EEC is estimated by using state-of-the-art exposure models such as EXAMS (exposure analysis modelling system) at the third level. All the major routes of pesticide transport are considered in this model simulation. The EEC is then estimated with the model for both lentic and lotic scenarios. Field residue monitoring may be necessary to resolve disagreements in parameter validation and interpretation. Output from the computer model is expressed as the number of days that the EEC exceeds the critical risk criteria, and the day is identified on which the minimum and maximum EECs occurred. The latter aids in determining if there should be a concern for risk to non-target aquatic organisms, which of these organisms are at risk, and the duration of this risk. If a risk is identified, the review proceeds to Level 4. On Level 4, actual field residue-monitoring studies may be requested and no modelling is completed.

Pesticide terrestrial releases. Efforts are concentrated on profiling residue levels of pesticides on vegetation or invertebrate (insect) surfaces or both. Maximum expected residues occur on day zero of application, or immediately after pesticide application. Actual residue data as supplied by the registrant or found in the literature are used. Next, the likelihood of exposure and hazard is determined by correlating the residue information, food items consumed, and wildlife utilization of crop areas with the available mammalian and avian toxicity data (Urban & Cook 1986).

With these toxicity data, the estimated residues in or on mammalian and avian food items is related to the number of pesticide granules, number of seeds, or number of baits likely to be ingested, or the amount of pesticide that may directly contact non-target organisms. Estimates of potential risk are made. In the final analysis the quantity of product, seeds, baits or granules available in a square foot basis is correlated with the amount of product, seeds, baits or granules needed to produce an LD_{50} on a per animal body-weight basis. Acute toxicity data are correlated with estimated or actual residues. A theoretical LC_{50} value for a certain active ingredient rate is developed and this value is then correlated with residues estimated to be on available items of food (Urban & Cook 1986).

In the assessment of subchronic or chronic hazards, the toxicity data available are related to the estimated or actual field residue data. Often, residue and field dissipation data are lacking, but extrapolations are used to estimate the residues likely to be found over time.

3.6 THE HAZARD ASSESSMENT/ RISK CHARACTERIZATION OF ECOLOGICAL EFFECTS: DESCRIPTION AND GUIDANCE

3.6.1 Summary of the process

Table 3.8 summarizes the steps that would be followed in the process of hazard assessment or characterization of ecological effects for industrial chemicals and pesticides. Guidance on this process is offered from a regulatory perspective. This process is described somewhat generically so as to increase its utility. The process can be broken down into four steps, each with multiple parts. Only some of the parts are discussed in general terms here. Many of the methods associated with each step were discussed above; many of the parts of the steps are obvious from the table and require no further discussion.

Data aquisition/information gathering

The first step is familiarization with the regulatory requirements and laws associated with approval, notification or registration of industrial chemicals (new and existing) or pesticides in each country where these will be produced, marketed, imported/exported or disposed of. Sections 3.2.2 through to 3.2.5 have given a brief, world-wide overview of chemical laws, regulatory requirements and activities. Relevant information is also discussed in Chapters 9 and 10 of this volume. These requirements may include minimal toxicity testing information and submission of certain applications, notification forms or other documentation.

Information must be gathered and reviewed on the physical–chemical characteristics of the industrial chemical or pesticide (e.g. water solubility, octanol/water partition coefficient). Also, information on production, use and disposal should be gathered and reviewed.

Information on the inherent toxicity (aquatic and/or terrestrial) of the substance may also be gathered and reviewed. This information consists of any acute, subchronic, or chronic toxicity data from any test organism, but for aquatic exposures it is mainly from algae, invertebrates (usually daphnids) and fish. For terrestrial exposures, data mainly from plants, mammals and birds are used.

For some chemicals that have been in production for some time there will be a greater amount of toxicity test information available. For others, for example, new chemicals, there may be little or no toxicity data. Depending on the production, use and disposal pattern, these toxicity data may be restricted to test species occurring only in certain local ecological communities or ecosystems. For example, if the substance is to be produced and disposed of near, or solely into, estuaries, toxicity tests using estuarine species should be emphasized (others should not be excluded) in the review. If

Table 3.8 Steps taken in hazard assessment/identification of ecological effects, risk characterization and risk management processes. Emphasis is on industrial chemicals. Steps above the line in number 2 would be completed mainly by the chemical company, submitter, petitioner, or marketer seeking to produce or market the industrial chemical or pesticide. Regulators or regulatory scientists could also complete some of the steps above the line to verify information or validate tests. Steps below the line in numbers 2 and 3 would usually be completed only by regulators or regulatory scientists. Risk management steps (number 4) are taken by the chemical company, submitter, petitioner or marketer with advice and recommendations from the regulatory agency or authority.

---

1 Data acquisition/information gathering

- Become familiar with regulatory requirements and laws
- Contact regulatory officials/authorities, as needed
- Consult general ecotoxicological references and data bases
- Collect information on physical–chemical characteristics and ecotoxicology data specific to chemical or pesticide

2 Determine ecological effects from information and test results

- As all tiered testing is completed
- Review/validate available toxicity information
- Perform additional testing if needed, or testing to fill new identified data gaps
- Submit information (notification, petition, etc.) to regulatory authority, agency, body, etc.

- Use quantitative structure–activity relationship (QSAR) methods if necessary
- Review information and validate toxicity information if not already done
- Select most sensitive test species and measurement test end-point (if possible, from chronic tests)
- Derive a level of concern (LOC), or predicted no-effect concentration (PNEC), etc., by dividing test end-point by an assessment or uncertainty factor (of 1–1000)
- Identify as completely as possible the hazard and characterize as completely as possible the ecological effects

3 Risk characterization (risk assessment)

- Compare LOC, or PNEC, etc., with a predicted environmental concentration (PEC) or expected environmental concentration (EEC) derived from an exposure characterization
- Identify definite risk (LOC < PEC), possible or probable risk (LOC 'similar' to PEC), or possible absence of risk (LOC > PEC)

4 Risk management

- Take regulatory steps (ban, use restrictions, special review candidate, pollution prevention) or non-regulatory steps (as mandated by the applicable law(s)) to control, reduce, or possibly reverse and thereby manage and mitigate the identified risk for the industrial chemical or pesticide
- After risk assessment is completed for pesticides, proceed with registration/reregistration
- Monitoring activities may be necessary in affected environment(s)
- Review chemical again in future if something occurs, e.g. increased production, new information, spills, high monitored environmental concentration

---

such toxicity data are lacking, additional testing should be conducted, or if this is not possible, substitute toxicity tests can be considered.

The most sensitive test species from tests found to be scientifically valid and acceptable should be used to determine the toxicity of the substance.

General references and data bases should be consulted to identify useful toxicity (and fate) information. Many of those listed below are used in North America and may be unknown in Europe, Australasia or in other parts of the world. A very useful reference is the five-volume set on acute toxicity of industrial chemicals to fathead minnows (University of Wisconsin-Superior, 1984–1990). Important general references include NAS (1972), Verschueren (1983), Howard (1989–1993) and Ramamoorthy *et al.* (1995). Other aquatic references that deal with general or specific classes of industrial chemicals or pesticides, and background information on chemicals, include Wood

(1953), Hollis & Lennon (1954), Applegate *et al.* (1957), McKee & Wolf (1963), Marvin & Proctor (1964), MacPhee & Ruelle (1969), ADMI (1974), Hohreiter (1980), Buikema *et al.* (1979), Johnson & Finley (1980), LeBlanc (1980), Buccafusco *et al.* (1981), Curtis & Ward (1981), Heitmuller *et al.* (1981), Slooff & Canton (1983), Slooff *et al.* (1983), University of Wisconsin-Superior (1992), Mayer & Ellersieck (1986), Mayer (1987), Environment Canada (1987), and USEPA (1989a). Useful terrestrial toxicity references include Hill *et al.* (1975), Schafer *et al.* (1983), Hudson *et al.* (1984), Schafer & Bowles (1985) and Hill & Camardese (1986).

Databases such as HEDSET and ECOTOX are good sources of toxicity test data and specific references. The ECOTOX data base is a system that is being developed by the USEPA, National Health and Environmental Effects Research Laboratories, Duluth, Minnesota, and Corvallis, Oregon, USA. It consists of the AQUIRE (aquatic life), PHYTOTOX (terrestrial plants) and TERRETOX (wildlife) databases. Also, the Office of Pesticide Programs' Ecological Effects Database of aquatic and terrestrial effects data, which has been reviewed and categorized as acceptable to fulfil pesticide registration and reregistration guideline requirements, has been added to the ECOTOX system.

The AQUIRE database contains over 133000 individual test records for 5700 chemicals and 2800 freshwater and marine organisms (as of March 1996). Over 9600 publications were reviewed for AQUIRE. The PHYTOTOX database contains over 74000 toxicity tests from 3900 references for more than 2000 chemicals and 1300 terrestrial plant species. The TERRETOX database contains 29051 toxicity tests from 10000 references for more than 1641 chemicals and 250 terrestrial animal species.

The ECOTOX database is available only to USA Government users. Eventually the system may be made available to others through the Chemical Information System (CIS). The AQUIRE database is now available to anyone through CIS on a fee basis.

Ecological effects determined from information and test results

The main literature references from which the toxicity test data entries in these databases are derived, should be reviewed to confirm that the test procedures and methodology used are valid and reliable. If the toxicity data are from older, invalid or equivocal tests, these data should not be used, or in some cases some of the data may be modified or corrected and then used. Preference is always given to using valid test data. Additional testing may be necessary to fill data gaps or to repeat tests using valid, current test methods.

If no toxicity data are located or are available, QSAR procedures can be used for a substance (e.g. a new chemical), provided it is a member of a chemical class that already has a QSAR. The QSAR will give an estimate of the inherent toxicity of the chemical. A number of QSARs have been the subject of literature publications (see, e.g., Lipnick *et al.* 1985a,b; Newsome *et al.* 1987, 1993). Quantitative structure–activity relationships procedures were discussed in section 3.5.1.

The test end-point from the most sensitive test species is selected. This may be an $LC(EC)_{50}$ value from an acute laboratory test, a NOEC, LOEC, MATC or GmMATC from a chronic laboratory test, a value from a multispecies or microcosm test, or even a value from a field or mesocosm test. Examples of end-points include mortality (acute tests), changes in growth and reproduction (subchronic or chronic tests), and changes in population abundances, dissolved oxygen, photosynthesis, respiration and nutrient concentrations in both multispecies laboratory tests and field tests (Hill *et al.* 1994).

Chronic test end-points are preferred over acute end-points. The former are based on more sensitive, sublethal, subtle, sometimes reversible effects, and the second are irreversible lethal effects (Chapman 1995b). Use of chronic end-points allows some delay and manoeuvring room, in that regulatory or pollution prevention practices can be implemented before irreversible harm or toxicity occurs to organisms and ecosystems.

Risk characterization, risk assessment and risk management

The steps here have already been discussed in sections 3.4.3 and 3.5.2. The information relating to these topics in Table 3.8 is self-explanatory.

The risk management step should be completed independently of risk characterization (Rodier & Zeeman 1994).

Future of hazard/risk assessment for pesticides (USEPA)

Many recommendations were made in SETAC (1994) that will result in changes in the assessment of hazard and risk of pesticides, in addition to a de-emphasis of mesocosm and field tests (section 3.5.1). For example, the new paradigm recommends a shift from the single EEC quotient method to a probabilistic approach using distributions of EECs, and that risk assessment quotients be replaced by estimates of specific-effect levels (e.g. LC_5). A modified tier testing scheme and risk assessment process with new chronic test triggers is also proposed. Risk characterization should be presented as a distribution of toxicity test end-points instead of a single-point estimate.

It seems certain that if these recommendations are implemented they will have widespread influence over the hazard/risk assessment of pesticides and possibly even on pesticide assessment activities in other countries.

3.7 CONCLUDING REMARKS: THE FUTURE OF CHEMICAL RISK ASSESSMENT FOR ECOLOGICAL SYSTEMS

Ecotoxicology used for the assessment of industrial chemicals and pesticides is now at a crossroads. The bottom-up approach to determine the hazard and risk of these chemicals must 'meet' the top-down approach, for significant progress to occur next in chemical risk assessment of ecological systems.

At the present time multilevel, multiphase or tiered testing schemes (mainly aquatic) with emphasis on single-species acute laboratory tests remain the primary means of assessing the hazard and risk of industrial chemicals and pesticides in many countries. Researchers continue to fine-tune the schemes and discuss related topics such as the use of uncertainty factors, the adequacy of the NOEC, establishment of sediment and terrestrial schemes, and development of new tests. Important international efforts also continue on the revision, updating, standardization and harmonization of tests and testing schemes (e.g. the OECD *Daphnia magna* chronic test efforts beginning in the 1980s). These efforts are needed in terrestrial ecotoxicology especially, because progress up to now has been much more limited in this area than in aquatic ecotoxicology. Even though these efforts are time-consuming and 'progress' may appear to be slow, they should be supported and encouraged by countries throughout the world, because harmonized, useful test guidelines and testing schemes can result from these activities.

The inadequacies of single-species tests have been made obvious. They are over-simplified and arbitrary, minimally realistic and laboratory to field extrapolations are difficult.

Microcosm, mesocosm and other field tests were developed, in part, as a response to these perceived inadequacies. The advantages of these multispecies tests have been described earlier. Perhaps the most important one is that a realistic picture of processes occurring in natural eco-systems with respect to the fate and effects of substances can be obtained.

Even with these advantages and the large amount of research activity and development of many promising methods in the 1980s, these tests still have not replaced single-species tests, and if anything, the influence of microcosm, mesocosm and field tests has decreased. Single-species tests continue to have powerful offsetting advantages that have not been overcome. These include economy, ease of operation, familiarity, usefulness as a screening tool, and the availability of a large toxicity database for some test organisms. Also, single-species tests continue to supply simple, easily understood values to regulatory bodies, on large numbers of chemicals, that can be used to make regulatory decisions.

Comparisons and extrapolations between single-species tests and microcosm/mesocosm/field tests should continue to narrow as more research in this area is completed. Also, uncertainty factors can become more specific and certainty increased. Instead of using a current factor of 10, for example, eventually using a lower factor such as six or two, may be possible as laboratory to field predictability improves.

As indicated by the recent increasing activities in OECD's RAAB, interest in terrestrial ecotoxicology is expanding. Terrestrial testing schemes, decision criteria and other procedures for assessing the hazard to soil-dwelling organisms need to be developed and evaluated, and efforts should continue on achieving international consensus. Standardized/harmonized test methods using beneficial and saprotrophic arthropods, annelids, birds, mammals and plants should continue to be developed.

The ecosystem (ecological system) approach to assessment of substances has been gaining in importance recently, stimulated in part by increasing activities in ecological risk assessment. The ecological risk assessment efforts are attempting to deal with problem areas in ecotoxicology, for example, ecosystem 'health', uncertainty, effects and exposure characterization, and ecological significance. Also, large-scale efforts are underway to protect ecosystems, for example, at the USEPA (USEPA 1995a). International organizations (e.g. OECD) discuss protection of ecosystems with greater and greater frequency in their published and unpublished documents (e.g. OECD 1995a,b).

Ecological models used to quantify the hazard/risk of industrial chemicals and pesticides to populations of organisms have been discussed briefly in this chapter. Evaluation and development of these models continues, but they have not yet achieved widespread use because of several limitations. The models may be applicable to only very specific ecosystems, for example, a northern USA lake. Also, the results of the models are not easily translated from end-points to triggers for regulatory action. How can chemical exposure be lowered so as to reduce the probability that a certain percentage of a population will not be, in turn, reduced? How much reduction in chemical exposure should be recommended or required? Until these questions are answered, modelling will be an adjunct to hazard and risk assessment. We must know better how 'normal' populations, communities and ecosystems operate and respond to stressors, and how 'resilient' they are to stressors. A theoretical framework has to be developed that will address the structure and function (processes) of 'unstressed' or 'normal' ecosystems.

These efforts may increase to a point in the future where they finally serve as the stimulus needed for ecotoxicology to combine both bottom-up and top-down approaches. This integration will require increased resources. It will result in a de-emphasis on single-species tests, an increased emphasis on, and shift of resources to, microcosm/mesocosm/field tests (and in the process reinvigorate research to develop new microcosm methods and improve existing methods). More sophisticated microcosm and modelling methods are needed, so that biologically and statistically significant comparisons between these tests and stressed whole ecosystems can be made. As this development continues, predictions of toxic effect to a wider array of communities and ecosystems can be made.

Finally, this integration must be translated into regulatory or pollution prevention initiatives and mechanisms on a world-wide basis. Cooperative international efforts in assessing the hazard and risk of industrial chemicals and pesticides must continue and must even be expanded. Updating and harmonizing test methods and assessment tools to 'improve' the review of substances that are of world-wide interest and that will be used in the integrated approach must also continue. A commitment of adequate levels of available resources for these tasks by countries throughout the world is needed to assure this cooperation and to continue to improve the process of hazard and risk assessment of industrial chemicals and pesticides. Ultimately, the utility of hazard and risk assessments used in international environmental protection efforts will depend on the resources available to complete these tasks.

3.8 ADDENDUM

As this chapter was being completed, chemicals which can be endocrine disrupters or modulators (nonylphenol and bisphenol A are examples) have become the subject of increasing interest in North America and in Europe. Endocrine disruptive chemicals interfere with various processes associated with natural hormones (for example, synthesis, secretion, binding), which are responsible for the reproductive, developmental, homoeostatic and behavioural processes of organisms.

We are at the very beginning phases of developing methods and testing schemes to identify (screen) these chemicals and to assess their ecological effects/hazard and risk. Many of the assessment procedures, methods and approaches described in this chapter will probably also be directly applicable to endocrine disrupters/modulators. The specific test methods to assess their toxicity and to recommend will depend on the quantity and kinds of valid pre-existing information available from previous ecological effects characterization and hazard assessment. Eventually, a testing scheme applicable specifically to these chemicals will be developed. The steps listed in Table 3.8 will be useful and directly applicable to these chemicals.

3.9 REFERENCES

Adams, W.J., Kimerle, R.A. & Barnett, J.W., Jr. (1992) Sediment quality and aquatic life assessments. *Environmental Science and Technology*, **26**, 1864–1875.

Ahlers, J., Diderich, R., Klaschka, U., Marschner, A. & Schwarz-Schulz, B. (1994), Environmental risk assessment of existing chemicals. *Environmental Science and Pollution Research International*, **1**, 117–123.

Allaby, M. (1994) *The Concise Oxford Dictionary of Ecology*. Oxford University Press, Oxford, 415 pp.

ADMI (1974) *Dyes and the Environment: Reports on Selected Dyes and Their Effects*, Vol. II. American Dye Manufacturers Institute, New York.

APHA, American Water Works Association & Water Pollution Control Federation (1955) *Standard Methods for the Examination of Water, Sewage and Industrial Wastes*, 10th edn. American Public Health Association, New York, 522 pp.

APHA, American Water Works Association & Water Pollution Control Federation (1960) *Standard Methods for the Examination of Water and Wastewater*, 11th edn. American Public Health Association, New York, 626 pp.

APHA, American Water Works Association & Water Environment Federation (1995) *Standard Methods for the Examination of Water and Wastewater*, 19th edn. American Public Health Association, Washington, DC.

ASTM (1993) *ASTM Standards on Aquatic Toxicology and Hazard Evaluation*. American Society for Testing and Materials, Philadelphia, PA, 538 pp.

ASTM (1996) *Annual Book of ASTM Standards, Section 11, Water and Environmental Technology*, Vol. 11.05, *Biological Effects and Environmental Fate; Biotechnology; Pesticides*. American Society for Testing and Materials, Philadelphia, PA, 1399 pp.

Anderson, B.G. (1944) The toxicity thresholds of various substances found in industrial wastes as determined by the use of *Daphnia magna*. *Sewage Works Journal*, **16**, 1156–1165.

Anderson, B.G. (1946) The toxicity thresholds of various sodium salts determined by the use of *Daphnia magna*. *Sewage Works Journal*, **18**, 82–87.

Applegate, V.C., Howell, J.H., Hall, A.E., Jr. & Smith, M.A. (1957) *Toxicity of 4,346 Chemicals to Larval Lampreys and Fishes*. Special Scientific Report—Fisheries No. 207, U.S. Department of the Interior, Washington, DC, 157 pp.

Auer, C.M., Nabholz, J.V. & Baetke, K.P. (1990) Mode of action and the assessment of chemical hazards in the presence of limited data: use of structure–activity relationships (SAR) under TSCA, Section 5. *Environmental Health Perspectives*, **87**, 183–197.

Auer, C.M., Zeeman, M., Nabholz, J.V. & Clements, R.G. (1994) SAR—the US regulatory perspective. *SAR and QSAR in Environmental Research*, **2**, 29–38.

Balk, F., de Bruijn, J.H.M. & van Leeuwen, C.J. (1993) *Guidance Document for Aquatic Effects Assessment*. Report prepared for the Organization for Economic Cooperation and Development (OECD), Paris.

Barnthouse, L.W., DeAngelis, A.L., Gardner R.H. *et al.* (1982) *Methodology for Environmental Risk Analysis*. ORNL/TM 8167, Oak Ridge National Laboratory, Oak Ridge, Tennessee.

Barnthouse, L.W. (1992) The role of models in ecological risk assessment: a 1990s' perspective. *Environmental Toxicology and Chemistry*, **11**, 1751–1760.

Bayer, E. & Fleischhauer, G. (1993) Status report on the testing activities according to the German programme for existing chemicals. *Chemosphere*, **26**, 1789–1822.

Belanger, S.E. (1994) Review of experimental microcosm, mesocosm, and field tests used to evaluate the potential hazard of surfactants to aquatic life and the relation to single species data. In: *Freshwater Field Tests for Hazard Assessment of Chemicals* (eds I.R. Hill, F. Heimbach, P. Leeuwangh & P. Mattiessen), pp. 287–314. Lewis Publishers, Boca Raton, FL.

Beyers, R.J. & Odum, H.T. (1993) *Ecological Microcosms*. Springer-Verlag, New York, 557 pp.

Blok, J. & Balk, F. (1995) Environmental regulation in the European community. In: *Fundamentals of Aquatic Toxicology* (ed. G.M. Rand), 2nd edn, pp. 775–802. Taylor & Francis Publishers, Washington, DC.

Boyle, T.P. (ed). (1985) *Validation and Predictability of Laboratory Methods for Assessing the Fate and Effects of Contaminants in Aquatic Ecosystems*, STP 865. American Society for Testing and Materials, Philadelphia, PA, 233 pp.

Buccafusco, R.J., Ells, S.J. & LeBlanc, G.A (1981) Acute

toxicity of priority pollutants to bluegill (*Lepomis macrochirus*). *Bulletin of Environmental Contamination and Toxicology*, **26**, 446–452.

Buikema, A.L., Jr. & Cairns, J., Jr. (eds) (1980) *Aquatic Invertebrate Bioassays*, STP 715. American Society for Testing and Materials, Philadelphia, PA, 209 pp.

Buikema, A.L., Jr., McGinniss, M.J. & Cairns, J., Jr. (1979) Phenolics in aquatic ecosystems: a selected review of recent literature. *Marine Environment Research*, **2**, 87–181.

Buikema, A.L., Jr., Niederlehner, B.R. & Cairns, J., Jr. (1982) Biological monitoring. Part IV — toxicity testing. *Water Research*, **16**, 239–262.

Burton, G.A., Jr. (1991) Assessing the toxicity of freshwater sediments. *Environmental Toxicology and Chemistry*, **10**, 1585–1627.

Burton, G.A., Jr. (ed.) (1992) *Sediment Toxicity Assessment*. Lewis Publishers, Boca Raton, FL, 457 pp.

Burton, G.A., Jr. & MacPherson, C. (1995) Sediment Toxicity Testing Issues and Methods. In: *Handbook of Ecotoxicology* (eds D.J. Hoffman, B.A. Rattner, G.A. Burton, Jr. & J. Cairns, Jr.), pp. 70–103. CRC Press, Boca Raton, FL.

Burton, G.A., Jr. & Scott, K.J. (1992) Sediment toxicity evaluations. *Environmental Science and Technology*, **26**, 2068–2075.

Cairns, J., Jr. (1983) Are single species toxicity tests alone adequate for estimating environmental hazard? *Hydrobiologia*, **100**, 47–57.

Cairns, J., Jr. (ed.) (1985) *Multispecies Toxicity Testing*. Pergamon Press, New York, 253 pp.

Cairns, J., Jr. (ed.) (1986a) *Community Toxicity Testing*, STP 920. American Society for Testing and Materials, Philadelphia, PA, 349 pp.

Cairns, J., Jr. (1986b) The myth of the most sensitive species. *BioScience*, **36**, 670–672.

Cairns, J., Jr. (1986c) What is meant by validation of predictions based on laboratory toxicity tests? *Hydrobiologia*, **137**, 271–278.

Cairns, J., Jr. (1988) What constitutes field validation of predictions based on laboratory evidence? In: *Aquatic Toxicology and Hazard Assessment: Tenth Volume*, STP 971 (eds W.J. Adams, G.A. Chapman & W.G. Landis), pp. 361–368. American Society for Testing and Materials, Philadelphia, PA.

Cairns, J., Jr. (1989) Editorial: Will the real ecotoxicologist please stand up? *Environmental Toxicology and Chemistry*, **8**, 843–844.

Cairns, J., Jr. & Cherry, D.S. (1993) Freshwater multi-species test systems. In: *Handbook of Ecotoxicology*, Vol. 1 (ed. P. Calow), pp. 101–116. Blackwell Scientific Publications, Oxford.

Cairns, J., Jr. & Dickson, K.L. (eds) (1972) *Biological Methods for the Assessment of Water Quality*, STP 528. American Society for Testing and Materials, Philadelphia, PA, 256 pp.

Cairns, J., Jr. & Dickson, K.L. (1995) Ecological hazard/risk assessment: lessons learned and new directions. *Hydrobiologia*, **312**, 87–92.

Cairns, J., Jr. & Niederlehner, B.R. (1987) Problems associated with selecting the most sensitive species for toxicity testing. *Hydrobiologia*, **153**, 87–94.

Cairns, J., Jr. & Niederlehner, B.R. (1994) *Ecological Toxicity Testing*. Lewis Publishers, Boca Raton, FL, 256 pp.

Cairns, J., Jr. & Pratt, J.R. (eds) (1989) *Functional Testing of Aquatic Biota for Estimating Hazards of Chemicals*, STP 988. American Society for Testing and Materials, Philadelphia, PA, 242 pp.

Cairns, J., Jr. & Pratt, J.R. (1993) Trends in ecotoxicology. *Science of the Total Environment*, **Suppl. 1993**, 7–22.

Cairns, J., Jr., Dickson, K.L. & Maki, A.W. (eds) (1978) *Estimating the Hazard of Chemical Substances to Aquatic Life*, STP 657. American Society for Testing and Materials, Philadelphia, PA, 273 pp.

Cairns, J., Jr., Dickson, K.L. & Westlake, G.F. (eds) (1976) *Biological Monitoring of Water and Effluent Quality*, STP 607. American Society for Testing and Materials, Philadelphia, PA. 246 pp.

Cairns, J., Jr., McCormick, P.V. & Belanger, S. (1992) Ecotoxicological testing: small is reliable. *Journal of Environmental Pathology, Toxicology and Oncology*, **11**, 247–263.

Cairns, M.A., Nebeker, A.V., Gakstatter, J.H. & Griffis, W.L. (1984) Toxicity of copper-spiked sediments to freshwater invertebrates. *Environmental Toxicology and Chemistry*, **3**, 435–445.

Calabrese, E.J. & Baldwin, L.A. (1993) *Performing Ecological Risk Assessments*. Lewis Publishers, Boca Raton, FL, 257 pp.

Calow, P. (1992) Can ecosystems be healthy? *Journal of Aquatic Ecosystem Health*, **1**, 1–5.

Calow, P. (ed.) (1993a) *Handbook of Ecotoxicology*, Vol. 1. Blackwell Scientific Publications, Oxford, 478 pp.

Calow, P. (1993b) Hazards and risks in Europe: challenges for ecotoxicology. *Environmental Toxicology and Chemistry*, **12**, 1519–1520.

Calow, P. (1994) Overview with observations on risk assessment and management. In: *Handbook of Ecotoxicology*, Vol. 2 (ed. P. Calow), pp. 1–4. Blackwell Scientific Publications, Oxford.

Carpenter, K.E. (1924) A study of the fauna of rivers polluted by lead mining in the Aberystwyth district of Cardiganshire. *Annals of Applied Biology*, **11**, 1–23.

Carpenter, K.E. (1925) On the biological factors involved in the destruction of river-fisheries by pollution due to lead-mining. *Annals of Applied Biology*, **12**, 1–13.

Carpenter, K.E. (1927) The lethal action of soluble metallic salts on fishes. *British Journal of Experimental Biology*, **4**, 378–390.

Carpenter, K.E. (1930) Further researches on the action of metallic salts on fishes. *Journal of Experimental Zoology*, **56**, 407–422.

Chapman, P.M. (1995a) Do sediment toxicity tests require field validation? *Environmental Toxicology and Chemistry*, **14**, 1451–1453.

Chapman, P.M. (1995b) Ecotoxicology and pollution—Key issues. *Marine Pollution Bulletin*, **31**, 167–177.

Chapman, P.M., Caldwell, R.S. & Chapman, P.F. (1996) A warning: NOECs are inappropriate for regulatory use. *Environmental Toxicology and Chemistry*, **15**, 77–79.

Chemical Products Safety Division (1977) *The Chemical Substances Control Law in Japan*. Basic Industry Bureau, Ministry of International Trade and Industry, Tokyo, Japan.

Clark, J.R. & Cripe, C.R. (1993) Marine and estuaries multi-species test systems. In: *Handbook of Ecotoxicology*, Vol. 1 (ed. P. Calow), pp. 227–247. Blackwell Scientific Publications, Oxford.

Clements, R.G., Nabholz, J.V., Johnson, D.W. & Zeeman, M. (1993a) The use and application of QSARs in the Office of Toxic Substances for ecological hazard assessment of new chemicals. In: *Environmental Toxicology and Risk Assessment*, STP 1179 (eds. W.G. Landis, J.S. Hughes & M.A. Lewis), pp. 56–64. American Society for Testing and Materials, Philadelphia, PA.

Clements, R.G., Nabholz, J.V., Johnson, D.W. & Zeeman, M. (1993b) The use of quantitative structure–activity relationships (QSARs) as screening tools in environmental assessment. In: *Environmental Toxicology and Risk Assessment: 2nd Volume*, STP 1216 (eds J.W. Gorsuch, F.J. Dwyer, C.G. Ingersoll & T.W. LaPoint), pp. 555–570. American Society for Testing and Materials, Philadelphia, PA.

Clements, R.G., Nabholz, J.V., Zeeman, M.G. & Auer, C. (1995). The relationship of structure–activity relationships (SARs) in the aquatic toxicity evaluation of discrete organic chemicals. *SAR and QSAR in Environmental Research*, **3**, 203–215.

Crow, M.E. & Taub, F.B. (1979) Designing a microcosm bioassay to detect ecosystem level effects. *International Journal of Environmental Studies*, **13**, 141–147.

Curtis, M.W. & Ward, C.H. (1981) Aquatic toxicity of forty industrial chemicals: testing in support of hazardous spill substance spill prevention. *Journal of Hydrology*, **51**, 359–367.

de Bruijn, J.H.M. & van Leeuwen, C.J. (1996) No-effect concentrations in environmental policy. In: *The Analysis of Aquatic Toxicity Data* (eds S.A.L.M. Kooijman & J.J.M. Bedaux), pp. 1–7. VU University Press, Amsterdam.

Dickson, K.L., Maki, A.W. & Brungs, W.A. (eds) (1987) *Fate and Effects of Sediment-bound Chemicals in Aquatic Systems*. Pergamon Press, New York, 449 pp.

Dickson, K.L., Maki, A.W. & Cairns, J., Jr. (1982) *Modeling the Fate of Chemicals in the Aquatic Environment*. Ann Arbor Science Publishers, Ann Arbor, MI, 413 pp.

Di Toro, D.M., Zarba, C.S., Hansen, D.J. *et al.* (1991) Technical basis for establishing sediment quality criteria for nonionic organic chemicals by using equilibrium partitioning. *Environmental Toxicology and Chemistry*, **10**, 1541–1583.

Donker, M.H., Eijsackers, H. & Heimbach, F. (eds) (1994) *Ecotoxicology of Soil Organisms*. Lewis Publishers, Boca Raton, FL, 470 pp.

Doudoroff, P. (1977) Keynote address—reflections on pickle-jar ecology. In: *Biological Monitoring of Water and Effluent Quality*, STP 607 (eds J. Cairns, Jr., K.L. Dickson & G.F. Westlake), pp. 3–19. American Society for Testing and Materials, Philadelphia, PA.

Doudoroff, P., Anderson, B.G., Burdick, G.E. *et al.* (1951) Bio-assay methods for the evaluation of acute toxicity of industrial wastes to fish. *Sewage Industrial Wastes*, **23**, 1381–1397.

ECETOC (1993a) *Aquatic Toxicity Data Evaluation*. Technical Report No. 56, European Centre for Ecotoxicology and Toxicology of Chemicals, Brussels.

ECETOC (1993b) *Environmental Hazard Assessment of Substances*. Technical Report No. 51, European Centre for Ecotoxicology and Toxicology of Chemicals, Brussels, 92 pp.

ECETOC (1994) *Environmental Exposure Assessment*. Technical Report No. 61, European Centre for Ecotoxicology and Toxicology of Chemicals, Brussels.

Ellis, M.M. (1937) Detection and measurement of stream pollution. *Bulletin of the United States Bureau of Fisheries*, **48**, 365–437.

Emans, H.J.B., Van der Plassche, E.J., Canton, J.H., Okkerman, P.C. & Sparenburg, P.M. (1993) Validation of some extrapolation methods used for effect assessment. *Environmental Toxicology and Chemistry*, **12**, 2139–2154.

Environment Canada (1987) *Canadian Water Quality Guidelines*. Task Force on Water Quality Guidelines, Canadian Council of Resource and Environment Ministers. Environment Canada, Ottawa.

Environment Canada (1995) *Toxic Substances Management Policy*. Government of Canada, Environment Canada, Ottawa, 10 pp.

EC (1994) *Risk Assessment of Existing Substances*. Technical Guidance Document. Directorate-General, Environment, Nuclear Safety, and Civil Protection, European Commission, Brussels.

Ferson, S. & Akcakaya, H.R. (1990) *RAMAS/Age User Manual: Modeling Fluctuations in Age-structured Populations*. Exeter Software, Setauket, NY, 145 pp.

Forbes, T.L. & Forbes, V.E. (1993) A critique of the use

of distribution-based extrapolation models in ecotoxicology. *Functional Ecology*, **7**, 249–254.

Forbes, V.E. & Forbes, T.L. (1994) *Ecotoxicology in Theory and Practice*. Chapman and Hall, London, 247 pp.

Fujiwara, K. (1979). Japanese law on new chemicals and the methods to test the biodegradability and bioaccumulation of chemical substances. In: *Analyzing the Hazard Evaluation Process* (eds K.L. Dickson, A.W. Maki & J. Cairns, Jr.), pp. 50–57. Water Quality Section, American Fisheries Society, Washington, DC.

Gearing, J.N. (1989) The role of aquatic microcosms in ecotoxicologic research as illustrated by large marine systems. In: *Ecotoxicology: Problems and Approaches* (eds S.A. Levin, M.A. Harwell, J.R. Kelly & K.D. Kimball), pp. 411–472. Springer-Verlag, New York.

Giesy, J.P., Jr. (ed.) (1980) *Microcosms in Ecological Research*, CONF-781101. National Technical Information Service, Springfield, VA.

Giesy, J.P., Jr. (1985) Multispecies tests: research needs to assess the effects of chemicals on aquatic life. In: *Aquatic Toxicology and Hazard Assessment: Eighth Symposium*, STP 891 (eds R.C. Bahner D.J. Hansen), pp. 67–77 American Society for Testing and Materials, Philadelphia, PA.

Giesy, J.P., Jr. & Hoke, R.A. (1989) Freshwater sediment toxicity bioassessment: rationale for species selection and test design. *Journal of Great Lakes Research*, **15**, 539–569.

Giesy, J.P., Jr. & Hoke, R.A. (1991) Bioassessment of the toxicity of freshwater sediment. *Verhandlungen Internationale Vereinigung Limnologie*, **24**, 2313–2321.

Gillett, J.W. (1989) Terrestrial microcosms and mesocosms in ecotoxicologic research. In: *Ecotoxicology: Problems and Approaches* (eds S.A. Levine, M.A. Harwell, J.R. Kelly & W.D. Kimball), pp. 280–313. Springer-Verlag, New York.

Gillett, J.W. & Gile, J.P. (1976) Pesticide fate in terrestrial laboratory ecosystems. *International Journal of Environmental Studies*, **10**, 15–22.

Gillett, J.W. & Witt, J.M. (eds) (1977) *Terrestrial Microcosms: Symposium on Terrestrial Microcosms and Environmental Chemistry*. National Science Foundation, Washington, DC, 35 pp.

Gould, S.J. (1995) Spin doctoring Darwin. *Natural History*, **104**, 6–9, 70–71.

Grandy, N.J. (1995) Role of the OECD in chemicals control and international harmonization of testing methods. In: *Fundamentals of Aquatic Toxicology* (ed. G.M. Rand), 2nd edn, pp. 763–773. Taylor & Francis Publishers, Washington, DC.

Graney, R.L., Kennedy, J.H. & Rodgers, J.H. (eds) (1994) *Aquatic Mesocosm Studies in Ecological Risk Assessment*. Lewis Publishers, Boca Raton, FL, 723 pp.

Greig-Smith, P.W. (1991) Environmental risk assessment of plant protection products: an approach to the development of guidelines. *Bulletin OEPP (Organisation Europeene et Mediterraneene pour la Protection des Plantes*, **21**, 219–226.

Greig-Smith, P.W. (1992) A European perspective on ecological risk assessment, illustrated by pesticide registration procedures in the United Kingdom. *Environmental Toxicology and Chemistry*, **11**, 1673–1689.

Hammons, A.S. (ed.) (1981) *Methods of Ecological Toxicology: a Critical Review of Laboratory Multispecies Tests*. Ann Arbor Science Publishers, Ann Arbor, MI, 307 pp.

Handley, J.W. & Knight, D.J. (1995) EC environmental risk assessment of new and existing chemicals. In: *Environmental Toxicology Assessment* (ed. M. Richardson), pp. 345–362. Taylor & Francis, London.

Harrass, M. & Sayre, P. (1989) Use of microcosm data for regulatory decisions. In: *Aquatic Toxicology and Hazard Assessment: Twelfth Symposium*, STP 1027 (eds U. Cowgill & L. Williams), pp. 202–223. American Society for Testing and Materials, Philadelphia, PA.

Hart, W.B., Doudoroff, P. & Greenbank, J. (1945) *The Evaluation of the Toxicity of Industrial Wastes, Chemicals, and other Substances to Freshwater Fishes*. Waste Control Laboratory, Atlantic Refining Co., Philadelphia, PA, 317 pp.

Hedgecott, S. (1994) Prioritization and standards for hazardous chemicals. In: *Handbook of Ecotoxicology*, Vol. 2 (ed. P. Calow), pp. 368–393. Blackwell Scientific Publications, Oxford.

Heitmuller, P.T., Hollister, T.A. & Parrish, P.R. (1981) Acute toxicity of 54 industrial chemicals to sheepshead minnows (*Cyprinodon variegatus*). *Bulletin of Environmental Contamination and Toxicology*, **27**, 596–604.

Henderson, C. (1960) *Bioassay Procedures, Aims, and Equipment*. Technical Report W60-3, 246–248. US Department of Health, Education, and Welfare, Washington, DC.

Henderson, C. & Tarzwell, C.M. (1957) Bio-assays for control of industrial effluents. *Sewage and Industrial Wastes*, **29**, 1002–1017.

Hill, E.F. & Camardese, M.B. (1986) *Lethal Dietary Toxicities of Environmental Contaminants and Pesticides to Coturnix*. Fish and Wildlife Technical Report 2, US Department of Interior, Washington, DC, 147 pp.

Hill, E.F., Heath, R.G., Spann, J.W. & Williams, J.P. (1975) *Lethal Dietary Toxicities of Environmental Pollutants to Birds*. Special Scientific Report—Wildlife No. 191, US Department of the Interior, Washington, DC, 61 pp.

Hill, I.R., Matthiessen, P. & Heimbach, F. (eds) (1993) *Guidance Document on Sediment Toxicity Tests and*

Bioassays for Freshwater and Marine Environments. From the SETAC-Europe, 'Workshop on Sediment Toxicity Assessment', 8–10 November 1993, Renesse, The Netherlands, 105 pp.

Hill, I.R., Heimbach, F., Leeuwangh, P. & Matthiessen, P. (eds) (1994) *Freshwater Field Tests for Hazard Assessment of Chemicals*. Lewis Publishers, Boca Raton, FL, 561 pp.

Hoekstra, J.A. & van Ewijle, P.H. (1993) Alternatives for the no-observed-effect level. *Environmental Toxicology and Chemistry*, **12**, 187–194.

Hoffman, D.J., Rattner, B.A., Burton, G.A., Jr. & Cairns, J., Jr. (eds) (1995) *Handbook of Ecotoxicology*. Lewis Publishers, Boca Raton, FL, 768 pp.

Hohreiter, D.W. (1980) *Toxicities of Selected Substances to Freshwater Biota*. Report ANL/ES-94, Argonne National Laboratory, Argonne, IL, 71 pp.

Holl, K.D. & Cairns, J., Jr. (1995) Landscape indicators in ecotoxicology. In: *Handbook of Ecotoxicology* (eds D.J. Hoffman, B.A. Rattner, G.A. Burton, Jr. & J. Cairns, Jr.), pp. 185–197. Lewis Publishers, Boca Raton, FL.

Hollis, E.H. & Lennon, R.E. (1954) *The Toxicity of 1,085 Chemicals to Fish*. Fish Toxicity Report No. 2, US Department of the Interior, Kearneysville, WV, 66 pp.

Howard, P.H. (1989–1993) *Handbook of Environmental Fate and Exposure Data for Organic Chemicals*, Vol. I (Large Production and Priority Pollutants), Vol. II (Solvents), Vol. III (Pesticides), and Vol. IV (Solvents 2). Lewis Publishers, Boca Raton, FL.

Hudson, R.H., Tucker, R.K. & Haegele, M.A. (1984) *Handbook of Toxicity of Pesticides to Wildlife*, 2nd edn. Resource Publication No. 153, US Department of the Interior, Washington, DC, 90 pp.

Huggett, R.J., Kimerle, R.A., Mehrle, P.M., Jr. & Bergman, H.L. (eds) (1992a) *Biomarkers Biochemical, Physiological, and Histological Markers of Anthropogenic Stress*. Lewis Publishers, Chelsea, MI, 347 pp.

Huggett, R.J., Kimerle, R.A., Mehrle, P.M., Jr. & Bergman, H.L. (1992b) Introduction. In: *Biomarkers Biochemical, Physiological, and Histological Markers of Anthropogenic Stress* (eds R.J. Huggett, R.A. Kimerle, P.M. Mehrle, Jr. & H.L. Bergman), pp. 1–3. Lewis Publishers, Chelsea, MI.

Inform (1995) *Toxics Watch 1995*. Inform, New York, 816 pp.

Ingersoll, C.G., Ankley, G.T., Benoit, D.A. *et al.* (1995) Toxicity and bioaccumulation of sediment-associated contaminants using freshwater invertebrates: a review of methods and applications. *Environmental Toxicology and Chemistry*, **14**, 1885–1894.

IPCS (International Programme On Chemical Safety) (1978) *Principles and Methods for Evaluating the Toxicity of Chemicals*, Part I, Environmental Health Criteria 6, World Health Organization, Geneva.

IPCS (International Programme On Chemical Safety) (1990) *Principles for the Toxicological Assessment of Pesticide Residues in Food*. Environmental Health Criteria 104, World Health Organization, Geneva.

Johnson, W.W. & Finley, M.T. (1980) *Handbook of Acute Toxicity of Chemicals to Fish and Aquatic Invertebrates*. Resource Publication No. 137, US Department of the Interior, Washington, DC, 98 pp.

Jones, J.R.E. (1938) The relative toxicity of salts of lead, zinc, and copper to the stickleback (*Gasterosteus aculeatus* L.) and the effect of calcium on the toxicity of lead and zinc salts. *Journal of Experimental Biology*, **15**, 394–407.

Kevern, N.R. & Ball, R.C. (1965) Primary productivity and energy relationships in artificial streams. *Limnology and Oceanography*, **10**, 74–87.

Kimball, K.D. & Levin, S.A. (1985) Limitations of laboratory bioassays: the need for ecosystem-level testing. *Bioscience*, **35**, 165–171.

Klopffer, W. (1994). Environmental hazard assessment of chemicals and products. Part I. General assessment principles. *Environmental Science and Pollution Research International*, **1**, 47–53.

Kooijman, S.A.L.M. & Bedaux, J.J.M. (1996) *The Analysis of Aquatic Toxicity Data*. VU University Press, Amsterdam, The Netherlands, 149 pp.

Lander, L., Walterson, E., Lander, L., Niemela, J. & Olsen, L. (1994) *Classification of Chemical Preparations as Dangerous for the Environment*. Consequences of Different Criteria Models. 11994-05-20, Final draft, Nordic Council of Ministers, Stockholm, 61 pp.

LeBlanc, G.A. (1980) Acute toxicity of priority pollutants to water flea (*Daphnia magna*). *Bulletin of Environmental Contamination and Toxicology*, **24**, 684–691.

Lee, G.F. & Jones, R.A. (1980) Role of biotransformation in environmental hazard assessments. In: *Biotransformation and Fate of Chemicals in the Aquatic Environment* (eds A.W. Maki, K.L. Dickson & J. Cairns, Jr.), pp. 8–21. American Society of Microbiology, Washington, DC.

Lee, H., II, Boese, B.L., Pelletier, J., Winsor, M., Specht, D.T. & Randall, R.C. (1993) *Guidance Manual: Bedded Sediment Bioaccumulation Tests*. EPA/600/R-93/183, Office of Research and Development, Newport, OR, 231 pp.

Levin, S.A., Kimball, K.D., McDowell, W.H. & Kimball, S.F. (1984) New perspectives in ecotoxicology. *Environmental Management*, **8**, 375–442.

Levin, S.A., Harwell, M.A., Kelly, J.R. & Kimball, K.D. (eds) (1989) *Ecotoxicology: Problems and Approaches*. Springer-Verlag, New York, 548 pp.

Life Systems, Inc. (1983) *Testing Triggers Workshop: Workshop Report*. Project 1247, Contract No. 68-01-6554, Office of Toxic Substances, US Environmental Protection Agency, Washington, DC, 62 pp.

Lipnick, R.L., Bickings, C.K., Johnson, D.E. & Eastwood,

D.A. (1985a) Comparison of QSAR predictions with fish toxicity screening data for 110 phenols. In: *Aquatic Toxicology and Hazard Assessment: Eighth Symposium*, STP 891 (eds R.C. Bahner & D.J. Hansen), pp. 153–176. American Society for Testing and Materials, Philadelphia, PA.

Lipnick, R.L., Johnson, D.E., Gilford, J.H., Bickings, C.K. & Newsome, L.D. (1985b) Comparison of fish toxicity screening data for 55 alcohols with the quantitative structure–activity relationship predictions of minimum toxicity for nonreactive nonelectrolyte organic compounds. *Environmental Toxicology and Chemistry*, **4**, 281–296.

Lokke, H. (1994) Ecotoxicological extrapolation: tool or toy? In: *Ecotoxicology of Soil Organisms* (eds M.H. Donker, H. Eijsackers & F. Heimbach), pp. 412–425. Lewis Publishers, Boca Raton, FL.

Lundahl, P. (1979) Hazard assessment in schemes for new chemicals in France. In: *Analyzing the Hazard Evaluation Process* (eds K.L. Dickson, A.W. Maki & J. Cairns, Jr.), pp. 23–29. American Fisheries Society, Washington, DC.

MacPhee, C. & Ruelle, R. (1969) *Lethal Effects of 1888 Chemicals Upon Four Species of Fish from Western North America*. Forest, Wildlife, and Range Experimental Station Bulletin No. 3, University of Idaho, Moscow, ID, 112 pp.

Maki, A.W. (1979) An analysis of decision criteria in environmental hazard evaluation programs. In: *Analyzing the Hazard Evaluation Process* (eds K.L. Dickson, A.W. Maki & J. Cairns Jr.), pp. 83–100. American Fisheries Society, Bethesda, MD.

Maki, A.W., Dickson, K.L. & Cairns, J., Jr. (eds) (1980) *Biotransformation and Fate of Chemicals in the Aquatic Environment*. American Society of Microbiology, Washington, DC, 150 pp.

Marsh, M.C. (1907) *The Effect of Some Industrial Wastes On Fishes*. Water Supply and Irrigation Paper No. 192, pp. 337–348, US Geological Survey, Washington, DC.

Marvin, K.T. & Proctor, R.R., Jr. (1964) *Preliminary Results of the Systematic Screening of 4,306 Compounds As 'Red-Tide' Toxicants*. US Fish and Wildlife Services Data Report 2, US Department of the Interior, Washington, DC, 84 pp.

Matthews, R.A., Buikema, A.L., Jr., Cairns, J., Jr. & Rodgers, J.H., Jr. (1982) Biological monitoring. Part IIA—receiving system functional methods, relationships and indices. *Water Research*, **16**, 129–139.

Mayer, F.L., Jr. (1987) *Acute Toxicity Handbook of Chemicals to Estuarine Organisms*. EPA 600/8-87/017, Office of Research and Development, Gulf Breeze, FL.

Mayer, F.L., Jr. & Ellersieck, M.R. (1986) *Manual of Acute Toxicity: Interpretation and Database for 410 Chemicals and 66 Species of Freshwater Animals*. Resource Publication No. 160, US Department of the Interior, Washington, DC, 506 pp.

McKee, J.E. & Wolf, H.W. (1963) *Water Quality Criteria*, 2nd edn. Publication 3-A, California State Water Resources Control Board, Sacramento, CA, 549 pp. Available from National Technical Information Service, Springfield, VA.

Menzer, R.E., Lewis, M.A. & Fairbrother, A. (1994) Methods in environmental toxicology. In: *Principles and Methods of Toxicology*, 3rd edn (ed. A.W. Hayes), pp. 1391–1418. Raven Press, Ltd., New York.

Miller, S. (1993) Where all those EPA lists come from. *Environmental Science and Technology*, **27**, 2302–2303.

Mitchell, G.C. (1994) System design for field tests in flowing waters: design and use of outdoor artificial streams in ecotoxicology. In: *Freshwater Field Tests for Hazard Assessment of Chemicals* (eds I.R. Hill, F. Heimbach, P. Leeuwangh & P. Matthiessen), pp. 127–139. Lewis Publishers, Boca Raton, FL.

Moore, D.R.J. & Biddinger, G.R. (1995) The interaction between risk assessors and risk managers during the problem formulation phase. *Environmental Toxicology and Chemistry*, **14**, 2013–2014.

Morgan, E. & Knacker, T. (1994) The role of laboratory terrestrial model ecosystems in the testing of potentially harmful substances. *Ecotoxicology*, **3**, 213–233.

Mount, D.I. (1981) Margins of safety for aquatic communities. In: *Aquatic Toxicology and Hazard Assessment: Fourth Conference*, STP 737 (eds D.R. Branson & K.L. Dickson), pp. 5–9. American Society for Testing and Materials, Philadelphia, PA.

Mount, D.I. (1985) Scientific problems in using multispecies toxicity tests for regulatory purposes. In: *Multispecies Toxicity Testing* (ed. J. Cairns, Jr.), pp. 13–18. Pergamon Press, New York.

Mount, D.I. & Stephan, C.E. (1969) Chronic toxicity of copper to the fathead minnow (*Pimephales promelas*) in soft water. *Journal of the Fisheries Research Board of Canada*, **26**, 2449–2457.

Nabholz, J.V. (1991) Environmental hazard and risk assessment under the United States Toxic Substances Control Act. *Science of the Total Environment*, **109/110**, 649–665.

Nabholz, J.V., Miller, P. & Zeeman, M. (1993a) Environmental risk assessment of new chemicals under the Toxic Substances Control Act (TSCA) Section Five. In: *Environmental Toxicology and Risk Assessment*, STP 1179 (eds W.G. Landis, J.S. Hughes & M.A. Lewis), pp. 40–55. American Society for Testing and Materials, Philadelphia, PA.

Nabholz, J.V., Clements, R.G., Zeeman, M., Osborn, K.C. & Wedge, R. (1993b) Validation of structure–activity relationships used by the USEPA's Office of Pollution Prevention and Toxics for the environmental hazard assessment of industrial chemicals. In: *Environmental*

Toxicology and Risk Assessment, 2nd volume, STP 1216 (eds J.W. Gorsuch, F.J. Dwyer, C.G. Ingersoll & T.W. LaPoint), pp. 571–590. American Society for Testing and Materials, Philadelphia, PA.

NAS (1972) *Water Quality Criteria 1972* National Research Council, National Academy of Science/ Engineering, US Government Printing Office, Washington, DC, 594 pp.

NAS (1981) *Testing for Effects of Chemicals on Ecosystems*. National Research Council, National Academy of Science, National Academy Press, Washington, DC, 103 pp.

NAS (1983) *Risk Assessment in the Federal Government: Managing the Process*. National Research Council, National Academy of Science, National Academy Press, Washington, DC.

Naylor, C. & Rodrigues, C. (1995) Development of a test method for *Chironomus riparius* using a formulated sediment. *Chemosphere*, **31**, 3291–3303.

Nebeker, A.V., Cairns, M.A., Gakstatter, J.H., Malueg, K.W., Schuytema, G.S. & Krawczyk, D.F. (1984) Biological methods for determining toxicity of contaminated freshwater sediments to invertebrates. *Environmental Toxicology and Chemistry*, **3**, 617–630.

Newsome, L.D., Johnson, D.E., Cannon, D.J. & Lipnick, R.L. (1987) Comparison of fish toxicity screening data and QSAR predictions for 48 aniline derivatives. In: *QSAR in Environmental Toxicology—II* (ed. K.L.E. Kaiser), pp. 231–250. D. Reidel, Dordrecht.

Newsome, L.D., Johnson, D.E. & Nabholz, J.V. (1993) Validation and upgrade of a QSAR study of the toxicity of amines to freshwater fish. In: *Environmental Toxicology and Risk Assessment*, STP 1179 (eds W.G. Landis, J.S. Hughes & M.A. Lewis), pp. 413–426. American Society for Testing and Materials, Philadelphia, PA.

Nixon, S. (1969) A synthetic microcosm. *Limnology and Oceanography*, **14**, 142–145.

Noppert, F., Van der Hoeven, N. & Leopold, A. (1994) *How to Measure No Effect—Towards a New Measure of Chronic Toxicity in Ecotoxicology*. Workshop Report, Netherlands Working Group on Statistics and Ecotoxicology, The Hague, 44 pp.

Norton, S.B., Rodier, D.J., Gentile, J.H., van der Schalie, W.H., Wood, W.P. & Slimak, M.W. (1992) A framework for ecological risk assessment at the EPA. *Environmental Toxicology and Chemistry*, **11**, 1663–1672.

Okkerman, P.C., Van der Plassche, E.J., Slooff, W., van Leeuwen, C.J. & Canton, J.H. (1991) Ecotoxicological effects assessment: a comparison of several extrapolation procedures. *Ecotoxicology and Environmental Safety*, **21**, 182–193.

Okkerman, P.C., Van der Plassche, E.J., Emans, H.J.B. & Canton, J.H. (1993) Validation of some extrapolation methods with toxicity data derived from multiple species experiments. *Ecotoxicology and Environmental Safety*, **25**, 341–359.

OECD (1979) *Report on the Assessment of Potential Environmental Effects of Chemicals. The Effects on Organisms other than Man and on Ecosystems*. Chemicals Testing Programme, Ecotoxicology Group, Organization for Economic Cooperation and Development, Paris, 28 pp.

OECD (1984) *Guidelines for Testing of Chemicals. Section 2: Effects on Biotic Systems*. Organization for Economic Cooperation and Development, Paris.

OECD (1989) *Report of the OECD Workshop on Ecological Effects Assessment*. Environment Monograph No. 26, Organization for Economic Cooperation and Development, Paris, 67 pp.

OECD (1992a) *On the Extrapolation of Laboratory Aquatic Toxicity Data to the Real Environment*. Environment Monograph No. 59, Organization for Economic Cooperation and Development, Paris, 43 pp.

OECD (1992b) *Report of the OECD Workshop on Quantitative Structure Activity Relationships (QSARs) in Aquatic Effects Assessment*. Environment Monograph No. 58, Organization for Economic Cooperation and Development, Paris, 95 pp.

OECD (1993) *Report of the OECD Workshop on Effects Assessment of Chemicals in Sediment*. Environment Monograph No. 60, Organization for Economic Cooperation and Development, Paris, 56 pp.

OECD (1995a) *Guidance Document for Aquatic Effects Assessment*. Environment Monograph No. 92, Organization for Economic Cooperation and Development, Paris, 116 pp.

OECD (1995b) *Report of the OECD Workshop on Environmental Hazard/Risk Assessment*. Environment Monograph No. 105, Organization for Economic Cooperation and Development, Paris, 77 pp.

Pack, S. (1993) *A Review of Statistical Data Analysis and Experimental Design in OECD Aquatic Toxicology Test Guidelines*. Unpublished Report, Shell Research Ltd., Kent, 42 pp.

Pedersen, F., Tyle, H., Niemela, J.R., Guttmann, B., Lander, L. & Wedebrand, A. (1994) *Environmental Hazard Classification*. Tema Nord 1994: 589, Nordic Council of Ministers, Copenhagen, 160 pp.

Perry, D.A. (1994) *Forest Ecosystems*. The Johns Hopkins University Press, Baltimore, MD, 649 pp.

Persoone, G. & Janssen, C.R. (1994) Field validation of predictions based on laboratory toxicity tests. In: *Freshwater Field Tests for Hazard Assessment of Chemicals* (eds I.R. Hill, F. Heimbach, P. Leeuwangh & P. Mattiessen), pp. 379–397. Lewis Publishers, Boca Raton, FL.

Peters, R.H. (1991) *A Critique for Ecology*. Cambridge University Press, Cambridge, 384 pp.

Pimentel, D. (1995) Amounts of pesticides reaching target

pests: environmental impacts and ethics. *Journal of Agricultural and Environmental Ethics*, **8**, 17–29.

Ramamoorthy, S., Baddaloo, E.G. & Ramamoorthy, S. (1995) *Handbook of Chemical Toxicity Profiles of Biological Species*, Vol. I (Aquatic Species), Vol. II (Avian and Mammalian Species). Lewis Publishers, Boca Raton, FL.

Rand, G.M. (ed.) (1995) *Fundamentals of Aquatic Toxicology*, 2nd edn. Taylor & Francis Publishers, Washington, DC, 1125 pp.

Reckhow, K.H. (1994) Importance of scientific uncertainty in decision making. *Environmental Management*, **18**, 161–166.

Reynoldson, T.B. & Day, K.E. (1993) Freshwater sediments. In: *Handbook of Ecotoxicology*, Vol. 1 (ed. P. Calow), pp. 83–100. Blackwell Scientific Publications, Oxford.

Rodier, D.J. & Mauriello, D.A. (1993) The quotient method of ecological risk assessment and modeling under TSCA: a review. In: *Environmental Toxicology and Risk Assessment*, STP 1179 (eds W.G. Landis, J.S. Hughes & M.A. Lewis), pp. 80–91. American Society for Testing and Materials, Philadelphia, PA.

Rodier, D.J. & Zeeman, M.G. (1994) Ecological risk assessment. In: *Basic Environmental Toxicology* (eds L.G. Cockerham & B.S. Shane), pp. 581–604. CRC Press, Boca Raton, FL.

Schafer, E.W., Jr. & Bowles, W.A., Jr. (1985) Acute oral toxicity and repellency of 933 chemicals to house and deer mice. *Archives of Environmental Contamination and Toxicology*, **14**, 111–129.

Schafer, E.W., Jr., Bowles, W.A., Jr. & Hurlbert, J. (1983) The acute oral toxicity, repellency, and hazard potential of 998 chemicals to one or more species of wild and domestic birds. *Archives of Environmental Contamination and Toxicology*, **12**, 355–382.

Scherer, E. (ed.) (1979) *Toxicity Tests for Freshwater Organisms*. Canadian Special Publication, Fisheries and Aquatic Science 44, Freshwater Institute, Winnipeg, 194 pp.

Shannon, L.J., Harrass, M.C., Yount, J.D. & Walbridge, C.T. (1986) A comparison of mixed flask culture and standardized laboratory model ecosystems for toxicity testing. In: *Community Toxicity Testing*, STP 920 (ed. J. Cairns, Jr.), pp. 135–157. American Society for Testing and Materials, Philadelphia, PA.

Shelford, V.E. (1917) An experimental study of the effects of gas wastes upon fishes, with especial reference to stream pollution. *Bulletin of the Illinois State Laboratory for National History*, **11**, 381–412.

Slooff, W. & Canton, J.H. (1983) Comparison of the susceptibility of 11 freshwater species to 8 chemical compounds. II. (Semi) chronic toxicity tests. *Aquatic Toxicology*, **4**, 271–282.

Slooff, W., Canton, J.H. & Hermans, J.L.M. (1983) Comparison of the susceptibility of 22 freshwater species to 15 chemical compounds. I. (Sub) acute toxicity tests. *Aquatic Toxicology*, **4**, 113–128.

Slooff, W., Van Oers, J.A.M. & De Zwart, D. (1986) Margins of uncertainty in ecotoxicological hazard assessment. *Environmental Toxicology and Chemistry*, **5**, 841–852.

Smrchek, J.C., Zeeman, M. & Clements, R. (1995) Ecotoxicology and the assessment of chemicals at the US EPA's Office of Pollution Prevention and Toxics: current activities and future needs. In: *Making Environment Science* (eds J.R. Pratt, N. Boxers & J.R. Stauffer), pp. 127–158. Ecoprint, Portland, OR.

Smrchek, J.C., Clements, R., Morcock, R. & Rabert, W. (1993) Assessing ecological hazard under TSCA: methods and evaluation of data. In: *Environmental Toxicology and Risk Assessment*, STP 1179 (eds W.G. Landis, J.S. Hughes & M.A. Lewis), pp. 22–39. American Society for Testing and Materials, Philadelphia, PA.

Soares, A.M.V.M. & Calow, P. (eds) (1993a) *Progress in Standardization of Aquatic Toxicity Tests*. Lewis Publishers, Boca Raton, FL, 224 pp.

Soares, A.M.V.M. & Calow, P. (1993b) Seeking standardization in ecotoxicology. In: *Progress in Standardization of Aquatic Toxicity Tests* (eds A.M.V.M. Soares & P. Calow), pp. 1–6. Lewis Publishers, Boca Raton, FL.

SETAC (1994) *FINAL REPORT: Aquatic Risk Assessment and Mitigation Dialogue Group*. Society of Environmental Toxicology and Chemistry Press, Pensacola, FL, 220 pp.

SETAC (1995) Environment Canada finalizes toxic substances management policy. *Society of Environmental Toxicology and Chemistry News*, **15**, 22.

SETAC (1996) An overview of European risk assessment requirements for new and existing substances. *Society of Environmental Toxicology and Chemistry News*, **16**, 24.

Sprague, J.B. (1970) Measurement of pollutant toxicity to fish. II. Utilizing and applying bioassay results. *Water Research*, **4**, 3–32.

Sprague, J.B. (1971) Measurement of pollutant toxicity to fish—III. Sublethal effects and 'safe' concentrations. *Water Research*, **5**, 245–266.

Sprague, J.B. & Fogels, A. (1977) *Watch the Y in Bioassay*, pp. 107–118. Technical Report No. EPS-5-AR-77-1, Environmental Protection Service, Halifax, Canada.

Steele, R.L. & Thursby, G.B. (1983) A toxicity test using life stages of *Champia parvula* (Rhodophyta). In: *Aquatic Toxicology and Hazard Assessment: Sixth Symposium*, STP 802 (eds W.E. Bishop, R.D. Cardwell & B.B. Heidolph), pp. 73–89. American Society for Testing and Materials, Philadelphia, PA.

Stephan, C.E. & Rogers, J.W. (1985) Advantages of using regression analysis to calculate results of chronic toxicity tests. In: *Aquatic Toxicology and Hazard Assessment: Eighth Symposium*, STP 891 (eds R.L.

Bahner & D.J. Hansen), pp. 328–338. American Society for Testing and Materials, Philadelphia, PA.

Suedel, B.C., Deaver, E. & Rodgers, J.H., Jr. (1996) Formulated sediment as a reference and dilution sediment in definitive toxicity tests. *Archives of Environmental Contamination and Toxicology*, **30**, 47–52.

Suter, G.W., II (1990) Endpoints for regional ecological risk assessment. *Environmental Management*, **14**, 9–23.

Suter, G.W., II (1993a) A critique of ecosystem health concepts and indexes. *Environmental Toxicology and Chemistry*, **12**, 1533–1539.

Suter, G.W., II (ed.) (1993b) *Ecological Risk Assessment*. Lewis Publishers, Boca Raton, FL, 538 pp.

Suter, G.W., II (1995) Introduction to ecological risk assessment for aquatic toxic effects. In: *Fundamentals of Aquatic Toxicology*, 2nd edn (ed. G. Rand), pp. 803–816. Taylor & Francis Publishers, Washington, DC.

Suter, G.W., II, Vaughan, D.S. & Gardner, R.H. (1983) Risk assessment by analysis of extrapolation error: a demonstration for effects of pollutants on fish. *Environmental Toxicology and Chemistry*, **2**, 369–378.

Suter, G.W., II, Rosen, A.E., Linder, E. & Parkhurst, D.F. (1987) Endpoints for responses of fish to chronic toxic exposures. *Environmental Toxicology and Chemistry*, **6**, 793–809.

Tansley, A.G. (1935) The use and abuse of vegetational concepts and terms. *Ecology*, **16**, 284–307.

Taub, F.B. (1976) Demonstration of pollution effects in aquatic microcosms. *International Journal of Environmental Studies*, **10**, 23–33.

Taub, F.B. (1984a) Introduction: laboratory microcosms. In: *Concepts in Marine Pollution Measurements* (ed. H.H. White), pp. 113–116. Maryland Sea Grant Publication, University of Maryland, College Park, MD.

Taub, F.B. (1984b) Measurement of pollution in standardized aquatic microcosms. In: *Concepts in Marine Pollution Measurements* (ed. H.H. White), pp. 159–192. Maryland Sea Grant Publication, University of Maryland, College Park, MD.

Taub, F.B. & Crow, M.E. (1978) Loss of critical species in a model (laboratory) ecosystem. *Verhandlungen Internationale Vereinigung Limnologie*, **20**, 1270–1276.

Taub, F.B. & Crow, M.E. (1980) Synthesizing aquatic microcosms. In: *Microcosms in Ecological Research* (ed. J.P. Giesy, Jr.), pp. 69–104. National Technical Information Service, Springfield, VA.

Taub, F.B., Kindig, A.C. & Conquest, L.L. (1987) Interlaboratory testing of a standardized aquatic microcosm protocol. In: *Aquatic Toxicology and Environmental Fate: Tenth Volume*, STP 971 (eds W.J. Adams, G.A., Chapman & W.G. Landis), pp. 384–405. American Society for Testing and Materials, Philadelphia, PA.

Taub, F.B., Kindig, A.C., Conquest, L.L. & Meador, J.P. (1989) Results of interlaboratory testing of the standardized aquatic microcosm protocol. In: *Aquatic Toxicology and Environmental Fate: Eleventh Volume*, STP 1007 (eds G.W. Suter II & M.A. Lewis), pp. 368–394. American Society for Testing and Materials, Philadelphia, PA.

Tebo, L.B., Jr. (1986) Effluent monitoring: historical perspective. In: *Environmental Hazard Assessment of Effluents* (eds R.A. Kimerle & A.W. Maki), pp. 13–31. Pergamon Press, New York.

UK Department of the Environment (1993) *Guidance on Risk Assessment of Existing Substances*. Unpublished report, Department of the Environment, London, 69 pp.

USEPA (1975) *Methods for Acute Toxicity Tests with Fish, Macroinvertebrates, and Amphibians*. EPA-600/3-75-009, US Environmental Protection Agency, Office of Research and Development, Washington, DC, 61 pp.

USEPA (1977) *Quality Criteria for Water*. US Environmental Protection Agency, US Government Printing Office, Washington, DC, 256 pp.

USEPA (1978) *Bioassay Procedures for the Ocean Disposal Permit Program*. EPA-600/9-78-010, US Environmental Protection Agency, Office of Research and Development, Gulf Breeze, FL, 121 pp.

USEPA (US Environmental Protection Agency) (1980) Water quality criteria documents; availability. *Federal Register*, **45**, 79318–79379.

USEPA (1983a) *Testing for Environmental Effects under the Toxic Substances Control Act*. US Environmental Protection Agency, Office of Pollution Prevention and Toxics, Washington, DC, 24 pp.

USEPA (1983b) *Technical Support Document for the Environmental Effects Testing Scheme*. US Environmental Protection Agency, Office of Pollution Prevention and Toxics, Washington, DC, 31 pp.

USEPA (1984) *Estimating Concern Levels for Concentrations of Chemical Substances in the Environment*. US Environmental Protection Agency, Office of Pollution Prevention and Toxics, Washington, DC, 31 pp.

USEPA (1985a) *Hazard Evaluation Division, Standard Evaluation Procedure, Avian Dietary LC50 Test*. EPA-540/9-85-008, US Environmental Protection Agency, Office of Pesticide Programs, Washington, DC.

USEPA (1985b) *Hazard Evaluation Division, Standard Evaluation Procedure, Avian Single-dose Oral LD50*. EPA-540/9-85-001, US Environmental Protection Agency, Office of Pesticide Programs, Washington, DC.

USEPA (1985c) *Hazard Evaluation Division, Standard Evaluation Procedure, Wild Mammal Toxicity Test*. EPA-540/9-85-004, US Environmental Protection

Agency, Office of Pesticide Programs, Washington, DC.

USEPA (US Environmental Protection Agency) (1985d) Toxic substance control act test guidelines; Final rules. *Federal Register* **50**, 39252–39516 (Part 796—Chemical fate testing guidelines, Part 797—Environmental effects testing guidelines).

USEPA (US Environmental Protection Agency) (1986) Proposed rules, Part 797—Amended. *Federal Register*, **51**, 490–492 (Section 797.1310, Gammarid acute toxicity test).

USEPA (US Environmental Protection Agency) (1987a) Revision of TSCA test guidelines. *Federal Register*, **52**, 19056–19082 (Part 796—Amended; Part 797—Amended).

USEPA (US Environmental Protection Agency) (1987b) Toxic substances control act test guidelines, proposed rule. *Federal Register*, **52**, 36334–36371 (Part 797—Amended, Subpart D—Microcosm guidelines).

USEPA (1988) *Estimating Toxicity of Industrial Chemicals to Aquatic Organisms Using Structure Activity Relationships* (ed. R.G. Clements). EPA-560-6-88-001, US Environmental Protection Agency, Office of Toxic Substances, Washington, DC.

USEPA (1989a) *Fish Toxicity Screening Data. Part 1: Lethal Effects of 964 Chemicals Upon Steelhead Trout and Bridgelip Sucker. Part 2: Lethal Effects of 2,014 Chemicals Upon Sockeye Salmon, Steelhead Trout, and Threespine Stickleback.* Unpublished reports by C. MacPhee (1974) & F. F. Cheng (1975), EPA 560/6-89-001, US Environmental Protection Agency Washington, DC.

USEPA (1989b) *Sediment Classification Methods Compendium.* Draft Final Report, US Environmental Protection Agency, Office of Water, Washington, DC.

USEPA (1991a) *Manual for the Evaluation of Laboratories Performing Aquatic Toxicity Tests* (Authors: D.J. Klemm, L.B. Lobring & W.H. Horning). EPA/600/4-90/031, US Environmental Protection Agency, Office of Research and Development, Cincinnati, OH, 108 pp.

USEPA (1991b) *Summary Report on Issues in Ecological Risk Assessment.* EPA/625/3-91/018, US Environmental Protection Agency, Risk Assessment Forum, Washington, DC.

USEPA (1991c) *Technical Support Document for Water Quality-based Toxics Control.* EPA/505/2-90/001, US Environmental Protection Agency, Office of Water, Washington, DC, 387 pp.

USEPA (1992) *Framework for Ecological Risk Assessment.* EPA/630-R-92/001, US Environmental Protection Agency, Risk Assessment Forum, Washington, DC, 41 pp.

USEPA (1993a) *ECOSAR, a Computer Program for Estimating the Ecotoxicity of Industrial Chemicals Based on Structure Activity Relationships.* EPA-748-F-93-001, US Environmental Protection Agency, Office of Pollution Prevention and Toxics, Washington, DC.

USEPA (1993b) *Methods for Measuring the Acute Toxicity of Effluents and Receiving Waters to Freshwater and Marine Organisms* (ed. C.I. Weber), 4th edn. EPA/600/4-90/027F, US Environmental Protection Agency, Office of Research and Development, Cincinnati, OH, 293 pp.

USEPA (1994a) *ECOSAR: Computer Program and User's Guide for Estimating the Ecotoxicity of Industrial Chemicals based on Structure Activity Relationships.* EPA-748-R-93-002, US Environmental Protection Agency, Office of Pollution Prevention and Toxics, Washington, DC.

USEPA (1994b) *EPA's Contaminated Sediment Strategy.* EPA 823-R-94-001, US Environmental Protection Agency, Office of Water, Washington, DC, 130 pp.

USEPA (1994c) *Methods for Assessing the Toxicity of Sediment-associated Contaminants with Estuarine and Marine Amphipods.* EPA/600/R-94/025, US Environmental Protection Agency, Office of Research and Development, Narragansett, RI, 140 pp.

USEPA (1994d) *Methods for Measuring the Toxicity and Bioaccumulation of Sediment–associated Contaminants with Freshwater Invertebrates.* EPA/600/R-94/024, US Environmental Protection Agency, Office of Research and Development, Duluth, MN, 133 pp.

USEPA (1994e) *Pesticide Reregistration Rejection Rate Analyses. Ecological Effects.* EPA 738-R-94-035, US Environmental Protection Agency, Office of Prevention, Pesticides, and Toxic Substances, Washington, DC, 167 pp.

USEPA (1994f) *Short-term Methods for Estimating the Chronic Toxicity of Effluents and Receiving Waters to Freshwater Organisms* (eds P.A. Lewis, D.J. Klemm, J.M. Lazorchak, T.J. Norberg-King, W.H. Peltier & M.A. Heber), 3rd edn. EPA/600/4-91/002, US Environmental Protection Agency, Office of Research and Development, Cincinnati, OH, 341 pp.

USEPA (1994g) *Short-term Methods for Estimating the Chronic Toxicity of Effluents and Receiving Waters to Marine and Estuarine Organisms* (eds D.J. Klemm, G.E. Morrison, T.J. Norberg-King, W.H. Peltier & M.A. Heber), 2nd edn. EPA/600/4-91/003, US Environmental Protection Agency, Office of Research and Development, Cincinnati, OH, 483 pp.

USEPA (1995a) *A Phase 1 Inventory of Current EPA Efforts to Protect Ecosystems.* EPA 841-S-95-001, US Environmental Protection Agency, Office of Water, Washington, DC, 357 pp.

USEPA (1995b) *World at a Glance: a Directory of International Chemicals Programs.* US Environmental Protection Agency, Office of Pollution Prevention and Toxics, Washington, DC, 20 pp.

USEPA (US Environmental Protection Agency) (1996) Proposed testing guidelines; notice of availability and request for comments. *Federal Register*, **61**, 16486–16488.

USFDA (1987) *Environmental Assessment Technical Handbook*. US Food and Drug Administration, Center for Food Safety and Applied Nutrition; Center for Veterinary Medicine, Rockville, MD.

USHHS (1992) *Toxicological Profile for Tin*. Report TP-91/27, US Department of Health and Human Services, Agency for Toxic Substances and Disease Registry, Washington, DC, 148 pp.

University of Wisconsin-Superior (1984–1990) *Acute Toxicities of Organic Chemicals to Fathead Minnows (Pimephales promelas)*, Vols. I–IV (eds L.T. Brooke, D.J. Call, D.L. Geiger, C.E. Northcote & S.H. Poirier). Center for Lake Superior Environmental Studies, University of Wisconsin-Superior, Superior, WI.

University of Wisconsin-Superior (1992) *Subchronic Toxicities of Industrial and Agricultural Chemicals to Fathead Minnows (Pimephales promelas)*, Vol. I (eds D.J. Call & D.L. Geiger). Center for Lake Superior Environmental Studies, University of Wisconsin-Superior, Superior, WI, 318 pp.

Urban, D.J. & Cook, N.J. (1986) *Hazard Evaluation Division, Standard Evaluation Procedure, Ecological Risk Assessment*. EPA 540/9-85-001, Office of Pesticide Programs, Washington, DC, 96 pp.

van den Berg, M. (1992/1993) Ecological risk assessment and policy-making in the Netherlands: dealing with uncertainties. *Network*, **6**, 8–11.

van Gestel, C.A.M. & Van Straalen, N.M. (1993) Soil invertebrates and micro-organisms. In: *Handbook of Ecotoxicology*, Vol. 1 (ed. P. Calow), pp. 251–277. Blackwell Scientific Publications, Oxford.

van Leeuwen, C.J. (1995) Ecotoxicological effects. In: *Risk Assessment of Chemicals. An Introduction* (eds C.J. van Leeuwen & J.L.M. Hermans), pp. 175–237. Kluwer Academic Publishers, Dordrecht.

van Leeuwen, C.J. & Hermans, J.L.M. (eds) (1995) *Risk Assessment of Chemicals. An Introduction*. Kluwer Academic Publishers, Dordrecht, 374 pp.

van Straalen, C.A.M. & van Gestel, N.M. (1994) Ecotoxicological test systems for terrestrial invertebrates. In: *Ecotoxicology of Soil Organisms* (eds M.H. Donker, H. Eijsackers & F. Heimbach), pp. 207–228. Lewis Publishers, Boca Raton, FL.

van Voris, P., Tolle, D.A. & Arthur, M.F. (1985a) *Experimental Soil-core Microcosm Test Protocol. A Method for Measuring the Potential Ecological Effects, Fate, and Transport of Chemicals in Terrestrial Ecosystems*. EPA/600/3-85/047, PNL-5400, Office of Research and Development, Corvallis, OR.

van Voris, P., Tolle, D.A., Arthur, M.F. & Chesson, J. (1985b) Terrestrial microcosms: applications, validation, and cost-benefit analysis. In: *Multi-species Toxicity Testing* (ed. J. Cairns, Jr.), pp. 117–142. Pergamon Press, New York.

Verschueren, K. (1983) *Handbook of Environmental Data on Organic Chemicals*, 2nd edn. Van Nostrand Reinhold Co., New York, 1310 pp.

Walker, P.M.B. (1989) *Cambridge Dictionary of Biology*. Cambridge University Press, Cambridge, 324 pp.

Warren, C.E. (1971) *Biology and Water Pollution Control*. W.B. Saunders Co., Philadelphia, PA, 434 pp.

Whittaker, R.H. (1957) Recent evolution of ecological concepts in relation to the eastern forests of North America. *American Journal of Botany*, **44**, 197–206.

Witt, J.M. & Gillett, J.W. (eds) (1977). *Terrestrial Microcosms and Environmental Chemistry*. National Science Foundation, Washington, DC, 147 pp.

Wood, E.M. (1953) *The Toxicity of 3,400 Chemicals to Fish*. Fish Toxicity Report No. 1, US Department of the Interior, Kearneysville, WV, 198 pp.

Zeeman, M.G. (1995a) Ecotoxicity testing and estimation methods developed under Section 5 of the Toxic Substances Control Act (TSCA). In: *Fundamentals of Aquatic Toxicology* (ed. G. Rand), 2nd edn, pp. 703–715. Taylor & Francis, Washington, DC.

Zeeman M.G. (1995b) EPA's framework for ecological effects assessment. In: *Screening and Testing Chemicals in Commerce*, pp. 69–78. OTA-BP-ENV-166, US Congress, Office of Technology Assessment, Washington, DC.

Zeeman, M. (1997) Aquatic toxicology and ecological risk assessment: US-EPA/OPPT perspective and OECD interactions. In: *Ecotoxicology: Responses, Biomarkers and Risk Assessment*, an OECD Workshop (ed. J. T. Zelikoff), pp. 89–108. SOS Publications, New Haven, NJ.

Zeeman, M. & Gilford, J. (1993) Ecological hazard evaluation and risk assessment under EPA's Toxic Substances Control Act (TSCA): an introduction. In: *Environmental Toxicology and Risk Assessment*, STP 1179 (eds W.G. Landis, J.S. Hughes & M.A. Lewis), pp. 7–21. American Society for Testing and Materials, Philadelphia, PA.

Zeeman, M., Auer, C.M., Clements, R.G., Nabholz, J.V. & Boethling, R.S. (1995) US EPA regulatory perspectives on the use of QSAR for new and existing chemical evaluations. *SAR and QSAR in Environmental Research*, **3**, 179–201.

Chapter 4
Evaluation of the Likelihood of Major Accidents in Industrial Processes

JOHN H. GOULD

4.1 INTRODUCTION

Undoubtedly the best method to estimate the likelihood of accidents is the statistical analysis of accidents that have already occurred. However, this is often difficult since, fortunately, major accidents rarely occur within the chemical industry. Even so, many organizations have spent considerable effort in collecting accident data, e.g. MHIDAS data base funded by the Health and Safety Executive (HSE). These databases can give a unique insight into what can go wrong and, to a limited extent, how often.

In reality, for many industries the lack of sufficient relevant data makes historical analysis inadequate, and the introduction of new or novel processes makes the application of historical data impossible. In most cases the frequency of accidents needs to be synthesized by a quantified risk assessment technique. Risk assessments based on experience are of limited use for novel processes or for rare accidents with such high consequences that society will not tolerate them. For these hazards a risk assessment using predictive techniques has been developed, initially in the nuclear industry, and successfully used in the petroleum and chemical industries. Similar techniques have also been successfully applied to wide areas of industry, from rail transport to pressure vessel design.

Risk assessment may appear to be a new technique with early applications outside the nuclear industry in the Canvey island study (HSE 1978) and the Rijmond study (Cremer & Warner Ltd 1981), but risk-based decisions have been made since 'the dawn of humanity'. Our environment has always been a risky place. People quickly learned that the benefit of eating tiger meat was far outweighed by the risks of trying to kill one. People made a risk-based decision, initially taking into account direct experience and later using historical data passed on to succeeding generations. A modern example of this is when we cross the road. Whilst most people do not have direct experience of being run over, we are aware of the consequences. We think we know the likelihood of being run over based on the estimate of the vehicle's speed and our own agility. Using this information we balance the risk of being run over against the benefit of crossing the road, and we make a risk-based decision.

Similar assessments have always been carried out within industry. These assessments have relied heavily on the experience of the operators and their supervisors. Much of this hard-earned experience is written down as plant rules, company standards and even regulations. A good example of industry's best practice incorporated into national guidance is the British Standard (British Standards Institute 1995) that controls ignition sources in flammable atmospheres.

These assessments have been formalized under recent UK legislation–Management of Health and Safety at Work Regulations (HSC 1992)—that requires companies to carry out risk assessments on all their activities and in many cases record their results. Similar developments for industrial safety have been carried out throughout the world and whilst this chapter focuses on risk assessment within British industry, the basic principles differ little world-wide.

In the nuclear, chemical and petrochemical industries, such risk assessments may be given a numerical solution. The nuclear industry refers

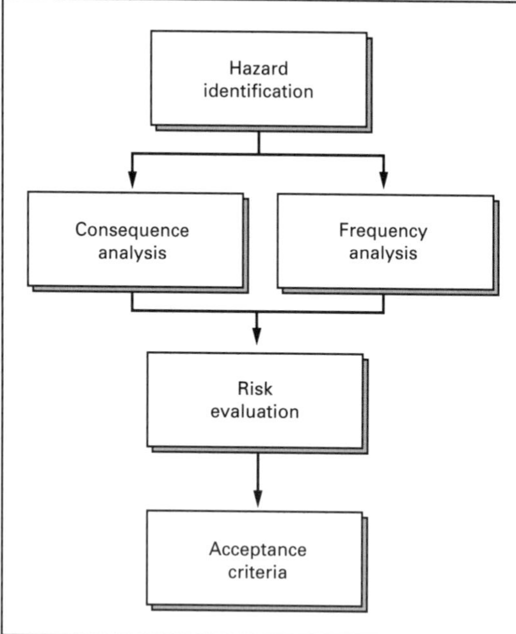

Fig. 4.1 The five stages of QRA.

to them as probabilistic safety assessments (PSA), whilst in the chemical and petrochemical industries they are called chemical plant quantified risk assessments (CPQRA), often shortened to QRA. All these techniques are basically the same and they are commonly divided into five stages: hazard identification, consequence analysis, frequency analysis, risk evaluation and acceptance criteria. These five stages are shown as a flow diagram in Fig. 4.1.

A significant point to note is that risk is a function of two parameters: the likelihood of occurrence of the undesired event and its consequences.

Risk = f (frequency × consequences)

Clearly there is scope for the frequency and the consequence analysis to be carried out in any order or even in parallel. There are also a considerable number of places where feedback loops can be and are utilized, where the results of one stage are reviewed and used to alter the basis of a previous stage in an iterative process. Once identified, a hazard can be excluded from the analysis on the grounds of its frequency or its consequence. Several flow diagrams have been produced to show how the different stages fit in with a risk management system, an example of which is shown in Fig. 4.2.

Whilst the above framework is often given, in reality activities that have a trivial risk are not usually assessed. Most of the risks are within the ALARP (as low as reasonably practicable) region and although the assessment may initiate risk reduction measures, reassessment will only be carried out if the assessment is easy to perform or if there is a need to show that the risk is below a set criterion.

The UK Regulatory Authority for health and safety (Health and Safety Executive; HSE) makes extensive use of quantified risk assessment (QRA) for the nuclear, offshore and onshore major chemical hazard industries. Although there is a thread of commonality within the QRA techniques used, there is a fundamental difference in approach that gives them an individual character. Significantly, none of the techniques resemble the procedures commonly shown in diagrams similar to Fig. 4.2. Many of the differences are due to the way the risk assessment is used within the three industries, as well as the assessment acceptance criteria set. These two points are discussed later.

4.2 DIFFERING USES OF QRA

But why carry out a QRA? Provided that all the risks are below the intolerable range, the frequency analysis can be disregarded. Identifying what can go wrong (hazard identification) and how bad the outcomes will be (consequence analysis) should be enough to initiate measures to help to prevent the accident occurring.

Should more information be needed on the relative ranking of the risks, in order to identify the best place to spend a limited amount of resources to achieve the maximum reduction in risk, then a semi-quantitative assessment of the relative likelihood of the hazardous events is sufficient.

There seems to be very little point in expending effort in evaluating the frequency of accidents unless there are criteria for comparing the risk levels calculated. Other than comparing the results against certain criteria, the main reason to quantify the risk assessment is when the law demands it. This is rarely the case, but there are

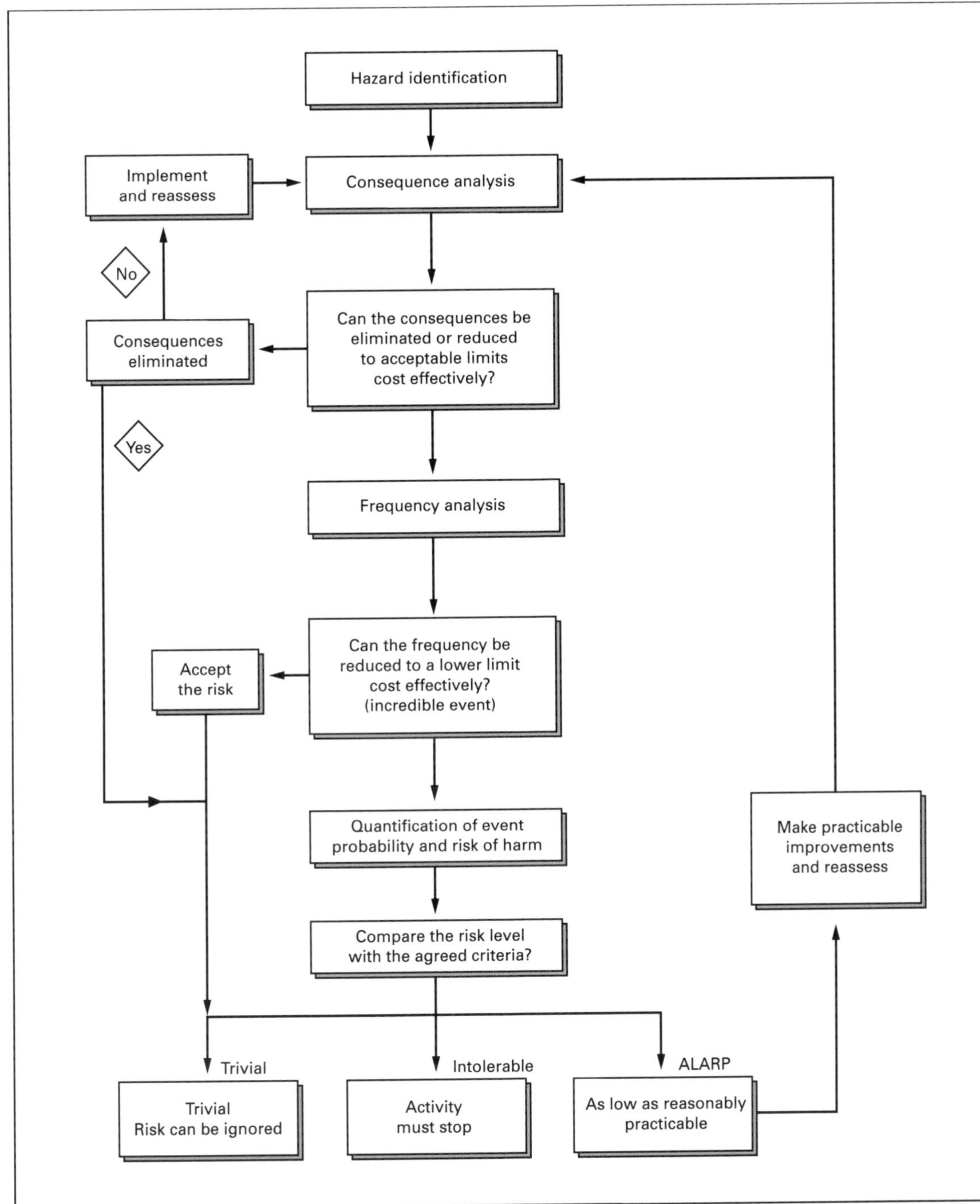

Fig. 4.2 A procedure for a risk assessment.

many situations where operators of industrial sites seek to justify that the risks from the operation are so low as to be acceptable or even trivial.

A brief review of the QRA requirements of the nuclear, offshore and onshore major chemical hazard industries show that while the absolute requirement to carry out a QRA is not always explicitly required, it is often the best or even the only way to show that the risks from an industrial activity are adequately mitigated, assessed and controlled.

4.2.1 Nuclear industry

In Britain the nuclear industry has to comply with the Nuclear Installation Act 1965 (as amended). Under this Act certain nuclear installations are required to obtain a licence before operating any nuclear site. To obtain a licence, the operator has to submit a safety case to the HSE. In assessing the safety case, the HSE has provided a framework to be used as a reference for the technical assessment—Safety Assessment Principles for Nuclear Power Plants (SAPs) (HSE 1992a). These SAPs are primarily aimed at the assessment of new plants. Quantified risk analysis is an important part of the safety case and it should also be an important part of identifying the critical safety factors during the design process.

The risk analysis consists of large fault and event trees with defined levels of plant damage or a large release of radioactive material as end-points and top events. The fault and event trees are drawn to show how the safety systems would need to fail in order for a plant damage state or a large release to occur. The fault tree is extended to include the failure of these safety systems and subsystems until the fault tree shows the system failure in terms of individual components. Campbell (1987) discusses the way a preconstruction safety report (PCSR) was used to modify the design of the Sizewell B pressurized water reactor. The PSA in the PCSR was not a comprehensive analysis and considered only a limited number of bounding initiating events. The 14 chosen were considered the most onerous events and the contributions from other events were not expected to significantly increase the total release frequency. Whilst this PCSR was adequate for preliminary design work, the HSE required a comprehensive PSA as part of the pre-operational safety report, after the design work was essentially complete, to demonstrate that the reactor could meet the criteria.

The SAPs detail the basis for regulatory acceptance of the PSA. The base events are used in the fault and event trees to establish the conditional failure probability of safeguard systems in response to a particular fault. The accident frequency is then evaluated by combining the fault frequency with the conditional probability, and summing all of the individual contributions to a particular end-point. For estimating the frequencies or probabilities of the base events on the fault tree, there are three principles detailed in the SAPs that are applicable to all risk assessments. These are listed below.

> P37 Where statistical data are used, they should be shown to be appropriate to the design and operating conditions of the plant and should relate to a relevant and sufficiently large population. The source of the data, the sample size and the uncertainty in the data should be specified. If changes to the source data are made to take account of differences between the available data and the plant conditions, these should be justified.
>
> P38 Where no relevant statistical data are available, judgements should be made and their basis stated. Particular attention should be paid to determining the sensitivity of the results of the PSA to such judgements.
>
> P39 For some fault sequences, it will not be possible to calculate the frequency of occurrence because the data are inadequate or no appropriate models are available. For example, for certain structural components such as pressure vessels where failure could lead to severe consequences, the failure frequency required to meet the accident frequency principles may be well below the values that can be justified by standard statistical estimation techniques. In all such cases a considered judgement should be made of the contribution to the predicted frequencies from such faults.

Whilst principle 37 is the correct way to use historical data, there are rarely enough data available and principles 38 and 39 are a pragmatic solution where the ideal is impossible. They allow expert judgement to be used to modify the estimated frequencies.

The risk assessments need only be as complex as is necessary to show that the plant will meet the criteria laid out in the SAPs. There is no sense in carrying out an overcomplicated and expensive PSA, to show that any plant is better than the pre-set criteria, when all that is really needed is to show that the criteria are met; a simple PSA will suffice. It is important to note that PSA is only one element of a nuclear safety case and it is complementary to the engineering and deterministic analysis.

4.2.2 Offshore industry

The inquiry carried out by Lord Cullen (Department of Transport 1990) following the *Piper Alpha* disaster produced a series of recommendations, which included the following.

> A demonstration by quantified risk assessment of major hazards that the acceptance of standards has been met in respect of risk to the integrity of the temporary safe refuge (TSR), escape routes and embarkation points and lifeboats from design accidental event and that all reasonably practical steps have been taken to ensure the safety of persons in the TSR and using escape routes and embarkation points.

Lord Cullen's recommendation on the use of QRA is implemented by the Offshore Installations (Safety Case) Regulations 1992 and HSE Guidance on the Regulations (HSE 1992c). These Regulations require a safety case to be submitted to, and accepted by, the HSE. The safety case requires that suitable use should be made of QRA as part of the demonstration of the adequacy of preventative and protective measures, particularly for the temporary refuge.

4.2.3 Onshore major chemical hazards

Quantified risk assessment is not a part of the UK regulatory regime for onshore major chemical hazard industries. An advisory committee was set up by the Health and Safety Commission in 1974 to consider the safety of large-scale hazardous industries. The committee published three reports (HSC 1976, 1979, 1984). These three reports suggested a framework of legislative and other methods within which the problems associated with the control of major hazards could be contained. The final report included recommendations on identification, control and mitigation of the risks from major hazards, stating that: 'The risks from a hazardous installation to an individual employee or member of the public should not be significant when compared with other risks to which he is exposed in every day life'.

Within the UK, the implementation of the European Seveso directive is the principal legislative control of major chemical hazards. The Control of Industrial Major Accidents Hazard 1984 Regulations (CIMAH) and the associated guidance (HSE 1990) requires operators to submit safety reports. As part of the safety report, operators are implicitly required to calculate the consequences of major accidents. This requirement is detailed in part 5a of schedule 6.

> Information relating to the potential major accidents, namely:
> (a) A description of the potential sources of a major accident and the conditions or events which could be significant in bringing one about;

The Guidance to the Regulations details the requirement of explicit quantification of the consequences of a major accident combined with relatively broad, but justifiable, *qualitative* predictions about the likelihood of occurrence. However, the requirement stops short of a full QRA approach.

However, the HSE does use QRA as a guide for its advice on land use planning around major chemical hazard installations. A zone is notified around each installation where the Local Planning Authority is required to consult the HSE on planning matters. The HSE uses QRA to assist it in its decision-making process.

This QRA is required to be consistent and transparent and both the criteria and the methodologies are openly published (HSE 1989).

4.3 RISK CRITERIA

All three industries have criteria set for their own

QRA. The nature and level of the criteria are partly due to history, but are mainly set so that they are appropriate for the purpose of the assessment. Comparison of the criteria may not be meaningful because of the different units and also the types of populations at risk.

All the criteria use the principles detailed in the HSE document The Tolerability of Risk from Nuclear Power Stations (ToR) (HSE 1992b). This document contains guidelines on the tolerable levels of risk from new nuclear power stations. These levels are used widely throughout the UK and elsewhere. More importantly, it puts into risk assessment terms the primary principle laid down in the Health and Safety at Work Act. This Act places a responsibility on employers to reduce the health and safety risks to their employees and the public as far as is reasonably practicable (SFAIRP). This, combined with the idea that there is a level of risk that is unacceptable and, possibly, a more contentious idea that there is a level of risk that is so small it is negligible, produces a framework on which risk criteria can be set. This is illustrated in Fig. 4.3, which shows an upper level of risk that is unacceptable and a lower level of risk that is broadly acceptable. The region in-between is where the risks need to be reduced to as low as reasonably practicable (ALARP).

The terms SFAIRP and ALARP are not interchangeable: SFAIRP is a term qualifying a legal duty as part of the Health and Safety at Work Act (1974); ALARP is a wider statement of principles, a risk management concept that would consider more factors than SFAIRP. Industrial

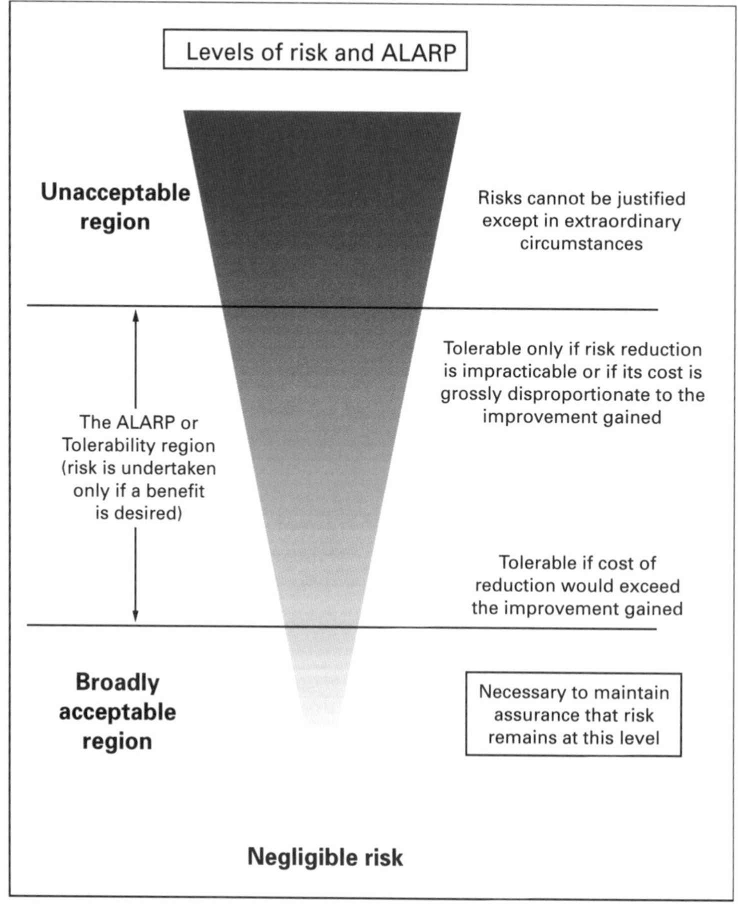

Fig. 4.3 ALARP triangle.

risks that are unquestionably ALARP should also satisfy or exceed the legal duties to reduce SFAIRP.

Other criteria commonly used with environmental risk assessments are described as the 'best available technology not entailing excessive cost' (BATNEEC) and the 'best practicable environmental option' (BPEO). Both of these criteria involve balancing the reduction in risk (to the environment) with the practicability of reducing that risk. Whilst there may be some difference in the legal interpretation of ALARP and BATNEEC or BPEO, for all practical purposes these criteria are unlikely to be significantly different because they all include the appraisal of the relevant costs and benefits.

Whilst this framework is almost universally accepted, the upper and lower values are harder to agree. The values proposed by the ToR document for frequency of death per year are 1×10^{-4} for the upper (just tolerable) level and 1×10^{-6} for the lower (broadly acceptable) level. These criteria are intended to apply to members of the public. The upper limit for a worker is set one order of magnitude higher than a member of the public, at 1×10^{-3}.

A brief summary of the acceptance criteria for the nuclear, offshore and the onshore major chemical hazard industries shows the differences between the QRAs. This summary does not give a complete picture and the author strongly recommends that the reference documents should be read thoroughly before trying to apply them.

4.3.1 Nuclear industry

It has already been stated that risk assessment is an important part of the nuclear safety case required to obtain a licence to install or operate a nuclear plant. The SAPs comprise a set of objectives, most of which need to be met as far as possible. The criteria for both qualitative and quantitative assessments are carried through from the HSE's ToR paper. The tolerability limit is taken as the basic safety limit (BSL) that must be reached for a licence to be issued. Each BSL is complemented by a best safety objective (BSO). This is the level at which the HSE does not ask for any further safety improvements.

A proposed nuclear plant must satisfy the BSL and improve safety according to the ALARP principle. The justification that an activity is ALARP has to be made on a case-by-case basis. Once the BSO is reached, then it is assumed that the risks are ALARP.

The SAPs have five principles that provide numerical criteria for the probabilistic safety analysis (PSA): doses to the public, risk to workers, large releases, plant damage and criticality accidents.

The criteria for doses to the public are given in Table 4.1. To provide consistency with assessments between nuclear installations, it is assumed that the effective dose is one that would be received by a person at typically 1 km downwind from the plant. These frequencies do not take into account the probability that a person will receive the dose and so do not take into account such factors as weather conditions, wind direction and the proportion of time a person spends away from the location.

The BSL is equivalent to an individual risk of death of 1×10^{-5} per year, and the BSO is equivalent to 1×10^{-7} per year. These criteria are one order of magnitude lower than the just tolerable and broadly acceptable levels proposed by the HSE's ToR document. However, taking into account

Table 4.1 Criteria for doses to the public.

	Total predicted frequency per year	
Maximum effective dose (mSv)	Basic safety limit	Best safety objective
0.1–1	1	10^{-2}
1–10	10^{-1}	10^{-3}
1–100	10^{-2}	10^{-4}
100–1000	10^{-3}	10^{-5}
>1000	10^{-4}	10^{-6}

society's particular aversion to nuclear risks, the SAPs conclude that these criteria are broadly consistent with the HSE's ToR document.

The criteria for a large release, i.e. a major accident where the dose to a person close to the site may cause prompt death, are set at 1×10^{-5} and 1×10^{-7} for the BSL and BSO, respectively.

4.3.2 Offshore industry

The Offshore Safety Case Regulations (HSE 1992c) treat QRA in two parts: the temporary safe refuge (TSR) and the assessment of the rest of the installation. The TSR is given specific criteria in the Safety Case Regulations and Guidance that give details of what constitutes a failure of the TSR. The HSE expects the QRA to show that the loss of integrity of the TSR within the minimum endurance time does not exceed 1×10^{-3} per year. The loss of integrity includes the ability of the refuge to protect personnel and essential equipment and provide a means of evacuation, escape and rescue. The endurance time is usually 1 h but the Guidance and Regulation allows for this time to be varied depending on the conditions. The Guidance and Regulations do not give any specific value for the maximum tolerable risk to the workforce but relies on the ToR document (HSE 1992b), which gives an upper limit of likelihood of death of 1×10^{-3} per year and the broadly acceptable region of $< 1 \times 10^{-6}$. This gives an extremely large ALARP range. There is no need to set a risk criterion for members of the public for the offshore industry.

4.3.3 Onshore major hazards

The risk criteria for land use planning in Britain are set out in the HSE's document 'Risk Criteria for Land Use Planning in the Vicinity of Major Industrial Hazards' (HSE 1989). In this document not only are the criteria given but also the methods for calculating the risks. Further information on the risk assessment methodology used with the HSE's risk assessment computer tool RISKAT is given in technical papers as they are developed (Pape & Nussey 1985). When considering the criteria it is important to remember that both the criteria and methodology are a package. Transferring the criteria to a different assessment methodology may make the criteria meaningless.

Importantly, the document discusses the use of death as an injury criterion and concludes that it was not appropriate for two reasons.

1 Society is concerned about risks of serious injury or other damage, as well as death.

2 There are technical difficulties in calculating the risks of death from a hazard to which individual members of a population may have widely differing vulnerabilities.

To overcome these difficulties, the criteria are given not for a risk of death per year but for a risk of receiving a 'dangerous dose or worse' per year. Dangerous dose is defined as a harmful effect that causes the following.

1 Severe distress to almost everyone.

2 A substantial fraction of people requiring medical attention.

3 Some people seriously injured, requiring prolonged treatment.

4 Any highly susceptible people might be killed.

The HSE developed principles given in the Advisory Committee on Major Hazards, Third Report (HSC 1984), as a basis for setting the criteria. The HSE developed the criteria from the ToR document and from a Royal Society Study Group Report (RSSG 1983). The one chance per million years (1 cpm) is kept as the lower bounds risk but uses the 'dangerous dose or worse injury' criteria rather than death. The HSE also assumes that for a vulnerable population the risk of receiving a 'dangerous dose or worse' equates to the risk of death. For development where there is a high proportion of vulnerable people, for example hospitals, the HSE proposed a more stringent lower bound of a 1/3 chance per million years of a dangerous dose or worse.

The upper limit was not taken from the RSSG because they assumed that a high element of risk was voluntary. The HSE proposed an upper limit of 10 chances per million years of receiving a dangerous dose or worse.

Although they are based on individual risk, the criteria have taken into account societal risk by adopting different criteria for differing sizes of developments and by producing equivalencies for other types of developments. For a development of less than 10 houses the HSE proposed a limit

Table 4.2 Housing equivalence for land use planning risk assessment.

House	Retail	Leisure (daytime)	Holiday/hotel accommodation
10 houses	100 people	100 people	25 people
30 houses	300 people	300 people	75 people

of 10 chances per million years of receiving a dangerous dose or worse and 1 cpm for a development of 30 houses. Table 4.2 shows the hypothetical housing size equivalence for various developments.

4.4 EVALUATING ACCIDENT FREQUENCIES

While risk assessment needs hazard identification, consequence analysis and a relative frequency estimation, it does not need to evaluate the absolute frequency of accidents. Accurate and precise frequency analysis is only required if the results of the risk assessment are to be compared with another different activity or an established criterion, or used in a cost–benefit analysis. Having established the requirement for a QRA rather than just a risk assessment, then the methodologies for evaluating the frequency of accidents are the same for nuclear, offshore and the onshore major chemical hazard industries.

There are three basic ways to determine the likelihood of an industrial accident.

1 Analysis of historical data.
2 Synthesized analysis using techniques such as fault and event trees.
3 Expert judgement.

None of these is a truly separate methodology, as there is always some overlap among all three. It is worthwhile considering them separately to highlight their strengths and weaknesses. From this it is possible to see how the likelihood of accidents can be estimated on a pragmatic basis, taking advantage of every appropriate methodology.

4.4.1 Analysis of historical data

The analysis of relevant historical data is the nearest we can get to actually measuring the likelihood of an accident. However, in order to do this there needs to be a statistically significant number of accidents. This methodology is possible for events like road accidents. The UK Department of Transport has produced statistical analyses of road accidents in Great Britain every year. These reports give the number of casualties from road accidents and an estimate on the traffic. From this information it is easy to calculate the casualty rate due to road traffic accidents. The casualty rate of car users in 1990 was 58 per 100 million vehicle-kilometres.

This actuarial approach to evaluating accidents breaks down for major industrial accidents. Data for major accidents are sparse or even non-existent because the number of major accidents is thankfully small. There are also many differences between what may appear to be similar industrial plants in the way that they are designed, constructed and operated. These differences make the direct correlation of data from plant to plant difficult. In addition to these problems, rapidly changing technology introduces the difficulty of applying yesterday's data to today's industrial process. Assuming that the data exist, and are available, then historical analysis is fraught with difficulties.

There are many databases that record accidents and these are often used as part of historical data analysis. These include statutory reporting of major accidents under the European Seveso directive and the UK reporting of injuries, diseases and dangerous occurrences regulations (HSE 1995). There are also several other databases available worldwide that record information on major accidents. Three major databanks that contain accident data are described below.

1 MHIDAS (Major Hazard Incident Data Service) developed by the Safety and Reliability Directorate of the United Kingdom Atomic Energy Authority (SRD) on behalf of the Major Hazards Assessment Unit of the United Kingdom Health and Safety

Executive. The system has been created to record details of those incidents involving hazardous materials that resulted in, or had the potential to produce, a significant impact on the public at large. It has established a database of major incidents that can be used for validating assumptions and judgements in safety assessment.

MHIDAS contains incidents from over 95 countries throughout the world, and particularly the USA, UK, Canada, Germany, France and India. The database was started in the early 1980s, but there are references to incidents going back to the early years of the twentieth century, and the database is continuously being updated. The data included in MHIDAS have been selected and indexed by professional engineers.

2 FACTS (Failure and Accident Technical Information System) is a databank containing technical data about incidents with dangerous substances operated by TNO (Organization for Applied Scientific Research) in Holland. The database stores over 14000 incidents and includes both accidents and near misses. The criterion for inclusion of an incident in the database is the presence of danger and/or damage to the nearby population and the environment.

3 WOAD (World Offshore Accident Database) is operated by VERITEC and records information on offshore accidents from blowouts to groundings. The database includes damage to fixed and mobile platforms, sub-sea production systems, lay barges and pipelines. Importantly, WOAD also contains data on the number of platforms, wells, persons on board, etc. Statistical analysis of the data is published regularly and the data base is accessible.

Details of accidents collected by regulatory authorities are not normally available to the public, although they may produce regular statistical analyses. An exception to this is the hydrocarbon database held by the Offshore Safety Division (OSD) of the UK HSE. The database was produced in response to one of Lord Cullen's recommendations into the *Piper Alpha* disaster (Department of Transport 1990). This database contains information on hydrocarbon leaks, spill and ignition in the UK sector of the offshore industry. As part of the inquiry's recommendation the operators have access to the data, particularly for the purpose of carrying out QRA.

Whilst there appear to be plenty of sources of data, none of them is complete. This is shown by considering data from chlorine installations. Chlorine failure data should be comprehensively documented, as chlorine is very noticeable and there should be plenty of information available considering that chlorine has been in use for many decades. A search of the MHIDAS database records for incidents involving releases of chlorine over the last 50 years shows 192 incidents, of which 60 occurred in Europe and 81 in the USA.

Information from the Industry Association indicates that there have been 175 incidents involving chlorine in Europe since 1953 at user premises alone. The difference between the number of incidents may be due to the definition of an incident from the two sources, a mismatch between the population reporting incidents or a failure of MHIDAS to obtain all the data on chlorine.

Clearly an Industry Association will have the opportunity to record all incidents involving chlorine, whilst MHIDAS can only record information from the public domain. Industry's data are used as a tool by which operators can learn from past incidents to prevent similar occurrences in the future. The Association can include events that may not be classed as an incident. Industry Associations will be able to collect data from all its association members, whilst MHIDAS is taking its information from the public domain. This is bound to cause some underreporting of incidents in MHIDAS. This is demonstrated when the number of incidents per year is plotted. Figure 4.4 shows that the number of incidents is increasing. This is the opposite to what is expected because, generally, the number of accidents in the chemical industry is falling as safety is improved. This anomaly is most certainly due to the under-reporting of older incidents, which are not recorded on MHIDAS.

When there is some confidence that all the data on the incidents have been collected, it is only possible to produce a failure rate provided that the populations of the installations are known using the simple formula:

$$\text{Failure rate} = \frac{\text{Number of incidents}}{\text{Population}}$$

Estimating the population from which the incidents

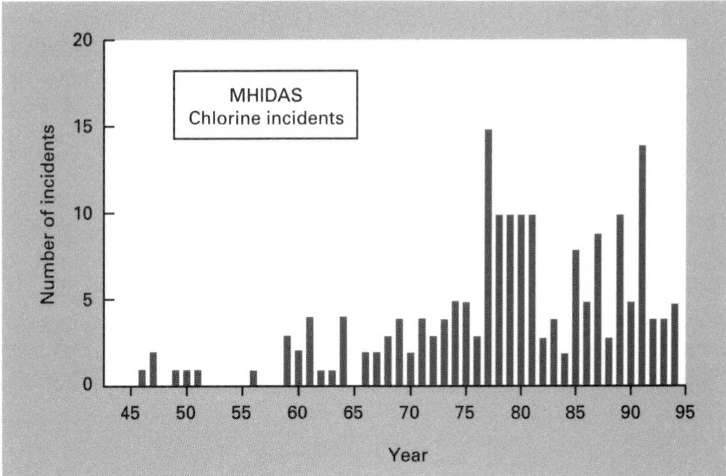

Fig. 4.4 Number of chlorine incidents over the past 50 years.

derive presents numerous difficulties. Continuing with chlorine as the example, the number of fixed chlorine storage tank failures is seven. This is confirmed by the MHIDAS database, the Health and Safety Commission's Second Report of the Advisory Committee on Major Hazards (HSC 1979) and an analysis of chlorine incidents (Harris 1978). The population of chlorine tanks is thought to average around 10000 over the last 70 years. This gives a failure rate according to the formula above of 1×10^{-5} per year.

$$\text{Failure rate} = \frac{7}{10000 \times 70}$$

Obviously the failure rate is highly dependent on the estimated number of vessels. Whilst 10000 is a good value based on expert judgement, the actual number must have varied from year to year as the production rate of chlorine has changed. No one has counted the actual number of chlorine vessels over this period and as many of the older vessels will have been taken out of operation, it is impossible to get an accurate estimate of the chlorine tank population over the past 70 years.

Where data are available, the nature of the information is not always obvious. Information on the population from which the data were obtained, and the details of the type of failures recorded (commonly referred to as the failure mode), are both required to ensure proper application of the derived failure rates.

Only when considering the possible causes of failure does the importance of detailing the specification of the population become clear. No two items will have identical failure rates; it will depend on the vulnerability of the item to the various initiating events. A study on pipework failures (Blything & Parry 1988) discusses 18 causes of pipework failures from corrosion to impact, with five basic root causes. This results in 90 root/failure cause combinations. The susceptibility of pipework to internal corrosion will depend on the fluid being carried and material of construction, whilst its susceptibility to impact will mainly depend on its location and physical protection. The relative importance of the causes will be affected by different site-specific factors. Because of this, each individual section of pipe will have a unique failure rate. Figure 4.5 shows two sets of radically different failure rate data for pipework derived from two different industries (Hawksley 1984).

Information on the material being carried and the standard of construction may be available for the historical data, but location and physical protection will not. The data collected will be an average failure rate for the population from which it was derived. However, all the details that will affect the relative importance of the different causes of failure are unlikely to be specified.

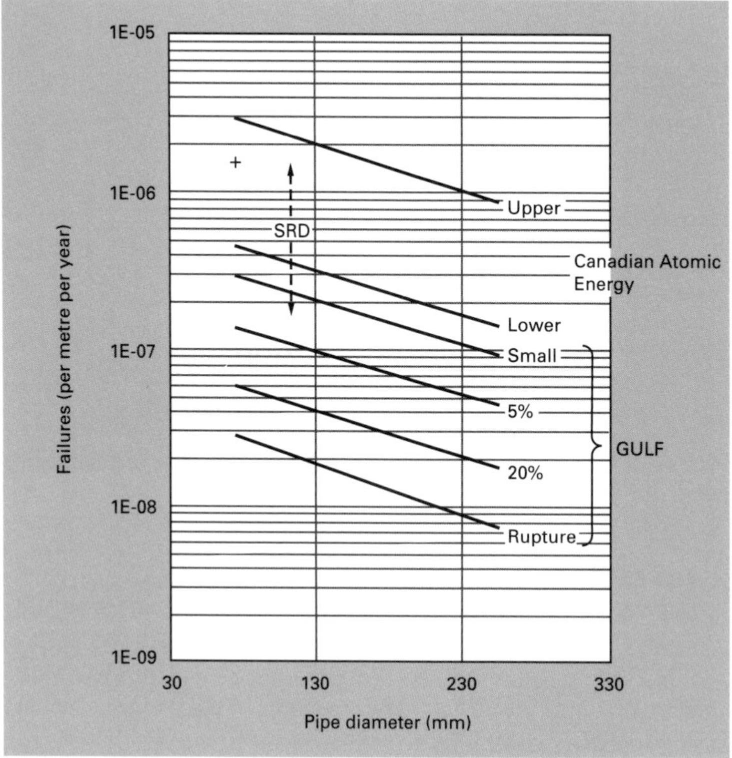

Fig. 4.5 Summary of some pipework failure rates.

Without these details it is difficult to be confident that the historical data used to derive the failure rate will match the population in the analysis.

4.4.2 Synthesized analysis

If there are insufficient historical data to carry out a statistical analysis then the frequency of the event can be calculated by considering the causes of the incidents. There are two commonly used methodologies for this—fault trees and event trees. An advantage of using synthesized analysis is that it can be made highly site specific, with synthesis tailored to the system being analysed. A distinct advantage over historical analysis is that it can identify accident scenarios that may never have occurred and provide a structure to allow risk reduction measures to be put into place to ensure that they never do.

A major difficulty with synthesized frequencies is that it still requires data to quantify the identified contributory causes. Whilst it is intended to resolve the synthesis to a point where the data are known, they will come from a system that is bound to differ slightly from the real situation. In many cases data will have to be modified or generated to quantify the contributory causes. There is often a problem in justifying the results from frequency analysis when the synthesis contains many judgements and is not supported by historical analysis.

Fault trees

Fault trees are by far the best known and most widely used technique for developing failure logic. Fault tree analysis was first used in 1962 by the Bell Telephone Laboratories in connection with a US Air Forces contract, to study a rocket launch control system. This was used to study unlikely events in a complex system. In risk assessment a fault tree is constructed to identify the contributing causes for any one event.

An undesired event must be selected and this is usually called the top event. This may be the product from the hazard identification stage. Most of the undesired events are obvious: release of toxic or flammable material, failure of a reactor trip system or failure of a water spray system to start. The fault needs to be defined precisely. Issues such as the size of the release of material and duration for it to be defined a failure must be resolved. It would be useless to calculate how often a water spray would not work, if 90% of the time it works but fails to provide sufficient water to fulfil its designed task.

It is important to define the system that is to be analysed. Attempts to construct quantified fault trees for generic systems often produce results with large uncertainties because the quantification of many of the causes will depend on specific factors.

The main task in constructing a fault tree is to reason backwards from the causes of the top event. These causes are related using simple logical relationships of AND, OR, etc. that allow the construction of a logical structure that models the failure modes of the system. Whilst the technique will eventually produce a list of the events that could lead to the fault (top event) coming about, it is important that analysis does not produce a straight list of causes. Lists, by their very nature, are limited and can very easily miss out contributory factors. The events are most often broken down into two to four contributory factors. Whilst the tree is being constructed there is a need for fairly free thinking based on a good knowledge of the system being modelled if all the contributing events are to be included.

These contributory factors are further broken down into their causes and the process is repeated until it is terminated at a base event. A base event is a cause that does not require further breaking down because either:

1 the event is quantifiable, i.e. there are some failure rate data available; or

2 the event can clearly be seen to have an insignificant effect on the top event.

When compiling a fault tree, certain base events often appear several times. This does not cause any difficulty when quantifying the tree. The list of causes (cut sets) is produced using Boolean algebra, which takes into account duplicate (common) base events. Attempts at producing a fault tree that does not have common base events are usually a result of preconceived ideas about the causes of failure, and may lead to branches of the tree being missed.

The set theory and the Boolean algebra used in fault-tree analysis are usually studied as part of undergraduate studies or an engineering training course. There are books (Lees 1980) available but it is probably better studied formally. Several computer programs are available that not only assist in drawing the fault tree but will also produce the cut sets and evaluate the tree.

As an example of how fault trees are constructed, consider the transfer of liquid chlorine from a road tanker via a flexible hose. To ensure the integrity of the system, two tests are carried out before starting the transfer: the hose is pressurized with compressed air to find out if it will hold pressure and then the hose is pressurized with chlorine vapour; and any leaking chlorine vapour can be detected with a solution of ammonia. When these two tests have been successfully completed, the liquid chlorine valve is opened and the transfer is started. Figure 4.6 shows the valve arrangement.

The sequence of operations is as follows.

1 Connect the road tanker to the system.

2 Open up the compressed-air valve to pressurize the system.

3 Close the compressed-air valve to isolate the system.

4 Check that the pressure does not drop over 2 min.

5 Open the chlorine vapour line to fill the system with chlorine vapour.

6 Test for chlorine with ammonia solution.

7 Provided that the two tests show no leaks, open the liquid chlorine valve to start the transfer.

The top event is the failure of the operator to detect a leak in the system that would lead to release of liquid chlorine.

Firstly, the tree considers that the top event could be caused by the failure of the pressure drop test and the leak test. The test could fail by the test not being carried out or by the test being inadequate. The fault tree is shown in Fig. 4.7. The base events have been quantified using either a human factor prediction technique or expert judgement. These are shown in Table 4.3.

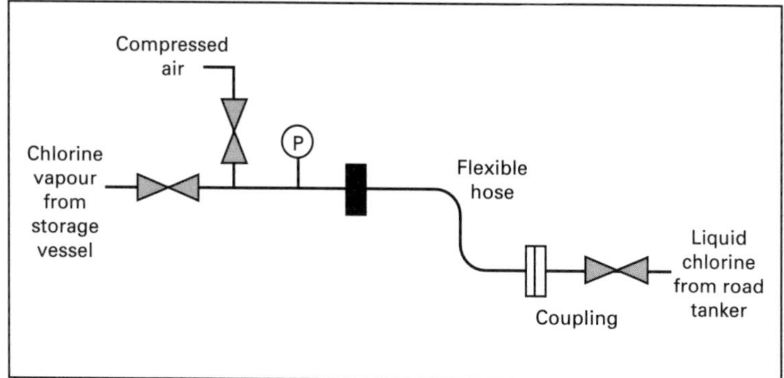

Fig. 4.6 Valve arrangement for transfer option.

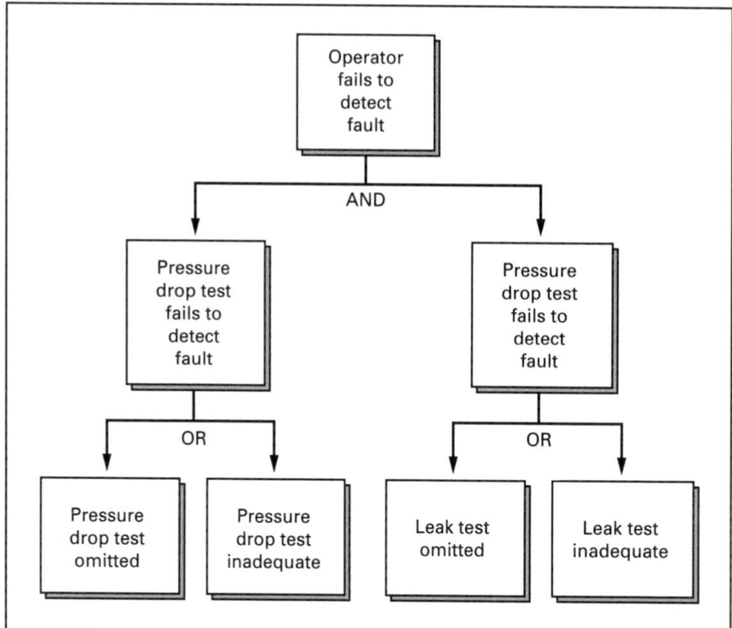

Fig. 4.7 Fault tree for testing procedure.

Table 4.3 Base event probabilities.

Base event	Probability
Pressure drop test omitted	0.01
Pressure drop test inadequate	0.0015
Leak test omitted	0.06
Leak test inadequate	0.008

The minimum cut sets, i.e. the possible causes of failure and their percentage contribution to the final probability for the fault tree, are as follows.

1 Pressure drop test omitted *and* leak test omitted, 66.5%.

2 Pressure drop test omitted *and* leak test inadequate, 17.7%.

3 Pressure drop test inadequate *and* leak test omitted, 12.5%.

4 Pressure drop test inadequate *and* leak test inadequate, 3.3%.
The overall probability of the operator failing to detect a fault is 7.2×10^{-4} per filling operation.

The fault tree in the example is only a trivial use of the methodology and does not need the rigorous analysis carried out by a formal system, but most applications are larger and more complex than the example. One such example of a fault tree analysis is a study that was carried out by the author (Gould 1993). In this study the frequency of catastrophic failure of a chlorine vessel was determined. The fault tree stretched over seven pages with 30 gates and 57 base events. The quantification of the base events themselves requires relevant historical data. In the above study a major contributor to the overall failure frequency is a construction defect. There are insufficient data on the number of defects found in chlorine vessels and no indication of the proportion of those defects that may lead to a vessel failure. Generic pressure vessel failure data (Smith & Warwick 1981) supplemented by more recent data (Davenport 1991) were used in the quantification of this base event. Most of these data are taken from boilers, with additional information from air receivers.

These data are not directly relevant to chlorine vessels because air receivers have a less rigorous inspection scheme than chlorine vessels and construction defects would more likely be identified and rectified before they would lead to failure. Additionally, air receivers have a more arduous duty with significant temperature and thermal cycling, which could lead to the growth of construction defects to failure. The estimated catastrophic failure frequency for a chlorine vessel ranged from 1.6 to 3.1×10^{-6} per year. This value is about one order of magnitude smaller than the failure frequency referred to in the section on historical data calculated from known failures and estimated population.

This fault tree is dwarfed by those used within the nuclear industry, where there is a need for detailed analysis of very complex safety systems that are broken down to their component levels.

Event trees

Whilst fault tree analysis applies backward reasoning, event trees use forward reasoning. The starting point is an initiating event and is often used to model how safety systems or other mitigating systems will work. The usual convention is for the event tree to work from left to right, although this is not mandatory. The tree is divided into columns, each column or node being an event.

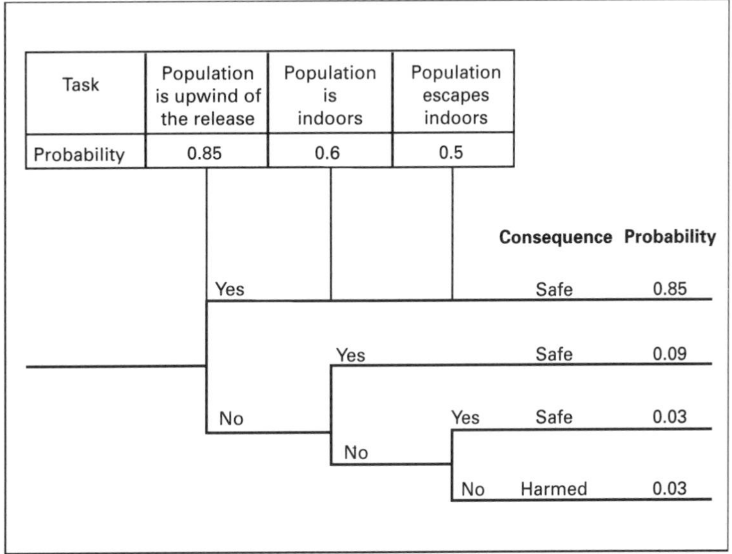

Fig. 4.8 Example event tree.

A simple event tree could consider the effect of a chlorine release on the surrounding population. For the chlorine to affect a particular person the release must be downwind of the individuals and they must be outdoors. If they are outside they may have a chance to 'escape' indoors before they are affected by the toxic gas. The event tree is shown in Fig. 4.8.

The likelihood of the individuals being affected by the chlorine release can be evaluated given the probabilities shown in Table 4.4. This gives the probability of harm occurring to individuals given a release of chlorine as 0.03. The probabilities will depend on the local weather condition, size of release, location being considered and the concentration of chlorine at that location. However, this type of application does not make best use of the analytical power of event trees. Event trees are very powerful for modelling series of events that are time-dependent, such as operator actions.

An event tree can be used as an alternative method to analyse the earlier fault tree example; probability of failure for a transfer hose test. Figure 4.9 shows this operation modelled as an event tree. The base events have been evaluated using a human factor prediction technique. The result is a slightly higher value than the fault tree result (7.2×10^{-4}), with the probability of a fault being undetected of 3×10^{-3} per operation.

Significantly, the event tree shows that if the compressed-air valve is left open, both the pressure drop test and the leak test become ineffective because the pressure will not drop, even if there is a leak and with the air supply on, and the chlorine vapour cannot enter the hose.

The event tree has identified a part of the operation that is critical and will allow effective action to be taken to improve the reliability of the system.

Table 4.4 Probabilities for example event tree.

Event	Probability
Upwind	0.85
Indoors	0.6
Escape	0.5

4.4.3 Expert judgement

Whilst relevant historical data are rarely (if ever) available for the frequency analysis of a QRA, the use of various methodologies to synthesized frequencies is very time consuming and they are not without their own difficulties.

Many risk assessments require a simple method for evaluating the frequency of events as an essential part of the quantification. This has led to the application of expert judgement to failure rates. There are two methods by which expert judgement is used to produce a failure rate.

1 Expert review of a series of failure rates to produce a weighted average either as a generic rate or one specific to a system.

2 Modification of an existing failure rate.

The first method has been used extensively within the loss prevention professions. All the failure rates associated with a system are collected. These are then weighted, depending on the deviation from the system being assessed and other factors thought to be significant. A best estimate is then produced by combining all the available information and applying expert judgement. A good example of this is the failure rates used within the HSE risk assessment tool RISKAT. The pipework failure frequency was derived by a review of 22 references.

The difficulty with this approach is the reconciliation of widely varying failure rates. Three of the failure rate references used by the HSE to produce the pipework failure rate are shown in Fig. 4.5. To derive the guillotine failure rate from the wide spread of data shown on the graph needs a great deal of detailed knowledge of how each reference was produced. Many of the references had partly derived the failure rate from common sources. Such references were weighted to remove the double counting of the original data.

This method of producing failure rates can be very time consuming and expensive. A simpler method for estimating failure rates is to modify an existing value. This can be a generic failure rate or one derived from historical data. An established failure rate that is as applicable as possible to the system being analysed is modified by applying values estimated by expert judgement to take into account the differences between the system in question and the original failure rate.

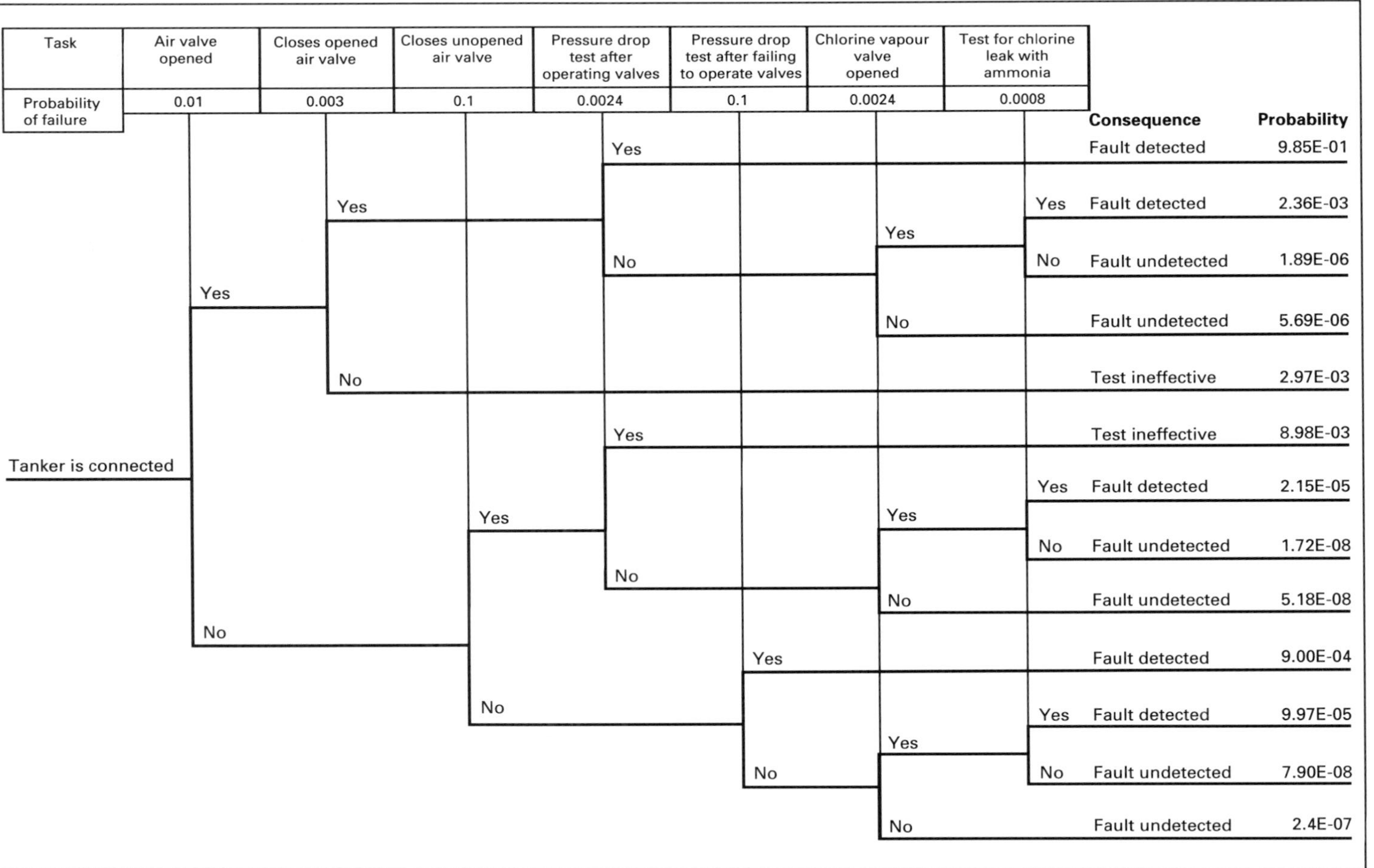

Fig. 4.9 Event tree for testing procedure.

Application of expert judgement needs to be done carefully to ensure that it does not turn into a random guess.

These derivations are part of the risk assessment methodology and should be presented with similar detail as the rest of the QRA. Organizations deriving failure rates in this manner need to keep a record including how they were derived, what factors were taken into consideration and how many times they have been used in any particular circumstances.

4.5 CONCLUSIONS

Risk assessment is a tool that has provided many benefits to industry. Quantified risk assessment is a powerful application of risk assessment that has the potential to benefit industry and society as a whole. Considerable effort is required to move from a risk assessment to a QRA and much of the work is in estimating the likelihood of the events being considered. The decision to undertake a QRA should consider not only the potential benefits but also the cost in personnel and expertise to carry out the assessment. The nuclear industry uses QRA to set a benchmark so that all the new nuclear sites are as safe or safer than the benchmark. The QRA is a major input into the design of a new reactor. Quantified risk assessment in the offshore industry is a product from the inquiry after the *Piper Alpha* incident. With the exception of the temporary safe refuges, it has been up to the industry to set its own assessment criteria. Much of the information produced by the QRA has been used in cost–benefit analysis to aid decisions on risk reduction work. The HSE uses QRA to assist in land use planning decisions around major chemical hazard sites in the UK. Quantified risk assessment is a unique tool that provides a method to compare the risks from different sites in order to make consistent decisions on land use planning around different sites.

Whilst concentrating on the quantification of accident frequencies, it is easy to miss the most important benefit from the work. The analysis of accident frequencies should give a powerful insight into what can go wrong and, more importantly, what can be done to reduce the likelihood of the accident. Both historical and synthesized frequencies can contribute to this. The likelihood of occurrence may not only reveal an idea of the magnitude of such factors, but also the ways and means to reduce them.

The views expressed in this paper are those of the author and not necessarily those of the HSE.

4.6 REFERENCES

Blything, K.W. & Parry, S.T. (1988) *Pipework Failures—a Review of Historical Incidents*, HSE/SRD Report R441, HMSO, London. ISBN 0-853-56300-4.

British Standards Institute (1995) *Electrical Apparatus for Potentially Explosive Atmospheres* (BS EN 50020), HMSO, London. ISBN 0-580-225533-X.

Campbell, J.F. (1987) *The Role of Probabilistic Safety Analysis in the Licensing of Sizewell 'B': Implications of Probabilistic Risk Assessment.* Elsevier Applied Science, London.

Cremer & Warner Ltd (1982) *Risk Analysis of Six Potentially Hazardous Industrial Objects in the Rijmond Area, a Pilot Study*, D. Reidel Publishing Company, Dordrecht, ISBN 90-277-1393-6.

Davenport T. (1991), *Proceeding of Reliability 91. A Further Survey of Pressure Vessel Failures in the UK.* Elsevier Applied Science, London.

Department of Transport (1990) *The Public Inquiry into the Piper Alpha Disaster by the Hon Lord Cullen*, HMSO, London, ISBN 0-1011310-2-X.

Gould, J.H. (1993) *The Fault Tree Analysis of a Catastrophic Failure of a Chlorine Storage Vessel.* HSE/SRD Report R603; HMSO, London. ISBN 0-85356401-9.

Harris, C. (1978) *Analysis of Chlorine Accident Reports*, The Chlorine Institute's 21st Plant Managers Seminar, Houston, Texas.

Hawksley, J.L. (1984) Some social, technical and economical aspects of risks of large chemical plants. In: *CHEMDRAWN III (chemical research applied to worlds needs conference, June 1984)*, The Hague.

Health and Safety at Work Act (1974) HMSO, London. ISBN 0-10-543774-3.

HSC (1976) *First Report of the Advisory Committee on Major Hazards*, Health and Safety Commission, HMSO, London. ISBN 0-11-88-0884-2.

HSC (1979) *Second Report of the Advisory Committee on Major Hazards*, Health and Safety Commission, HMSO, London. ISBN 0-11-88-3299-9.

HSC (1984) *Third Report of the Advisory Committee on Major Hazards*, Health and Safety Commission, HMSO, London. ISBN 0-11-88-3753-2.

HSC (1992) *Management of Health and Safety at Work Regulations 1992, Approved Code of Practice*, Health and Safety Commission, HMSO, London. ISBN 0-11-886330-4.

HSE (1978) *Canvey, an Investigation of Potential Hazards from Operations in the Canvey Island/Thurrock area*, Health and Safety Executive, HMSO, London. ISBN 0-11-883200-X.

HSE (1989) *Risk Criteria for Land Use Planning in the Vicinity of Major Industrial Hazards*, Health and Safety Executive, HMSO, London. ISBN 0-11-885491-7.

HSE (1990) *A Guide to the Control of Industrial Major Accident Hazard Regulations 1984 HS(R)21(rev)*, Health and Safety Executive, HMSO, London. ISBN 0-11-885579-4.

HSE (1992a) *Safety Assessment Principles for Nuclear Power Plants*, Health and Safety Executive, HMSO, London. ISBN 0-11-882043-5.

HSE (1992b), *The Tolerability of Risks from Nuclear Power Stations*, Health and Safety Executive, HMSO, London. ISBN 0-11-86368-1.

HSE (1992c) *A Guide to the Offshore Installations (Safety Case) Regulations: L30*, Health and Safety Executive, HMSO, London. ISBN 0-11-882055-9.

HSE (1995) *A Guide to the Reporting of Injuries, Diseases, and Dangerous Occurrences Regulations*, Health and Safety Executive, HMSO, London. ISBN 0-7176-1012-8.

Lees, F.P. (1980) *Loss Prevention in the Process Industries*, Butterworth Heinemann, London. ISBN 0-7506-1529-X.

Pape, P.R. & Nussey, C. (1985) A basic approach for the analysis of risks from major hazards. *Institution of Chemical Engineers Symposium, The Assessment and Control of Major Hazards*, Manchester, April 1985.

RSSG (1983) Royal Society Study Group Report, Risk Assessment. The Royal Society, London ISBN 0-85-403208-1.

Smith, T.A. & Warwick, R.G. (1981) *A Survey of Defects in Pressure Vessels in the UK for the Period 1962–1978 and its Relevance to Nuclear Primary Circuits*, SRD/HSE Report R203, United Kingdom Atomic Energy Agency, Harwell (HMSO, London).

Chapter 5
Assessing Risks to Ecosystems and Human Health from Genetically Modified Organisms

RAMON J. SEIDLER, LIDIA S. WATRUD AND S. ELIZABETH GEORGE

5.1 INTRODUCTION

Since the 1970s, scientists have been altering the genetic make-up of living creatures (Tzotzos 1995). Techniques in molecular biology have made it possible to incorporate genes from one organism into virtually any other organism's genetic composition to create a broad array of unique life forms. These unique organisms may have traditional or totally new uses for applications ranging from agriculture, food and beverage and pharmaceutical production, to environmental clean-up and energy production. Through this technology industry anticipates promises of unique economic opportunities from patented 'new' life forms. These rapid developments in molecular biotechnology are forcing society to think about new concepts in biology and to contemplate the potential effects that recombinant DNA technology can have on ecosystems and on human health (Levin *et al.*, 1987; Levin & Strauss 1990; Regal 1990). Questions about proprietary rights to novel germplasm, cloned genes, and the patenting of 'new' life forms have also been raised.

Genetically modified organisms (GMOs) have already been released experimentally into the environment. Essentially, these releases have been small-scale tests, designed to examine primarily the product efficacy, not usually to evaluate potential environmental risks. Some negative environmental perturbations have been noted in some studies where risk assessment has been an experimental objective. However, since the reported effects were limited in scale and communicated largely through technical channels, few adverse public reactions to these results have been noted (Leopold 1995).

It has been over 20 years since the first recombinant microorganism was constructed and probably half as many years since they were suggested for use in the environment. There have already been more than 1000 applications to allow the testing of genetically engineered organisms, most of which have or will be released in the USA (Mellon & Rissler 1993). Once released, interactions with biotic and abiotic factors can disperse microbial agents over considerable distances (Seidler & Hern 1988; Lighthart & Kim 1989; Seidler *et al.* 1994) and a desire for properly conducted risk assessments is supported by most who are involved with this technology.

The widespread environmental experimentation with GMOs raises important questions. Can the risks of GMOs be assessed without actually releasing the agents directly into the natural environment? Can the impacts be anticipated and is knowledge of ecology sufficient to predict with confidence the fate and survival of a microbial agent and how it might interact with ecosystems and humans? Concern has been expressed by numerous scientists regarding the release of GMOs before such risks are understood (Halvorsen *et al.* 1985; Levin & Strauss 1990; Sharples 1990). Clearly, the possible risks that GMOs may pose to the natural environment must be estimated and assessed before decisions are made as to their release.

In order to ensure public health and environmental safety, the United States Government mandated that release of genetically engineered microorganisms be regulated by *Federal Register* statutes (1986). The United States Environmental Protection Agency (USEPA) regulates microbial pesticides and microbial 'chemicals' through

Subdivision M (Microbial Pesticides) of the Federal Insecticide, Fungicide, and Rodenticide Act (FIFRA) and the Toxic Substances Control Act (TSCA). In the European Union, release of genetically engineered microorganisms and the microbial plant protection products are regulated through Directive 90/220/EEC and Directive 91/414/EEC, respectively. Under both of these directives, risk assessments of health and environmental effects of the products are determined.

In response to the diversity of product opportunities as well as risk assessment concerns raised through many organizations and individuals around the world, a massive research and development effort is under way in Europe to address risk assessment issues. Designated the 'biotechnology programme', the research is funded through the European Commission and designated as part of the fourth European Union research and technological development framework programme for 1994–1998. Currently there are about 528 laboratories from every Union country sharing a budget of 73 million euros.

The basic features of general risk assessment of GMOs are understandably different from those associated with chemicals. Genetically modified organisms are living organisms and therefore, unlike chemicals that may become diluted, GMOs have the potential to disperse to new habitats, colonize those sites, and multiply. Their novel activities, including the production of metabolic products, enzymes and toxins, will occur as long as the GMOs remain metabolically active. Once established, living organisms cannot be recalled. The last 12 years have seen a new era of environmental research devoted in part to the risk assessment issues associated with the environmental release of GMOs (Bourquin & Seidler 1986; Levin *et al.*, 1987). This research seemed justified considering the anticipated widespread development and application of GMOs and their multitude of capabilities.

The understanding of basic ecological and evolutionary concepts as to why organisms may survive or flourish in natural ecosystems has changed compared to only 15–20 years ago (Regal 1990; see also Chapter 6). Many of the risk assessment decisions on safety issues will depend upon whether and how one weighs new concepts within ecology and evolution relative to older theories. For example, conventional wisdom is that if a new trait were engineered, selection by the environment would return the organisms to their original state. Exotic species introductions mostly fail to establish in nature because the new conditions are not conducive to their survival (Williamson 1994). However, the engineering of an already established host begins with an ecological advantage. Genetic engineering may also enhance that competitive advantage. When exotic species go out of control it is not only because they have escaped natural constraints, but because they sometimes find new ecological opportunities (Regal 1990). A thorough discussion of the ecological concepts involved in biological introductions, excluding GMOs, along with reasons for their success or lack of success in establishment is given in Chapter 6. Most of the concepts discussed in Chapter 6 are directly applicable to GMOs as well.

The purposes of this chapter are to describe and characterize primarily genetically engineered microorganisms and the relevant tools used for their risk assessment, provide examples of observed and possible environmental effects induced by GMOs, summarize methods available for detecting effects (both human and environmental), and provide discussions on possible hazards associated with their repeated, large-scale environmental releases that is anticipated in the near future. Many unknowns still exist about the environmental and human health significance of risks from GMOs, despite approximately 11 years of research. However, much has also been learned about how to detect experimentally the possible risks, and the nature of the risks has now been documented; both subjects are topics to be discussed in this chapter. The sections below will: (i) illustrate specific examples of genetically engineered microorganisms; (ii) define the risk-associated issues; (iii) report known ecological and health effects of GMOs; and (iv) highlight continuing and future research needs for the use of GMOs.

5.2 EXAMPLES OF GMO PRODUCTS

5.2.1 Bacteria

With the advent of modern molecular methods,

the identification, isolation and movement of genes from one organism to another, even between species or taxa which may traditionally have breeding barriers, has been greatly facilitated. Consequently, the variety of microbes that can produce novel products or have novel and useful activities is increasing dramatically. The scope of microbial activities being developed ranges from crop protection and environmental restoration, to food, beverage, pharmaceutical and specialty product production (Harlander 1992; Edgington 1994; Valigra 1994; Watrud & Seidler 1997). The examples given below will illustrate a variety of specific applications for genetically modified bacteria developed for either environmental or contained applications.

Crop protection (pest control for insects and diseases)

Intentional selection and development of bacteria intended for large-scale environmental release have primarily been made for agricultural purposes, particularly for insect control. Between 1948 and 1995, 25 bacterial products were registered for use in the USA as microbial pesticides (W. Schneider, pers. comm.). Eighteen of those registrations occurred after 1992. In part, it has been this increased interest in microbials for pest control that has led to the recent creation of a Division of Biological Pesticides within the United States Environmental Protection Agency. The most widely used bacterial species for biological control of insect pests is *Bacillus thuringiensis* subsp. *kurstaki* (*B. t. kurstaki*), which has a high degree of specificity for lepidopteran insects. The active pesticidal ingredients in *B. t. kurstaki* and in the related species *B. t.* subsp. *tenebrionis*, which has a high degree of activity against certain coleopteran insect pests, are proteins (McPherson *et al.* 1988; Hofte & Whiteley, 1989). Using both classical breeding and recombinant methods, geneticists have tried to create strains that express higher levels of one or both of the desired insecticidal protein activities (Gawron-Burke & Baum 1991; Bosch *et al.* 1994). Several engineered strains of *B. thuringiensis* have recently been registered (W. Schneider, pers. comm.). Additional *Bacillus* species registered in the USA for the control of coleopteran insect pests include *B. popilliae* and *B. lentimorbus* (W. Schneider, pers. comm.). In addition to efforts to improve microbial formulations of *Bacillus* species, pesticidal genes isolated from *Bacillus* species have been, or are being, cloned into a number of plant species, including major food and fibre crops such as corn, cotton, potatoes and woody species (reviewed in Watrud *et al.* 1996). A concern resulting from the anticipated millions of acres to be planted in crops expressing *B. thuringiensis* proteins is an increase in the rate of development of resistance in insect pests (Tabashnik 1994). In terms of human health, a potential concern is whether allergenic reactions may develop in people exposed to higher levels of the insecticidal proteins, due to increased environmental and dietary exposures. Non-target effects of microbial forms of *B. t. kurstaki* on insects and on microbial biota in soil have recently been reviewed (Watrud & Seidler 1997).

Bacillus species and *Pseudomonas* species have each been proposed and evaluated for use as biological control agents for *Rhizoctonia* and *Pythium*, fungi which cause damping-off diseases of seedlings (Weller 1988; Whipps 1992; Lumsden *et al.* 1995). Bacterial treatments for disease control have been evaluated as seed treatments, as root drenches for transplants and as dry or liquid soil treatments. Although short-term ecological and toxicological effects and other safety considerations of microbial pesticides typically have been addressed (Levin 1995), studies on long-term non-target effects of bacterial pesticides have often been limited (Watrud & Seidler 1996).

Both fungi and bacteria pathogenic or phytotoxic to specific types of plants have been proposed for weed control. Bacterial genera reported to have herbicidal activity include isolates of *Streptomyces*, *Bacillus* and *Pseudomonas* (Charudattan 1990; Hoagland 1990; Stonard & Miller-Wideman 1995). As with the fungal biocontrol agents which are described below, a major concern of using biocontrol agents that are themselves pathogens is their actual degree of host specificity; i.e. will non-target crop or native species also be affected? Increases in the competitiveness of existing plant pathogens due to the acquisition of traits such as tolerance to pesticides, heavy metals or temperature extremes, by genetic recombination, are also of potential concern.

Plant growth-promoting biofertilizers/rhizobacteria

Use of commercial inocula of nitrogen-fixing strains of *Rhizobium* or *Bradyrhizobium* for seed treatment of legumes has become a standard agronomic practice. Attempts have also been made to boost the nitrogen-fixing capacity of the inocula by the use of recombinant methods (Paau 1991; Hall 1995). For graminaceous species such as barley and wheat, non-symbiotic nitrogen fixers (*Azospirillum lipoferum, A. brasiliensis, A. chroococcum*), have been tested in greenhouses and in the field for a number of years, often yielding positive effects on plant growth and yield, particularly in warmer climates (Dobereiner *et al.* 1988; Bhattarai & Hess 1993; Di Ciocco *et al.* 1994; Zaady *et al.* 1994). Effects of environmental conditions on efficacy of inocula have also been studied; however, ecological studies on non-target effects of *Azospirillum* inocula have not been reported.

Several bacterial preparations (*P. cepacia, P. fluorescens, B. subtilis*) have been examined as seed treatments which may permit or enhance plant growth by inhibiting damping-off fungi (Weller 1988; Whipps 1992; Lumsden *et al.* 1995). The distinction between biological control agents used for controlling soil-borne plant diseases and plant growth-promoting rhizobacteria may be largely a semantic one. To date, there are no registrations in the USA under the Toxic Substances Control Act (TSCA) for plant growth-promoting bacteria *per se*. However, approximately one-quarter of the bacteria registered as microbial pesticides in the USA have plant diseases as their targeted pests.

Biomass conversion agents

In addition to starch or sugar crops such as corn and sugarcane, lignocellulosic wastes can also be used as substrates for the production of ethanol by fermentation. Pretreated mechanically and chemically, e.g. by milling and acid or alkaline hydrolysis, complex wastes then can be subjected to fermentation. Traditional sources of lignocellulosic biomass are the byproducts of logging, wood and paper processing operations. Short-rotation woody or herbaceous species such as poplars and switchgrass and agricultural residues, including corn stover, seed hulls, straws and peels, are also potential commercial sources of lignocellulose. Several species of engineered bacteria (*Klebsiella planticola, K. oxytoca, Zymomonas mobilis, Streptomyces lividans*) have each been considered for production of ethanol from agricultural residues or for degradation of lignocellulosic residues from paper milling operations (Helsot 1990; Trotter 1990; McCarthy & Williams 1992; Wood & Ingram 1992; Crawford *et al.* 1993; Sprenger 1993). Anaerobic digestion of lignocellulosic and municipal wastes also has been evaluated as a means of producing energy in the form of methane (Wyman & Goodman 1993). The feasibility of producing lipids as a result of metabolism of sugars produced by photosynthetic activities of marine microalgae such as diatoms also has been proposed (de la Noue & de Pauw 1988; Wyman & Goodman 1993). Specialty chemical production (e.g. for solvents such as acetone) by engineered isolates of *Clostridium acetobutylicum* has also been studied (Mermelstein *et al.* 1993).

Bioremediation and mining applications

A number of isolates of *Pseudomonas* species (*P. putida, Burkholderia cepacia*) have been studied biochemically, genetically and, to a lesser degree, ecologically for their potential use and non-target effects in the remediation of polluted soils, sediments and surface and ground waters (Mulbry & Kearney 1991; Fan & Scow 1993; Edgington 1994; Krumme *et al.* 1994; Ramos *et al.* 1994). Laboratory selection and genetic engineering for customized degradation of given aromatic compounds by individual isolates has been actively pursued for a number of years (Singleton 1994; Daubaras *et al.* 1995). However, intentional environmental applications of engineered organisms for bioremediation purposes have been few, limited in part by environmental and public opinion concerns. Those issues, along with cost considerations, often have led to management of both marine and terrestrial spills by physical or chemical methods which enhance desired indigenous microbial activities. These methods include the use of absorbents (straw or sawdust or biopolymers), agitation to increase aeration and oxidation, pH modification, and addition of fertilizers (organic and inorganic) and organic substrates such as molasses or starch (King *et al.* 1992; Rouchaud *et al.* 1993). Less frequently, non-engineered inocula have been added to achieve the desired degradation and biotransformations. In 1990, a polyaromatic hydrocarbon-degrading strain of *Pseudomonas fluorescens*, HK44, containing an engineered bioluminescent reporter plasmid for naphthalene catabolism (King *et al.* 1990) was approved for field testing in the USA under the TSCA (P. Sayre, pers. comm.). To date, however, engineered inocula have been used primarily in contained laboratory facilities or microcosms, or in bioreactors that have been brought to contaminated field sites or to centralized treatment centres.

Recently, progress has been reported in the development of transformation systems for *Thiobacillus* and *Leptospirillum*, genera which may prove to be useful in the management of leachates from mining operations (Rawlings & Silver 1995). Other proposed applications for bacteria in mining include the bioconcentration of valuable metals and the detoxification of waste streams by the adsorption or complexation of heavy metals by microorganisms (Francis 1990; Wales & Sagar 1990; Sahoo *et al.* 1992; Volesky & Holan 1995). Potential roles for using microbially based processes to desulphurize coal with isolates of *Thiobacillus, Leptospirillum*, or *Rhodococcus* have also been proposed (Merretig *et al.* 1989; Kilband & Jackowski 1992; Mannivannan *et al.* 1994).

Food biotechnology/speciality products

Historically, the longest term *de facto* or intentional applications of bacterial activities by humans have been the use of fermented foods. Traditional and ethnic foods such as sausages and sauerkraut, soy sauce and miso, salami and pepperoni, yogurt, cheese and buttermilk have a long history of consumption world-wide. In addition to creating characteristic flavours, aromas or textures, the microbial activities often have enhanced the storage life of the meats, milk, fruit and grains which have served as the substrates for the bacterial activities. Dairy products produced as a result of controlled fermentations with known bacterial inocula include yogurt, buttermilk, sour cream and Swiss cheese (Harlander 1992). As the sciences of microbiology, biochemistry and genetics have advanced, the ability to select, improve and create strains with desired activities has also increased. Specific examples of the applications of biotechnology to the dairy industry include strain improvement and development of deoxyribonucleic acid (DNA)-based diagnostics for both inocula and contaminants in foods (Ramos & Harlander 1990; McIntyre & Harlander 1993). Additional useful products resulting from bacterial fermentations include organic acids such as citric acid, which can be used as a food preservative or flavouring agent, vitamins, and amino acids such as lysine and methionine, which are used as food or feed supplements. Organic acids and amino acids produced as a result of fermentations also may serve as substrates for subsequent modifications, for example, the chemical combination of microbially produced phenylalanine with aspartic acid to produce the sweetener, aspartame (de Boer & Dijkhuizen 1994). Enzymes used in food processing or detergent industries (i.e. proteases, invertases, lipases and amylases), also may be derived from bacterial fermentations (Harlander 1992). In addition to the production of fermented foods, bacterial activities have also been utilized to produce food ingredients ranging from thickeners, sugars and vitamins, to pigments, flavours and amino acids (Harlander 1992; Gableman 1994). Enzymes derived from bacteria (e.g. amylase, glucose isomerase, pullulanase) are used in food and beverage processing for starch liquefaction, high-fructose corn syrup production, and beer production. Biotechnological methods also are being used to select and modify strains of *Lactobacillus* to improve the digestibility and preservation of silage (Flores 1991; Wallace 1994).

Most examples given above have been taken from processes that generally have been carried out indoors, in vats or fermentors. As the use of recombinant strains increases, some concerns may arise as to the health or environmental effects of those strains, or novel products produced by them, in the event of spills or leaks. Potential adverse effects include short-term fish kills resulting from lowered oxygen levels in contaminated lakes or streams. Additional concerns include health risks to susceptible individuals who may be immunocompromised or allergic to recombinant proteins expressed in the released microbes.

Pharmaceuticals

Many of the antibiotics in common use today are fermentation products or synthetic derivatives of bacterial metabolites produced during closely controlled fermentations. Prominent among the medically useful bacterial species that have served as sources of numerous commercial antibiotics are the genera *Streptomyces*, *Actinomyces*, and *Bacillus*. The well-known antibiotics streptomycin, actinomycin and bacitracin are produced, respectively, by *Streptomyces*, *Actinomyces* and *Bacillus* (Glasby 1992). As techniques for transforming key antibiotic-producing species of microbes have become available, cloning of genes controlling metabolic pathways involved in antibiotic synthesis have permitted the development of strains which may serve as the source of new, novel and useful antibiotics (Malpartida & Hopwood 1992; Hopwood 1993; McDaniel *et al.* 1993; Valigra 1994; Bedford *et al.* 1995; Kakinuma *et al.* 1995). The availability of technologies to produce new types of antibiotic compounds is becoming increasingly important as new antibiotic-resistant strains of human and veterinary pathogens continue to develop and spread in clinical and outdoor environments. In addition to the production of novel bacterial compounds in bacterial hosts well suited for growth in fermentors, recombinant technologies are being applied to develop microbially derived vectors to optimize the production of mammalian compounds in bacteria, bacterial compounds in plant cell cultures, and monoclonal antibodies in animal cell lines. Recombinant pharmaceuticals used for human and veterinary growth enhancement, metabolic regulation, or for therapeutic purposes include human and animal (bovine and porcine) growth hormones, insulin, inteferons, blood clotting factors, tissue plasminogen activator and vaccines (Ladisch & Kohlmann 1992; Thayer 1992, 1994; Borman 1994; Munn 1994). One concern of biotechnology applications for the production of pharmaceuticals is allergenicity of susceptible individuals to recombinant proteins. Environmental concerns resulting from pharmaceutical applications of biotechnology are limited primarily to events resulting from spills or leaks, which might result in fish kills or other non-target effects.

5.2.2 Fungi

Crop protection

Twelve of the 45 microbial pest control agents registered in the USA are fungi. The agents include *Beauveria* and *Metarrhizium*, used for the control of insects, *Colletotrichum*, used to control the weed joint vetch in rice, *Cronartium*, used to control water hyacinths and *Puccinia*, used for control of nutsedge. Several fungi (*Trichoderma* and *Gliocladium*) are registered for the control of plant pathogenic fungi *Pythium* and *Rhizoctonia*, which cause damping-off diseases of seedlings. In France and Italy, hypovirulent strains of the causal organism of chestnut blight, *Endothia (Hyponectria) parasitica*, have been successfully used in the field to treat the disease (MacDonald & Fulbright 1991; Nuss 1992). Some of the major ecological concerns which arise with the increasing use of biological agents include non-target effects of the released agents on beneficial organisms (James & Lighthart 1994; Watrud & Seidler 1997) and the potential for increased host range or competitiveness resulting from genetic recombination between released and indigenous strains (Cisar *et al.* 1994). Health concerns centre largely on potential allergenicity of the inocula to sensitive individuals.

Mycorrhizae

Each of the major classes of mycorrhizal fungi can enhance inorganic nutrient uptake in host plants and help alleviate the effects of stressors ranging from drought, disease organisms and heavy metals to agricultural chemicals (Gildon & Tinker 1983; Heggo & Angle 1990; Hetrick *et al.* 1994). Among the three major classes of mycorrhizae, i.e. the ectomycorrhizal fungi, the ericoid mycorrhizal fungi, and the vesicular–arbuscular mycorrhizal (VAM) fungi, laboratory culture has become routine primarily only for members of the ectomycorrhizal group. Accordingly, it is the group of ectomycorrhizal fungi which includes the basidiomycete genera *Pisolithus* and *Laccaria* and the ascomycete genera *Tuber* and *Morchella* which has received the most commercial attention in terms of inoculum production and genetic improvement.

Basidiomycete ectomycorrhizal inocula are used to inoculate conifer species used in restoration of areas damaged by mining and smelting operations. Among the ericoid mycorrhizal fungi that typically form symbiotic associations with ericaceous plants, only two genera (*Hymenoscyphus* and *Oidiodendron*) can be cultured routinely in the laboratory (R.G. Linderman, pers. comm.). The VAM fungi, which include the phycomycetous genera *Glomus* and *Gigaspora*, can associate with most herbaceous species, and at some life stages with some woody species as well. However, these fungi evade extended axenic culture in the laboratory. Accordingly, the VAM fungi routinely are maintained by being grown in association with the roots of host plants such as grasses, onions or legumes. Most biotechnological work to date with mycorrhizal fungi has focused on the development of molecular probes to identify effective strains, particularly of ectomycorrhizal and VAM fungi (Simon *et al.* 1993; Lanfranco *et al.* 1995; Paolocci *et al.* 1995; Tommerup *et al.* 1995). As axenic culture methods become improved or available for the various types of mycorrhizal fungi, the potential to transform mycorrhizal fungi to select or create more competitive strains is expected to increase (Hall 1995). However, given the relative lack of specificity of ectomycorrhizal fungi among conifers and of VAM fungi among herbaceous and some hardwood species, the ability to mitigate the spread of (undesired) strains, which may displace or outcompete desired strains, could be environmentally or economically problematic.

Biomass conversion

Yeasts, especially *Saccharomyces* spp., are the traditional organisms of choice for production of ethanol from fruits and grains, potatoes and sugarcane. Other yeast species, including *Yarrowia* isolates, also have been evaluated for industrial purposes, including the biotransformation of lipids (Helsot 1990). Isolates of the filamentous fungus *Trichoderma*, noted above as a biological agent for plant disease control and a producer of enzymes, also have been proposed for use in the conversion of lignocellulosic wastes for fuel production and for the treatment of cotton-based textiles. This industrial interest in *Trichoderma* accordingly has resulted in research on the molecular biology and safety of the organism (Penttila *et al.* 1991; Nevalainen *et al.* 1994). The oyster mushroom *Pleurotous ostreatus*, which is edible, also has been evaluated as a commercial candidate to degrade ligninocellulosic wastes (Kerem & Hadar 1993). In addition to their degradative activities, bacteria and fungi may also serve in biomass conversions, biopulping, or biobleaching, or as pollution prevention agents by minimizing the accumulation of toxic metabolites. However, the possibility of toxic metabolite production by remediating agents also needs to be considered in risk evaluations.

Bioremediation

Utilization of fungal activities in bioremediation may occur by management of indigenous inocula or by the addition of exogenous inocula. For example, composting may effect a remediation due to activities of indigenous flora on straw or sawdust added to polluted soils. The white rot fungus *Phanaerochaete chrysosporium* has been reported to degrade a number of compounds, including 2,4,5-trichlorophenol (2,4,5-T), the BTEX complex (benzene, toluene, ethylbenzene and xylene), and pentachlorophenol (PCP) (Joshi & Gold 1993; Yadav & Reddy 1993). Additional filamentous fungi evaluated as polyaromatic hydrocarbon bioremediating agents include *Acremonium* and *Cunninghamella* (Pothuluri *et al.* 1995). Filamentous fungi such as *Rhizopus* (the common bread mould) and *Absidia* each have been proposed for use in biosorption of heavy metals for detoxification of waste streams (Volesky & Holan 1995). As transformation methods and vectors for fungal systems become more available, increased efforts in engineering industrial strains are anticipated. One concern associated with the environmental release of degradative fungal strains, many of which may be plant pathogens or saprophytes in nature, is the development of more competitive pathogens as the result of genetic recombination with indigenous pathogens or other compatible strains. Another concern is the potential for adverse effects on the quality and storage life of plant-based food and fibre crops. Allergenic reactions in sensitive individuals, either to the fungi or to recombinant proteins produced by them, also are of concern.

Food/beverage industries

Whether added intentionally as pure cultures or inadvertently as indigenous flora on fruits and grains, historically yeasts (especially *Saccharomyces* species) have been used to produce alcoholic beers, ales and wines. In more recent times, genetic selection and improvement of yeasts have been explored to attain more consistently desired aromas and flavours (Rank & Xiao 1991; Colagrande *et al.* 1994). They are also being used to produce light beers, with reduced starch, sugar and calorie contents, and to help regulate the level of ethanol in final products. The composition of high-fructose corn syrups used in non-alcoholic beverages, such as carbonated sodas and juices, and in products such as jams, jellies and pancake syrups, also can be adjusted by the use of genetically improved strains or enzymes such as invertases, amylases, glucose isomerases and pullanases derived from the selected strains (Harlander 1992).

Human and veterinary pharmaceuticals

Prominent among the strains of filamentous fungi which have served as sources of antibiotics or as models for chemical synthesis are the genera *Penicillium* and *Cephalosporium*, known for their synthesis of the broadly used penicillins and cephalosporins (Weil *et al.* 1995). Strains of the phycomycete *Rhizopus* have served as gene sources, precursors or models for the production of steriods, such as the cortisones used as anti-inflammatory agents and in the manufacture of steroid hormones used in birth-control pills (Breskvar *et al.* 1991; Vidyarthi & Nagar 1994). Genetic engineering of yeasts as hosts for the production of heterologous proteins has also been explored (Buckholz & Gleeson 1991).

5.2.3 Viruses

Seven of the 45 microbial pest control agents currently registered with the USEPA are viruses used to control lepidopteran insect pests of forest and agronomic species. Efforts to select and modify viral insecticides include using genetic engineering to broaden the host range of viruses (Vlak 1993). Recent approvals for field testing in the USA for engineered microbial pesticides include recombinant insecticidal baculoviruses which express insect-specific toxin genes derived from scorpions (W. Schneider, pers. comm.). With the advent of genetic engineering of higher plants, viral elements including the genes encoding viral coat proteins of plant pathogens have been cloned into a number of food crops, including tomatoes, potatoes and cucurbit species. Other viral elements, such as the cauliflower mosaic virus (CaMV35S) promoter, also have been instrumental in obtaining or optimizing expression of diverse genes in engineered plants (Sanders *et al.* 1987). Engineered plants expressing viral coat proteins have been demonstrated to confer protection to one or more types of viruses, particularly the types which served as the source of the viral coat proteins (reviewed in Watrud *et al.* 1996). In addition to plant protection by expression of viral genes (coat proteins, ribozymes, viral satellite ribonucleic acids (RNAs), some plants may be protected from certain viruses by treatment with either closely related viral strains or with avirulent strains. The concept of cross-protection using intact viruses has been used to control tristeza, an aphid-transmitted virus disease of citrus crops (Fulton 1986). Aphids exposed to avirulent strains of the virus are used to transmit the avirulent strains to citrus trees, thereby minimizing the effects of virulent viral strains on the citrus trees. Other crops for which the cross-protection approach of using viral inocula to protect crops against viral diseases is being evaluated include cacao (for swollen shoot), papaya (for ringspot), avocado (for sun blotch), and stone and pome fruits, such as peaches and apples (Whipps 1992). Hypovirulent strains of the fungal causal agent of chestnut blight (*Cryphonectria (Endothia) parasitica*), reported to have been successfully used to control the disease, have consistently shown a correlation with the presence of virus-like double-stranded RNAs (MacDonald & Fulbright 1991; Nuss 1992). An engineered hypovirulent strain of *Cryphonectria* received approval for field testing in the USA in 1994 (J. Rissler, pers. comm.). Mammalian pathogens, including poliomyelitis, influenza and sindbai viruses, currently are being evaluated as expression vectors in animal cell systems for the production of heterologous proteins (Schlesinger 1995). The use of transgenic plants as factories to produce viral components such as coat proteins or other antibodies, which might be used in the production of vaccines for human or veterinary applications, also has been proposed (Hiatt *et al.* 1989). Environmental concerns which emanate from the use of virus-resistant transgenic plants include the selection for novel, more competitive viruses, potentially with broader host range arising as the result of genetic recombination. Another concern related to the use of transgenic virus-resistant plants is that they may serve as reservoirs of inoculum for viral transmission, via biological or mechanical means, to non-protected plants. Allergenicity and pathogenicity to sensitive individuals may cause potential health concerns following exposure to viral expression systems and expression of foreign proteins.

5.3 ISSUES TO CONSIDER IN CONDUCTING RISK ASSESSMENTS OF GMOs

A document written in 1984 (Bourquin & Seidler 1986) developed and guided the earliest strategies to provide methods for conducting risk assessments of GMOs released to the environment. In that report, research plans were developed that addressed the need for increased research in microbial ecology in order to provide knowledge on how to conduct GMO environmental risk assessments. The focus of basic research needed to conduct appropriate risk assessments is summarized as follows.

1 Methods are needed to detect, identify and enumerate released GMOs.

2 Methods are needed to determine the fate and transport of the GMOs.

3 Genetic stability, i.e. propensity for gene transfer, must be assessed.

4 Hazard Assessments (pathogenicity to non-target spp., including humans and disruption of environmental processes) must be thoroughly investigated.

5 A means to mitigate possible effects would be prudent.

Environmental risk of GMOs has been described as a function of the exposure and the hazard posed by the GMO (Bourquin & Seidler, 1986; Levin & Strauss 1990). This relationship has helped conceptually to develop strategies for identifying relevant research needs to address the ecological concerns of releasing GMOs to the environment. Thus, exposure is a general function of organism survival and multiplication (cell density), which in turn influences GMO transport in the environment. Hazard is a function of ecological effects, including competitiveness and metabolic activities, that perturb the habitat, as well as pathogenicity, virulence, toxicity and allergenicity. Therefore it is essential to understand the fundamental ecology and physiology of the organism to be released and the basic character of the release environment in order to address questions concerning risk assessment of the GMO (USEPA 1992). These evaluations to date largely have been conducted on a case-by-case basis because most products are unique and previously have not been encountered by regulators, ecologists and the environment.

In order to conduct a risk assessment of a GMO, information is needed on at least five technical issues that comprise the exposure and hazard subcomponents for a microorganism (Bourquin & Seidler 1986; Levin *et al.* 1987; Hall 1995). These include: (i) knowledge of how to detect and enumerate the GMO; (ii) estimates on organism survival, multiplication and potential competitiveness; (iii) transport and subsequent colonization; (iv) possible transfer of genetic traits; and (v) the ability to cause adverse environmental effects, including disruption of environmental processes and pathogenicity, toxicity and infectivity. Useful technical guidance applicable to further risk assessment issues can be found in the July 1994 and November 1994 publications of the USEPA Risk Assessment Forum (USEPA 1994a,b) and in Sayler & Sayre (1995). These technical issues are now characterized in further detail and are presented as a central theme for conducting risk assessment of GMOs.

5.3.1 The need to detect, identify and enumerate

The most fundamental requirement for GMO risk assessments is the necessity to have an absolutely specific means to detect and enumerate the target GMO from environmental samples (Bourquin & Seidler 1986; Levin *et al.* 1987). This makes it possible to detect and document any hazard, should one exist, and evaluate the components of the 'exposure' side of the risk assessment equation (survival, multiplication, transport and competitiveness). The enumeration procedure must be so specific that the GMO can be readily distinguished from indigenous microbes of the same species. This simple requirement may be complex in practice because of the diversity of habitats that must be sampled and the lack of any single methodology applicable to recovery of GMOs from all habitats (Donegan *et al.* 1991; Angle *et al.* 1994).

5.3.2 Conditions impacting survival, multiplication and competitiveness in the environment

By having a general understanding of the ecology of the GMO and knowledge of the release habitat, certain generalities may be developed regarding expectations for the survival and potential for regrowth of the test organism. For example, if the GMO strain was isolated from the intestine of an insect, it may be expected to better colonize that habitat when reintroduced into nature than to colonize soil (Armstrong *et al.* 1989). Such a microorganism obtained from an insect also might be more readily transported off-site through colonization of other insects. Knowledge of the maximum growth temperature of an organism will also be useful in estimating hazards. For example, a recombinant organism isolated from soil and having a maximum growth temperature of 32°C would be most unlikely to colonize the gastrointestinal tract of a warm-blooded animal, let alone cause a disease in such a host. Scanferlato *et al.* (1989) studied the survival of a genetically engineered *Erwinia carotovora* in aquatic microcosms. This plant pathogen did poorly, did not become established and did not survive beyond 32 days.

Understanding the capabilities and conditions that may lead to population growth in the environment might be conveniently accomplished through pilot studies in contained environments, such as in microcosms or greenhouses (Fredrickson & Seidler 1989; Cripe *et al.* 1992; Hood & Seidler 1995; Krimisky *et al.* 1995). Gillett *et al.* (1978) have defined a microcosm as '... a controlled, reproducible laboratory system which attempts to simulate the situation in a portion of the real world' (see also Chapter 3). The intent is that the microcosm duplicate, as faithfully as possible, environmental conditions in order to simulate the proposed field site. Perhaps the most relevant practical use of microcosms is that they can provide results that can be used to recommend how often to sample, where to sample, and what general concentrations of the GMO may be anticipated in the field environment. Furthermore, research has demonstrated that studies in these contained environments may provide a first approximation of anticipated fate/survival in the field (Bentjen *et al.* 1988; Armstrong 1989; Bolten *et al.* 1991; Angle *et al.* 1995; Wagner-Dobler *et al.* 1992). Other natural factors that influence survival may otherwise be difficult to anticipate and may include an ability to persist under starvation conditions and grow in association with other organisms (intestines of animals, insects, rhizosphere of plants) or with decaying tissue of plants or animals. All the latter situations also may be simulated in microcosms.

A valid question has been raised in risk assessment discussions: 'what cell densities are important in assessing risks and is there a threshold response level below which any potential risks will diminish?' For example, in order to exhibit detectable metabolic activities, it is predicted that ca. 10^6–10^7 cells g^{-1} (or cells ml^{-1}) will be required, whether it be to induce production of transconjugants, to cause change in a substrate concentration, or to enable outcompeting of indigenous microflora, etc. (Olson *et al.* 1990; Doyle *et al.* 1991; Short *et al.* 1991; Walter *et al.* 1991; De-Leij *et al.* 1994; Ripp *et al.* 1994). If the GMO has declined to less than 10^3 cells g^{-1} it is very unlikely that an undesirable metabolic or ecological event effect would occur, let alone be detected. It is conceivable, however, that a critical environmental substrate could allow the GMO to multiply and achieve significant levels, but there are few known examples where this has occurred except for disease processes in animals and plants. For documented situations where GMOs have multiplied significantly following their release into the environment, the reader is referred to Armstrong *et al.* (1989), Raaijmakers *et al.* (1995) and De-Leij *et al.* (1995). Whatever the GMO is designed for, it is likely that large cell densities (in excess of 10^7 cells g^{-1}) will be necessary for it to carry out its intended function in the environment. Periodic sampling in microcosm habitats over several weeks to months will provide useful data for anticipating cell numbers that a specific GMO might achieve.

5.3.3 Fate (dispersal and transport) of GMOs

It has been demonstrated with early field experiments, and on a smaller scale with microcosms, that wind, water and insects can transport GMOs from the original site of deposition (Seidler & Hern 1988; Armstrong *et al.* 1989, 1990; Lighthart & Kim 1989; Seidler *et al.* 1994). These studies revealed that once a GMO is released into the environment, it can spread. This is not an unexpected observation, but few earlier studies failed to investigate the transport of microbes with such specificity and detail. Transport with GMOs has been illustrated not only through a simple physical change of location in soil, but also through a change in the habitat as well. Thus, it has been shown that bacteria can travel from leaves to insects to soil (Armstrong *et al.* 1989, 1990). Therefore, when planning a field release locale, investigators, regulators and the general public should realize that dissemination may occur to a certain extent. Movement of the GMO will not necessarily be a risk *per se* but mobility should be taken into consideration as a component of the risk assessment relationship. Mobility could lead to GMO establishment at new sites, perhaps in new habitats, and therefore may lead to new exposures.

There is obviously a need to understand the fate and rate of transport to evaluate exposures in the environment. Test procedures are available for assessing the fate of GMOs in microcosms. These model ecocystems can provide very useful information that deals with GMO dispersal, transport, survival and potential effects, but obviously they are scale limited (Angle *et al.* 1995).

It is presumed that there are interactions of biotic and abiotic components, as well as fundamental ecological processes, in microcosms that reflect processes in the environment. Studies clearly have demonstrated that a key to the successful application of microcosms is the necessity of controlling and monitoring environmental conditions during the course of the experimental period (Armstrong 1989; Wagner-Dobler *et al.* 1992). An advantage of using microcosms is that they can be constructed to contain fewer complexities compared to the natural environment. However, this simplification in itself can be a disadvantage because the natural environment is not simple. Obviously microcosms are closed systems, while natural ecosystems are open.

5.3.4 Gene flow and recombination

Gene transfer has been documented to occur within more than two dozen bacterial genera and numerous additional species (Levin *et al.* 1992). In order for gene transfer to occur between microbes, certain physiological, metabolic and cell density requirements must be met. Gene transfer under artificial situations in the laboratory or in microcosms usually requires millions of cells per unit volume of material in order to obtain detectable transfers (Knudsen *et al.* 1988; van Elsas & Smit 1994). Because these metabolic and cell density conditions are not thought to occur with any regularity in the natural environment, some believe that gene transfer among GMOs and the indigenous bacteria is an extremely infrequent event. However, more recent data lead to other conclusions. The three well-characterized mechanisms by which bacteria may undergo genetic exchange (transformation, transduction and conjugation) are, in all likelihood, operational in soil (Smit *et al.* 1991; van Elsas & Smit 1994).

Recently, studies from two laboratories have rigorously demonstrated that gene transfer from experimentally released bacteria into indigenous bacteria has occurred in a watershed and an aquifer (Fulthorpe & Wyndham 1991; Zhou & Tiedje 1995). Transduction and conjugation also have been demonstrated experimentally to take place in the phylloplane, rhizosphere and epilithon of rocks in a stream (Knudsen *et al.* 1988; Farrand 1992; Hill *et al.* 1992; Kidambi *et al.* 1994).

Soil and its component parts have been demonstrated to be both a hindrance (compartmentalization) and a stimulator (nutrients, rhizosphere effect) of gene transfers. It also has been confirmed that when the recombinant DNA is maintained in the chromosome as opposed to on a transmissible plasmid, the incidence of gene transfer by conjugation generally drops to undetectable levels, as anticipated (Smit *et al.* 1991; 1992). Thus, if one is prudent in the molecular design of the recombinant organism and can avoid habitats that promote cell contact and high cell densities and provides low available substrates for growth, the transfer of recombinant DNA into indigenous bacteria is not a detectable event.

Recently, evidence also has accumulated that bacteriophage-mediated transduction plays a significant role in the transfer of both plasmid and chromosomal DNA in aquatic ecosystems (Ripp & Miller 1995). Using lakewater microcosms incubated in a freshwater lake, investigators demonstrated that particulate matter increased transduction up to 100-fold. Furthermore, up to 40% of the *Pseudomonas aeruginosa* in natural systems contain DNA sequences homologous to phage genomes, indicating probable prior interaction with a transducing bacteriophage (Ogunseitan *et al.* 1992).

It is well documented that gene transfer between bacteria requires specific physiological, nutritional and environmental conditions. These conditions can be met in a variety of habitats where natural cell densities are sufficient to promote necessary cell, bacteriophage or naked DNA interactions to allow genetic exchange to occur at detectable rates. Deoxyribonucleic acid transfers of recombinant genes have been facilitated by naturally occurring mobilizing plasmids in river epilithon where bacterial numbers are very high on rock surfaces (Hill *et al.* 1992).

5.3.5 Characterizing ecological effects

The experimental search for ecological effects from GMOs has taken a tortuous and intellectually challenging route because of the potential political, social, economic and scientific consequences and challenges of finding such an organism. Uncertainties over possible ecological effects are probably derived from a combination of our uncertain and limited knowledge of microbial ecology and the documented 'horror' stories of significant and adverse effects from introductions of certain higher species of plants and animals (Halvorson *et al.* 1985; Sharples 1990). Scientists have tended to speculate that certain undesirable consequences brought about by higher plant and animal introductions (see Chapter 6) might be extended to GMOs. This has perhaps sensationalized the anticipated discovery of GMO-induced effects and made the documented effects seem less important.

Sharples (1990) has compiled a list of characteristics that probably assist in the establishment of new organisms in new niches. An inspection of these traits reveals their applicability to micro-organisms as well as to higher organisms:

> '... the organisms with generalized requirements and broad tolerances seem to make better invaders ...'
>
> '... exploiting a used or underused resource will facilitate establishment ...'
>
> '... unique physiological features, as in the ability to metabolize unusual compounds, may facilitate establishment ...'
>
> '... an organism may survive better if it is preadapted to survive when the new habitat is like the one in which it evolved ...'

It may prove valuable to keep these concepts in mind as new GMOs are constructed and examined for possible environmental perturbations.

Despite the verifications that GMOs may cause undesirable changes in a habitat (Doyle *et al.* 1991; Short *et al.* 1991; Seidler 1992; Doyle *et al.* 1995), questions remain as to how best to evaluate and detect these potential detrimental effects with new GMOs. Another equally perplexing issue is a criticism as to the significance of an effect that is 'transient' in nature as opposed to one that is longer in duration. It must, of course, be clear that the recovery from a 'transient' effect actually leaves the habitat in a 'recovered' state that is unaltered from its original structure and function. Most investigations have used only broad, general assays to monitor habitat changes induced by GMOs. However, when detailed methodologies are used to track perturbations, evidence for taxonomic changes in bacterial populations has been revealed, despite the return in culturable bacterial numbers to control levels (Donegan *et al.* 1995).

Many end-points have been considered in evaluating possible effects from GMOs (Table 5.1). Initially, it was not obvious which end-points would be the most logical to investigate. A 'shotgun approach' was used in early research with hopes of capturing GMO-induced alterations (Seidler 1992; Doyle *et al.* 1995). This approach resulted in the detection of effects documented in changes of respiration, cell numbers, and certain soil enzymes (Doyle *et al.* 1991; Short *et al.* 1991; Wang *et al.* 1991). Monitoring these population changes and other risk assessment experiments can be conducted readily, and perhaps should be conducted in microcosms. Population trends, biochemical activities, mineralization or leaching of soil enzymes, and plant biomass trends also have been investigated and documented to be useful when monitoring for GMO perturbations (Bolten *et al.* 1991).

Table 5.1 Experimental end-points relevant to measuring possible ecological risks associated with genetically modified organisms (GMOs) in the environment (After Seidler 1992.)

Impact of GMO on numbers or kinds of viable bacteria
Metabolic activity (respiration, kinetics of CO_2 evolution, substrate-induced respiration)
Numbers of Gram-negative bacteria present
Bacterial populations in the rhizosphere and rhizoplane
Species diversity
Biomass of colonized plants
Microbial biomass
Biomass of plant shoots
Nitrogen content of wheat shoots
Total viable fungi
Numbers of cellulose utilizers
Number of chitin utilizers
Numbers of denitrifiers
Numbers of nitrifiers
Numbers of protozoans
Nutritional groups of bacteria present
Activity of soil enzymes
Kinetics of nitrogen transformations
Numbers of non-symbiotic dinitrogen fixers
Competitive ability of GMOs
Effects of microbial pest control agents (MPCAs) on mycorrhizal colonization of roots
Effects of MPCAs on mycorrhizal colonization and plant growth
Numbers of soil nematodes, mites, Collembola
Trophic groups of nematodes
Transfer of recombinant DNA to indigenous microorganisms

The day may be rapidly approaching when experimental trials of GMOs will be routinely conducted in the open environment with little or no preliminary microcosm investigations. This certainly has become the trend in testing transgenic plants (Mellon & Rissler 1993). Another approach to initial or preliminary assessments of GMOs may be to conduct field trials with the unaltered parental strain. In this case there may be doubt if the parental strain would accurately mimic key ecological traits that pertain to risk assessment aspects of the GMO. It seems likely that differences in key metabolic characteristics of the GMO may significantly alter its survival and this may influence its characteristics for exposure or hazard assessment. Therefore, if experiments are to be conducted in microcosms, this should be done using the (recombinant) GMO strain.

5.3.6 Health effects

The deliberate environmental release of GMOs has initiated investigations of potential health effects associated with exposure to these organisms. Humans may come into contact with these microorganisms or their products in the agricultural or industrial setting during production or application (Grunnet & Hansen 1978; Olenchock 1988). Because soil, water or air may contain high concentrations of the microbial product, the primary routes of exposure are through inhalation, ingestion or skin penetration (Levy 1986).

The new *Escherichia coli* host–vector systems constructed during the initial recombinant DNA studies in the 1970s raised several human health issues (Gorbach 1978a,b). Colonization of the intestinal tract and transfer to, and expression of, foreign genes by the intestinal microbiota were of particular concern. Researchers studied colonization of the human gastrointestinal tract by *E. coli* strains K12 and B (and their derivatives) and found that the colonization of normal healthy adults did not occur (Anderson 1975; Smith 1975, 1978; Levy *et al.* 1980). However, *E. coli* strains were able to establish in the intestines of germ-free and antibiotic-treated mice (Cohen *et al.* 1979; Levy *et al.* 1980; Laux *et al.* 1981; Levy & Marshall 1981). Gene transfer to the normal human intestinal flora was plausible, but *E. coli* strains harbouring plasmids were demonstrated to have a selective disadvantage in their colonizing ability (Anderson 1978; Laux *et al.* 1982; Duval-Iflah & Chappuis 1984).

Aside from the pesticide products that require health effects testing as described in Subdivision M of FIFRA, little health effects research has been done on biotechnology agents and their products or byproducts. To date, the general consensus has been that because the strains used for biotechnology applications are of little risk to 'normal' people, more research emphasis has been placed on environmental effects. However, as research into potential health effects is conducted, the health effects of some biotechnology strains may be less certain than previously thought. As summarized in Table 5.2, the majority of the published research has dealt with microorganisms used for biodegradation or the byproducts of this process. Recently, Health Canada has completed an extensive study on the pathogenicity of environmental and clinical *P. aeruginosa* (because of its potential application for biodegradation) isolates looking at a variety of end-points which will be addressed in Section IIF(1) (Godfrey *et al.* 1996).

In general, there are two fundamental questions that should be answered when considering health effects associated with biotechnology agents and products: (i) Does the microorganism(s) involved impact health; and (ii) does the microbial product or byproduct result in an increased health risk? In order to address these questions, three general areas will be explored and their relevance discussed: pathogenicity (comprised of mortality, morbidity and infectivity); allergenicity of both the product of interest and the microbial component; and product (or byproduct) toxicity.

Pathogenicity

The likelihood of an industrial release of a known

Table 5.2 Examples of biotechnology agents and their products where health effects studies have been reported in the scientific literature.

Application	Microbial strain or product	References
Pest control	*Bacillus thuringiensis* subsp. *israelensis*	Held *et al.* 1986; Mayes *et al.* 1989; Kawanishi *et al.* 1990
	B. thuringiensis subsp. *kurstaki*	Sherwood *et al.* 1991a,b
	Pseudomonas syringae	Olenchock 1988; Goodnow *et al.* 1990
	Autographa californica	Hartig *et al.* 1989, 1992
Biodegradation	*Pseudomonas* sp.	Kaiser *et al.* 1981
	P. aeruginosa	George *et al.* 1989, 1990, 1993
	P. putida	Kaiser *et al.* 1981
	Burkholderia (Pseudomonas) cepacia	George *et al.* 1990, 1991
	Xanthomonas (Pseudomonas) maltophilia	George *et al.* 1989, 1990, 1993
	Mycobacterium sp.	Kaiser *et al.* 1981
	Flavobacterium sp.	Kaiser *et al.* 1981
	Norcardia sp.	Kaiser *et al.* 1981
Product	Alcalase/subtilisin	Sarlo *et al.* 1991; Ritz *et al.* 1993
Byproduct	2,4,5-Trichlorophenoxyacetate metabolites	George *et al.* 1992
	Pentachlorophenol metabolites	DeMarini *et al.* 1990
	Crude oil degradation metabolites	Claxton *et al.* 1991a,b

pathogen is very low. In fact the *Federal Register* statutes (1986) regulating the release of GMOs mandate that the organism used not be pathogenic. However, because of the increased exposure potential and the possibility for the use of identified or unknown opportunistic pathogenic agents, it is imperative to determine if an organism has the probability to infect workers or the general public. Therefore, several *in vivo* and *in vitro* systems have been used to determine if biotechnology agents and their products are potentially hazardous. Many of these use rodent models to determine morbidity and establish LD_{50} values. For example, Roe *et al.* (1991) have demonstrated that the solubilized parasporal crystalline protein of *Bacillus thuringiensis* subsp. *israelensis* has an LD_{50} of 1.3 mg kg^{-1} following i.p. injection of the Swiss-Webster mouse. The LD_{50} in the CD rat is 9 mg kg^{-1}. Following intranasal exposure, George *et al.* (1993) have established an LD_{50} of 1.05 × 107 CFU for *P. aeruginosa* strain AC869 in the C3H/HeJ mouse.

In order to cause disease, an organism must evade or overwhelm the defence barriers of its potential host and multiply. Therefore, survival temperature range plays an important role in the infection process. As stated earlier, if an organism cannot survive above 30°C, it is unlikely that it will persist and multiply in a mammalian host. However, if the microorganism can persist at the host temperature, it can attach to, or invade, host cells. A list of temperature ranges for selected biotechnology agents is presented in Table 5.3. Additionally, in order to achieve colonization, an organism must avoid the host immune systems. Once inside the cell, an organism may multiply and be disseminated throughout the host. Pathogenicity factors, such as adhesins, extracellular enzymes and cytotoxins, may enhance disease progression. However, disease is an inadvertent outcome of infection (Finley & Falkow 1989).

Production of pathogenicity factors by an invasive microorganism may facilitate infection (Finley & Falkow 1989). Significant attention has been given to pathogenicity factors attributed to *P. aeruginosa* and other pseudomonads (Table 5.4). Members of the genus *Pseudomonas* are used for environmental applications because of their effectiveness in biocontrol and versatile substrate utilization abilities. Even though most of these microorganisms generally are considered harmless soil isolates, they may cause opportunistic infections following exposure of susceptible individuals to high concentrations. *Pseudomonas aeruginosa* is recognized as an opportunistic human pathogen and is capable of causing disease, especially in immunosuppressed hosts and patients with leukaemia or cystic fibrosis (Bodey *et al.* 1978; Schimpff 1980; Guiot *et al.* 1981). Grimwood *et al.* (1993) have concluded that exoenzyme expression correlates with lung deterioration in cystic fibrosis disease progression. Some isolates of *Burkholderia cepacia* (formally *Pseudomonas cepacia*) are involved in nosocomial infections (Martone *et al.* 1981) and can colonize and cause pneumonia and septicaemia in cystic fibrosis patients (Rosenstein & Hall 1980; Sajjan *et al.* 1992). *Pseudomonas putida*, *P. fluorescens*, and *Xanthomonas maltophilia* (formally *Pseudomonas maltophilia*) have been

Table 5.3 Temperature range for survival of some bacterial species of biotechnology relevance.

Biotechnology agent	Temperature optimum (°C)*
Pseudomonas aeruginosa	37
P. putida	25–30
P. fluorescens	25–30
P. aureofaciens	~30
P. syringae	~25–30
Burkholderia cepacia	~30–35
Xanthomonas maltophilia	35
Bacillus thuringiensis	30–40 (10–40)†
B. subtilis	30–40 (10–50)
B. popliae	~30
B. lentimorbis	~30
Thiobacillus ferroxidans	30–35 (10–37)
Sulfolobus sp.	>50–87
Rhizobium meliloti	25–30 (4–42.5)†
Bradyrhizobium japonicum	25–30 (25–42)†

* Bergeys' Manual of Systematic Bacteriology, volume 1 (Kreig & Holt 1984), volume 2 (Sneath *et al.* 1986), and volume 3 (Staley *et al.* 1989).
† Numbers in parentheses indicate temperature range.

Table 5.4 Pathogenicity factors identified in *Pseudomonas* and *Burkholderia* spp (denoted by X).

Trait	*P. aeruginosa*	*B. cepacia*	*P. fluorescens*	*P. putida*
Protease	X	X	X	
References	Liu 1974; Woods & Sokol 1986; Grimwood *et al.* 1989	Bevivino *et al.* 1994	Wilson & Miles 1975; Tan & Miller 1992	
Exotoxin A	X			
References	Woods *et al.* 1982; Woods & Sokol 1986; Grimwood *et al.* 1989; Woods *et al.* 1991			
Elastase	X	X	X	
References	Liu 1974; Woods *et al.* 1982	McKevitt *et al.* 1989	Oakley & Bonnerjee 1963	
Haemolysin	X		X	
References	Liu 1974; Woods & Sokol 1986		Wilson & Miles 1975	
Phospholipase C (lecithinase)	X		(X)	
References	Grimwood *et al.* 1989; Woods *et al.* 1991		Esselmann & Liu 1961; Liu 1974; Wilson & Miles 1975	
Collagenase	X			
Reference	Liu 1974			
Pyochelin	X	X		
References	Woods *et al.* 1991	Bevivino *et al.* 1994		
Lipase	X	X	X	X
References	Jørgensen *et al.* 1991; Wolfarth *et al.* 1992	Jørgensen *et al.* 1991	Tan & Miller 1992; Wolfarth *et al.* 1992	Wolfarth *et al* 1992
Enterotoxin	X			
Reference	Liu 1974			
Adhesins				
Alginate/mucoid Polysaccharide	X	X		
References	Liu 1974; Marcus *et al.* 1989; Pedersen *et al.* 1992	Straus *et al.* 1989		
Exoenzyme S	X			
References	Woods & Sokol 1986; Woods & Que 1987; Grimwood *et al.* 1989; Woods *et al.* 1991			
Pili/fimbriae	X	X		
References	Ramphal *et al.* 1984; Farinha *et al.* 1994; Tang *et al.* 1995	Kuehn *et al.* 1992; Goldstein *et al.* 1995		

shown to cause secondary infections in cancer patients and are involved in other opportunistic infections (Gilardi 1991; Anaissie *et al.* 1986).

In order to establish regulations for biotechnology products, Health Canada has investigated several end-points for use in the risk analysis of *P. aeruginosa*, a potential isolate of bioremediation products (Godfrey *et al.* 1996). Clinical and environmental isolates have been screened for exoenzyme production (Toxin A, protease, phospholipase C, exoenzyme S, elastase), cytotoxicity, serotype, serum sensitivity, phagocytic killing, opsonic phagocytosis, A and B band lipophysaccharide (LPS) probe reactivity, pilus phage sensitivity and restriction fragment length polymorphism type. Of all of the end-points measured, exoenzyme S is the only one to demonstrate a good correlation to mouse virulence (LD_{50} in the neutropenic mouse model), with elastase production showing possible correlation to LD_{50}. Because the mechanism of pathogenicity by *P. aeruginosa* is multifactorial, the elimination of one pathogenicity factor, such as exoenzyme S, may not eliminate pathogenicity (Godfrey *et al.* 1996). Others have suggested that, with the exception of alginase production by *P. aeruginosa* cystic fibrosis isolates, there is no one factor that differentiates between clinical and environmental isolates of *P. aeruginosa* and *B. cepacia* (Nicas & Iglewski 1986). However, Bevivino *et al.* (1994) have shown that rhizosphere isolates do have several unique traits not found in their clinical counterparts, such as wider temperature range for growth, nitrogen fixation ability, and indole acetic acid production. In this study, clinical isolates produced protease and pyochelin, and adhered to human cells.

Allergenicity

A second category of deleterious outcomes following human exposure to biotechnology agents and their products is an allergic reaction. This can culminate in a skin reaction or a more serious respiratory response to the allergen, such as bronchitis and/or asthma. Because a variety of fungi can cause severe respiratory reactions, it is especially important to consider the potential allergenicity of fungal pest control agents such as *Metarhizium anisophilae* and *Beauvaria bassiana*. Additionally, bacterial strains that secrete extracellular enzymes and/or polysaccharides may be allergens.

An extensively studied enzyme product that elicits an allergic reaction is alcalase (subtilisin). Following its introduction into the detergent industry in the late 1960s, this alkaline protease, produced by *Bacillus subtilis* and supplied by Novo Industri A/S (Bagsvaerd, Denmark), was identified as a Type I (IgE) respiratory allergen (Flindt 1969; Pepys *et al.* 1969). Franz *et al.* (1971) reported respiratory symptoms in detergent industry workers 3–8 h following their work shift. Another group at risk, the consumer, also has been shown to develop respiratory symptoms from exposure to the enzyme component in the detergent (Shapiro & Eisenberg 1971).

Detergent industry workers were closely monitored once alcalase was identified as the allergen. Several studies of workers have been reported that correlate exposure to a positive skin prick test and total serum immunoglobulin E (IgE) levels (Shapiro & Eisenberg 1971; Weill *et al.* 1971; How *et al.* 1978). Once exposure control was implemented in the detergent plant (e.g. dust levels reduced, protective clothing worn by enzyme handlers), fewer instances of respiratory allergenicity were reported. Additionally, because acceptable exposure levels have been established for alcalase, it is now used as the benchmark for comparison of new enzyme products (Sarlo *et al.* 1991).

A third area of potential allergen exposure arises from the consumption of transgenic plants. The United States Food and Drug Administration has conceded that it is possible for genetically modified plants to express foreign proteins that could cause the food to be allergenic and now has a protocol to help regulate potentially allergenic transgenic foods (Nestle 1996). Interestingly, Nordlee *et al.* (1996) identified a Brazil nut allergen in an engineered soybean, which expressed brazil nut albumin. Therefore, it is important to determine if an engineered food has an allergenic potential so that the consumer can be notified of potential risks.

Byproduct toxicity

Use of bioremediation to reduce or eliminate hazardous waste is economically important. The ultimate goal of waste management is to eliminate toxicity of the waste and reduce risk of exposure to the remaining hazardous components. There are many bacteria and fungi that can metabolize, or co-metabolize, many complex chemicals either alone or in consortia, both aerobically and anaerobically. However, the scenario at the hazardous waste site or biodegradation facility may not be as controlled as the laboratory incubator, so a biodegradation process may not occur as readily or identically as that observed in the laboratory. Therefore, because the reduction of one compound does not necessarily indicate reduced toxicity of the resulting byproducts, a variety of methods can be used to determine if a chemical degradation process reduces the overall toxicity (cytotoxicity, genotoxicity, teratogenicity and neurotoxicity) of the target compound(s) or complex mixtures.

There are two general approaches to toxicity reduction: (i) if one or a few chemicals are involved and if the metabolic pathway(s) is known, then the toxicity of the metabolites can be established and the toxicity of the process predicted; and (ii) if the metabolic pathway is unknown or if the target is a complex mixture of chemicals, then extracts from bench-scale biodegradation processes can be analysed for a reduction in overall toxicity prior to large-scale use. An example of the first scenario is that of pentachlorophenol (PCP) degradation. The pathway for PCP degradation from anaerobic sewage sludge studies is shown in Fig. 5.1 (Mikesell & Boyd 1988). Pentachlorophenol induces hepatic carcinomas and adenomas in mice and chromosomal aberrations in Chinese hamster ovary cells (NTP 1989; DeMarini *et al.* 1990). The metabolites 2,3,4,5-tetrachlorophenol and 3,4,5-trichlorophenol are more genotoxic than the parental compound PCP in the prophage induction bioassay in *E. coli* (DeMarini *et al.* 1990). Tetrachlorohydroquinone, a major PCP metabolite identified in rodent studies, is genotoxic in V79 Chinese hamster cells and may be responsible for oxidative DNA damage *in vivo* (Jansson & Jansson 1991). A similar scenario occurs following 2,4,5-trichlorophenoxyacetic acid (2,4,5-T) metabolism by *B. cepacia* (George *et al.* 1992). One of the metabolites, 2,4,5-trichlorophenol, is 100-fold more genotoxic than the parental compound in the prophage induction bioassay. Therefore, if the parental compounds PCP and 2,4,5-T are metabolized, the chlorinated intermediates of these compounds may accumulate, and the resulting toxicity may be considerably higher than that observed for the unremediated waste. Because the metabolic pathway is known for PCP and 2,4,5-T and the resulting intermediates are more toxic than the parental compounds, strains can be constructed that do not accumulate the more toxic intermediates or another microorganism with alternative degradative pathways can be employed for the bioremediation process.

Toxicity tests have been used to monitor the toxicity of hazardous waste sites and bioremediation (Table 5.5). Claxton *et al.* (1991a,b) used the *Salmonella* assay to assess the genotoxicity during the bioremediation efforts following the Alaskan oil spill in March 1989 in Prince William Sound, Alaska. The efficacy of two fertilizers, one oil soluble and the other water soluble, was studied. The reduction or generation of toxicants during bioremediation was of particular interest. The results concluded that the mutagenicity

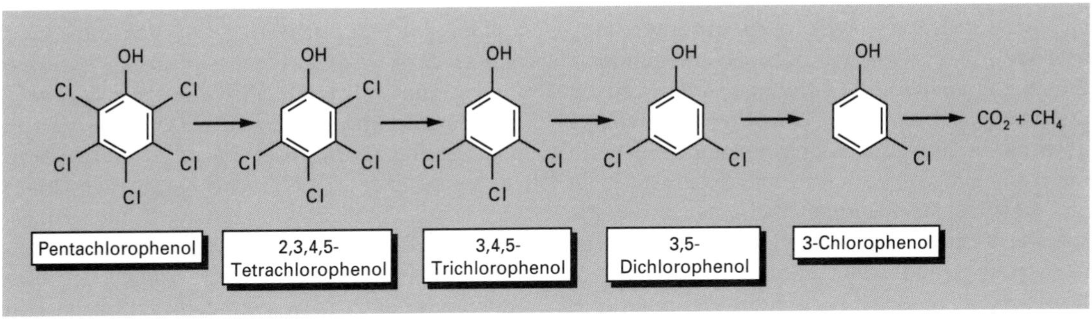

Fig. 5.1 Degradation pathway for pentachlorophenol. (After Mikesell & Boyd 1988.)

Table 5.5 Use of bioassays to determine the toxicity of bioremediation processes and hazardous waste.

Chemical	Matrix	Bioassay	References
PCB metabolites*	–	Sister chromatid exchange *Salmonella* reversion	Sayler *et al.* 1982
Chlorophenols	–	Prophage induction (*E. coli*)	DeMarini *et al.* 1990
2,4,5-T	–	*Salmonella* reversion Prophage induction (*E. coli*)	George *et al.* 1992
Pesticide waste	Soil	*Salmonella* reversion	Somich *et al.* 1990
Benzo[*a*]pyrene	Soil and sludge	*Salmonella* reversion	Miller *et al.* 1988
Wood-preserving waste	Soil and water	Microtox™ *Salmonella* reversion *Aspergillus nidulans* *Bacillus subtilis* DNA repair	Donnelly *et al.* 1987a,b; Aprill 1990
Petroleum waste	Soil	Microtox™ *Salmonella* reversion Seed germination Plant	Aprill 1990; Wang & Bartha 1990
Hazardous waste complex mixture	Soil and water	*Salmonella* reversion SOS (*E. coli umu*) Prophage induction (*E. coli*)	Donnelly *et al.* 1988; Houk & DeMarini 1988; McDaniel *et al.* 1993

* PCB, polychlorinated biphenyl; 2,4,5-T, 2,4,5-trichlorophenoxyacetic acid.

declined over time and that both the naturally occurring processes and fertilizer supplement contributed to the reduction in toxicity.

5.4 EXAMPLES AND METHODS FOR DETECTING EFFECTS CAUSED BY GMOs

5.4.1 Environmental perturbations from GMOs

What kinds of environmental perturbations might be anticipated in a worst-case scenario involving a GMO? Are there any precedents in the literature that might provide answers to this question? There are now many examples of documented effects on beneficial organisms or ecosystem processes induced following exposure to a GMO (Table 5.6). Overall, the documented effects include the transfer of genetic material to indigenous bacteria, production of toxic metabolites, changes in metabolic activity of the soil community, competition resulting in the suppression of indigenous microorganisms and changes in the biodiversity of organisms present in the habitat. These changes represent exactly the kinds of perturbations that were speculated as possible occurrences, early in the risk assessment of GMOs (Halvorson *et al.* 1985; Bourquin & Seidler 1986). Genetically modified organisms that are effectors of ecosystem perturbations also represent a large and diverse type of taxonomic and metabolic group.

Perhaps more studies have been conducted with a plant/soil ecosystem to explore possible GMO effects than with any other system. The first field studies in the USA involved the release of ice–*Pseudomonas syringae* on to strawberries and potatoes to control frost damage (Seidler & Hern 1988; Seidler *et al.* 1994). In those experiments it was learned that despite reasonable precautions, the aerosol spray of the GMO resulted in considerable drift off the main plot. Although elaborate monitoring devices were employed, it transpired

Table 5.6 Genetically engineered microorganisms documented to have caused an ecological effect and/or transmitted DNA to another organism. (From Seidler 1992; Doyle *et al.* 1995.)

Microorganism	Comment	References
Alcaligenes	Degraded chlorinated phenols and transferred plasmid to indigenous bacteria; natural selection favours recipients	Fulthorpe & Wyndham 1991; Nakatsu *et al.* 1995
Alcaligenes eutrophusa EO106 (pRO101)	Transiently altered the bacterial diversity of aquatic microcosm compared to uninoculated lakewater	Leser 1995
Enterobacter cloacae	Gene transfer inside insect gut	Armstrong *et al.* 1990
Pseudomonas putida	Degraded 3-chlorobenzoate and transferred degradative capacity to indigenous *Pseudomonas* sp.	Pertsova *et al.* 1984
Pseudomonas sp. RC1	Reduced colonization of wheat rhizoplane by indigenous fluorescent pseudomonads, which help in suppressing take-all disease of wheat by the fungus *Gaumannomyces graminis* var. *tritici*	Fredrickson & Seidler 1989
Pseudomonas sp. B13*	Gene transfer of 3-chlorobenzoate activity into *Alcaligenes* in a natural aquifer	Zhou & Tiedje 1995
Pseudomonas putida PPO301 (pRRO103)	Accumulated 2,4-dichlorophenol from 2,4-dichlorophenoxyacetic acid degradation, which reduced CO_2 evolution from soil and decreased number of fungal propagules in soil to undetectable levels	Doyle *et al.* 1991; Short *et al.* 1991
Pseudomonas cepacia AC100	Degraded 2,4,5-trichlorophenoxyacetic acid; caused changes in taxonomic diversity of indigenous microbiota, possibly as a result of metabolic activity by the GMO	Bej *et al.* 1991
Pseudomonas fluorescens	Plasmid DNA transferred to several genera of indigenous bacteria	Smit *et al.* 1991
P. fluorescens lux-modified denitrifier	Caused short-term reduction in overall microbial diversity in very small pore-size soil	White *et al.* 1994
Pseudomonas aureofaciens Kmr, xylE, lacZY chromosomal modification	Caused large perturbation (up to 100-fold) changes in microbial populations on seeds and root systems of spring wheat. Changes were transient but could still be detected in colony morphology differences when cultured from maturing plant roots	De-Leij *et al.* 1994
Plant growth-promoting fluorescent pseudomonads	Unexpected dispersal to plant shoots following seed inoculation	Raaijmakers *et al.* 1995
Streptomyces lividans TK23-3651 (pSE5)	Increased transiently the CO_2 from soil microcosms when lignocellulose was added (pSE5 codes for enhanced production of extracellular lignin peroxidase and H_2O_2)	Wang *et al.* 1991
Lactobacillus plantarum	Recombinant strains containing a *Clostridium thermocellum*-derived cellulose gene inoculated into grass mini-silos successfully proliferated and competed with epiphytic lactic acid bacteria in silage	Sharp *et al.* 1992

* This strain is not genetically engineered.

that large (150 mm diameter) Petri dishes could adequately serve as indicators of the extent of microbial deposition on and off the plot. Only 0.001% of the sprayed *P. syringae* drifted some 30 m off the central plot area. This still translated into a considerable number of viable cells (estimated 10^7). Although this increased transport occurred, no known undesirable consequences resulted. This may not always be the case with other GMOs. Therefore, it is prudent to anticipate dispersal of a GMO beyond the boundaries of a test, especially in aerosol releases. Monitoring may easily be conducted as described above to document drift for cases where legal or sensitive environmental issues may be involved.

Two studies assessing similar potential GMO effects were carried out independently in different parts of the world and the results merit special mention. The studies reported by Fredrickson & Seidler (1989) and by White *et al.* (1994) both described the possible impacts caused by a different root-colonizing GMO on the bacterial flora associated with wheat. In both cases a *Pseudomonas* species was the GMO and in both cases the introduced organism resulted in substantial reductions in indigenous microbial populations found naturally on wheat seeds and root systems. At least in one case the introduced organisms suppressed other fluorescent pseudomonads, organisms that help to reduce the incidence of take-all disease, a fungal disease of wheat.

In other studies, GMOs placed on seeds or inoculated directly into soil were found to be capable of moving or being transported to the aerial parts of plants. For example, insects could transport GMOs between soil and plant parts and even on to new insects via contaminated leaves (Armstrong *et al.* 1989, 1990). In a recent study, plant growth-promoting fluorescent pseudomonads colonized the surface and interior tissues of cotyledons in row crops following initial inoculation on to seeds (Raaijmakers *et al.* 1995). It is clear that at least with some applications of GMOs, transport may be the rule rather than the exception to their ecological fate.

Two documented cases have been described of GMO effects resulting from the production of metabolites. These studies validate the potential concern that an introduced taxon can significantly alter its surroundings. *Streptomyces lividans* has been genetically modified to enhance its production of extracellular lignin peroxidase and hydrogen peroxide to facilitate the degradation of lignin (Wang *et al.* 1989). Whenever the recombinant strain was added to native soil supplemented with lignin, there was an increase in the amount of carbon dioxide respired over all control treatments. Because GMO strains did not cause an increased evolution in carbon dioxide in sterile soil, the researchers proposed that the enzyme-induced lignin breakdown by the GMO provided fresh substrate for indigenous soil microflora and that this synergistic relationship caused the enhanced mineralization of carbon. Thus, the lignin peroxidase produced by the GMO had a significant short-term impact on carbon turnover rates that peaked out by day 6–9, presumably when the added lignin had been biodegraded.

The synergism established between an added bacterium (GMO) and components of the indigenous microbiota illustrate dramatically how a GMO may interact with, and influence the metabolic characteristics of, the indigenous community. It would be of great interest to determine whether the breakdown of natural or added lignin had any impact on other soil parameters, such as fertility, moisture-holding capacity, pH, etc., and what proportion of the resident lignin can be decomposed by this GMO. In this case, microcosm studies could provide answers to both sets of questions.

The second case of a deleterious GMO metabolite resulted from an unanticipated combination of ecological and biochemical circumstances. The use of GMOs as agents of bioremediation to facilitate environmental clean-up is a most worthy goal. However, the metabolic limitations and capabilities of any altered metabolic traits should be well documented prior to open field trials to evaluate efficacy. A case in point that illustrates this caution is the *P. putida* PPO301 (pRO103) organism that is capable of degrading 2,4-dichlorophenoxyacetic acid (2,4-D). When this GMO was added to a soil microcosm containing added 2,4-D, the first metabolic product of 2,4-D breakdown, 2,4-dichlorophenol (2,4-DCP), accumulated to 70–90 $\mu g\,g^{-1}$ of soil. The 2,4-DCP is more toxic than 2,4-D, and fungal propagules in the experimental treatment decreased from over 3×10^6 to undetectable levels by day 18 (Short *et al.* 1991). Toxicity of the 2,4-DCP also was demonstrated on Petri dish assays, where relative fungal counts were determined in the presence of increasing concentrations of 2,4-DCP. Growth of all soil fungi was reduced by as little as 10 $\mu g\ ml^{-1}$ of media of 2,4-DCP and by 25 $\mu g\ g^{-1}$ of soil in pure culture assays. Interestingly, the 2,4-DCP did not reduce the numbers of soil bacteria. The accumulation of 2,4-DCP was the result of two factors. First, the bacterium could not degrade 2,4-D completely so the 2,4-DCP accumulated. Furthermore, the arid soil used in these studies did not contain any indigenous microbes capable of degrading 2,4-D or 2,4-DCP. Thus, in this soil with this GMO, 2,4-DCP accumulated. These observations are also valuable because they validate the use of microcosms for investigating ecological perturbations and the results were easy to measure since changes in microbial populations were the impacted end-point.

5.4.2 Toxicological considerations

In this section, *in vivo* models and tier testing approaches will be presented for use in determining the potential pathogenicity, infectivity and allergenicity of biotechnology agents and their products. End-points that are characteristic for pathogenic strains have been described above so pathogenicity and infectivity factors will not be discussed further in this section.

In vivo infectivity and pathogenicity models

Several animal models have been used to detect potential health effects of biotechnology agents and their products, including conventional mice (with and without fasting) and rats (Table 5.7). From these animal models, LD_{50} values can be established. Aside from mortality, morbidity end-points including body and tissue weight loss, heart rate, hypothermia, vasodilation, jejunal haemorrhaging and liver centrilobular congestion have been detected (George *et al.* 1991, 1993; Roe *et al.* 1991). Additionally, clearance (from the lungs, small and large intestine, caecum), colonization (gastrointestinal tract), translocation (spleen, liver, mesenteric lymph nodes), and pulmonary inflammatory response have been determined following oral or intranasal exposure (George *et al.* 1989, 1991, 1993).

Other animal models that may be useful for determining the mortality and morbidity of GMOs are: (i) pulmonary exposure of rats to agar- or agarose-encapsulated *P. aeruginosa* or other organisms (Cash *et al.* 1979; Woods *et al.* 1980; Starke *et al.* 1987); (ii) neutropenic mouse model to determine LD_{50} (Godfrey *et al.* 1996); and (iii) the *Streptococcus zooepidemicus*/influenza virus pulmonary exposure mouse model (Sherwood *et al.* 1988). The latter has been used to determine health effects of *P. aeruginosa, B. cepacia, X. maltophilia, B. thuringiensis* subsp. *israelensis, B. thuringiensis* subsp. *kurstaki*, and *B. thuringiensis* subsp. *aizawai* (Kawanishi *et al.* 1990; George *et al.* 1991, 1993).

Allergenicity

Sarlo & Clark (1992) have described a tier approach to determine the respiratory allergenicity potential of low-molecular-weight chemicals. This strategy has application for biotechnology chemical products and may be suitable for protein products. The tier approach is shown in Fig. 5.2. If a chemical can modify a protein, then there is a potential to yield an immunogenic chemical–carrier conjugate (Sarlo & Clark 1992). The ability of a biotechnology product to react with a protein demonstrates its potential as an allergen. The level 3 guinea pig injection model involves subcutaneous injection of the potential allergen in an olive oil carrier over a period of 6 weeks. At the end of the treatment regiment, respiratory reactivity (following intratracheal challenge of the potential allergen), active cutaneous anaphylaxis testing (intradermal injection with the chemical protein conjugate using Evans blue dye intracardially injected as the indicator), and allergen-specific IgE levels are measured by enzyme-linked immunosorbent assay.

Guinea pig inhalation tests constitute the final tier (level 4). Animals are sensitized by exposure to aerosols for 5 days (3 h day^{-1}) and respiratory rate and breath peak height are monitored continuously. Two weeks following sensitization, animals are exposed to an aerosol, and the respiratory rate and breath peak height are determined over a

Table 5.7 *In vivo* models for potential health effects of biotechnology agents and their products.

Agent/product	Species	Strain	Route*	References
28 K Polypeptide of *Bacillus thuringiensis* subsp. *israelensis*	Mouse	Swiss-Webster	i.p.	Roe *et al.* 1991
			Oral	Mayes *et al.* 1989
	Rat	Sprague-Dawley	i.p	Mayes *et al.* 1989
	Rat	CD	i.p.	Roe *et al.* 1991
			s.c.	Roe *et al.* 1991
			i.v.	Roe *et al.* 1991
			Oral	Roe *et al.* 1991
	Japanese quail	–	i.p.	Roe *et al.* 1991
			s.c.	Roe *et al.* 1991
			i.v.	Roe *et al.* 1991
			i.n.	Roe *et al.* 1991
B. thuringiensis subsp. *israelensis*	Mouse	CD-1	i.n. i.t.	Kawanishi *et al.* 1990
B. thuringiensis subsp. *kurstaki*	Mouse	CD-1	i.n. i.t.	Kawanishi *et al.* 1990
B. thuringiensis subsp. *aizawai*	Mouse	CD-1	i.n. i.t.	Kawanishi *et al.* 1990
B. thuringiensis subsp. *wuhanensis*	Mouse	CD-1	i.n.	Sherwood *et al.* 1991
Pseudomonas syringae	Rat	Sprague-Dawley	Inh.	Goodnow *et al.* 1990
P. putida	Rat	Sprague-Dawley	Oral i.c. i.p. Inh.	Kaiser *et al.* 1981
P. aeruginosa	Mouse	CD-1	Oral	George *et al.* 1989
			i.n.	George *et al.* 1991
		C3H/HeJ	i.n.	George *et al.* 1993
Burkholderia cepacia	Mouse	CD-1	Oral	George *et al.* 1990
			i.n.	George *et al.* 1991
Xanthomonas maltophilia	Mouse	CD-1	Oral	George *et al.* 1989
		C3H/HeJ	i.n.	George *et al.* 1993

*i.c., intracutaneous; i.n., intranasal; Inh., inhalation ; i.p., intraperitoneally; i.t., intratracheal; i.v., intravenous; s.c., subcutaneous.

30-min period. Readers should refer to Sarlo & Clark (1992) for a more-detailed description of the level 1–4 procedures.

Skin testing of detergent factory workers has been used to identify atopic individuals (Newhouse *et al.* 1970; Franz *et al.* 1971; Juniper *et al.* 1977; How *et al.* 1978). Subjects were injected cutaneously with different concentrations of alcalase protein in physiological saline and the skin response observed (Sarlo *et al.* 1991; How *et al.* 1978).

Tier testing approach for microbial and biochemical pest control agents

The United States Environmental Protection Agency regulates new microbial and biochemical pest control products using the approach presented in Subdivision M of the Pesticide Testing Guidelines (Anderson *et al.* 1989). The Tier I toxicology guidelines, which address acute oral, dermal, pulmonary, and intravenous toxicity and pathogenicity,

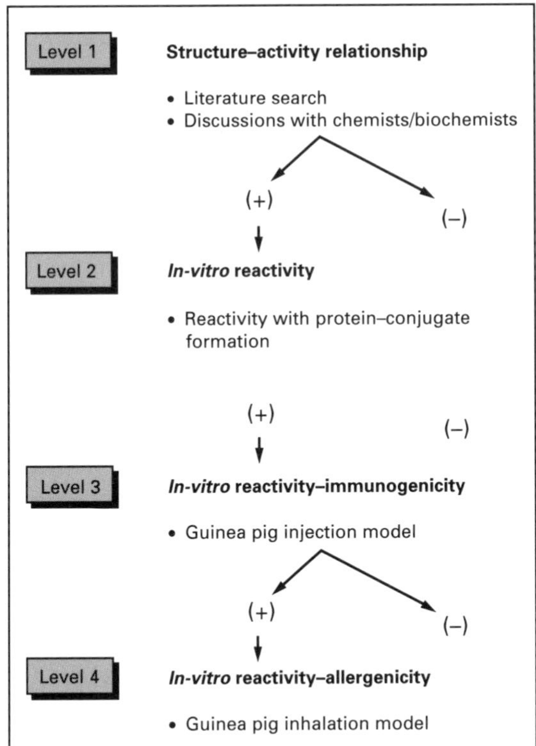

Fig. 5.2 Proposed multilevel approach for assessment of new chemicals as respiratory allergens. Positive results at one level should lead to further testing at the next level with the final test using the inhalation model. If a negative result is obtained at one level, then no further testing should be done with the chemical in question. (From Sarlo & Clark 1992 with permission.)

as well as eye irritation and infection, hypersensitivity and cell culture tests (viral agents), are suitable for GMOs. Tier II testing describes methods for acute and subchronic toxicity/pathogenicity studies, and Tier III focuses on: (i) reproductive and fertility effects; (ii) oncogenicity studies, (iii) immunodeficiency studies; and (iv) a primate infectivity and pathogenicity study. Even though the guidelines are established for regulation of pest control agents, they have some applicability for determining health effects of GMOs.

5.4.3 Reliability of end-point measurements

In order to assure workers and the population at large, the safety of environmentally released GMOs must be established. If a potential environmental release is not from a pathogenic genus, health or environmental effects testing on this particular microbial strain may not have been done. Even though the probability is low, there is a potential for GMOs to harbour genes that are harmful to human health or the environment. By developing, validating and using methods that examine the potential effects of the newly generated strains, such as those described in the previous sections, researchers and industry can reassure the workers and the general public that the isolates are safer.

In order to evaluate reliably the measurements conducted on environmental aspects of GMO risk assessments, properly designed ecologically relevant experiments are essential. It is crucial that any pre-release studies be conducted with environmental materials that will ultimately receive the GMO. Thus, local water, soil, plants, etc., should be employed. This common-sense approach to evaluating survival, transport and effects will help to ensure that relevant and hopefully reliable information on the ecology of the organism will be obtained. As indicated in Table 5.1, numerous experimental end-points have been tested or proposed to measure risks associated with GMOs. A recent publication discusses many of these end-points (Stotzky *et al.* 1993) and the reader is referred to that report for a more detailed analysis.

One of the keys to taking reliable end-point measurements is the ability to control the environment in which microcosm analyses may be evaluated. Without proper environmental controls to measure and control light, temperature, soil moisture, etc., there is little hope of either simulating natural conditions or obtaining repeatable data on basic ecological parameters such as GMO survival. How and when to sample the environment to measure end-points can best be obtained from published data on other GMOs and from trial-and-error experiences. Basic microbiological techniques to ensure that sample integrity is maintained free from extraneous contaminants, a reliable procedure to detect and enumerate the exact strain under test, and a reasonable culturing milieu are all necessities. As stated earlier, we also have found when conducting field trials that a quality assurance plan must be in place to ensure the quality and repeatability of the data collected (Seidler & Hern 1988).

The determination of potential health effects of GMOs is not a precise science. The selection of an animal model to determine morbidity and mortality of all relevant GMOs is difficult. It is important to consider the clearance rate of the GMO from the animal tissues because human pathogens do not necessarily cause disease in animal models. Exoenzyme and toxin production, in combination with the toxicology data, are helpful in determining if there is a potential for adverse health effects. However, the pathogenicity factors described previously are from the *Pseudomonas* species because of the widespread use of the pseudomonads in bioremediation and other environmental applications (predominantly *P. aeruginosa* has been studied). Additionally, Godfrey *et al.* (1996) have shown that only the pathogenicity factor exoenzyme S correlates with LD_{50} in *P. aeruginosa* isolates tested, even though Grimwood *et al.* (1993) have demonstrated a correlation between several exoenzymes and cystic fibrosis lung deterioration. Therefore, the possession of pathogenicity factors by a biotechnology agent may have an impact on the potential risk associated with exposure.

However, if the GMO of interest does cause mortality, morbidity, or it colonizes a selected animal, and if it has some of the recognized pathogenicity factors (a pest control agent, for example), then the potential to cause adverse human health effects may increase. This information, along with the exposure data, should be considered prior to release of the GMO.

Generally, healthy adults should not experience any adverse reactions following exposure to GMOs. However, because some of the GMOs proposed for environmental release may cause opportunistic infections, there are several groups of individuals that may be at a greater risk following exposure due to their immune status. These include: (i) young children and infants; (ii) the elderly; (iii) chemotherapy patients; (iv) individuals with an active illness; (v) convalescents using antibiotics or other medications; and (vi) atopic persons. For example, isolates of *B. cepacia*, an organism found in the soil, cause disease in cystic fibrosis victims (Rosenstein & Hall 1980). *Clostridium difficile* can colonize the healthy, human adult intestinal tract in low numbers but causes pseudomembranous colitis in patients on antibiotic therapy (Bartlett 1979). Green *et al.* (1990) have examined the public health implications of a *B. thuringiensis* var. *kurstaki* spray in Oregon and have found that *B. thuringiensis* var. *kurstaki* is culturable from clinical specimens from exposed individuals. Three patients with prior medical problems harboured *B. thuringiensis* var. *kurstaki* which could not be ruled out as a causative disease agent. However, detrimental effects from controlled human exposure (oral and inhalation) to *B. thuringiensis* have produced no adverse effects (Fisher & Rosner 1959).

It is important to identify potential allergens so that atopic individuals can be identified and their exposure reduced. The detergent industry experience with alcalase is a case in point (Franz *et al.* 1971). Once researchers recognized alcalase as an allergen and identified atopic individuals, industrial exposures were reduced in the following ways: (i) workers who handled the raw materials were outfitted with protective clothing; (ii) dust levels were reduced in the plants so that employee exposures were lessened; and (iii) atopic individuals were placed in jobs where exposure was reduced or eliminated. The industry continues to closely monitor its workers, even though the allergenicity-related complaints have declined.

Otherwise healthy individuals may have some risk following exposure to GMOs or their products. One point to consider is the transfer of antibiotic resistance genes from the environmentally released GMOs to other microorganisms and mammals, including humans. The spread of the antibiotic-resistant genes has the potential to reduce the effectiveness of the presently used antibiotics to fight disease. One case in point is multidrug-resistant *Mycobacterium tuberculosis* (Pearson *et al.* 1992). Because the organism is resistant to a host of currently used antibiotics, it is more difficult to eradicate the disease. Several groups have demonstrated the flow of antibiotic-resistance genes from free-living microorganisms to microorganisms associated with humans. Marshall *et al.* (1990) described the spread of an *E. coli* strain that harboured antibiotic-resistance markers on a plasmid from the microflora of animals to humans. A study by Linton (1986) reported a similar finding, suggesting that *E. coli*-resistance traits can transfer from animal microflora to human microflora, ultimately rendering potential pathogens resistant to drugs and therefore much more difficult to treat. The probability of other (non-antibiotic) introduced genes to transfer from the GMOs to humans may result in some unexpected transfer of the engineered traits to humans. The long-term effects of this transfer are unknown; however, there is the potential for expression of these traits.

Finally, the toxicity of both the byproducts and end-products should be considered. This is especially important in terms of bioremediation, where hazardous waste reduction and elimination may result in production and accumulation. Therefore, it is important to have an understanding of the remediation process and the ultimate biological activity (toxicity) of the finished product, not just the chemical analysis that measures elimination of the parental compound(s).

5.5 FUTURE RISK ASSESSMENT CONCERNS AND RESEARCH NEEDS

The use of microorganisms and management of microbial activities long predates the use of pure cultures as inocula. However, it has not been until the last 10–20 years, with the advent and growth of recombinant techniques that allow the development of novel genotypes and activities, that public concerns have been raised about the safety of using microbes, particularly for intentional environmental releases. To date, three of the four engineered microbes approved for commercial use in the USA, under FIFRA, are non-viable formulations of engineered microbes (*P. fluorescens* containing toxin genes from various *Bacillus thuringiensis* subspecies). Engineered microbes approved for small-scale experimental environmental release under the TSCA include *P. fluorescens (P. aureofaciens)* containing the *lac zy* gene from *E. coli*, *P. fluorescens* strain HK44 for polyaromatic hydrocarbon degradation, and *R. meliloti* containing antibiotic resistance and nitrogen fixation (*nif*) genes. A request for commercial registration approval under the TSCA of an *R. meliloti* strain containing an enhanced nitrogen fixation (*nif*) capability is currently under review by the United States Environmental Protection Agency. To date, with the exception of widely used agronomic inocula such as nitrogen-fixing bacteria (*Rhizobium* spp. and *Bradyrhizobium* spp.), silage inocula such as *Lactobacillus* spp. and the microbial insecticide *B. t. kurstaki*, relatively few inocula have been intentionally released to the environment on a large-scale basis. Most of the non-agronomic applications noted above have either been experimental ones, performed under semi-contained conditions of laboratory microcosms or greenhouses, or they are being used under the contained conditions of fermentation tanks or other types of bioreactors. Where field data are available, numerous examples have been reported for presumably transient effects of both biologicals and chemicals on populations and activities of soil biota, nutrient cycling and diversity (Watrud & Seidler 1997). As the size of field tests and commercial uses increase, questions about the effects of multiple applications and long-term use of biotechnology products also arise. These include questions of both fate of the biologicals (organisms, nucleic acids, proteins) and ecological effects on non-target organisms and communities, nutrient cycling processes, and on below-ground and above-ground biological diversity. However, in the absence of long-term monitoring studies and the lack of consensus on suitable biological indicators or methodology, long-term impacts remain unknown.

Scientific issues which repeatedly have been raised regarding the release of recombinant microbes include their fate and their potential adverse effects on non-target species. In traditional chemical risk or ectoxicological assessments, single species have typically been tested for short periods of time under laboratory conditions. Pending results with the first 'tier' of indicator organisms, additional organisms may be tested or longer-term tests may be carried out. With the current increased interest in the use of biologicals, traditional ways of measuring persistence, transport, and effects (for chemicals) need to be re-evaluated for their suitability in assessing the effects and risks of biologicals. Unlike chemicals, bioengineered organisms can reproduce; nucleic acids, enzymes or toxins associated with biological agents may retain biological activity in natural environments for extended periods of time. Genetic exchange between introduced and indigenous flora can result in transport and persistence of the biological activity beyond the site and time of introduction. Experience and data from large-scale, repeated or long-term uses of engineered microbes is limited. A need for research therefore continues in order to answer questions regarding fate and effects of the introduced microbes and microbial products, not only at the points of release but also in areas adjacent, downwind or downstream from the points of application or release. Methods need to be developed not only to look at single-species or single-site effects, but to examine effects on communities of organisms in different ecosystems in diverse geographic areas. For example, if a product is designed for agronomic use, such as crop protection, non-target effects perhaps should be examined not only within the agronomic system or on the target pest, but also on adjacent agronomic, native, weedy and horticultural species in the surrounding area. Studies on the effects of engineered microbes on soil foodweb components (plant symbionts, nematodes, litter arthropods, protozoans and microbes), nutrient cycling, plant biomass, nutrient status and plant community diversity, may be useful to identify early indicators of adverse effects to the diversity, functioning, and sustainability of ecosystems. To facilitate the capability to handle large sample numbers, automated methods of sample preparation and analysis may be needed. Given a suitable database, development of predictive models to permit estimation of risks from novel biologicals may also be useful. A need for longer-term monitoring of effects on local and geographic bases should be considered since, as we have learned from persistent chemicals such as dichlorodiphenyltricholoroethane (DDT) and the polychlorinated biphenyls (PCBs), ecological effects may not be apparent until decades after their initial introduction.

In summary, we have presented an overview of the complex interactions of genetically engineered microorganisms with the human and natural environments. It has become clear from this information that risk assessments of these interactions are complex and at present at a very preliminary stage of development. Additional model systems are needed involving both humans and the natural environment in order to provide reassurances that properly focused risk assessments can indeed provide reliable predictions about the potential risks of these novel life forms. Currently available data from short-term experiments point toward minimal risks associated with GMOs. However, there is neither sufficient commercial production nor consumer utilization of these diverse new products to definitively draw any conclusions about the human safety and ecological effects resulting from the long-term use and repeated applications of recombinant organisms or products derived from them.

5.6 REFERENCES

Anaissie, E., Fainstein, V., Pitlik, S., Kassamali, H., Bodey, G.P. & Rolston, K. (1986) *Pseudomonas putida*: an emerging pathogen in cancer patients. In: *Abstracts of the Annual Meeting of the American Society for Microbiology* (ed. P.A. Hartman), 86 pp., abstract 417. American Society for Microbiology, Washington, DC.

Anderson, E.S. (1975) Viability of, and transfer of a plasmid from *E. coli* K12 in the human intestine. *Nature*, **255**, 502–504.

Anderson, E.S. 1978. Plasmid transfer in *Escherichia coli*. *The Journal of Infectious Diseases*, **137**, 686–687.

Anderson, J., Edwards, D.F., Hazel, W.J. *et al.* (1989) *Subdivision M of the Pesticide Testing Guidelines: Microbial and Biochemical Pest Control Agents*. US Environmental Agency, Office of Pesticide and Toxic Substances, Washington, DC.

Angle, J.S., Levin, MA., Gagliardi, J.V., McIntosh, M.S. & Glew, J.G. (1994) *Pseudomonas aureofaciens* in soil: survival and recovery efficiency. *Microbiological Releases*, **2**, 247–254.

Angle, J.S., Levin, M.A., Gagliardi, J.V. & McIntosh, M.S. (1995) Validation of microcosms for examining the survival of *Pseudomonas aureofaciens* (lac ZY) in soil. *Applied and Environmental Microbiology*, **61**, 2835–2839.

Aprill, W., Sims, R.C., Sims, J.L. & Matthews, J.E. (1990) Assessing detoxification and degradation of wood preserving and petroleum wastes in contaminated soil. *Waste Management Research*, **8**, 45–65.

Armstrong, J.L. (1989) Assessing the persistence of recombinant bacteria in microcosms. In: *Evaluation of terrestrial microcosms for detection, fate, and survival analysis of genetically engineered microorganisms and their recombinant genetic material* (eds J.K. Fredrickson & R.J. Seidler) US Environmental Protection Agency, EPA, 600/3-89/043, Corvallis, OR.

Armstrong, J.L., Porteous, L.A. & Wood, N.D. (1989) The cutworm *Peridroma saucia* (Lepidoptera: Noctuidae) supports growth and transport of pBR322-bearing bacteria. *Applied and Environmental Microbiology*, **55**, 2200–2205.

Armstrong, J.L., Porteous, L.A. & Wood, N.D. (1990) Transconjugation between bacteria in the digestive tract of the cutworm *Peridroma saucia*. *Applied and Environmental Microbiology*, **56**, 1492–1493.

Bartlett, J.G. (1979) Antibiotic-associated pseudomembranous colitis. *Reviews of Infectious Diseases*, **1**, 530–539.

Bedford, D.J., Schweizer, E., Hopwood, D.A. & Khjosla, C. (1995) Expression of a functional fungal polyketide synthase in the bacterium *Streptomyces coelicolor* A3 (2). *Journal of Bacteriology*, **177**, 4544–4548.

Bej, A.K., Perlin, M., Atlas, R.M. (1991) Effect of introducing genetically engineered microorganisms on soil microbial community diversity. (*Federation of European Microbiological Societies.) Microbiology Ecology*, **86**, 169–176.

Bentjen, S.A., Fredrickson, J.K., vanVoris, P. & Li, S.W. (1988) Intact soil-core microcosms for evaluating the fate and ecological impact of the release of genetically engineered microorganisms. *Applied and Environmental Microbiology*, **55**, 198–202.

Bevivino, A., Tabacchioni, S., Chiarini, L., Carusi, M.V., Gallo, M.D. & Visca, P. (1994) Phenotypic comparison between rhizosphere and clinical isolates of *Burkholderia cepacia*. *Microbiology*, **140**, 1069–1077.

Bhattarai, T. & Hess, D. (1993) Yield responses of Nepalese spring wheat (*Triticum aestivum* L.) cultivars to inoculation with *Azospirillum* spp. of Nepalese origin. *Plant and Soil*, **151**, 67–76.

Bodey, G.P., Rodriguez, V., Chang, H.-Y. & Narboni, G. (1978) Fever and infection in leukemic patients. *Cancer*, **41**, 1610–1622.

Bolten, H., Fredrickson, J.K., Thomas, J.M., Li, S.W., Workman, D.J., Bentjen, S.A. & Smith, J.L. (1991) Field calibration of soil-core microcosms: fate of a genetically engineered rhizobacterium. *Microbial Ecology*, **21**, 163–173.

Borman, S. (1994) Platelet growth factor cloned and characterized. *Chemical and Engineering News*, **72**, 9.

Bosch, D., Schipper, B., van der Kieij, H., de Maagd, R.A. & Stiekema, W.J. (1994) Recombinant *Bacillus thuringiensis* crystal proteins with new properties: possibilities for resistance management. Biotechnology, **12**, 915–918.

Bourquin, A. & Seidler, R.J. (1986) Research plan for test methods development for risk assessment of novel microbes released into terrestrial and aquatic ecosystems. In: *Biotechnology and the Environment: Research Needs* (ed. G.S. Omen), pp. 18–61. Noyes Data Corporation, Park Ridge, NJ.

Breskvar, K., Cresnar, B., Plaper, A. & Hudnik-Plevnik, T. (1991) Localization of the gene encoding steroid hydroxylase cytochrome P-450 from *Rhizopus nigricans* inside a *Hin*dll fragment of genomic DNA. *Biochemical and Biophysical Research Communications*, **178**, 1078–1083.

Buckholz, E.F. & Gleeson, M.A.G. (1991) Yeast systems for the commercial production of heterologous proteins. *Biotechnology*, **9**, 1067–1072.

Charudattan, R. (1990) Pathogens with potential for weed control. In: *American Chemical Society Symposium Series No. 439. Microbes and Microbial Products as Herbicides*, pp. 132–154. Washington, DC.

Cash, H.A., Woods, D.E., McCullough, B., Johanson, W. & Bass, J.A. (1979) A rat model of chronic respiratory infection with *Pseudomonas aeruginosa*. *American Review of Respiratory Disease*, **119**, 453–459.

Cisar, C.R., Spiegel, F.W., TeBeest, D.O. & Trout, C. (1994) Evidence for mating between isolates of *Colletotrichum gloeosporioides* with different host specificities. *Current Genetics*, **25**, 330–335.

Claxton, L.D., Houk, V.S., Williams, R. & Kremer, F. (1991a) Effect of bioremediation on the mutagenicity of oil spilled in Prince William Sound, Alaska. *Chemosphere*, **23**, 643–650.

Claxton, L.D., Houk, V.S., Williams, R. & Kremer, F. (1991b) Oil spill cleanup. *Nature*, **353**, 24.

Cohen, P.S., Pilsucki, R.W., Myhal, M.L., Rosen, C.A., Laux, D.C. & Cabelli, V.J. (1979) Colonization potentials of male and female *E. coli* K12 strains, *E. coli* B, and human fecal *E. coli* strains in the mouse GI tract. *Recombinant DNA Technical Bulletin*, **2**, 106–113.

Colagrande, O., Silva, A., Fumi, M.D. (1994) Recent applications of biotechnology in wine production. *Biotechnology Progress*, **10**, 2–18.

Crawford, D.L., Doyle, J.D. Wang, Z., Hendricks, C.W., Bentjen, S.A., Bolton, H.J., Fredickson, J.K. & Bleakley, B.H. (1993) Effects of a lignin peroxidase-expressing recombinant *Streptomyces lividans* TK23.1 on biogeochemical cycling and microbial numbers and activities in soil microcosms. *Applied and Environmental Microbiology*, **59**, 508–518.

Cripe, C.R., Pritchard, P.H. & Stern, A.M. (1992) *Workshop: Application of Microcosms for Assessing the Risk of Microbial Biotechnology Products.* US Environmental Protection Agency, EPA/600/R-92/066, Gulf Breeze, FL, 141 pp.

Daubaras, D.L., Hershberger, C.D., Kitano, K. & Chakrabarty, A.M. (1995) Sequence analysis of a gene cluster involved in metabolism of 2,4,5-trichlorophenoxyacetic acid by *Burkholderia cepacia* AC1100. *Applied and Environmental Microbiology*, **61**, 1279–1289.

de Boer, L. & Dijkhuizen, L. (1994) Microbial and enzymatic processes for L-phenylalanine production. *Advances in Biochemical Engineering/Biotechnology*, **42**, 1–27.

de la Noue, J. & de Pauw, N. (1988) The potential of

microalgal biotechnology: a review of production and uses of microalgae. *Biotechnology Advances*, **6**, 725–770.

De-Leij, F.A.A.M., Sutton, E.J., Whipps, J.M. & Lynch, J.M. (1994) Effect of a genetically modified *Pseudomonas aureofaciens* on indigenous microbial populations of wheat. (*Federation of European Microbiological Societies*) *Microbiology Ecology*, **13**, 249–257.

De-Leij, F.A.A.M., Sutton, E.J., Whipps, J.M., Fenlon, J.S. & Lynch, J.M. (1995) Field release of a genetically modified *Pseudomonas fluorescens* on wheat: establishment, survival, and dissemination. *Biotechnology*, **13**, 1488–1492.

DeMarini, D.M., Brooks, H.G. & Parkes, D.G., Jr. (1990) Induction of prophage lambda by chlorophenols. *Environmental and Molecular Mutagenesis*, **15**, 1–9.

Di Ciocco, C.A., Rodriguez-Caceres, E.A. & Di Ciocco, C.A. (1994) Field inoculation of *Setaria italica* with *Azospirillum* spp in Argentine humid pampas. *Field Crops Research*, **37**, 253–257.

Dobereiner, J., Reis, V. M. & Lazarini, A.C. (1988) New N_2 fixing bacteria in association with cereals and sugar cane. In: *International Congress on Nitrogen Fixation* (eds H. Bothe, F.J. de Bruijn & W. E. Newton), pp. 717–722. Gustav Fischer Verlag GmbH & Co. KG, Stuttgart.

Donegan, K., Matyac, C., Seidler, R.J. & Porteous, A. (1991) Evaluation of methods for sampling, recovery, and enumeration of bacteria applied to the phylloplane. *Applied and Environmental Microbiology*, **57**, 51–56.

Donegan, K.K., Palm, C.J., Fieland, V.J. *et al.* (1995) Changes in levels, species, and DNA fingerprints of soil microorganisms associated with cotton expressing the *Bacillus thuringiensis* var. *kurstaki* endotoxin. *Applied Soil Ecology*, **2**, 111–124.

Donnelly, K.C., Brown, K.W. & DiGiullio, D.G. (1988) Mutagenic characterization of soil and water samples from a superfund site. *Nuclear and Chemical Waste Management*, **8**, 135–141.

Donnelly, K.C., Brown, K.W. & Kampbell, D. (1987a) Chemical and biological characterization of hazardous industrial waste. I. Prokaryotic bioassays and chemical analysis of a wood-preserving bottom-sediment waste. *Mutation Research*, **180**, 31–42.

Donnelly, K.C., Brown, K.W. & Scott, B.R. (1987b) Chemical and biological characterization of hazardous industrial waste. II. Eukaryote bioassay of a wood-preserving bottom sediment. *Mutation Research*, **180**, 43–53.

Doyle, J.D., Short, K.A., Stotzky, G., King, R.J., Seidler, R.J. & Olsen, R.H. (1991) Ecologically significant effects of *Pseudomonas putida* PPO301 (pRO103), genetically engineered to degrade 2,4-dichlorophenoxyacetate, on microbial populations and processes in soil. *Canadian Journal for Microbiology*, **37**, 682–691.

Doyle, J.D., Stotzky, G., McClung, G. & Hendricks, C.W. (1995) Effects of genetically engineered microorganisms on microbial populations and processes in natural habitats. *Advances in Applied Microbiology*, **40**, 237–287.

Duval-Iflah, Y. & Chappuis, J.P. (1984) Influence of plasmids on the colonization of the intestine by strains of *Escherichia coli* in gnotobiotic and conventional animals. In: *Current Perspectives in Microbial Ecology* (eds M.J. Klug & C.A. Reddy), pp. 264–272. American Society for Microbiology, Washington, DC.

Edgington, S.M. (1994) Environmental biotechnology. *Biotechnology*, **12**, 1338–1342.

Esselmann, M.T. & Liu, P.V. (1961) Lecithinase production by gram-negative bacteria. *Journal of Bacteriology*, **81**, 939–945.

Fan, S. & Scow, K.M. (1993) Biodegradation of trichloroethylene and toluene by indigenous microbial populations in soil. *Applied and Environmental Microbiology*, **59**, 1911–1918.

Farinha, M.A., Conway, B.D., Glasier, L.M. *et al.* (1994) Alteration of the pilin adhesin of *Pseudomonas aeruginosa* PAO results in normal pilus biogenesis but a loss of adherence to human pneumocyte cells and decreased virulence in mice. *Infection and Immunology*, **62**, 4118–4123.

Farrand, S.K. (1992) Conjugal gene transfer on plants. In: *Microbial Ecology: Principles, Methods, and Applications* (eds M.A. Levin, R.J. Seidler & M. Rogul), 945 pp. McGraw-Hill, New York.

Federal Register (1986) Coordinated framework for regulation of biotechnology; announcement of policy and notice for public comment. *Federal Register*, **51**, 23302–23393.

Finley, B.B. & Falkow, S. (1989) Common themes in microbial pathogenicity. *Microbiological Reviews*, **53**, 210–230.

Fisher, R. & Rosner, L. (1959) Toxicology of the microbial insecticide, thuricide. *Agriculture and Food Chemistry*, **7**, 686–688.

Flindt, M.L.H. (1969) Pulmonary disease due to inhalation of derivatives of *Bacillus subtilis* containing enzyme. *Lancet*, **1**, 1177–1181.

Flores, D.A. (1991) Biotechnology and the improvement of silage (tropical and temperate) rumen digestion: a mini-review. *Applied and Microbiological Biotechnology*, **35**, 277–282.

Francis, A.J. (1990) Microbial dissolution and stabilization of toxic metals and radionuclides in mixed wastes. *Experientia*, **46**, 840–851.

Franz, J., McMurrain, K.D., Brooks, S. & Bernstein, I.L. (1971) Clinical, immunological and physiologic observations in factory workers exposed to *Bacillus subtilis* enzyme dust. *Journal of Allergy*, **47**, 170–179.

Fredrickson, J.K. & Seidler, R.J. (eds) (1989) Evaluation of terrestrial microcosms for detection, fate, and survival analysis of genetically engineered microorganisms and their recombinant genetic material. US Environmental Protection Agency, EPA600/3-89/043, Corvallis, OR.

Fulthorpe, R.R. & Wyndham, R.C. (1991) Transfer and expression of the catabolic plasmid pBRC60 in wild bacterial recipients in a freshwater ecosystem. *Applied and Environmental Microbiology*, **57**, 1546–1553.

Fulton, R.W. (1986) Practices and precautions in the use of cross protection for plant virus disease control. *Annual Review of Phytopathology*, **24**, 67–81.

Gableman, A. (1994) *Bioprocess Production of Flavor, Fragrance and Color Ingredients*, 361 pp. Wiley Press, New York.

Gawron-Burke, C. & Baum, J.A. (1991) Genetic manipulation of *Bacillus thuringiensis* insecticidal crystal proteins in bacteria. *Genetic Engineering Principles and Methods*, **13**, 237–263.

George, S.E., Kohan, M.J., Walsh, D.B., Stead, A.G. & Claxton, L.D. (1989) Polychlorinated biphenyl-degrading pseudomonads: survival in mouse intestines and competition with normal flora. *Journal of Toxicology and Environmental Health*, **26**, 19–37.

George, S.E., Kohan, M.J., Whitehouse, D.A., Creason, J.P. & Claxton, L.D. (1990) Influence of antibiotics on intestinal tract survival and translocation of environmental *Pseudomonas* species. *Applied and Environmental Microbiology*, **56**, 1559–1564.

George, S.E., Kohan, M.J., Whitehouse, D.A. *et al.* (1991) Distribution, clearance, and mortality of environmental pseudomonads in mice upon intranasal exposure. *Applied and Environmental Microbiology*, **57**, 2420–2425.

George, S.E., Whitehouse, D.A. & Claxton, L.D. (1992) Genotoxicity of 2,4,5-trichlorophenoxyacetic acid biodegradation products in the *Salmonella* reversion and lambda prophage-induction bioassays. *Environmental Toxicology and Chemistry*, **11**, 733–740.

George, S.E., Kohan, M.J., Gilmour, M.I. *et al.* (1993) Pulmonary clearance and inflammatory response in C3H/HeJ mice after intranasal exposure to *Pseudomonas* spp. *Applied and Environmental Microbiology*, **59**, 3585–3591.

Gilardi, G.L. (1991) *Pseudomonas* and related genera. In: *Manual of Clinical Microbiology* (eds A. Balows, W.J. Hausler, Jr., K.L. Herrmann, H.D. Isenberg & H.J. Shadomy, 5 edn, pp. 429–441. American Society for Microbiology, Washington, DC.

Gildon, A. & Tinker, P.B. (1983) Interactions of vesicular-arbuscular mycorrhizal infections and heavy metals in plants II. The effects of infection on uptake of copper. *New Phytologist*, **95**, 263–268.

Gillett, J.W., Witt, J.M. & Wyatt, C.J. (1978) Symposium on Terrestrial Microcosms and Environmental Chemistry, June 1977, Corvallis, Oregon, NSF/RA79-0026. National Science Foundation, Washington, DC.

Glasby, J.S. (1992) *Encyclopedia of Antibiotics*, 467 pp. John Wiley and Sons, Chichester.

Godfrey, A., Campbell, M., Lam, J., Paranchych, W., Speert, D. & Woods, D.E. (1996) Current research priorities pertaining to health requirements for biotechnology products under the Canadian Environmental Protection Act. In: *Proceedings of the Seventh Annual Symposium on Environmental Releases of Biotechnology Products: Risk Assessment Methods and Research Progress* (eds M. Levin, C. Grimm & J.S. Angle), pp. 211–227. US Environmental Protection Agency/US Department of Agriculture/Environment Canada University of Maryland Biotechnology Institute, College Park, MD.

Goldstein, R., Sun, L., Jiang, R.-Z., Sajjan, U., Forstner, J.F. & Campanelli, C. (1995) Structurally variant classes of pilus appendage fibers coexpressed from *Burkholderia (Pseudomonas) cepacia*. *Journal of Bacteriology*, **177**, 1039–1052.

Goodnow, R.A., Katz, G., Haines, D.C. & Terrill, J.B. (1990) Subacute inhalation toxicity study of an ice-nucleation-active *Pseudomonas syringae* administered as a respirable aerosol to rats. *Toxicology Letters*, **54**, 157–167.

Gorbach, S.L. (1978a) Recombinant DNA: an infectious disease perspective. *Journal of Infectious Diseases*, **137**, 615–623.

Gorbach, S.L. (1978b) Risk assessment protocols for recombinant DNA experimentation. *Journal of Infectious Disease*, **137**, 704–714.

Green, M., Heumann, M., Sokolow, R., Foster, L.R., Bryant, R. & Skeels, M. (1990) Public health implications of the microbial pesticide *Bacillus thuringiensis*: an epidemiological study, Oregon, 1985–1986. *American Journal of Public Health*, **80**, 848–852.

Grimwood, K., To, M., Rabin, H.R. & Woods, D.E. (1989) Inhibition of *Pseudomonas aeruginosa* exoenzyme expression by subinhibitory antibiotic concentrations. *Antimicrobial Agents and Chemotherapy*, **33**, 41–47.

Grimwood, K., To, M., Semple, R.A., Rabin, H.R., Sokol, P.A. & Woods, D.E. (1993) Elevated exoenzyme expression by *Pseudomonas aeruginosa* is correlated with exacerbations of lung disease in cystic fibrosis. *Pediatric Pulmonology*, **15**, 135–139.

Grunnet, K. & Hansen, J.C. (1978) Risk of infection from heavily contaminated air. *Scandinavian Journal of Work Environmental Health*, **4**, 336–338.

Guiot, H.F.L., van der Meer, J.W.M. & van Furth, R. (1981) Selective antimicrobial modulation of human microbial flora: infection prevention in patients with decreased host defense mechanisms by selective elimination of potentially pathogenic bacteria. *Journal of Infectious Diseases*, **143**, 644–654.

Hall, G. (1995) Environmental release of genetically modified rhizobia and mycorrhizas. In: *Genetically Modified Organisms. A Guide to Biosafety* (ed. G.T. Tzotzos), pp. 64–92. CAB International, Wallingford.

Halvorson, H.O., Pramer, D. & Rogul, M. (eds) (1985) *Engineered Organisms in the Environment: Scientific Issues*, 239 pp. American Society for Microbiology, Washington, DC.

Harlander, S. (1992) Food biotechnology. In: *Encyclopedia of Microbiology*, pp. 191–207. Academic Press, Inc., San Diego, CA.

Hartig, P.C., Chapman, M.A., Hatch, G.G. & Kawanishi, C.Y. (1989) Insect virus: assays for toxic effects and transformation potential in mammalian cells. *Applied and Environmental Microbiology*, **55**, 1916–1920.

Hartig, P.C., Cardon, M.C. & Kawanishi, C.Y. (1992) Effect of baculovirus on selected vertebrate cells. *Development of Biological Standards*, **76**, 313–317.

Heggo, A. & Angle, J.S. (1990) Effects of vesicular-arbuscular mycorrhizal fungi on heavy metal uptake by soybeans. *Soil Biology and Biochemistry*, **22**, 865–869.

Held, G.A., Huang, Y-S. & Kawanishi, C.Y. (1986) Effect of removal of the cytolytic factor of *Bacillus thuringiensis* subsp. *israelensis* on mosquito toxicity. *Biochemical and Biophysics Research Communications*, **141**, 937–941.

Helsot, H. (1990) Genetics and genetic engineering of the industrial yeast *Yarrowia lipolytica. Advances in Biochemical Engineering/Biotechnology*, **43**, 43–73.

Hetrick, B.A.D., Wilson, G.W.T. & Figge, D.A.H. (1994) The influence of mycorrhizal symbiosis and fertilizer amendments on establishment of vegetation in heavy metal mine soil. *Environmental Pollution*, **86**, 171–179.

Hiatt, A., Cafferkey, R. & Bowdish, K. (1989) Production of antibodies in transgenic plants. *Nature*, **342**, 76–78.

Hill, K.E., Weightman, A.J. & Fry, J.C. (1992) Isolation and screening of plasmids from the epilithon which mobilize recombinant plasmid pD10. *Applied and Environmental Microbiology*, **58**, 1292–1300.

Hoagland, R.E. (1990) Microbes and microbial products as herbicides. In: *American Chemical Society Symposium Series No. 439. Microbes and Microbial Products as Herbicides*, pp. 2–52. Washington, DC.

Hofte, H. & Whiteley, H. R. (1989) Insecticidal crystal proteins of *Bacillus thuringiensis. Microbiological Reviews*, **53**, 242–255.

Hood, M.A. & Seidler, R.J. (1995) Design of microcosms to provide data reflecting field trials of germs. In: *Molecular Microbial Ecology Manual* (eds A.D.L. Akkermans, J.D. van Elsas & F.J. de Bruijn). Kluwer Academic Publishers, Dordrecht.

Hopwood, D.A. (1993) Genetic engineering of *Streptomyces* to create hybrid antibiotics. *Current Opinions in Biotechnology*, **4**, 531–537.

Houk, V.S. & DeMarini, D.M. (1988) Use of the microscreen phage-induction assay to assess the genotoxicity of 14 hazardous industrial wastes. *Environmental and Molecular Mutagenesis*, **11**, 13–29.

How, M.J., Goodwin, B.F.J., Juniper, C.P. & Kinshott, A.K. (1978) Comparative serological and clinical findings in subjects exposed to environmental allergens. *Clinical Allergy*, **8**, 347–360.

James, R.R. & Lighthart, B. (1994) Susceptibility of the convergent lady beetle (Coleoptera: Coccinellidae) to four entomogenous fungi. *Environmental Entomology*, **23**, 190–192.

Jansson, K. & Jansson, B. (1991) Induction of mutation in V79 chinese hamster cells by tetrachlorohydroquinone, a metabolite of pentachlorophenol. *Mutation Research*, **260**, 83–87.

Jørgensen, S., Skov, K.W. & Diderichsen, B. (1991) Cloning, sequence, and expression of a lipase gene from *Pseudomonas cepacia*: lipase production in heterologous hosts requires two *Pseudomonas* genes. *Journal of Bacteriology*, **173**, 559–567.

Joshi, D.K. & Gold, M.H. (1993) Degradation of 2,4,5-trichlorophenol by the lignin-degrading basidiomycete *Phanerochaete chrysosporium. Applied and Environmental Microbiology*, **59**, 1779–1785.

Juniper, C.P., How, M.J., Goodwin, B.F.J. & Kinshott, A.K. (1977) *Bacillus subtilis* enzymes: a 7-year clinical, epidemiological and immunological study of an industrial allergen. *Journal of the Society for Occupational Medicine*, **27**, 3–12.

Kaiser, A., Classen, H.-G., Eberspächer, J. & Lingens, F. (1981) Acute toxicity testing of some herbicides-, alcaloids-, and antibiotics-metabolizing soil bacteria in the rat. *Zentralblatt für Bakteriologie, Mikrobiologie und Hygiene, Abteilung I: Originale, Reihe B, Hygiene*, **B173**, 173–179.

Kakinuma, S., Ikeda, H., Takada, Y., Tanaka, H., Hopwood, D.A. & Omura, S. (1995) Production of the new antibiotic tetrahydrokalafungin by transformants of the kalafungin producer *Streptomyces tanashiensis. Journal of Antibiotics (Tokyo)*, **48**, 484–487.

Kawanishi, C.Y., George, S.E. & Sherwood, R.L. (1990) Health effects assessment of pulmonary exposure to biotechnology agents. In: *Review of Progress in the Biotechnology—Microbial Pest Control Agent Risk Assessment Program*, pp. 214–215. US Environmental Protection Agency. EPA/600/9-90/029, Office of Research and Development, Corvallis, OR.

Kerem, Z. & Hadar, Y. (1993) Effect of manganese on lignin degradation by *Pleurotus ostreatus* during solid-state fermentation. *Applied and Environmental Microbiology*, **59**, 4115–4120.

Kidambi, S.P., Ripp, S. & Miller, R.V. (1994) Evidence for phage-mediated gene transfer among *Pseudomonas aeruginosa* strains on the phylloplane. *Applied and Environmental Microbiology*, **60**, 496–500.

Kilband, J.J.I. & Jackowski, K. (1992) Biodesulfurization of water-soluble coal-derived material by *Rhodococcus. Biotechnology and Bioengineering*, **40**, 1107–1114.

King, J.M.H., DiGrazia, P.M., Applegate, B., Burlage, R.,

Sanseverino, J., Dunbar, P., Larimer, F. & Sayler, G.S. (1990) Rapid, sensitive bioluminescent reporter technology for naphthalene exposure and biodegradation. *Science*, **249**, 778–780.

King, R.B., Long, G.M. & Sheldon, J.K. (1992) *Practical Environmental Bioremediation*, 149 pp. Lewis Publishers, Boca Raton, LA.

Knudsen, G.R., Walter, M.V., Porteous, L.A., Prince, V.J., Armstrong, J.L. & Seidler, R.J. (1988) A predictive model of conjugative plasmid transfer in the rhizosphere and phyllosphere. *Applied and Environmental Microbiology*, **54**, 343–347.

Kreig, N.R. & Holt J.G. (eds) (1984) *Bergey's Manual of Systematic Bacteriology*, Vol. 1. Williams and Wilkins, Baltimore, MD.

Krimisky, S., Wurbel, R.P., Naess, I.G., Levy, S.B., Wetzler, R.E. & Marshall, B. (1995) Standardized microcosms in microbial risk assessment. *BioScience*, **9**, 590–599.

Krumme, M.L., Smith, R.L., Egestorff, J., Theim, S.M., Tiedje, J.M., Timmis, K.N. & Dwyer, D.F. (1994) Behavior of pollutant degrading microorganisms in aquifers: predictions for genetically engineered organisms. *Environmental Science and Technology*, **28**, 1134–1138.

Kuehn, M., Lent, K., Haas, J., Hagenzieker, J., Cervin, M. & Smith, A.L. (1992) Fimbriation of *Pseudomonas cepacia*. *Infection and Immunity*, **60**, 2002–2007.

Ladisch, M.R. & Kohlmann, K.L. (1992) Recombinant human insulin. *Biotechnology Progress*, **8**, 469–478.

Lanfranco, L., Wyss, P., Marzachi, C. & Bonfante, P. (1995) Generation of RAPD-PCR primers for the identification of isolates of *Glomus mosseae*, an arbuscular mycorrhizal fungus. *Molecular Ecology*, **4**, 61–68.

Laux, D.C., Myhal, M.L. & Cohen, P.S. (1981) Relative colonization potentials of *E. coli* K-12 and human fecal strains in streptomycin-treated mice. In: *Molecular Biology, Pathogenicity, and Ecology of Bacterial Plasmids* (eds S.B. Levy, R.C. Clowes & E.L. Koenig), p. 624. Plenum Press, New York.

Laux, D.C., Cabelli, V.J. & Cohen, P.S. (1982) The effect of plasmid gene expression on the colonizing ability of *E. coli* HS in mice. *Recombinant DNA Technical Bulletin*, **5**, 1–5.

Leopold, M. (1995) Public perception of biotechnology. In: *Genetically Modified Organisms. A Guide to Biosafety* (ed. G.T. Tzotzos), pp. 8–16. CAB International, Wallingford.

Leser, T.D. (1995) Validation of microbial community structure and ecological functional parameters in an aquatic microcosm designed for testing genetically engineered microorganisms. *Microbial Ecology*, **29**, 183–201.

Levin, M. (1995) Microbial pesticides: safety considerations. In: *Genetically Modified Organisms: A Guide to Biosafety* (ed. G.T. Tzotzos), pp. 93–109. CAB International, Wallingford.

Levin, M. & Strauss, H.S. (1990) Introduction: overview of risk assessment and regulation of environmental biotechnology. In: *Risk Assessment in Genetic Engineering* (eds M. Levin & H. Strauss), pp. 1–17. McGraw-Hill, Inc., New York.

Levin, M.A., Seidler, R.J., Bourquin, A.W., Fowle, J.R. & Barkay, T. (1987) EPA developing methods to assess environmental release. *Biotechnology*, **5**, 38–45.

Levin, M.A., Seidler, R.J. & Rogul, M. (eds) (1992) *Microbial Ecology: Principles, Methods, and Applications*, 945 pp. McGraw-Hill, New York.

Levy, S.B. (1986) Human exposure and effects analysis for genetically modified bacteria. In: *Engineered Organisms in the Environment: Scientific Issues* (eds J. Fiksel & V. Covello), pp. 23–25. American Society for Microbiology, Washington, DC.

Levy, S.B. & Marshall, B. (1981) Risk assessment studies of *E. coli* host-vector systems. *Recombinant DNA Technical Bulletin*, **4**, 91–98.

Levy, S.B., Marshall, B., Rowse-Eagle, D. & Onderdonk, A. (1980) Survival of *Escherichia coli* host-vector systems in mammalian intestine. *Science*, **209**, 391–394.

Lighthart, B. & Kim, J. (1989) Simulation of airborne microbial droplet transport. *Applied and Environmental Microbiology*, **55**, 2349–2355.

Linton, A.H. (1986) Flow of resistance genes in the environment and from animals to man. *Journal of Antimicrobials and Chemotherapy*, **18** (Suppl. C), 189–197.

Liu, P.V. (1974) Extracellular toxins of *Pseudomonas aeruginosa*. *Journal of Infectious Diseases*, **130**, S94–S99.

Lumsden, R.D., Lewis, J.A. & Fravel, D.R. (1995) Formulation and delivery of biocontrol agents for use against soilborne plant pathogens. In: *Biorational Pest Control Agents* (eds F.R. Hall & J.W. Barry), pp. 166–182. American Chemical Society, Washington, DC.

MacDonald, W.L. & Fulbright, D.W. (1991) Biological control of chestnut blight: use and limitation of transmissible hypovirulence. *Plant Disease*, **75**, 656–661.

McCarthy, A.J. & Williams, S.T. (1992) Actinomycetes as agents of biodegradation in the environment—a review. *Gene*, **115**, 189–192.

McDaniel, R., Ebert-Khosla, S., Hopwood, D.A. & Khosla, C. (1993) Engineered biosynthesis of novel polyketides. *Science*, **262**, 1546–1557.

McIntyre, D.A. & Harlander, S.K. (1993) Construction of first generation lactococcal integrative cloning vectors. *Applied Microbiology and Biotechnology*, **40**, 348–355.

McKevitt, A.I., Bajaksouzian, S., Klinger, J.D. & Woods, D.E. (1989) Purification and characterization of an extracellular protease from *Pseudomonas cepacia*. *Infection and Immunity*, **57**, 771–778.

McPherson, S.A., Perlak, F.J., Fuchs, R.L., Marrone, P.G., Lavrik, P.B. & Fischhoff, D.A. (1988) Characterization of the coleopteran-specific protein gene of *Bacillus thuringiensis* var. *tenebrionis*. *Biotechnology*, **6**, 61–66.

Malpartida, F. & Hopwood, D.A. (1992) Molecular cloning of the whole biosynthetic pathway of a *Streptomyces* antibiotic and its expression in a heterologous host. *Biotechnology*, **24**, 342–433.

Mannivannan, T., Sandhya, S. & Pandey, R.A. (1994) Microbial desulfurization of coal by chemoautotrophic *Thiobacillus ferrooxidans*—an iron mine isolate. *Journal of Environmental Science and Health—Part A. Environmental Science and Engineering*, **29**, 2045–2061.

Marcus, H., Austria, A.L. & Baker, N.R. (1989) Adherence of *Pseudomonas aeruginosa* to tracheal epithelium. *Infection and Immunity*, **57**, 1050–1053.

Marshall, B., Petrowski, D. & Levy, S.B. (1990) Inter- and intraspecies spread of *Escherichia coli* in a farm environment in the absence of antibiotic usage. *Proceedings of the National Academy of Science USA*, **87**, 6609–6613.

Martone, W.J., Osterman, C.A., Fisher, K.A. & Wenzel, R.P. (1981) *Pseudomonas cepacia*: implications and control of epidemic nosocomial colonization. *Reviews of Infectious Diseases*, **3**, 708–715.

Mayes, M.E., Held, G.A., Lau, C. *et al.* (1989) Characterization of the mammalian toxicity of the crystal polypeptides of *Bacillus thuringiensis* subsp. *israelensis*. *Fundamental and Applied Toxicology*, **13**, 310–322.

Mellon, M. & Rissler, J. (eds) (1993) *The Gene Exchange*, Vol. 4, No. 2. Union of Concerned Scientists, Washington, DC.

Mermelstein, L.D., Papoutsakis, E.T., Petersen, D.J. & Bennett, G.N. (1993) Laboratory engineering of *Clostridium acetobutylicum* ATCC 824 for increased solvent production by enhancement of acetone formation enzyme activities using a synthetic acetone operon. *Biotechnology and Bioengineering*, **42**, 1053–1060.

Merretig, U., Wiotzka, P. & Onken, U. (1989) The removal of pyritic sulphur from coal by *Leptospirillum*-like bacteria. *Applied Microbiology and Biotechnology*, **31**, 626–628.

Mikesell, M.D. & Boyd, S.A. (1988) Enhancement of pentachlorophenol degradation in soil through induced anaerobiosis and bioaugmentation with anaerobic sewage sludge. *Environmental Science and Technology*, **22**, 1411–1414.

Miller, R.M., Singer, G.M., Rosen, J.D. & Bartha, R. (1988) Photolysis primes biodegradation of Benzo[*a*]pyrene. *Applied and Environmental Microbiology*, **54**, 1724–1730.

Mulbry, W. & Kearney, P.C. (1991) Degradation of pesticides by microorganisms and the potential for genetic manipulation. *Crop Protection*, **10**, 334–346.

Munn, C.B. (1994) The use of recombinant DNA technology in the development of fish vaccines. *Fish and Shellfish Immunology*, **4**, 459–473.

Nakatsu, C.H., Fulthorpe, R.R., Holland, B.A., Peel, M.C. & Wyndham, R.C. (1995) The phylogenetic distribution of a transposable dioxygenase from the Niagara River watershed. *Molecular Ecology*, **4**, 593–603.

Nestle, M. (1996) Allergies to transgenic foods—questions of policy. *New England Journal of Medicine*, **334**, 726–728.

Nevalainen, H., Suominen, P. & Taimisto, K. (1994) On the safety of *Trichoderma reesei*. *Journal of Biotechnology*, **37**, 193–200.

Newhouse, M.L., Tagg, B., Pocock, S.J. & McEwan, A.C. (1970) An epidemiologic study of workers producing enzyme washing products. *Lancet*, **1**, 689–693.

Nicas, T.I. & Iglewski, B.H. (1986) Production of elastase and other exoproducts by environmental isolates of *Pseudomonas aeruginosa*. *Journal of Clinical Microbiology*, **23**, 967–969.

Nordlee, J.A., Taylor, L.L., Townsend, J.A., Thomas, L.A. & Bush, R.K. (1996) Identification of a brazil-nut allergen in transgenic soybeans. *New England Journal of Medicine*, **334**, 688–692.

NTP (1989) *Toxicology and carcinogenesis studies of pentachloropenol (CAS No. 87-86-5) in B6C3F$_1$ mice (feed studies)*. National Toxicology Program Technical Report, No. 349, National Institutes of Health Publication No. 88-2804, Research Triangle Park, NC.

Nuss, D.L. (1992) ACS Symposium Series No. 595. Biological control of chestnut blight: an example of virus-mediated attenuation of fungal pathogenesis. *Microbiological Reviews*, **56**, 561–576.

Oakley, C.L. & Bonerjee, N.G. (1963) Bacterial elastase. *Journal of Pathology and Bacteriology*, **85**, 489–506.

Ogunseitan, O.A., Sayler, G.S. & Miller, R.V. (1992) Application of DNA probes to analysis of bacteriophage distribution patterns in the environment. *Applied and Environmental Microbiology*, **58**, 2046–2052.

Olenchock, S.A. (1988) Quantitation of airborne endotoxin levels in various occupational environments. *Scandinavian Journal of Work Environmental Health*, **14**, 72–73.

Olson, B.H., Ogunseitan, O.A., Rochelle, P.A, Tebbe, C.C. & Tsai, Y.L. (1990) The implications of horizontal gene transfer for the environmental impact of genetically engineered microorganisms. In: *Risk Assessment in Genetic Engineering* (eds M. Levin & H. Strauss), pp. 163–168. McGraw-Hill, Inc., New York.

Paau, A.S. (1991) Improvements of *Rhizobium* inoculants by mutation, genetic engineering and formulation. *Biotechnology Advances*, **9**, 173–184.

Paolocci, F., Angelini, P., Cristofari, E., Granetti, B. & Arcioni, S. (1995) Identification of *Tuber* spp and corresponding ectomycorrhizae through molecular markers. *Journal of Science/Field/Agriculture*, **69**, 511–517.

Pearson, M.L., Jereb, J.A., Frieden, T.R. *et al.* (1992) Nosocomial transmission of multidrug-resistant *Mycobacterium tuberculosis*: a risk to patients and health care workers. *Annals of Internal Medicine*, **117**, 191–196.

Pedersen, S.S., Høiby, N., Espersen, F. & Koch, C. (1992) Role of alginate in infection with mucoid *Pseudomonas aeruginosa* in cystic fibrosis. *Thorax*, **47**, 6–13.

Penttila, M., Teeri, T.T., Nevalainen, H. & Knowles, J.K.C. (1991) The molecular biology of *Trichoderma reesei* and its application to biotechnology. *Symposium Series of the British Mycological Society*, **18**, 85–102.

Pepys, J., Hargreave, T.E., Longbottom, J.L. & Faux, J.A. (1969) Allergic reactions of the lungs to enzymes of *Bacillus subtilis*. *Lancet*, **1**, 1181–1184.

Pertsova, R.N., Kunc, S. & Golovleva, L.A. (1984) Degradation of 3-chlorobenzoate in soil by pseudomonads carrying biodegradative plasmids. *Folia Microbiology (Prague)*, **29**, 242–247.

Pothuluri, J.V., Selby, A., Evans, F.E., Freeman, J.P. & Cerniglia, C.E. (1995) Transformation of chrysene and other polycyclic aromatic hydrocarbon mixtures by the fungus *Cunninghamella elegans*. *Canadian Journal of Botany*, **73**, 1025–1033.

Raaijmakers, J.M., Vander Sluis, I., vanden Hout, M., Bakker, P.A.H.M. & Schippers, B. (1995) Dispersal of wild-type and genetically-modified *Pseudomonas* spp. from treated seeds or soil to aerial parts of radish plants. *Soil Biology and Biochemistry*, **27**, 1473–1478.

Ramos, J.L., Diaz, E., Dowling, D., de Lorenzo, V., Molin, S., O'Gara, F., Ramos, C. & Timmis, K.N. (1994) The behavior of bacteria designed for biodegradation. *Biotechnology*, **12**, 1349–1356.

Ramos, M.S. & Harlander, S.K. (1990) DNA fingerprinting of lactococci and streptococci used in dairy fermentations. *Applied Microbiology and Biotechnology*, **34**, 386–374.

Ramphal, R., Sadoff, J.C., Pyle, M. & Sillpigni, J.D. (1984) Role of pili in the adherence of *Pseudomonas aeruginosa* to injured tracheal epithelium. *Infection and Immunity*, **44**, 38–40.

Rank, G.H. & Xiao, W. (1991) Alteration of industrial food and beverage yeasts by recombinant DNA technology. *Annals of New York Academy of Science*, **646**, 155–171.

Rawlings, D.E. & Silver, S. (1995) Mining with microbes. *Biotechnology*, **13**, 773–778.

Regal, P. (1990) Gene flow and adaptability in transgenic agricultural organisms: long-term risks and overview. In: *Risk Assessment in Agricultural Biotechnology: Proceedings of the International Conference* (eds J.J. Marois & G. Bruening), pp. 102–110. University of California, Davis, CA.

Ripp, S. & Miller, R.V. (1995) Effects of suspended particulates on the frequency of transduction among *Pseudomonas aeruginosa* in a freshwater environment. *Applied and Environmental Microbiology*, **61**, 1214–1219.

Ripp, S., Ogunseitan, O.A. & Miller, R.V. (1994) Transduction of a freshwater microbial community by a new *Pseudomonas aeruginosa* generalized transducing phage, UT1. *Molecular Ecology*, **3**, 121–126.

Ritz, H.L., Evans, B.L.B., Bruce, R.D., Fletcher, E.R., Fisher, G.L. & Sarlo K. (1993) Respiratory and immunological responses of guinea pigs to enzyme-containing detergents: a comparison of intratracheal and inhalation models of exposure. *Fundamental and Applied Toxicology*, **21**, 31–37.

Roe, R.M., Kallapur, V.L., Dauterman, W.C. *et al.* (1991) Vertebrate toxicology of the solubilized parasporal crystalline proteins of *Bacillus thuringiensis* subsp. *israelensis*. *Reviews of Pesticide Toxicology*, **1**, 119–130.

Rosenstein, B.J. & Hall, D.E. (1980) Pneumonia and septicemia due to *Pseudomonas cepacia* in a patient with cystic fibrosis. *Johns Hopkins Medical Journal*, **147**, 188–189.

Rouchaud, J., Gustin, F., Roisin, C., Grevy, L. & Raimond, Y. (1993) Effects of organic fertilizers on aldicarb soil biodegradation in sugar beet crops. *Archives of Environmental Contamination and Toxicology*, **24**, 67–74.

Sahoo, D.K., Kar, R.N. & Das, R.P. (1992) Bioaccumulation of heavy metal ions by *Bacillus circulans*. *Bioresource Technology*, **41**, 177–179.

Sajjan, U.S., Corey, M., Karmali, M.A. & Forstner, J.F. (1992) Binding of *Pseudomonas cepacia* to normal human intestinal mucin and respiratory mucin from patients with cystic fibrosis. *Journal of Clinical Investigation*, **89**, 648–656.

Sarlo, K. & Clark, E.D. (1992) A tier approach for evaluating the respiratory allergenicity of low molecular weight chemicals. *Fundamental and Applied Toxicology*, **18**, 107–114.

Sarlo, K., Polk, J.E. & Ritz, H.L. (1991) Guinea pig intratracheal (IT) test to assess respiratory allergenicity of detergent enzymes: comparison with the human data base. *Journal of Allergy and Clinical Immunology*, **87**, 343.

Sayler, G.S. & Sayre, P. (1995) Risk assessment for recombinant pseudomonads released into the environment for hazardous waste degradation. In: *Bioremediation: the Tokyo 1994 Workshop*, pp. 263–272. Organization of Economic Cooperation and Development, Paris.

Sayler, G.S., Reid, M.C., Perkins, B.K. *et al.* (1982) Evaluation of the mutagenic potential of bacterial polychlorinated biphenyl biodegradation products.

Archives of Environmental Contaminant Toxicology, **11**, 577–581.

Scanferlato, V.S., Orvos, D.R., Cairns, J., Jr. & Lacy, G.H. (1989) Genetically engineered *Erwinia carotovora* in aquatic microcosms: survival and effects on functional groups of indigenous bacteria. *Applied and Environmental Microbiology*, **55**, 1477–1482.

Schimpff, S.C. (1980) Infection prevention during profound granulocytopenia. New approaches to alimentary canal microbial suppression. *Annals of Internal Medicine*, **93**, 358–361.

Schlesinger, S. (1995) RNA viruses as vectors for the expression of heterologous proteins. *Journal of Molecular Biotechnology*, **3**, 155–165.

Seidler, R.J. (1992). Evaluation of methods for detecting ecological effects from genetically engineered microorganisms and microbial pest controls agents in terrestrial systems. *Biotechnology Advances*, **10**, 149–178.

Seidler, R.J. & Hern, S. (1988) *Special report: release of ice minus recombinant bacteria.* US Environmental Protection Agency, EPA/600/3-88/060, ERL-Corvallis-473, Environmental Research Laboratory, Corvallis, OR.

Seidler, R.J., Walter, M.V., Hern, S., Fieland, V., Schmedding, D. & Lindow, S. (1994) Measuring the dispersal and reentrainment of recombinant *Pseudomonas syringae* at California test sites. *Microbial Releases*, **2**, 209–216.

Shapiro, R.S. & Eisenberg, B.E. (1971) Sensitivity to proteolytic enzymes in laundry detergents. *Journal of Allergy*, **47**, 76–79.

Sharp, R., O'Donnell, A.G., Gilber, H.G. & Hazlewood, G.P. (1992) Growth and survival of genetically manipulated *Lactobacillus plantarum* in silage. *Applied and Environmental Microbiology*, **58**, 2517–2522.

Sharples, F.E. (1990) Ecological aspects of hazard identification for environmental uses of genetically engineered organisms. In: *Risk Assessment in Genetic Engineering* (eds M. Levin & H. Strauss), pp. 1–17. McGraw-Hill, Inc., New York.

Sherwood, R.L., Thomas, P.T., Kawanishi, C.Y. & Fenters, J.D. (1988) Comparison of *Streptococcus zooepidemicus* and influenza virus pathogenicity in mice by three pulmonary exposure routes. *Applied and Environmental Microbiology*, **54**, 1744–1751.

Sherwood, R.L., Byrne, M.J., Kawanishi, C.Y. & Sjoblad, R. (1991a) Comparison of hemolytic and toxic components of *Bacillus cereus*, *B. thuringiensis* var. *israelensis* (BTI) and *B. thuringiensis* var. *kurstaki* (BTK). In: *Abstracts of the Annual Meeting of the American Society for Microbiology* (eds J.A. Morello & J.E. Domer), p. 322 (abstract Q-274). American Society for Microbiology, Washington, DC.

Sherwood, R.L., Mega, W.M., Kawanishi, C.Y. & Sjoblad, R. (1991b). Murine toxicity of vegetative preparations of *Bacillus thuringiensis*. In: *Abstracts of the Annual Meeting of the American Society for Microbiology* (eds J.A. Morello & J.E. Domer), p. 322 (abstract Q-273). American Society for Microbiology, Washington, DC.

Short, K.A., Doyle, J.D., King, R.J., Seidler, R.J., Stotzky, G. & Olsen, R.H. (1991) Effects of 2,4-dichlorophenol, a metabolite of a genetically engineered bacterium, and 2,4-dichlorophenoxyacetate on some microorganism-mediated ecological processes in soil. *Applied Environmental Microbiology*, **75**, 412–418.

Simon, L., Levesque, R.C. & Lalonde, M. (1993) Identification of endomycorrhizal fungi colonizing roots by fluorescent single-strand conformation polymorphism-polymerase chain reaction. *Applied and Environmental Microbiology*, **59**, 4211–4215.

Singleton, I. (1994) Microbial metabolism of xenobiotics: fundamental and applied research. *Journal of Chemistry, Technology and Biotechnology*, **59**, 9–23.

Smit, E., van Elsas, J.D., van Neen, J.A. & DeVos, W.M. (1991) Detection of plasmid transfer from *Pseudomonas fluorescens* to indigenous bacteria in soil by using bacteriophage vphi-R2f for donor counterselection. *Applied and Environmental Microbiology*, **57**, 3482–3488.

Smit, E., van Elsas, J.D. & van Neen, J.A. (1992) Risks associated with the application of genetically modified microorganisms in terrestrial ecosystems. *(Federation of European Microbiologal Societies) Microbiology Reviews*, **88**, 263–278.

Smith, H.W. (1975) Survival of orally administered *E. coli* K12 in alimentary tract of man. *Nature*, **255**, 500–502.

Smith, H.W. (1978) Is it safe to use *Escherichia coli* K12 in recombinant DNA experiments? *Journal of Infectious Diseases*, **137**, 655–660.

Sneath, P.H., Mair, N.S., Sharpe, M.E. & Holt, J.G. (eds) (1986) *Bergey's Manual of Systematic Bacteriology*, Vol. 2. Williams and Wilkins, Baltimore, MD.

Somich, C.J., Muldoon, M.T. & Kearney, P.C. (1990) On-site treatment of pesticide waste and rinsate using ozone and biologically active soil. *Environmental Science and Technology*, **24**, 745–749.

Sprenger, G.A. (1993) Approaches to broaden the substrate and product range of the ethanolagenic bacterium *Zymomonas mobilis* by genetic engineering. *Journal of Biotechnology*, **27**, 225–237.

Staley, J.T., Bryant, M.P., Pfennig, N. & Holt, J.G. (eds) (1989) *Bergey's Manual of Systematic Bacteriology*, Vol. 3. Williams and Wilkins, Baltimore, MD.

Starke, J.R., Edwards, M.S., Langston, C.L. & Baker, C.J. (1987) A mouse model of chronic pulmonary infection with *Pseudomonas aeruginosa* and *Pseudomonas cepacia*. *Pediatric Research*, **22**, 698–702.

Stonard, R.J. & Miller-Widerman, M.A. (1995) Herbicides and plant growth regulators. In: *Agrochemicals from Natural Products* (ed. C.R.A. Godfry), pp. 283–310. Marcel Dekker, Inc., New York.

Stotzky, G., Broder, M.W., Doyle, J.D. & Jones, R.A. (1993) Selected methods for the detection and assessment of ecological effects resulting from the release of genetically engineered microorganisms to the terrestrial environment. *Advances in Applied Microbiology*, **38**, 1–98.

Straus, D.C., Lonon, M.K., Woods, D.E. & Garner, C.W. (1989) Production of an extracellular toxic complex by various strains of *Pseudomonas cepacia*. *Journal of Medical Microbiology*, **30**, 17–22.

Tabashnik, B.E. (1994) Evolution of resistance to *Bacillus thuringiensis*. *Annual Review of Entomology*, **39**, 47–79.

Tan, U. & Miller, K.J. (1992) Cloning, expression, and nucleotide sequence of a lipase gene from *Pseudomonas fluorescens* B52. *Applied and Environmental Microbiology*, **58**, 1402–1407.

Tang, H., Kays, M. & Prince, A. (1995) Role of *Pseudomonas aeruginosa* pili in acute pulmonary infection. *Infection and Immunity*, **63**, 1278–1285.

Thayer, A. (1992) Recombinant blood clot factor cleared for sale. *Chemical and Engineering News*, **70**, 6.

Thayer, A. (1994) Bioengineered cystic fibrosis drug approved. *Chemical and Engineering News*, **72**, 5.

Tommerup, I.C., Barton, J.E. & O'Brien, P.A. (1995) Reliability of RAPD fingerprinting of three basidiomycete fungi *Laccaria, Hydnangium* and *Rhizoctonia*. *Mycological Research*, **99**, 179–186.

Trotter, P.C. (1990) Biotechnology in the pulp and paper industry: a review. 1. Tree improvement, pulping and bleaching, and dissolving pulp applications. *Technical Association of the Pulp and Paper Industry*, **73**, 198–204.

Tzotzos, G.T. (ed.) (1995) *Genetically Modified Organisms. A Guide to Biosafety*, 213 pp. CAB International, Wallingford.

USEPA (1992) *Monitoring small-scale field tests of microorganisms. Prevention, pesticides, and toxic substances* (TS-788). US Environmental Protection Agency, EPA/700/R-92/008, Washington, DC.

USEPA (1994a) *A Review of Ecological Assessment Case Studies from a Risk Assessment Perspective*, Vol. 2. US Environmental Protection Agency, EPA/630/R-94/003, Risk Assessment Forum, Washington, DC.

USEPA (1994b) *Ecological Risk Assessment Issue Papers*. US Environmental Protection Agency, EPA/630/R-94/009, Washington, DC.

Valigra, L. (1994) Engineering the future of antibiotics. *New Scientist*, **142**, 25–27.

van Elsas, J.D. & Smit, E. (1994) Some considerations on gene transfer between bacteria in soil and rhizosphere. In: *Molecular Ecology of Rhizosphere Microorganisms* (eds F. O'Gara, D.N. Dowling & B. Boesten), pp. 151–164. Verlagsgesellschaft mbH, Weinheim.

Vidyarthi, A.S. & Nagar, S.J. (1994) Immobilized fungal spores for microbial transformation of steroids: 11 alpha-hydroxylation of progesterone. *Biological Memoirs*, **20**, 15–19.

Vlak, J.M. (1993) Genetic engineering of viruses for insect control. In: *Molecular Approaches to Fundamental and Applied Entomology*, pp. 90–127. Springer-Verlag, New York, NY.

Volesky, B. & Holan, Z.R. (1995) Biosorption of heavy metals. *Biotechnology Progress*, **11**, 235–250.

Wagner-Dobler, I., Pipke, R., Timmis, K.N. & Dwyer, D.F. (1992) Evaluation of aquatic sediment microcosms and their use in assessing possible effects of introduced microorganisms on ecosystem parameters. *Applied and Environmental Microbiology*, **58**, 1249–1258.

Wales, D.S. & Sagar, B.F. (1990) Recovery of metal ions by microfungal filters. *Journal of Chemical Biotechnology*, **49**, 345–355.

Wallace, R.J. (1994) Ruminal microbiology, biotechnology, and ruminant nutrition: progress and problems. *Journal of Animal Science*, **72**, 2992–3003.

Walter, MV., Porteous, L.A., Prince, V.J., Ganio, L. & Seidler, R.J. (1991) A microcosm for measuring survival and conjugation of genetically engineered bacteria in rhizosphere environments. *Current Microbiology*, **22**, 117–121.

Wang, X. & Bartha, R. (1990) Effects of bioremediation on residues, activity and toxicity in soil contaminated by fuel spills. *Soil Biology and Biochemistry*, **22**, 501–505.

Wang, Z., Crawford, D.L., Pometto A.L., III & Rafii, F. (1989) Survival and effects of wild-type, mutant, and recombinant *Streptomyces* in a soil ecosystem. *Canadian Journal of Microbiology*, **35**, 535–543.

Wang, Z., Crawford, D.L. Magnuson, T.S., Bleakley, B.H. & Hertel, G. (1991) Effects of bacterial lignin peroxidase on organic carbon mineralization in soil, using recombinant Streptomyces strains. *Canadian Journal of Microbiology*, **37**, 287–294.

Watrud, L.S. & Seidler, R.J. (1997) Non-target ecological effects of plant, microbial and chemical introductions to soil systems. In: *Soil Chemistry and Ecosystem Health* (ed. P.M. Huang). Soil Science Society of America, Madison, WI (in press).

Watrud, L.S., Metz, S.G. & Fischhoff, D.A. (1996) Engineered plants in the environment. In: *Engineered Organisms in Environmental Settings: Biotechnological and Agricultural Applications* (eds M. Levin & E. Israeli), pp. 165–189. CRC Press, Boca Raton, FL.

Weill, H. Waddell, L.C. & Ziskind, M. (1971) A study of workers exposed to detergent enzymes. *Journal of the American Medical Association*, **217**, 425–433.

Weil, J., Miramonti, J. & Ladisch, M.R. (1995) Biosynthesis of cephalosporin C: regulation and recombinant technology. *Enzyme and Microbial Technology*, **17**, 88–90.

Weller, D.M. (1988) Biological control of soilborne plant

pathogens in the rhizosphere with bacteria. *Annual Review of Phytopathology*, **26**, 379–407.

Whipps, J.M. (1992) Status of biological disease control in horticulture. *Biocontrol Science and Technology*, **2**, 3–24.

White, D., Crosbie, J.D., Atkinson, D. & Killham, K. (1994) Effect of an introduced inoculum on soil microbial diversity. *(Federation of European Microbiological Societies) Microbiology Ecology*, **14**, 169–178.

Williamson, M. (1994) Community response to transgenic plant release: predictions from British experience of invasive plants and feral crop plants. *Molecular Ecology*, **3**, 75–79.

Wilson, G.S. & Miles, A. (1975) *Topley and Wilson's Principles of Bacteriology, Virology and Immunity*, pp. 811–812. The Williams and Wilkins Co., Baltimore, MD.

Wolfarth, S., Hoesche, C., Strunk, C. & Winkler, U.K. (1992) Molecular genetics of the extracellular lipase of *Pseudomonas aeruginosa* PAO1. *Journal of General Microbiology*, **138**, 1325–1335.

Wood, B.E. & Ingram, L.O. (1992) Ethanol production from cellobiose, amorphous cellulose, and crystalline cellulose by recombinant *Klebsiella oxytocia* containing chromosomally integrated *Zymomonas mobilis* genes for ethanol production and plasmids expressing thermostable cellulase genes from *Clostridium thermocellum*. *Applied and Environmental Microbiology*, **58**, 2103–2110.

Woods, D.E. & Que, J.U. (1987) Purification of *Pseudomonas aeruginosa* exoenzyme S. *Infection and Immunity*, **55**, 579–586.

Woods, D.E. & Sokol, P.A. (1986) Role of *Pseudomonas aeruginosa* extracellular enzymes in lung disease. *Clinical and Investigative Medicine*, **9**, 108–112.

Woods, D.E., Straus, D.C., Johanson, W.G., Jr., Berry, V.K. & Bass, J.A. (1980) Role of pili in adherence of *Pseudomonas aeruginosa* to mammalian buccal epithelial cells. *Infection and Immunity*, **29**, 1146–1151.

Woods, D.E., Cryz, S.J., Friedman, R.L. & Iglewski, B.H. (1982) Contribution of toxin A and elastase to virulence of *Pseudomonas aeruginosa* in chronic lung infections of rats. *Infection and Immunity*, **36**, 1223–1228.

Woods, D.E., Sokol, P.A., Bryan, L.E. *et al.* (1991) *In vivo* regulation of virulence in *Pseudomonas aeruginosa* associated with genetic rearrangement. *Journal of Infectious Diseases*, **163**, 143–149.

Wyman, C.E. & Goodman, B.J. (1993) Biotechnology for production of fuels, chemicals, and materials from biomass. *Applied Biochemistry and Biotechnology*, **39**, 41–59.

Yadav, J.S. & Reddy, C.A. (1993) Degradation of benzene, toluene, ethylbenzene, and xylenes (BTEX) by the lignin-degrading basidiomycete *Phanerochaete chrysosporium*. *Applied and Environmental Microbiology*, **59**, 756–762.

Zaady, E., Okon, Y. & Perevolotsky, A. (1994) Growth response of Mediterranean herbaceous swards to inoculation with *Azospirillum brasilense*. *Journal of Range Management*, **47**, 12–15.

Zhou, J.Z. & Tiedje, J.M. (1995) Gene transfer from a bacterium injected into an aquifer to an indigenous bacterium. *Molecular Ecology*, **4**, 613–618.

Chapter 6
Assessing Risks to Ecological Systems from Biological Introductions (Excluding Genetically Modified Organisms)

DANIEL SIMBERLOFF AND MARTIN ALEXANDER

6.1 INTRODUCTION AND SCOPE

Several factors must be considered in assessing the risk posed by an introduced species. First, the species must invade a particular ecosystem. Yet even if it survives, it need not have a discernible impact on that ecosystem. The probability of either successful invasion or disruptive effect is highly uncertain. This uncertainty is the basis for some panic concerning the outcome of the release of non-indigenous species. Conversely, despite the uncertainty, some resource managers and biologists are unconcerned about novel introductions. Both groups muster evidence to support their contentions, but rarely have members of either camp attempted a formal risk assessment for introduced species. That some introduced organisms have become established and caused major disruptions proves that a risk exists, especially for poorly known organisms—i.e. most organisms. On the other hand, the likely failure of many—and probably most—introduced species to become established or to have a detrimental impact (Simberloff 1981) suggests that the risk is small for most species. Data do not exist, however, to assess how small is small or to determine for most introduced organisms the likelihood of being problematic. Further, some of those relatively few introduced species that have survived have generated huge costs, leading to the enormous concern about introductions.

Thus the risk assessor concerned with biological introductions probably will encounter two disparate groups of experts. The first will emphasize the evidence for problems and will cite acknowledged instances of major ecological impact. The other will downplay such evidence as idiosyncratic and will present plausible arguments that the average introduced species is very unlikely to create an environmental problem. Adherents of the first camp would respond that the effect of the average introduced species is irrelevant; one should be interested not in the mean effect but in that of the outliers. This expert background noise characterizes the setting for risk assessment of introduced species. We begin in section 6.2 by outlining key differences between introduced species and other types of environmental threats, including differences that produce the uncertainty mentioned above. The main differences are that: (i) introduced species reproduce and multiply; (ii) they disperse by myriad means, often in large 'jumps' that are inherently very difficult to predict; (iii) their interactions with other living entities are extremely difficult to predict; and (iv) they evolve, and evolution has large elements of chance.

The risk presented by an introduced species rests on four factors: survival, multiplication, dispersal and ecological effects. The first three produce the risk of exposure, while the fourth constitutes the hazard in the typical risk assessment dichotomy (e.g. Suter 1993). In section 6.3 we discuss what general information is available on survival, proliferation and dispersal, with an emphasis on what is not known generically about these processes. We also show that, with respect to dispersal, a distinction can be drawn between risks posed by a particular species, inadvertently or deliberately introduced or whose introduction is contemplated, and those associated with processes that can introduce many species as an unintended by-product of some activity. Thus, we consider briefly the assessment of risks posed by particular pathways or commodities that are not themselves introduced species but may carry large numbers of them. For example, ballast water used in international shipping carries living organisms of many species all over the globe (Carlton & Geller 1993). Both living commodities (e.g. ornamental plants, aquaculture stock) and non-living commodities (e.g. unprocessed logs) may be agents of inadvertent introduction of species. Then, in section 6.4, we consider the myriad effects that introduced species can have on populations, communities and ecosystems. In sections 6.5 and 6.6 we discuss how the uncertainties associated with introduced species, and various characteristics of introduced species as opposed to chemicals and physical processes, complicate the choice of assessment end-points.

To a large extent, effects at the community and ecosystem levels that might be expected of an introduced species, and therefore the risks that are considered, rest on how communities and ecosystems are conceived. There is a spectrum of opinion about the nature of communities and their importance as a level of biological organization. The *Framework for Ecological Risk Assessment* (USEPA 1992; see also Chapters 2 and 3) represents one extreme of this spectrum: 'Community—An assemblage of populations of different species within a specified location in space and time'. The other extreme is the superorganismic community (Clements 1936), which has an integrity and organic nature quite comparable to that of an individual, with the different populations interacting in stylized ways and with various functions, such as energy flow, fully analogous to the physiological functions of individuals.

The controversy over the nature of communities is very old (Simberloff 1980; Underwood 1986), remains a burning issue, and will not be resolved here. Suffice it to say that both ends of the spectrum are still represented. Thus, Wilson & Sober (1989) write about holistic communities and ecosystems in a Clementsian vein, while the definition in the *Framework*, though extreme, is currently held by some ecologists. Ecologists hold a similar range of views on ecosystems, with some perceiving ecosystem processes like nutrient cycling as analogues to the physiology of organisms (Rambler *et al.* 1988) and others viewing them as the inevitable by-product of the co-occurrence of multiple populations. The debate about emergent properties of communities and ecosystems (Simberloff 1980; Wilson & Sober 1989) is simply another aspect of this longstanding controversy: can every aspect of the structure and function of communities and ecosystems be comprehended by a full understanding of their component populations and the physical setting, or do certain features emerge only by virtue of the complex interactions among populations, so that these features are intractable to reductionist study?

Clearly, if the community or ecosystem is seen simply as an epiphenomenal consequence of the co-occurrence of populations, then the effects of introduced species on traits at these levels are unlikely to be of concern. Thus assessment end-points are unlikely to be community or ecosystem features. Rather, the effects of an introduced species will be considered, and sought, as the sum of its effects on the various component populations. If, on the other hand, the community or ecosystem is considered a major level of organization, the effects of introduced species are likely to be sought on traits at this level, particularly emergent traits. A new biological entity might well be expected to sunder completely the previous community and, through its interactions with other populations and with the physical environment, to generate an entirely new community: i.e. a small change in the species composition could lead to a very large change in the functioning of the community and ecosystem. In summary, very different effects would be considered and very different assessment end-points selected, depending on the definition of communities and ecosystems.

Here we have chosen not to restrict ourselves to the definition of a community as simply a co-occurring group of populations, and we shall consider effects that span the spectrum from effects on individuals and populations through effects on communities and ecosystems. Regardless of whether communities and ecosystems are holistic, they have features that can be measured only at those levels, and even if they are only compound properties (e.g. various diversity measures) rather than emergent properties, the public is concerned with some of these features, and laws even address them. Further, it is entirely possible that there are great differences among communities and ecosystems, with some highly organized and others simply collections of coexisting populations.

Finally, we consider introduced species to include any species introduced to regions not contiguous with its original range, thus encompassing organisms moved between regions of a single nation. A recent US government document, *Harmful Non-indigenous Species in the United States* (US Congress 1993), has numerous examples of the effects we discuss in section 6.4, plus an extended consideration of risk assessment of various introduced species. However, the report considers impacts only of species originating outside the USA. There is no logical or scientific justification for this narrowed scope. For example, the rainbow trout (*Oncorhynchus mykiss*), though indigenous to the continental USA, has been introduced widely beyond its natural range to the detriment of many native species. Even more limited movement can be problematic. The yellow bush lupine (*Lupinus arboreus*), introduced from San Francisco, has greatly modified the entire dune community at Humboldt Bay in northern California by its survival in the nutrient-poor dune mat and subsequent modification of many features of the environment (references in Simberloff 1991). In any event, it would be perverse to exclude intracontinental introductions from our purview in light of the great concern in many nations about the movement of foreign organisms even after they establish a beachhead. For example, the spread and risks in the USA posed by such Eurasian scourges as the zebra mussel (*Dreissena polymorpha*) and purple loosestrife (*Lythrum salicaria*) are targets of major assessment efforts.

We will show throughout this paper that random aspects of the biology of living organisms (particularly evolution), the complexity of various two-species and multispecies interactions, and a dearth of crucial information on the autecology and synecology of almost all species that do not directly affect agriculture or human health, render risk assessment for introduced species vastly more difficult than for chemicals or physical processes. Thus, as in Chapter 5 on genetically modified organisms (GMOs), we emphasize the various kinds of effects wrought by introduced species and we rely heavily on examples. We believe that, in most instances, this approach is probably the best currently possible for introduced species.

6.2 KEY DIFFERENCES BETWEEN INTRODUCED SPECIES AND CHEMICAL OR PHYSICAL FACTORS AFFECTING THE ENVIRONMENT

First, living organisms reproduce and, in the course of reproduction, may multiply (see Chapter 5). With a chemical pollutant, simply ceasing to add it to the environment inevitably lessens the risk of any ecological effect as the molecule breaks down naturally or various mitigation procedures facilitate the breakdown of the molecule and/or its breakdown products or their transport from the site. With a living organism, arresting its introduction to a site need not substantially lower the risk of an ecological effect—even if the organism dies—so long as it can reproduce there.

For some introduced species, a population increases only slightly initially, then remains stable for many generations as individuals just replace themselves. In 1870, about 12 pairs of Old World tree sparrows (*Passer montanus*) were released in St Louis, Missouri. They quickly established a persistent population, estimated at 2500–25 000 individuals, which has been restricted for over a century to St Louis and adjacent areas (Lever 1987). Stragglers are occasionally reported as far away as Wisconsin, but the population never expands in range or size. The Mediterranean fruit fly (*Ceratitis capitata*), a major pest associated with fruit and vegetables, appears to have established a persistent but generally small population in southern California despite repeated, extensive eradication campaigns by the US Department of Agriculture. Sporadic outbreaks in the Los Angeles area are probably not the result of repeated introductions, but manifestations of a sparse population that has persisted for at least 5 years and possibly much longer (Carey 1991). For microbial risk assessment, the situation may be considerably worse than these metazoan examples indicate. Microorganisms can persist for long periods by various means (see Chapter 5 and section 6.3.1), at such low numbers as to be undetectable, yet can grow rapidly and spread under certain circumstances. Chemicals normally would not have such abilities, though a sequestered chemical could be liberated by some physical or chemical perturbation.

Reproduction and low-level, long-term maintenance at one site, however, are probably not the most common trajectory of a surviving introduced species. Most introduced species that survive at all probably increase in numbers and spread, at least initially. Both the tree sparrow and the Mediterranean fruit fly increased, although the initial increase was limited and there was almost no geographical spread. A more probable result, if an introduced species survives, is substantial population increase associated with at least some geographical spread. Although maximum rate of increase is limited (and can be estimated) by such life-history characteristics as generation length and fecundity, all species have the capacity for exponential increase in an appropriate environment.

From the standpoint of potential ecological effects, the possibility of geographical spread is of even greater concern than the likelihood of local increase. Indeed, all species have means of dispersal. As for GMOs (Chapter 5), autonomous dispersal of introduced species is a characteristic that distinguishes living organisms from chemical pollutants. Because the nature of this dispersal process and of range changes generally has been extensively studied (e.g Johnson 1969; Pielou 1979), some assessment of risk of spread can be attempted. Yet this process of spread, particularly for species that move of their own volition rather than passively (e.g. by wind or water currents, or phoresy), is qualitatively different from that of chemicals. Indeed, many living organisms are capable of dispersal over extremely long distances, and such dispersal is likely to make risk assessment particularly difficult, a problem discussed in more detail in section 6.3.3. It is also worth noting that even species that depend on either physical processes or other species for transport can increase their ranges substantially through increments because they can reproduce and multiply as each generation reaches a new suitable site.

For example, although the Asian fungus that causes Dutch elm disease (*Ophiostoma ulmi*) requires two bark beetles for dispersal in North America, the fungus spread rapidly from its initial site of introduction in the early twentieth century to infect most elms in the eastern USA (Elton 1958; von Broembsen 1989).

Interestingly, one of its beetle vectors, introduced from Europe at approximately the same time as the fungus itself, spread more rapidly than the fungus (Elton 1958); nevertheless, the fungus ultimately reached almost the entire range of its host.

Another characteristic difference between some introduced species and chemical or physical factors is the range of organisms that can be affected. For example, pathogens affecting animals, plants, and microorganisms can be particularly host-specific, as can parasitoids of insects and some phytophagous insects. Such limited ranges of susceptible species are not characteristic of chemicals or physical processes, which typically act on a wide range of species. Some introduced species similarly affect many other species (see section 6.4.1).

A key difference between chemicals and introduced species is that the latter can evolve, and this evolution can either increase or decrease the risk of ecological effects. For example, the Dutch elm fungus was introduced to North America from Europe on infected logs. In North America, it evolved more pathogenic strains that have been implicated in a recent new outbreak of the disease in Europe (von Broembsen 1989). On the other hand, numerous introduced pathogens and their hosts have co-evolved such that initially virulent diseases have become relatively benign (Ewald 1983). For the myxoma virus introduced to control introduced rabbits in Australia, the virus has become attenuated and the rabbits have become resistant (Krebs 1985; Williamson 1992). Yet evolution itself has numerous unpredictable aspects. For example, it may be severely constrained by which mutations arise during a particular period or by which chromosomal crossovers occur. Such chance components of evolution vastly complicate risk assessment and differ qualitatively from any processes associated with chemical or physical factors.

Another way in which introduced species characteristically differ from chemical and physical factors affecting the environment is that, once established, they can be either impossible or very difficult to remove. Extremely expensive and controversial attempts to eradicate even small insect pest populations have rarely been successful (Dahlsten 1986). Eradications of the coypu (*Myocastor coypus*) in Great Britain (Gosling 1989) and the giant African snail (*Achatina fulica*) in Florida (Simberloff 1997) stand out among many failures to eliminate introduced pests (see section 6.7).

An introduced species can generate another kind of bizarre problem—a development completely different from anything encountered with chemical pollutants—it may become a valued ecosystem component. For example, although eucalyptus introduced to Angel Island in San Francisco Bay provides a much less suitable habitat than the native plants for resident native animal populations, a California State Department of Parks and Recreation plan to remove them generated a storm of protest from people who found eucalyptus aesthetically pleasing (Azevedo 1990). Similarly, numerous game mammals introduced by hunters destroy native plants; feral pigs are of particular concern. But their effects cannot be contained because of political pressure and they are often spread by hunters.

6.3 SURVIVAL, PROLIFERATION AND DISPERSAL

For an introduced species at a particular site, four factors must be considered in a risk assessment: (i) probability that the organism will survive; (ii) probability that the survivor will multiply; (iii) probability that the population will disperse from the initial site to another location at which establishment is possible; and (iv) probability that the species will be harmful. If probabilities for the first or fourth factors are zero (i.e. the species does not survive or has no harmful effects), the probability of a deleterious environmental impact is zero. If the probability for the third factor is zero (i.e. the organism fails to spread), the probability of environmental impact beyond the initial site need not be zero. There is also the probability that genetic information associated with environmental impact will be transferred to another species; this is discussed in section 6.4.1. If such a 'lateral transfer' occurs, a similar analysis of survival, multiplication, dispersal, and harmful effects must be conducted for the organism receiving the genetic information (Alexander 1985a,b). This approach to risk analysis is also used by the US Department of Agriculture (Orr *et al.* 1993).

6.3.1 Survival

The number of propagules of an introduced species will probably be critical to its initial reproduction and multiplication and thus to the risk of a potential ecological effect. Every species has a 'minimum viable population size' (Shaffer 1981; Simberloff 1988)—or better, range of population sizes—such that, when the population falls below this range, rapid extinction is likely because of a number of forces, all of which affect small populations more heavily than large ones. For example, some dioecious species experience difficulties finding mates at low densities, while other species may require group stimulation of ovarian development or mating (Simberloff 1986a, 1988). These problems concerning breeding at low population size, collectively termed the 'Allee effect', have been the target of some modelling efforts (Dennis 1989). For a number of classes of introduced species, an increase in the number of propagules contributes to an increased probability of survival, although many other factors also contribute (Smallwood 1990).

Survival and increase also depend on the environment at the site of introduction. The percentage of surviving propagules varies with the particular organism and the environment (Liang *et al.* 1982). Some species can survive in one environment for many years with no replication but disappear rapidly in other environments. The critical environmental differences may be subtle, as in the case of microorganisms that persist in one soil but decline in another. None the less, every species has a range of habitats compatible with survival and reproduction; often the range for survival is wider than it is for reproduction. For example, many plant species' geographical ranges are limited not by the ability of adults to survive, but by their ability to reproduce at the margins of their range (e.g. Nielson & Wullstein 1983). In fact, many species may exist as metapopulations (i.e. loosely connected sets of populations) in which a few 'source' sites with ideal habitat produce dispersing individuals that colonize lower quality 'sinks' (Pulliam 1988). Whereas such a species would not be capable of maintaining itself in the sinks, these may constitute the bulk of the range.

The vast majority of propagules of plants, animals and microorganisms almost certainly die without issue because they end up in unsuitable habitats (e.g. terrestrial seeds land in water, microorganisms colonize too acidic or alkaline an environment, parasitic organisms fail to find a suitable host). Determining exactly what constitutes a receptive environment for a species often requires intensive research. Information on the survival of microorganisms comes chiefly from the public health and agricultural literature. Abundant data exist, but only for a few organisms that are important in diseases of humans, livestock, and agricultural crops, for a few bacteria used as indicators of faecal contamination, and for several bacteria of agronomic importance (Alexander 1985a). This information shows clearly that certain microorganisms are able to persist, often for long periods, in environments in which they are not indigenous (Liang *et al.* 1982). Although the relative frequency among microorganisms of species that survive in alien habitats is unknown, we know that the rate of death of such introduced microorganisms is affected by a number of physical and chemical factors, including drying, moisture level, pH, solar radiation, salinity and the presence of organic and inorganic toxins (see also Chapter 5).

Moreover, although the survival of many microorganisms is not greatly affected by these stresses, they do not persist. For most potential microbial invaders, it is not the physical and chemical properties of the environments (at least of soils and waters) that preclude establishment. In laboratory tests, samples of soil or water that are sterilized (to render them free of other microorganisms) are readily colonized by species that cannot—as well as those that can—survive or proliferate in non-sterile samples of the same soil or water. Undoubtedly, the activities of some of the resident populations resist the establishment of an invader. The activities of concern may be competition for limited resources between the recent arrival and indigenous populations, parasitism or grazing by protozoans or other predators. Although grazing pressure and the degree of competition and parasitism vary by habitat, it is usually impossible to predict which community or habitat will be suitable for invasion, except for environments that have such adverse conditions because of physical or chemical properties that only species tolerant of these factors will flourish.

Thus many microbial species are precluded from establishment because they fail to compete with the indigenous microflora and as a result die from starvation or because of predation. As in metazoans and plants, however, competition between introduced and native species is extremely difficult to document in the field (see section 6.4.1). Although microorganisms have parasites, parasitism is not known to eliminate introduced microorganisms. None the less, parasites of microorganisms may be important when a species achieves substantial population size or biomass; an impact of parasites on a low-density population of a particular host species is unlikely. Again, the survival of a microorganism in one environment but not in another is probably the result of differences in the competitive and/or predatory regimes. Data to support this assertion are sparse, however.

To survive, an introduced microorganism must tolerate abiotic stresses. It must also avoid, evade or cope with many competing microorganisms, the grazing activity of predators, and possibly attack by parasites. Starvation is a major stress, and a survivor must be able to endure periods of nutrient deprivation. Some microorganisms overcome these obstacles by forming resting structures (i.e. the endospores of certain bacterial genera, cysts of many protozoans, and sclerotia, chlamydospores, conidia and other structures of fungi). These resting structures can persist for many years. Even some bacteria and fungi with no specialized structures will persist for months, years and even decades. Also, certain bacteria enter a dormant state in which they appear to be injured in some manner. Injured bacteria found in natural waters (Roszak & Colwell 1987) that have undergone some physiological stress such that they will not multiply in the usual media used for their enumeration can be recovered in appropriately supplemented media. These 'injured' bacteria generally would not be detected if one used conventional procedures; yet they may endure for long periods and ultimately give rise to a population that grows and has a deleterious impact. In many instances, the number of surviving microorganisms may be below the level of detection and deemed to be absent. Given suitable conditions, however, the few survivors will multiply to large and possibly harmful populations. A potentially major impact from an introduced species initially below detection limits, however, is not characteristic of chemical or physical factors.

Some metazoans and plants also have resting stages in which they can remain viable for long periods, even in a stressful environment. The seeds of many species may remain dormant, buried in the soil for many years before germination — this is a typical trait of annuals and other short-lived species. Thus one cannot assume that a plant has disappeared from an area simply because no seedlings are observed. It is possible to assay soil for viable seeds but the procedure can be onerous, especially if the seed density is low. Among animals, resistant eggs of nematodes, fairy shrimp, rotifers, mosquitoes and other species correspond to the resting stages of microorganisms and seeds of plants. Similarly, sponge gemmules, bryozoan statocysts, and other life-history stages constitute persistent resting stages and often are characteristically produced when the environment becomes harsh.

Therefore, considerable uncertainty exists in predicting the probability of survival of introduced microorganisms, plants and animals, except for the very few groups that have been intensively studied. Methods exist for assessing survival, but in view of the frequent lack of understanding of the reasons for elimination or endurance and of the contribution of abiotic and biotic factors to death, few generalizations are possible.

6.3.2 Proliferation

For most microorganisms, although the term 'growth' is often used, it actually refers to multiplication (i.e. increases in cell number or population size). Filamentous microorganisms (i.e. fungi, many algae and some bacteria) may increase in biomass without a concomitant increase in countable numbers; thus, they grow without necessarily multiplying. Some specialists refer to increase in numbers or biomass as 'colonization' or 'establishment'. To avoid semantic difficulties in this discussion, growth, multiplication and colonization are all considered proliferation. For plants and metazoans, 'population growth' and 'multiplication' both mean an increase in number, while 'colonization' refers to the establishment of propagules as well as a subsequent increase in numbers.

After an introduction, a propagule that does not diapause can multiply rapidly. For an unlimited environment, numerous models of local population growth predict the initial trajectory (Krebs 1985). For plants and metazoans, depending on available knowledge, these models may incorporate such features as age and sex structure. If interactions with other species do not intervene, local population growth can often be modelled quite accurately. However, the main stresses imposed by an introduced species would be experienced at a much broader geographical scale as the initial population spreads to form a metapopulation or completely separate populations. Metapopulation dynamics are just beginning to be modelled (e.g. Hanski & Gilpin 1991), and few empirical data are available to test the models. Moreover, the simple trajectories of single-species population growth are greatly complicated when interactions come into play. Prediction may be possible in such instances, but many data are needed and detailed observations and often experiments are necessary to establish which interactions are crucial in limiting a population's growth.

Proliferation of microorganisms is manifested by the outbreak and spread of diseases of humans, other animals, and plants as well as the development of phytoplankton blooms, the spoilage of foods, and the appearance of large bacterial numbers on roots that emerge early. The issue in risk assessment is not whether proliferation of any microorganism will occur but whether population density or biomass of a particular species will increase in a given environment. Currently, the data bearing on this issue are scarce.

Proliferation is essential for any species to have an environmental effect because the number of initial propagules is nearly always too small to be of ecological concern. Proliferation requires that nutrients that can be used by the particular organism be available. For most microorganisms, the limiting nutrient is carbon (C) because the supply of inorganic nutrients is generally not limiting. In those instances in which much readily degradable organic matter is present with a high C:N or C:P ratio, the limiting nutrient may be nitrogen or phosphorus. For algae and photosynthetic bacteria, the limiting nutrient is inorganic—usually nitrogen or phosphorus. The episodic increase in the supply of limiting nutrients, however, is not sufficient to result in an increase in the abundance of an individual species requiring that nutrient. Many coexisting species may use the same nutrients. Which of these species respond and proliferate cannot generally be predicted. The sole exception is an environment in which a truly unique nutrient exists, and, apart from a few organic materials that support a limited range of microorganisms, the truly unique nutrients are host organisms. The host for a parasite constitutes a unique nutrient, although the uniqueness requires that the parasite overcome many barriers to infection (i.e. skin, cutin, phagocytes, lignified tissues and antibodies).

From the viewpoint of exposure analysis, the magnitude of the increase in microbial population size or biomass can be enormous. A few bacterial cells can multiply to yield 10^6, 10^9, 10^{12}, or more cells, and the exposure rises in parallel with this increase in abundance. Similarly, a biomass of less than 1.0 ng may increase to 1 kg or more, as in some algal blooms. The ultimate exposure in the worst case is the maximum population size or biomass to which susceptible populations or communities are exposed.

Microbial proliferation rates vary enormously (Alexander 1985b). At one extreme are many soil bacteria; because of the slow turnover of organic matter in many soils, no more than a few cell divisions may occur each year. At the other extreme are bacteria, fungi, algae or protozoans for which a doubling in cell numbers or biomass may require less than 1 h. Unrestricted proliferation of a single cell with a doubling time of 1 h would yield a population 8 million times larger after 1 day, which is a phenomenal increase in exposure. A key word is 'unrestricted', because rarely do conditions permit unrestricted growth. Yet few of the restrictions on microbial proliferation are characterized, other than nutrient limitations and host responses. Even these two restrictions have qualifications, since an environment in which microorganisms are limited by a particular organic nutrient often receives episodic inputs of that nutrient. Similarly, a host species that typically exhibits resistance to parasites contains compromised or genetically more susceptible individuals, or it undergoes modifications because of environmental changes that permit rapid proliferation of a parasitic microorganism.

Maximum plant and metazoan reproductive rates also span an enormous range; generally the rate is inversely correlated with body size (Bonner 1965; Fenchel 1974). Reproductive rates are useful in assessing risk from an introduced species, but other factors often outweigh them. If a species enters an ecosystem during or soon after a disturbance that greatly reduces potentially competing populations, a high reproductive rate may increase the probability of initial establishment and even dominance. Weedy plants with high reproductive rates often dominate locally after a hurricane or trail-clearing in a forest. Maximum rates measured under ideal conditions, however, probably bear little relation to realized rates under most field conditions, except perhaps at the outset of an introduction. Also, many species that are common in nature and ecologically important have low maximum reproductive rates, while many species with high reproductive potential are uncommon except in sparsely distributed disturbed areas. Thus the effects on a system by a quickly reproducing introduced species may be temporary.

Studies of individual microorganisms in the laboratory provide hints of the traits that may result in rapid proliferations in nature, but no more than hints. Species with the fastest multiplication rates might be those that proliferate most readily in nature. As with plants and metazoans, however, this assumption is tenuous because many of the dominant species in nature do not multiply quickly. In fact, for plants and metazoans, many simple models of species interactions fail because they entail this assumption. The intrinsic rate of increase of a species in isolation under ideal conditions must be considered along with the limitations imposed by physical and chemical constraints in the environment, the concentration and turnover of limiting nutrients, grazing pressure, and the effects of other organisms. Given the lack of knowledge about the impact of these limitations on proliferation, it must be concluded that rate of multiplication and, indeed, whether a particular introduction will proliferate at all usually cannot be predicted. Among the few microorganisms that are exceptions are pathogens of humans and economically important plant and animal species, as well as microorganisms of environments that are so harsh (e.g. solar salt ponds and hot springs) that few species are able to tolerate the abiotic stresses.

The main attempt to avoid the intensive research needed to fill the lacunae described above, and thus to produce a shortcut to predicting survival and initial proliferation, is the hypothesis of 'biotic resistance' (Simberloff 1986b), which states that introduced species are less likely to survive in more diverse, complex communities because of the increased 'resistance' of various sorts from resident species. Among possible forces opposing the insertion of an introduced species into a community are competition, predation and parasitism. Sometimes the survival and effects of an introduced species seem obviously attributable to release from biological resistance. For example, the invasion of Lake Huron by the alewife (*Alosa pseudoharengus*), along with its subsequent proliferation and the numerous accompanying ecological effects, depended on the prior introduction of the sea lamprey (*Petromyzon marinus*), which greatly reduced populations of native fishes such as the lake trout (*Salvelinus namaycush*) and burbot (*Lota lota*) that would have competed with the alewife (Smith 1968).

However, it is unlikely that the simple criteria of size and complexity of the resident community will greatly help predict what constitutes a receptive environment. For example, one avatar of the biotic resistance hypothesis is that disturbed habitats are more easily invaded than undisturbed ones. Yet a close examination of records of invasions into various habitat types casts doubt on this view. The habitats, such as agricultural ones, that appear particularly invasible are generally anthropogenic ones — new groups of species in a highly human-modified physical setting. Disturbance *per se* does not seem to lead automatically to invasibility. For example, naturally disturbed habitats such as fire disclimax forests or high-energy beaches do not appear to have more introduced species than do other pristine habitats (Simberloff 1986b). Similarly, another version of the biotic resistance hypothesis is that prior invasion of a system by one or more species hinders successful subsequent invasion. In the systems in which this effect has been studied, however, it appears that the species introduced earlier were a priori more likely to survive independently of which other and how many other species were present, while later invaders were poorer colonists that would in any circumstances have had a low probability of survival (Keller 1984; Washburn 1984; Simberloff & Boecklen 1991). For microorganisms, except for some highly stressed or nutrient-poor environments, there is little evidence that species-poor communities are more easily invaded than species-rich communities.

A complication in assessing risk from a novel biological entity is that even if a habitat is adequate for survival and reproduction of a species, colonization has a stochastic element such that 'replicate' introductions do not have the same trajectory. For example, a few pairs of the Old World house sparrow (*Passer domesticus*) were released in Brooklyn, New York, in 1851, never to be seen again. A larger number of propagules were released there in 1852, with the same result. In 1853, at the same time of year and in the same place, a similar number of propagules were released. The population multiplied enormously, the species spread through North America, and it is now one of the most common birds on the continent (Long 1981), displacing native martins, swallows and wrens (Sharples 1982). Similar tales abound among insect introductions for biological control. It is often assumed that such differences in the outcomes of 'replicate' releases were caused by genetic differences among the propagules, but such differences are probably unmeasurable for most propagules of potential introduced species (Simberloff 1985). Thus, at least some fraction of the apparent stochasticity of biological introductions will always be present.

6.3.3 Dispersal

Once a species has survived at its site of introduction, its potential spread must be assessed. Many introduced species are mobile, and their dispersal often does not follow the trajectories associated with the movement of chemical pollutants. The probability of dispersal from the point of introduction is critical to risk assessment. The species' impact at the original location may be negligible. If it is transported to a more hospitable site, however, it may cause major harm. Even if its impact on the environment of the initial introduction site is substantial, this could be highly localized unless the species spread to new areas. Moreover, the issue of dispersal is not merely one of physical dissemination. Because the propagule must reach the new locale alive, dispersal must be assessed together with an evaluation of factors that could kill the propagule during dispersal.

Organisms disperse by many means (Mackenzie *et al.* 1985; Upper & Hirano 1991). Microorganisms use one or more of several modes: (i) through the air; (ii) in association with currents or by mixing in streams, rivers, lakes or marine waters; (iii) over the soil surface with runoff after precipitation; (iv) through the soil with vertical movement of water; (v) through groundwater in aquifers; (vi) by splashes or raindrops falling on foliage or the soil surface; (vii) in connection with animal movement; and (viii) as a consequence of human activity or implements. In addition, microorganisms can move by ballistic discharges as well as by hyphal growth by fungi, phototaxis by algae in surface waters, or motility by bacteria. The distances achieved by such means, however, rarely exceed a few centimetres—a few metres at most—and thus such movement is unlikely to be important in risk assessment.

Aerial dispersal is often the chief or only dispersal mode of a microorganism (Upper & Hirano 1991). Individuals may be lifted into the air when winds dislodge propagules from plant foliage; by ballistic discharges that allow fungal spores to move away from the plant or soil surface with small particles of soil carried aloft by the wind; as aerosols from surface irrigation waters as well as from sewage treatment plants; or as dust from farming, construction or other human activities. Aerial dispersal depends on the nature of aerial dissemination of particles, the factors that move the microbial propagules into the air, and the death rate of propagules during such transport.

Many microorganisms require water to move (Alexander 1971). Such movement may be tied to lateral currents in both freshwater and marine environments, vertical mixing in many bodies of water, or the lateral transport of water and suspended soil particles following rainfall or snowmelt. The extent of movement depends on the physical transport of the water, the factors that place the organisms in the path of flow, and survival of the organisms as they move. Although bacteria and viruses also pass through soil with percolating waters, few propagules move far because they are sorbed to soil surfaces or retained by the physical filtration associated with small pores in soil. None the less, many microorganisms can enter the underlying aquifer by passing through channels. While lateral dissemination of bacteria and viruses may occur with the moving stream of groundwater, distances traversed are short.

Biological vectors often move microorganisms. The vectors may be birds, large terrestrial animals, insects, rodents, fishes, earthworms, growing roots and probably zooplankton. The number of microbial propagules borne by such vectors may be small, but the dissemination is often remarkably efficient because the vector may carry the propagule unerringly to a suitable habitat, as is common with insect, bird and rodent transmission.

Metazoans and plants use most of the dispersal means of microorganisms. They are transported passively by wind and water, carried phoretically by animals, and are often moved by human activities. Movement through soil is relatively unimportant and quite local, in any event. Even passive movement by animals and plants is often initiated, just as in microorganisms, by a behaviour or process that places the propagule in the vicinity of the transport agent (e.g. plants explosively release their seeds to wind and water currents, spiderlings assume postures and spin silk threads that facilitate ballooning). Virtually all animal and plant species have characteristic lifecycle stages during which dispersal is likely, and these stages are often associated with behaviour or structures that enhance dispersal (Johnson 1969).

Many animals, however, have more active, sustained behaviours that enhance the probability of great dispersal. Some are migratory, for example, and others have innate behaviour that causes them to travel far from the natal area. Further, many animals and some plants time their movements to increase the likelihood of dispersing effectively and terminating the dispersal in a habitat suitable for existence and reproduction.

Many factors governing the transport of inanimate materials, especially particles, would likewise govern the transport of microorganisms and small plants and metazoans. Indeed, much of the modelling of such dispersal is based on particle transport or related models, such as smoke-stack diffusion. A critical difference, however, is that living organisms can die (or viruses can lose infectivity). Passive dispersal models that do not account for the decline of viability or infectivity will overestimate the distance likely to be dispersed or the number of viable propagules that will arrive at a site. Unfortunately, there are not substantial data on death during dispersal in various media.

Distances traversed by microorganisms, plants and metazoans range from a few centimetres to thousands of kilometres. In particular, active or passive aerial dispersal (including transport by birds) can move organisms vast distances. The literature on agricultural and forest entomology, public health, veterinary medicine and plant pathology provides considerable data on distances and means of spread.

Faced with the bewildering array of dispersal means and their varying efficiencies, Pielou (1979) distinguished between two distinct forms of spread. In the first, which she called diffusion (cf. Hastings 1996), a species' spread approximates increasing concentric circles for which the circumferences become progressively more warped. The Colorado potato beetle (*Leptinotarsa decemlineata*) in Europe (Nowak 1975) and the Japanese beetle (*Popillia japonica*) in the USA (Elton 1958) are good examples. The rate at which such species' ranges expand is a function of their biology. Various models, beginning with simple diffusion models (e.g. Anderson & May 1991; Strauss & Levin 1991; Upper & Hirano 1991) describe the process. These typically consider the initial density of organisms and rate and efficacy of the means of dissemination. The larger the source and the hardier the propagule, the greater the probability of successful dispersal. The most extensively developed models are epidemiological, for microbial disease agents. Such models typically predict where microorganisms will go and how many will arrive alive. Those that relate to plant pathogens are particularly useful for studying aerial dispersal of spores that lead to disease in economic crops, while those for human infectious agents primarily concern host-to-host transmission and movements of individual affected hosts.

While some recent diffusion models (e.g. Hengeveld 1989) seem to simulate some observed spreads of introduced species strikingly closely, it is too early to tell if their predictions will continue to be accurate. The warping of the range circumference as diffusion proceeds is probably caused by heterogeneities in the physical environment (Simberloff 1986b), and it is quite possible that, if one knew enough about the habitat requirements of a particular species, a diffusion model could be modified to reflect these heterogeneities. For example, the Atlantic Ocean and Chesapeake Bay seem to have prevented the Japanese beetle from spreading evenly in all directions from its point of introduction in New Jersey. The unsuitability of ocean as a habitat for a terrestrial beetle is easy to deduce, but the influences of other habitat gradients on diffusion dispersal will be more subtle.

Cellular automata are a more recent kind of model that, like diffusion models, depict the local, gradual spread of introduced species (Hastings 1996). In these computer-intensive models, each spatial location is in one of a number of states (e.g. high or low population density). At each time increment, the state of each location changes according to the states of nearby locations and rules based on the biology of the species. A stochastic version of a cellular automaton, a 'Richardson model', allows for chance events. Cellular automata are so recent that one cannot yet say if their predictions are more accurate than those of modern diffusion models.

Models for aerial dispersal of microorganisms are probably useful in risk assessment of non-pathogens. The paucity of information on death, however, will probably affect their utility because different species' death rates differ greatly owing to radiation, desiccation and temperature stresses encountered during aerial transmission. Models for waterborne dispersal have received less attention for microorganisms, plants and metazoans. Among aquatic introduced plants and animals, the spread of many of the most problematic species (e.g. white amur (*Ctenopharyngodon idella*), European carp (*Cyprinus carpio*), zebra mussel, purple loosestrife, water hyacinth (*Eichhornia crassipes*), and Eurasian watermilfoil (*Myriophyllum spicatum*)) has not been modelled in more than a cursory fashion despite extensive empirical data. It may seem reasonable to use models for the transport of chemicals in water to study microbial or small plant or metazoan dissemination; however, the attachment of microorganisms to particles, the likely different death rates of different species, and differences in death rates between sorbed and free cells suggest that transport models of non-living materials might not apply well to organisms. Thus, predicting even the gradual, diffusive spread of living organisms is difficult, despite a number of possible models.

By contrast to more or less regular diffusion, some introduced species have spread irregularly from the outset or after a brief period of circular range expansion. Often several foci arise simultaneously by long-distance 'jumps' (Pielou 1979), each subsequently serving as a base for slower circular growth or yet another long-distance leap. The aphid *Hydaphis tatarica* was restricted to a small area of southern Russia, apparently by the limited range of its host, Tatarian honeysuckle (*Lonicera tatarica*). It was in the process of spreading gradually westward and had just reached the Moscow region when it was described in 1935. As the host honeysuckle was planted as an ornamental throughout much of central and southern Europe, however, the aphid's range increased greatly and irregularly, often to areas not contiguous with the original range.

Predicting jump-dispersal seems a far more formidable task than modelling diffusion dispersal. Probably many more propagules actually undergo jump-dispersal than are recorded. Most of them never establish ongoing populations because they either fall in unsuitable habitat or fail to increase for the various reasons outlined above. Yet it is clear that many suitable sites are not reached by adequate propagules of the myriad of introduced species that have survived and often increased dramatically in many areas of the globe. Because physical forces such as upper air currents can sometimes be identified as likely agents of jump-dispersal, at least direction may be predictable. For other modes, such as transport of seeds on birds' feet (Simpson 1952), generating useful probabilities would be much more difficult. Nevertheless, long-distance transport by birds was certainly important in establishing ranges of many invertebrates and freshwater algae, as well as flowering plants (Carlquist 1974; Pielou 1979), and so it must be a common event.

Human movement of living organisms, both deliberate and inadvertent, usually constitutes jump-dispersal. Recreational vehicles have transported gypsy moth (*Lymantria dispar*) egg masses on land and zebra mussel juveniles in freshwater. Innumerable introductions of terrestrial, freshwater and marine species have occurred in ship's ballast. The ornamental plant industry spread many important species, such as purple loosestrife, while individuals seeking attractive flowers dispersed others (e.g. water hyacinth (*Eichhornia crassipes*)), and forestry activities yet others (e.g. *Melaleuca quinquenervia*). Fish and game departments as well as individual fishermen have completely changed the ichthyofaunas of many areas, including the American West, where in several drainages most of the native species are threatened by invaders (Moyle 1986). Many species have been introduced through deliberate or inadvertent release of pets. For certain taxa, introduction societies have changed entire faunas. In the Hawaiian islands, at least 70 species of passeriform and columbiform birds have been introduced; many survived and completely dominate lowland areas, while native birds are now almost restricted to upland native forests (Simberloff & Boecklen 1991). Where an introduction is deliberate, an initial jump-dispersal can be predicted. For many inadvertent introductions, detailed prediction will be impossible, although heavy use of certain transportation routes will likely generate correspondingly large numbers of jump-dispersals.

Unplanned introductions: dispersal en masse

To this point, we have discussed assessing risks for specific species either already introduced or whose introduction is proposed. Certain activities, however, pose a different sort of risk in that they may lead to the inadvertent introduction of many species whose identities can be predicted only partly at best. For example, in the USA alone, 2.5 million gallons of ballast water, originating all over the globe, are released every hour (Carlton *et al.* 1994). Although this practice was originally thought to be innocuous regarding the introduction of species relative to those carried by the soil previously used as ballast, it has become apparent that ballast water is completely changing marine biogeography (Carlton & Geller 1993) and that many major effects are likely. For example, several scores of species introduced in this fashion can be found together fouling floating objects in San Francisco Bay and Los Angeles Harbor. Detecting even major ecological effects is often difficult, however, and predicting which organisms will cause what effect is extremely difficult (sections 6.4 and 6.6). None the less, as over 3000 marine species are in transit in the ballast of ships on the oceans on any given day (Carlton & Geller 1993) and at least 367 species of plants and animals have been found in ballast water arriving in Oregon just from Japan, one can be sure that ballast water is a major pathway for introductions and generates major effects.

Based on information about marine shipping routes, where ballast-borne species are most likely to come from and go to may be predictable, although this sort of analysis has been conducted only for the Pacific Ocean. Once an organism arrives in a new biogeographical region by such means, the sorts of models discussed in section 6.3.3 might be used to predict its further spread; however, such initial long-distance movement of an organism clearly would involve an enormous component of chance.

Activities that can introduce species *en masse* greatly complicate risk assessment, as the risk of inadvertent dispersal to various sites must be concatenated with risks of various effects (section 6.4) for each potential invader. None the less, an assessment of risk posed by an activity like the pet and ornamental plant industries or a pathway for dispersal *en masse* may be a worthwhile exercise. The US Department of Agriculture *Generic Non-Indigenous Pest Risk Assessment Process* (Orr *et al.* 1993) encompasses just such phenomena. For example, the risks of introducing forest pests in unprocessed logs imported into the USA were estimated, with the subsequent decision to permit the importation of Monterey pine (*Pinus radiata*) and two other tree species from Chile (USDA 1993) but to bar larch (*Larix* spp.) from Siberia and the Soviet Far East (USDA 1991). For each assessment, first the species that might be transported in this way were identified and an estimate of the likelihood of such movement was developed (i.e. the dispersal part of the assessment). Then, for each species judged potentially harmful, the assessment team estimated the likelihood of various effects (see section 6.4) as well as the potential costs if they were to occur. In addition to alerting the public and appropriate authorities to the potential problems, the exercise formalized what is known about the risks and indicated key uncertainties.

In general, risk assessments for unplanned introductions tend to differ from those for planned introductions in three ways (Ruesink *et al.* 1995). First, when an introduction is planned, species tend to be selected specifically so that they will be able to survive in the target environment, and so the probability of initial survival is far lower for unplanned introductions. Thus, assessment procedures for planned introductions tend to focus more on the likelihood of survival and effects in unintended places within the general target region, while procedures for assessment of unplanned introductions emphasize the probabilities that various species traverse a certain path and survive at all upon arrival. Second, the number of individuals in a propagule is likely, in a planned introduction, to suffice for initial survival and increase, while propagules may be too small and/or infrequent for unplanned introductions to get past this initial step (cf. section 6.3.1). Third, risk assessment for unplanned introductions emphasizes procedures for detecting propagules, a factor that need not be considered for planned introductions. Ruesink *et al.* (1995) summarize the operating procedures of various USA agencies for planned and unplanned introductions.

6.4 EFFECTS

An introduced species can affect an ecosystem in numerous ways. Often, deliberate introductions of game animals are said to 'fill an empty niche', and inadvertently introduced species that establish especially large populations are also occasionally said to have 'occupied a vacant niche'. It is notoriously difficult, however, to show that a niche truly is empty (Herbold & Moyle 1986), and in one sense it seems impossible that there is such an entity. Because ultimately resources are metabolized in some way, if only by bacteria, it is difficult to imagine that a surviving introduced species would not at least affect the topology of energy-flow webs and nutrient cycles (Simberloff 1991). In any event, even a species that fills a classically empty niche can have an enormous ecosystem impact. In 1788, the first English settlers to Australia brought seven cattle, seven horses, and 44 sheep. By 1974, cattle alone numbered 30 million and were producing 300 million cowpats daily, which are not removed by native dung beetles and so modify existing plant communities, provide breeding habitat for numerous insect species, and in some areas they dry and blanket the ground. Three beetle species introduced in 1967 filled this 'empty niche' by establishing over large areas and removing vast numbers of cowpats, working the dung into the soil. Their full impact, yet to be calculated, must be enormous.

Perhaps, rather than using terminology related to niches, it would be useful simply to consider whether newly introduced species with radically different ways of 'making a living' than any resident species are particularly likely to have a major impact. For example, the destruction of ground-nesting endemic birds by predatory mammals introduced to many oceanic islands could easily have been predicted, because these islands had supported no mammals other than bats and no other species 'substituting' for predatory mammals so as to affect the populations of these birds (Simberloff 1994). Perhaps the profound impact of the zebra mussel in the USA (Garton *et al.* 1993) can be at least partially explained by the fact that this species has two traits that are extremely rare among freshwater molluscs in North America: planktotrophic larvae and byssal threads for attachment (L. Johnson, pers. comm.).

The sense that the major effects of some invasions might have been predicted from careful consideration of the traits of the invader has led to persistent attempts to predict the effects of biological introductions by compiling lists of traits that predispose species to invasiveness. Although the results of such attempts, which were popular in the 1960s (e.g. several papers in Baker & Stebbins 1965), are still occasionally published, this approach has largely been discarded (Mack 1996). One problem is that such lists, or at least readers who sought to use them, often confounded the traits favouring effective dispersal, traits predisposing toward establishment, and traits conducive to major impact on the resident community once established.

However, the major reason why this approach is no longer popular is that there were always exceptions—plants with 'weedy' traits that are not weeds, and plants lacking weedy traits that are problematic weeds (Mack 1996). Such exceptions abound because the kinds of interactions that an introduced species can potentially have with the resident community are so numerous. Moreover, many of these rest so heavily on idiosyncrasies of one or more species that they would never have been considered with respect to invasion effects. Thus it is not feasible with a simple listing to make more than a first stab at predicting invasion results (references in Simberloff 1989). An analogous approach in public health would be to attempt to predict the nature and effects of a pathogen by studying its structure and molecular make-up without looking at infected individuals.

The biotic resistance hypothesis, described in section 6.3.2 with respect to establishment and proliferation of an invading organism, has been equally applied to attempts to predict effects, at least partially, in response to the numerous problems with the list approach. However, although laudable in that it focuses on the recipient community, it too has proven to be not especially predictive, doubtless because the single trait of species richness is simply too crude a characterization of a community (Simberloff 1986b, 1989).

In general, attempts to predict the ecological consequences of introduced species are confounded by the enormous number of possible effects, whether on particular species or on entire communities. An introduced species can have a major impact on a single population (e.g. if it is a host-specific pathogen or parasite), or it may alter one or more processes important to ecosystem function. As noted in section 6.1, expectations of how important such alterations will be rest heavily on views about the significance of communities and ecosystems as units of biological organization. Although biologists and environmental scientists often have favourite species or processes that they feel should be examined in any assessment of ecological effects of introduced species, many of the favoured species or processes do not seem always to have a role so critical as to warrant concern. Unfortunately, no consensus has been reached on which species or processes should be singled out for assessment, nor has research been designed to facilitate such decisions. It is almost platitudinous to say that the perceived importance of the ecological consequences of an introduced species depends on the perceived importance of the species or processes affected. Yet the problem is often lack of agreement on which are the important species or processes.

6.4.1 Types of effects

Effects of introduced species can be divided into direct and indirect effects. A species can affect another directly by, for example, killing it, eating it or displacing it. Or the effect can be indirect—for example, a species can modify the habitat of another species or reduce its prey. Although such indirect effects are often subtle and difficult to elucidate, they are extremely important (Strauss 1991).

An introduced species can affect another as a pathogen or parasite. Myxoma virus introduced from South America into Australia in 1951 initially killed over 99% of the huge European rabbit population, although subsequent evolution of both virus and rabbit has allowed some recovery (Krebs 1985; Williamson 1992). Frequently, various epidemiological models, often of the diffusion type, can predict more or less accurately the geography and time course of initial spread of a pathogen (Anderson & May 1982; Dobson & May 1986). The evolution of benignity and resistance, however, can severely limit the longer-term predictive value of these models in terms of effects. Further, jump-dispersal of infected hosts can be a more important mode of spread than diffusion.

The impact of an introduced pathogen can range from insignificant to devastating on its host plants or animals. A particular microorganism may multiply on or within a host plant and cause little harm, or it may cause overt damage to a small or large percentage of individuals of a susceptible host species. If that host fills a 'keystone' role (Simberloff 1991) in the community, this microorganism could disturb the entire community. Such major disruptions are well documented for several fungi (e.g. those causing chestnut blight and Dutch elm disease). The diseases caused by these fungi have had major impacts on community structure and ecosystem function, even though each acted on a single host species. The affected host was a critical component of plant communities, and the effects rippled through populations that interacted with the hosts. For example, American chestnut (*Castanea dentata*) comprised 25% of the trees of many forests (Elton 1958), more than 40% of overstory trees and 53% of basal area (Krebs 1985). Several insect species that are host-specific to American chestnut are either endangered or extinct because of the destruction of their host by chestnut blight (Opler 1978). Subsequent effects of the loss of these species on their predators and parasites are unstudied. Oak wilt disease (*Ceratocystis fagacearum*) has increased on many native oak species because populations of red oak (*Quercus rubra*), which is particularly susceptible, increased greatly when the chestnut disappeared (Quimby 1982). It seems inconceivable that the loss of such a dominant tree could not have affected many other ecosystem features, but no substantial studies have been performed. Whether the consequences of the disappearance of chestnut qualify as important effects depends on how significance is assessed.

Such impacts will be especially devastating if the host population has had no contact with the pathogen or parasite and thus has no resistance. The literature of human and veterinary medicine attests to the marked decline in host populations following exposure to novel viruses or bacteria. The impact is also affected by the genetic heterogeneity of the potential host population, its density, and the stage of its lifecycle at which exposure to the pathogen occurs. The severity of disease is greater if plants in a community are not diverse and if genotypes in a population are few, as in the devastation caused by fungi in agricultural monocultures. Furthermore, if the population of susceptible individuals is dense, the effect of a parasitic microorganism is likely to be far greater than if the host population is sparse (Pimentel 1985; Levin & Harwell 1986). The stage in the lifecycle of host plants or the age of animal hosts may also alter pathogen impact.

Many dramatic examples illustrate damage to natural ecosystems by introduced pathogenic microorganisms. In less than 50 years following its introduction from Asian nursery material, the fungus *Endothia parasitica* became established on 91 million ha of USA forestland and almost totally eliminated the American chestnut, as noted above. Introduction of the fungus *Ophiostoma ulmi* had a devastating impact on forests containing large populations of American elms. *Phytophthora cinnamomi*, a fungus pathogenic to nearly a thousand plant species, has similarly affected large wooded areas following its introduction into new regions of Australia and the USA (von Broembsen 1989). The introduction of the rinderpest virus into parts of Africa resulted in massive death of cattle and wild ungulates (Holmes 1982) with subsequent ripple effects on other components of the community, such as vegetation and predators (Barbault 1992). Similar effects of myxoma virus and avian disease organisms are discussed elsewhere in this section.

Competition generally is notoriously difficult to document in the field, and the decline of native species said to be due to competition from introduced species often can be ascribed also to other forces. Without experiment, it is impossible to establish the causes with certainty, For example, the decline of the otter (*Lutra lutra*) in Britain and Sweden in the 1950s was long believed to be caused by competition for space with the introduced American mink (*Mustela vison*), which spread rapidly at approximately this time. More recently, the decline of the otter has been attributed to organochlorine pesticides (Chanin & Jefferies 1978).

Nevertheless, sometimes observations convincingly implicate competition from an introduced species in the decline of a native species. For example, the introduction of barn owls (*Tyto alba*) to the Seychelles coincided with the decline of the endemic Seychelles kestrel (*Falco araea*), probably because of nest site competition (Penny 1974). The limiting resource does not even have to be used in the same way for the introduced species to harm the native species. For example, because nest boxes on poles for eastern bluebirds (*Sialia sialis*) in Bermuda are used as perches by the introduced great kiskadee (*Pitangus sulphuratus*), the bluebirds cannot nest in them (Samuel 1975). In addition to such 'resource competition', in which a species detrimentally affects another by pre-empting a limiting resource, an introduced species can engage in 'interference competition' by aggressive behaviour. On the Japanese island of Oshima, the introduced grey-bellied squirrel (*Callosciurus caniceps*) chases the native oriental white-eye (*Zosterops palpebrosa*) from camellia flowers. This interference affects both the bird and the plant because the white-eye pollinates it (de Vos *et al.* 1956).

The prevalence of interspecific competition in nature and the nature of the evidence required to demonstrate it have been controversial subjects in ecology for over a decade (e.g. Lewin 1983). So it is not surprising that it is difficult to estimate quantitatively the fraction of introductions that result in interspecific competition and the importance of that competition. Ebenhard (1988), admitting that the evidence is often sketchy, found reports of potential competition in a selection of the literature for 9% of mammal introductions and 18% of bird introductions. However, reports may tend to be published where there is a possibility of competition, so these figures may exaggerate the prevalence of competition among introductions as a whole.

An introduced species can eat a native animal or plant. Bird species all over the world have been eliminated by introduced rats, mustelids, and feral dogs, cats and pigs (King 1984; Atkinson 1989). The most famous case is that of the lighthouse keeper's cat, which arrived on Stephen Island, New Zealand, in 1894 and eliminated the entire population of the Stephen Island wren (*Xenicus lyalli*) within 1 year (Greenway 1967).

Numerous introductions of insect predators and parasitoids for biological control of insect pests and weeds associated with agriculture have lowered the populations of the latter to insignificant levels (Krebs 1985). Although such an approach is widely touted as 'environmentally friendly', particularly in comparison to chemical control, in many cases introduced biological control agents harmed or even eliminated non-target native species (Howarth 1991; Simberloff & Stiling 1996). Sometimes the resulting impact is far from the point of introduction. For example, the cactus moth (*Cactoblastis cactorum*) was introduced to the island of Nevis in the West Indies in 1957 to control various species of prickly pear (*Opuntia*). It island-hopped, probably by its own flight, throughout the West Indies, reaching the Florida Keys by 1990 and infecting and threatening the entire remaining wild population of the endemic semaphore cactus (*Opuntia spinosissima*), a candidate endangered species (Simberloff 1992). Because of the mobility of living organisms, it will be hard to establish a reasonable scope for assessing the risk of potential ecological effects. Clearly, the host range of a parasite or predator is a key component in this assessment.

Given the difficulty of determining the population controls of species of plants and animals that do not stand out in any way (the vast majority), and the remarkable sets of circumstances that revealed the existence of a few local extinctions caused by biological control introductions, there is every reason to think that predators and parasitoids introduced for biological control have caused numerous local extinctions and possibly global extinctions as well (Howarth 1991; Simberloff 1992). Similarly, it is not possible to estimate accurately how many previous introductions have inimically affected native species by predation. Ebenhard (1988) found the possibility of such an effect for about one-third of 290 mammal species introduced world-wide, but this can be only the roughest estimate. Moreover, the effects of mammals are far more likely to be noticed than those of most other taxa.

An introduced plant can affect native plants by allelopathy. For example, the introduced African ice plant (*Mesembryanthemum crystallinum*) devastated native vegetation in California (Vivrette & Muller 1977; Macdonald *et al.* 1989). It is an annual that accumulates salt throughout its life. When it dies, rain and fog leach the salt into the soil, where it suppresses growth and germination of native species. An effect similar to allelopathy can be induced by introduced nitrogen-fixing plants. For example, the Atlantic shrub *Myrica faya* was introduced to young, nitrogen-poor volcanic regions of the island of Hawaii where there are no native nitrogen-fixers. The shrub forms nearly monospecific stands, to the detriment of native plants adapted to nitrogen-poor soils and to the benefit of other exotic plants (Vitousek 1986).

Introduced species can be vectors or reservoirs of disease to which they are resistant but native species are susceptible. The major reason that so many native bird species of Hawaii have gone extinct and that so many of the remainder are threatened is habitat destruction (discussed below). Key contributors, however, are avian pox and malaria vectored by birds introduced from Asia in the late nineteenth and twentieth centuries (van Riper *et al.* 1986). These diseases may prevent native species from colonizing otherwise suitable upland forest in which densities of introduced species are high. An introduced species may even serve as a reservoir for a disease that was not introduced with it. In Puerto Rico, the introduced small Indian mongoose (*Herpestes auropunctatus*) carries rabies but did not introduce the disease (Silverstein & Silverstein 1974). It is well known that human diseases brought by Europeans to North and South America, Australia and New Zealand, and various small islands around the world, devastated many native peoples (e.g. Crosby

1986). There is every reason to think that other animals were equally affected by species introduced by Europeans, and the same may be true of plants as well. Establishing that a native species is limited in range or numbers by a disease is extremely difficult, and determining the origin and reservoir of that disease is even more so.

Probably the most important ecological effect of introduced species is modification of the habitat, because this can affect entire communities of species and frequently whole ecosystems. In the eighteenth and nineteenth centuries, much of the northeastern North American coast consisted of mud flats and salt marshes, not the current rocky beaches. This change was wrought by the European periwinkle snail (*Littorina littorea*) (Bertness 1984; Dean 1988), which was introduced to Nova Scotia in about 1840 and has slowly expanded its range southward. The periwinkle eats algae on rocks and rhizomes of marsh grasses. When it is experimentally excluded, algae and then mud quickly cover the rocks, after which grasses invade the mud. Thus the physical nature of the entire intertidal zone of a large region has changed, with consequent change of the entire community. Introduced feral pigs (*Sus scrofa*) have similarly modified entire ecosystems by rooting and selectively feeding on plant species with starchy bulbs, tubers and rhizomes (references in Simberloff 1991). Further, they have greatly modified soil characteristics by thinning the forest litter, mixing organic and mineral layers, and creating bare ground. In turn, these changes increased concentrations of nitrogen and potassium in soil solution and accelerated the leaching of many minerals (Singer *et al.* 1984). In some areas, the changes have greatly aided the invasion of exotic plants (Loope & Scowcroft 1985).

An introduced species can devastate an entire ecological community through habitat modification even if it does not reproduce. For example, grass carp are widely used to control rooted aquatic plants, often with great success from the standpoint of the target species alone (Ashton & Mitchell 1989). However, fish introductions for this purpose always have unintended side-effects (Courtenay & Williams 1992). Grass carp are often so voracious that they destroy a large fraction of the vegetation, even non-target species, thereby affecting the fishes and invertebrates that inhabit this vegetation. These effects are sufficiently detrimental that environmental managers generally do not want grass carp permanently established, leading to the use of sterile triploids (Thorgaard & Allen 1992). However, no experimental evidence is available on the recovery of communities devastated by grass carp, and the grazing is often so severe that one might expect extremely persistent effects even after disappearance of the carp. It is also noteworthy that supposedly 'sterile' triploid fishes may not be completely precluded from reproducing (Thorgaard & Allen 1992).

An introduced plant can modify an entire plant community by various means. We discuss above such impacts by a nitrogen-fixer in nitrogen-poor soil and by allelopathy. Fire enhancement can have equally great effects. In 1906, *Melaleuca quinquenervia* was introduced from Australia to south Florida, where it has displaced less fire-resistant species, such as cypress, over thousands of hectares (Ewel 1986). Introduced plants similarly enhance fire in Hawaii (Vitousek 1986). Because the fire regime can determine the composition of a plant community, and the plant community in turn constitutes the habitat for the animal community, such species that effect great changes in the fire regime can have an enormous impact.

Introduced plant species can modify the habitat by constituting new forests. Along rivers in the arid southwestern USA, salt-cedar (*Tamarix* spp.) and Russian olive (*Eleagnus angustifolia*) have had far-reaching effects because they form new forests (Knopf & Olson 1984; Vitousek 1986). Salt-cedars were introduced in the nineteenth century, and their deep roots allow them to maintain themselves in situations where other plants cannot, such as on the floodplain of the Colorado River. These forests provide habitat for animal species. Also, salt-cedars transpire so heavily around desert springs that they have replaced entire marsh communities with a monospecific forest. The effects of Russian olive, found further north, are similar. Similarly, although mangroves cover intertidal soft substrates in sheltered tropical bays and estuaries in most of the world, they were unknown in Hawaii, where such sites were unforested. In 1902, seedlings of red mangrove (*Rhizophora mangle*) were planted on Molokai.

This species has since spread to other islands by natural dispersal, and perhaps by deliberate plantings, and now forms forests up to 20 m high. Although the consequences of this new habitat have not been studied, they must be enormous. For example, healthy mangrove swamps drop about 4000 kg of leaves annually per hectare, and the roots form critical habitat for fishes and shrimp (Carey 1982) and accumulate sediment (Holdridge 1940).

Finally, introduced species can hybridize with native species, potentially modifying the native species in undesirable ways or even changing it so much that it would not be regarded as the same species (Rhymer & Simberloff 1996). Similar threats from GMOs are cited by Seidler *et al.* (Chapter 5). For introduced species, several examples are already known. Introgression from cultivated sorghum (*Sorghum vulgare*) has rendered shattercane (*Sorghum bicolor*) and Johnson grass (*S. halepense*) more serious pests (Harlan 1982). Numerous plants endemic to islands suffer hybridization by pollen influx from related introduced species (Ellstrand 1992). Introduction of the fish *Gambusia affinis* and *G. holbrooki* for mosquito control has led to their hybridization with a restricted endemic (*G. heterochir*) and threatens the existence of the latter as a separate species (Courtenay & Meffe 1989). In the Seychelles, the local subspecies of the Madagascar turtle dove (*Streptopelia picturata rostrata*) has been destroyed by hybridization with the nominate subspecies (*S.p. picturata*) (Penny 1974).

For plants and metazoans, the circumstances under which such hybridization is most likely can be predicted. Of course, subspecies can exchange genes. Species within groups with mainly behavioural means of reproductive isolation are probably more likely to be subject to such a process, as the hybrids are at least fertile and prior selection for behavioural avoidance would not have occurred (Ebenhard 1988). Wild congeners of feral domestic species are likely to hybridize; for example, few 'pure' polecats (*Mustela putorius*) probably remain in Britain because of widespread release of the domestic ferret, *M. 'furo'* (Lynch 1995). However, it would be difficult to generate quantitative predictions about the risk of such hybridization or its effects on the species concerned. Many closely related species do not exchange genes, either because of a postzygotic isolating mechanism, such as chromosomal incompatibility, or because of a subtle prezygotic isolating mechanism. Moreover, some species that are not closely related have hybridized in nature, particularly among plants. The particular genes exchanged can have such varied and unpredictable effects as to defy generalization.

Even in the absence of introgression, hybridization can threaten a native species. The European mink (*Mustela lutreola*) is rapidly declining in most of its range. One key reason is hybridization with the more numerous introduced American mink (*M. vison*). The larger American mink males mate with European mink females, but the embryos resorb and no gene flow occurs. However, these females do not reproduce that year, while American mink females are mated by conspecific males (Rozhnov 1993).

'Lateral transfer' of genes is also possible for microorganisms (Stotzky *et al.* 1991; see also Chapter 5). It may occur even if the original introduced population dies out. Thus, the possibility of hazard remains in the form of genetic information. The species now bearing genetic information coding for deleterious traits may survive, it may be widely dispersed, and it may multiply. Indeed, the recipient microorganism may be more ecologically fit than the originally introduced individual and thus constitute a higher risk. Most of the information on gene transfer comes from laboratory studies using highly artificial test conditions or environmental samples that do not mimic nature well. Many of these studies suggest that bacteria are engaging in a wild gene-exchange orgy. Some of the extrapolation by microbial geneticists is excessive, but it is likely that gene exchange does occur in nature. However, its frequency, particularly of genes potentially harmful to the environment, is unknown (references in Regal 1986). The movement of genes into a population need not require introduction of the entire living organism. Rather, components like pollen and plasmids can serve this function.

Most of the effects that we have discussed are direct. As noted above, myriad indirect effects are possible and could be difficult to recognize, much less to predict. For example, the large blue butterfly (*Maculina arion*) was inadvertently extinguished in Britain by an unregulated biological control introduction (Ratcliffe 1979). Caterpillars of the blue must develop in underground nests of the ant *Myrmica sabuleti*. The ant, in turn, cannot nest in overgrown areas. Changing land-use patterns and reduced livestock grazing left primarily rabbits maintaining the habitat. Then an *ad hoc* attempt to control rabbits with myxoma virus reduced their populations and, consequently, reduced ant populations to the point where the butterfly disappeared.

Since species can interact through shared prey or hosts, shared predators, parasites and pathogens, many types of habitat modification, and a variety of tritrophic interactions, the possible sorts of indirect effects are enormous, and there seems no way to do more than to list some obvious possibilities in each case. Certainly a quantitative estimate of the probability of various effects is currently impossible.

We have focused primarily on effects of introduced species on resident populations, with some examples of community-wide effects. This emphasis reflects the biases of the invasion literature. We have touched on some effects on ecosystem processes, such as nutrient cycles and fire regimes. As we noted in section 6.1, an abiding controversy in ecology is whether the functioning of ecosystems can be understood by analysis of its components or whether ecosystems are holistic entities that must be studied at the ecosystem level. Perhaps the attention paid to various levels of ecological organization by students of biological invasions reflects their stance toward this controversy. In any event, additional consideration of ecosystem-level consequences of introduced species is summarized by Ramakrishnan & Vitousek (1989).

6.4.2 Stochasticity, inherent unpredictability and dose–response

As noted in section 6.3.2, there is a stochastic element to whether a propagule survives and reproduces at a site. Similarly, there is variation in the effect that an introduced species has on a community once it has successfully dispersed. For example, different influenza invasions in the same human populations have different effects (Regal 1986). Some fraction of this variation may be explained by even minor genetic differences among the individuals in the different propagules. Data on such differences are not usually available.

Further, even aside from truly stochastic variation (e.g. meteorological conditions during the arrival of a group of propagules) and uncontrolled genetic variation, the dynamics of some very simple models of interactions among species are so 'chaotic' that the interactions appear to be random and certainly cannot be predicted with much precision (May 1987). Even with identical initial conditions, several alternative equilibrium states are predicted as possible outcomes of introduction of the same species to a patchy landscape (Drake *et al.* 1993). All these features of introduced species and their dynamics greatly complicate the ability to predict their effects, though they by no means prevent the assignment of great risk to certain species and lower risk to others.

For microorganisms, another source of variation is quite unpredictable at present. The relationship between the dose of a toxicant and the response of a test population is important for risk assessment of chemicals. A simple dose–response relationship is not evident for introduced microorganisms. It is evident, however, that large populations of some species will produce an effect, while small populations cause no demonstrable change. A simple test system might allow the establishment of a mathematical relationship between population size of a species and the response of some susceptible species, permitting, for example, an analogue of an LD_{50}. Despite the many parasites, pathogens and hosts that have been investigated, either such simple dose–response curves do not characterize microorganisms, or perhaps there are heterogeneities (e.g. genetic differences) among the propagules. Nevertheless, the outbreak of plant disease sometimes can be predicted from knowledge of the number of fungal propagules present at a site (Shrum 1978). Complicating the establishment of quantitative relationships between exposure and response is the fact that many potentially harmful propagules can be viable but inactive. The density of viable propagules is not necessarily the concentration that poses a threat, because the percentage of these organisms that are active or will be active in the near future depends on the particular environment.

The establishment of a microorganism in an environment or the creation of an undesirable effect often requires large populations, and a small introduction will not survive for long or will have no impact. In effect, a threshold exists for many microorganisms below which the population is not maintained and a deleterious impact is not observed. Although few microorganisms have been studied in this regard, the threshold varies greatly with the particular organism. The clearest evidence comes from studies of bacteria that infect humans: for ingested bacteria, 180 cells of *Shigella flexneri*, 10^5 cells of *Salmonella typhi*, but 10^8 cells of *Vibrio cholerae* and *Escherichia coli* are needed (Collins, cited by Levy 1986). Reasons for these differences are unknown. In some environments, they may result from predation that eliminates a small population but leaves sufficient survivors of a large population to cause an effect (Ramadan *et al.* 1990). In other words, whether the species has an impact boils down to whether it survives long enough in large enough numbers.

On the other hand, an organism that is frequently introduced into an environment in small numbers may cause an effect, whereas the same population size introduced infrequently will not. This difference may result from changes in the suitability of the environment and the propitiously timed proliferation of the small numbers. In short, there may be stochastic elements in this phenomenon similar to those already discussed with respect to survival.

6.5 DEFINING END-POINTS

End-points for some potential effects of introduced species are straightforward and do not differ in kind from those associated with some chemical pollutants and physical processes. For example, an introduced species that preys on a native species might be expected to reduce the prey population, and an appropriate assessment end-point would be this population decline. The measurement end-point would then be identical to the assessment end-point—one would monitor the population. If the assessment end-point were construed in this instance as some change in community structure or function, again the measurement end-point might well be population size, or it might be some nutrient or energy flow believed to be associated with the prey species. Many ecological effects discussed in section 6.4.1 similarly would not present unique problems in defining end-points.

The evolution of both the introduced and the native species, however, and the possibility of gene exchange (discussed in sections 6.2 and 6.4.1) complicate greatly the definition of justifiable end-points, as with GMOs (see Chapter 5). The evolution of resistance, virulence and benignity can be very quick, particularly for species with short lifecycles. At other times, however, such processes can be lengthy, with an end effect that is important none the less. Even the co-evolution in the myxoma-rabbit system in Australia, which seems rapid from a biological control standpoint, would be viewed as slow in the context of a typical risk assessment. There were about six epizootics in the quarter century after viral introduction, and the rabbit mortality rate did not fall substantially until the third one (Krebs 1985).

The acquisition of novel hosts by introduced insects will bedevil risk analyses. For example, the American fruit fly *Rhagoletis pomonella* originally was almost wholly a pest of hawthorn, although apples introduced from Europe had been within its range and habitat for almost two centuries. In 1865 it was first recorded as attacking apples in the Hudson River Valley, and this 'apple race' spread rapidly (Bush 1975). Currently it is an important apple pest in much of eastern and mid-western USA. Many successful biological control programmes entail new associations between host and natural enemy, and these associations include species that had been thought to be monophagous or oligophagous (Hokkanen & Pimentel 1984, 1989). Indeed, the very conditions in which most species are introduced (small population size and novel environment) might be expected to lead to rapid evolution, which may include a new host range (Roderick 1992); moreover, a single gene mutation can change host specificity (Williamson 1992). Such host range expansions occur in nature, as with the Australian gall wasp (Dennill *et al.* 1993). Genetically modified organisms entail a similar threat of increased host range resulting from genetic change (see Chapter 5).

Similarly, species can evolve an expanded tolerance of physical factors that would greatly increase the probability of important ecological effects, but this might take a long time. The evolution of resistance to insecticides and herbicides is well known (references in Begon *et al.* 1990), sometimes occurring quickly and other times much more slowly. Similarly, plants evolve tolerance of novel soil contaminants at varying rates (e.g. Walley *et al.* 1974). Although the cause of sudden outbreaks in which available resources seem unchanged is uncertain, genetic changes such as a mutation or a recombinant event are often likely (Simberloff & Colwell 1984). Because both mutations and chromosomal crossover are random events, it is not possible to predict their nature and very difficult to specify even the probability of a detectable change of this sort. Further, these events can happen at any time, including many generations after introduction.

Sudden dramatic increases in geographical range are also documented, such as that for the collared dove (*Streptopelia decaocto*) in Europe (Simberloff & Colwell 1984). While none of these has been linked conclusively to genetic as opposed to habitat change, it is not impossible that such changes have occurred.

An introduced species can hybridize with a native species to produce a new pest. For example, North American cordgrass (*Spartina alterniflora*) brought to England in shipping ballast hybridized with the non-invasive native *S. maritima*. The hybrids were sterile, but after many years a hybrid individual underwent a doubling of chromosome number to produce a fertile, invasive species, *S. anglica* (Thompson 1991).

Perhaps the key aspect of possible genetic and evolutionary change that bears on choice of end-points is that much of it is not incremental, but would arise abruptly and might begin to generate effects almost instantaneously. Thus, whereas the addition of a chemical might be expected gradually to lower the size of one or more populations or gradually to affect a population statistic such as fertility, a biological introduction or a GMO (see Chapter 5) might have little or no effect for a period, until a specific recombinant event or mutation produces a different sort of organism with different effects.

6.6 UNCERTAINTY OF INTRODUCED SPECIES AS OPPOSED TO CHEMICAL POLLUTANTS AND PHYSICAL PROCESSES

Biological introductions, then, pose a series of unique challenges to risk assessment. These are largely associated with the complexity of communities of interacting organisms and the likelihood of some degree of evolutionary change, which is at least partly random. Such complexity forces a risk assessor to consider a myriad of possible effects—many of them probably important to community function—as assessment end-points. Additionally, the complexity of community organization means that there will be some uncertainty in associating measurement end-points with assessment end-points, particularly assessment end-points at the community and ecosystem levels. Even if a set of measurement end-points can be agreed upon as relevant, sufficient monitoring will probably entail a heavy economic and work burden because there are so many factors to monitor. The evolutionary change complicates end-point choice in two ways: it makes temporal limits questionable; and it means that change need not be incremental.

Moreover, quantitative risk assessment models are more difficult to devise for introduced species than for chemical pollutants or physical processes, because living organisms are quirky and communities are complex. Characteristics of living organisms, such as evolution, dispersal behaviour and dormancy, are all difficult to model. Even static community models are highly idiosyncratic and their predictions not well borne out; accurate dynamic community ecological models simply do not exist.

Forecasts have been attempted on the basis of ecological effects of similar genotypes or species. There is reasonable doubt, however, that such predictions can be useful as anything more than a 'shopping list' of potential effects to anticipate. For example, the distinctly different trajectories of closely related introduced species suggest that the experience of one could not have helped predict the effects of the other. Consider the tree sparrow, restricted to the vicinity of St Louis, and the house sparrow, which spread over most of North America (see sections 6.2 and 6.3.2). These congeners are extremely similar in both morphology and habits in their native Old World ranges. There is currently no convincing explanation for the dramatic difference in their effects. Similarly, numerous aphelinid wasps of the genus *Aphytis* have been introduced to California to control California red scale (*Aonidiella aurantii*), with widely differing results (Simberloff 1986b). Some have quickly disappeared, others have become established but were severely limited in range or density; a few established substantial populations over wide areas but had very different effects on the scale. In this rare instance, intensive research explained some of the differences in these trajectories (Luck *et al.* 1982), and the results indicate that no superficial effort would have predicted the different outcomes.

Further, the literature on introductions is highly biased. Almost certainly most introductions fail to establish ongoing populations; of those that do establish, almost certainly most have rather small effects on the target community or ecosystem (Simberloff 1991). Introduced species that have survived and had a major impact, however, are much more likely to have been noticed. Probably the more substantial the impact, the greater the probability that the species will be studied and the study published.

6.6.1 Action in the face of uncertainty

Risk assessment has a cachet of infallibility among the public because it is seen as scientific (Freudenberg 1988). O'Brien (1994) argues that, for many chemical pollutants, the levels of uncertainty surrounding various key estimates are so great that a quantitative risk assessment is meaningless—worse, it is misleading because of the apparent precision of a quantitative probability statement, and it discourages consideration of very different alternatives. This argument can be raised a fortiori against risk assessment for introduced species. As we have seen, the uncertainties are far greater than for chemicals.

Nevertheless, decisions will be made even in the absence of good science and adequate predictive capacity. It seems certain that risk assessments will be conducted. Further, they can be useful heuristic devices even if the probabilities err by orders of magnitude. The risk assessment for black carp (*Mylopharyngodon piceus*) introduction into the USA as a potential biological control agent for zebra mussel and other aquatic pests (Nico & Williams 1996) is a good example (see below). Thus, guidance must be provided to the risk assessor, but the enormous uncertainties must be noted. As pointed out in section 6.3, predictive ability or knowledge of four major factors is needed to perform a risk assessment: probabilities that the introduced species will survive, that survivors will multiply, that propagules will disperse, and that the organism will be harmful. With some introduced organisms, information about one or more of these factors will be extensive; more often, such information will be scanty. It may be possible to draw on analogies with closely related organisms. For example, the US Department of Agriculture and the US Fish and Wildlife Service rely on such analogies in their risk assessments of introduced species and genetically engineered organisms. Such information should be used, however, only with an explicit acknowledgement that related organisms often behave in vastly different fashions with respect to one or more of the four factors.

Often a Delphic process is used to provide essential information on introduced species. This approach calls on committees of experts on related organisms in an attempt to assess the nature of risks from new species. Members of such committees are generally 'experts' in the sense that they know a lot about related organisms. Even though they are not able to muster evidence beyond what is available in the scientific literature, their 'feel' for a group of organisms makes the Delphic approach the correct one for risk assessment. However, because the Delphic approach requires a team that must be composed of experts, risk assessment for introduced species using this approach is burdensome. None the less, no other current method provides adequate insight into these risks. The risk assessments for importation of Siberian and Chilean logs (section 6.3.3) give an indication of the size of the team and the depth of expertise required. Table 6.1 gives an example, for the risk assessment of the proposed black carp introduction (Nico & Williams 1996), of the way in which a Delphic procedure operates and the algorithms that are established for estimating probabilities for various processes, combining them into an overall statement of risk (Anonymous 1996).

Of course, an expert's 'feel' cannot substitute for experimentally tested knowledge. When asked to deal with applied environmental problems, ecologists typically claim that further research is needed. While this need is perhaps greater with regard to risk from biological introductions than for most other environmental problems, a vast literature already exists on several factors

Table 6.1 Generic risk assessment and algorithms for assigning values for introduction of non-indigenous aquatic organisms, US Department of Agriculture (Anonymous 1996), with assignments and total assessment (in **bold**) for proposed introduction of black carp. (From Nico & Williams 1996.)

(a) Generic uncertainty codes.

Uncertainty code	Symbol	Description
Very certain	VC	As certain as I am going to get
Reasonably certain	RC	Reasonably certain
Moderately certain	MC	More certain than not
Reasonably uncertain	RU	Reasonably uncertain
Very uncertain	VU	A guess

(b) Determination of establishment probability. Each component probability is high (H), medium (M) or low (L). Total probability is to be assigned value of lowest component probability.

Total probability of establishment	=	Probability that species disperses	×	Probability that species survives dispersal	×	Probability that species establishes population	×	Probability that species spreads
M*		**H** **VC**		**H** **VC**		**M** **RC**		**H** **RC**

* Nico & Williams incorrectly calculate total probability as H.

(c) Determination of establishment consequence. Perceptual consequence comprises aesthetic, political, social and other factors.

Total consequence of establishment	=	Economic consequence	×	Environmental consequence	×	Perceptual consequence
H		H		L,M,H		L,M,H
H		L,M,H		H		L,M,H
M		M		M		L,M,H
M		M		L		L,M,H
M		L		M		L,M,H
M		L		L		M,H
L		L		L		L
H		**L**		**H**		**M**
		MC		**VC**		**MC**

Table 6.1 *Continued.*
(d) Organism risk potential. Low risk defined as acceptable risk, organism of little concern, not justifying mitigation. Medium risk defined as unacceptable risk, organism of moderate concern, mitigation justified. High risk defined as unacceptable risk, organism of major concern, mitigation justified.

Total organism risk potential	=	Establishment probability	×	Consequence of establishment
H		H		H
H		M		H
M		L		H
H		H		M
M		M		M
M		L		M
M		H		L
M		M		L
L		L		L
H		**M**		**H**

important for risk assessment, and an expert team would assess this literature. Unfortunately, much of that information does not deal directly with the organisms of likely concern to risk assessors, except for pathogens of humans, livestock or crops.

Further, most potentially relevant genus-wide or family-wide generalizations are riddled with exceptions. Thus, one research need is information on the specific morphological or physiological characteristics that are the basis for survival, multiplication, dispersal and ecological effects. Currently available information on a number of species will have to serve for now as a generic basis for predictions of risk. Thus, until the amount of such information increases significantly, risk assessment for introduced species will be speculative at best.

For risk assessments of introduced species to be scientifically defensible and transparent, it is crucial that assessors acknowledge the levels of uncertainty. For example, the US Department of Agriculture generic risk assessment process (Orr *et al.* 1993) mandates declaring how certain the expert team is about each step in an assessment, and it provides an algorithm for combining the uncertainties to give an indication of the uncertainty of the project as a whole. This approach is illustrated in the assessment for Chilean logs (USDA 1993).

6.7 RECOVERY

Recovery from an introduced species can differ greatly from recovery from chemicals or physical processes because of several irrevocable changes that a new organism can cause (Simberloff 1990a,b). This will be particularly true if one seeks a full restoration, on a species-by-species basis, rather than simply recovery of ecosystem function. Full recovery will be impossible if the introduced species has caused global extinction of a resident species, or it can take a very long time if local extinction must be redressed by natural immigration from a distant source. The opposite effect is probably even more problematic—it is tremendously difficult to eliminate an introduced species once it has become established, as noted above. Even where a sustained effort has succeeded in eradicating an introduced species, restoration has been slow at best. In some cases, it will be impossible. For example, virtually all endemic forest birds of Guam have been eliminated by the introduced Australian brown tree snake, *Boiga irregularis* (Savidge 1987). Even if the snake were now removed, in what sense could the island community ever be 'restored'? As observed above, a genetic change in a resident species caused by an introduced species, even if the latter does not persist, can further complicate recovery.

Hybridization, lateral transfer, and even 'ordinary' evolution, such as resistance to a new disease or pest, will change resident species permanently.

Non-evolutionary effects of introduced species, such as habitat modification or various forms of population suppression (e.g. predation) resemble effects of chemical pollutants. Yet, because introductions are usually irreversible, assessment of recovery from introduced species usually differs from assessment of recovery from chemical pollutants.

6.8 SUMMARY

Four factors — survival, multiplication, dispersal and ecological effects — determine the risk posed by an introduced species. For some species, information exists for one or more of these factors, but little or no relevant information exists for many introduced species. Thus risk assessors often examine species closely related to the species introduced, an exercise that requires recognition of significantly different ways in which closely related species behave.

The fact that introduced species reproduce and may multiply renders risk assessment for such species more difficult than for chemical pollutants or physical processes. The assessment can be further complicated by the myriad of ways in which introduced species disperse, both actively and passively. Particularly noteworthy is the frequency of jump-dispersal in comparison to simple diffusion. Although the exact trajectory of such discontinuous dispersal is extremely difficult to predict, it may be possible to assess the probability that it will occur and to assign greater probabilities to certain geographical routes than to others. One factor to consider is that jump-dispersal is characteristic of, but not restricted to, transport of living organisms by humans.

Once an introduced species reaches a site, its potential effects on the target ecosystem are numerous. Since many types of effects have been studied extensively by ecologists, the literature can be useful for beginning to assess potential risks. Studies may indicate, for example, special vigilance about certain types of effects. None the less, for most species that have the potential of being introduced, the literature on effects is deficient.

Perhaps the key difference between introduced species and chemical pollutants or physical processes — the factor that most affects risk assessment — is that biological entities evolve. Because evolution can affect survival, multiplication, dispersal and effects, it can profoundly influence the risk that a species poses. Because it is a stochastic process, however, and because many aspects of evolution remain quite unpredictable, performing a risk assessment for an introduced species can be complicated. At present, a Delphic process remains the most reasonable starting point, once the uncertainties involved are clearly stated.

6.9 REFERENCES

Alexander, M. (1971) *Microbial Ecology*. Wiley, New York.

Alexander, M. (1985a) Ecological consequences: reducing the uncertainties. *Issues in Science and Technology*, **1**(3), 57–68.

Alexander, M. (1985b) Survival and growth of bacteria. In: *Potential Impacts of Environmental Release of Biotechnology Products: Assessment, Regulation, and Research Needs* (eds J.W. Gillett, A.M. Stern, S.A. Levin, M.A. Harwell, M. Alexander & D. Andow), pp. 63–67. Ecosystems Research Center, Cornell University, Ithaca, NY.

Anderson, R.M. & May, R.M. (eds) (1982) *Population Biology of Infectious Diseases*. Springer-Verlag, Berlin.

Anderson, R.M. & May, R.M. (1991) *Infectious Diseases of Humans: Dynamics and Control*. Oxford University Press, Oxford.

Anonymous (1996) *Report to the Aquatic Nuisance Species Task Force. Generic Nonindigenous Aquatic Organisms Risk Analysis Review Process*. US Department of Agriculture, Riverdale, MD.

Ashton, P.J. & Mitchell, D.S. (1989) Aquatic plants: patterns and modes of invasion, attributes of invading species and assessment of control programmes. In: *Biological Invasions: a Global Perspective* (eds J.A. Drake, H.A. Mooney, F. di Castri, R.H. Groves, F.J. Kruger, M. Rejmanek & M. Williamson), pp. 111–154. Wiley, Chichester.

Atkinson, I. (1989) Introduced animals and extinction. In: *Conservation for the Twenty-First Century* (eds D. Western & M. Pearl), pp. 54–75. Oxford University Press, New York.

Azevedo, M. (1990) Of eucalyptus and ecology. *California Waterfront Age*, **6**(1), 16–20.

Baker, H.G. and Stebbins, G.L. (1965) *The Genetics of Colonizing Species*. Academic Press, New York.

Barbault, R. (1992) *Ecologie des peuplements*. Masson, Paris.

Begon, M., Harper J.L. & Townsend, C.R. 1990. *Ecology: Individuals, Populations, and Communities*, 2 edn. Blackwell Scientific Publications, Boston.

Bertness, M.D. (1984) Habitat and community modification by an introduced herbivorous snail. *Ecology*, **65**, 370–381.

Bonner, J.T. (1965) *Size and Cycle: an Essay on the Structure of Biology*. Princeton University Press, Princeton, NJ.

Bush, G.L. (1975) Sympatric speciation in phytophagous parasitic insects. In: *Evolutionary Strategies of Parasitic Insects and Mites* (ed. P.W. Price), pp. 187–206. Plenum, New York.

Carey, J. (1982) Mangroves: swamps nobody likes. *International Wildlife*, **12**(5), 19–28.

Carey, J.R. (1991) Establishment of the Mediterranean fruitfly in California. *Science*, **253**, 1369–1373.

Carlquist, S. (1974) *Island Biology*. Columbia University Press, New York.

Carlton J.T. & Geller, J.B. (1993) Ecological roulette: the global transport of nonindigenous marine organisms. *Science*, **261**, 78–82.

Carlton, J.T., Reid, D. & van Leeuwen, H. (1994) *The role of shipping in the introduction of nonindigenous aquatic organisms to the coastal waters of the United States (other than the Great Lakes) and an analysis of control options*. US Coast Guard and the National Sea Grant Program/Connecticut Sea Grant Project R/ES-6. New London, CT.

Chanin, P.R.F. & Jefferies, D.J. (1978) The decline of the otter *Lutra lutra* L. in Britain: an analysis of hunting records and discussion of causes. *Biological Journal of the Linnean Society*, **10**, 305–328.

Clements, F.E. (1936) Nature and structure of the climax. *Journal of Ecology*, **24**, 252–284.

Courtenay, W.R., Jr. & Meffe, G.K. (1989) Small fishes in strange places: a review of introduced poeciliids. In: *Ecology and Evolution of Livebearing Fishes (Poeciliidae)* (eds G.K. Meffe & F.F. Snelson), pp. 319–331. Prentice-Hall, Englewood Cliffs, NJ.

Courtenay, W.R., Jr. & Williams, J.D. (1992) Dispersal of exotics species from aquaculture sources, with emphasis on freshwater fishes. In: *Dispersal of Living Organisms into Aquatic Systems* (eds A. Rosenfield & R. Mann), pp. 49–81. Maryland Sea Grant Program, College Park, MD.

Crosby, A.W. (1986) *Ecological Imperialism: the Biological Expansion of Europe, 900–1900*. Cambridge University Press, Cambridge.

Dahlsten, D.L. (1986) Control of invaders. In: *Ecology of Biological Invasions of North America and Hawaii* (eds H.A. Mooney and J.A. Drake), pp. 275–302. Springer-Verlag, New York.

Dean, C. (1988) Tiny snail is credited as a force shaping the coast. *New York Times*, **August 23**, 15, 19.

Dennill, G.G., Donnelly, D. & Chown, S.L. (1993) Expansion of host-plant range of a biocontrol agent *Trichilogaster acaciaelongifoliae* (Pteromalidae) released against the weed *Acacia longifolia* in South Africa. *Agriculture, Ecosystems and Environment*, **43**, 1–10.

Dennis, B. (1989) Allee effect: population growth, critical density and the chance for extinction. Natural Resource Modelling **4**, 481–538.

de Vos, A., Manville, R.H. & Van Gelder, R.G. (1956) Introduced mammals and their influence on native biota. *Zoologica: New York Zoological Society*, **41**, 163–194.

Dobson, A.P. & May, R.M. (1986) Disease and conservation. In: *Conservation Biology: the Science of Scarcity and Diversity* (ed. M.E. Soule), pp. 345–365. Sinauer, Sunderland, MA.

Drake, J.A., Flum, T.E., Witteman, G.J. *et al.* (1993) The construction and assembly of an ecological landscape. *Journal of Animal Ecology*, **62**, 117–130.

Ebenhard, T. (1988) Introduced birds and mammals and their influence on native biota. *Viltrevy (Swedish Wildlife Research)*, **13**(4), 1–106.

Ellstrand, N.C. (1992) Gene flow by pollen: implications for plant conservation genetics. *Oikos*, **63**, 77–86.

Elton, C.S. (1958) *The Ecology of Invasions by Animals and Plants*. Chapman and Hall, London.

Ewald, P.W. (1983) Host–parasite relations, vectors, and the evolution of disease severity. *Annual Review of Ecology and Systematics*, **14**, 465–485.

Ewel, J.J. (1986) Invasibility: lessons from south Florida. In: *Ecology of Biological Invasions of North America and Hawaii*, (eds H.A. Mooney & J.A. Drake), pp. 214–230. Springer-Verlag, New York.

Fenchel, T. (1974) Intrinsic rate of natural increase: the relationship with body size. *Oceanologia*, **14**, 317–326.

Freudenberg, W.R. (1988) Perceived risk, real risk: social science and the art of probabilistic risk assessment. *Science*, **242**, 44–49.

Garton, D.W., Berg, D.J., Stoeckmann, A.M. & Haag, W.R. (1993) Biology of recent invading species in the Great Lakes: the spiny water flea, *Bythotrephes cederstroemi*, and the zebra mussel, *Dreissena polymorpha*. In: *Biological Pollution: the Control and Impact of Invasive Exotic Species* (ed. B.N. McKnight), pp. 63–84. Indiana Academy of Science, Indianapolis.

Gosling, M. (1989) Extinction to order. *New Scientist* **121**(1654), 44–49.

Greenway, J.C. (1967) *Extinct and Vanishing Birds of the World*. Dover, New York.

Hanski, I. & Gilpin, M. (1991) Metapopulation dynamics: brief history and conceptual domain. *Biological Journal of the Linnean Society*, **42**, 3–16.

Harlan, J.R. (1982) Relationships between weeds and crops. In: *Biology and Ecology of Weeds* (eds W. Holzner & M. Numata), pp. 91–96. Dr. W. Junk, Boston.

Hastings, A. (1996) Models of spatial spread: a synthesis. *Biological Conservation*, **78**, 143–148.

Hengeveld, R. (1989) *Dynamics of Biological Invasions*. Chapman and Hall, London.

Herbold, B. & Moyle, P.B. (1986) Introduced species and vacant niches. *American Naturalist*, **128**, 751–760.

Hokkanen, H. & Pimentel, D. (1984) New approach for selecting biological control agents. *Canadian Entomologist*, **116**, 1109–1121.

Hokkanen, H. & Pimentel, D. (1989) New associations in biological control: theory and practice. *Canadian Entomologist*, **121**, 829–840.

Holdridge, L.R. (1940) Some notes on mangrove swamps of Puerto Rico. *Caribbean Forester*, **1**, 19–29.

Holmes, J.C. (1982) Impact of infectious disease agents on the population growth and geographical distribution of animals. In: *Population Biology of Infectious Diseases* (eds R.M. Anderson & R.M. May), pp. 37–51. Springer-Verlag, Berlin.

Howarth, F.G. (1991) Environmental impacts of classical biological control. *Annual Review of Entomology*, **36**, 485–509.

Johnson, C.G. (1969) *Migration and Dispersal of Insects by Flight*. Methuen, London.

Keller, M.A. (1984) Reassessing evidence for competitive exclusion of introduced natural enemies. *Environmental Entomology*, **13**, 192–195.

King, C.M. (1984) *Immigrant Killers: Introduced Predators and the Conservation of Birds in New Zealand*. Oxford University Press, Auckland.

Knopf, F.L. & Olson, T.E. (1984) Naturalization of Russian-olive: implications to Rocky Mountain wildlife. *Wildlife Society Bulletin*, **12**, 289–298.

Krebs, C.J. (1985) *Ecology: the Experimental Analysis of Distribution and Abundance*, 3 edn. Harper & Row, New York.

Lever, C. (1987) *Naturalized Birds of the World*. Longman, London.

Levin, S.A. & Harwell, M.A. (1986) Potential ecological consequences of genetically engineered organisms. *Environmental Management*, **10**, 495–513.

Levy, S.B. (1986) Human exposure and effects analysis for genetically modified bacteria. In: *Biotechnology Risk Assessment* (eds J. Fiksel & V.T. Covello), pp. 56–74. Pergamon, New York.

Lewin, R. (1983) Santa Rosalia was a goat. *Science*, **221**, 636–639.

Liang, L.N., Sinclair, J.L., Mallory, L.M. & Alexander, M. (1982) Fate in model ecosystems of microbial species of potential use in genetic engineering. *Applied and Environmental Microbiology*, **44**, 708–714.

Long, J.L. (1981) *Introduced Birds of the World*. Universe Books, New York.

Loope, L.L. & Scowcroft, P.G. (1985) Vegetation response within exclosures in Hawaii: a review. In: *Hawaii's Terrestrial Ecosystems: Preservation and Management* (eds C.P. Stone & J.M. Scott), pp. 377–402. University of Hawaii Cooperative National Park Resources Studies Unit, Honolulu.

Luck, R.F., Podoler, H. & Kfir, R. (1982) Host selection and egg allocation behavior by *Aphytis melinus* and *A. lingnanensis*: comparison of two facultatively gregarious parasitoids. *Ecological Entomology*, **7**, 317–408.

Lynch, J.M. (1995) Conservation implications of hybridisation between mustelids and their domesticated counterparts: the example of polecats and feral ferrets in Britain. *Small Carnivore Conservation*, **13**, 17–18.

Macdonald, I.A.W., Loope, L.L., Usher, M.B. & Hamann, O. (1989) Wildlife conservation and the invasion of nature reserves by introduced species: a global perspective. In: *Biological Invasions: a Global Perspective* (eds J.A. Drake, H.A. Mooney, F. di Castri *et al.*), pp. 215–255. Wiley, Chichester.

Mack, R.N. (1996) Predicting the identity and fate of plant invaders: emergent and emerging approaches. *Biological Conservation*, **78**, 107–121.

Mackenzie, D.R., Barfield, C.S., Kennedy, C.G., Berger, D.R. & Taranto, D.J. (1985) *The Movement and Dispersal of Agriculturally Important Biotic Agents*. Claitor's Publishing Division, Baton Rouge, LA.

May, R.M. (1987) Nonlinearities and complex behavior in simple ecological and epidemiological models. Annals of the New York Academy of Science, **504**, 1–15.

Moyle, P.B. (1986) Fish introductions to North America: patterns and ecological impact. In: *Ecology of Biological Invasions of North America and Hawaii* (eds H.A. Mooney & J.A. Drake), pp. 27–43. Springer-Verlag, New York.

Nico, L.G. & Williams, J.D. (1996) *Risk Assessment on Black Carp (Pisces: Cyprinidae)*. US Department of the Interior, National Biological Service, Gainesville, FL.

Nielson, R.P. & Wullstein, L.H. (1983) Biogeography of two southwest American oaks in relation to atmospheric dynamics. *Journal of Biogeography*, **10**, 275–297.

Nowak, E. (1975) *The Range Expansion of Animals and its Causes*. Smithsonian Institution and National Science Foundation, Washington, DC.

O'Brien, M.H. (1994) The scientific imperative to move society beyond 'just not quite fatal'. *Environmental Professional*, **16**, 356–365.

Opler, P.A. (1978) Insects of American chestnut: possible importance and conservation concern. In: *The American Chestnut Symposium* (ed. W. McDonald), pp. 83–85. American Chestnut Association, Morgantown, WV.

Orr, R.L., Cohen, S.D. & Griffin, R.L. (1993) *Generic Non-Indigenous Pest Risk Assessment Process*. US Department of Agriculture, Animal and Plant Health Inspection Service, Beltsville, MD.

Penny, M. (1974) *Birds of Seychelles and the Outlying Islands*. Collins, London.

Pielou, E.C. (1979) *Biogeography*. Wiley, New York.
Pimentel, D. (1985) Using genetic engineering for biological control: reducing ecological risks. In: *Engineered Organisms in the Environment: Scientific Issues*. (eds H.O. Halvorson, D. Pramer & M. Rogul), pp. 129–140. American Society for Microbiology, Washington, DC.
Pulliam, H.R. (1988) Sources, sinks, and population regulation. *American Naturalist*, **132**, 652–661.
Quimby, P.C. (1982) Impact of diseases on plant populations. *Biological Control of Weeds with Plant Pathogens* (eds R. Charudattan & H.L. Walker), pp. 47–60. Wiley, New York.
Ramadan, M.A., El-Tayeb, O.M. & Alexander, M. (1990) Inoculum size as a factor limiting success of inoculation for biodegradation. *Applied and Environmental Microbiology*, **56**, 1392–1396.
Ramakrishnan, P.S. & Vitousek, P.M. (1989) Ecosystem-level processes and the consequences of biological invasions. In: *Biological Invasions: a Global Perspective* (eds J.A. Drake, H.A. Mooney, F. di Castri *et al.*), pp. 281–300. Wiley, Chichester.
Rambler, M., Margulis, L. & Fester R. (1988) *Global Ecology*. Academic Press, New York.
Ratcliffe, D. (1979) The end of the large blue butterfly. *New Scientist*, **8**, 457–458.
Regal, P.J. (1986) Models of genetically engineered organisms and their ecological impact. In: *Ecology of Biological Invasions of North America and Hawaii* (eds H.A. Mooney & J.A. Drake), pp. 111–129. Springer-Verlag, New York.
Rhymer, J.M. & Simberloff, D. (1996) Extinction by hybridization and introgression. *Annual Review of Ecology and Systematics*, **27**, 83–109.
Roderick, G.K. (1992) Postcolonization evolution of natural enemies. In: *Selection Criteria and Ecological Consequences of Importing Natural Enemies* (eds W.C. Kauffman & J.E. Nechols), pp. 71–86. Entomological Society of America, Lanham, MD.
Roszak, D.B. & Colwell, R.R. (1987) Survival strategies of bacteria in the aquatic environment. *Microbiological Reviews*, **51**, 365–379.
Rozhnov, V.V. (1993) Extinction of the European mink: an ecological catastrophe or a natural process? *Lutreola*, **1**, 10–16.
Ruesink, J.L., Parker, I.M., Groom, M.J. & Kareiva, P.M. (1995) Reducing the risks of nonindigenous species introductions. *BioScience*, **45**, 465–477.
Samuel, D.E. (1975) The kiskadee in Bermuda. Newsletter of the Bermuda Biological Station, **4**.
Savidge, J. (1987) Extinction of an island avifauna by an introduced snake. *Ecology*, **68**, 660–668.
Shaffer, M.L. (1981) Minimum population sizes for species conservation. *BioScience*, **31**, 131–134.
Sharples, F.E. (1982) Spread of organisms with novel genotypes: thoughts from an ecological perspective. *Recombinant DNA Technical Bulletin*, **6**, 43–56. (June 1983.)
Shrum, R.D. (1978) Forecasting of epidemics. In: *Plant Disease* (eds J.G. Horsfall & E.B. Cowling), Vol. 2, pp. 223–238. Academic Press, New York.
Silverstein, A. & Silverstein, V. (1974) *Animal Invaders: the Story of Imported Wildlife*. Atheneum, New York.
Simberloff, D. (1980) A succession of paradigms in ecology: essentialism to materialism and probabilism. *Synthese*, **43**, 3–39.
Simberloff, D. (1981) Community effects of introduced species. In: *Biotic Crises in Ecological and Evolutionary Time* (ed. M.H. Nitecki), pp. 53–81. Academic Press, New York.
Simberloff, D. (1985) Predicting ecological effects of novel entities: evidence from higher organisms. In: *Engineered Organisms in the Environment/Scientific Issues* (eds H.O. Halvorson, D. Pramer & M. Rogul), pp. 152–161. American Society for Microbiology, Washington, DC.
Simberloff, D. (1986a) The proximate causes of extinction. In: *Patterns and Processes in the History of Life* (eds D.M. Raup & D. Jablonski), pp. 259–276. Springer-Verlag, Berlin.
Simberloff, D. (1986b) Introduced insects: a biogeographic and systematic perspective. In: *Ecology of Biological Invasions of North America and Hawaii* (eds H.A. Mooney & J.A. Drake), pp. 3–26. Springer-Verlag, New York.
Simberloff, D. (1988) The contribution of population and community biology to conservation science. *Annual Review of Ecology and Systematics*, **19**, 473–511.
Simberloff, D. (1989) Which insect introductions succeed and which fail? In *Biological Invasions: A Global Perspective* (eds J.A. Drake, H.A. Mooney, F. di Castri *et al.*), pp. 61–75. Wiley, Chichester.
Simberloff, D. (1990a) Reconstructing the ambiguous: can island ecosystems be restored? In: *Ecological Restoration of New Zealand Islands* (eds D.R. Towns, C.H. Daugherty & I.A.E. Atkinson), pp. 37–51. Department of Conservation, Wellington.
Simberloff, D. (1990b) Community effects of biological introductions and their implications for restoration. In: *Ecological Restoration of New Zealand Islands* (eds D.R. Towns, C.H. Daugherty & I.A.E. Atkinson), pp. 128–136. Department of Conservation, Wellington.
Simberloff, D. (1991) Keystone species and the community effects of biological introductions. In: *Assessing Ecological Risks of Biotechnology* (ed. L.R. Ginzburg), pp. 1–19. Butterworth-Heinemann, Boston.
Simberloff, D. (1992) Conservation of pristine habitats and unintended effects of biological control. In: *Selection Criteria and Ecological Consequences of Importing Natural Enemies* (eds W.C. Kauffman & J.E. Nechols), pp. 103–117. Entomological Society of America, Lanham, MD.

Simberloff, D. (1994) Why do introduced species appear to devastate islands more than mainland? *Pacific Science*, **49**, 87–97.

Simberloff, D. (1997) Eradication. In: *Strangers in Paradise: Impact and Management of Nonindigenous Species in Florida* (eds D. Simberloff, D. Schmitz & T. Brown), pp. 221–228. Island Press, Washington, DC.

Simberloff, D. & Boecklen, W.J. (1991) Patterns of extinction in the introduced Hawaiian avifauna: a reexamination of the role of competition. *American Naturalist*, **138**, 300–327.

Simberloff, D. & Colwell, R.K. (1984) Release of engineered organisms: a call for ecological and evolutionary assessment of risks. *Genetic Engineering News*, **4**(8), 4.

Simberloff, D. & Stiling, P.D. (1996) Risks of species introduced for biological control. *Biological Conservation*, **78**, 185–192.

Simpson, G.G. (1952) Probabilities of dispersal in geologic time. *Bulletin of the American Museum of Natural History*, **99**, 163–176.

Singer, F.G., Swank, W.T. & Clebsch, E.E.C. (1984) Effects of wild pig rooting in a deciduous forest. *Journal of Wildlife Management*, **48**, 464–473.

Smallwood, K.S. (1990) *Turbulence and ecology of invading species.* PhD dissertation, University of California, Davis.

Smith, S.H. (1968) Species succession and fisheries exploitation in the Great Lakes. *Journal of the Fisheries Research Board of Canada*, **25**, 667–693.

Stotzky, G., Zeph, L.R. & Devanas, M.A. (1991) Factors affecting the transfer of genetic information among microorganisms in the soil. In: *Assessing Ecological Risks of Biotechnology* (ed. L.R. Ginzburg), pp. 95–122. Butterworth-Heinemann, Boston.

Strauss, H.S. & Levin, M.A. (1991) Use of fate and transport (dispersal) models in microbial risk assessment. In: *Risk Assessment in Genetic Engineering* (eds M.A. Levin & H.S. Strauss), pp. 240–271. McGraw-Hill, New York.

Strauss, S.Y. (1991) Indirect effects in community ecology: their definition, study, and importance. *Trends in Ecology and Evolution*, **6**, 206–210.

Suter, G.W., II (1993) *Ecological Risk Assessment.* Lewis Publishing, Chelsea, MI.

Thompson, J.D. (1991) The biology of an invasive plant. *BioScience*, **41**, 393–401.

Thorgaard, G.H. & Allen, S.K. (1992) Environmental impacts of inbred, hybrid, and polyploid aquatic species. In: *Dispersal of Living Organisms into Aquatic Ecosystems* (eds A. Rosenfield & R. Mann), pp. 281–288. Maryland Sea Grant Program, College Park, MD.

Underwood, A.J. (1986) What is a community? In: *Patterns and Processes in the History of Life* (eds D.M. Raup & D. Jablonski), pp. 351–367. Springer-Verlag, Berlin.

US Congress (1993) *Harmful Non-Indigenous Species in the United States.* Office of Technology Assessment, US Government Printing Office, Washington, DC.

USDA (1991) *Pest Risk Assessment of the Importation of Larch from Siberia and the Soviet Far East.* Forest Service Miscellaneous Publication 1495, US Department of Agriculture, Washington, DC.

USDA (1993) *Pest Risk Assessment of the Importation of Pinus radiata, Nothofagus dombeyi, and Laurelia philippiana Logs from Chile.* Forest Service Miscellaneous Publication 1517, US Department of Agriculture, Washington, DC.

USEPA (1992) *Framework for ecological risk assessment.* Risk Assessment Forum, US Environmental Protection Agency, EPA/630/R-92/001, Washington, DC.

Upper, C.D. & Hirano, S.S. (1991) Aerial dispersal of bacteria. In: *Assessing Ecological Risks of Biotechnology* (ed. L.R. Ginzburg), pp. 75–93. Butterworth-Heinemann, Boston.

van Riper, C., van Riper, S.G., Goff, M.L. & Laird, M. (1986) The epizootiology and ecological significance of malaria in Hawaiian landbirds. *Ecological Monographs*, **56**, 327–344.

Vitousek, P.M. (1986) Biological invasions and ecosystem properties: can species make a difference? In: *Ecology of Biological Invasions of North America and Hawaii* (eds H.A. Mooney & J.A. Drake), pp. 163–176. Springer-Verlag, New York.

Vivrette, N.J. & Muller, C.H. (1977) Mechanism of invasion and dominance of coastal grassland by *Mesembryanthemum crystallinum. Ecological Monographs*, **47**, 301–318.

von Broembsen, S.L. (1989) Invasions of natural ecosystems by plant pathogens. In: *Biological Invasions: A Global Perspective* (eds J.A. Drake, H.A. Mooney, F. diCastri *et al.*), pp 77–83. Wiley, Chichester.

Walley, K., Khan, M.S.I. & Bradshaw, A.D. (1974) The potential for evolution of heavy metal tolerance in plants. I. Copper and zinc tolerance in *Agrostis tenuis. Heredity*, **32**, 309–319.

Washburn, J.O. (1984) The gypsy moth and its parasites in North America: a community in equilibrium? *American Naturalist*, **124**, 288–292.

Williamson, M. (1992) Environmental risks from the release of genetically modified organisms (GMOs): the need for molecular ecology. *Molecular Ecology*, **1**, 3–8.

Wilson, D.S. & Sober, E. (1989) Reviving the superorganism. *Journal of Theoretical Biology*, **136**, 337–356.

Chapter 7
Retrospective Assessment, Ecoepidemiology and Ecological Monitoring

GLENN W. SUTER II

7.1 INTRODUCTION

The topic of this chapter is retrospective ecological risk assessment. Retrospective assessments address risks from actions that began in the past and therefore can be assessed on the basis of measurements of the state of the environment. Retrospective assessments most commonly are driven by a causal factor such as an oil spill, an existing effluent or a hazardous waste site. However, the chapter is particularly concerned with those retrospective assessments that attempt to identify the reality and causes of apparent effects. That practice has been termed ecological epidemiology, or ecoepidemiology, by analogy to epidemiology, the estimation of risks to humans based on measurement of morbidity and mortality (Bro-Rasmussen & Lokke 1984; Suter 1990, 1993a; Fox 1991). Prominent recent examples include assessments of the causes of forest decline in central Europe, of fishless lakes in northeastern USA, and of declines in populations of piscivorous and raptorial birds during the 1950s and 1960s.

A premise of the chapter is that environmental monitoring, which has often been treated as an end in itself, is more properly treated as the data-collection component of an environmental risk assessment.* Until recently, this premise has not generally been accepted by those who design and conduct monitoring studies. More commonly, the goal of monitoring has been the elucidation of environmental status and trends. While this goal is important, it is not sufficient to justify the effort and expense of environmental monitoring. Monitoring is justified only when it supports decision-making by providing the means to determine whether there is a need for action to protect or remediate the environment. The discipline that provides the technical basis for such decisions is ecological risk assessment.

This chapter explains how to perform retrospective ecological risk assessments, how to make inferences about the reality and causes of observed effects in ecoepidemiology, and how to design monitoring studies to be relevant to risk assessments. It does not address specific techniques for sampling, analysis and testing. Such topics are beyond the scope of this chapter and are discussed in numerous other volumes, including a companion to this Handbook, *Handbook of Ecotoxicology*. Finally, the diversity of assessment problems, types and amounts of data with which an assessor may work, and assessment tools that may be applied in retrospective ecological risk assessments, are so diverse that a book chapter can only provide an approach for organizing and conducting such assessments and an entrée to the literature. Table 7.1 provides a set of examples of retrospective ecological risk assessments which not only provide an indication of the diversity of problems but also provide a set of case studies to supplement this chapter.

* The term monitoring is used here to indicate any activity that measures or counts biological, chemical or physical properties of the ambient environment. It does not necessarily imply collection of a time-series of repeated measurements.

7.2 RETROSPECTIVE ECOLOGICAL RISK ASSESSMENT PARADIGM AND METHODS

Within the last decade, ecological risk assessment

Table 7.1 Examples of retrospective ecological risk assessments.

Effects-driven assessments: ecoepidemiology
The attribution of the crash of peregrine falcon populations to DDT metabolites is explained by one of the principal investigators (Peakall 1993) and presented in terms of Koch's postulates (Suter 1993a). Evidence included population surveys; residue analyses, toxicity testing, and identification of characteristic symptoms.

The apparent decline of forests in Europe and North America has been the source of much confusion and debate due to the potentially complex and indirect causal chains, the differences among sites, and the difficulty of performing realistic tests on large and long-lived organisms. Reviews of the evidence concerning the cause of forest decline, including application of Koch's postulates, can be found in Woodman & Cowling (1987), Skelly (1989), Bernard *et al.* (1990) and Shriner *et al.* (1990).

Ecoepidemiological inference was applied to the possibility that reproductive failure of lake trout is caused by contaminant effects. The inference was based principally on the hatchability of eggs and subsequent fry survival, and contaminant levels in eggs and adult fish (Mac & Edsall 1991).

Ecoepidemiological inference was applied to the possibility that mink and otter populations have declined in the Great Lakes drainage. Because of the difficulty of characterizing these populations, the inference was as much concerned with the reality of the putative decline as whether it was caused by contaminants (Wren 1991).

Deformities and associated reproductive failures of piscivorous birds in the Great Lakes basin have been causally associated with chlorinated hydrocarbons (Colborn 1991; Gilbertson *et al.* 1991; Giesy *et al.* 1994a,b).

The causes of mass mortalities and population declines of seals in Europe have been a subject of investigation and controversy. Toxicity tests feeding contaminated fish to seals and equivalent polychlorinated biphenyl (PCB) doses to mink provided the basis for concluding that the reproductive failures and associated declines in the common seal population in the western Wadden Sea were due to PCBs (Feijnders 1986). Other halogenated organic chemicals have been implicated in the declines of Baltic grey and ringed seals, but without controlled studies the conclusions are tentative (Olsson 1994, 1995).

Source-driven assessments
The *Exxon Valdez* oil spill was followed by numerous and extensive studies of the fate of the oil and the effects of the spill on wildlife, fisheries and intertidal communities. The results of studies by Exxon and other non-government scientists are summarized by Wells *et al.* (1995).

Waste sites provide the impetus for many ecological risk assessments. Both aquatic and terrestrial risks were assessed at the pesticide-contaminated Baird McGuire site (Burmaster *et al.* 1991; Menzie *et al.* 1992).

Exposure-driven assessments
The fact that the herbicide atrazine has been frequently found in surface waters in North America has led to concerns about potential ecological effects. Because effects have not been documented in the field, the risks from this exposure were assessed on the basis of exposure estimates from both runoff models and measured concentrations, and effects estimates were derived from various tests of atrazine toxicity (Solomon *et al.* 1996).

The high levels of lead in game bird habitats and in birds led to concerns about effects on game bird populations. Risks to mourning doves and their predators were assessed based on measured lead levels, toxicity studies and incidents of avian mortality (Kendall *et al.* 1996).

Estimates of regional acidification of lakes and streams were the impetus for estimating regional effects on fisheries based on observed effects in acidified waters, laboratory tests, and models of organismal and population responses (Baker *et al.* 1990a,b).

has become the predominant paradigm for assessment of the effects of human actions on the environment (Risk Assessment Forum 1992; Suter 1993a). Ecological risk assessment is not always distinct from hazard assessment, environmental impact assessment, and other styles of environmental assessment, but it differs in organization and emphasis (Suter 1993a). Most importantly, it differs in having a standard conceptual paradigm that tends to compel greater rigour and

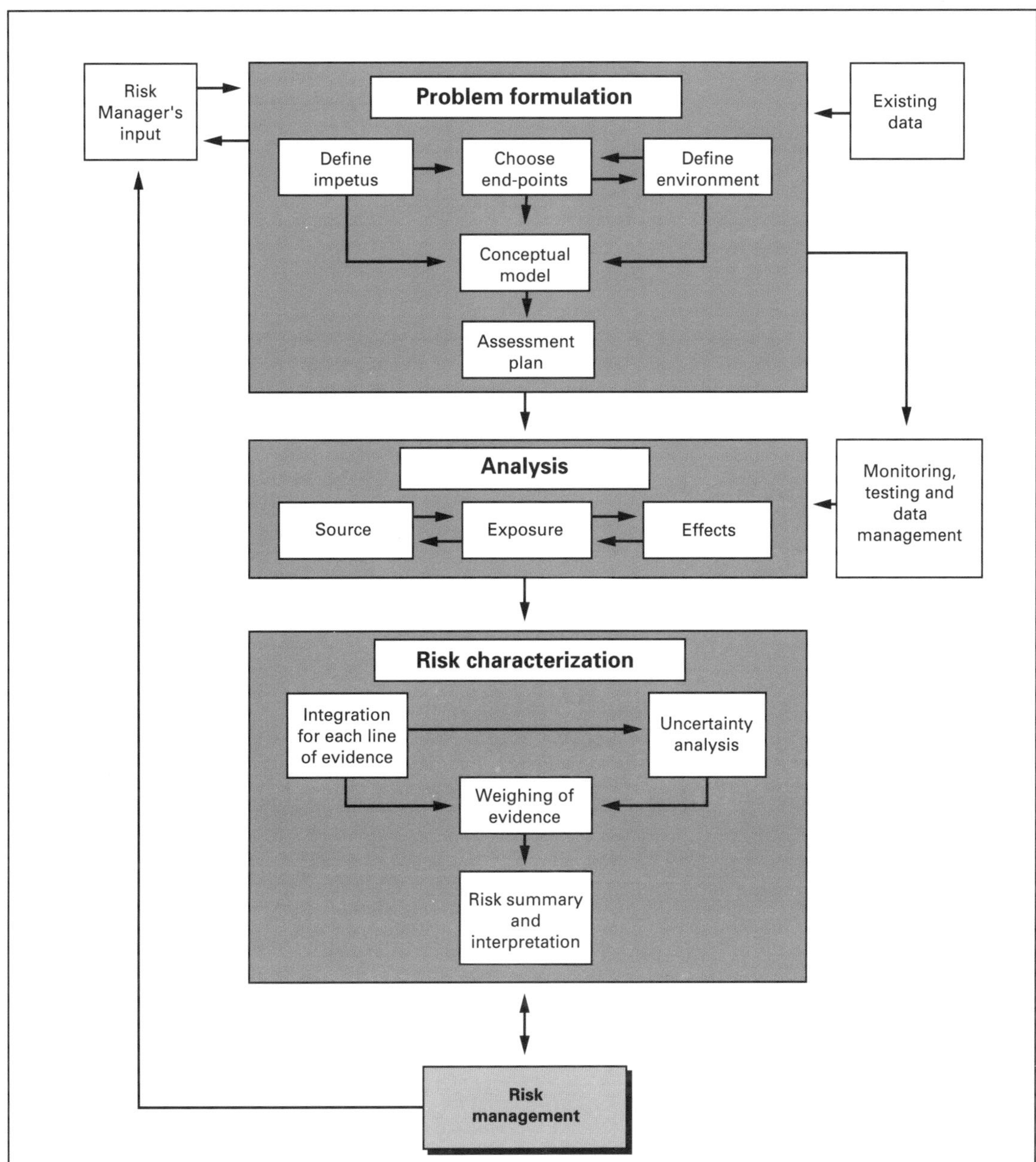

Fig. 7.1 A paradigm for retrospective ecological risk assessments. It is based on the author's previous retrospective risk paradigm but integrates appropriate USEPA concepts and terminology (Risk Assessment Forum 1992; Suter 1993a).

thoroughness than other assessment traditions. Most versions of the ecorisk paradigm are designed for the assessment of risks posed by a proposed future release of a chemical, effluent, bioengineered organism (and introduced species), or other potentially hazardous agent (i.e. predictive assessments). Predictive ecological risk assessments are discussed in other chapters. Retrospective assessments that are based solely on estimates of the source and on modelling of exposure and effects are conceptually and procedurally equivalent to predictive assessments. However, when measurements of exposure or effects are used, retrospective assessments can become much more complex but also can be more accurate and reliable. Figure 7.1 presents a version of the ecorisk paradigm that is applicable to all retrospective ecological risk assessments. Its principal difference from the USEPA's paradigm (Risk Assessment Forum 1992) is that it allows for cases in which the source is unknown and must be determined rather than being prescribed during the problem formulation.

7.2.1 Problem formulation in retrospective assessments: what are we assessing?

Problem formulation is the critical but often neglected initial stage in ecological risk assessment that determines what decisions are to be made, what information concerning ecological risks is needed to make the decision, and how that information is to be generated.

Impetus definition: why are we doing this assessment?

Problem formulation for retrospective studies must begin by identifying the impetus for the assessment. There are three general alternatives (Table 7.1).

1 Source-driven assessments address releases that began in the past and have uncertain consequences, such as hazardous waste dumps, industrial effluents, and oil spills.

2 Effects-driven assessments address apparent ecological effects, such as fish kills and declines in populations, that have uncertain magnitudes and causes.

3 Exposure-driven assessments address observed

Table 7.2 Properties of the source, effect or exposures that constitute the impetus for a risk assessment, and general temporal and spatial properties that should be characterized during the problem formulation.

Property	Examples
Sources	
Form	Spill, aqueous effluent, atmospheric deposition, buried waste, etc.
Composition	Chemicals, hazardous physical properties such as temperature and pH, complex materials such as petroleum fractions, etc.
Receptor	Media contaminated or to which releases occur
Intensity	Concentrations, radiation flux, volumes, release rates, etc.
Effects	
Entity	Affected population, species, assemblage, ecosystem, etc.
Nature	Mass mortality, temporal decline in abundance, reduction in range, etc.
Magnitude	Number killed, proportional decline, etc.
Exposure	
Entity	Exposed population, species, assemblage, ecosystem, etc.
Type	Chemicals in water, chemicals in organs, biomarkers, etc.
Intensity	Concentration, dose, proportion of individuals exposed, etc.
Spatial and temporal properties	
Duration	Duration of release event, time since initiation of continuous release, duration of population decline, etc.
Frequency	Interval between events, number of events, etc.
Timing	Occurrence relative to biological events such as spawning, relative to environmental factors such as low flows, etc.
Scale	Area contaminated, area within which effects occurred, heterogeneity, etc.

exposures, such as high chemical concentrations in fish, acidified lakes, or oil on birds, that have uncertain causes and consequences.

Properties that should be addressed in the definition of the impetus are listed in Table 7.2.

Environment definition: what is the context of the assessment?

Once the impetus for the assessment has been characterized, the problem formulation proceeds with the definition of the environment to be assessed, the assessment end-points, the measures of exposure and effects, and the conceptual model. These steps are conceptually interlinked and must in effect be performed simultaneously. That is, knowledge of the ways in which the released materials move through the environment, their fate in media, the modes of exposure of organisms, the responses of organisms to exposure, and the interactions among organisms all go into defining the physical and temporal scope of the assessment, the assessment end-points, and the processes that will be included in the conceptual model.

The environmental definition must be sufficient to allow the assessors and risk manager to perform the problem formulation and to justify their decisions to readers and reviewers. However, the environmental definition should be limited to a description of the features of the environment that are relevant to understanding and performing the assessment. It should not be an occasion for cataloguing all of the species and physical features of the area, as has been done in many environmental assessments. For example, all of the species of plants in a region need not be enumerated to assess risks of soil contamination to plant production. In fact, because the relative sensitivity of plants to chemicals other than herbicides is poorly characterized, a plant species list would be of little use. Rather, it is sufficient to describe the major vegetation types, any plant species that are of special concern because they are rare or have particular cultural values (e.g. medicinal plants), and any plants that have properties that are relevant to estimation of exposure or effects (e.g. metal accumulators).

Assessment end-point definition: estimate risks to what?

Assessment end-points are the explicit expression of the environmental value to be protected (Suter 1989, 1993a; Risk Assessment Forum 1992). It is the ecological equivalent of the lifetime cancer risk to a reasonable, maximally exposed individual in human health risk assessments, or the annual probability of a Class 9 accident (fuel melt with breach of containment) in risk assessments for nuclear power plants. Therefore, it must be something that is important and can be estimated, not some vague desire like healthy ecosystems. Criteria for selection of assessment end-points are presented in Table 7.3. The selection of the assessment end-points depends on a knowledge of the receiving environment and of the contaminants provided by the assessment scientists, as well as the values that will drive the decision (provided by the risk manager). A completely specified assessment end-point for any ecological risk assessment that measures effects includes: an entity such as a Caspian tern population; a property of that entity, such as young fledged per nest; a level of effects to be detected, such as 15% reduction relative to reference populations; and a desired degree of statistical confidence, such as 20%. Clearly, the outcome of the assessment and the decisions that follow from it depends on the end-points selected.

Conceptual model definition: what do we think is going on?

The conceptual model summarizes the results of the problem formulation and guides the analytical phase of the assessment. It is a working hypothesis about how the hazardous agent or action may affect the end-point entities (Risk Assessment Forum 1992; Barnthouse & Brown 1994; Suter *et al.* 1994). Typically, it includes a graphical representation of the entities involved and the processes that link them. It also includes a narrative that describes those entities and processes plus aspects of the problem formulation that are not included in the graphic, such as the spatial and temporal limits. An example of a graphical conceptual model is presented in Fig. 7.2.

Table 7.3 Criteria for selection of assessment end-points for ecological risk assessments. (After Suter 1989, 1990a; Risk Assessment Forum 1992.)

Policy goals and societal values Because the risk to the assessment end-point is the basis for decision-making, it is important that it reflects the policy goals and societal values that the risk manager is expected to protect
Ecological relevance Entities and properties that are significant determinants of the properties of the system of which they are a part are more worthy of consideration than those that could be added or removed without significant system-level consequences. Examples include a keystone predator species or the process of primary production
Susceptibility Susceptible entities are those that are potentially highly exposed and responsive to the exposure
Operationally definable Without an unambiguous operational definition, it is not possible to determine what must be measured and modelled in the assessment, and the results of the assessment are too vague to be balanced against costs of regulatory action or against countervailing risks

Assessment plan: How are we going to estimate risks?

The output of the problem formulation is a plan for conducting the assessment, including what sampling, analysis, testing and measurement will be conducted and how the *ad hoc* data and existing data will be used to estimate risks. It should include specific sampling, analysis and measurement methods that will be used to characterize effects, sources and exposure, as well as the models that will be used to relate those measures to each other and estimate risks. The assessment plan should include a plan for assuring the quality of the data, including plans for data management. It should also specify why the measurements are needed and how they will be used in the assessment. Therefore, the development of the assessment plan requires that the assessors plan the analytical and risk characterization phases of the risk assessment so that data needs are specified.

Measures of effects or measurement end-points are statistical or arithmetical summaries of observations used to estimate the effects of exposure on the assessment end-point (Suter 1989; Risk Assessment Forum 1992; Suter 1993a). They include test end-points, such as LD_{50} values or dose–response functions, and summaries of field measurements, such as catch per unit effort or mean density. The distinction between assessment and measurement end-points is needed because, for the following reasons, the end-point for a set of measurements should not simply be adopted as the end-point of the assessment.

1 The property measured is, at best, an estimate of the property to be protected (e.g. a mean fledging rate from a sample of tern nests) and often it is a related effect that must be extrapolated to the assessment end-point (e.g. a chronic no-observed-adverse-effect level (NOAEL) for chickens).

2 The results of a particular series of observations can be summarized in a number of ways, resulting in alternative measurement end-points which may be more or less useful for estimating risks to the assessment end-point.

3 Multiple measurement end-points may be used to estimate risks to a single assessment end-point.

The utility of a measurement end-point depends not so much on its inherent properties, such as cost or variance (although these are always important), but on its ability to contribute to an estimate of the risk to the assessment end-point given the available extrapolation models and other available information. A highly precise measurement obtained at low cost is worthless unless it can be used to estimate a property that someone cares about enough to make the basis for a decision.

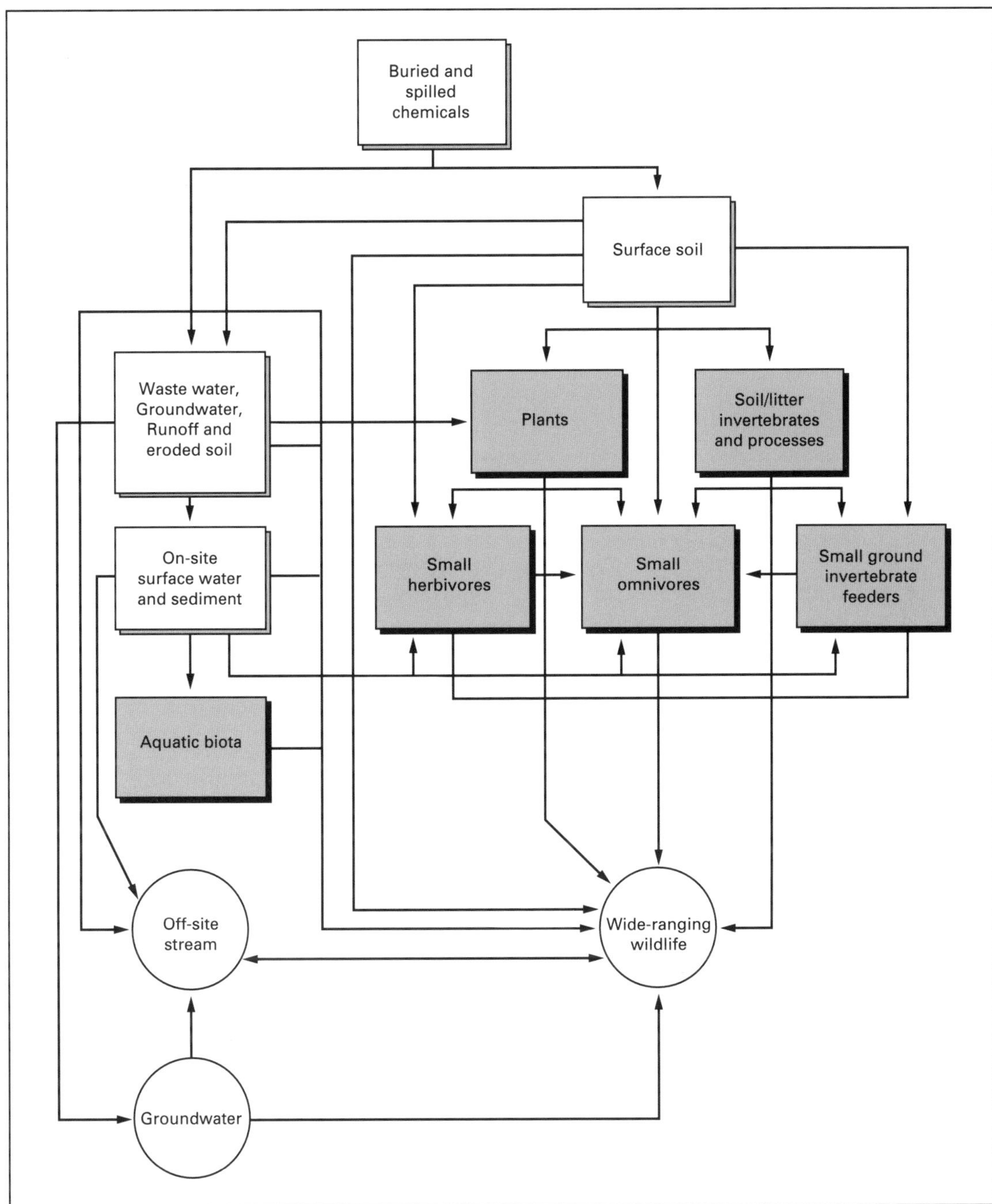

Fig. 7.2 A generic conceptual model for ecological risk assessment for a waste disposal site. The shaded boxes represent the classes of receptors for which assessment end-points will be identified. Circles represent receptors that are outside the scope of the assessment because they are more appropriately included in a larger scale multi-site risk assessment.

Measures of exposure must also be specified during the problem formulation. Most commonly in ecoepidemiological studies these are summaries of concentrations of contaminants in ambient media (e.g. means over time at each sampling location, or hectares of soil with concentrations greater than some prescribed value). Body burdens of chemicals and biomarkers of exposure potentially provide measures of internal exposure. If measures of concentrations are not available, exposure levels may be modelled from source and receiving environment characteristics. As with measures of effects, there may be policy considerations as well as technical restraints on the measures of exposure. For example, the best estimate of the exposure of aquatic biota to aqueous contaminants may be concentrations in the dissolved phase, but regulators often prefer the conservatism of using total concentrations (dissolved plus suspended particulate). Specification of the measures of exposure should include the media, constituents, limits of detection and enough information about the needed spatial and temporal coverage and desired level of precision to allow the statistical design of the sampling and analysis plan.

If sources are known or postulated, the measurements that will be used to characterize them must be planned. Since mathematical models of transport and fate are typically used to associate sources with exposures, the models must be prescribed so that all needed source parameters can be measured or estimated from measurements.

The risk manager's input

The role of the risk manager in the problem formulation is important and controversial. In nearly all cases, the individual scientists who perform the assessment are not the individuals with the authority to make a decision about removing a product from the market or spending millions of pounds on remediation. Therefore, the assessment scientists should involve the risk manager in the problem formulation to ensure that the assessment will provide the information needed to make the decision and to ensure that there is support for the schedule and level of effort implicit in the problem formulation. In addition it is desirable to distinguish clearly the roles of the risk managers and assessment scientists in the process, so that the biases of the risk manager do not improperly influence the selection of data, models and assumptions. The risk manager must deal with values (i.e. what should be) and the assessment scientist should deal with facts (i.e. what is). Although the distinction between these roles is not perfect (Schrader-Frechette & McCoy 1994), it is possible to make a working distinction in nearly all cases.

The relationship of risk management to problem formulation can be illustrated by a hypothetical example of contaminant deposition in a forest. The assessment scientists might inform the risk manager that tree growth and soil community properties are likely to be affected. The risk manager's response will be determined by the situation and the political and cultural context. A risk manager in the USA might determine that the growth of trees is an assessment end-point but soil community properties are not because he cannot defend a regulatory decision based on effects on soil organisms. However, a Dutch risk manager might consider soil invertebrates to be important as well as the trees. The American assessment scientists might, based on their conceptual model, determine that the risks to trees are in large part mediated by changes in soil properties resulting from effects on soil organisms. In that case, it is incumbent upon them to include effects on soil organisms in the ecological risk assessment, but as an intermediate causal factor and not as an end-point. Although the American and Dutch assessment scientists would both include the soil biota in their field studies, differences in risk management would influence the design of the studies. The Americans would measure effects on the processes performed by the soil community that mediate effects on trees, while the Dutch scientists would measure changes in species composition of the soil invertebrates.

7.2.2 Analysis phase: what do we know about sources, exposure and effects?

The analytical phase of the risk assessment uses the conceptual model, literature data, and the results of any *ad hoc* monitoring and testing to characterize the source, exposure and effects. Although the analytical phase should be guided by the problem formulation, the conceptual model should be treated as a working hypothesis and modified as information becomes available.

Analysis of effects

In retrospective ecological risk assessments, effects data might be available from field monitoring, from toxicity testing of the contaminated media, and from traditional single-chemical laboratory toxicity tests. The analysis of effects summarizes and analyses the data concerning effects in such a way that they can be related to the exposure estimates, thereby allowing characterization of the risks to each assessment end-point during the risk characterization phase. The effects analysis phase is primarily devoted to analysis of the data from single-chemical toxicity tests. This is because, being drawn from the literature, they are relatively poorly related to the conditions being assessed and therefore require the most analysis and interpretation. Even when they are performed *ad hoc*, single-chemical toxicity tests usually use standard protocols and species exposed to pure chemicals in laboratory media.

Single chemical toxicity tests. The analysis of effects must determine which of the available data are relevant to each assessment end-point, and reanalyse and summarize them as appropriate to make them useful for risk characterization. This requires consideration of two issues. The first issue—what form of measurement end-point best approximates the assessment end-point—should have been considered during the problem formulation. However, the availability of unanticipated data and better understanding of the situation after data collection will often require reconsideration of this issue. For example, better understanding of the distribution of contaminants may cause the data to be spatially aggregated differently.

The second issue in analysis of effects is expression of the effects data in a form that is consistent with expressions of exposure. Integration of exposure and effects defines the nature and magnitude of effects given the spatial and temporal pattern of exposure levels. Therefore, the relevant spatial and temporal dimensions of effects must be defined and used in the expression of effects. For example, if the exposures are episodic and brief, then effects that are induced in that time period must be extracted from the effects data for the chemicals of concern and analysis of effects-monitoring data should focus on biological responses such as mass mortalities that could occur rapidly rather than long-term average properties.

Most of the work of effects analysis is devoted to determination of the relationship between exposure and effects for each chemical of concern. In conventional risk assessments, this involves deriving an exposure–response model from laboratory toxicity tests. This requires analysis of the test data to derive a test end-point and extrapolation from the test end-point to the assessment end-point. The extrapolation may be performed in various ways, including the following (Suter 1993a).

Selection It can be assumed that the end-point species, life stages and responses are equal to those in the most sensitive reported test or in the test that is most similar in terms of taxonomy or other factors.

Safety factors A test end-point can be divided by 10, 100 or 1000 to estimate a safe level.

Species sensitivity distributions A percentile of the distribution of test end-point values for various species can be used to represent a level that would be protective of that percentage of the exposed community.

Regression models Regressions of one taxon on another, one life stage on another, one test duration on another, etc., can be used to extrapolate among taxa, life stages, durations, etc.

Mathematical models Toxicodynamic models can be used to estimate effects on organisms from physiological responses and population or ecosystem models to estimate effects on populations or ecosystems from organism responses.

Media toxicity tests. A considerable increase in realism can be obtained by testing the contaminated media rather than individual chemicals in laboratory media. This can be done in at least three ways. The most direct approach is to cage, pen or plant organisms along a gradient of contamination or at contaminated and reference sites. This approach is relatively easy for immobile organisms such as plants and bivalve molluscs and difficult for organisms that are mobile and forage for food. It is highly realistic in that the organisms are subject to realistic conditions and variation in exposure. However, such studies are subject to the effects of variation among sites in conditions other than contamination and to loss of the study due to vandalism, predation or extreme conditions. In addition, cage effects may modify the sensitivity of the organisms.

A more common approach is to bring contaminated and reference media into the laboratory for toxicity testing. This is a very active area of ecotoxicology and tests have been developed for ambient waters, sediments, soils and biota (Mount & Norberg 1985; Norberg & Mount 1985; Schimmel *et al.* 1989).

Finally, the least used technique is to bring contaminated biota into the laboratory and test them. This technique is appropriate if the contaminant is persistent and bioaccumulated, or if it is known to cause persistent injury. For example, herring eggs from areas exposed to spilled oil and from unexposed areas were brought into the laboratory and their hatching rates and frequencies of abnormalities were recorded (Pearson *et al.* 1995). Similarly, eggs from adult fish that have accumulated persistent pollutants in the Great Lakes have been tested for hatchability and fry survival (Mac & Edsall 1991).

Because the exposure component of the exposure–response analysis for these tests is not different from the exposure analysis for the assessment, the discussion of exposure–response analysis for these tests is deferred to the risk characterization. However, as part of the analysis of effects, it is important to consider whether some qualitative or quantitative extrapolation model needs to be applied to the ambient media toxicity tests to make them relevant to the assessment end-point. The types of extrapolation models used with single chemical toxicity tests are potentially useful for these tests as well.

Biosurveys. Biological surveys of effects include a wide variety of techniques for enumeration and characterization of biological populations, communities and ecosystems. In the simplest case, the measurement end-point for the biological survey is an estimate of the assessment end-point. In such cases, the effects analysis consists of summarizing the data in such a way as to reveal the relationship of effects to exposure. Examples would include plotting the species richness of a fish community on exposure axes, such as kilometres upstream and downstream of a source, proportional dilution of the source, or concentrations of a particular chemical.

If the measurement end-points do not directly estimate the assessment end-point, then the relationship between them must be characterized. For example, if data are available for stream macroinvertebrates and the assessment end-point is some property of the fish community, then the relationship between them must be characterized in terms of the trophic dependence of fish on invertebrates, the relative sensitivity of fish and invertebrates, the similarity of their exposure, and other relevant properties.

Indirect effects. Ecological risk assessments have followed human health risk assessments in emphasizing direct toxic effects. However, non-human organisms are much more subject to indirect effects, such as habitat modification and reductions in the abundance of food species, competitors or predators. This is particularly the case for pesticides and other chemicals that have highly selective toxicity. Modern herbicides have low toxicity to animals but they reduce primary production and alter habitat structure, thereby modifying entire ecosystems. Insecticides reduce the abundance of insects and crustaceans, thereby reducing resources for fish, birds, bats and other insectivores. These indirect effects, which should have been identified in the conceptual model, should be quantified as far as possible in this component of the assessment. Biological surveys of areas contaminated by use or disposal of chemicals can potentially reveal indirect effects, but, because the exposures are uncontrolled and unreplicated, indirect effects are difficult to distinguish in such studies. When they are available, the results of microcosm, mesocosm or field tests can be used to quantify empirically the indirect effects, or, for less-selective chemicals, the combined direct and indirect effects. Alternatively, simple assumptions can be made, such as an *x*% loss of seagrass beds will result in an x% reduction in the abundance of species that depend on seagrass for any of their life stages. Less commonly, but more rigorously, ecosystem models may be used to estimate the consequences for all end-point taxa of toxic effects on all modelled components of the exposed ecosystem (Bartell *et al.* 1992; O'Neill *et al.* 1982; Emlen 1989; Suter 1993a).

Exposure–response profile. The output of the analysis of effects is the exposure–response profile. For individual contaminants of concern, this should indicate how the effects increase with increasing duration and concentration of exposure. It should also, to the extent that such information is available and relevant, indicate the effects of environmental variables such as temperature and pH on toxic effects. It should indicate the mode of action and the variation among taxa, life stages, and processes in sensitivity.

For ambient media toxicity tests, the exposure–response profile should summarize the results in terms of the spatial and temporal distribution, the nature and magnitude and the consistency of toxicity. If more than one test is performed on a contaminated medium, the relative sensitivities of the tests should be explained as far as possible in terms of the relative sensitivities of the species and life stages involved, in terms of the nature and duration of the exposure in the test system, or other relevant factors.

Analysis of exposure

Organisms must be exposed before there can be a risk. In risk assessments, the primary purpose of the analysis of exposure is to provide an appropriate estimate of exposure to parameterize the exposure–response model. This purpose is relatively clear and simple in predictive risk assessments which use mathematical models to simulate exposure based on source and receiving environment parameters. However, in retrospective studies this traditional exposure analysis is supplemented by analysis of contaminant concentrations in ambient media and in the receptor organisms and by analysis of biomarkers of exposure.

Whether based on measurements or modelling, the exposure estimate must have appropriate dimensionality. Three dimensions must be considered: intensity, time and spatial scales (Suter 1993a; Suter *et al.* 1994). For chemical contaminants, intensity is most commonly expressed as concentration, but dose may be more appropriate. Time includes duration of exposure, frequency and timing relative to biological events such as hatching or physical events such as snowmelt. Space should be defined in biologically relevant terms. For example, dietary exposures should be averaged over areas corresponding to foraging ranges.

External exposure can be estimated from analyses of food items or of the contaminated media (water, sediment and soils) to which organisms are directly exposed. Analyses of ambient media are nearly always the most abundant data type available to the ecological risk assessor. The analysis of exposure must deal primarily with the question: to what media and what components of the media are the receptors exposed? Examples include the following.

1 To what depth in the soil should exposure of plants or invertebrates to chemicals be assumed to occur? Plant roots may extend several metres in depth, but most roots are in the A horizon.

2 Are aquatic organisms exposed to contaminants associated with suspended particles or to only the dissolved fraction? Dissolved concentration is the best estimate of exposure, but most analyses are of total recoverable concentrations.

3 Should wildlife be assumed to drink waste waters, even when the waters are highly saline or have extreme pH values?

These questions can be answered by research, but are more often answered by judgement or policy. For example, it is possible to either determine the distribution of plant roots in a particular combination of vegetation and soil type, or to assume a default value for rooting depth, such as 30 cm.

Internal exposure is measured either as body burdens or biomarkers. Measures of internal exposure have the considerable advantage that they are better approximations of the effective exposure, which is the exposure at the site of action (McCarty & Mackay 1993). In particular, they avoid the issues of avoidance, bioavailability and uptake, because assimilated chemicals were not avoided, were bioavailable and were taken up. Biomarkers of exposure are biochemical changes in an organism that are indicative of exposure (Huggett *et al.* 1992). They indicate that the chemical or class of chemicals has been taken up and is in a biologically active form. For example, blood or brain cholinesterase levels can be used to estimate exposure to organophosphate pesticides even though the pesticide does not form persistent residues in the organism. Biomarkers may be particularly useful when the interest is in estimating total exposure to a group of chemicals that have a common mode of action. For example, the H4IIE bioassay provides a toxicity-normalized measure of exposure to chlorinated diaromatic hydrocarbons (Tillit *et al.* 1991; Giesy *et al.* 1994b). The greatest impediment to use of internal exposure metrics in ecological risk assessment is the general lack of exposure–response data that would allow estimation of effects. A related problem is the lack of agreement about how the body burdens should be measured: Whole body or specific organs? Depurated or not? Washed or not? Finally, many of the biomarkers of exposure are not very specific. They may respond to a variety of contaminants and to natural variables.

Exposure analysis based on modelling of transport and fate is useful when analyses of contaminants in the environment cannot be performed or when the data must be supplemented or confirmed by modelled values. Analytical data may be inadequate because of low quality, because sampling did not cover all locations or conditions of concern, because sample sizes are small, because the limits of detection were not low enough, or because the analyses did not include the fractions of the media or the forms of the chemical that are relevant to ecological exposures. In addition, exposures that involve multiple routes of uptake, such as the consumption of various food items, water and soil by wildlife, must be modelled even when extensive chemical analyses have been performed (Sample & Suter 1994).

Many models are available for describing the processes of transport, transformation and exposure (Calamari 1993), and standard sets of models have been developed by the USEPA and many other environmental agencies. The selection of a model should be based on the following criteria.

Suitability Does it include the exposure pathways, represent the receiving environment, have appropriate temporal and spatial scales and estimate the appropriate exposure parameters?

Acceptability Has the model been verified or validated; is it based on defensible scientific principles; and is it accepted by the scientific and regulatory community?

Practicality Can the model be parameterized and implemented given the limits of time, resources and scientific understanding?

In retrospective assessments, many of the model parameters can, and should be, obtained from the specific site or region being modelled. As more data become available, the need for simulation declines. For example, if concentrations of chemicals in invertebrates are adequately measured, it is not necessary to model their exposure in order to estimate exposure of their predators.

The output of the analysis of exposure is an exposure profile. It summarizes the exposure estimates in terms of the intensity, time and extent dimensions. It explains the measurements, models and assumptions used to generate the estimates. Finally, it presents the uncertainties associated with the estimates, explains their origins, and suggests how further iterations of the assessment might reduce them.

Analysis of sources

For source-driven assessments, the source is described in the problem formulation. For effects- and exposure-driven assessments, the source is not defined a priori and must be characterized in the analysis phase of the assessment. That is, in attempting to assess the cause of particular effects or exposures, it is necessary to seek out potential sources and characterize them. The relevant characteristics are the same as discussed above (section 7.2.1), and the same problem exists of defining what constitutes a source for purposes of the assessment. The differences are that the source is undefined, there may be many candidate sources, and more than one source may contribute. Therefore, on the basis of chemical and physical measurements and analyses one must characterize each candidate source in sufficient detail so that it will be possible, in the risk characterization, to infer their contributions to the exposure and effects (see p. 191).

7.2.3 Risk characterization: what is our risk estimate and how good is it?

Risk characterization consists of integration of the available information about exposure and effects, integration of information about sources and exposures, analysis of uncertainty, weighing of evidence, and presentation of conclusions in a form that is appropriate to the risk manager and stake-holders (Fig. 7.1). The integration process should be carried out for each line of evidence independently so that the implications of each are explicitly presented. This makes the logic of the assessment clear and allows independent weighting of the evidence. For each line of evidence, it is necessary to evaluate the relationship of the measurement end-point to the assessment end-point, the quality of the data, and the relationship of the exposure metrics in the exposure–response data to the exposure metrics for the site.

This section deals with risk characterization for retrospective ecological risk assessments in general, and includes assessments of contaminated sites and spills and other assessments for which sources are known. The problem of determining causation when apparent effects are reported is related, but is sufficiently distinct to be treated separately (section 7.3).

Integration of environmental monitoring data: what risks are inferred from biological surveys?

The first line of evidence is supplied by the environmental monitoring programme. This line of evidence is the most realistic in that it represents the actual state of the environment, but it provides relatively poor evidence of causation. In many cases, the measurement end-point (e.g. number of fish species in 10 seine hauls) will be an estimate of the assessment end-point property (e.g. fish species richness) so that no extrapolation will be necessary. It is only necessary to estimate the uncertainty associated with the measurement end-point. However, in other cases it is necessary to consider the relationship of the measurement end-point (e.g. fledging success) to the assessment end-point (e.g. likelihood of population extinction). In this case, the extrapolation could be performed using a demographic model, but in many cases the extrapolation is performed simply by professional judgement. In some cases, monitoring data do not provide a basis for estimating risks, but they provide supporting evidence. For example, small-mammal trapping is seldom sufficient to estimate effects on populations, but if all animals trapped at a site were young of the year or if many animals had pathologies, the findings would tend to support other evidence suggesting the occurrence of toxic effects.

The quality of monitoring data must be evaluated. This includes not only the usual quality assurance and quality control (QA/QC) issues, such as whether procedures were followed and whether detection limits were adequate, but also more fundamental questions about the appropriateness of the procedures (e.g. if seining is an appropriate sampling technique for this fish community). At this point it is important to talk with the individuals who performed the monitoring in order to learn about strengths and weaknesses of the data (e.g. few fish were collected on a particular date because flows were extremely high). The quality of biological survey data strictly limits their interpretation. However, even minimal or qualitative information is potentially useful because it constrains the judgements that can be made concerning the state of the system. For example, a visual survey of a terrestrial site can serve to indicate that it is vegetated, that the dominant species are the same as on a nearby uncontaminated site, that the density is similar, and that the plants are not visibly injured. Such a finding does not mean that significant phytotoxic effects have not occurred, but they put a cap on the potential severity of the effects. If such conditions exist on a site where an analysis of the soil indicates that chemicals occur at severely phytotoxic concentrations, then one might re-examine the relevance of the phytotoxicity data or conduct additional studies to determine whether speciation or some other factor in the toxicity tests is not applicable to the site.

The risk characterization for the biological survey data must estimate the level of apparent effects and evaluate whether they are real and associated with the contaminants rather than other environmental factors. If the exposure is treated as categorical (i.e. uncontaminated versus contaminated or discrete classes of contamination), the estimate of effects is the difference in the levels of the end-point properties between or among the categories. If the exposure is treated as continuous (e.g. a gradient downstream from an outfall), then the effects can be derived from an exposure–response curve, as in toxicity tests. Results may be expressed as the estimated level of effects, or the likelihood that effects exceed some prescribed threshold. In any case, the issue of the reality of apparent effects must be considered in terms of the possibility that apparent effects are due to sampling error or confounding variables such as habitat difference (section 7.4). The association of the effects with the contaminant consists of defining contaminant concentrations and temporal dynamics for each exposure category or for the exposure gradient. The relationship of the exposure metrics to the effects must not be taken for granted. Measurements of chemical concentrations may be inadequate not only because of poor quality but also because of site-specific considerations such as the possibility that effects are imposed by episodes of high concentrations that were not sampled.

Integration of media toxicity data: what is the cause of observed toxicity?

The second line of evidence is the testing of the contaminated media for toxicity. Most of these are tests of the toxicity of media that have been brought into the laboratory. The relationship of the test end-points to assessment end-points depends on the modes of exposure in the tests being relevant to the field exposures and on the test organisms, life-stages and responses being relevant to the assessment end-point. These questions are answered in a generic sense by the validation of media tests against field surveys. For standard tests of ambient water used in the USA it has been shown that where toxicity is detected, the species richness of aquatic communities is diminished (Dickson *et al.* 1992; Hartwell *et al.* 1995a,b). For tests that have not been validated against the field, inferences must be made, such as reduction in seedling growth in the laboratory is equivalent to reductions in primary production of the plant community or of growth of sensitive plant species (depending on the form of the assessment end-point).

The quality of these tests is often limited by the performance of test organisms in reference media. Diseases, upstream contamination or unsuitable physical–chemical properties of the media may cause organisms to die or perform poorly in the tests. In general, it is advisable to use both control media (e.g. laboratory water) to determine that the test procedures are adequate and reference media (e.g. upstream water) for determination of the incremental toxicity of the contaminants of concern. However, appropriate control soils and sediments have proved difficult to formulate because of the importance of texture, nutrient levels, organic matter content and other variables.

The relationship of effects to exposure is relatively straightforward if analyses are performed on the tested media; i.e. the effects are caused by the constituents of the tested media and those media are the ones to which organisms in the field are exposed. The principal issue to be addressed is the degree to which the test media represent the variance over space and time of contamination levels in the field. The assessors should use information about the contaminant concentrations in the test media relative to the full range of concentrations reported for the site, and their understanding of the factors that control variance in exposure (e.g. hydrology and sediment texture), to determine whether tests are representative of the worst conditions that persist for a sufficient period of time or over a sufficient area to cause significant effects on the assessment end-point.

Interpretation of media test results is relatively simple: if toxicity is observed, the medium is toxic, and, for most currently used tests, the responses measured are clearly relevant to population or community-level end-points. If toxicity is not observed it cannot be concluded that there have been no significant toxic effects in the field. As discussed above, the sensitivity and relevance of the test must be evaluated to avoid false-negative conclusions. If toxicity is observed in any tests, the distribution of the effects relative to sources or to contamination levels should be demonstrated; i.e. the relationship of exposure expressed as categories (e.g. areas of a site with different levels or types of soil contamination), as concentrations of a marker contaminant (a chemical that is believed to be primarily responsible for toxicity or is representative of the major contaminants), or as spatial gradients (e.g. distance from a source) to the frequency or intensity of toxic effects should be tabulated and plotted.

Media toxicity tests are usually used simply to indicate whether and to what extent the contaminated media are toxic. However, if a sufficient number of tests are performed over a sufficient range of contaminant concentrations, multivariate statistics may be used to identify individual contaminants or groups of contaminants that are responsible for the effects and, with particularly good data sets, concentrations that correspond to thresholds for toxicity. Alternatively, chemical fractionation and treatment techniques can be used to identify constituents that are responsible for toxicity. Techniques for water are described in Chapter 12 of the *Handbook of Ecotoxicology* (a companion to this Handbook) and techniques for sediment are described by Ankley *et al.* (1992).

Integration of chemical toxicity data

Single-chemical toxicity data provide the third major line of evidence. They are in general much more tenuously related to the assessment end-point and to events in the field than the other lines of evidence. The tenuous relation of the standard test end-points to assessment end-points should have been dealt with in the analysis of effects by applying extrapolation models (see p. 185).

The relationship to exposure is commonly expressed as a quotient of the ambient exposure concentration (*AEC*) divided by the toxicologically effective concentration (*TEC*):

$$HQ = AEC/TEC$$

The *TEC* may be a test end-point, a test end-point corrected by a factor or other extrapolation model, or a regulatory criterion or other benchmark value. This type of analysis is commonly used for screening purposes, to determine whether particular chemicals are credible contributors to risk and therefore worthy of further assessment. In that case, conservative *AEC* values, such as 95% upper confidence limits or maxima of measured ambient concentrations, are used and $HQ > 1$ is treated as evidence that the chemical is worthy of concern.

If exposure, effects or both are expressed as distributions, the results of risk integration can be expressed as the probability that $HQ \geq 1$ (Suter *et al.* 1983). Distributions of *TEC* can be derived from the variance on regression models that relate test species and life stages to species and stages of interest, distributions of test end-points, or probabilistic population or ecosystem models (O'Neill *et al.* 1982; Suter *et al.* 1983; Sloof *et al.* 1986; Barnthouse *et al.* 1987; Bartell *et al.* 1992).

Integration of exposure and sources: where did it come from?

This question may be peripheral to a retrospective risk assessment, or central. For example, immediately following the *Exxon Valdez* oil spill there was little concern for determining the source of oil on marine birds and mammals because the source was manifest. However, years after the spill it was necessary to consider whether ongoing exposures to petroleum hydrocarbons were due to mobilization of old spilled oil or due to ongoing relatively small releases of oil from tankers and other boats. The principal approaches to determining sources of exposure are empirical and modelling.

Empirical approaches to determining sources may be qualitative or quantitative. Qualitative approaches may be as simple as identifying the sole source that could be responsible for a particular exposure, given the distribution of contamination. A more typical qualitative approach is to analyse for contamination upstream and downstream of the proposed source in order to verify that the source is in fact responsible. Other qualitative approaches are more ingenious. For example, atmospheric deposition was demonstrated to be a significant source of exposure to persistent chlorinated chemicals in the Laurentian Great Lakes by analysing them in waters that receive them only by atmospheric deposition, in the lakes on Isle Royale, an undeveloped island in Lake

Superior (Czuczwa & Hites 1986). For complex materials such as petroleum and polychlorinated biphenyls (PCBs), the contribution of different sources can be estimated quantitatively by fingerprinting. For example, following the *Exxon Valdez* spill, the contributions of that spill, spills of diesel fuel, and other sources were estimated from the relative proportions of polyaromatic hydrocarbons (PAHs) in biological samples (Bence & Burns 1995).

Modelling for source identification and characterizing, termed receptor modelling, is the logical inverse of modelling to predict the transport and fate of releases of chemicals (Gordon 1988). That is, rather than using release rates to predict exposures, concentrations in biota and media are used to estimate the contributions of potential sources. Such modelling cannot only identify sources, but can apportion exposures among multiple sources.

Weight of evidence: what does it all mean?

This section addresses the general issues of weighing the evidence concerning the nature and magnitude of ecological risks. Inference in ecological risk assessments is made by weight of evidence rather than traditional scientific standards of proof (Risk Assessment Forum 1992; Suter & Loar 1992). The traditional standard for inference in science is, in effect, proof beyond a reasonable doubt. That standard is embodied in the use of a 95% confidence requirement before a hypothesized phenomenon is deemed to be demonstrated and in scepticism concerning results that have not been replicated in an independent study. This is appropriate for pure science, which is engaged in adding to the body of reliable knowledge concerning the nature of the world. However, risk assessors do not have the luxury of suspending judgement until a scientific standard of confidence is met. Decisions are made on schedules that are not within the control of scientists and will be made on other bases if scientific input is not available. Suspension of judgement constitutes an abrogation of responsibility.

Given that risks have been estimated based on each line of evidence, the process of weighing the evidence amounts to determining what estimate of risks is most consistent with those results. If the assessment end-point is defined in terms of some threshold for significance, then the process can be conducted in two steps. First, for each line of evidence determine whether it is consistent with exceedence of the threshold, inconsistent with exceedence, or ambiguous. Second, determine whether the results as a whole indicate that it is likely or unlikely that the threshold is exceeded. If the results for all lines of evidence are consistent or inconsistent, the result of the weighing of evidence is clear. Assuming that there is no consistent bias in the assessment, agreement among multiple lines of evidence is strong evidence to support a conclusion. However, if there are inconsistencies, the true weighing of evidence must occur. The weights are determined based on the following analyses.

Relevance List the ways in which the results could be wrong because they are fundamentally inappropriate or so inaccurate as to nullify the results and evaluate the likelihood that they are occurring in this case. For example, single-chemical toxicity tests could be performed with the wrong form of chemical, in media that differ from the site media in ways that significantly affect toxicity, or the tests may be insensitive due to short duration, a resistant species, or the lack of measures of sublethal effects.

Scope Determine whether the data encompass the relevant range of conditions. For example, if streams are not sampled following a rainstorm that occurs shortly after a pesticide application, pesticide analyses are unlikely to detect concentrations that could be associated with episodes of acute lethality in aquatic organisms.

Quality Evaluate the quality of the data in terms of the protocols for sampling and analysis, the expertise of the individuals involved in the data collection, the adequacy of the quality control during sampling, sample processing, analysis and recording of results, and any other issues that are known to affect the quality of the data for purposes of risks assessment.

Quantity Evaluate the adequacy of the data in terms of the number of observations taken. Results based on small sample sizes are always given less weight.

These and other considerations can be used as

points to consider in forming an expert judgement or consensus about which way the weight of evidence tips the balance. Table 7.4 presents an example of a summary of the results of weighing evidence based on this sort of process. Alternatively, the considerations can be used to assign a grade to each line of evidence (e.g. high, moderate or low weight). This still leaves the inference to a process of expert judgement or consensus but makes the bases clearer to readers and reviewers. Finally, a scoring system could be developed that would formalize the weighing of evidence. For example, a numerical weight could simply be assigned to each line of evidence, a plus or minus sign assigned depending on whether the evidence is consistent or inconsistent with the hypothesized risk and the weights summed across lines of evidence. Such systems are not currently used, although one is being developed by the state of Massachusetts. They would have the advantage of being open and consistent, but may not give as reasonable a result in every case as a careful *ad hoc* weighing of the evidence would. However the weighing of evidence is performed, it is incumbent on the assessment scientist to make the basis for the judgement as clear as possible to readers and reviewers.

Uncertainty in the risk characterization

Uncertainties should be listed and, as far as possible, quantified during all phases of the risk assessment.

Table 7.4 A summary of risk characterization by weight of evidence for a fish community in a stream contaminated by industrial waste.

Evidence	Result*	Explanation
Biological surveys	+	Fish community abundance and species richness are both low, but habitat quality may be a causal factor
Bioindicators	±	Frequencies of histopathologies were elevated in largemouth bass and bluegill, and levels of fecundity and other indicators of reproductive condition were reduced in largemouth bass, relative to reference fish
Toxicity tests	+	Water was toxic to medaka and redbreast sunfish embryos, but apparently not to fathead minnows and *Ceriodaphnia*. Tests were not conducted during high-contamination episodes
Fish analyses	±	Concentrations in maximally contaminated channel catfish exceed the concentration causing reduced growth and liver pathology in that species, but most concentrations are much lower and the toxic body burdens were for larval fish
Media analyses	+	Copper, Hg, and possibly Ni and Ag were detected episodically at toxic concentrations, but average concentrations are non-toxic
Weight of evidence	+	Although the evidence is not strong, this reach appears to pose a significant risk to the fish community from episodes of high metal concentrations and from polychlorinated biphenyls, which are mostly undetectable in water

* +, indicates that the evidence is consistent with the occurrence of a 20% reduction in species richness of abundance of the fish community. –, would indicate that the evidence is inconsistent with the occurrence of a 20% reduction in species richness of abundance of the fish community. ±, indicates that the evidence is too ambiguous to interpret.

Although the concept of uncertainty is central to risk assessment, and much has been written on the subject of risk and uncertainty, risk assessments in the regulatory arena have made little use of formal quantitative uncertainty analysis methods (Morgan & Henrion 1990; NRC 1994). The uncertainty in the various components of risk assessments must be estimated using techniques that are appropriate to the data and models. The types and sources of uncertainties have been catalogued in various ways (Suter 1990b). They include natural environmental variability, sampling variance, measurement error, extrapolation error and model structural uncertainty.

The analysis of uncertainty should present the uncertainties associated with each line of evidence and the uncertainty associated with the final risk estimate. The estimation of uncertainty associated with a weight-of-evidence analysis is not straightforward. In general, the uncertainty should be less than that of any of the lines of evidence. That is, for a case with an assessment end-point of a 10% reduction in fish species richness, if biological surveys show an 80% reduction in the number of fish species relative to the average of reference sites, if more than half of fathead minnows die in all toxicity tests of site waters, and if five chemicals are found in site waters at concentrations exceeding chronic values for at least one fish toxicity test, the uncertainty concerning the conclusion that effects are significant is small. Given that consistency and the fact that effects are estimated to be well above the threshold for significance, it matters little that some component uncertainties are large. For example, the sensitivity of other fish species may differ by more than an order of magnitude from that of fathead minnows, but the biological survey data indicate that the fish in this community were not so much less sensitive as to negate the results of the tests.

In less ideal cases, the evidence is inconsistent and indicated effects are near the threshold for significance. In such cases, the estimation of uncertainty must be based on the logic used to weigh the evidence. If the conclusion about risk is based on one line of evidence that is judged to be much stronger than the others, then the uncertainty associated with that analysis is the overall uncertainty. If there are conflicting lines of evidence and no one line is strong enough to provide a clear basis for the conclusion, then the uncertainty is clearly high. In any case, one should at least provide an estimate of uncertainty on a qualitative scale (low, moderate, high, etc.) and identify the most important sources of uncertainty.

7.2.4 The risk assessment–risk management interface

The risk assessor's task does not end with the production of a document containing the risk assessment. The next critical step is the process of educating the risk manager. Depending on the relationship between the risk assessors and manager, this may be an informal and friendly process or a formal process of exchanging written comments and responses or a series of formal and potentially adversarial meetings. However, it is always desirable to view this interaction as an educational one, in which the assessors ensure that the risk manager understands the results, including the uncertainties, and the risk manager is sufficiently open about his level of understanding and his technical and policy concerns to allow the assessors to expand, or clarify, the risk characterization in order to make it more useful.

Presentation of the results of the risk assessment to the public is an extension of the risk assessment–risk management interchange in that a democratic public is the ultimate risk manager. Although there is an extensive literature of communicating risks to the public, it does not specifically address the presentation of ecological risks. However, the same advice about an open process of mutual education is applicable to the public as well as the designated risk manager.

Because of the potential complexity of retrospective assessments, this process requires more creativity and effort than simply condensing the risk assessment into an executive summary. Maps, graphs, graphical conceptual models, and other presentations can be helpful. If there is a long-term relationship between a set of risk assessors and risk managers, it may be worthwhile developing interactive software that allows the risk managers to examine the consequences of alternative assumptions or decisions.

7.2.5 Phasing and iteration of assessments

It is often efficient and technically advisable to perform ecological risk assessment iteratively. This allows assessors to make efficient use of data and to focus sampling and analysis on critical data gaps. In general, three sorts of assessments are used in an iterative strategy: screening, interim and conclusionary.

First, before any sampling and analysis is performed, existing data and the results of site reconnaissance are used to perform a screening assessment. In screening assessments, chemicals in media are screened to determine whether their concentrations are sufficiently low that they can be eliminated from further consideration. This is done by comparing conservative estimates of exposure concentrations to conservative ecotoxicological benchmarks, by comparing concentrations to background or by comparing the list of detected chemicals to waste inventories. Chemicals that cannot be eliminated because their concentrations are too high, because their concentrations are unknown, or because their toxicities, background concentrations or occurrence in the waste are unknown must be retained.

Interim assessments are performed at an intermediate stage in a sampling and analysis programme to use the early results to focus and guide further studies. They are performed like screening assessments and have the same purpose of eliminating some chemicals, media and receptors from further study so that resources can be focused on the greatest potential hazards. In addition, they provide better estimates of variance so that sampling designs can be adjusted. Finally, they indicate which techniques are inadequate. For example, detection limits may be too high due to interferences, or biological sampling techniques may not be collecting enough organisms.

Conclusionary assessments are those that are intended to provide the basis for reaching a decision about remediation or regulation. They should have data of adequate quality and quantity to provide a good basis for risk estimation and management. However, even when data are meagre, they should provide the best possible description of the risks and associated uncertainties as a basis for the decision.

7.3 CAUSAL INFERENCE IN ECOLOGICAL EPIDEMIOLOGY

Inference in ecological epidemiology is primarily concerned with demonstrating the credibility of the idea that certain effects have occurred and that they resulted from exposure to a particular agent. Once those inferences have been made, the current and future implications of that causal relationship can be estimated.

7.3.1 Inference by weight of evidence

As discussed above (section 7.2.3), inference in ecological risk assessment is made by weight of evidence. The need for this approach is clear in epidemiology because direct evidence of causation is usually unavailable, and, even with infinite time and resources, epidemiological and ecoepidemiological studies do not allow decisive study designs and clear results (Hill 1965; Susser 1986a,b). The nature and cause of events that are observed in the field are always questionable, because controlled, replicated, and randomized studies are not possible. Fish kills, forest declines, oil spills, etc., are unique events. Multiple samples taken at exposed and unexposed sites or sites where effects were, and were not, observed are pseudoreplicates because the treatments are not replicated (Eberhardt 1976; Hurlbert 1984). For example, multiple samples taken at sites above and below a waste-water outfall are pseudoreplicates because the outfall is not replicated. The differences between upstream and downstream samples may be due to the stream gradient, or due to anthropogenic effects other than the contamination that is thought to be the cause (siltation, flow depletion or augmentation, etc.). In addition, the sites are not randomly assigned to treatments. Therefore, a difference between two sites may be due to some bias in the assignment of sites, including contamination from other sources, habitat differences, geochemical factors, etc. Since observational studies cannot be conclusive, they must be supported by other lines of evidence and the results of the multiple lines of evidence must be weighed to support a conclusion.

In the absence of clear and complete information, one must weigh all available information to reach a conclusion. The result of such a process is not as simple as a statement that a hypothesis has been rejected. Rather, it is typically a statement about the nature of an effect and associated cause and the associated uncertainty. For example, one might simply indicate that the number of birds killed by a spill was estimated to be 1000 but may be as high as 5000, or the likelihood that as many as 100 birds were killed is estimated to be 0.85. The remainder of this section is concerned with how this weighing of evidence is performed in ecoepidemiological studies.

7.3.2 Is there a real effect?

The first question that must be answered in an ecoepidemiological study is: has a real effect occurred? In human epidemiology this question is answered by comparing the incidence of a disease in a particular population with the expected incidence. An improbably high incidence is indicative of a real effect of some causal factor. By analogy, an ecological effect is a change in the state of an ecological system that is sufficiently anomalous to require some explanation rather than being attributed to natural variability. The cause may be anthropogenic or due to some extraordinary natural phenomenon. The attribution of effects may result from reported incidents (e.g. dead geese on a golf course), the accumulation of observations (e.g. the decline in abundance of peregrine falcons in the 1950s), or results of studies designed to elucidate effects (i.e. environmental monitoring). In any case, it is necessary to somehow infer that the apparent effect is real and not simply a result of normal variability in mortality or abundance before assigning a cause or working out the implications.

The simplest type of effect to demonstrate is the occurrence of conditions such as mass mortalities or fishless streams that are so anomalous that they are manifestly not a result of random variance. Use of this argument depends on the degree of scepticism of the risk manager and stakeholders. A single dead bird in a 10-ha cornfield does not constitute a manifest effect but a thousand dead birds does. There are no medical records or actuarial statistics that could be used to determine what constitutes an abnormal mortality rate for birds. Rather, the threshold number of mortalities for manifest effects depends on the experience and perspective of the individuals making the judgement. If possible, such thresholds should be defined a priori. For example, many governments have definitions of reportable fish kills.

A second approach is to compare the state of a system to the state predicted by a mechanistic or empirical model. That is, if the factors that control the parameter of concern are known, then those parameters can be measured and used to predict what the state of the system should be. For example, one might use habitat models to determine whether a species should be present at a location and its approximate abundance (Bovee 1982; Orth 1987; Bovee & Zuboy 1988; Wright *et al.* 1989). In general, this approach is severely limited by the availability of reliable models and the knowledge needed to develop them. Most existing models are relatively simple and empirically derived. However, they can provide increased objectivity in judging the reality of large apparent effects relative to the expert judgements used to determine that effects are manifestly real.

Most commonly, the effects are not manifest and the system is not well enough understood to allow modelling of the unexposed state. In such cases one may attempt to determine whether effects are occurring by comparing the putatively affected system with some reference system. Comparisons can be performed in various ways, and defining the appropriate comparison is difficult and often controversial. Although field studies cannot definitively establish causation because the causal factors are not controlled, they are appropriate to discriminating effects from natural patterns or random variance. That is, the goals of field study designs are to establish the likelihood that sites are not different, the likelihood that a site has changed over time, etc. This suits them more to answering the questions of the likelihood that real effects have occurred than questions about causation. An improbable difference in a response parameter between a contaminated site and a set of reference sites is weak evidence that the contamination is responsible for the difference but is stronger evidence that something is affecting the contaminated site that is worthy of investigation. Good references on field study design include Green (1979, 1984), Eberhardt & Thomas (1991), Underwood (1994), and Wiens & Parker (1995).

Observational studies such as those used in ecoepidemiology are always subject to confounding variables. Decreasing the likelihood that they will either cause false apparent effects or hide real effects requires effort and ingenuity. First, the influence of confounding variables can be diminished by increasing the types and numbers of comparisons performed. As is discussed below, confounding variables are unlikely to occur and act similarly at multiple times and locations. Second, the environment at the study sites and potential reference sites can be characterized in detail so as to ensure as far as possible that the sites are ecologically similar and therefore less likely to have important confounding variables. For example, the USEPA's procedure for assessing effects on stream communities exposed to aqueous effluents begins with a habitat analysis of the exposed sites and reference sites (Plafkin *et al.* 1989). Third, the relationship between the measured parameter and environmental variables that potentially confound the analysis of anthropogenic effects can be studied. The information concerning environmental variables can then be used in site-specific models, such as multivariate analysis of covariance to identify residual variance that may be due to real effects.

The simplest, most common, and least reliable comparison is of a pair of sites (i.e. an effects parameter at a pair of nominally affected and unaffected sites are compared). The comparison will inevitably show a difference and that difference may be large relative to within-site variance. Without more information, however, it is not possible to determine whether the difference is due to natural between-site variance or a real effect that requires some explanation.

A significant increase in inferential power is obtained by comparing the apparently affected site to multiple reference sites (Underwood 1994). This allows the assessor to gain some idea of what constitutes normality by examining the variance among reference sites. For example, a study of the decline in abundance of San Joaquin kit foxes on an oil field began by comparing contaminant levels in foxes from the oil field to levels in foxes from the immediately surrounding area (Suter *et al.* 1992). The results showed that concentrations of several metals were consistently and statistically significantly higher in the oil-field foxes. However, when the study was expanded to include foxes from other sites, it was found that the levels of nearly all metals were not exceptionally high in foxes from the declining population on the oil field but were exceptionally low in foxes from the surrounding area.

An expansion of this idea of multiple reference sites is comparison of apparently affected sites to a regional reference established by studying numerous sites in a region (Plafkin *et al.* 1989). For example, the state of Ohio has been divided into regions that are reasonably uniform ecologically. For each region, the range of properties of the stream macroinvertebrate and fish assemblages has been established for a set of reasonably uncontaminated and undisturbed streams (Whittier *et al.* 1988; Ohio Environmental Protection Agency 1990). The fauna of any stream in the region can be characterized as affected if its properties lie outside the regional reference range.

Before–after comparisons (i.e. comparison of measurements taken at a site before and after an apparent effect is observed or an event occurs that may cause effects) are conceptually more appealing than comparisons of sites because one expects a site to be more comparable to itself than to other sites. However, temporal variation due to ecological succession, weather or other factors are the norm. Therefore, before–after comparisons are, like site-to-site comparisons, likely to be confounded. In addition, such comparisons are often impossible because there are no before measurements. The locations of effects of interest often are unpredictable and useful baseline studies are seldom available even when actions might reasonably be expected to have effects. Therefore, before–after comparisons are useful when the location and time of an action are known ahead of time or when a long-term monitoring programme includes the location of an apparent effect.

Much of the recent effort in statistical design of ecological field studies has gone into the development of designs that combine spatial and temporal comparisons. These before–after control-impact (BACI) studies require that prior to occurrence of some event, potentially affected and unaffected sites should be identified and characterized and then characterized again after the event (Green 1979). The BACI design allows the assessor to determine whether the similarity of exposed and reference sites is changed by the exposure. This more elaborate type of study design is less likely to be confounded than other comparative designs, but it has been shown to be confounded in practice when there is temporal variation in the degree of similarity of the control and impact sites (Smith *et al.* 1993). Further elaborations of the BACI design include temporal replication of before and after samples (Stewart-Oaten *et al.* 1986) and spatial replication of the control sites (Underwood 1994).

All of these comparisons are categorical. That is, it is assumed that ecosystems can be divided into those that are potentially affected and reference systems (e.g. oiled beaches versus unoiled beaches) or multiple clear categories of exposure (e.g. fields with zero, one, two or three pesticide treatments). However, many ecological systems vary along habitat gradients, many human actions result in gradients of exposure rather than presence/absence or other categorical exposures, and many effects are gradually imposed over time. It is appropriate to use gradient analyses for cases of continuous variation in space, time series analyses for cases of continuous variation in time, or a combination (e.g. exposure level-by-time interactions). Gradient analysis involves regressing or simply plotting the response parameter against environmental variables. For example, streams exhibit natural downstream gradients in fish species richness and abundance, and a deviation from that gradient would indicate a potential effect. Gradient analysis is much less subject to the influence of pseudoreplication than the categorical comparisons of sites discussed above (Wiens & Parker 1995).

Time series analyses are analogous to gradient analyses in that they provide more of a context for inference. For example, an apparent decrease in invertebrate species richness shown by a before–after comparison may be shown by time series analysis to be part of a long-term trend, part of a pattern of year-to-year fluctuation, or a real deviation from background temporal dynamics. In many cases a pre-impact time series is not available, but time series analysis can still be useful. If the response parameter shows apparent improvement over time to some background level, that pattern is suggestive of a real effect from which the system is recovering.

Confidence that a comparison represents a real effect depends on: (i) the magnitude of the observed difference; (ii) the natural variability in the parameter being measured; and (iii) information concerning the appropriateness of the comparison. The first two criteria are the stuff of routine sample statistics. For example, if the magnitude of difference between the exposed and reference systems is large, it is more likely that an effect has occurred; if the variance within a system is small relative to that difference, it is more likely that an effect has occurred. Similarly, if the slope of a gradient or the magnitude of the deviation from a gradient is large relative to the variance, it is more likely that an effect has occurred. The third criterion is more complex. In general it is more likely that a difference represents a real effect if the natural environmental factors that control the parameter of interest are known and have been shown not to account for the apparent effect. For example, knowing that nitrogen limits plant production in the absence of contaminants, one could determine with greater confidence whether observed differences are due to natural variability by analysing soil nitrogen than by elaborate sampling designs. In any case, any analysis of the likelihood that a real effect has occurred should include a list of the natural factors that are known or expected to influence the response and a qualitative, if not quantitative, analysis of the potential influence of those factors.

The presentation of results of the statistical analysis of field studies for the elucidation of effects has received little attention. The use of appropriate plots of response parameter values and confidence intervals serves to show both the magnitudes of differences and the associated uncertainties. A common solution is to display the magnitude of the parameter with indicated confidence intervals (Fig. 7.3). McDonald has recommended presenting the ratio of the mean response at the assessment site to the mean at reference sites plotted with various confidence intervals (Fig. 7.4). In the example depicted, it is clear that not only is there little confidence that the sites are different, but the magnitude of the difference (4%) is small, so it is unlikely that there is a real effect. This provides much more information than simply testing the null hypothesis and stating that the differences are not statistically significant.

The above discussion has assumed that a limited number of assessment end-points have been identified that are represented by a limited number of response parameters (measurement end-points) which are individually statistically analysed. However, there are cases in which multiple parameters may be analysed in aggregate. For example, if there are no clear assessment end-points, assessors may measure a variety of response parameters in the hopes that in aggregate they will capture the important responses of the system. These parameters may be arithmetically combined into an index which can be compared among sites, over time, or to a regional reference value (Karr *et al.* 1986). Although these indices are commonly used, they have many practical and conceptual limitations (Suter 1993b). In this context, the most salient problem is that indices of heterogeneous variables obscure the nature of differences between sites. For example, two sites on a stream may have the same Index of Biotic Integrity (IBI) but have very different species compositions or even species number (Karr *et al.* 1986).

A more informative approach to using multiple parameters to determine whether effects have occurred is to use multivariate statistical techniques. A conceptually direct approach is to determine the likelihood that the putatively affected system belongs to the same portion of an n-dimensional-state space as the reference system, where n is the number of measured parameters and the reference state space is the volume within the n-dimensional Cartesian space within which observed values of the parameters for the reference systems lie (Johnson 1988). A low probability of belonging to the reference state space implies a real effect on the system, as defined by the measured set of parameters. Other multivariate statistical approaches include similarity metrics, principal components analysis, and multivariate discriminant analysis (Gauch 1982; Boyle *et al.* 1984; Smith *et al.* 1989).

Finally, the reality of an apparent effect is supported by its association with a causal agent. That is, an observation of dead birds or low plant diversity is more likely to be a real effect rather than a random event if there is evidence for an anthropogenic cause. Therefore, a relatively weak answer of 'yes' to the question 'is there a real

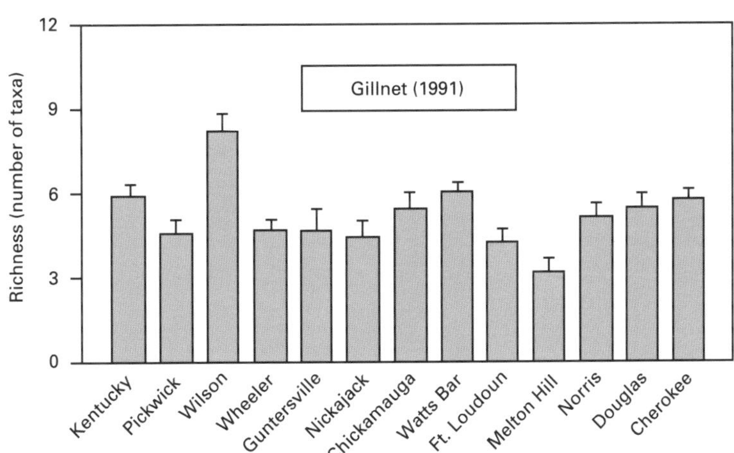

Fig. 7.3 A comparison of the species richness (number of species in all samples) of fish communities in reservoirs with 95% confidence intervals.

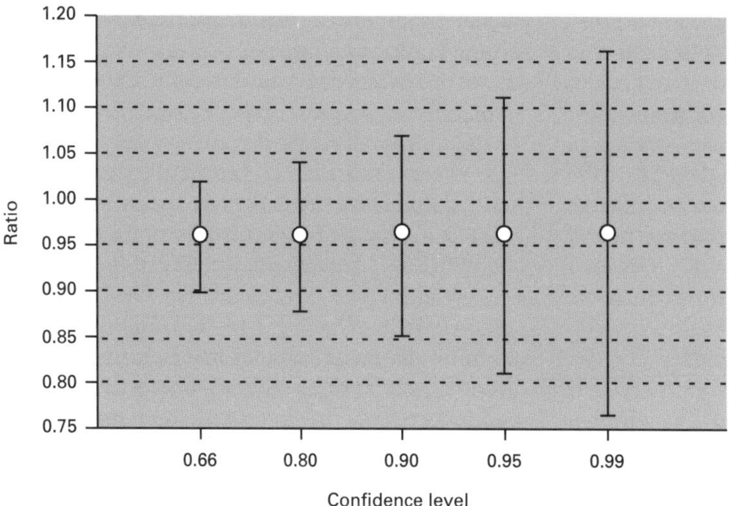

Fig. 7.4 Hypothetical 66%, 80%, 90%, 95% and 99% confidence intervals on the ratio of the mean value of a response parameter in a contaminated area relative to a reference area. Graphical presentation by L. McDonald, WEST Inc., Cheyenne, WY.

effect?' may be a strong answer of 'yes' once the question 'what is the cause?' is answered (section 7.3.3).

These approaches to determining whether there is a real effect are not mutually exclusive. One might, for example, find that the abundance of a species is much lower at one site compared to reference sites. The argument that this apparent effect is real might be more convincing if habitat models are used to show that the apparently affected site is as suitable as reference sites.

In most ecoepidemiological studies, the question of the reality of effects is treated as equivalent to the question of causation. That is, if an environmental change can be shown to be caused by a particular contaminant, the effect is real. However, in many cases sufficient information is available to address the question of reality but not causation. In addition, many studies that claim to demonstrate causation have done no more than demonstrate the reality of an effect and its association with a possible cause. In those cases, it can be very useful to treat the question of reality as distinct and resolvable, even if causality is unknown or unclear.

7.3.3 What is the cause?

The nature of causation is an old metaphysical problem that has been satisfactorily solved in the limited context of scientific experiments but not in observational sciences such as epidemiology that deal with the uncontrolled conditions of the real world. Anyone who follows the news, learns of cases in which some cause for a human disease or degenerative condition is proposed and then contested. These range from the debates about whether electromagnetic fields from electrical transmission lines cause cancer to arguments about whether people in Mediterranean Europe have low rates of coronary artery disease because of wine, olive oil, garlic or some other factor. In general, there are three approaches to determining causation in epidemiology that are potentially applicable to ecoepidemiology: employ a standard of proof such as Koch's postulates; evaluate the epidemiological evidence against a set of criteria such as Hill's or Susser's criteria; or apply a hypothetico-deductive approach to eliminate potential causes. The following summary of these approaches is based primarily on more extensive discussions (see Fox 1991; Suter 1993a).

Koch's postulates

Koch's postulates are a standard of proof for demonstrating that a particular pathogen causes a disease. It has been adapted for demonstrating that particular contaminants cause a human disease (Yerushalmy & Palmer 1959; Hackney & Kinn

1979) or an ecological effect (Adams 1963; Woodman & Cowling 1987; Suter 1990c). The following adaptation of Koch's postulates for ecological epidemiology of pollution effects is from Suter (1993a).

1 The injury, dysfunction or other putative effects of the toxicant must be regularly associated with exposure to the toxicant and any contributory causal factors.

2 Indicators of exposure to the toxicant must be found in the affected organisms.

3 The toxic effects must be seen when healthy organisms are exposed to the toxicant under controlled conditions, and any contributory factors should contribute in the same way during the controlled exposures.

4 The same indicators of exposure and effects must be identified in the controlled exposures as in the field.

The strength of this standard of proof results from the concordance of uncontrolled studies of the real world and controlled laboratory studies. Field observations tend to be highly variable and ambiguous and do not provide clear evidence of causation, but they indicate the actual state of nature. The toxicity tests or other controlled studies carefully control causal factors, but the resulting confidence about causation is limited to the test system. The fact that an effect is associated with a contaminant in the field and also in the laboratory (rules 1 and 3) is strong evidence for causation, but the proof comes from demonstrating that the results are associated with the same types and levels of indicators of exposure in both the field and laboratory (rules 2 and 4).

Application of rule 1 requires that we define 'regular association'. It may be defined as consisting of Hume's criteria for causation: (i) spatial and temporal contiguity; (ii) temporal succession; and (iii) consistent conjunction (Suter 1993a). The first two criteria are common sense and relatively easily applied. That is, a cause must be associated with the effect in space and time, and the effect must occur after the cause. However, consistent conjunction is conceptually more complex. In its strict form, it requires that every case of effect x should be associated with cause y, and every instance of cause y should result in effect x. This is clearly an impossibly strict criterion for many ecoepidemiological studies. However, ecoepidemiological studies can be designed to facilitate the demonstration of consistent conjunction.

First, many ecological effects of interest, such as mass mortalities or reductions in plant production, are broadly defined and have many potential causes (i.e. effect x is associated with causes y_1, y_2, y_3, ...). For that reason, it is desirable to define the effect as clearly and narrowly as possible. That is, rather than defining the effect as mass mortality, define it as mortality with certain characteristic symptomology. For example, reproductive failure of fish-eating birds in the Great Lakes has been associated with a 'Great Lakes embryo mortality, edema, and deformities syndrome' or GLEMEDS (Gilbertson *et al.* 1991). This is an important role for the histopathological and physiological measures that are not sufficiently important in themselves to constitute assessment end-points, but tend to be more specific to particular chemicals or chemical classes than the more important but less specific responses of populations and communities. However, even higher-level effects can be defined in such a way as to be more specific. For example, mass mortalities of fish typically do not include all species, trophic groups or life stages in the affected community, and the composition of the set of dead fish may be characteristic of certain contaminants, routes of exposure or other causal factors (Meyer & Barklay 1990).

A different set of problems arises when attempting to demonstrate that cause y is always associated with effect x. Although y may be the true cause, it may not always cause effect x because the subjects are not always in a susceptible state. That is, the occurrence of an effect may require that the organisms be stressed by parasites, malnutrition, physical conditions, or some other factor; the effect may occur only when certain environmental conditions prevail (e.g. low hardness and metal toxicity to fish), or the effect may occur only in certain genotypes or life stages. The solution in this case is to define the cause more specifically. That is, metal concentrations $>y$ are not consistently associated with fish mortality, but metal concentrations $>y_1$ in conjunction with pH $<y_2$ or hardness $<y_3$ are consistently associated with fish mortality.

Rule 1 requires well-designed field studies that demonstrate association of effects with the putative causal factors. This requires, in turn, that the studies be designed in such a way as to minimize the likelihood that false associations are perceived or true causal associations are obscured. Many of the issues of study design that are discussed in the previous section are relevant here as well. That is, study designs that show an association of the effect with the cause in both space and time are likely to be more reliable than a comparison in either dimension alone, comparisons to multiple reference sites are more reliable than comparisons to a single site, studies that demonstrate gradients in response along gradients of exposure are more reliable than those that make categorical comparisons, and studies that match sites or otherwise control for potentially confounding factors are more reliable than those that do not. The fundamental difference from the study designs for demonstrating the reality of apparent effects lies in the requirement that the causal factors be measured along with the effects parameters so that they can be associated. The need to associate cause and effects provides an additional argument for favouring gradient studies over categorical comparisons. That is, not only are gradient designs less subject to inferential errors due to pseudoreplication (Wiens & Parker 1995), but also they have the potential to generate exposure–response relationships which, in themselves, strengthen the case for causation (see below).

Rule 2 requires that elevated concentrations of the toxic chemical or its metabolites be detected in the affected organisms. This requires that the chemical be persistent, that it be detectable at all concentrations that can cause effects, and that the measurements be relevant to the effective internal exposure. For example, copper may cause acutely lethal effects by damaging the gill epithelia of fish without increasing whole-body concentrations relative to the normal physiologically required levels. An alternative to analysis of contaminant chemical concentrations is analysis of specific biomarkers of exposure. Biomarkers are potentially useful for chemicals that cannot be analysed in biota, but relatively few are sufficiently specific to a chemical or class of chemicals to be useful. Examples include cholinesterase inhibition as a marker for organophosphate insecticides, elevated δ-aminolevulinic acid dehydrogenase (ALAD) as a marker for lead, and benzo[*a*]pyrene (BaP) adducts as markers for BaP (McCarthy & Shugart 1990; Huggett *et al.* 1992).

Rule 3 requires appropriate toxicity tests. The appropriateness of a test depends on the degree to which the test physically simulates the field conditions associated with the effects. This includes species, life stages, durations, physical/chemical conditions and chemical forms. This requirement may be met by existing test data or by *ad hoc* tests. In particular, if the chemicals of concern are persistent, the contaminated media may be collected and used in toxicity tests. Even if the chemicals are not persistent, spiking site-specific media can give greater relevance to the tests than using standard test media.

Rule 4 requires that the body burdens, biomarkers, characteristic symptomology, and other measures that characterize the exposure of the test organisms to the chemicals be shown to match those in the field. It is not sufficient to satisfy rule 3 by finding, for example, that the chemical associated with death in the field causes death in the laboratory. The indicators of exposure in the field (rule 2) must also occur in the laboratory. This requirement is the least commonly satisfied component of Koch's postulates and is left out of some formulations (Woodman & Cowling 1987). Body burdens, biomarkers of exposure or other measures of internal exposure are seldom reported in published toxicity tests and can significantly add to the cost of *ad hoc* tests. However, satisfying rule 4 significantly enhances the evidence of causation. In effect, satisfying rule 3 shows that the contaminant can cause the effect, while demonstrating the concordance of exposure in the laboratory with that in the field (rule 4) shows that it did cause the effect.

All of the components of Koch's postulates have been satisfied for numerous ecoepidemiological studies (Suter 1993a). However, they should not be considered a requirement for determining causation in ecoepidemiological studies. In many cases, resources are not available to complete the necessary studies or the studies are not possible. In addition, because of variation in the quality of both field and laboratory studies and the possibility of disagreement about interpretation of results, particularly about whether field and laboratory results are concordant, inference based on Koch's postulates may not be definitive. Finally, in a strict philosophical sense, Koch's postulates are not an absolute standard of proof. Satisfying Koch's postulates is probably the best goal for those designing ecoepidemiological studies, but causation can also be satisfactorily demonstrated by weighing the evidence based on a set of criteria.

Criteria for causation

Most epidemiological studies do not employ a set of firm criteria for causation. Rather, they evaluate the standards for causation with respect to various factors which constitute a check list or set of points to consider. The most commonly cited list is Hill's (1965) (Table 7.5). Unlike Koch's postulates, there is no expectation that all of these factors will be satisfied or that causation will be proved through their use.

There is some obvious overlap of these factors with Koch's postulates. Consistency and temporality are two of the three criteria for regular association, and specificity of association tends to facilitate the detection of a regular association. Also, Hill's 'experiment' is analogous to Koch's rule 3. Those factors need not be discussed again.

Strength of association. This criterion is not necessary to prove causation. True causes may be weak or may appear weak because of poor measures of exposure or effects. Therefore, it has no negative effect on the causal hypothesis. However, if the effect is strongly associated with the putative cause, this greatly strengthens the argument. Epidemiologists typically discuss strength of association in absolute terms such as the relative risk (i.e. the frequency of disease in the subject population relative to reference) or the slope of the dose–response function. However, in terms of demonstrating causation, it is better to demonstrate that the association is strong relative to alternative causes or confounding factors rather than in absolute terms.

Biological gradient. Since Paracelsus, the idea that effects are a function of exposure is the central premise of toxicology. Therefore, the demonstration of biological gradients with respect to a causal factor that varies in either space or time can provide strong evidence of causation. In general, this criterion can be judged in the same way as any other correlation or regression analysis. High correlation coefficients are indicative of good evidence for causation. However, poor correlations may be due to poor measurements, few measurements, or measurements over a small portion of the relevant range of exposure. Therefore, a lack of evidence of gradients only weakly negates the evidence for causation. Monotonic exposure–response relationships are most convincing, but complex forms may result from complex biology or, in field studies or tests of contaminated ambient media, from variance in confounding factors.

Plausibility. The plausibility criterion addresses the reasonableness of the putative causal association given current knowledge of the biology, chemistry and physics of the causal agent and receptors. The best indication of plausibility is the existence of a well-defined mechanism of action of the causal agent that would result in the observed effect. Because the causal mechanisms or the circumstances of the exposure may be truly novel, true causes may be implausible a priori. Therefore, plausibility is more likely to strengthen than negate causal hypotheses unless laws of physics are violated.

Coherence. Hill defined coherence as the consistency of available evidence. If a variety of evidence has been obtained relative to the putative causal relationship and it is all consistent, this constitutes powerful supporting evidence. A few inconsistencies make for weak support, and numerous inconsistencies among lines of evidence, particularly if the quality of the inconsistent evidence is high, can strongly negate a causal hypothesis.

Analogy. Analogy refers to the existence of independent cases in which the putative cause or a similar cause (e.g. a related chemical) was shown to have the same or a similar effect. Analogy is strong if there are many analogous cases and if they are highly similar to the case in question. An

Table 7.5 Hill's (1965) factors for evaluating the likelihood of causal associations in epidemiological assessments (Suter 1993a). Those in **bold** type are also among Susser's criteria (Susser 1986a).

Criterion	Explanation
Strength	A stronger response to a putative cause is more likely to indicate true causation. This means either a severe effect or a large proportion of organisms responding in the exposed areas relative to reference areas, and a large increase in response per unit increase in exposure. In other words, a steep exposure–response curve situated low on the exposure scale.
Consistency	A more consistent association of an effect with a putative cause is more likely to indicate true causation. This is a weak form of Hume's third criterion: consistent conjunction. Hill's discussion implies that the case for causation is stronger if the number of instances of consistency is greater, if the systems in which consistency is observed are diverse, and if the methods of measurement are diverse.
Specificity	The more specific the effect, the more likely it is to have a single consistent cause. This is equivalent to our suggestion that regular association is more readily established if a characteristic effect is identified. Also, the more specific the cause, the easier it is to associate it with an effect. For example, it is easier to demonstrate that localized pollution caused an effect than that a regional pollutant caused an effect.
Temporality	A cause must always precede its effects. This is Hume's second criterion and Susser's 'time-order'.
Biological gradient	The effect should increase with increasing exposure. This is the classic requirement of toxicology that effects must be shown to increase with dose. (Susser treats the demonstration of a monotonic dose–response relationship as a component of coherence.)
Plausibility	Given what is known about the biology, physics, and chemistry of the putative cause, the receiving environment, and the affected organisms, is it plausible that the effect resulted from the cause? Susser treats this as a component of coherence.
Coherence	Is the hypothesized relationship between cause and effect consistent with all available evidence?
Experiment	Because Hill was concerned with effects on humans, he emphasized 'natural experiments' rather than the controlled exposures required by Koch's third postulate. An ecological example would be observations of recovery of a receiving community following abatement of an effluent. However, toxicity tests and other controlled studies would also fall under this criterion.
Analogy	Is the hypothesized relationship between cause and effect similar to any well-established cases?

absence of analogous cases does not affect the evidence for causation.

Susser (1986b) added probability and predictive performance to Hill's and his own earlier lists of criteria. Probability is the conventional statistical criterion of estimated probability that the difference between exposed and control groups is due to chance. If it is unlikely that the effect is due to chance, the difference is said to be statistically significant. Both Susser, in his inclusion of this criterion, and Hill, in his rejection of it, emphasized that this potential criterion is different in kind from the others in that it is not a logical basis for rejecting or accepting a causal relationship but simply a quantitative one (Hill 1965; Susser 1986b). Statistical significance does not indicate biological significance. It is redundant with 'strength of association' but depends on the degree of replication, variance in sampling and analysis, and natural variability as well. On the other hand, it is conventionally a major component of arguments concerning causation.

Predictive performance is the testing of the deduction that if *x* causes *y* then it should be possible to observe some previously unobserved consequence of *y*. For example, if a particular chemical is the cause, one might predict that a particular histopathology should be observed in the affected population. Subsequent observations of the predicted histopathology would support the inference of causation.

Although this criteria set outnumbers Koch's postulates, it does not explicitly include Koch's rules 2 and 4. That is, the criteria do not distinguish between a regular association with a cause that is based on inferred exposure and those that are based on evidence of internal exposure, or at least clear evidence of external exposure. In some cases, this distinction is not critical. However, if there is some uncertainty as to whether the affected organisms engaged in behaviour that would lead to exposure, whether they spent sufficient time in a contaminated area to be significantly exposed, whether the contaminant is in a bioavailable form, or if there was any other serious question about the actual occurrence of exposure, Koch's rules 2 and 4 may be critical to determining causation. Therefore, one might add evidence of internal exposure to Hill's and Susser's criteria (Table 7.5).

The results of a causal analysis based on these factors are typically presented as a narrative. However, as an aid to communication with reviewers and the public, it would be desirable to present results in terms of a table that checks off factors that are consistent with the causal hypothesis or that scores factors based on their influence on the causal inference. Table 7.6 presents one possible form of a table. For each criterion the assessor would assign a result and a corresponding rating in terms of the number of plus or minus signs designating strength of positive or negative effect on the causal hypothesis, zero for a neutral effect, or NA if not applicable. The sets of possible ratings presented in Table 7.6 represent the author's judgement as to the possible importance of the factor based in part on the judgements of Susser (1986a) and Fox (1991). For example, incompatible time order would strongly negate the causal hypothesis but implausibility would only weekly negate it. Table 7.6 could be modified to be more relevant to a particular assessment, but this should be done during the problem formulation rather than during the analysis of causation in order to minimize the potential to bias the results. One might add a fourth column to the table for scoring the quality of the information that led to the result.

Process of elimination

The classical basis for inference in science is the disproof of hypotheses. This is because it is possible to show that a general proposition is untrue by showing a contrary case, but it is not possible to demonstrate that it is true by showing that there are no contrary cases unless the universe of possible cases has been examined (Platt 1964; Popper 1968). This is true for epidemiology as well as pure science (Susser 1986b). An example is the process of elimination that was used by the National Acid Precipitation Assessment Programme (NAPAP) to determine which lakes had been acidified by deposition of mineral acids (Baker *et al.* 1990b). If an acidic lake had chemistry characteristic of acid mine drainage or of being in a watershed dominated by sulphate minerals, or if the acidity was not due to organic acids, then acidification by atmospheric deposition was eliminated for the lake, otherwise it was demonstrated (Fig. 7.5).

It is very difficult to prove ecological causation by eliminating all alternative causes because the alternatives are nearly always numerous and it is not possible in most cases to demonstrate that all possible alternatives have been considered. Nevertheless, in addition to building a case for the most likely cause, as discussed above, it is advisable to address the most plausible alternative causes. Elimination of any plausible alternative strengthens the case for the proposed cause. Incompatible temporality, inconsistency on replication, and incoherence with established scientific facts can serve to eliminate a proposed cause (Susser 1986a).

Table 7.6 Format for a table to summarize results of an inference concerning causation in ecoepidemiology. In an application, one result and a corresponding effect on hypothesis rating would be selected for each criterion.

Criterion	Results	Effect on hypothesis
Strength of association	Strong, moderate, weak, none	+++, ++, +, 0
Consistency of association	Invariate, regular, most of the time, seldom (preferably, present numerical results)	+++, ++, +, 0
Specificity of cause	High, moderate, low	++, +, 0
Specificity of effect	High, moderate, low	++, +, 0
Temporality	Compatible, incompatible, uncertain	++ −−−, 0
Biological gradient	Clearly monotonic, weak or other than monotonic, none found	+++, +, −
Plausibility	Plausible, implausible	+, −
Coherence	Evidence: all consistent, most consistent, many inconsistencies	+++, +, −−−
Experiment	Experimental studies: concordant, ambiguous, inconcordant, absent	+++, +, −−−, 0
Analogy	Analogous cases: many or few but clear, few or unclear, none	++, +, 0
Probability	Probability association occurred by chance: very low, low, high (or present numerical results)	++, +, −
Predictive performance	Prediction: confirmed, failed	+++, −
Internal exposure	Detected, undetected, undetermined	++, −−, 0

7.4 MONITORING

In an ecological risk assessment context, monitoring focuses on providing source, exposure and effects data needed for risk characterization. This is a departure from the more traditional practice of selecting environmental indicators and measuring them on some spatial or temporal pattern. Traditional monitoring programmes provide information about environmental status and trends that may be quite useful. However, the history of such programmes is bleak because they typically have not proved to be sufficiently useful to sustain the interest of environmental managers. Monitoring to support a decision by providing information for a risk assessment is more likely to be sustained until useful results are obtained.

7.4.1 What to monitor

Rather than monitoring indicators that are in some undefined sense indicative of something important, when monitoring to support risk assessment one monitors things that are important. In a risk assessment context, things may be important because they are assessment end-points, because they provide approximations or components of assessment end-points, or because they are thought to be causally connected to assessment end-points. In other words, one should monitor in order to generate good measurement end-points and good exposure metrics as discussed above (section 7.2).

In addition, one should design monitoring programmes to contribute to the development of causal relationships and not just to describe one component of the environment. While it is usually not necessary to monitor all components of a causal chain, without some linkage of the moni-

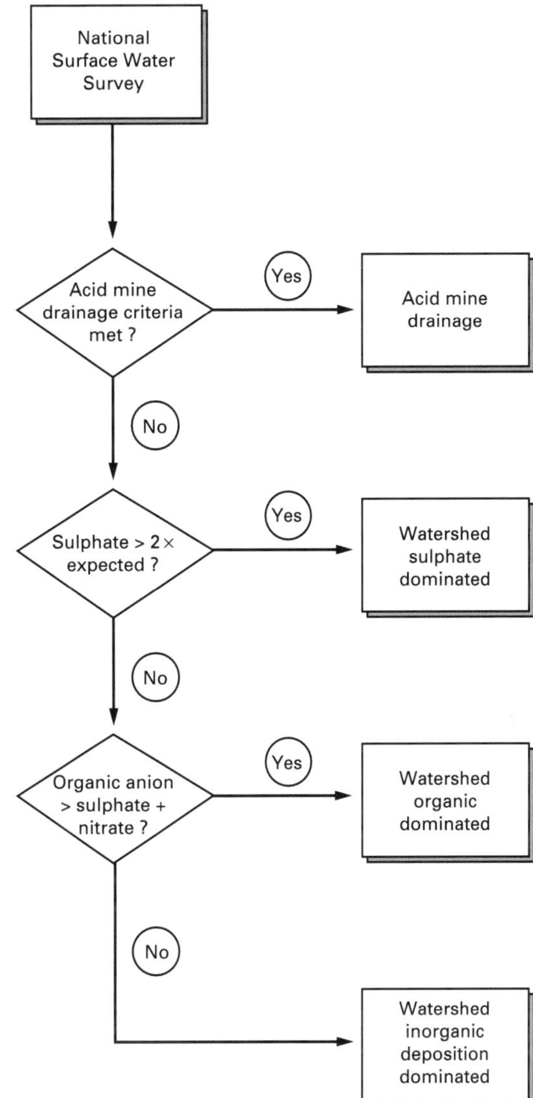

Fig. 7.5 A process of elimination for establishing that lake acidity is due to acid deposition. (After Thornton *et al.* 1994.)

toring programme to causes and effects the results will often be inconsequential. This is because environmental management decisions are nearly always decisions about changing a causal relationship. Many monitoring programmes have shown that the concentrations of some contaminant are increasing but have not been able to say what sources are responsible or what effects are occurring (Suter 1993a). Similarly, many programmes have suggested that effects are occurring without being able to indicate the cause (Olsson 1995). Therefore, the monitoring framework should be causal (USEPA 1995). Such a conceptual framework is presented in Fig. 7.6. It is an amalgamation of the pressure/state/response framework which is widely used for national environmental accounting

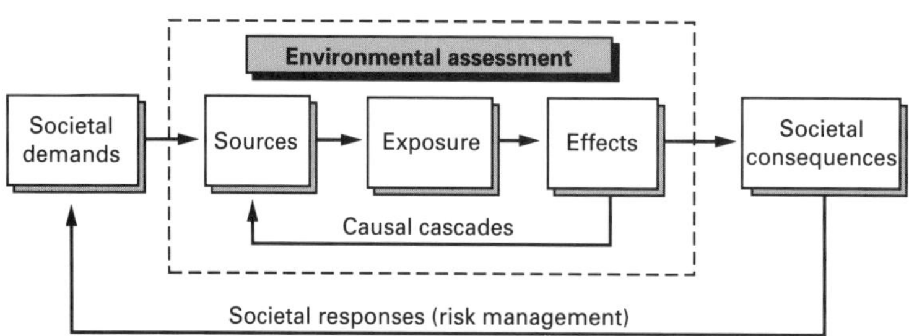

Fig. 7.6 A causal framework for environmental monitoring based on the pressure/state/response framework (OECD 1991, 1993; Adriaanse 1993; USEPA 1995).

(OECD 1991, 1993; Adriaanse 1993; USEPA 1995) and a framework for causal inference in retrospective ecological risk assessments (Suter 1990c, 1993a).

This framework serves to emphasize the socioeconomic context of environmental monitoring and assessment, as well as the causal linkages that result in environmental degradation or enhancement. The box in the centre, which is equivalent to the analysis box in Fig. 7.1, represents the domain of ecological risk assessment and therefore of environmental monitoring. The area outside the box is the socioeconomic arena in which demands for resources and economic efficiency create pressures on the environment and in which the effects of those pressures cause degradation in market and non-market values of environmental resources, which result in changes in demand due to changes in public behaviour or regulatory decisions to modify the pressures. This monitoring framework is, in effect, a generic conceptual model of the environmental system that is analysed in the risk assessment and risk management paradigm.

In this framework, the process of designing a monitoring programme consists of deciding what needs to be known about the system in order to understand the risks that are being imposed on it and intervene properly. As discussed above, these decisions are part of the problem formulation. Any monitoring programme that is not prompted by a prior risk assessment should begin with a problem formulation process. That is, it should identify the impetus for the programme, define the environment to be monitored, define the spatial and temporal scope of the programme, and define a conceptual model of the system being monitored, including the sources and processes of exposure and effects induction that are considered relevant to the programme.

Deciding what to monitor is more difficult when the goal is to provide information relevant to the assessment of ecological risks in general rather than to support a specific assessment with a definable impetus. In such cases, it is tempting, but ultimately counterproductive, to resort to monitoring of things that are claimed to be environmental indicators without reference to what they indicate. However, even in the absence of defined assessment end-points, those things that are in some sense important should be monitored. Importance is a property that flows in the opposite direction from causation (Fig. 7.7). People's values determine what properties of the environment are important, but public values are poorly defined. The properties of the environment that are valued can be determined from people's behaviour (e.g. recreational use or purchases of resources), by social surveys, by reviewing policies and decisions of environmental protection agencies, or by consulting elected representatives, appointed officials, and spokesmen for constituencies such as environmental organizations (see Part 3). Valued environmental properties (equivalent to assessment end-points) must be translated into things that are technically susceptible to measurement (equivalent to measurement end-points). Detected changes in these measures of effects will be important

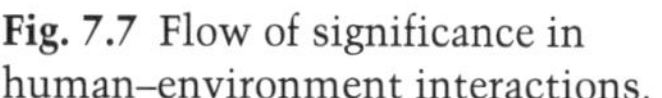
Fig. 7.7 Flow of significance in human–environment interactions.

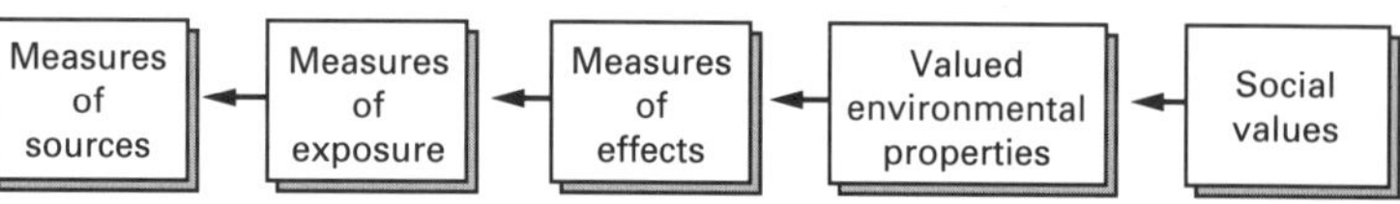

in themselves because they are linked to values. By extension, exposures that are potentially responsible for those effects are important, and sources of the exposures are important.

Monitoring susceptible systems is an intermediate strategy between monitoring to support a specific assessment with an identifiable impetus and monitoring out of general concern for status and trends in the environment. That is, one may not know enough about a situation to formulate an assessment problem and design a programme to address it, but one may have ideas about what systems and components are susceptible to injury given anticipated human activities or knowledge of biological or ecological properties that lead to susceptibility to many contaminants or activities. The trap to be avoided in this approach is designing a monitoring programme for the problems of the past. For example, patterns of relative sensitivity for new chemicals are likely to be different from those for chemicals that caused problems in the past.

Given that one has decided what to monitor based on the importance of a property for a particular risk assessment, general importance or susceptibility, one must decide exactly what to measure, how, when and where. The answers to these questions require a combination of technical judgement and pragmatic judgement about what sorts of activities are practical and can be funded and sustained. A recent review of potential parameters to be monitored in a USA national environmental monitoring programme can provide a starting point for the selection process (Hunsaker & Carpenter 1990).

7.4.2 What to do with monitoring results

Unfortunately, much more attention has been devoted to designing monitoring programmes than to determining how the results are to be used. As a result, reports of many monitoring studies serve primarily to take up shelf space. Alternative uses of monitoring results can be arranged in a hierarchy based on the amount of interpretation derived from the data.

The highest level of interpretation is the use of monitoring results for ecological risk assessment. This requires that the monitoring data and any related information be sufficient to estimate the risks due to some source of exposure or determine the reality and cause of some apparent effect (sections 7.2 and 7.3). This is most likely to be feasible when the monitoring programme has been designed for the assessment by the problem formulation process. When monitoring is conducted to support a risk assessment, the monitoring programme does not have a distinct data analysis and interpretation function. Rather, it is incumbent on the individuals involved in the sampling and analysis to assure the quality of the data and, when data are questionable, to aid in their proper use and interpretation. Not all data collected in violation of procedures are useless and not all data collected according to procedures are useful.

A somewhat lower level of interpretation is the use of monitoring data to suggest relationships between causes and effects for further investigation and assessment. This requires that a variety of potentially causal agents be monitored so that potential causes or sets of interacting causes can be extracted through multivariate statistics.

Such analyses are referred to as synoptic and should include a variety of factors, not simply levels of anthropogenic contamination. For example, potential causes of forest decline included climatic factors, site factors (soil nutrients, pH, etc.), stand factors (age, density, etc.), and air pollution levels (Wallace 1978; Schulze 1989).

The lowest level of interpretation is the documentation of environmental status and trends. This is the simplest analysis to perform but the least informative. That is, one can say as a result of a single year of monitoring that a particular stream has 36 species (status), and after a number of years of monitoring one can say that the number has declined at a mean rate of 0.2 species per year (trend). While such relatively undigested data may be of some interest in themselves, their primary utility is as resources for those who use them to answer their own assessment questions. If the monitoring results are to be a resource to others, it is important to document the quality and ensure the integrity of the data. It is also important to provide the data in a disaggregated form so that assessors with a variety of interests (e.g. effects in a particular river reach, river basin or region) can make use of the data by aggregating them appropriately.

Status and trends reports can provide more useful information if results can be related to some standard. Chemical concentrations are commonly compared to regulatory standards or criteria to determine where and with what frequency exceedences occur.

When there are no standards or criteria, concentrations of naturally occurring chemicals can be compared to the range of background concentrations. An example of such results might be: at 19% of stations in a river basin, one or more metals exceed water quality standards, and one or more of the metals for which there are no standards exceed background at 36% of stations.

For biological monitoring, something equivalent can be done by comparing to an acceptable ecological state, for example, the USEPA's Environmental Monitoring and Assessment Programme proposed to categorize sites as nominal, marginal or subnominal with respect to a set of alternate criteria (Thornton *et al.* 1994). Criteria for acceptability may be based on: (i) whether the ecosystem provides a valued resource; (ii) whether the ecosystem is perceived to be degraded by the public; (iii) whether the ecosystem is degraded based on expert judgement; (iv) whether the ecosystem differs from undisturbed ecosystems with respect to the monitored properties; or (v) whether the ecosystem differs from its historic condition based on palaeoecological comparisons. This type of analysis would allow one to state that, for example, 28% of the wetland area in a region is in a degraded (i.e. subnominal) state.

An intermediate strategy between presenting status and trends without interpretation and normalizing them to some criterion or acceptable state is to normalize them with respect to some base state, such as the first year of the monitoring programme. That is, the proportional change in the monitored parameters in the current year could be presented relative to year one or to some other base state. If directions of change can be associated with policy or public values, a plus or minus sign should be added to indicate whether the direction of change is judged to be an improvement or decrement in the state of the system.

Because monitoring programmes typically measure and report results for numerous environmental parameters at numerous sites, presentation of the results of status and trends analyses to the public and policy makers is problematic. The problem of numerous sites is most commonly resolved by spatial aggregation. That is, results for all sites in a region or political unit are presented in terms of mean values, proportions of values exceeding some criterion, or other appropriate summary. An alternative that is becoming increasingly practical is to generate custom reports. That is, rather than providing a dump of all data (which is not interpretable by the requester) or summaries based on standard aggregations (which may not be appropriate to the requester), a monitoring programme could develop the capability of readily generating custom reports that aggregate data for any particular set of parameters and area of interest. Finally, rather than presenting numerical or graphical results, maps may be generated. Sets of spatially associated data points can be converted to mapped areas using a variety of techniques that generate isopleth contours (e.g. kriging) or polygons (e.g. Thieson polygons).

The problem of summarizing multiple parameters in a useful manner is more difficult. There are two general strategies for dealing with this problem: selecting a limited set of parameters and aggregating parameters. Parameter selection is based on the realization that some parameters are of interest to decision makers and the public and others are primarily of interest to the assessment scientists who perform risk assessments and other interpretations of the data. Hence, status and trends results for those environmental parameters that are, in some sense, valued by the public and its representatives who manage the environment can be reported. A related basis for selection is to use parameters that are in some sense integrative. For example, species abundances integrate anthropogenic factors such as contamination, physical disturbance, and harvesting rates, as well as natural habitat variables such as flooding intensity, because the abundance of species depends on the entire array of habitat properties, including those that are not monitored. Often, these selection criteria will be complementary.

Aggregation of parameters is an obvious solution to the problem of presenting monitoring results.

A set of heterogeneous parameters may be aggregated by adding them, taking their mean, or in some other manner numerically combining them into an index. The index may then be given a name like 'ecosystem health'. However, proper aggregation requires that the aggregate property be comprehensible and of interest to decision makers and the public. Most indices of heterogeneous variables are not transparent or comprehensible and the audience must accept the pronouncement of its creator that it represents ecosystem health or whatever other name is applied. In addition, these indices may obscure important effects, have arbitrary and usually undefined statistical properties, and possess a variety of other properties that make them difficult to use and interpret (Suter 1993b). Also, recent results suggest that indices of heterogeneous variables like the IBI are less sensitive to toxic effects than simpler metrics such as species number or diversity (Dickson *et al.* 1992; Hartwell *et al.* 1995a,b).

Aggregation can be performed in a more defensible manner by using the component parameters to calculate some higher level property of the monitored system. For example, monitoring of fish in a lake could generate measurements of the number and size of fishes by species. These data may be aggregated into community properties such as number of species, species evenness or standing biomass. Appropriate higher-level properties are less obvious for larger scale systems and for more heterogeneous parameter sets. However, one commonly useful strategy is to normalize all parameters to some benchmark condition. For example, all habitat areas can be reduced to proportion of a region occupied by its potential natural vegetation and all species numbers can be reduced to the proportion of potentially naturally occurring species (e.g. in the New World, species occurring in the region in pre-Columbian conditions). This sort of aggregation is more understandable but must still be done in a way that generates properties of interest.

The sort of aggregation discussed above requires that the aggregated properties be natural components of a common ecological property. Alternatively, ecological properties may be aggregated on the basis of common economic metrics. Resource economists and ecological economists have developed a variety of techniques for expressing the state of the environment in monetary units (see Part 3). This is possibly the least informative aggregation in terms of informing the audience about the nature of the problems, and it is extremely controversial. However, when non-market values are included, it can impress the magnitude of environmental change on some audiences in a way that purely ecological parameters cannot.

The best currently used solution to the problem of presenting status and trends results is the use of Dutch amoeba diagrams (ten Brink 1989). As shown in Fig. 7.8, the various parameters are rays on the diagram. The reference situation is the circle, which represents the recent historical situation, and the abundances are normalized to that value so that rays outside the circle indicate species that have increased and those within the circle that have decreased. The management objective for each species is presented as a dashed line across each ray, and the anticipated effects of current policies are represented as dark overlays (decreases) or pale extensions (increases) of the rays. This diagram incorporates many of the points about presenting status and trends data discussed above. It reduces the parameter set to selected important properties (vertebrate abundances and a few other important parameters), it normalizes data to a reference state, and also relates them proportionally to a criterion, the ecological management objective.

7.5 CONCLUSIONS

This chapter addresses the assessment of risks due to events that were initiated in the past and have ongoing potential consequences (retrospective risk assessments), the specific problems of determining the reality and cause of apparent effects on the environment (ecological epidemiology), and the gathering of environmental data to support ecological assessment and management (monitoring). The reader should have detected some common themes throughout these sections. Environmental data should be gathered either to supply the needs of a planned assessment or to fit into a general cause–effect framework which could support performance of an assessment when issues arise. Assessments, in turn, should be carried out in

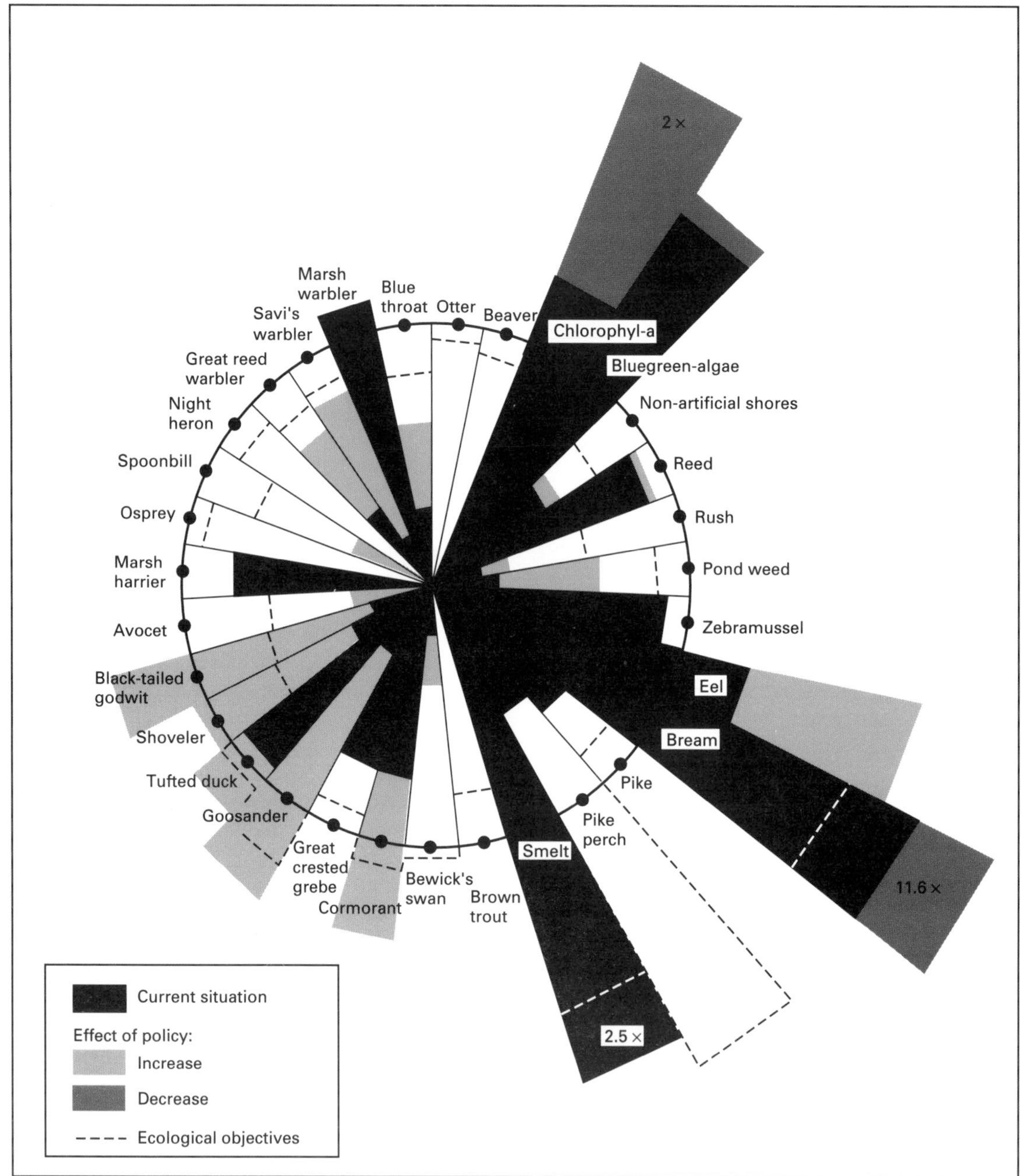

Fig. 7.8 An example of an amoeba diagram (Botterweg 1995). The rays on the diagram are the end-point parameters. The circle is the reference situation and the abundances are normalized to that value so that rays outside the circle indicate parameters that have increased and those within the circle those that have decreased. The management objective is presented as a dashed line across each ray, and the anticipated effects of current policies are represented as dark overlays (decreases) or pale extensions (increases) of the rays.

Table 7.7 Actions suggested by lessons learned in communicating between decision makers and scientists through the global climate programme (SPA 1992).

Action	Lesson learned
Supply interim information	Decision makers need interim information on relevant findings, in addition to periodic assessments. Decisions do not conform to the dates for assessment reports. Interim information is required for informed decisions driven by short time-frames
Tie assessment to important policy time-frames	Assessments should be keyed to important dates, such as international policy discussions, reauthorization of environmental legislation (e.g. Clean Water Act Amendments, Wetland Protection Act Amendments), policy formulation and other decision-making activities
Provide useful information before reliable predictions	Assessments can provide useful information even if the 'penultimate' model predictions of the effects of stressors are not yet available. Existing information can be synthesized and integrated with results from simple empirical models, statistical analyses and similar analytical approaches to provide useful, interim information for decision makers
Know decisions are not either/or	Decision-makers' choices are not simply either to pursue research or to implement management alternatives. The challenge is to define the appropriate levels of each over time. Researchers need to provide a broad array of information to address the complex and interacting decisions. Decision makers, for their part, need to recognize the long time scales involved in research and, thus, the importance of continuity of funding and programme goals.
Place comparative risk in relative context	Assessing change in a relative risk context is difficult, but extremely important
Address urgent need for education	A concerted effort is needed to educate decision makers on the facts and uncertainties of any environmental issue. Considering that public concern is often the impetus for formulating policy, scientists need to communicate technical information to the public more effectively, as well as more frequently. In addition, scientists need to learn more about the decision-making process and the types of information most useful for policy. Frequent two-way communication between decision makers and researchers is fundamental if research is to play an effective role in the decision-making process
Manage uncertainty	There are more ways to manage uncertainties than simply trying to reduce them. For example, building resilient institutions and methodologies would provide a flexible response to any future change, albeit at potentially significant costs. Contingency plans could allow decision makers to prepare for possible climate outcomes through R&D technologies, without needing to employ them
Know research does not always provide the answer	Decision makers need to realize that additional research actually could increase the amount of uncertainty in some areas. Researchers should enquire about how much certainty decision makers need in order to take a specific action. To this end, uncertainties that do not matter for decision making should be so identified

(Continued on p. 214)

Table 7.7 *Continued.*

Action	Lesson learned
Develop an ongoing assessment process for research	To improve communication and better inform decision makers, research efforts should include an iterative assessment process. These assessments would not only help to identify the relevant questions, but would also serve to structure the research results and, thus, facilitate clearer communication between the two communities. Furthermore, the assessment process would provide valuable input to the planning of policy-relevant research

such a manner as to support real environmental decisions in a timely manner. The importance of this linkage between decision making, assessment, and data collection is a hard-won lesson of the environmental management community over the last decade. The conclusions of one group of scientists and decision makers involved in the issue of global climate change are summarized in Table 7.7. Although they are framed in terms of a long-term and large-scale problem, they are relevant to all programmes that address an existing environmental problem. Even relatively short-term monitoring and assessment programmes benefit from procedures to ensure that the assessment science is driven by the needs of the decision makers and that the decisions and policies are fully informed by the science.

7.6 REFERENCES

Adams, D.F. (1963) Recognition of the effects of fluorides on vegetation. *Journal of the Air Pollution Control Association*, **13**, 360–362.

Adriaanse, A. (1993) *Environmental Policy Performance Indicators*. Ministry of Housing, Physical Planning, and Environment, Bilthoven, The Netherlands.

Ankley, G.T., Schubauer-Berigan, M.K. & Hoke, R.A. (1992) Use of toxicity identification techniques to identify dredged material disposal options: a proposed approach. *Environmental Management*, **16**, 1–6.

Baker, J.P., Bernard, D.P., Christensen, S.W. & Sale, M.J. (1990a) *Biological Effects of Changes in Surface Water Acid-base Chemistry*. National Acid Precipitation Assessment Program, Washington, DC.

Baker, L.A., Kaufmann, P.R., Herlihy, A.T. & Eilers, J.M. (1990b) *Current Status of Surface Water Acid-base Chemistry*. National Acid Precipitation Assessment Program, Washington, DC.

Barnthouse, L.W. & Brown, J. (1994) Conceptual model development. *Ecological Risk Assessment Issue Papers*, US Environmental Protection Agency, EPA/630/R-94/009, Washington, DC.

Barnthouse, L.W., Suter, G.W., II, Rosen, A.E. & Beauchamp, J.J. (1987) Estimating responses of fish populations to toxic contaminants. *Environmental Toxicology and Chemistry*, **6**, 811–824.

Bartell, S.M., Gardner, R.H. & O'Neill, R.V. (1992). *Ecological Risk Estimation*. Lewis Publishers, Ann Arbor, MI.

Bence, A.E. & Burns, W.A. (1995) Fingerprinting hydrocarbons in the biological resources of the *Exxon Valdez* spill area. In: *Exxon Valdez Oil Spill: Fate and Effects in Alaskan Waters* (eds P.G. Wells, J.N. Butler & J.S. Hughes), pp. 84–140. American Society for Testing and Materials, Philadelphia.

Bernard, J.E., Lucier, A.A. & Johnson, A.H. (1990) *Changes in Forest Health and Productivity in the United States and Canada*. National Acid Precipitation Assessment Program, Washington, DC.

Botterweg, J. (1995) Use of bioindicators in ecological effects assessment for water management purposes in The Netherlands. In: *Comparative Risk Analysis and Priority Setting for Air Pollution Issues* (eds S.D. Lee and T. Schneider), pp. 162–168. Air and Waste Management Association, Pittsburgh, PA.

Bovee, K.D. (1982) *A Guide to Stream Habitat Analysis using the Instream Flow Incremental Methodology*. US Fish and Wildlife Service, FWS/OBS-82/26, Fort Collins, CO.

Bovee, K.D. & Zuboy, J.R. (1988) *Proceedings of a Workshop on the Development and Evaluation of Habitat Suitability Criteria*, US Fish and Wildlife Service, Fort Collins, CO.

Boyle, T.P., Sebaugh, J. & Robinson-Wilson, E. (1984) A hierarchical approach to the measurement of changes in community induced by environmental stress. *Journal of Environmental Testing and Evaluation*, **12**, 241–245.

Bro-Rasmussen, F. & Lokke, H. (1984) Ecoepidemiology—a casuistic discipline describing ecological disturbances and damages in relation to their specific causes;

exemplified by chlorinated phenols and chlorophenoxy acids. *Regulatory Toxicology and Pharmacology*, **4**, 391–399.

Burmaster, D.E., Freshman, J.S., Burris, J.A., Maxwell, N.I. & Drew, S.R. (1991). Assesment of methods for estimating aquatic hazards at Superfund-type sites: a cautionary tale. *Environmental Toxicology and Chemistry*, **10**, 827–842.

Calamari, D. (ed.) (1993) *Chemical Exposure Prediction*. Lewis Publishers, Boca Raton, FL.

Colborn, T. (1991) Epidemiology of Great Lakes bald eagles. *Journal of Toxicology and Environmental Health*, **33**, 395–453.

Czuczwa, J. & Hites, R. (1986) Airborne dioxins and dibenzofurans: sources and fates. *Environmental Science and Technology*, **25**, 1619–1627.

Dickson, K.L., Waller, W.T., Kennedy, J.H. & Ammann, L.P. (1992) Assessing the relationship between ambient toxicity and instream biological response. *Environmental Toxicology and Chemistry*, **11**, 1307–1322.

Eberhardt, L.L. (1976). Quantitative ecology and impact assessment. *Journal of Environmental Management*, **4**, 27–70.

Eberhardt, L.L. & Thomas, J.M. (1991) Designing environmental field studies. *Ecological Monographs*, **61**, 53–73.

Emlen, J.M. (1989) Terrestrial population models for ecological risk assessment: a state-of-the-art review. *Environmental Toxicology and Chemistry*, **8**, 831–842.

Feijnders, P.J.H. (1986) Reproductive failure in common seals feeding on fish from polluted coastal waters. *Nature*, **324**, 456–457.

Fox, G.A. (1991) Practical causal inference for eco-epidemiologists. *Journal of Toxicology and Environmental Health*, **33**(4), 359–374.

Gauch, H.G. (1982) *Multivariate Analysis in Community Ecology*. Cambridge University Press, Cambridge.

Giesy, J.P., Ludwig, J.P. & Tillitt, D.E. (1994a). Deformities in birds of the Great Lakes region: assigning causality. *Environmental Science and Technology*, **28**, 128–135A.

Giesy, J.P., Ludwig, J.P. & Tillitt, D.E. (1994b) Dioxins, dibenzofurans, PCBs and colonial fish-eating water birds. In: *Dioxins and Health* (ed. A. Schecter), pp. 249–307. Plenum Press, New York.

Gilbertson, M., Kubiak, T., Ludwig, J. & Fox, G. (1991) Great Lakes embryo mortality, edema, and deformities syndrome (GLEMEDS) in colonial fish-eating birds: similarity to chick-edema disease. *Journal of Toxicology and Environmental Health*, **33**, 455–520.

Gordon, G.E. (1988) Receptor models. *Environmental Science and Technology*, **22**, 1132–1142.

Green, R.H. (1979) *Sampling Design and Statistical Methods for Environmental Biologists*. John Wiley & Sons, New York.

Green, R.H. (1984) Some guidelines for the design of biological monitoring programs in the marine environment. In: *Concepts in Marine Pollution Measurements* (ed. H.H. White), pp. 647–655. Maryland Sea Grant College, College Park, MD.

Hackney, J.D. & Kinn, W.S. (1979) Koch's postulates updated: a potentially useful application to laboratory research and policy analysis in environmental toxicology. *American Review of Respiratory Diseases*, **1119**, 849–852.

Hartwell, S.I., Dawson, C.E., Durell, E.Q. & Jordahl, D.H. (1995a) Integrated measures of ambient toxicity and fish community diversity. Maryland Department of Natural Resources, Chesapeake Bay Research and Monitoring Division, Annapolis, MD.

Hartwell, S.I., Dawson, C.E., Jordahl, D.H. & Durell, E.Q. (1995b) Demonstrating a method to correlate measures of ambient toxicity and fish community diversity. Maryland Department of Natural Resources, Chesapeake Bay Research and Monitoring Division, Annapolis, MD.

Hill, A.B. (1965) The environment and disease: association or causation. *Proceedings of the Royal Society of Medicine*, **58**, 295–300.

Huggett R.J., Kimerle, R.A., Mehrle, P.M. & Bergman, H.L. (eds) (1992) *Biochemical, Physiological, and Histological Markers of Anthropogenic Stress*. Lewis Publishers, Boca Raton, FL.

Hunsaker, C.T. & Carpenter, D.E. (1990) *Ecological Indicators for the Environmental Monitoring and Assessment Program*. US Environmental Protection Agency, Research Triangle Park, NC.

Hurlbert, S.H. (1984) Pseudoreplication and the design of ecological field experiments. *Ecological Monographs*, **54**, 187–211.

Johnson, A.R. (1988) Diagnostic variables as predictors of ecological risk. *Environmental Management*, **12**, 515–523.

Karr, J.R., Fausch, K.D., Angermeier, P.L., Yant, P.R. & Schlosser, I.J. (1986) Assessing biological integrity in running waters; a method and its rationale. Illinois Natural History Survey Special Publication 5, Champaigne, IL.

Kendall, R.J., Lacher, T.E., Bunck, C. *et al.* (1996) An ecological risk assessment of lead shot exposure in non-waterfowl avian species: upland game birds and raptors. *Environmental Toxicology and Chemistry*, **15**, 4–20.

Mac, M.J. & Edsall, C.C. (1991) Environmental contaminants and the reproductive success of lake trout in the Great Lakes: an epidemiological approach. *Journal of Toxicology and Environmental Health*, **33**, 375–394.

McCarthy, J.F. & Shugart, L.R. (eds) (1990) *Biomarkers of Environmental Contamination*. Lewis Publishers, Bocas Raton, FL.

McCarty, L.S. & Mackay, D. (1993) Enhancing ecotoxicological modeling and assessment. *Environmental Science and Technology*, **27**(9), 1719–1728.

Menzie, C.A., Burmaster, D.E., Freshman, D.S. & Callahan, C. (1992) Assessment of methods for estimating ecological risk in the terrestrial component: a case study at the Baird and McGuire Superfund site in Holbrook, Massachusetts. *Environmental Toxicology and Chemistry*, **11**, 245–260.

Meyer, F.P. & Barklay, L.A. (1990) *Field Manual for the Investigation of Fish Kills.* US Fish and Wildlife Service, Resource Publication 177, Washington, DC.

Morgan, M.G. & Henrion, M. (1990) *Uncertainty: a Guide to Dealing with Uncertainty in Quantitative Risk and Policy Analysis.* Cambridge University Press, Cambridge.

Mount, D.I. & Norberg, T. (1985) A seven-day life-cycle toxicity test. *Environmental Toxicology and Chemistry*, **3**, 425–434.

NRC (1994). *Science and Judgement in Risk Assessment.* National Research Council, National Academy Press, Washington, DC.

Norberg, T.J. & Mount, D.I. (1985) A new fathead minnow (*Pimephales promelas*) subchronic toxicity test. *Environmental Toxicology and Chemistry*, **4**, 711–718.

OECD (1991) *Environmental indicators: a preliminary set.* Organization for Economic Cooperation and Development, Paris.

OECD (1993) *OECD Core Set of Indicators for Environmental Performance Reviews.* Organization for Economic Cooperation and Development, Paris.

Ohio Environmental Protection Agency (1990) *The Use of Biocriteria in the Ohio EPA Surface Water Monitoring and Assessment Program.* Division of Water Quality Planning and Assessment, Columbus, OH.

Olsson, M. (1994) Diseases and environmental contaminants in seals from the Baltic and Swedish west coast. *Science of the Total Environment*, **154**, 217–227.

Olsson, M. (1995) Ecological effects of airborne contaminants in arctic aquatic ecosystems: a discussion on methodological approaches. *Science of the Total Environment*, **160/161**, 619–630.

O'Neill, R.V., Gardner, R.H., Barnthouse, L.W., Suter, G.W., Hildebrand, S.G. & Gehrs, C.W. (1982) Ecosystem risk analysis: a new methodology. *Environmental Toxicology and Chemistry*, **1**, 167–177.

Orth, D.J. (1987) Ecological considerations in the development and application of instream flow-habitat models. *Regulated Rivers: Research and Management*, **1**, 171–181.

Peakall, D.B. (1993) DDT-induced eggshell thinning: an environmental detective story. *Environmental Review*, **1**, 13–20.

Pearson, W.H. Moksness, E. & Skalski, J.R. (1995) A field and laboratory assessment of oil spill effects on survival and reproduction of Pacific herring following the *Exxon Valdez* spill. In: *Exxon Valdez Oil Spill: Fate and Effects in Alaskan Waters* (eds P.G. Wells, J.N. Butler & J.S. Hughes), pp. 626–661. American Society of Testing and Materials, Philadelphia.

Plafkin, J.L., Barbour, M.T., Porter, K.D., Gross, S.K. & Hughes, R.M. (1989) *Rapid Bioassessment Protocols for Use in Streams and Rivers: Benthic Macroinvertebrates and Fish.* US Environmental Protection Agency, Washington, DC.

Platt, J.R. (1964) Strong inference. *Science*, **146**, 347–353.

Popper, K.R. (1968) *The Logic of Scientific Discovery.* Harper and Row, New York.

Risk Assessment Forum (1992) *Framework for Ecological Risk Assessment.* US Environmental Protection Agency, Washington, DC.

Sample, B.E. & Suter, G.W. (1994) *Estimating exposure of terrestrial wildlife to contaminants.* Oak Ridge National Laboratory, Oak Ridge, TN.

Schimmel, S.C., Morrison, G.E. & Heber, M.A. (1989) Marine complex effluent toxicity program: test sensitivity, repeatability, and relevance to receiving water toxicity. *Environmental Toxicology and Chemistry*, **8**, 739–746.

Schrader-Frechette, K.S. & McCoy, E.D. (1994) How the tail wags the dog: how value judgements determine ecological science. *Environmental Values*, **3**, 107–120.

Schulze, E.-D. (1989) Air pollution and forest decline in a spruce (*Picea abies*) forest. *Science*, **244**, 776–783.

Shriner, D.S., Heck, W.W. *et al.* (1990) *Response of Vegetation to Acid Deposition and Air Pollution.* National Acid Precipitation Assessment Program, Washington, DC.

Skelly, J.M. (1989) Forest decline versus tree decline—the pathological considerations. *Environmental Monitoring and Assessment*, **12**, 23–27.

Sloof, W., van Oers, J.A.M. & de Zwart D. (1986) Margins of uncertainty in ecotoxicological hazard assessment. *Environmental Toxicology and Chemistry*, **5**, 841–852.

Smith, E.P., Pontasch, K.W., & Cairns, J., Jr. (1989) Community similarity and the analysis of multispecies environmental data: a unified statistical approach. *Water Research*, **24**, 507–514.

Smith, E.P., Orvos, D.R. & Cairns, J., Jr. (1993) Comments and concerns in the use of the before–after-control-impact (BACI) model for impact assessment. *Canadian Journal of Fisheries and Aquatic Science*, **50**, 627–637.

Solomon, K.R., Baker, D.B., Richards, R.P. *et al.* (1996) Ecological risk assessment for atrazine in North American surface waters. *Environmental Toxicology and Chemistry*, **15**, 31–76.

SPA (1992) *Report of Findings: Joint Climate Project to*

Address Decision Maker's Uncertainties. Science and Policy Associates, Electric Power Research Institute, Pleasant Hill, CA.

Stewart-Oaten, A., Murdoch, W.W. & Parker, K.R. (1986) Environmental impact assessment: 'pseudoreplication' in time? *Ecology*, **67**, 929–940.

Susser, M. (1986a) Rules of inference in epidemiology. *Regulatory Toxicology and Pharmacology*, **6**, 116–186.

Susser, M. (1986b) The logic of Sir Carl Popper and the practice of epidemiology. *American Journal of Epidemiology*, **124**, 711–718.

Suter, G.W., II (1989). Ecological endpoints. *Ecological Assessment of Hazardous Waste Sites: A Field and Laboratory Reference Document* (eds W. Warren-Hicks, B.R. Parkhurst & J.S.S. Baker), pp. 2-1-2-28. US Environmental Protection Agency, EPA/600/3-89/013, Corvallis Environmental Research Laboratory, Corvallis, OR.

Suter, G.W., II (1990a) Endpoints for regional ecological risk assessments. *Environmental Management*, **14**, 9–23.

Suter, G.W., II (1990b) Uncertainty in environmental risk assessment. In: *Acting Under Uncertainty: Multidisciplinary Conceptions* (ed. G.M. von Furstenberg), pp. 203–230. Kluwer Academic Publishers, Boston.

Suter, G.W., II (1990c) Use of biomarkers in ecological risk assessment. In: *Biomarkers of Environmental Contamination* (eds J.F. McCarthy & L.L. Shugart), pp. 419–426. Lewis Publishers, Ann Arbor, MI.

Suter, G.W., II (1993a). *Ecological Risk Assessment*. Lewis Publishers, Boca Raton, FL.

Suter, G.W., II (1993b). A critique of ecosystem health concepts and indices. *Environmental Toxicology and Chemistry*, **12**, 1533–1539.

Suter, G.W., II & Loar, J.M. (1992) Weighing the ecological risks of hazardous waste sites: The Oak Ridge case. *Environmental Science and Technology*, **26**(3), 432–438.

Suter, G.W., II, Vaughan, D.S. & Gardner, R.S. (1983) Risk assessment by analysis of extrapolation error, a demonstration for effects of pollutants on fish. *Environmental Toxicology and Chemistry*, **2**, 369–378.

Suter, G.W., II, Rosen, A.E., Beauchamp, J.J. & Kato T.T. (1992). *Results of Analysis of Fur Samples from the San Joaquin Kit Fox and Associated Water and Soil Samples from the Naval Petroleum Reserve No. 1, Tupman, California*. Oak Ridge National Laboratory, Oak Ridge, TN.

Suter, G.W., II, Gillett, J.W. *et al.* (1994) Characterization of exposure. *Ecological Risk Assessment Issue Papers*. US Environmental Protection Agency, EPA/630/R-94/009, Washington, DC.

ten Brink, B.J.E. (1989) *Amoeba-approach: A Method for Description and Assessment of Ecosystems*. Ministry of Transportation, Public Works and Water Management, Tidal Water Division, The Hague.

Thornton, K.W., Saul, G.E. & Hyatt, D.E. (1994) *Environmental Monitoring and Assessment Program Assessment Framework*. US Environmental Protection Agency, Research Triangle Park, NC.

Tillit, D.E., Giesy, J.P. & Ankley, G.T. (1991) Characterization of the H4IIE rat hepatoma cell bioassay as a tool for assessing toxic potency of planar halogenated hydrocarbons in environmental samples. *Environmental Science and Technology*, **25**, 87–92.

Underwood, A.J. (1994) On beyond BACI: sampling designs that might reliably detect environmental disturbances. *Ecological Applications*, **4**(1), 3–15.

USEPA (1995) *A Conceptual Framework to Support Development and use of Environmental Information in Decision-Making*. Environmental Statistics and Information Division, US Environmental Protection Agency, Washington, DC.

Wallace, H.R. (1978) The diagnosis of plant diseases of complex etiology. *Annual Review of Phytopathology*, **16**, 379–402.

Wells, P.G., Butler, J.N. & Hughes, J.S. (eds) (1995) *Exxon Valdez Oil Spill: Fate and Effects in Alaskan Waters*. American Society for Testing and Materials, Philadelphia, PA.

Whittier, T.R., Larson, D.P., Highes, R.M., Rohm, C.M., Gallant, A.L. & Omernick, J.M. (1988) *The Ohio Stream Regionalization Project: A Compendium of Results*. US Environmental Protection Agency, EPA/600/3-87/025, Corvallis, OR.

Wiens, J.A. & Parker, K.R. (1995) Analyzing the effects of accidental environmental impacts: approaches and assumptions. *Ecological Applications*, **5**(4), 1069–1083.

Woodman, J.N. & Cowling, E.B. (1987) Airborne chemicals and forest health. *Environmental Science and Technology*, **21**, 120–126.

Wren, C.D. (1991) Cause–effect linkages between chemicals and populations of mink (*Mustella vison*) and otter (*Lutra canadensis*) in the Great Lakes basin. *Journal of Toxicology and Environmental Health*, **33**, 549–585.

Wright, J.F., Armitage, P.D., Furse, M.T. & Moss, D. (1989) Prediction of invertebrate communities using stream measurements. *Regulated Rivers: Research and Management*, **4**, 147–155.

Yerushalmy, J & Palmer, C.E. (1959) On the methodology of investigations of etiologic factors in chronic disease. *Journal of Chronic Disease*, **10**(1), 27–40.

Chapter 8
Epidemiology and Environmental Risk Assessment

JOSEPH V. RODRICKS AND THOMAS A. BURKE

8.1 INTRODUCTION

Risk assessment is a systematic means for organizing and evaluating scientific information relevant to the question of whether, and with what likelihood, individuals exposed to agents in their environments will suffer harm. Here the term 'agents' is used broadly, and includes any of the physical, biological and chemical entities found in the environment and with which individuals may come into contact. Almost all such agents can, under some conditions, cause some type of harm. The purpose of risk assessment in this context is to evaluate the likelihood that harm has occurred or could occur in humans under their actual or expected conditions of contact with (or exposure to) those agents. It is recognized that, to be maximally useful, risk assessments need to include careful evaluations both of what is known and what is unknown about the risks under study (NRC 1983, 1994).

There are, of course, many possible uses of risk assessment results; for example, to guide regulatory and public health decisions; to assist in the identification of public health priorities; to inform the courts in cases of injuries alleged to be caused by consumer products or environmental exposures; to provide businesses with information regarding the possible dangers associated with their facilities and their products; and even to help establish environmental and public health research agendas. It seems that the effective functioning of modern societies increasingly demands understanding of health and safety risks that may have been introduced by past and current technologies, and predictions of the likely consequences of the introduction of new ones.

Epidemiology is the core science of public health, providing a basis for evaluating and describing risks in the human population. Historically, much of our understanding of health risks from environmental and occupational exposures has been gained from epidemiological studies of exposed populations (i.e. smoking, radiation, lead, benzene). While toxicology-based animal studies have supplanted human studies in the quantitative estimation of risks for many regulatory applications, epidemiology remains essential to the validation of risk assessment approaches and the understanding of disease causality in humans. For example, well-conducted positive human studies are essential to the determination that an agent is a human carcinogen.

There are several advantages of epidemiological data over animal data. Because they describe risks in the human population, epidemiological studies avoid the uncertainties of interspecies extrapolation. They also allow the evaluation of a wider range of adverse health outcomes which may be related to environmental exposures. This is particularly important as the practice of risk assessment continues to expand to include diverse, non-cancer end-points such as reproductive, neurological or immunological effects. Epidemiological studies also provide information based on actual levels of human exposure, thereby reducing the uncertainties associated with high-to-low dose extrapolation required for animal studies.

There are also disadvantages in current epidemiological methods that have limited the utility of many epidemiological studies for quantitative risk assessment. Two major difficulties are introduced here. First, there are limitations on the ranges of risk detectable using epidemiology.

Several factors contribute, notably the practical problems involved in the assembly of populations sufficiently large to detect relatively small or even moderately large risks. This same type of limitation applies to all methodologies available to measure the risks associated with environmental agents. Thus, while the epidemiological method can provide significant and direct information regarding large risks (e.g. from smoking or from inadequately controlled occupational exposures), neither it nor other (experimental) methods can, at the present time, provide direct, reliable measures of the many small-to-moderate risks to which humans are exposed, especially through environmental media. Risk assessment was introduced as a means for developing some insight into such currently immeasurable risks, and its most difficult (and controversial) aspects arise when it is necessary to *extrapolate* from the range of measurable into the range of non-measurable risks. Unless such extrapolation is undertaken, we are forced to remain silent regarding the very large number of relatively small risks to which human populations are exposed. The notion driving the introduction of risk assessment is that practical decision-making cannot await the development of highly certain scientific data, that it is better to make decisions under uncertainty (as long as the uncertainties are reasonably well understood) than to ignore public health consequences (Paustenbach 1989; NRC 1994).

The second significant limitation to the epidemiological method concerns the fact that the potentially harmful consequences of human exposures to environmental agents cannot be discovered until after, sometimes long after, exposures have occurred. Of course, the results from epidemiological research can be useful in predicting the consequences of future exposures, but, in an ideal world, it would be highly desirable to understand the possible risks associated with environmental agents before human exposures were allowed to occur. This desire prompted the development of experimental animal studies, and results from such studies (which, of course, have their own limitations) are now widely used to assess possible human risks (NRC 1983).

Notwithstanding these limitations (and others to be reviewed later), epidemiology remains the most powerful tool available for understanding how, and to what extent, agents in the environment affect human health. The purpose of this chapter is to review the various methods used by epidemiologists to study the aetiologies of disease and to establish causal relationships, and to show how results from epidemiology studies in certain limited populations can be used to assess risks in other populations exposed under different conditions. Specific examples involving occupational and environmental exposures are used to illustrate these methodologies and applications of epidemiology.

This chapter is limited in scope to the risks of latent diseases or toxic injuries caused by chemical agents that come to be present in occupational settings and in the general environment. Epidemiological studies have made, and continue to make, major contributions to our understanding of the health risks associated with drugs, medical devices and consumer products (Rothman 1986). They also are the principal means by which we have come to understand the many roles of biological pathogens in human diseases. We do not discuss here these well-known aspects of epidemiology, but rather devote our attention to emerging uses of epidemiology to understand and help solve public health problems created by chemical exposures found in occupational settings and in environmental media. We note here the many ways cited by Suter (Chapter 7) in which the emerging discipline of ecoepidemiology has began to draw upon some of the methodological tools applied by epidemiologists.

8.2 MEASURES OF DISEASE AND RISK

The measurement of illness, injury and death is essential to the understanding of population risk. This section describes the metrics used in epidemiology to examine disease and mortality in populations, and to compare the risks in different populations. Sources of information on mortality and illness are also discussed.

8.2.1 Measures of morbidity

Perhaps the most fundamental measure of risk in

a population is the incidence of a disease, or morbidity. Incidence is defined as the number of new cases of a disease that occur in a population at risk for developing the disease during a specified period of time. Incidence has provided public health professionals with a valuable measure of population risk, enabling them to track trends, compare populations, and identify increases in population risk of disease. From lung cancer to human immunodeficiency virus (HIV) infection, annual incidence data have identified those at high risk and provided a metric for measuring the effectiveness of prevention strategies. For the incidence rate to provide an accurate measure of risk, it is important that the denominator reflects all those who have the potential for developing the disease.

$$\text{Incidence rate per 1000} = \frac{\text{Number of new cases in the population over a specified time}}{\text{Number of persons at risk during that time period}} \times 1000$$

Prevalence is also an important measure of population health status in occupational and environmental health. Prevalence is defined as the number of persons in the population with a disease at a specified time divided by the total number of persons in the population at that time.

$$\text{Prevalence per 1000} = \frac{\text{Number of cases in the population at a specified time}}{\text{Number of persons in the population at that time}} \times 1000$$

The measure of prevalence for a single point in time is called the point prevalence (i.e. the prevalence of influenza on 26 February). The number of cases of a disease that exist over a specified period of time (month, year) is called the period of prevalence. Prevalence does not measure when the disease developed or take into account the duration of the disease. It is therefore not as useful as incidence in the investigation of causality. However, disease prevalence is an important consideration in many risk management decisions. For example, the prevalence of childhood asthma in a population may be an important consideration in developing air pollution control strategies. Similarly, identifying a high prevalence of musculoskeletal disorders in certain occupational settings may be important in shaping efforts to reduce worker risks.

While prevalence provides an important measure of the extent of a disease in a population, it does not provide an estimate of the risk of developing a disease. For example, the prevalence of breast cancer may be higher in a wealthy population than in a poor population because of improved survival due to better access to medical care. If the incidence rates of the same populations are compared, it is likely that the poor community would be shown to have a higher incidence and therefore would have a higher risk of developing the disease. There is an interrelationship between incidence and prevalence. Prevalence is proportional to the incidence multiplied by the duration of the disease.

8.2.2 Measures of mortality

Mortality rates quantify the risk of dying from given diseases or injuries in a population and allow the comparison of risks across populations. Perhaps the most often used measure of mortality is the annual mortality rate for all causes.

$$\text{Annual mortality rate for all causes per 1000} = \frac{\text{Total deaths from all causes in 1 year}}{\text{Population at mid-year}} \times 1000$$

The annual mortality rate is an example of a crude rate; i.e. it is a summary rate based upon the number of events in a population over a specified time. While crude rates are useful, specific rates are often more informative in the investigation of risk. For example, age- and sex-specific mortality rates present the number of deaths in the specified group and limit the denominator to the population at risk in that age and sex group. Disease- or cause-specific rates are most useful in describing population risks of mortality from specific diseases. The annual mortality rate from lung cancer in white females is an example of a specific rate.

$$\text{Annual mortality rate for lung cancer in white females per 1000 population} = \frac{\text{Lung cancer deaths in white females per year}}{\text{Number of white females in mid-year population}} \times 1000$$

Epidemiologists often use other measures of mortality that do not provide a quantification of risk, but are useful in characterizing the impacts of disease. For example, the case fatality rate is the percentage of people diagnosed with a disease who die within a specified period. Proportionate mortality is the percentage of total deaths due to a specific cause, which is useful for identifying the major causes of death in a population. Years of potential life lost represents another mortality measure which calculates productive years of life lost due to mortality from disease or injury.

8.2.3 Comparing rates in different populations

Understanding and comparing the rates of morbidity and mortality in different populations are fundamental to epidemiology. Simple ratios of incidence or mortality can reveal significant differences in risk between populations. For example, the ratio of cancer mortality in males to females in the USA is 585.5/383.8 or 1.53; therefore the risk for males is 53% higher than females. A similar comparison of mortality between blacks and whites of both sexes (624.3/457.2 or 1.37) reveals that the risk of mortality from cancer is 37% higher for blacks than for whites (USDHHS 1994a).

In comparing populations it is important to recognize that there may be fundamental differences in the population composition that may affect the risks of morbidity and mortality. Age is probably the most important variable that can influence the rates of disease and mortality in a population. Simply put, for many causes of death and disease, an older population is at higher risk than a younger population. Suppose that a study of the relationship between air pollution and cardiovascular mortality is being designed to compare an inner city population with high pollution to a suburban community with low pollution. Given the higher proportion of elderly in the inner city and the higher risk of heart disease in the elderly, it is essential to consider the differences in age distribution when comparing mortality rates. To address this problem, age-adjusted rates are calculated using statistical procedures to remove the effect of differences in the age composition of the populations. Similar approaches can also be used to adjust for other population factors, such as sex, race and socioeconomic status.

There are two methods of age adjustment: direct and indirect. Direct adjustment applies the observed age-specific rates to a standard population to calculate an expected number of deaths. Indirect adjustment for age can also be accomplished by calculation of the standardized mortality ratio (SMR). This approach is often used in epidemiological studies of occupational risks. The SMR is calculated by totalling the observed number of deaths and dividing it by the expected number of deaths based upon the age-specific mortality rate for the general or comparison population.

$$\text{SMR} = \frac{\text{Observed number of deaths per year}}{\text{Expected number of deaths per year}}$$

This approach is also used in descriptive environmental epidemiological studies to determine if communities with suspected exposures have higher than expected rates of morbidity or mortality. Disease incidence rates can be compared using an analogous approach through the calculation of a standardized incidence ratio (SIR). For example, cancer incidence in a town can be compared with the number of cases that would be expected based upon state or national incidence rates to determine if incidence is elevated.

8.2.4 Sources of data on incidence and mortality

There are numerous approaches to the collection of incidence and mortality data which form the basis of public health surveillance systems and provide the quantitative basis for epidemiological research. Historically, vital statistics have provided important insights into the causes of mortality,

Table 8.1 Examples of data sources in epidemiology.

Vital statistics	Birth and mortality records
Reportable disease statistics	New cases of notifiable diseases
Disease registries	Data bases on case registries, such as cancer registries
Population surveys	Interview and diagnostic surveys of the population
Health and life insurance statistics	Morbidity and mortality of insured populations
Hospital admission and discharge data	Cases treated in hospitals
Absenteeism data	Missed days of work or school
Armed forces and veterans statistics	Morbidity and mortality
Labour statistics	Worker illness, injury, and mortality
Social security statistics	Disability benefit data
Census data	Counts and characteristics of populations by location

and reportable diseases have been tracked to provide early warning systems and shape prevention approaches. In recent years the amount of data on disease incidence and population health status has increased dramatically as public health agencies, health-care providers, employers and insurance companies have developed expanded data collection and analysis capabilities. Table 8.1 lists a number of sources of data on disease incidence, mortality and risk factors. These databases are an invaluable resource for epidemiological research, the identification of emerging health issues, the validation of current risk assessment approaches, and the epidemiological evaluation of population risks.

8.3 MEASURES OF EXPOSURE AND DOSE

An essential determinant of the risk posed by a chemical is its dose–response relationship. The latter refers to the relationship between the magnitude, frequency and duration of the dose of a chemical received by exposed populations and the incidence, prevalence or severity of the resulting injury or disease. Evidence that toxic responses increase with increasing doses of a chemical contributes significantly to the determination of whether the chemical actually *caused* the response (section 8.5). Knowledge of dose–response relationships is also essential to the conduct of risk assessments involving exposures not identical to those at which risks were actually measured; if the exposures being assessed fall within the range of measured dose–response relationships, then some form of *interpolation* is needed to estimate risk, and if (as is the far more common case) exposures fall below that range, then the more difficult problem of dose–response *extrapolation* needs to be faced (USEPA 1986a).

In the preceding section we discussed various measures of morbidity and mortality, and of methods for comparing risks of different populations. Here we turn to the problem of measuring doses. As suggested in the discussion of the two major uses of dose–response relationships, the problem of identifying doses received by exposed individuals has two manifestations. First, the successful conduct of an epidemiological investigation depends heavily upon identifying the doses of the agent under study, or some surrogate measure, received by the populations under study. Second, the successful conduct of a risk assessment depends upon estimating the doses received by the individuals who are the subject of that assessment. The discussion here does not further distinguish these two manifestations of the problem, although it should be recognized that not all of the methods discussed are equally applicable in the conduct of epidemiology studies and in the conduct of risk assessments.

8.3.1 Basic concepts

The term *exposures* is ordinarily used to describe the contact between individuals and an environmental medium containing a chemical(s). Thus, consumption of foods and beverages containing deliberately introduced chemicals, inhalation of contaminated air, direct skin contact with cosmetics or industrial solvents, and inadvertent ingestion of contaminated soils are examples of chemical exposures. Such contacts create opportunities for exposed individuals to receive a dose; the latter can be measured in several different ways, but all reflect attempts to capture the dose metric that is the most relevant determinant of a chemical's toxic action (USEPA 1986a). With respect to human populations (as against experimental settings) there are often serious practical limitations in obtaining such measures, and it is often difficult to acquire knowledge of what is the most relevant measure of dose. In theory, it appears that the most relevant dose metric would be the concentration of the toxic substance (which is very often not the chemical to which individuals are exposed but rather a metabolite(s) thereof), at the site of toxic action, integrated with the duration of time over which such contact occurs (USEPA 1986a, 1988a). As might be imagined, such a measure is difficult to acquire and various surrogate measures of dose are usually sought. Perhaps the closest surrogates are body burdens, which might include blood levels (now the common measure of lead exposure) and lipid concentrations of fat-soluble, persistent organic substances such as chlorinated dioxins and polychlorinated biphenyls (PCBs). In recent years, measurements of protein and deoxyribonucleic acid (DNA) adducts have shown promise for a few substances as what have come to be called 'biomarkers' of exposure. However, practical applications of such dose measures are still in their infancies (NRC 1994).

The most common environmental dose measures are, as might be expected, those that are easiest to measure. For substances whose toxic effects are not systemic, but rather at initial points of contact (e.g. respiratory and dermal toxicants), the usual 'dose' measures are the concentrations of the chemical in the medium of contact multiplied by the duration of that contact. If the duration is intermittent, a decision has to be made regarding the appropriate determinant of toxic risks—which could be represented either by the 'peak' value of medium concentration multiplied by the actual duration of medium contact, or by the long-term, average, concentration–time profile. Thus, the health risk associated with a carcinogen exposure could, depending upon its mode of biological action, be better correlated with intermittent, relatively high exposures, or with long-term, cumulative exposures, irrespective of whether those exposures were actually received intermittently or continually. Often epidemiologists attempt to develop several different measures of population exposures and seek to determine which correlates most closely with excess risks (Rothman 1986).

When the toxic effects of interest are systemic, and no theories or practical tools are available to identify 'target site doses', attempts are usually made to develop estimates of 'absorbed dose'. This is usually expressed as mass of chemical systemically absorbed (mg day^{-1}, for example), normalized to body mass; its typical units are mg kg^{-1} day^{-1}, and this is perhaps the most common measure of dose in risk assessment (USEPA 1988a).

Epidemiological investigations are often dependent upon highly indirect and often qualitatively expressed surrogates for exposure and dose. These include educated guesses regarding past exposure levels, time spent in various exposure situations, duration and frequency of exposure, and even guesses regarding the extent of dermal contact, especially with substances encountered occupationally. Often epidemiologists are able to develop rough, semi-quantitative ranges of exposure experienced by occupationally and environmentally exposed individuals, and are thus able to ascertain whether risk increases with dose. Such semi-quantitative dose–response data can often add substantially to an understanding of whether causal relations between exposure and toxic response are likely, but may not be of much value in quantitatively evaluating risks, especially those that may exist outside the range of observation (see Paxton *et al.* 1994 for an example of how such estimates can be derived).

8.3.2 Determinants of exposure and dose

Whether the context is the epidemiological investigation of exposed populations or the attempt to develop estimates for other exposed populations that are to be the subjects of risk assessments, a systematic attempt to identify all information relevant to dose estimation needs to be undertaken. In most cases the ultimate estimations of dose will depend upon whatever data happen to be available pertinent to the specific population of interest, but a systematic inquiry into all relevant aspects of the question is advisable. At least the following areas of inquiry are ordinarily investigated.

The population

A careful delineation of the populations to be investigated is the typical starting point for exposure and dose assessment. Individuals exposed in occupational settings have been the subject of most epidemiological investigations involving chemical exposures. (Patient populations have been central to the study of the effects of drugs and medical devices.) In recent years epidemiologists have also attempted to measure risks in populations exposed environmentally, but, except in the case of health effects observable relatively soon after the initiation of exposures, clear successes have been relatively uncommon, because of difficulties in obtaining relevant health and exposure data (NRC 1991). In the context of predictive risk assessment, the population to be evaluated might be individuals exposed occupationally, or through air, water, foods or soils. Among the latter, for example, might be individuals exposed to emissions from a manufacturing facility, or groups of people consuming fish from waters that have been the subject of chemical contamination (USEPA 1988a). The absence of a clear definition of the population to be evaluated can completely undermine an epidemiological study and can, at the very least, muddle the outcome of a risk assessment (Rothman 1986).

Sources of population exposure

Once the population of interest is defined it becomes possible to identify each of the sources of the chemical(s) that might ultimately lead to an exposure. In the occupational setting each operation that might create a worker exposure is identified. Each environmental release that might eventually reach members of the non-occupational population of interest requires similar elucidation. In some instances, care must be taken to avoid focus only on obvious sources, because those that might escape immediate attention could, in the end, be more significant. The example of lead exposures, presented in section 8.8.3, reveals some of the pitfalls of a too-narrow focus.

Pathways to the population

In some instances, individuals may be directly exposed to the chemical at its source; the typical case involves dermal exposures of workers to chemicals encountered on the job (Paxton *et al.* 1994). But the more common phenomenon includes the movement of a chemical from its source into one or more environmental media and, in many cases, from one medium into another, before individuals come into contact (Swann & Eschenroeder 1983). The movement of a chemical, through, for example, volatilization, solubilization or uptake into plants, describes the *pathways* from source to individuals who become exposed. The example of lead is, again, useful to describe this phenomenon, because it can travel many pathways (section 8.8.3).

Media of exposure

The environmental media through which individuals contact the chemical are the so-called media of exposure. Irrespective of whether, in the final analysis, each contributes significantly to an individual's exposure, it is useful to attempt to identify all potentially relevant exposure media. It is usually considered necessary for a risk assessor to be all-inclusive until it is clear (and clearly documented) that particular media can be disregarded as of trivial importance to the overall exposure and dose (USEPA 1989).

Concentrations of chemicals in exposure media

In the ideal it would be desirable to acquire data on the distributions of chemical concentrations in the media of exposure. Such information can be obtained through sampling and analysis (measurement) or through the application of various models that have been developed to account for the movement of a chemical from its source to the ultimate medium of exposure (see, for example, Swann & Eschenroeder 1983; McKone & Layton 1986). Whether measurement or modelling is used, significant uncertainties in this area cannot be avoided, and will often be among the most important in a risk assessment. These uncertainties arise for many reasons: the difficulty of obtaining statistically representative samples, both in time and space, and the inherent limitations of any attempt to model physical reality, are perhaps the two central problems (Millard 1987). It is often the case that information on past exposures will be completely absent, and attempts to 'reconstruct' past exposures through modelling will often fail for lack of data on 'source terms' and other conditions that determine exposure. In such cases it may be possible to obtain only crude, qualitative measures of exposure. When long-term, average exposures are of interest it is often possible to obtain reasonable estimates, but obtaining estimates of short-term 'peaks' may be quite problematic (Rodricks 1996).

Population behaviour influencing exposure media contact

The duration and frequency of contact of individuals in the population with the media of exposure are determinants of the dose(s) received. The number of hours per day, days per week, weeks per year, and years per lifetime that an individual breathes contaminated air or drinks contaminated water will determine the intake of the medium per day, week, year or lifetime. While it is reasonable to assume that workers are on the job 8 hours per day and 5 days per week for a 40–50-year working lifetime, it may be that for some workers only a portion of those times is spent working in contact with exposure media. For individuals exposed outside the workplace, behaviour patterns are far more variable. How many hours per week, for example, is it likely that nearby residents will be in specific areas in which contact with contaminated air from an industrial facility will be experienced? How many fish containing dioxin will individuals living downstream from a paper-mill consume per week, and how much fish will they consume per serving? Information useful for answering such questions can be developed, but to do so may require significant research. In the context of predictive risk assessments, broad, generic assumptions are often adopted to deal with data gaps (Paustenbach 1989; USEPA 1989).

Media contact rates

How much air does an individual breathe per hour or per day? How much water or food does an individual consume per day? How do these media intake rates vary among individuals in a population? Without answers to these questions it will not be possible to answer questions about either the amount of chemical exposure occurring at each route into the body (airways, gastrointestinal tract, skin) or the amounts absorbed through the lungs, gastrointestinal tract and skin into the bloodstream. The latter estimates of systemic dose require, of course, information on rates of absorption across these three barriers. It must also be remembered that, depending upon the circumstances, individuals may be exposed to a chemical simultaneously through more than one medium, and that the *total dose* received through all media is the determinant of risk. Of course, the media may also be contaminated with more than one chemical. The USEPA has developed data on media contact rates for different segments of the population in the USA (Pao *et al.* 1982; ICRP 1984; USEPA 1989).

Body burdens

In some instances it may be possible to obtain a measure of body burdens (see the example of lead, below), either through direct measurements in exposed individuals or through modelling of the pharmacokinetic behaviour of a chemical once it is absorbed. Such measures are becoming increasingly important, and we shall have more to say about them in the closing section.

8.3.3 Dose estimates

The goal of dose estimation is to understand the distribution of received doses, and its variation over time, in the population of interest. Obtaining complete distribution estimates, except in those relatively uncommon cases in which direct body burden measures can be developed in all or a statistically representative number of the population of interest, is usually a practical impossibility. Broad population averages, or estimates of doses somewhere on the tail of a distribution (e.g. near the 90th or 95th percentile), are usually the best that can be done, and even these estimates are subject to substantial uncertainties (NRC 1994). And, as has already been noted, sometimes only semi-quantitative estimates (e.g. years on a job in

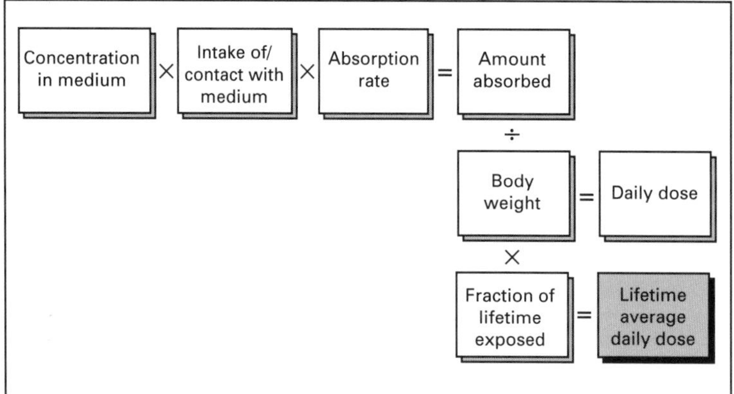

Fig. 8.1 Outline of procedure for calculating maximum and lifetime average daily doses. The latter measure is typically used in risk assessment when cumulative lifetime exposure is the critical determinant of risk, as in the case of carcinogens. In many cases the dose metric most relevant to risk estimation is not well established.

which exposures were 'high') can be developed. Because no useful statements regarding risks can be made unless some type of exposure estimate is developed, it is usually the practice that assessors develop the best possible estimates, point out the significant uncertainties in them, and offer some perspective on the likely influence of those uncertainties on the ultimate assessment of risk. Figure 8.1 summarizes the calculations typically made to estimate two common dose measures (lifetime average and maximum daily). As noted, other dose metrics may be appropriate depending upon the chemical toxicant and its mode of action.

Table 8.2 Major types of epidemiological studies conducted to evaluate environmental risks (see section 8.4).

Experimental
Clinical trials
Community trials
Observational
Ecological
Analytical
Cohort
Case–control
Cross-sectional

8.4 EPIDEMIOLOGICAL METHODS

This section describes the major types of epidemiological studies conducted to evaluate environmental risks. There are two major classes of study design: experimental and observational. Within these classes there are a number of different study methods. Table 8.2 is a listing of the methods most often applied to the evaluation of environmental risk factors. The potential for bias and confounding that may affect study outcomes are important considerations in the design and evaluation of studies. These issues are also discussed in this section.

8.4.1 Experimental studies

Experimental epidemiological studies are those in which the investigator has the ability to control the conditions under which the study is conducted. There are two kinds of experimental studies: clinical trials and community trials. Clinical trials evaluate a therapeutic agent or procedure in individual subjects. In this approach the investigator randomly assigns volunteer subjects to an experimental or control group, attempting to make the two groups as similar as possible except for the factor being studied. The health status of the two groups is observed and can be compared to evaluate the efficacy and safety of the treatment, and to identify long-term effects. Randomization is important to avoid selection bias in clinical trials. For example, if healthier participants are more likely to be assigned to the experimental group, the observed results may not be due to the treatment being studied. Clinical trials are important in the evaluation of new drugs and medical procedures. Since they are ethically limited to the evaluation of potentially beneficial agents, clinical trials have had limited use in environmental risk assessment.

Community trials are studies of populations to evaluate risk management strategies. In this design, a community is identified to receive a 'treatment', usually a public health intervention or prevention strategy. The health status of the community is then observed for trends or compared with other communities without the 'treatment'. Community trials were instrumental in determining the beneficial effects of water fluoridation in the prevention of dental caries. This approach might also be used to evaluate the effectiveness of risk management strategies such as smoking prevention measures, handgun regulation or lead poisoning prevention.

Experimental studies must be carefully designed to ensure that the experimental groups are representative of the populations of interest, to avoid or limit potential adverse effects, and to inform participants of all known risks or potential adverse effects.

8.4.2 Observational studies

Most of the epidemiological basis for risk assessment has been derived from observational studies. There are two types of observational studies: descriptive and analytical. Descriptive studies evaluate the distributions of disease or injury in populations. Descriptive studies also include investigations of temporal and geographical associations between risk factors or exposures and disease rates in populations. These studies are also called correlational or ecological studies. Analytical studies focus on the disease and exposure experience of individuals to investigate causal relationships between risk factors and disease in populations. There are three major kinds of analytical studies: cohort studies, case–control studies and cross-sectional studies.

8.4.3 Ecological studies

Ecological studies examine the relationship between disease rates and exposures in population groups. These studies do not consider information on individuals. The units of analysis are usually the rate of disease and the population level of exposure. Ecological studies have been used, for example, to demonstrate the correlation between the age of housing and the rate of lead poisoning. Ecological trend studies examine the relationship between temporal changes in exposure levels and disease. An investigation demonstrating the association between the increasing percentage of women who smoke and the subsequent increase in female lung cancer mortality is an example of an ecological trend study.

Ecological studies have played an important role in environmental risk assessment. For example, early ecological studies of the relationship between cancer mortality and drinking water quality (Cantor *et al.* 1978) were instrumental in the development of regulatory approaches to reduce chemical contamination and investigate the health effects of disinfection byproducts. More recently, ecological trend studies of the relationship between particulate air pollution and mortality have led to a reconsideration of national air pollution control strategies in the USA (HEI 1995).

It must be emphasized that because ecological studies do not consider the exposure or disease status of individuals, they cannot demonstrate causality. In fact, the correlation of two population level variables may be totally unrelated to causality. This type of correlation is known as the ecological fallacy. (One far-out example of the ecological fallacy is the reported association between the sale of polyester leisure suits and increase in cancer mortality in the early 1970s.) The primary value of ecological studies is in hypothesis generation, or as a component of an overall weight of evidence evaluation in environmental risk assessment (see Chapter 7).

8.4.4 Cohort studies

Cohort studies are also called prospective or follow-up studies. The approach is to compare the disease experience of exposed and non-exposed individuals. The investigator identifies a group of exposed individuals and a group of non-exposed individuals and both groups are followed to compare the incidences of disease or death. Figure 8.2 presents the approach to a cohort study. If a greater

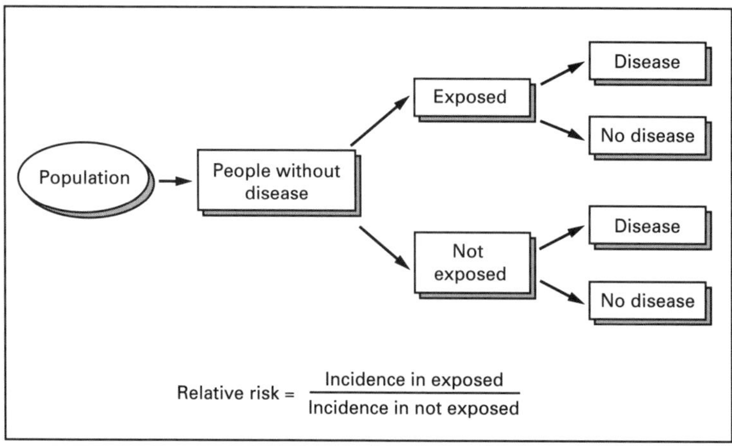

Fig. 8.2 Design of a cohort study.

proportion of the exposed group develops the disease than the non-exposed group, a positive association is said to exist. The measure of association is the relative risk, defined as:

$$\text{Relative risk} = \frac{\text{The incidence in exposed}}{\text{The incidence in non-exposed}}$$

If the relative risk is greater than 1, the risk is greater in exposed persons than non-exposed and there is a positive association which may be causal. (Determination of causality is discussed later.) A relative risk of less than 1 indicates the absence of an association between the risk factor and the disease.

A prospective or longitudinal cohort study identifies the study population at the beginning of the study and follows it for an appropriate time until the disease develops or not. This can be problematic for studying diseases with long latency periods between exposure and illness; many cancers, for example, have latency periods of more than 15–20 years. Length of follow-up is a critical consideration in designing appropriate cohort studies. For example, a negative cohort study of cancer in occupationally exposed workers must be carefully examined to determine if the length of time between exposure and follow-up is appropriate to allow for the latency period of the disease. In order to shorten the time necessary to conduct a cohort study, historical or retrospective studies are often appropriate. A historical cohort study uses data on past exposures and compares disease rates at the time the study is conducted. This approach is most often used in evaluating occupational groups where employment records allow ascertainment of past exposures. Such cohort studies of occupationally exposed groups have been essential to the identification and quantification of many environmental risks, including those associated with chromium, asbestos and radium (Langard 1990).

8.4.5 Case–control studies

Case–control studies examine the relationship between exposure and disease by comparing people with the disease (cases) with people without the disease (controls). The method compares the proportion of cases exposed to the risk factor of interest to the proportion of controls who were exposed. Figure 8.3 presents the approach to a case–control study. Because a case–control study does not measure disease incidence, it does not provide a direct measure of relative risk. The measure of association in a case–control study is called the odds ratio, or relative odds, defined as:

$$\text{Odds ratio} = \frac{\text{The odds of exposure in the case group}}{\text{The odds of exposure in the control group}}$$

If the odds ratio is greater than 1, there is a greater proportion of exposed subjects in the case group than in the controls, and a positive association

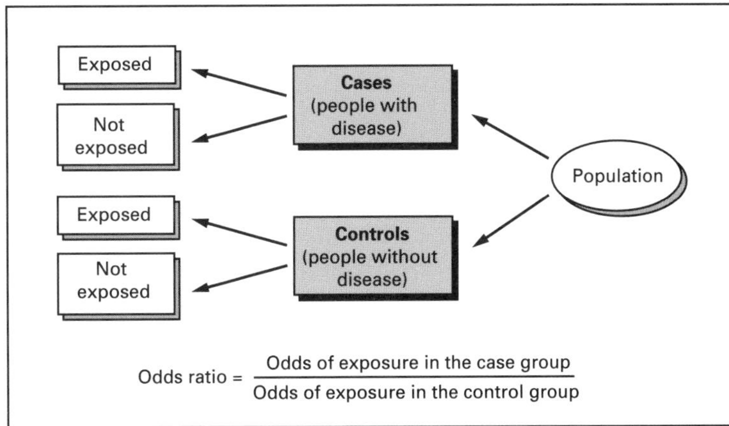

Fig. 8.3 Design of a case–control study.

is said to exist between the risk factor and disease. The greater the odds ratio, the stronger the association.

Case–control studies are particularly appropriate for the study of relatively rare diseases. The case–control design has been useful in establishing the association between lung cancer and smoking, and has been widely used in cancer epidemiology. For such diseases it is often impractical, or impossible, to conduct cohort studies with enough participants to generate an adequate number of cases to achieve meaningful results. Case–control studies must be carefully designed to ensure that the results are generalizable to other populations. Cases should be selected to be representative of all cases, and controls to be representative of the general population. One particular difficulty with case–control studies is the challenge of reconstructing past exposure information. Often accurate information on past workplace or environmental exposures, or even smoking habits, is not available (Gordis 1996).

8.4.6 Cross-sectional studies

A third type of analytical study often used in the evaluation of environmental risks is the cross-sectional study, or prevalence study. Such studies may be descriptive in approach; for example, a study of the prevalence of asthma in school-age children. This approach may also measure exposure and disease simultaneously in a study population. For example, cross-sectional studies have been conducted to compare the prevalence of disease in communities with hazardous waste sites with the prevalence of disease in communities without such sites. Two measures of association can be calculated. The prevalence of disease in the exposed can be compared with its prevalence in the non-exposed, or the prevalence of exposure in those with disease can be compared with the prevalence of exposure in those without the disease.

Cross-sectional studies may reveal associations between risk factors and disease; however, they cannot establish causal relationships. Because they are studies of prevalent cases, they do not include those who have died from the disease. Also, because they establish disease and exposure status simultaneously, they do not consider the temporal relationship between exposure and disease. This is an important limitation in studies of environmental cancer risks, given the long latency of most cancers. For example, current environmental exposures of cancer patients may have no relationship to past exposures that contributed to the development of the disease. As in the case of ecological studies, cross-sectional studies are primarily used to generate hypotheses for further study. They may be very useful in addressing questions regarding the health status of a population, or as part of an overall weight-of-evidence evaluation of risk.

8.4.7 Bias and confounding in epidemiological studies

In evaluating the results of epidemiological studies,

and weighing the evidence for causal relationships, it is important to examine potential sources of bias and confounding that may result in erroneous estimates of risk. Bias has been defined by Schlesselman as 'any systematic error in the design, conduct or analysis of a study that results in mistaken estimates of the exposure's effect on the risk of disease' (Schlesselman 1982). Major types of bias include selection bias, misclassification bias and information bias.

Selection bias is encountered when cases and controls, or exposed and non-exposed individuals, are selected in a way that leads to an inaccurate measurement of the degree of association between exposure and disease. This may occur when study subjects are not representative of the general population at risk. One classic example of selection bias that has been seen in occupational studies is the 'healthy worker' effect. This bias occurs when the risks of exposure observed in worker populations underestimate the risk to the general population. For example, mortality studies of exposed workers often find lower numbers of deaths than would be expected when compared to the general population mortality rates. Such a finding may not reflect true risks of exposure, but may be biased due to job selection factors which require that workers are often healthier than the general population.

Misclassification bias occurs when the disease or exposure status of study subjects is incorrect or misclassified; this may occur, for example, when persons presumed to be non-exposed were in fact exposed to the agent being studied. Such situations may arise in case–control studies when individuals may not be aware of past exposures. Exposure misclassification can be particularly troublesome in environmental studies when the exposures of concern are low-level, ubiquitous pollutants.

Information bias results from measurement error in the assessment of exposure and disease. One type of information bias is recall bias, which describes the phenomenon that cases may be more likely to recall past exposures than are healthy controls. Parents of a sick child, for example, are more likely to recall past exposures than parents of a healthy child. Information bias can also be introduced through errors in abstracting data from records, or through interviewer error. When studies depend upon surrogates for exposure information (i.e. when the subject is deceased), family members may not have accurate information concerning past exposures, such as smoking habits or workplace exposures. Non-response bias may also be an important source of error in epidemiological studies, because no information is obtained from people who do not respond. Moreover, non-responders may not be randomly distributed with respect to exposure or disease status.

Confounding may introduce a serious problem in evaluating causality in epidemiological studies. Confounding factors are risk factors that are related to both the disease of interest and the risk factor being studied. Smoking is a confounding variable that must be controlled for in many epidemiological studies. An ecological study of air pollution and lung cancer mortality may provide an erroneous measure of association if the investigator does not consider the possible impact of smoking prevalence on community lung cancer mortality.

There are a number of analytical techniques that can control for the influence of confounders. Study subjects can be matched by smoking status or age. The analyses can also be stratified to assess the impact of the confounder. Examination of the relative risks of both smokers and non-smokers is an example of such an analysis. Confounding is a particularly difficult issue in many environmental studies when risks being evaluated are relatively small and the potential number of confounding exposures is quite large (Rothman 1986).

In the establishment of causality in epidemiological studies, it is essential to assess the potential effects of bias and confounding. Bias represents a design flaw or inadequacy that can seriously affect measures of association. It must be recognized, however, that confounding is not an error in study design, but is rather, as Gordis describes, 'a valid finding that describes the nature of the relationship between several factors and the risk of disease' (Gordis 1996). While no study can be totally free of bias or potential confounding, it is important for investigators to consider these sources of error in designing studies and to describe them when reporting study results.

8.5 ESTABLISHING CAUSAL RELATIONSHIPS

The typical result of an epidemiological investigation consists of a set of statistical evaluations of the associations between exposures/doses and risks of disease or death. As noted earlier, the epidemiologist is seeking to determine whether any observed associations are not due to chance, and a number of statistical tools are available to make such determinations. Although it has become common to designate as 'statistically insignificant' associations that have greater than a 5% likelihood of being due to chance, it is important to recall that any such definition of 'statistical significance' is based on convention, and is not a 'law of nature'. Thus, while it is important that scientists follow some common conventions when making such determinations, it should always be kept in mind that, when other factors are considered, statistical associations having a greater than 5% probability of being due to chance might be considered important and those having less than this level might turn out to be unimportant (Rothman 1986).

Associations between exposures and risks of disease or death that are found to be 'statistically significant' by any single criterion do not, by themselves, establish the existence or absence of a causal link between the exposure and the disease. Epidemiological studies (except those that involve randomized, controlled trials) cannot be strictly controlled, in the sense that laboratory experiments can be controlled, and so the observation that a statistical association exists between the measured values (exposure and risk of disease or death) is hardly sufficient to establish causation.

Epidemiologists have adopted 'working guidelines' for examination of the issue of causation. The usual starting point is a determination of whether any observed association could be the result of measurement or selection bias (see section 8.4.7). If such biases can be reasonably ruled out, the important issue of confounding is assessed (see section 8.4.7). If confounding is not a likely explanation, and the observed association is not likely due to chance, the so-called *Hill criteria* are applied to judge the possibility that association reflects a cause–effect relationship. These criteria were first suggested in a paper published by A. Bradford Hill in 1965 (Hill 1953, 1965) and have become widely accepted by epidemiologists. The criteria are (with one exception, see Table 8.3) not strict 'rules' for judgement, but rather are general guidelines meant to be applied by experts to the epidemiological evidence (see also Chapter 7).

The discussion of causation presented here is limited in scope, and is devoted only to a determination of whether causation can be established between an exposure (which may consist of a single chemical or a complex mixture) and the risk of a particular disease outcome. Epidemiologists and other public health professionals are generally interested in understanding all of the factors that may contribute to the development of a disease state (Doll & Peto 1978). The fact that, in certain occupational settings, it has been shown that excessive benzene exposures can increase leukaemia risks is, without doubt, important but such an observation is of limited utility in understanding the factors that contribute to overall leukaemia rates. Epidemiologists seek to identify all factors that contribute to the development of a disease, and recognize a definition of causation that is broader than the one we have discussed here. The goal of identifying all factors that increase the risk of a disease, whether or not they are, in a strict scientific sense, causal, is within the scope of epidemiology. Thus, for example, an understanding that malnutrition is a 'risk factor' for a number of diseases caused by pathogens, even though it is by no means either a sufficient or necessary cause of the disease, assists in the design of prevention programmes. The search for such risk factors, and for an understanding of their relative importance, is a major element of epidemiological science (Rothman 1986).

8.6 EPIDEMIOLOGICAL EVIDENCE REGARDING CHEMICAL CARCINOGENS

Expert judgements resulting from the application of the Hill criteria may take several forms. Perhaps the most well recognized are those used by the International Agency for Research on Cancer (IARC) to judge epidemiological evidence regarding chemical carcinogens. The IARC periodically

Table 8.3 Criteria for judging causation based on epidemiological evidence (Hill 1965).

Criteria	Questions to be examined	Evidence for causation* increases when:
Temporal relationship	Does the cause precede the effect?	Causation not possible unless this criterion is met
Plausibility	Is the association consistent with other knowledge?	Evidence from animal and other experimental studies consistent with epidemiology findings
Consistency	Are the results consistent among studies?	Consistent pattern of results from several studies, especially when different methods are used
Strength of association	What is the strength of the observed association between cause and effect (increased risk)?	Association is very strong
Dose–response relationship	Does risk increase with increasing dose?	Risk clearly increases with increasing dose (magnitude or duration of exposure, or both)
Reversibility	Does removal of the possible cause reduce risk?	Risk declines when cause is removed

* The evidence for a causal relationship increases with increasing adherence to the Hill criteria. Expert judgement is required for a final conclusion. *Note.* The evidence of causation is applicable to populations, and applies to individuals only in a probabilistic manner.

convenes working groups of experts to examine both epidemiological and experimental evidence on individual chemicals, chemical mixtures and occupational settings. The results of these deliberations are presented in monographs in which the evidence is discussed and evaluated, and judgements are made regarding the existence of causal relationships (IARC 1987). With respect to epidemiological evidence, the IARC working groups classify substances into one of the following categories.

1 *Sufficient evidence* of carcinogenicity—there is sufficient evidence from epidemiological studies to support a causal relationship between exposures to the agent(s) and some type(s) of human cancer.

2 *Limited evidence* of carcinogenicity—there is some evidence from epidemiological studies to support a causal relationship, but alternative explanations, such as bias and confounding variables, cannot be excluded.

3 *Inadequate evidence* of carcinogenicity—there are few or no relevant data, or the available data do not support a causal relationship.

The IARC also evaluates evidence of carcinogenicity from animal studies and similarly categorizes the evidence (sufficient, limited, inadequate). The agency then classifies agents into one of five groups, based on a combined evaluation of human and animal evidence. The IARC classification scheme is presented in Table 8.4, with selected examples of Group 1 substances presented in Table 8.5.

8.7 EPIDEMIOLOGICAL EVIDENCE AND HEALTH EFFECTS OTHER THAN CANCER

The tools of epidemiology have made significant contributions to our understanding of how chemical agents in the environment cause diseases other than cancer. No system comparable to that developed by IARC for carcinogens has been developed to

Table 8.4 The IARC classification of evidence of carcinogenicity for humans (see Table 8.5 for example of Group 1 agents).

Group	Evidence*	
	Human data	Animal data
1 Agent is carcinogenic	Sufficient	
2A Agent is probably carcinogenic	Limited	Sufficient
2B Agent is possibly carcinogenic	Limited or inadequate	Sufficient
3 Agent is not classifiable as to carcinogenicity	Inadequate	Limited or inadequate
4 Agent is probably not carcinogenic	Inadequate or negative	Clearly negative

* See text for definitions of IARC terms. Some substances may be placed into Group 1 even when human evidence is not sufficient; if animal and mechanistic data are convincing, IARC is increasingly considering mechanistic data in all classifications.

Table 8.5 Some of the chemicals and occupational exposures listed by IARC as carcinogenic to humans (Group 1) (IARC 1987).*

Some occupational exposures
Boot and shoe manufacturer (certain exposures)
Furniture manufacturer (wood dusts)
Nickel refining
Rubber industry (certain occupations)
Underground haematite mining, when radon exposure exists
Some chemicals
Arsenic and arsenic compounds
Asbestos (when inhaled)
Chromium and certain chromium compounds (when inhaled)
Benzene
Diethylstilboestrol (DES)
2-Naphthylamine, benzidine (starting materials for manufacture of certain dyes)
Vinyl chloride (starting material for PVC plastic manufacture)
Mustard gas
Some chemical mixtures
Tobacco smoke
Smokeless tobacco products
Soots, tars, mineral oils†
Analgesic mixtures containing phenacetin

* Note that in many cases data on cancer rates were collected under exposure conditions that no longer exist.
† Mineral oils now in commercial production generally do not have the polyaromatic hydrocarbon content they had at the time the evidence of carcinogenicity was gathered.

weigh and evaluate evidence of other adverse health effects, but a number of environmentally important chemicals have been the subject of extensive and revealing epidemiological investigations. As in the case of carcinogens, most of the opportunities for investigation have arisen in occupational settings, and application of those results to non-occupational environments requires some degree of extrapolation (sections 8.3 and 8.8.2). In Table 8.6 are collected several substances of considerable environmental and public health importance, the toxic hazards of which have been revealed by epidemiological study; experimental investigations have, of course, also contributed heavily to our understanding.

In recent years epidemiologists have attempted to investigate populations outside the occupational setting, particularly those in groups of people residing close to major industrial and hazardous waste facilities (Upton *et al.* 1989; Rodricks 1996). As noted in section 8.4, such investigations are inherently subject to major methodological difficulties, yet, as reported by the National Research Council of the National Academy of Sciences (NRC 1991), numerous efforts are under way to determine whether adverse health effects can be detected in such populations. Many such studies have uncovered associations between living in proximity to, for example, hazardous waste sites and excess rates or prevalences of a variety of symptoms and diseases, but in most cases it has been impossible to draw firm conclusions regarding causation (NRC 1991). Thus,

Table 8.6 Some environmentally important chemicals that have been the subject of extensive epidemiological study.

Chemical	Significant non-cancerous toxic hazards detected in humans	Reference
Ozone	Pulmonary function impairment	USEPA, 1986b
Sulphur oxides	Bronchoconstriction Pulmonary function impairment	USEPA, 1996
Nitrogen oxides	Bronchoconstriction Pulmonary function impairment	USEPA, 1982
Carbon monoxide	Neurological and cardiovascular effects (due to hypoxia)	USEPA, 1994
Lead	Encephalopathy, anaemia	USEPA, 1986c
Mercury	Neurological effects (tremor) Acrodynia	USEPA, 1995
Polychlorinated biphenyls	Chloracne	USEPA, 1988b

for example, many such investigations have had to rely upon self-reported health problems or on highly incomplete medical information. Exposures have had to be assessed using surrogate measures (e.g. residence near an industrial site for a given period to time) that are of uncertain relationships to actual exposures. Although the results from such studies can mislead if their limits are not carefully characterized, it is clear that the advancement of epidemiological methods will require continuing efforts in these directions, with improvements in the tools for measuring exposures and disease outcomes.

8.8 THE ROLE OF EPIDEMIOLOGY IN RISK ASSESSMENT—TWO EXAMPLES

8.8.1 Introduction

So far we have been concerned with the problem of direct measurement of health risks in human populations. Such measurements, using the methods of epidemiology, are of obvious public health importance. However, we have yet to deal with the possible applications of such measurements to populations different from those in which risks were measured directly. Thus, for example, the discovery by epidemiologists that benzene exposures in certain occupational environments increases the risk of leukaemia does not, in and of itself, answer questions about whether individuals exposed environmentally, at very much lower levels, might also be at risk. Nor do data regarding the cardiovascular effects of carbon monoxide in relatively healthy workers directly reveal the size of such risks in populations, including much older individuals, that might also exhibit higher rates of underlying heart disease. We undertake a *risk assessment* whenever we seek to estimate the probability that adverse health effects will occur in populations in which direct measurement of risks has not been, or cannot be, made (NRC 1994).

In this section we present two examples of the use of epidemiology data as the basis for such risk assessments. The cases selected involve a non-carcinogenic, toxic pollutant, *lead*, that may reach people through several environmental media, and a chemical carcinogen, hexavalent *chromium*, that is important in both occupational and non-occupational settings. We begin with some general background information on risk assessment.

8.8.2 Content of risk assessment

Risk assessment is a systematic way to organize and evaluate data and knowledge, and their associated uncertainties, pertaining to the health risks that might arise in populations exposed to hazardous agents. In the present case we are concerned with chemical toxicities for humans. Risk assessment proceeds in steps, and asks the following questions (NRC 1983, 1994).

Step 1. What types of *hazards* are known to be associated with the chemical of interest? Chemical hazards include flammability, explosivity, radioactivity and any of a large number of forms of toxicity. Any of these could be the subject of a risk assessment. Here, we are concerned with toxicity.

Step 2. What is the relationship between the risk of toxicity and exposure—the *dose–response* relation? Typical measures of toxic response include the frequency of occurrence (incidence) and severity of effect.

Step 3. What *exposure*, or dose of chemical, is or could be experienced by the population of interest?

Step 4. What *risk of toxicity* exists for the population of interest? This question is answered by combining the results of Steps 1 and 2 (risk of toxicity as a function of dose) with the results from Step 3 (the dose experienced by the population of concern).

Information for the evaluation of hazards and dose–response for a specific chemical derives from the scientific literature. Information for the exposure assessment derives from an evaluation of how the chemical is used, enters the environment, and reaches the population of concern, and is estimated as described in section 8.3. Exposure assessment might involve, for example, identifying how much of a pesticide reaches people because of its use on food crops, or the amount of intake of a pollutant in individuals ingesting water. It could also involve estimation of the intake of an air pollutant by individuals living near, or working in, a manufacturing facility, or that received by people living near a hazardous waste site. In the latter case, several chemicals may reach people through air, their drinking water, and even through contaminated soils. Exposure assessment may take place prior to the introduction of a chemical into commerce, or only after exposure has already occurred. Before a risk assessment is embarked upon, the assessor needs to have a clear understanding of the population group or subgroup of interest (McKone & Layton 1986; USEPA 1988a).

Toxic hazard data derive primarily from two sources: epidemiology studies and animal toxicity experiments. The comparative advantages and disadvantages of epidemiology and animal studies are summarized in Table 8.7.

Dose–response evaluation (Step 2) is concerned with understanding what the available scientific literature reveals about the probability of a toxic effect occurring (the risk) as a function of the size and duration of the dose. At the present time, scientists are capable of measuring dose–response relations over a limited range of doses and risks in both epidemiology and animal studies. Only relatively large risks, usually occurring at high doses, can be readily measured. No means are available to measure risks at low doses, such as those that arise in many environmental settings (NRC 1994). This limitation is primarily the result

Table 8.7 Comparison of epidemiology and animal studies for identifying toxic properties.

	Epidemiology studies	Animal studies
Opportunity to conduct study	Often not possible	Always possible
Opportunity to obtain information prior to human exposure	No	Yes
Time requirements	Years to decades after exposure begins	Weeks to years after exposure initiated
Species of interest	Yes*	No
Cause–effect determination	Difficult	Usually easy
Opportunity to obtain quantitative dose–response data	Not frequently	Always

* Note that epidemiology studies may not provide data on both sexes or on all relevant subgroups of the human population.

of the fact that all studies necessarily involve limited numbers of subjects. This is a statistical limitation. Unfortunately, the principal concerns of many risk assessments relate to exposures at relatively low doses, occurring in large populations of people. Epidemiology, as we have emphasized, can rarely measure such low probability events (Moolgavkar 1986).

To say anything at all about low-dose risks, the assessment must adopt certain untested hypotheses about the nature of the dose–response relationship in the low-dose–low-risk region. Various mathematical models can be applied to the measured dose–response data, obtained either from human or animal studies, and parameters are derived that can be used to create a mathematical relationship that might hold in the low-dose region. Several such models are available. Each is based on certain biological or statistical principles that are thought to apply to dose–response phenomena. None is based on a thorough empirical test and all must therefore be considered to be hypotheses about low-dose risks (NRC 1983; Moolgavkar 1986; Upton *et al.* 1989).

In the regulatory context, at least in the USA, two classes of models are typically used. For all toxic effects except carcinogenicity, the dose–response model assumes that a *threshold dose* must be exceeded before any toxicity occurs. Several procedures exist for deriving an estimate of the threshold dose for large human populations when the available dose–response relations reveal the threshold dose for only a relatively small human population or for a test animal population (Rodricks 1992). There is widespread use of these procedures for specifying 'acceptable daily intakes' (or toxicity reference doses, RfDs, as they are called by USEPA) for chemicals causing non-carcinogenic forms of toxicity.

Carcinogens, at least in the USA, are treated differently. Based on various biological theories regarding the carcinogenic process, it is assumed that carcinogenic effects do not require a threshold dose to be exceeded before they become manifest. Any dose greater than zero is assumed to increase the probability (risk) of a carcinogenic process; as the dose increases, so does the risk. Borrowing from radiation biology, chemical risk assessors further assume that, at low doses, the risk increases in direct proportion to dose—this is the so-called *linear, no-threshold* model (Moolgavkar 1986; USEPA 1986a). The USEPA has derived cancer potency estimates for several carcinogens based on dose–response data obtained from epidemiological investigations (Table 8.8). The agency also derives potencies (which are usually called slope factors, see Table 8.8) based on animal data.

Alternative models of carcinogen dose–response models exist, including threshold models. Most show equal or lower risks at a given dose than does the linear model. The linear, no-threshold model thus yields what the USEPA terms the *upper*

Table 8.8 The USEPA slope (potency) factors* for carcinogens based on dose–response data developed from epidemiological studies.

Chemical	Oral route SF $(\text{mg kg}^{-1}\ \text{day}^{-1})^{-1}$	Inhalation route SF $(\mu\text{g m}^{-3})^{-1}$
Arsenic	1.5	4.2×10^{-3}
Asbestos	—	2.3 $(\text{fibres ml}^{-1})^{-1}$
Benzene	2.9×10^{-2}	8.3×10^{-6}
Cadmium	—	1.8×10^{-3}
Chromium(VI)	—	1.2×10^{-2}
Nickel (refinery dust)	—	2.4×10^{-4}

* Slope factors (SF) are estimates of excess lifetime cancer risks per unit of lifetime average daily dose (in $\text{mg kg}^{-1}\ \text{day}^{-1}$ for oral exposures) or per unit of continuous average lifetime exposure (in $\mu\text{g m}^{-3}$ for inhalation). The SFs were estimated by the USEPA based on dose–response data derived from epidemiology studies and the application of a *linear, no-threshold* model for extrapolation to low doses (see text, section 8.8.2). The absence of SFs for some oral exposures is explained by the lack of evidence of carcinogenicity by that route and the observation that the substance produces cancer by inhalation only at contact sites and not systemically.

bound on risk; actual risks almost certainly do not exceed the upper bound, are probably lower, and could for some carcinogens be zero (NRC 1994). In the face of uncertainty regarding the true dose–response relationship, and because of the concern that risk should not be underestimated, regulatory officials usually choose the upper bound estimate of risk as a basis for decision-making (NRC 1994). Much controversy surrounds this regulatory choice, and other such choices in the risk assessment process, which tend to yield the highest plausible estimates of risk. There are specific cases, however, in which data become available to provide a basis for departure from the usual, 'worst-case' approach. A recent report from the NRC urges the USEPA to incorporate models for risk assessment that are based on knowledge of the biological processes underlying the production of toxicity (NRC 1994). Such models are under development, but have not yet been widely used in risk assessment (OTA 1993).

For carcinogens for which a no-threshold model is used, risks may be presented as an upper bound on the excess probability of developing cancer over a lifetime. These probabilities are obtained from combining the dose–response evaluation (which yields values for upper bound on lifetime risk per unit of dose, as in Table 8.8) with the dose estimates for the exposed population. The probability is a unitless fraction; most environmental risks carry probabilities below 1/100, and many seem to fall in the one 1/1000 to 1/1 000 000 range. Recall that these are upper bounds on risks, at least as long as they are based on the linear, no-threshold model and other assumptions that reflect the adoption of a policy of caution in the face of scientific uncertainty (Rodricks 1992).

For non-carcinogenic forms of toxicity, the ratio of the estimated dose experienced by the exposed population to the estimated population threshold dose is calculated. This so-called *hazard quotient* (sometimes called *risk quotient*; Chapter 1), while not a direct measure of risk of the type used for carcinogens, nevertheless provides decision makers a guide to the potential public health impact of an exposure. As the hazard quotient rises above 1.0, it is expected that human risk (i.e. the fraction of the population experiencing doses greater than the threshold dose) also rises, while values less than 1.0 suggest the absence of a significant risk (USEPA 1988a; NRC 1994).

Risk assessments typically conclude with a description of all relevant data and their limitations, a restatement of the critical assumptions used, and the quantitative outcomes. This is the information sought by risk managers to help them decide whether, in specific contexts, actions should be taken to reduce risks and, if so, the degree of risk reduction needed to ensure public health protection.

It will be useful now to discuss specific applications of risk assessments based primarily upon epidemiological data.

8.8.3 Hazards and risks of lead exposure

Introduction

It is reasonably well documented that the public health burden in societies in which blood lead levels commonly exceed ca. $25\,\mu g\,dl^{-1}$ (the usual measure of lead exposure) is substantial. In such cases significant numbers of individuals are likely to suffer at least moderately serious clinical effects of several types. In societies faced with such conditions it would seem that efforts to reduce lead exposures should be directed at identifying the most significant sources of those exposures and finding ways to reduce their impact. This section presents a brief sketch of the various sources of lead exposure that most societies face, and of the principal health concerns that arise at different levels of exposure. It ends with a summary of the health issues that arise once exposures reach 'low' levels.

Sources of lead and pathways leading to human exposures

Although the natural level of lead in the earth's crust contributes to human exposure, it is the mining, smelting, processing, use and disposal of lead and lead compounds that are the primary sources. The major sources of lead exposure and the various pathways that lead can travel to reach humans are depicted in Fig. 8.4. As Fig. 8.4 shows, the primary environmental media through which humans become exposed are air, dusts, food and

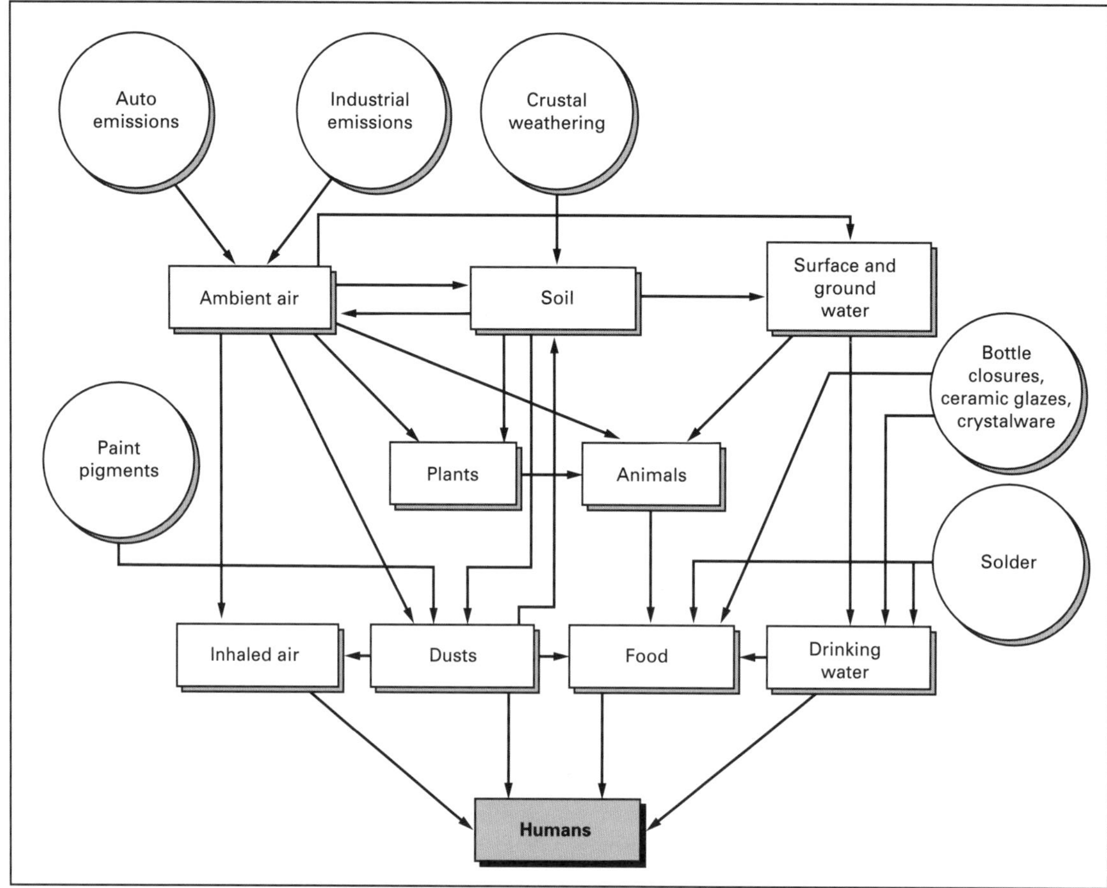

Fig. 8.4 Primary sources of lead and pathways leading to human exposures. (From USEPA 1986c.)

water. The relative contributions of these four media to total human exposure depend on many factors, including the concentrations of lead in each, the physical and chemical forms of the lead, the rates of human contact with each, and the relative rates of lead absorption through the human lungs and gastrointestinal tract and into the blood. It can also be seen from Fig. 8.4 that the concentrations of lead in each medium and the physical and chemical nature of the lead present are themselves dependent upon many factors that depend upon the source of lead and the pathway(s) by which it reaches the medium.

Several generalizations emerge from the information in Fig. 8.4. First, the relative contributions of the various lead sources to overall human exposure will vary both geographically and over time. Within specific countries and regions there will be some areas where industrial emissions predominate and areas where paints or vehicle emissions are of greater importance. These patterns can, of course, change over time as societies take action to limit one or more sources. Relative source contributions among countries can also vary widely. Wherever lead-containing solders or ceramic glazes are in widespread use, food is likely to be a more important source than in countries where this use has been reduced or abandoned. Some knowledge of the relative source contributions in countries considering programmes to reduce lead exposures would seem critical to ensuring that the most cost-effective and least risky means of achieving those reductions are adopted.

Uptake of lead from the environment and its distribution in the body

Once humans inhale air and dusts containing lead and ingest foods, waters, paints and dusts containing lead, some of the chemical is absorbed into the blood and distributed in the body. The fraction of lead that is absorbed depends upon many factors, including the medium containing the lead, its chemical and physical forms, and the nutritional status and age of the exposed person. As is widely known, there is a well-documented concern over the uptake and absorption of lead by young children. Children are far more prone to exposures through dusts and paint chips than are adults and, moreover, the rate of lead absorption through the gastrointestinal tract of a young child typically exceeds that in an adult. Thus, for a given level of lead in the environment, young children (perhaps up to the age of 7 years) develop higher body burdens than do adults (IPCS 1995).

Once absorbed lead is distributed in the body, there is a relatively rapid distribution in blood and in soft tissues, and a slower distribution to bone. Bone continually accumulates lead and, under some conditions, the metal may be released from this storage site. The half-life of lead in blood is 28–36 days, but it is much longer in bone. Transfer of lead to the fetus occurs throughout gestation; some of the studies of the effect of lead on IQ involve attempts to correlate levels of lead in fetal cord blood with learning ability. Blood lead levels (PbB, in $\mu g\,dl^{-1}$) are typically used as the measure of lead body burden, and most health effects data involve this measure of exposure.

Major health effects

It has been documented through many case reports and epidemiological studies of diverse types that lead exposures can result in a range of biological effects, depending upon dose. These effects include, at the low end, interference in the functioning of certain enzymes to, at the high end, devastating effects on the central nervous system and on renal function. A brief listing of some of the principal health effects associated with various levels of lead exposure is given in Table 8.9 (adults) and Table 8.10 (children). Note that as PbB levels decline, the degree of scientific certainty regarding the listed effect tends to decline. As levels reach 10–15 $\mu g\,dl^{-1}$, there is substantial controversy in the scientific community regarding either the certainty with which a causal connection has been established or the size of the health risk. Some observers suggest that levels below 10 $\mu g\,dl^{-1}$ are also of concern; this conclusion is based not so much on actual evidence of biological effects at or below 10 $\mu g\,dl^{-1}$, but on the possibility that no clear threshold of effect has been established based on observations at levels above that figure (Schwartz 1994). Other scientists question whether effects occur at levels as low as 10–15 $\mu g\,dl^{-1}$, and there is substantial debate in the literature regarding the magnitude of reported effects at these low levels (IPCS 1995).

It seems clear that PbB levels greater than 25–30 $\mu g\,dl^{-1}$ are of significant concern. Effects of lead on certain enzymes and on other biochemical parameters have been demonstrated at somewhat lower levels. According to a recent review by the International Programme on Chemical Safety (IPCS), a unit of the World Health Organization (WHO), only at PbB levels greater than 20 $\mu g\,dl^{-1}$ are the affected parameters of possible clinical significance; some enzyme changes occurring at lower PbB levels are of uncertain clinical significance (IPCS 1995). The IPCS also notes that some but not all epidemiology studies show an association between some indices of fetal growth and maturation at PbB levels of 15 $\mu g\,dl^{-1}$ and above.

The two major health concerns that emerge at low (less than 20–25 $\mu g\,dl^{-1}$) PbB levels pertain to the reported effects of lead exposure on blood pressure in adults and on IQ in children.

Decrements in IQ, obtained by several measures, have been observed in both cross-sectional and prospective studies to be associated with increases in PbB levels. The consistency of the observed associations suggests a causal relationship, but, according to a recent WHO review (IPCS 1995), definitive evidence of causation below PbB levels of 25 $\mu g\,dl^{-1}$ is lacking. The WHO experts note, however, that 'The size of the apparent IQ effect, as assessed at 4 years (of age) and above, is a deficit between 0 and 5 points (on a scale with a standard deviation of 15) for each ... 10 $\mu g\,dl^{-1}$ increment

Table 8.9 Some of the major health effects of lead in adults and their relationship to blood lead (PbB) levels.

PbB ($\mu g\,dl^{-1}$)	Effect*	Degree of scientific certainty†
>80	Irreversible brain damage	High
	Kidney damage	
	Infertility	
60–80	Anaemia	High
	Testicular dysfunction	
	Pregnancy complications	
	Neuropathies	
40–50	Reduced haemoglobin synthesis	High
	Neuropathies	Moderate
	Cognitive deficits	Moderate
	Premature birth	Moderate
15–30	Minor blood effects (enzyme inhibition)	High–moderate
	Interference in vitamin D	High–moderate
	Metabolism interference	
	Elevated blood pressure	Moderate
<10–15	Elevated blood pressure	Controversial
	Decreased fetal growth	Controversial

* Observations are based on many types of epidemiological studies.
† Our judgements regarding the database (see text).

Table 8.10 Some of the major health effects of lead in children and their relationship to blood lead (PbB) levels.

PbB ($\mu g\,dl^{-1}$)	Effect*	Degree of scientific certainty†
>80	Irreversible brain damage	High
	Kidney damage	
60–80	Anaemia	High
	Renal dysfunction	High–moderate
>40	Reduced haemoglobin synthesis, other blood effects	High–moderate
15–30	Cognitive, behavioural deficits	Moderate
<10–15	Cognitive deficits	Controversial

* Observations are based on many types of epidemiological studies.
† Our judgements regarding the data base (see text).

in PbB level, with a likely apparent effect size of between 1 and 3 points'. Such estimates are based on population studies and can be applied only in a probabilistic manner to individual children. The associations observed thus far do not provide definitive evidence of a threshold for this effect (although it is extremely difficult to make such a demonstration on such evidence alone); nevertheless, the evidence for an IQ effect below PbB levels of 10–15 $\mu g\,dl^{-1}$ is uncertain and subject to considerable controversy. These findings raise the highly problematic issue of individual versus population risks: a small shift in IQ may have small consequences for an individual child, but is a small shift in IQ distribution within a very large population similarly inconsequential? No answer to this question has been fully explored, yet public health officials have generally relied upon the findings of IQ decrements to support programmes for further reductions in lead exposures.

8.8.4 Carcinogenic risks of hexavalent chromium

Data on the toxicity of various forms of chromium emerged during the nineteenth century in the form of case reports of health problems among individuals exposed to very high levels of chromium in occupational settings. Among the effects reported in workers from the mid-nineteenth century until about 1960 were nasal preformations, various forms of skin and respiratory tract irritation, and gastro-intestinal disorders (PHS 1953; USEPA 1984). Most of these case reports concerned workers exposed to high levels of soluble, hexavalent chromium compounds, often in the form of airborne mists. Such mists were produced in the chrome-plating and other related industries where hexavalent chromium compounds were used. A few reports exist of gastrointestinal problems in workers involved in chromite ore processing, and skin and respiratory tract irritation were sometimes reported in such workers (PHS 1953).

Several reports appeared in the 1930s suggesting that chromic acid mists and soluble chromates—all forms of hexavalent chromium—might increase the risk of lung cancers in workers exposed to high levels of these airborne materials (Langard 1990). Although certain investigations were not able to reproduce the original findings of Machle & Gregorius (1948), additional studies during the 1940s and early 1950s were confirmatory (Langard 1990). Most authors pointed to various soluble chromium compounds as the hazardous agents. In 1951, Mancuso and Hueper, on the basis of findings of excess lung cancer cases in a chromite ore processing facility, suggested that insoluble chromium compounds might be involved, although they offered no explanation of how a highly insoluble material could be biologically active. In a 1953 Public Health Service report, and in subsequent studies, the evidence increasingly pointed to airborne forms of hexavalent chromium as the carcinogenically active form of chromium (Langard 1990).

Thus, the weight of available evidence implicates inhaled forms of soluble hexavalent chromium as a cause of lung cancer among workers, and also suggests that such risks might occur in workers exposed to chromite ores during processing, where both dust levels and total chromium levels were high. No evidence exists that either trivalent or hexavalent chromium is carcinogenic when ingested or dermally absorbed. The current views of the Public Health Service and of the USEPA support this position (USEPA 1984; USDHHS 1994b).

Of the various epidemiology studies available showing the relationship between the magnitude of chromium exposure and risk of lung cancer, the USEPA selected one as providing the best information for risk assessment. The study selected by the USEPA (Mancuso 1975) summarizes many years of observation of a group of workers at a chromate plant, and indeed provides a reasonably complete picture of the chromium exposures experienced by the workers. As is standard practice in such studies, Mancuso examined the incidence rates of lung cancers in workers exposed to various levels of chromium.

No excess of lung cancer was observed in workers unless they had accumulated at least 1000 $\mu g\,m^{-3}$ years of exposure to chromium. These units, $\mu g\,m^{-3}$ years, are a standard measure of cumulative exposure. The units of '$\mu g\,m^{-3}$' reflect the concentration of chromium in the air breathed by the workers, and 'years' reflect the total number of working years the chromium exposure was experienced. Thus, for example, a 1000 $\mu g\,m^{-3}$ years exposure might mean that a worker exposed on the job for 10 years experienced an average airborne chromium concentration of 100 $\mu g\,m^{-3}$ throughout that time. A worker experiencing 50 $\mu g\,m^{-3}$ would have to have worked 20 years to achieve the same cumulative chromium exposure.

A cancer potency for hexavalent chromium was derived by the USEPA by fitting a mathematical model to dose–response data that then allows extrapolation from the relatively high doses experienced by workers in an occupational setting to the lower environmental exposures that might be experienced by other population groups.

The USEPA-derived potency for hexavalent chromium for assessing lifetime carcinogenic health risks is 1.2×10^{-2} $(\mu g\,m^{-3})^{-1}$. This inhalation potency is based on data from the Mancuso (1975) worker study that we described earlier and the assumption of low-dose linearity. According to the USEPA, inhalation is the only route by which hexavalent chromium has demonstrated carcinogenicity. An estimate of the upper bound on lifetime lung cancer risk (LCR) is obtained by multiplying the lifetime average airborne concentration (LAAC in $\mu g m^{-3}$) that might be experienced by populations multiplied by the potency to yield an incremental lifetime cancer risk, LCR, as follows:

$$LCR = LAAC \times Potency$$

This LCR is an upper-bound estimate of the probability over a lifetime of acquiring an additional cancer as a result of being exposed at the levels assumed. For example, an incremental upper-bound lifetime cancer risk of one in a million (i.e. 10^{-6}) indicates that no more than one additional case of cancer per lifetime might be incurred for every one million people exposed at the assumed level of exposure, with the recognition that actual risk is likely to be less and could be zero. Such a very low risk is much smaller than any risk that scientists are capable of measuring. Note that most environmental exposures are many hundreds or thousands of times less than the minimum chromium exposure level that was observed by Mancuso to increase lung cancer risk in workers, yet all such exposures would, under the USEPA model for risk assessment, be associated with some increased risk of lung cancer.

8.9 TRENDS IN EPIDEMIOLOGY AND SURVEILLANCE

This chapter has presented an overview of the methods of epidemiology and their applications and limitations in environmental health risk assessment. Although epidemiological data are considered the 'gold standard' for understanding risks to the human population, there are many limitations in the application of epidemiology to risk assessment. Historically, epidemiological studies have been aimed at identifying hazards and often did not provide adequate information on exposure or dose to derive quantitative estimates of population risk. With the increasing applications of risk assessment to understand public health problems, there is an increasing appreciation of the role of epidemiology in validating animal models, weighing evidence of risks to humans, and establishing causal relationships between exposures and human disease.

In recent years there have been a number of advances that have the potential greatly to enhance the role of epidemiology in environmental health risk assessment. These advances include: (i) improved surveillance of exposures and diseases in human populations; (ii) development of biomarkers of exposure, susceptibility and disease; and (iii) an increased recognition of the complementary roles of toxicology and epidemiology in evaluating risks to human health.

Surveillance of mortality, disease and exposure is increasingly recognized as an essential tool in identifying hazards and evaluating environmental health impacts. Improved collection and management of data on mortality and disease will continue to enhance the ability of epidemiologists to identify populations at risk, and to improve the design of future studies. The expansion of cancer and other disease registries, along with improved tracking and availability of medical records due to changes in health care, will also provide important information resources for future epidemiological investigations.

Exposure surveillance, as demonstrated by the case example of blood lead screening, will improve the identification of populations at risk and provide refined exposure data for epidemiological studies. For example, there are now registries for workers exposed to hazardous agents ranging from heavy metals to dioxin. As analytical techniques continue to improve, there will be an increasing amount of environmental exposure surveillance information on chemical pollutants in air, water, the indoor environment and the food supply. The National Human Exposure Assessment Survey (NHEXAS) currently under way in the USA will provide new insights into the pathways and distributions of population exposures to a wide range of environmental pollutants (Burke & Sexton 1995).

Although current applications are limited, rapid advances in the development of biological markers, or biomarkers, will undoubtedly lead to refinements in the application of epidemiology to risk assessment. Biomarkers of exposure have the potential to improve exposure classification, thereby reducing a major source of uncertainty, and to increase understanding of dose–response relationships in human studies. Biomarkers of early health effects may help reduce methodological problems such as non-response and recall bias associated with the study of cancer and other diseases with long latency periods between exposure and clinical disease. Such markers may also make it possible to evaluate preclinical effects of low-dose exposures, and reduce the uncertainties currently associated with high- to low-dose extrapolation of epidemiological findings. In addition, biomarkers of susceptibility are shedding new light on the role of genetics in the development of disease, and have important implications for the identification and study of populations at high risk.

Perhaps the most important trend in epidemiology and risk assessment is the increasing recognition that epidemiology is an essential component of a multidisciplinary approach to understanding environmental risks. Over the past 15 years, the role of epidemiology has been supplanted by an increasing dependence upon animal models to assess many environmental health risks, particularly cancer. Unfortunately, during this time there has not been a great deal of interdisciplinary cooperation between epidemiologists and other risk professionals. Recently, however, fundamental questions concerning the relevance of animal data to the human experience have underscored the unique role of epidemiology in validating animal models and establishing causality. As a result, there has been growing recognition of the importance of epidemiology in deriving estimates of risk, and epidemiologists have acknowledged the need for improving the utility of epidemiology for risk assessors (Burke 1995). Some have suggested the development of guidelines for the use of epidemiological information in quantitative risk assessment (Graham *et al.* 1995), and the USEPA recently released revised cancer risk assessment guidelines that rely heavily upon the use of epidemiological data in the classification of carcinogens (USEPA 1996). These efforts represent important steps toward closing the gap between toxicology and epidemiology in risk assessment, and recognizing the complementary role of animal and human studies in evaluating and preventing risks to public health.

8.10 REFERENCES

Burke, T.A. (1995) The proper role of epidemiology in regulatory risk assessment: reactions from a regulator's perspective. In: *The Role of Epidemiology in Regulatory Risk Assessment* (ed. J. Graham), pp. 149–154. Elsevier Science, Amsterdam.

Burke, T.A. & Sexton, K. (1995) Integrating science and policy in a national human exposure assessment survey. *Journal of Experimental and Analytical Environmental Epidemiology*, **5**(3), 283–296.

Cantor, K.P., Hoover, R., Mason, T.J. & McCabe, L.J. (1978) Associations of cancer mortality with halomethanes in drinking water. *Journal of the National Cancer Institute*, **61**(4), 979–985.

Doll, R. & Peto, R. (1978) Cigarette smoking and bronchial carcinoma. *Journal of Epidemiology and Community Health*, **32**, 303–313.

Gordis, L. (1996) *Epidemiology*, p. 89. W.B. Saunders Company, Philadelphia, PA.

Graham, J.D., Paustenbach, D.J. & Butler, W.J. (1995) Epidemiology and risk assessment: divorce or marriage? In: *The Role of Epidemiology in Regulatory Risk Assessment*. (ed. J. Graham), pp. 1–14, Elsevier Science, Amsterdam.

HEI (1995) *Particulate Air Pollution and Daily Mortality. Replication and Validation of Selected Studies*. Health Effects Institute, Cambridge, MA.

Hill, A.B. (1953) Observation and experiment. *New England Journal of Medicine*, **248**, 995–1001.

Hill, A.B. (1965) The environment and disease: association and causation? *Proceedings of the Royal Society of Medicine*, **58**, 295–300.

IARC (1987) *IARC Monographs on the Evaluation of Carcinogenic Risks to Humans. Overall Evaluations of Carcinogenicity: an Updating of IARC Monographs, Volumes 1 to 42*. Supplement 7, International Agency for Research on Cancer, World Organization, Lyon.

ICRP (1984) *Report of the Task Group on Reference Man*. International Commission on Radiation Protection Publication No. 23, Pergamon Press, New York.

IPCS (1995) *Environmental Health Criteria 165: Inorganic Lead*. International Programme on Chemical Safety, World Health Organization, Geneva.

Langard, S. (1990) One hundred years of chromium and cancer: a review of epidemiological evidence and selected case reports. *American Journal of Industrial Medicine*, **17**, 189–215.

Machle, W. & Gregorius, F. (1948) Cancer of the respiratory system in the United States chromate-producing industry. *Public Health Reports*, **63**, 1114–1127.

Mancuso, T.F. (1975) Consideration of chromium as an industrial carcinogen. *Presented at the International Conference on Heavy Metals in the Environment*, Toronto, Ontario, 27–31 October 1975, pp. 343–356.

McKone, T.E. & Layton, D.W. (1986) Screening the potential risk of toxic substances using a multimedia compartment model: estimation of human exposure. *Regulatory Toxicology and Pharmacology*, **6**, 359–380.

Millard, S. (1987) Environmental monitoring, statistics, and the law: room for improvement. *American Statistician*, **41**(4), 249–259.

Moolgavkar, S. (1986) Carcinogenesis modeling: from molecular biology to epidemiology. *American Review of Public Health*, **7**, 151–169.

NRC (1983) *Risk Assessment in the Federal Government—Managing the Process*. National Research Council, National Academy of Sciences, National Academy Press, Washington, DC.

NRC (1991) *Environmental Epidemiology*. National Research Council, National Academy Press, Washington, DC.

NRC (1994) *Science and Judgment in Risk Assessment*. National Research Council, National Academy of Sciences, National Academy Press, Washington, DC.

OTA (1993) *Researching Health Risks*. Office of Technology Assessment, OTA-BBS-570, US Congress, Washington, DC.

Pao, E.M., Fleming, K.H., Guenther, P.M. & Mickle, S.J. (1982) *Foods Commonly Eaten by Individuals: Amount Eaten per Eating Occasion*. Home Economics Research Report 44, US Department of Agriculture, Washington, DC.

Paustenbach, D.J. (1989) Health risk assessments: opportunities and pitfalls. *Columbia Journal of Environmental Law*, **14**, 379–410.

Paxton, M.B., Rodricks, J.V., Chinchilli, V. & Brett, S.M. (1994) Leukemia risk associated with benzene exposure in the pliofilm cohort. *Risk Analysis*, **14**, 147–154.

PHS (1953). *Health of Chromate Workers* (ed. W.M. Gafafer). US Public Health Service publication No. 192, US Department of Health, Education and Welfare, Washington, DC.

Rodricks, J.V. (1992) *Calculated Risks*. Cambridge University Press, Cambridge.

Rodricks, J.V. (1996) Assessing carcinogenic risks associated with indoor air pollutants. In: *Indoor Air and Human Health* (eds R.B. Gammage and B.A. Berven), 2 edn, pp. 289–307. CRC Lewis Publishers, Boca Raton, FL.

Rothman, K.J. (1986) *Modern Epidemiology*. Little, Brown and Company, Boston, MA.

Schlesselman, J.J. (1982) *Case–Control Studies: Design, Conduct and Analysis*. Oxford University Press, New York.

Schwartz, J. (1994) Low-level lead exposure and children's IQ: a meta-analysis and search for a threshold. *Environmental Research*, **65**, 42–55.

Swann, R.L. & Eschenroeder, A. (1983) *Fate of Chemicals in the Environment: Compartmental and Multimedia Models for Predictions*. American Chemical Society, Washington, DC.

Upton, A.C., Kneys, T. & Toniolo, P. (1989) Public health aspects of toxic chemical disposal sites. *American Revue of Public Health*, **10**, 1–25.

USDHHS (1994a) *Health United States 1993*, p. 105. US Department of Health and Human Services Publication No. (PHS) 94-1232. Hyattsville, MD.

USDHHS (1994b) *Seventh Annual Report on Carcinogens*. US Department of Health and Human Services, National Toxicology Program. Pursuant to Section 301(b) (4) of the Public Health Service Act as Amended by Section 262, PL 95-622, Washington, DC.

USEPA (1982) *Review of the National Ambient Air Quality Standards for Nitrogen Oxides: Assessment of Scientific and Technical Information*. OAQPS Staff Paper. U.S. Environmental Protection Agency, EPA-450/5-82-002, Office of Air Quality Planning and Standards, Research Triangle Park, NC.

USEPA (1984) *Health assessment document for chromium*. US Environmental Protection Agency, EPA-600/8-83-014F, Environmental Criteria and Assessment Office, Research Triangle Park, NC.

USEPA (1986a) *Guidelines for Estimating Exposure*. US Environmental Protection Agency, **51** Federal Register 34042–34054. US Government Printing Office, Washington, DC.

USEPA (1986b) *Air Quality Criteria for Ozone and other Photochemical Oxidants*, Vol. V. US Environmental Protection Agency, EPA/600/8-84/020eF, Environmental Criteria and Assessment Office, Research Triangle Park, NC.

USEPA (1986c) *Air Quality Criteria for Lead*, Vol. 4. US Environmental Protection Agency, EPA/600/8-83-028dF, Environmental Criteria and Assessment Office, Research Triangle Park, NC.

USEPA (1988a) *Superfund Exposure Assessment Manual*. US Environmental Protection Agency, EPA/540/1-88/001, Office of Remedial Response, Washington, DC.

USEPA (1988b) *Drinking Water Criteria Document for Polychlorinated Biphenyls (PCBs)* US Environmental Protection Agency, ECAO-CIN-414, Environmental Criteria and Assessment Office, Cincinnati, OH.

USEPA (1989) *Exposure Factors Handbook*. US Environmental Protection Agency, EPA/600/8-89/043, Office of Health and Environmental Assessment, Washington, DC.

USEPA (1994) *National Ambient Air Quality Standards for Carbon Monoxide—Final Decision*. US Environmental Protection Agency, **59**(146) Federal Register 38906.

USEPA (1995) *Integrated Risk Information System (IRIS)*. U.S. Environmental Protection Agency, Office of Health and Environmental Assessment, Environmental Criteria and Assessment Office, Cincinnati, OH.

USEPA (1996) *National Ambient Air Quality Standards for Sulfur Oxides (Sulfur Dioxide)—Final Decision*. US Environmental Protection Agency, **61**(100) Federal Register 25566.

Part 2
Risk Assessment in Legislation

Chapter 1 made the point that much of the current policy and legislation involved with environmental protection is risk-, not just hazard- or suspicion-based. Within this, and in line with the recognition of a need to distinguish assessment procedures from management decisions, it is often possible to recognize separate information-gathering instruments and controlling instruments.

The two chapters in this Part are largely concerned with the information-gathering and assessment procedures, especially in the context of chemicals, in EU (and UK) and USA legislation. They draw together many of the references to particular items of legislation that will be found scattered through both Parts 1 and 4.

Chapter 9
Application of Risk Assessment in Policy and Legislation in the European Union and in the United Kingdom

NORMAN J. KING

9.1 HISTORICAL PERSPECTIVE

The UK was the first country to develop a modern industrial economy. Great wealth was created but unrestrained and unsustainable industrial development also led to: pollution of the air, water and land; urban squalor; much poor health; and the over-exploitation of resources. The roots of current environmental legislation in the UK go back to the middle of the nineteenth century when these problems of industrialization began to be tackled by government. The experience of most other countries in Europe was similar as they followed the model of industrial development pioneered by the UK. Had current techniques for risk assessment been developed and applied earlier, some of the environmental damage in Europe over the last two centuries might have been prevented. Instead, early environmental legislation was concerned with correcting damage caused by the failure to identify and manage risks.

Risk assessment as a tool in environmental protection had to await the development of the concept and its application in other fields, such as engineering and insurance, where it was used to handle uncertainty. However, in the UK, some aspects of what would now be recognized as risk assessment were incorporated into the concepts of 'Best practicable means' and 'As far as reasonably practical' developed, respectively, in legislation to control air pollution and hazards in the workplace. These concepts require industry and the relevant enforcement agencies to consider the risks posed by a process or practice and how these might be reduced at an acceptable cost. As a result of this experience, it has been relatively easy to incorporate risk assessment in modern environmental and safety practice and legislation in the UK.

Environmental risks can range from local (e.g. the redevelopment of an old gasworks site) to global (e.g. depletion of the ozone layer). This diversity does not necessarily require a change in risk assessment methodology, although the information requirements can be very different. The risk management strategy based on the risk assessment, however, will reflect the scale of the problem—a global problem requires a global solution.

9.2 RISK ASSESSMENT IN THE UK AND OTHER MEMBER STATES OF THE EUROPEAN UNION (EU)

The Health and Safety Executive (HSE), which is responsible for enforcing legislation on workplace safety in the UK, has adopted risk assessment as a pro-active tool for improving safety and to set priorities for the use of its resources in all areas of its responsibilities. Much of the legislation which the HSE enforces also requires the use of risk assessment by the industries being regulated. Examples include the Control of Substances Hazardous to Health (COSHH) Regulations (HSE 1994a) and the Control of Industrial Major Accident Hazards (CIMAH) Regulations (HSE 1984). As a result of this experience, the HSE has played a leading role in promoting and extending the use of risk assessment in government and industry in the UK.

Some of the HSE's responsibilities include aspects of environmental protection, for example, the CIMAH Regulations (HSE 1984) and the Control of Pesticides Regulations (MAFF 1986) and an assessment of environmental risks is required. However, the Department of the Environment

(DOE) generally leads on policy to control risks to the environment and it has been concerned that the assessment of environmental risks has lacked the consistency of approach apparent in the field of occupational safety. In an effort to rectify this, the DOE in 1995 published a guide aimed specifically at those concerned with assessing environmental risks (DOE 1995a). Figure 9.1 is taken from that guide and shows diagrammatically the steps advocated for an environmental risk assessment and how it can lead into risk management. The guide includes a few worked examples as annexes.

The DOE has encouraged the use of the guidance and used it itself in the report on dioxins published in 1996 (DOE 1996a). Her Majesty's Inspectorate of Pollution (HMIP; now part of the Environment Agency) also used the guidance in its report on dioxin emissions from municipal incinerators published in 1996 (DOE 1996b). The DOE undertook to evaluate the use and effectiveness of this guidance at the end of 2 years.

Legislation in most other countries in the EU placed a greater emphasis on technological standards to achieve improvement, and risk assessment is more difficult to fit within such a regulatory framework. However, most recognize its value and use it in aspects of their policies to protect man and the environment. The Netherlands, in particular, introduced the concept of risk assessment into its environmental policy in the mid-1980s and has played a leading role in developing some of the risk assessment procedures incorporated into EU legislation (see below). Sweden, Finland and Austria, on the other hand, based much of their approach to the control of risks from chemicals on hazard assessment. Because these countries have a more extensive list of banned or restricted chemicals than EU countries and they have been particularly concerned about exposure to dangerous substances contained in imported products, the problem of conflicts with the 'Single market' arose during negotiations to extend membership of the EU to these countries. This difficulty was not resolved during the negotiations on membership but must be resolved within 4 years of the accession of these countries. As yet, it is unclear how this difficulty will be resolved.

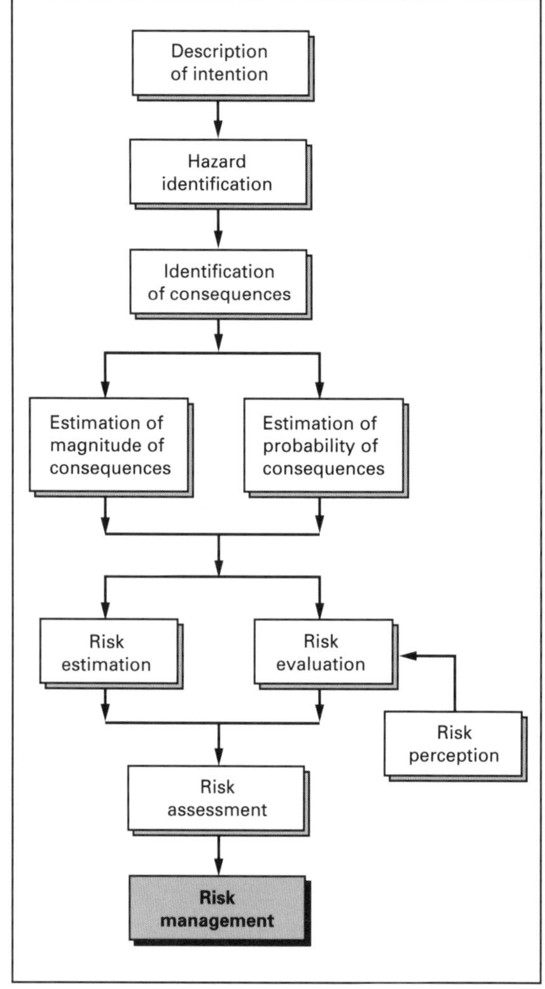

Fig. 9.1 From intention to risk management.

9.3 DEVELOPMENTS IN THE EU

The Treaty of Rome, agreed in 1957 and which led to the creation of the European Community and ultimately to the European Union, does not include any provisions relating directly to environmental protection. In the early years of the Community, Article 100 of the Treaty was used to enact some legislation which had health, safety or environmental benefits, although this article was concerned principally with the harmonization of trade within the Community: see, for example, the Directive on the classification, packaging and labelling of dangerous substances (EC 1967).

In 1972 the Member States set up an Environmental Action Programme. This was essentially an agreement by the Member States to work together on measures to protect their national environments and that of the Community. Over the next 15 years, more than 200 environmental measures were adopted at Community level, usually using Article 100 or Article 235 (a general provision covering cooperation between the Member States).

In 1987 changes were made to the original Treaty which introduced objectives aimed at protecting humans and the environment, and clear legal powers (Articles 130r, s and t) to take action at the Community level. The objectives set out in these Treaty revisions can be summarized as follows.

1 To preserve, protect and improve the quality of the environment.

2 To contribute to the protection and improvement of human health.

3 To ensure the prudent utilization of natural resources.

The Maastricht Treaty in 1993 added requirements to promote 'Sustainable and non-inflationary growth respecting the environment' and to take account of subsidiarity. 'Subsidiarity' has been variously interpreted. One definition, favoured particularly in the UK, is that action at the level of the European Union should only be considered when there are clear environmental or economic advantages from action at that level. The Maastricht Treaty also stated that the 'Precautionary Principle' should guide environmental policy. Indeed it has been argued, particularly by environmental interest groups, that the Treaty gives precedence to this principle over scientific evidence of risk and hence over risk assessment.

Since the Maastricht Treaty, the term 'European Union' has tended to replace 'European Community' as the collective title for the Member States and the various European institutions.

Early environmental legislation in the EU was not based on a systematic assessment of risk. Following the pattern already set in the Member States, legislation was usually aimed at correcting environmental problems created by poorly controlled industrial processes. Additional important considerations were the improvement and harmonization of standards of environmental protection across the EU. Priorities for action and the actions themselves were often seen to be self-evident. Although elements of risk assessment are implicit in some earlier legislation (e.g. Directives on environmental impact assessment and major industrial accidents), it has emerged as a policy tool as the EU has adopted a more preventive approach to environmental issues and as governments came to accept the need for a framework within which the costs, direct and indirect, and the benefits of taking (and the consequences of not taking) action can be assessed.

The Treaty texts, which underpin environmental legislation in the EU, do not mention risk assessment. However, action in the EU to protect and improve the environment generally takes place within the framework of the Environmental Action Programme. This sets out broad objectives and proposals for achieving these through new or revised legislation. Each programme is for a set number of years. The current Action Programme is the Fifth (EC 1993a) and it runs from 1993 until 2000. It identifies risk assessment and risk reduction as important activities.

9.4 INTERNATIONAL INTEREST IN RISK ASSESSMENT AS A LEGISLATIVE TOOL

Much work has been done in the USA to develop risk assessment and to widen its scope to areas such as environmental protection. For a comprehensive account of the development and use of risk assessment in that country, see Chapter 10.

The USA has also done much to encourage the use of risk assessment in the programmes of international agencies. Some of these agencies, such as the International Commission for Radiological Protection and the Organization for Economic Cooperation and Development (OECD), have developed widely accepted methodologies or guidance for risk assessment and promoted their application to problems at the international level.

9.5 THE PRECAUTIONARY PRINCIPLE

This is sometimes claimed to be an alternative to risk assessment, particularly by environmental interest groups. Several versions of this principle are in circulation and no single version has universal support. However, it is generally accepted that the original source of all of them is the 'Vorsorgeprinzip' set out by the German Federal Government in 1976. The following translation of the German text was published by the Royal Commission on Environmental Pollution (RCEP 1988):

> Environmental policy is not fully accomplished by warding off imminent hazards and the elimination of damage which has occurred. Precautionary environmental policy requires furthermore that natural resources are protected and demands on them made with care.

This is a broad statement which includes the idea of conserving resources. Subsequent interpretations have tended to concentrate on the theme of not delaying action until damage has occurred. One example of such an interpretation was included in the 'Rio Declaration' adopted in 1992 by governments at the United Nations Conference on Environment and Development (UN 1992):

> Where there are threats of serious or irreversible damage, lack of full scientific certainty shall not be used as a reason for postponing cost-effective measures to prevent environmental degradation.

This is similar to the interpretation in the UK's Environment White Paper (DOE 1990). Such interpretations are not in conflict with the scientific assessment of risk but they imply a lesser burden of proof in order to justify action in situations where a risk assessment indicates consequences which are potentially serious and long term. The DOE guidance on environmental risk assessment (DOE 1995a) suggests that it would be appropriate to apply the principle to the control of substances which are persistent and build up in the environment. The guide also proposes that the principle should apply pro-actively in situations where dealing with any likely damage will be costly.

The principal features in the policies and legislation of other EU Member States, e.g. Denmark, Germany and Sweden, and is regularly cited in EU legislation, including some where risk assessment is a requirement. It features in North Sea Declarations in the context of the control of pollution from substances which are persistent, toxic and liable to bioaccumulate: see, for example, the Esjberg Declaration (Danish EPA 1995).

9.6 EXAMPLES OF THE APPLICATION OF RISK ASSESSMENT IN THE UK AND IN THE EU

What follows is not a comprehensive account. It is intended to illustrate the growing use of environmental risk assessment in a few important areas of policy and to provide a few specific and more-detailed examples of its application. A comprehensive account of the use of risk assessment in Government in the UK was published recently by an interdepartmental group (HSE 1996). This covers many types of risk, including environmental risks, and should be consulted for a wider range of examples.

9.6.1 Risk assessment and standards

Risk assessment is increasingly being used in the UK and in the EU as an aid in standard setting. However, it is only an aid and a range of factors which are not strictly part of risk assessment are often considered before a standard is set (e.g. the economic impact, including the impact on trade, of any proposed measure; public expectations; social equity).

Risk assessment is often the starting point when a standard is being considered. If there is sufficient information, it can determine the magnitude of risk and, if a dose–response relationship can be established, it can derive a level of exposure for any 'acceptable' level of risk. Even if a dose–response relationship cannot be established, risk assessment will identify and indicate the relative significance of each source of exposure and the options for controlling exposure from each source. It can also provide a consistent basis for comparing risks, e.g. from alternative processes or substances.

Compliance with standards should be enforced and monitored and risk assessment can be used to help design effective enforcement and monitoring strategies. Conversely, the results of monitoring can be used to assess the effectiveness of a standard and, if necessary, trigger a further risk assessment.

Some standards, while intended to reduce environmental risks, are technology-based and risk assessment plays little part in the standard setting process. Examples in UK legislation include standards based on the concept that exposure should be controlled to 'as low as reasonably practical' or that emissions should be minimized by applying the concept of 'best available technology not entailing excessive cost' (BATNEEC). The former concept is a feature of radiation protection in the UK and in many other countries: the latter is the basis of integrated pollution control in the UK and of the EU Directive on integrated pollution prevention and control (EC 1993b). Integrated pollution control is applied to industrial processes which are considered to have the greatest potential to cause harm to humans or the environment. Hence, an assessment of risk lies behind the decision to require BATNEEC. In the UK there is also a requirement after emissions have been minimized that the remaining wastes be disposed of by the 'best environmental option'. It is difficult to see how this can be achieved, except through a thorough assessment of the risks associated with each disposal option.

'Best available technology' is also the basis of some kinds of product standards, e.g. for emissions from motor vehicles, or in other situations where there is limited scope for enforcement, e.g. discharge standards for ships or for off-shore oil production platforms. Even technology-based standards are often underpinned by environmental quality standards which are based on an assessment of risk. Thus, the technology-based emission standards for vehicles and industrial processes are set with the intention of meeting quality standards for ambient air.

9.6.2 Chemicals

Since chemicals are traded products, their control at the national level is likely to lead to barriers to trade within the EU. The environmental problems which they create may also extend beyond national boundaries. For these reasons, controls tend to be introduced at EU level and the national requirements are therefore relatively uniform across the EU. Descriptions of the regimes in the UK for new chemicals and for existing chemicals are given below. Similar systems, with some differences, to reflect national administrative arrangements, exist in the other Member States.

The Notification of New Substances Regulations 1993 (HSE 1993) implement in the UK the 'Seventh Amendment' (EC 1992a). New chemicals are defined as those not appearing in the European Inventory of Existing Commercial Chemical Substances (EINECS) (EC 1990a). The EINECS lists substances on the EU market in quantities greater than 1 tonne prior to 1981.

An earlier Directive, 'The Sixth Amendment' (EC 1979) required manufacturers and importers of new chemicals to notify the 'Competent Authority' (in the UK, the HSE and the DOE acting jointly) prior to placing a chemical on the market and to provide information on its properties. The amount of information required increased as the amount of the chemical placed on the market exceeded fixed tonnage thresholds. The notifier also had to propose appropriate classification and labelling and precautions for safe use. Finally, the notifier was required to disclose any likely adverse effects identified from an assessment of the information. These requirements imply that the notifier had to carry out at least a hazard assessment. They have been subsumed in the 'Seventh Amendment'.

In the UK, the Competent Authority also routinely carried out such an assessment. Although this was not required by the Directive, it was seen as a logical use of information to give early warning of possible risks arising from the manufacture, handling, transport, use or disposal of a substance. Figure 9.2 shows the steps in this UK hazard assessment procedure. It employed realistic worstcase scenarios to generate the exposure information (estimated doses and predicted environmental concentrations). The assessments also provided reassurance to other Member States, an important point since a notification accepted in one Member State was sufficient for marketing to take place throughout the EU. Because each Member State acts on behalf of the whole EU, operating to common standards, efficient systems for the exchange of information and for resolving any disputes between countries are essential features of the EU scheme. Test methods for generating information on the properties of new chemicals and procedures for ensuring the quality of the information are set out in annexes to other EC instruments (EC 1988, 1992b). In the UK these have been declared statutory codes of practice which support the regulations.

The 'Seventh Amendment' extended the period

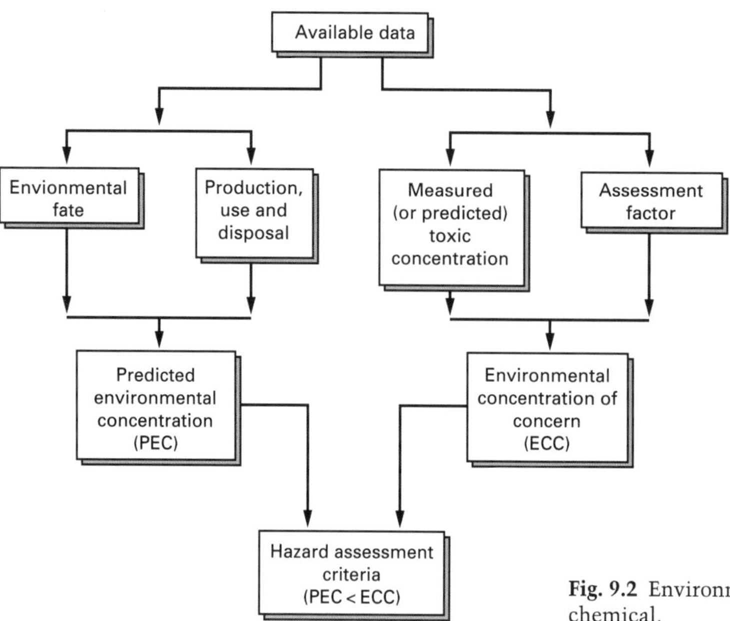

Fig. 9.2 Environmental hazard assessment of a new chemical.

of notification prior to marketing from 45 to 60 days and introduced the new requirement that the competent authority provide a risk assessment based on the information received. Guiding principles for this risk assessment are set out in EC 1992b. More detailed guidance is included in the supporting material to the UK regulations (HSE 1994b) and in technical guidance published by the Commission of the European Community (EC 1996). The latter covers the risk assessment of existing chemicals (see below) as well as new chemicals. The UK regulations also offer a reduced fee to notifiers who provide a risk assessment which the Competent Authority agrees complies with the requirements of the Directive. In most respects, the new requirement for risk assessment simply formalized existing UK practice and ensured that similar assessments were carried out in other Member States.

In common with many other countries, *ad hoc* risk assessment of 'existing chemicals' has been a regular feature of policy in the UK for many years. Published examples include several of the publications in the DOE's Pollution Paper series and HSE's Toxicology Reviews (both available from HMSO). Such assessments have often led to regulatory action or other forms of control. Risk assessment of chemicals is also an established feature of specific safety schemes, in the UK and elsewhere, such as those for pesticides, food, medicines and veterinary products.

The need for *ad hoc* risk assessment will continue in order to deal with new problems as they emerge but many countries, including the UK, have begun to move to a more pro-active, precautionary and systematic approach. With many thousands of chemicals to consider, any such change in approach has the potential to consume scarce expert resources on a scale which is beyond the capability of any single country. Recognizing the benefits which would come from a cooperative effort, the countries of the OECD agreed to work together and, in 1987, the Council of the OECD adopted a Decision/Recommendation (OECD 1987) to formalize this agreement. The Member Countries agreed to strengthen their programmes or to establish new ones to investigate existing chemicals more systematically in order to identify those posing risks to humans or the environment and to enable such risks to be controlled. This is the basis of the existing chemicals programme in the OECD. A list of 'High production volume chemicals' was agreed and responsibility for gathering risk-related information on each one was shared out among the Member Countries. The chemical industries in the OECD countries have agreed to cooperate and, in particular, to provide the information which they hold.

In parallel with the work in the OECD and as a contribution to it, the EU countries agreed, in 1993, to set up a scheme for existing chemicals to complement that for new chemicals. This took the form of an EU Regulation on the evaluation and control of the risks of existing substances (EC 1993c). As an EU Regulation, its requirements apply directly in the Member States, but implementing legislation in the UK (HSE 1994c) was needed to cover enforcement. The Competent Authority is the DOE and HSE acting jointly. The DOE/HSE publication 'How to report data on existing chemical substances' (HSE 1994d) explains the requirements of the Regulation and what companies need to do to comply with it. It includes the full text of the EU Regulation.

The Regulation requires companies producing or importing more than 10 tonnes per year of any substance listed in EINECS to notify the European Commission. They must then submit all the information which they hold on the properties of the substance and which is relevant to assessing risks to humans or the environment. This has to be submitted in electronic form using a software package (HEDSET) developed by the Commission for this purpose. There is a phased programme for the submission of information, high-tonnage chemicals taking precedence. The final phase of reporting will not be completed until 1998. Based on these submissions and using agreed selection criteria, a priority list is drawn up and each substance is allocated to a Member State. The first list was published in 1994 by the EC Commission (EC 1994a) and the Competent Authority must identify a rapporteur to carry out the risk assessment. Based on this risk assessment, the rapporteur may then put forward a strategy for consideration by the Member Countries for reducing any risks identified.

In 1994 the European Commission issued a Commission Regulation (EC 1994b) setting out guiding principles for the rapporteur to follow. This is supported by much more technical guidance issued by the European Commission in 1996 (EC 1996). The guidance covers new as well as existing chemicals. Between them, these set out how the rapporteur is to carry out a risk assessment. A worked example of the kind of approach advocated in the guidance is included as an annex in the DOE guidance on environmental risk assessment (DOE 1995a). Although it covers only one specific use scenario, it illustrates the application of the principles of risk assessment to an existing substance.

No guidance is available at EU level on developing a risk reduction strategy. To fill this gap in the UK, 'Risk reduction for existing substances' was produced by a Government/Industry working group in 1995 (DOE 1995b). Figure 9.3 is taken from that publication. It sets out in diagrammatic form how a risk reduction strategy for a substance may be derived from a risk assessment. It is also a reminder that a risk assessment of a chemical is not an end in its own right but rather a step towards greater safety throughout the 'lifecycle' of the chemical. Guidance on 'Risk–benefit analysis of existing substances' was developed similarly in 1995 (DOE 1995c). Both publications are available from the DOE.

As stated previously, Sweden has adopted a somewhat different approach to the control of chemicals. This is described in a publication from the Swedish Chemical Inspectorate (KEMI), 'A Common Strategy for International Chemicals Control' (KEMI 1995). While accepting the concept of risk assessment, this emphasizes the Precautionary Principle and the need for a substitution policy which requires that hazardous substances be replaced with less-hazardous substances wherever possible. Swedish accession to the EU makes such a policy difficult, as its implementation will create barriers to trade which will require notification under the technical standards Directive (EC 1983) and agreement in the EU for action under the Marketing and Use Directive. This Directive requires a risk assessment of the substance and of alternatives before substitution is agreed at the EU level.

9.6.3 Genetically modified organisms

The UK legislation in this field implements the EU Directives on contained use (EC 1990b) and on deliberate release of GMOs to the environment (EC 1990c). These are implemented in the UK by the Genetically Modified Organisms (GMOs) (Contained Use) Regulations (HSE 1992) and the Genetically Modified Organisms (Deliberate Release) Regulations (DOE 1992, 1995d), respectively. Both Directives require that the risks

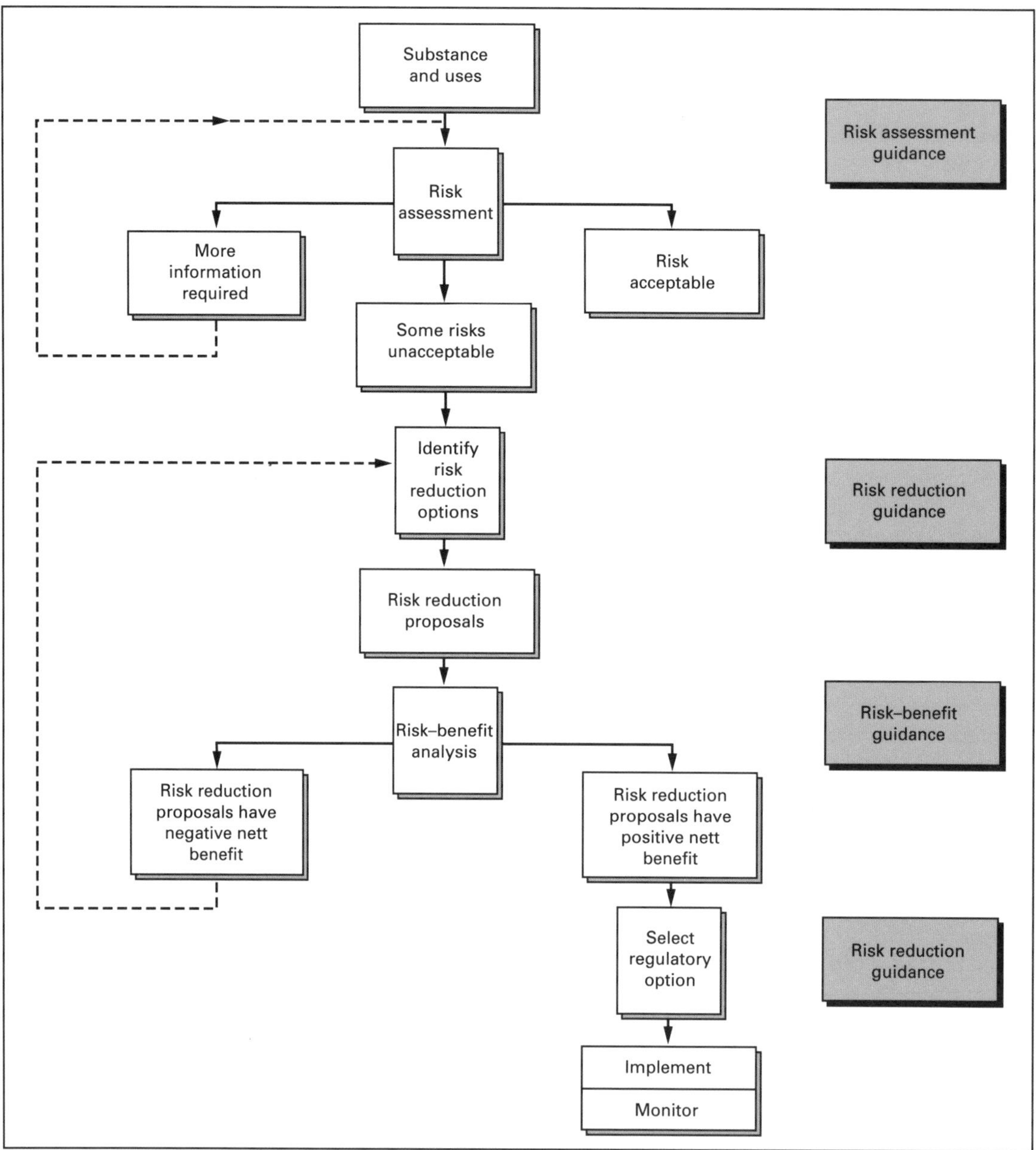

Fig. 9.3 Building a risk-reduction strategy from a risk assessment.

to human health and to the environment be evaluated and reported. What follows describes the approach to risk assessment adopted in the UK on proposals to release GMOs to the environment. It is based on the DOE's Guidance Note 1 (DOE 1993), but similar approaches have been adopted in other EU countries (see also Chapter 5).

The six main steps are as follows.

1 Identify the hazards associated with the GMO.
2 Identify how each hazard might be realized in the particular circumstances of the proposed release.
3 Estimate the magnitude of harm caused by each hazard if realized.
4 Estimate the likelihood and frequency that each

hazard will lead to harm.

5 On the basis of steps 3 and 4, estimate for each hazard the risk of harm.

6 Modify the proposal, if necessary, to reduce any risks to the minimum.

These steps apply equally to effects on health as well as to wider environmental effects.

It is recognized that quantitative information will not be available in many cases. A reasoned evaluation based on the best available quantitative or qualitative measure of risk to the environment or to health is what is required.

The DOE guide to risk assessment (DOE 1995a) includes a worked example of this approach as an annex.

9.6.4 Contaminated land

Land contamination emerged as an environmental issue in the UK in the mid-1970s. It is generally associated with industrial activities on a site, often past activities, or with poor waste disposal practices. In 1977 the DOE issued a circular (DOE 1977) to local authorities drawing attention to the problems which could arise when contaminated sites are redeveloped, announcing the creation of a central source of advice (the Interdepartmental Committee on the Redevelopment of Contaminated Land (ICRCL)). Public awareness rose sharply soon afterwards, with the publicity given to the large-scale contamination discovered at Love Canal (USA), Lederkerke (The Netherlands) and Shipham (UK).

In the UK, there is no legal definition of contaminated land but the term is often used to mean that contamination is present to an extent that it has the potential to affect the present or future use of the site. On that basis it has been estimated that between 100 000 and 200 000 ha are contaminated in the UK.

There is no specific legislation requiring that contaminated land be cleaned up. In general, the owner or the occupier of a site is responsible for its safety, and contamination can affect safety. However, local authorities have powers to intervene if this is the only way to ensure safety. The UK policy was reviewed in 1994 and set out in 'Framework for Contaminated Land' (DOE 1994). The review stated that decisions on clean-up should be based on an assessment of risk which took account of the current or proposed use of the land (the concept of 'suitability for use').

Most clean-up of contaminated sites in the UK takes place when a site is redeveloped. The owner or developer of a site is responsible for the costs of any remedial treatment, but such costs should be reflected in the market value of the site before and after it is developed. In general, the cost of any remedial treatment is accommodated within the overall cost of the development.

Much experience has been gained over the last 20 years and extensive guidance material is now available. For example, see the ICRCL Guidance Note 59/83 (DOE 1987) and 'Framework for contaminated land' (DOE 1994). Nevertheless, this is not an area of exact science and decision-making on individual sites will usually involve considerable uncertainties. It is against this background that risk assessment is advocated as a tool.

The starting point is generally whether or not the site is likely to be contaminated and, if it is, with what and to what extent. Much can be learned from a desk study of the previous uses of the site. This may confirm that contamination is unlikely to be a problem in the context of the planned development. However, if the desk study points to the likelihood of significant contamination, a site survey involving sampling and analysis will be necessary to provide information on which an assessment of any risks can be based (BSI 1988).

A range of hazards may have to be considered depending on the type and degree of contamination, the physical features of the site and the proposed use of it. The hazards might include, for example, effects on building materials, plants, groundwater, worker safety or human health. 'Safe levels' for these effects can be established in many cases from the scientific literature, by testing or by modelling, and the results compared with those from the site survey.

Possible effects on human health are often the most critical consideration, and a probabilistic computer model is being developed for the DOE to predict the risk to human health (see DOE 1995e). This allows assumptions about the physical characteristics of a site, proposed use and aspects of human behaviour within that use, and pathways to humans to be modelled in order to assess human intake of a contaminant from the soil. After taking account of intakes from other sources, the total intake can then be compared with values for tolerable daily intakes. The model can be used to establish, for any set of exposure assumptions, the level of a contaminant in the soil that might lead to an intake in excess of the tolerable daily intake.

At present there is no EU legislation in this field and it is arguable, on grounds of subsidiarity, that harmonization would not be appropriate. Nevertheless, the problems are common to most of the Member States and there is much sharing of experience and of the results of research.

In The Netherlands, a rather different approach to that used in the UK has developed. This places less emphasis on risk assessment and more on the inherent properties of the contaminants; i.e. hazard assessment is the basis for decision-making on the need for remedial action. However, the major difference is that policy in The Netherlands is not based on the concept of 'suitability for use' but is to clean up contaminated sites so that they are suitable for any purpose. This approach has led to the development of standards for a range of soil contaminants based on 'background' or 'normal' levels in soil. Policy in The Netherlands places a high value on land and on precaution and, to a greater extent than in the UK, it has led to the development of techniques for cleaning up soil to remove contaminants. On the other hand, it generally leads to more costly forms of clean-up and it has generated some problems in disposing of the residues from the treatment processes.

An account of policy and practice in The Netherlands is given in 'Reclaiming Contaminated Land' (Cairney 1987).

9.6.5 Ecolabelling

The concept of official ecolabelling schemes emerged as it was realized that consumers could, and were willing to, play a part in stimulating the production and marketing of more 'environmentally friendly' products. Unjustifiable claims for the environmental merits of their products by some manufacturers indicated the need to develop official schemes. The German 'Blue Angel' scheme set up in 1978 was followed by similar schemes in several other countries.

The objectives of such schemes are as follows.

1 To provide information to prospective purchasers on the potential impact of products on the environment and give them guidance on choosing products with least impact.

2 To encourage the development and marketing of products with reduced impact on the environment but which still perform effectively.

3 To use the purchasing power of the consumer to complement regulatory and other controls.

The consumer is ultimately helped and encouraged to make an informed choice on risks to the environment but, before the ecolabel is awarded, the product will have had to meet agreed criteria and have undergone an appraisal which has much in common with risk assessment.

In 1992 an EU Regulation (EC 1992c) was agreed to harmonize ecolabelling across the European Community. All consumer products are in principle eligible for the award of the ecolabel, with the exceptions of food, drink and pharmaceuticals. Products falling within the scheme are divided into product categories, e.g. adhesives, detergents, washing machines, etc., and criteria to evaluate them are developed and published. The evaluation is based on a lifecycle analysis of the product, which includes many elements of an environmental risk assessment as shown in the assessment matrix shown in Table 9.1. However, currently the process places greater emphasis on inherent properties, i.e. on hazard than on the likelihood of these properties being realized in the environment.

9.7 CONCLUSIONS

Risk assessment is gaining favour as a tool in environmental protection in the UK and in the EU. Its use ensures that the available scientific and technical information is assembled and used in a rigorous, coherent and transparent manner. It is used to identify and compare the options for control and it provides a sound basis for comparing risks. It is the usual starting point for assessing the costs of possible actions and the benefits arising from such actions. It also provides a basis for assessing the consequences and costs of failing to act.

While risk assessment is often the first step in developing a control strategy, factors other than those included in the risk assessment may be relevant in deciding what level of risk to tolerate. Nevertheless, it provides a sound base to which consideration of these other factors can be added when drawing up and implementing a programme to manage the risks.

Table 9.1 Assessment matrix used in the ecolabelling process.

	Product lifecycle				
Environmental fields	Pre-production	Production	Distribution (including packaging)	Utilization	Disposal
Waste relevance					
Soil pollution and degradation					
Water contamination					
Air contamination					
Noise					
Consumption of energy					
Consumption of natural resources					
Effects on ecosystems					

In recent years, much progress has been made in the UK in harmonizing the approaches to risk assessment used or advocated by government. Progress in the EU is less advanced. While the principle of risk assessment is accepted in many areas of EU policy, its exact role in specific cases often remains a subject of lively debate. Thus the need for harmonized approaches to preserve and foster the 'Single market' may be in conflict with existing national legislation and national views on the tolerability of risks. More recent environmental and safety legislation in the EU often includes requirements for risk assessment, but much of the earlier legislation does not, and it is not a simple matter to introduce consistent approaches in such a situation: difficulties remain in reconciling risk assessment and the Precautionary Principle, and subsidiarity and the harmonization of standards of safety; there is no consensus on how to link risk, benefits and costs.

Continuing efforts are needed nationally and internationally to extend and improve the utility and acceptability of risk assessment. In this context, a particular need is to develop the interface with the economic analysis of costs and benefits and to reach a consensus, nationally and internationally, on methodologies and on when it is appropriate to apply them.

9.8 REFERENCES

BSI (1988) *Draft Code of Practice for the Identification of Potentially Contaminated Land and its Investigation.* DD175:1988, British Standards Institution, London.

Cairney, T. (1987) *Reclaiming Contaminated Land.* Blackie, London.

Danish EPA (1995) *Esjberg Declaration; Ministerial Declaration from the Fourth Ministerial Conference on the Protection of the North Sea.* Environmental Protection Agency, Copenhagen.

DOE (1977) *The Redevelopment of Contaminated Land.* Department of the Environment, Joint Circular 49/77, HMSO, London.

DOE (1987) *Guidance on the Assessment and Redevelopment of Contaminated Land*, 2 edn. Interdepartmental Committee on the Redevelopment of Contaminated Land Guidance Note 59/83, Department of the Environment, London.

DOE (1990) *This Common Inheritance: Britain's Environmental Strategy.* Department of the Environment, HMSO, London.

DOE (1992) *Genetically Modified Organisms (Deliberate Release) Regulations.* Department of the Environment, SI 1972 No.3280, HMSO, London.

DOE (1993) *Guidance Note 1. Assessing and control-ling risks from the deliberate release of genetically modified organisms.* Department of the Environment, London.

DOE (1994) *Framework for Contaminated Land.* Department of the Environment, London.

DOE (1995a) *A Guide to Risk Assessment and Risk Management for Environmental Protection.*

Department of the Environment, HMSO, London. ISBN 0 11 753091 3.

DOE (1995b) *Risk Reduction for Existing Substances.* Department of the Environment, London.

DOE (1995c) *Risk Benefit Analysis of Existing Substances.* Department of the Environment, London.

DOE (1995d) *Genetically Modified Organisms (Deliberate Release) Regulations.* Department of the Environment, SI 1995 No.304, HMSO, London.

DOE (1995e) *Draft guidance on determination of whether land is contaminated under the provisions of the Environmental Protection Act 1990.* Department of the Environment, London.

DOE (1996a) *Dioxins in the environment.* Department of the Environment, HMSO, London.

DOE (1996b) *Release of dioxins from municipal incinerators.* Department of the Environment, HMSO, London.

EC (1967) *Classification, packaging and labelling of dangerous substances.* Directive 67/548/EEC; OJ L196 16.08.67, European Community, Brussels.

EC (1979) *Sixth amendment to the 1967 Directive on the classification, packaging, and labelling of dangerous substances.* Directive 79/831/EEC; OJ L259 15.10.79, European Community, Brussels.

EC (1983) *On the provision of information in the field of technical standards and regulations.* Directive 83/189/EEC; OJ L109 26.04.93, European Community, Brussels.

EC (1988) *Testing new chemicals.* Directive 88/302/EEC; OJ L133 30.05.88, European Community, Brussels.

EC (1990a) *European inventory of existing chemical substances* (EINECS). OJ C146 15.06.90, European Community, Brussels.

EC (1990b) *Contained use of genetically modified microorganisms.* Directive 90/219/EEC; OJ L117 08.05.88, European Community, Brussels.

EC (1990c) *Deliberate release of genetically modified organisms.* Directive 90/220/EEC; OJ L117 08.05.90, European Community, Brussels.

EC (1992a) *Seventh amendment to the 1967 Directive on the classification, packaging and labelling of dangerous substances.* Directive 92/32/EEC; OJ L154 05.06.92, European Community, Brussels.

EC (1992b) *Testing of new chemicals.* Directive 92/67/EEC; OJ 383 29.12.92, European Community, Brussels.

EC (1992c) *Community ecolabel award scheme.* Regulation 880/92/EEC; OJ L99 11.04.92, European Community, Brussels.

EC (1993a) Fifth environmental action programme: towards sustainability. Cm. (92) 23/Final/11, European Community, Brussels.

EC (1993b) *Proposal for Directive on integrated pollution protection and control.* OJ C311 17.11.93, European Community, Brussels.

EC (1993c) *The evaluation and control of the risks of existing substances.* Regulation R793/93/EEC; OJ L84 05.04.93, European Community, Brussels.

EC (1994a) *First list of priority substances as foreseen under Council Regulation 793/93.* Commission Regulation R1179/94; OJ L131 26.05.94, European Community, Brussels.

EC (1994b) *Principles for the assessment of risk to man and the environment of existing substances.* Commission Regulation R1488/94; OJ L161 29.06.94, European Community, Brussels.

EC (1996) *Technical guidance on assessing the risks of new and existing chemicals.* Commission of the European Communities, Brussels.

HSE (1984) *Control of Industrial Major Accident Hazards Regulations.* Health and Safety Executive, SI 1984 No.1902 (subsequently amended by SI 1988 No.142 and SI 1990 No.2325), HMSO, London.

HSE (1992) *Genetically Modified Organisms (Contained Use) Regulations.* Health and Safety Executive SI 1992 No. 3217, HMSO, London.

HSE (1993) *Notification of New Substances Regulations.* Health and Safety Executive, SI 1993 No. 3050, HMSO, London.

HSE (1994a) *Control of Substances Hazardous to Health Regulations.* Health and Safety Executive, SI 1994 No.3246, HMSO, London.

HSE (1994b) *Risk Assessment of Notified Substances.* Health and Safety Executive, London. ISBN 0 71 760758 5.

HSE (1994c). *The Notification of Existing Substances (Enforcement) Regulations.* Health and Safety Executive, SI 1994 No.1806, HMSO, London.

HSE (1994d) *How to report data on existing chemical substances.* Health and Safety Executive, HMSO, London. ISBN 0 11 752812 9.

HSE (1996). *Use of risk assessment within Government Departments.* Health and Safety Executive, London.

KEMI (1995) *A common strategy for international chemicals control work in Sweden.* KEMI (National Chemicals Inspectorate), Stockholm.

MAFF (1986) *Control of Pesticides Regulations.* Ministry of Agriculture Fisheries and Food, SI 1986 No.1510, HMSO, London.

OECD (1987) *Decision/Recommendation on the systematic investigation of existing chemicals.* C(87) 90 (Final), Organization for Economic Cooperation and Development, Paris.

RCEP (1988) *Royal Commission on Environmental Pollution, Twelfth Report, Appendix 3.* HMSO, London. ISBN 0 10 103102 5.

UN (1992) *Rio Declaration on Environment and Development.* United Nations, New York. ISBN 9 21 100509.

Chapter 10
Application of Risk Assessment in Policy and Legislation in North America

RANDALL S. WENTSEL

10.1 INTRODUCTION

Advances of science and technology give us new tools to address environmental issues. As we face up to indoor air pollution issues like radon, regional issues like non-point sources of water pollution, or global issues like ozone depletion, biodiversity and climate change, we must use risk analysis and other new tools more frequently. However, risk assessment cannot be applied indiscriminately to all environmental issues.

Policy makers who support the use of risk analysis believe that federal programmes to protect public health and safety and the environment should be carefully targeted to address the worst risks first, and that the risk reduction achieved should be worth the cost. Politicians also hear complaints about 'unfunded federal mandates' to state and local governments and the growing cost of compliance of environmental requirements by industries.

Opponents of this use of risk assessment argue that exclusive reliance on risk assessment and costs to define problems and evaluate solutions ignores other equally important facets of policy decisions, such as timeliness, fairness to all segments of society and practicality. They also state that risk assessment and cost–benefit analysis undervalue environmental and health benefits, exaggerate costs and focus on relatively small costs and risks spread throughout the entire population.

10.2 HISTORY

10.2.1 Application of risk to environmental issues

The history of the field of risk analysis is the history of the development and use of various techniques for gathering and analysing information about hazards. Many different analytical techniques have been employed, both qualitative and quantitative. Most are borrowed from other disciplines, including actuarial accounting, economics, biology, geology, geography and engineering.

Agreement on terminology in the field of risk analysis has often been difficult to do. However, as will have been apparent from the previous chapters, there is general agreement on the following terms.

Risk analysis a term that encompasses various risk activities, such as: risk assessment, risk management, risk communication and comparative risk.

Risk assessment 'the use of the factual base to define the health effects of exposure of individuals or populations to hazardous materials and situations' (NRC 1983). Risk assessment is a process which evaluates scientific data. It has four basic steps of analysis: hazard identification, dose–response relationships, exposure analysis and risk characterization.

Risk management uses the results of the risk assessment along with economic analysis, societal issues and concerns, policy issues, and technology and engineering input to determine the management of the health or environmental issue of concern.

Relative risk relative (quantitative) magnitudes of the risks estimates.

Comparative risk relative (subjective) magnitudes of the concerns after the risk management factors have been included.

10.2.2 Impact of federal statutes

Rapid development of environmental risk analysis

methods was stimulated in the last two decades, in part by the enactment of major federal environmental laws, such as: the Clean Air Amendments of 1970, the Federal Water Pollution Control Act Amendments of 1972, the Toxic Substances Control Act in 1976, the Resource Conservation and Recovery Act of 1976, and the Comprehensive Environmental Response, Compensation, and Liability Act of 1980. Implementation of these laws required the development of criteria, standards, limits and other numerical descriptors of environmental quality. Some also required agencies to weigh social and economic impacts of regulations against the benefit of improved safety that the regulations were intended to achieve. Risk analysis was adapted to satisfy both requirements.

These statutes to protect human health or the environment vary in specificity: some grant federal agencies broad discretionary power, while others limit the flexibility of the regulators. Factors such as costs, benefits of use, social impacts or technical feasibility are not used consistently in environmental statutes. Risk analysis methods were developed to measure the risks avoided, the major benefit of most environmental regulations. The US Environmental Protection Agency (USEPA) began in the late 1970s to estimate both costs and benefits under two environmental statutes: the Toxic Substances Control Act (TSCA) and the Federal Insecticide, Fungicide and Rodenticide Act (FIFRA) (Schierow 1994).

10.2.3 Initial activities of federal agencies

Federal agencies began to use chemical risk assessment in the 1970s to estimate the cancer-causing potential of chemicals in commerce. Their efforts in the field of risk analysis were encouraged when, in 1980, the US Supreme Court issued an opinion that required the Occupational Safety and Health Administration (OSHA) to perform risk assessments of toxic chemicals as a basis for regulating occupational exposures. The use of risk assessment in the development of regulations continued to be controversial, in part because authorizing legislation often provided little guidance about how risks should be balanced against other factors. Thus, for example, in 1987, a federal appeals court nullified a USEPA rule restricting air emissions of vinyl chloride because the agency had not used risk assessment properly (Rosenthal *et al.* 1992). Other risk assessment decisions also have been overturned by federal courts because the assessment was judged to be of insufficient technical quality.

By the mid-1970s, agencies had begun efforts to improve coordination among programmes and to ensure consistent use of uniform risk assessment procedures within and across agencies of the federal government. Formal procedures for extrapolating research results to estimate human health effects—i.e. chemical risk assessment—were adopted in the late 1970s. In 1976, the USEPA established an internal working group, the Carcinogen Assessment Group (CAG), which published the first interim guidelines for assessing risks of suspected carcinogens. In 1977, the USEPA, the Consumer Product Safety Commission (CPSC), the Food and Drug Administration (FDA), and OSHA formed the Interagency Regulatory Liaison Group (IRLG). The IRLG met as a forum for voluntary coordination and information exchange until 1980. One of its products was a 'cancer policy' that attempted to present the scientific basis for determining the carcinogenicity of a substance.

10.2.4 The NAS framework

The agencies' efforts to systematize risk analysis were criticized by some scientists and industrial representatives who were concerned that policy matters were overly influencing scientific judgements, and thus the risk assessment process. In response, congress requested a study by the National Academy of Sciences (NAS) of institutional arrangements to improve agencies' use of risk assessment. This led to the landmark report entitled 'Risk Assessment in the Federal Government: Managing the Process' (NRC 1983). This report presented the framework for conducting risk assessment. It discussed the need to separate risk assessment from risk management and recommended that uniform risk assessment guidelines be adopted by the federal government. While a uniform guide for the federal government has not been adopted, in 1985 the Office of Science and Technology Policy (OSTP) in the office of the President produced a report on chemical carcinogens and proposed a method to assess their hazards. This report described the state of the science upon which decisions could be made.

In the Clean Air Act Amendments of 1990, Congress asked the NAS National Research Council to assess the current state of the USEPA risk analyses. The 1994 NAS report *Science and Judgment in Risk Assessment* (NRC 1994) listed their concerns and identified major themes for their review. The concerns have included the following.

1 The lack of scientific data quantitatively relating chemical exposure to health risks.

2 The divergence of opinion within the scientific community on the merits of the underlying scientific evidence.

3 The lack of conformity among reported research results needed for risk characterization.

4 The uncertainty of results produced by theoretical modelling, which is used in the absence of measurements.

5 In response to its mandates, the USEPA has traditionally adopted risk assessments that, for the most part, incorporate conservative default options (i.e. those that are more likely to overstate than to understate human risk).

6 Better data and increased understanding of biological mechanisms should enable risk assessments that are less dependent on conservative default assumptions and more accurate as predictions of human risk.

Major questions addressed in the report were as follows.

Default options Is there a set of clear and consistent principles for modifying and departing from default options?

Data needs Is enough information available to the USEPA to generate risk assessments that are protective of public health and are scientifically plausible?

Validation Has the USEPA made a sufficient case that its methods and models for carrying out risk assessments are consistent with current scientific information available?

Uncertainty Has the USEPA taken sufficient account of the need to consider, describe and make decisions in light of the inevitable uncertainty in risk assessment?

Variability Has the USEPA sufficiently considered the extensive variation among individuals in their exposures to toxic substances and in their susceptibilities to cancer and other health effects?

Aggregation Is the USEPA appropriately addressing the possibility of interactions among pollutants in their effects on human health, and addressing the consideration of multiple exposure pathways and multiple adverse health effects? (NRC 1994).

The report contains 70 recommendations for improving risk assessment (NRC 1994). The major recommendations from the 1994 report are as follows.

• The USEPA should generally retain its conservative approach to risk assessment in the initial phase of setting standards, but the USEPA should more clearly state its principles.

• The USEPA should develop and use an iterative approach to risk assessment, beginning with relatively inexpensive screening techniques and moving on to more resource-intensive levels of data gathering, model construction and model application as each situation warrants.

• The USEPA should work to continually improve the models and data used in risk assessments and develop a standard procedure for deviating from its conservative approach to risk assessment when warranted by scientific considerations.

• In its reports to decision makers and the public, the USEPA should present information about the resources and magnitudes of uncertainty as well as point estimates of risk.

• Risk assessment is a set of tools, not an end in itself. The limited resources available should be spent to generate information that helps risk managers to choose the best possible course of action among the available options.

10.2.5 Cancer and non-cancer guidelines

Following the 1983 NAS framework, agencies such as the USEPA, have promulgated their own risk assessment guidelines. The USEPA guidelines for cancer risk assessment were finalized in 1986 and updated draft guidelines were released for comment in late 1995. In a related report, the USEPA in 1986 published a classification guide for carcinogens. Revised proposed guidelines for carcinogen risk assessment were released for comment in August 1995 (USEPA 1995a).

The risks of non-cancer effects of chemical exposure to humans also have been recognized and analysed. As early as 1980, just before it was disbanded, the IRLG was developing guidelines for the risk assessment of effects on reproduction and human development. To date, the USEPA has promulgated final guidelines for risk assessment of genetic mutations and adverse effects on human development. Draft guidelines for neurotoxicity risk assessment were recently released for comment (USEPA 1995b). Guidelines have yet to be established for assessing risks of reproductive failure, respiratory effects or damage to the immune system.

The USEPA has also provided more guidance on risk characterization and the risk assessor/risk manager interface (Habicht 1992). Concerning discussions between the risk assessor and risk management, the risk assessment information must be clearly presented, separate from any non-scientific risk management considerations. Discussion of risk management options should follow, based on consideration of all relevant factors, both scientific and non-scientific. Key scientific information on data and methods in both risk characterization (e.g. use of animal or human data for extrapolating from high to low doses, use of pharmacokinetics data) must be highlighted. The assessor should write a statement of confidence in the assessment that identifies all major uncertainties, along with comment on their influence on the assessment. The USEPA established policy to present information on the range of exposures derived from exposure scenarios and on the use of multiple risk-descriptors (i.e. central tendency, high end of individual risk, population risk, important subgroups, if known) consistent with terminology in agency guidelines.

Guidelines are necessary to ensure that risk assessments are conducted consistently and are, therefore, more easily evaluated by independent experts. However, guidelines do not ensure that scientists will agree with resulting risk estimates. In fact, some criticize the USEPA's risk estimates for carcinogens because they do not like the guidelines or think different rules should apply to certain chemicals. Controversy surrounds risk assessment, partly because the field is so young that its methods have not been studied thoroughly and validated.

10.2.6 Council, committee and societal actions

Other groups are also active in risk assessment and risk management. The National Research Council (NRC) of NAS has a permanent Committee on Risk Assessment Methodology (CRAM). They are working to consider changes in the scientific foundation of risk assessment that have occurred since the 1983 report. They issued a report entitled 'Issues in Risk Assessment' in 1993 (NRC 1993).

A Commission on Risk Assessment and Management was also authorized by the Clean Air Act Amendments. The Commission became active in 1994. They are mandated to do the following.

1 Consider issues related to how human health risk assessments are performed and, most importantly, how the results of human health risk assessments are used in regulatory decision-making.

2 Focus on the five aspects of their mandate:
- (a) uses and limitations of risk assessment in decision-making;
- (b) appropriate exposure scenarios;
- (c) uncertainty and risk communication;
- (d) risk management policy issues; and
- (e) consistency across agencies.

3 Consider the 1994 NRC report 'Science and Judgment in Risk Assessment'.

4 Report to Congress in March and September 1996.

Within the White House National Science and Technology Council, a Committee on Environment and Natural Resources (CENR) was formed. The mission of CENR is to develop a strategic guidance for environmental research and development activities across the federal government. Each of 10 subcommittees under the CENR is charged with defining critical policy questions or issues relevant to their environmental issue area (e.g. air quality, toxic substances, water resources) or cross-cutting methodological (e.g. risk assessment, social and economic sciences) area (Fig. 10.1). The scientific knowledge and corresponding research necessary to meet those policy challenges are then to be identified. Within CENR, a subcommittee on risk assessment has been established (OSTP 1995).

The subcommittee on risk assessment focuses on methodological research topics and upon the development of means for effective translation

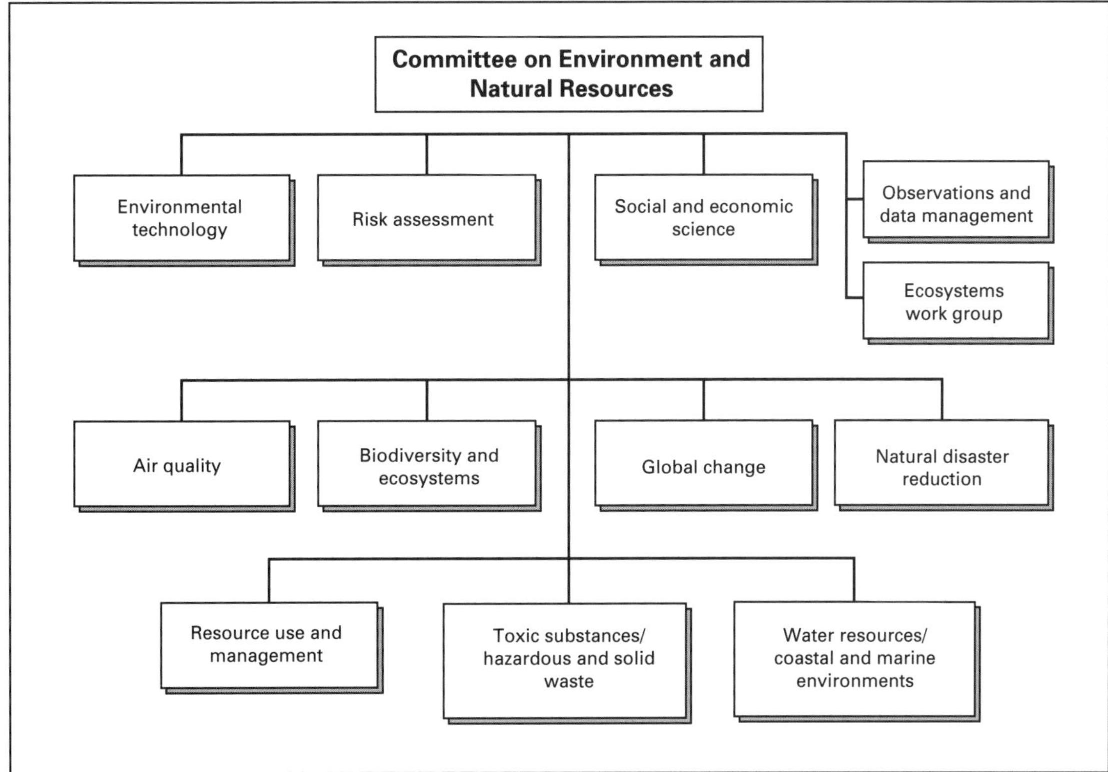

Fig. 10.1 The structure of the US Committee on Environment and Natural Resources.

and transmittal of scientific knowledge to policy makers. They have a dual role: to understand how key data gaps in each issue area influence (and perhaps inhibit) risk assessment; and to identify high-priority needs for methodological research for assessment, characterization and communication of risks. The subcommittee was established to review federal risk assessment research programmes and to develop ways to improve the enhancement and coordination of federal activities. In examining the current state of risk assessment research across the federal agencies and proposing research priorities that might enhance the *performance of risk assessments*, the subcommittee is considering issues such as the following.

1 Gaps in scientific knowledge that influence assessments for different types of risks and end-points.

2 Data needs for the performance of risk assessments and the potential for sharing data within and across agencies.

3 Alternative approaches to risk assessment across agencies, across federal statutes and internationally (CENR 1995).

10.2.7 Ecological risk assessment

Ecological risk assessments were initiated by the USEPA in order to develop water quality criteria required under the Clean Water Act. The first ecological risk assessment was a 1981 Synfuels assessment (Suter 1993). Ecological risk assessments initially followed the 1983 NAS framework, but in 1991 several workshops were held by the USEPA, NAS and others to reassess the procedures for ecological risk assessment. In 1992 the USEPA published a report entitled 'Framework for Ecological Risk Assessment' (USEPA 1992; Chapter 2). This new framework, while similar to the human health framework, recognizes the difference between ecological and human processes. Differences between them include the various end-points in ecological risk assessments, such as protection of bald eagles or maintaining species diversity in an aquatic system. This framework appears to have been generally accepted as the most appropriate procedure for ecological assessments. In December 1995, the USEPA released a draft guidance document on ecological risk assessment for peer review and comment (USEPA 1995c). The Departments of Energy and Defense also have guidance documents providing additional details on ecological risk assessment (USDOE 1994; Wentsel *et al.* 1994). In addition, the Canadians have published ecological risk assessment guidance for remediation efforts (Gaudet 1994).

10.2.8 Key USEPA reports

In 1987, the USEPA published 'Unfinished Business: A Comparative Assessment of Environmental Problems' (USEPA 1987). This effort was the first time that environmental problems were compared to each other without regard to programmatic context. Much of the Congressional activity and interest in comparative risk or risk ranking was generated from this report. In this report, 31 environmental problems were ranked by USEPA scientists based on the relative risk that each posed in four categories: cancer risks, non-cancer health risks, ecological effects and welfare effects (which included such effects as impairment of visibility and materials damage). It is important to note that the risks were ranked assuming that current regulatory controls were in place. So the low relative risk of hazardous waste sites was, in part, due to the existence of regulations and funds to treat that problem. Other problem areas (e.g. indoor air pollution) were ranked as a relatively high risk because they have never been regulated by the USEPA or any other federal agency. In addition, the report emphasized that the rankings were based on information available at the time, which, according to the scientists, contained 'substantial gaps'. The results indicated that no environmental problem ranked relatively high or relatively low in all four categories of risk. Problems that were ranked as relatively high or moderate in three or more categories included: air pollutants, stratospheric ozone depletion, pesticide residues on food and other pesticide risks. Other problems that were ranked as relatively high risks to human health included: hazardous air pollutants; indoor air pollution; indoor radon; pesticide application; exposure to consumer products; and worker exposures to chemicals. The other problems that ranked high in risk to ecology or human welfare included: global warming; surface water pollution; physical alteration of wetlands, estuaries, and other aquatic habitats; and mining wastes. One of the conclusions of the report was that the scientists' ranking of risks to human health differed from the ranking implied by national polls of public concerns. The public reported greatest concern about chemical waste disposal, water pollution, chemical plant accidents and air pollution. They had moderate concern about oil spills, worker exposure, pesticides and drinking water. The public had a relatively low concern about indoor air pollution, consumer products, radiation (other than nuclear power) and global warming. The USEPA budget appeared to correlate better with the priorities assigned by the public than with the risks thought to be greatest by the USEPA scientists. Interpretation of these results requires caution. For example, the low relative risk of hazardous waste sites (as indicated by data available in 1987) was due, in part, to the existence of regulations and availability of funds to treat the problem. Problems such as indoor air pollution were characterized as relatively risky, at least in part, because they were not regulated by any federal agency. Additional concerns on the report have centred on the methods used to rank the environmental problems. Ranks were assigned using qualitative professional judgements, incomplete data sets, and other unvalidated methods. The feasibility of controlling risks, the nature of the activities that posed risks, and the distribution of risks and benefits were not considered.

In a follow-up study the USEPA Science Advisory Board, in 1990, produced a well-regarded report entitled 'Reducing Risk: Setting Priorities and Strategies for Environmental Protection' (USEPA 1990). The purpose of the report was to determine strategies for reducing risks and recommend approaches for ranking and reducing risk in the future. Two committees were formed, one to study ecological and welfare risks and a second committee to address human health risks. The committees were to rank risks within their areas of concern. However, the report pointed out that the data gaps, uncertainty and judgement issues made the process of ranking risk tenuous and unreliable. The authors of the 'Reducing Risk' report also recommended 10 actions for the USEPA to take to reduce risk and improve the risk assessment process. While these recommendations have been well received, they have not been implemented by the USEPA through a top-down approach. The report also addressed problems with current cost–benefit analysis of environmental resources. Current methods assume that the future value of an ecological resource must be less than its present value. This policy leads to depletion of irreplaceable natural resources. The 'willingness to pay' technique is also misused in economic analysis. The public may not care about wetlands, but the contribution that wetlands make to the larger ecosystem is of value to the public now and in the future. Techniques need to be developed to assess the real long-term value of ecosystems (see Chapters 11 and 13).

10.2.9 Comparative risk assessment

As a result of the two reports on risk by the USEPA Science Advisory Board and their members, comparative risk assessment, a procedure for ranking environmental problems by their relative risk, has received support from risk managers and policy makers. It is used to prioritize by risk alone, risk and cost–benefit, or using all of the risk management components. Due to the federal regulatory mandates, which impact the budgets of state and local governments, they are in favour of comparative risk assessment. No standard methods are available for the ranking of risks. The USEPA Office of Policy, Planning and Evaluation (OPPE) has, however, funded comparative risk projects for States and has supported human health risk rankings and ecological risk rankings for the USEPA regions (USEPA 1993). These studies have had a mixed success. However, the involvement of various environmental offices, policy makers and the public in priority setting of environmental issues is very important. Van Houtven & Cropper (1993) reported on the use of risk ranking within an environmental category. Benefit–cost analysis was used to evaluate three USEPA regulatory actions. One example on the cost–benefit of banning asbestos gave results in cost per cancer case avoided and it was a good example of how relative risk assessment, within a category, could be used. Comparative risk assessment is an attractive use of risk assessment, cost–benefit analysis, and involving the public. The unfunded federal regulatory mandates to state and local governments and the limits in the federal budget will push decision makers toward this process. However, data gaps, subjective issues and uncertainty in any ranking of risks must be recognized. The USEPA Science Advisory Board report on reducing risk ranked ecological risks in three broad categories. Groupings of environmental problems in broad categories may minimize the imprecision of the process. The OPPE efforts in this area should be evaluated for use as a framework for ranking environmental issues. Improvements in methods for comparative risk assessment will increase its use in budgeting and prioritization of environmental problems.

Some groups argue against priority setting based on risk because it focuses on the annual probability of death or disease. They believe that the process ignores other equally important factors, such as the degree of force used to implement the risk reduction strategy and the fairness of the result to all segments of society (Schierow 1995).

The California Comparative Risk Project (CCRP 1994) differed from previous comparative risk projects by developing three lists of environmental issues instead of just one list of environmental problems for analysis. The CCRP developed 'environmental topic lists', which compared and ranked environmental hazards from three distinct and relatively consistent subsets of environmental issues.

List I included the traditional statutory division of environmental problems into the media (air, water and land) that are impacted by the release of toxic substances from different sources.

List II addressed major environmental stressors.

List III considered overarching categories of human activity which impact the environment.

The topic areas are presented in Table 10.1. Human health risk rankings are presented in Table

Table 10.1 Environmental topic areas (CCRP 1994).

LIST I: ENVIRONMENTAL RELEASES TO MEDIA BY SOURCES	Microbiological contamination
Water	New chemicals
Industrial releases to surface water	Non-native organisms
Municipal releases to surface water	Oil/petroleum
Non-point source releases	Ozone
Releases to groundwater	Particulate matter
Storage tank releases	Pesticides—agricultural use
	Pesticides—non-agricultural use
Land	Persistent/bioaccumulative organochlorines
Active hazardous waste generators	Radionuclides
Inactive hazardous waste sites	Radon
Solid waste disposal sites	SO_x and NO_x
Treatment, storage and disposal facilities	Stratospheric ozone depletors
	Substances that alter pH, salinity and hardness
Air	Thermal pollution
Mobile sources	Total suspended solids, biological oxygen demand and nutrients
Residential and consumer product sources	Volatile organics
Stationary and commercial area sources	
LIST II: ENVIRONMENTAL HEALTH STRESSORS	LIST III: POTENTIAL THREATS TO ENVIRONMENTAL INTEGRITY
Alteration of aquatic habitats	Agricultural practices
Alteration of terrestrial habitats	Commercial/industrial practices
Asbestos	Energy management practices
Carbon monoxide	Municipal/governmental practices
Electromagnetic fields	Natural resource practices
Environmental tobacco smoke	Recreational practices
Genetically engineered products or organisms	Residential/consumer practices
Greenhouse gases	Transportation systems
Inorganics	Water management practices
Lead	

10.2. The risk-rankings for Environmental Health Stressors (environmental topic list II) and ecological stressors are provided in Tables 10.3 and 10.4, respectively. These rankings should be considered in the context of the other decision-making factors discussed in the summary report. In using these rankings, the caveats to these tables are presented in the Appendix.

Slovic *et al.* (1995) presented the results of a comparative risk assessment study in Canada. The results showed that Canadian toxicologists had a lower perception of risk and were more supportive of the use of chemicals than the Canadian public. The attitude of the public was quite negative and showed the same lack of dose–response sensitivity found in the earlier USA studies. Both of the groups studied lacked confidence in the value of animal studies for predicting human health risks. Technical judgements of toxicologists were found to be associated with factors such as affiliation, gender and world views.

10.2.10 Uses of risk assessment

Many functions of the federal government are intended, either directly or indirectly, to reduce risks to public health and safety and the environment. In this function, risk analysis is used as a tool for evaluating what is known about things that cannot be known with certainty due to effects of hazards that are unpredictable, lack of scientific understanding of data, or uncertainty with models or other extrapolations. Risk analysis produces an estimate, and they vary due to the quality of

Table 10.2 Human health risk-rankings of environmental releases to media by sources (CCRP 1994). Populations at disproportionate risk of high impact are given in parentheses. Topics within each rank are ordered alphabetically.

Rank	Release source
High	Mobile source releases to air (children; people with respiratory or cardiac conditions; those living near transportation corridors) Natural source releases to groundwater (those drinking from contaminated water supplies) Residential and consumer product source releases to air (children; smokers; those living in regions with high radon sources) Stationary and commercial area source releases to air (children; people with respiratory or cardiac conditions; those living near emission sources)
Medium	Anthropogenic source releases to groundwater (infants; those drinking from contaminated water supplies) Inactive hazardous waste sites (those near undiscovered or uncontrolled sites) Non-point source releases to surface water (subsistence/sport fishers; those on private wells)
Low	Industrial releases to surface water (subsistence/sport fishers) Municipal releases to surface water (subsistence/sport fishers) Treatment, storage and disposal facilities (those near uncontrolled releases)
'Unrankable'	Topic area lacks sufficient toxicological or exposure data to reach a scientifically supportable evaluation Active hazardous waste generators Solid waste disposal sites Storage tank releases

information used by the assessor. Risk analysts can only discuss the likelihood of various outcomes, discuss the role of scientific judgement, and present risks as statistical probabilities. Agencies and departments evaluate risks, formally or informally, in most of their activities. Uses of risk assessment include the following (CENR 1995).

1 Defining problems and predicting risks.

2 Selecting risk avoidance or mitigation strategies and developing management programmes.

3 Setting standards to protect human or ecological health, and evaluating ongoing risk reduction activities.

4 Determining management and policy priorities.

Policy or non-scientific decisions in risk assessment include choices of assumptions and methods to address uncertainty in available scientific data. These decisions become significant when two federal agencies evaluate the same data and calculate different levels of risk. Risk assessment techniques typically rely on multiple assumptions which may be untested, or even untestable. In human health risk assessment, assumptions include: the relevance of animal data to humans, the choice of a dose–response model to extrapolate from high (experimental exposures) to the lower environmental doses, exposure pathways and protection levels, i.e. 95% of the affected population (Schierow 1995). The NRC report (1994) identified at least 50 decisions required in conducting a cancer risk assessment that cannot be made on a scientific basis, and many of these decisions have strong implications for public policy.

To improve risk assessment, the NAS National Research Council (NRC 1994) panel concluded that the most effective way would be by improving the quality and comprehensiveness of knowledge. The NAS recently evaluated the availability of data for risk analyses of 189 hazardous air pollutants and concluded that the USEPA did not have 'sufficient data to assess fully the health risks—within the time permitted by the Clean Air Act Amendments of 1990.'

Risk assessment has a major role in Proposition 65—The Safe Drinking Water and Toxic Enforcement Act of 1986—which became effective on 1

Table 10.3 Environmental health stressors (CCRP, 1994) (to be used only in conjunction with the caveats in Appendix 10.6).

Rank	Human health	Social welfare
High	Environmental tobacco smoke Inorganics Persistent organochlorines Ozone Particulate matter Radionuclides (natural sources) Radon Volatile organics	Alteration of aquatic habitats Alteration of terrestrial habitats Environmental tobacco smoke Greenhouse gases Lead Ozone Particulate matter Pesticides—agricultural use Pesticides—non-agricultural use Radionuclides Stratospheric ozone depletors Volatile organics
Medium	Carbon monoxide Lead Microbiological contaminants Pesticides—agricultural use Pesticides—non-agricultural use	Asbestos Inorganics Microbiological contaminants Non-native organisms Oil and petroleum products Persistent organochlorines Radon SO_x and NO_x
Low	Alteration of acidity, salinity, or hardness of water Radionuclides (anthropogenic) SO_x and NO_x Total suspended solids, biological oxygen demand or nutrients in water	Alteration of acidity, salinity or hardness of water Carbon monoxide Thermal pollution Total suspended solids, biological oxygen demand or nutrients in water

January 1997 (RPR 1995). The governor must publish a list, updated each year, of chemicals known by the state to cause cancer or reproductive toxicity. Businesses cannot 'knowingly and intentionally' expose any individual to the chemicals, 12 months after they are listed, without first giving 'clear and reasonable warning'. Exceptions are: no warning is required if exposures would result in a risk lower than 'no significant risk', defined as 'one excess case of cancer per 100 000 individuals exposed over a 70-year lifetime' for carcinogens, and as 'less than one-thousandth of the no-observed-effect level' for reproductive toxicants. Other exceptions also apply that are not risk-related (e.g. businesses with fewer than 10 employees), and the law prohibits discharges of chemicals into state drinking water sources 20 months after listing. Proposition 65 now lists approximately 500 chemicals, including some with no commercial use and 75–100 that are drugs whose potential health risks are told to users when they are prescribed. For commercially useful chemicals more than 250 numerical standards have been set, and none of these standards has been challenged in court.

There may be a greater role for risk assessment in the Endangered Species Act (ESA). A recent NAS National Research Council report (NRC 1995) stated—'Despite the major advances that have been made in models for predicting mean extinction times, the existing models still have substantial limitations. Most of the models deal with only one risk factor at a time and fail to incorporate the interactive effects of multiple

Table 10.4 Ecological 'health' stressors (CCRP 1994).

Rank	Stressor
High	Alteration of aquatic and wetland habitats Alteration of terrestrial habitats Inorganics Non-native organisms Ozone SO_x and NO_x
Medium	Alteration of acidity, salinity or hardness of water Greenhouse gases Lead Persistent organochlorines Oil and petroleum products Pesticides—agricultural use Pesticides—non-agricultural use Total suspended solids, biological oxygen demand or nutrients in water
Low	Microbiological contaminants Particulate matter Volatile organics

factors on reducing the time to extinction. This might result in a tendency for such models to underestimate the risk of extinction.'

The NRC offered several recommendations for strengthening the implementation of the ESA which deal with the use of risk assessment (NRC 1995).

1 Extinction models should play a more central role in ESA decisions, especially as guides to research.
2 Extinction models should be used to categorize species according to relative risk of extinction.
3 Levels of risk to trigger ESA decision must be framed as a probability of extinction during a specified time period, and it will be necessary to view extinction over longer time periods (i.e. of the order of hundreds of years).
4 An estimate of the number of individuals needed for long-term survival must be based on the biology of the organism.
5 To the degree that they can be quantified, the levels of risk associated with endangered status should be higher than those for threatened status.
6 The procedures used to make ESA decisions should be explicit and well documented.

The NRC Committee identified two main categories of risk: (i) the risk of extinction; and (ii) the risks associated with unnecessary expenditures in the face of substantial uncertainties about future events. They noted that policy decisions or public input had played a role in setting different levels of risk for different taxonomic groups. Since implementation of the ESA, numerous models have been developed for estimating the risk of extinction that provide valuable insights into potential impacts of management activities and recovery plans. However, despite major advances, most of these models still have substantial limitations, such as not taking into account multiple risks and the interaction of all the factors promoting extinction. They also noted that many ESA listing decisions and recovery plans have not met the guidelines suggested by current scientific thinking (i.e. population viability and risk assessment). Those favouring the use of extinction models and other risk assessment tools for listing and recovery planning state that they are scientifically sound and should lead to better rates of species survival. Their weakness is that extinction models are still relatively crude and have substantial limitations. In addition, such modelling is expensive and time consuming, especially when combined with the field research needed to produce the data needed for model input.

10.3 EXECUTIVE ORDERS AND LEGISLATIVE BACKGROUND

10.3.1 Executive orders

The USEPA has performed cost–benefit and risk analyses in support of its major regulatory initiatives and many other important environmental decisions for many years. The potential usefulness of cost–benefit analyses was discussed at the USEPA even before Presidents Ford and Carter began to encourage this activity in all regulatory agencies. Since February 1981, when President Reagan issued Executive Order 12291, cost–benefit analyses have been required for 'major rules'—i.e. rules likely to result in: 'an annual effect on the economy of $100 million or more'; a 'major increase in costs or prices for consumers, individual industries, federal, state, or local government agencies, or geographic regions'; or 'significant adverse effects on competition, employment, investment, productivity, innovation, or on the ability of United States-based enterprises to compete with foreign-based enterprises in domestic or export markets.' A requirement for risk analysis was not explicit in President Reagan's order but is implied by the need to quantify net benefits of environmental regulations—i.e. the monetary value of risk avoided (e.g. lives saved or environments protected) due to regulation less the cost of implementation and enforcement (Schierow 1993).

Studies have found that, between 1981 and 1990, 44 (3%) of the approximately 1480 regulations finalized by the USEPA qualified as major rules requiring cost–benefit–risk analysis. Between February 1981 and February 1986, the estimated cost of preparing a formal analysis ranged from $210 000 to $2380 000 and averaged $675 000 per analysis. The effect of the analyses on decisions often was not apparent, however. Other factors, such as health risks and potential economic impacts, generally determined the choice (Schierow 1993).

On 30 September 1993, President Clinton issued Executive Order (EO) 12866 (EO 1993). The order repeals EO 12291 and, for the first time, requires agencies to 'consider the degree and nature of the risks posed by various substances or activities within its jurisdiction' and to conduct cost–benefit analysis for all 'significant regulatory actions', including any substantive action expected to lead to promulgation of a 'major rule' (as defined in EO 12291) and other rules that may: adversely affect the economy, a sector of the economy, the environment, public health or safety, or state, local, or tribal governments or communities; create a serious inconsistency with an action taken or planned by another agency; alter the budgetary impact of entitlements, grants, user fees or loan programmes or the rights and obligations of recipients thereof; or raise novel legal or policy issues arising out of legal mandates, the President's priorities, or the principles for regulatory planning and review that are set out in the order and summarized below (EO 1993). Agencies are permitted to determine the significance of regulatory actions with oversight by the Office of Management and Budget (OMB). The order establishes 12 principles of regulatory planning and review. Executive Order 12866 requires that regulations, including environmental and health regulations, be assessed for cost and benefit to the public. It includes assessing the nature of the risk, seeking views of state and local governments, using comparative risk assessment in regulatory decision-making, and evaluating alternative approaches. The OMB Office of Information and Regulatory Affairs (OIRA) shall provide guidance to agencies in early 1996 (Table 10.5).

The White House, in 1996, released benefit–cost analysis guidance to support EO 12866 (RPR 1996). The document addresses three basic components for the economic analysis: statement of need for the proposed action, an examination of alternative approaches, and an analysis of benefits and costs. Economic analysis refers to any systematic procedure to evaluate real or anticipated resource expenditures and losses (costs) relative to real or anticipated gains (benefits). Cost–benefit–risk analysis is the quantification and monetary valuation of the expenditures, gains and losses, and the calculation of net benefits to society associated with the adoption of a particular regulation

Table 10.5 Executive Order 12866: risk-related provisions (OSTP 1995).

In September 1993, President Clinton signed Executive Order 12866 on Regulatory Planning and Review to guide federal agencies in the regulatory process. The Executive Order contains provisions directing agencies to consider risks in making regulatory decisions.

1 'In setting regulatory priorities, each agency shall consider, to the extent reasonable, the degree and nature of the risks posed by various substances or activities within its jurisdiction' [section 1(b)(4)].
2 In developing regulations, federal agencies should consider '... how the action will reduce risks to public health, safety, or the environment, as well as how the magnitude of the risk addressed by the action relates to other risks within the jurisdiction of the agency' [section 4(c)(1)(D)].

The Executive Order also establishes a Regulatory Working Group to 'serve as a forum to assist agencies in identifying and analysing important regulatory issues (including ... the methods, efficacy, and utility of comparative risk assessment in regulatory decision-making ...)' [section 4(d)].

(or alternative management strategy) to address an environmental hazard (Schierow 1994).

Problems with the guidance concern monetizing of non-monetary benefits; for example, values for ecosystems, future use of environmental resources, and discounting future health risks. Uncertainty issues are also controversial in that economists state that costs of a regulation are fairly certain, while benefits and risk calculations have a much larger uncertainty and should weigh less in the decision-making process. Risk reduction is typically a part of the benefit. When benefits are calculated into monetary terms to allow cost–benefit–risk assessment, inaccuracies, value judgements and lack of appropriate methods occur. The various benefit techniques consist of calculating the dollar values of health effects, which include studies of how much people are willing to pay to avoid exposure to a hazard or particular adverse effect, or savings of direct costs, such as health-care expenditures, salary loss for the duration of an illness, or the years of work lost to premature death. The intent is to estimate the gross monetary value of benefits to society and individuals. These numbers are then compared with the costs of the regulation.

The USEPA's Science Advisory Board (USEPA 1990) was critical of methods that assume that the future value of an ecological resource must be less than its present value. In the 1990 report they stated that this policy inevitably leads to depletion of irreplaceable natural resources. Reliance on measures such as the public's 'willingness to pay' exacerbates the problem because although the public may not care about wetlands, for example, they are none the less valuable now and in the future. They concluded that new techniques are needed to assess the real long-term value of ecosystems.

Environmental justice issues advocate that certain subgroups may be burdened with a disproportionate share of environmental risks. In addition, other groups may gain a disproportionate share of the risk reduction, while taxpayers and consumers bear the cost of implementation and compliance. Instead of, or in addition to, weighing a regulation's total costs against total benefits, they want inequities to be described and avoided (Schierow 1995).

10.3.2 Legislative activity

Congress can be separated into four groups on risk assessment (Wentsel 1994). One group wants to use risk assessment, cost–benefit analysis, and a comparison of the risk relative to other risks to which the public is exposed to evaluate all new USEPA regulations. They want to ensure that USEPA regulations will provide benefits to human health and the environment that justify the cost. While this may be a commonsense position, many of the supporters of this approach see it as a way to block environmental progress and get regulatory relief by slowing down the regulatory process by requiring an extensive analysis on even minor regulations. Many policy makers want the results of risk analysis to help identify economically efficient choices among environmental management strategies; in other words, how to get the 'biggest bang for the buck'. Spending for environmental protection should be managed more efficiently because it involves a considerable amount of money — too much to spend wastefully. They generally believe that spending for environmental protection is excessive because too many public and private resources are expended to produce small or uncertain gains in environmental protection and public health. Legislation to address these concerns would require federal agencies to conduct cost–benefit–risk analysis of regulations. The most severe critics in this group assert that environmental regulations have an adverse impact on the national economy and international competitiveness of American businesses. They want agencies to use cost–benefit–risk analysis to identify less expensive strategies to reduce risks (Schierow 1995).

A second group in Congress wants to use risk assessment as a scientific tool that is part of the regulatory decision-making process. They want risk assessment used to help identify problems, set regulatory priorities, compare effectiveness of risk management options, communicate to the public and identify research needs. These policy makers promote risk analysis as an objective scientific basis for environmental planning and decisions by federal agencies, Congress and the public. Risk analysis is needed to support or replace the piecemeal environmental policy that has grown in response to environmental issues over the last 25 years. They favour legislation mandating increased use of risk analysis by the USEPA and reports to Congress and the public to help policy makers evaluate and prioritize problems based on sound science.

A third group holds a position similar to many environmental groups that risk assessment causes delay, excludes the public from the decision-making process, implies that risk is acceptable, and obscures the role of judgement and values. They stress the limitations of risk analysis. Activists for environmental justice (i.e. avoidance of disproportionate risks to low-income and minority communities) oppose efforts to increase the influence of risk analysis on environmental decisions, because it tends to focus attention on relatively small risks to large populations, instead of multiple or large risks to smaller groups, such as workers, the economically disadvantaged or ethnic minorities (Schierow 1995).

The fourth group is largely uninformed.

Congressional proposals to develop and use risk analysis methods to evaluate regulatory proposals and to inform legislators were made as early as 1979. Their main thrust was to provide regulatory relief. In 1979, Congressman Ritter (Republican—Pennsylvania) introduced a bill to improve and promote the use of risk comparisons for regulatory alternatives. Following hearings in May 1980, Mr Ritter introduced a more comprehensive proposal, the 'Risk Analysis Research and Demonstration Act', to improve, coordinate and facilitate the use of risk analysis by federal agencies. On introducing the legislation, Mr Ritter argued '... that assessment and comparison risk are essential before regulations are promulgated to reduce those risks'. Mr Ritter introduced similar legislation each year throughout the 1980s. In 1982, Representative Ritter's bill (House Report 6159) was reported from the Committee on Science and Technology (House Report 97–625), passed the House, amended, and was referred to the Senate. Senator Schmitt (Republican—New Mexico) introduced companion legislation (Senate Bill 3006) which was reported by the Committee on Commerce, Science and Transportation with amendment but without a written report. The measure was not considered on the floor. Representative Jim Martin (Republican—North Carolina) proposed a bill to establish a Central Board of Scientific Risk Assessment to improve the scientific review and evaluation of risk assessments made by federal agencies. A subcommittee of the Committee on Science and Technology forwarded a 'clean bill' (House Report 4192) in lieu to the full committee, but there was no further action. Mr Ritter's bill of 1983 also was considered but rejected in favour of House Report 4192 by the subcommittee. Beginning with the 101st Congress, Mr Ritter promoted his ideas about the need for risk comparisons and analyses within the framework of legislation to reauthorize research at the USEPA.

The USEPA reports *Unfinished Business* and *Reducing Risk* influenced Senator Moynihan to become an active advocate of risk analysis during the 102nd Congress. His involvement initiated an intellectual evaluation of risk analysis as a decision-making tool. His proposal attempts to ensure that information about risks will be used to identify relatively more cost-effective approaches to environmental protection.

10.3.3 103rd Congress

Many legislative proposals in the 103rd Congress employed the concept of environmental risk, and several would require that risks, costs and relative risks be analysed by the USEPA. The Environmental Risk Reduction Act (Senate Bill 110), put forward by Senator Moynihan, directed the USEPA to use its resources to reduce risks through risk assessment and ranking of risks and options for management of risks. The Administration is directed to use the input from two technical committes on relative risks and on environmental benefits to support the ranking process. New research on environmental economics and risk analysis is also proposed. The Johnston amendment to the Department of the Environment Act (Senate Bill 171) required publication by the USEPA of an estimate of the risk addressed by each new regulation, the cost to implement and comply with the new regulation, and a comparative analysis of the risk relative to other risks to which the public is exposed. It also required the USEPA to certify that the analysis is supported by the best available scientific data. The regulation will 'substantially advance the purpose of protecting the human health and safety and the environment', and the regulation will produce benefits to health or the environment that justify the cost (Table 10.6). In a floor debate, Senator Johnston stated that 'the reason for this amendment ... is because we have seen instance after instance where unreasonable regulations have been adopted costing the tax-payer billions of dollars where this kind of analysis would have avoided that'. The Senate passed his amendment by 95 votes to 3. The thinking behind the amendment is to provide regulatory relief and to use risk management as a tool to produce practical, logical, and efficient regulations; using the best science available. This proposed amendment would apply to all final regulations and to preliminary regulatory actions such as Advanced Notices of Proposed Rule Making. The thrust of Senate Bill 110 complements the main points in the Johnston amendment. The ranking of risk is so that the most serious environmental issues are addressed first. The Bill also supports economic evaluation through cost–benefit analysis. The proper management of risk is the concern of both pieces of legislation. However, they do not directly address societal concerns, public perception, risk communication to the public, a framework for risk management, uncertainty in economic cost–benefit estimates, or the idea that risk analysis is a dynamic process and an imperfect science.

The impact of the Johnston amendment on environmental legislation in the 103rd Congress cannot be understated. A final vote in the House on the Department of the Environment Act was not taken because a majority of the House members wanted this amendment attached to the Act and the House leadership was opposed to the amendment. It impacted and delayed revisions of the Safe Drinking Water Act (SDWA) and Superfund legislation. When compromise language on the Johnston amendment was agreed to and passed by the Senate in the SDWA, the House would not compromise on their position. House leadership stated that risk–cost benefit language should not be attached to environmental bills.

Table 10.6 Main points of Johnston amendment.

Provide an estimate of the risk addressed by each new regulation
Determine the cost to implement and comply with the new regulation
Conduct a comparative analysis of the risk relative to other risks to which the public is exposed
Certify that the analysis is supported by the best available scientific data
Determine that the regulation will substantially advance the protection of human health and the environment
Determine if the regulation will produce benefits to health or the environment that justify the costs

The administration initially opposed the amendment. They viewed it as too costly to implement as well as causing delay in the implementation of regulations because the amendment applies to all regulations. They stated that EO 12866 addressed the issues of the amendment, except possibly comparative risk assessment. However, they eventually did not oppose the compromise language attached to the SDWA.

The Democrat House leadership opposed the amendment because it was seen as providing too much regulatory relief and delay. The refusal of the Democrats to compromise on the amendment led to delay and eventually the killing of all major environmental legislation. However, the Democrats are not solely to blame; Republican leaders determining, in August 1994, that they might gain control of the Senate, blocked the passage of any major environmental legislation. They did not want to give President Clinton any legislative victories before the election.

However, risk language was attached to the Federal Crop Insurance Reform and Department of Agriculture Reorganization Act of 1994 and passed by both Houses. An Office of Risk Assessment and Cost–Benefit Analysis was created with a role to assess risks to human health and the environment and prepare cost–benefit analyses for major regulations with an impact on the economy of $100 million or more.

10.3.4 104th Congress

Schierow (1995) presented an excellent review of progress of risk legislation half-way through the 104th Congress. At the start of the 104th Congress, legislative activity concerning risk analysis and cost–benefit analysis was linked to proposals for regulatory reform. Key regulatory reform issues are presented in Table 10.7. The Job Creation and Wage Enhancement Act of 1995 (House Report 9), which was introduced in association with the House Republican party 'Contract with America', passed the House, amended 3 March 1995. In the Senate, three Committees passed similar legislation on regulatory reform in the spring of 1995–Governmental Affairs, Senate Bill 291; Judiciary, Senate Bill 343; and Energy and Natural Resources, Senate Bill 333. The bill supported by Senator Dole, Senate Bill 343, and passed by the Judiciary Committee, was brought to the Senate floor. In July, the bill was debated and amended on the Senate floor; three votes to closet debate could not get the required 60 votes to enable the bill to be voted on, so it was not considered further. In the Senate, the SDWA was passed with compromise risk language in the setting of drinking water standards. Opponents of risk language in environmental legislation argue that risk analysis is not a pure science and not entirely objective. They claim that the science used in risk analysis is immature and is only reliable for assessing immediate threats or the risk of developing cancer. They maintain that for most chemicals, health effects and ecological effects, data do not exist, and without data, risk analysis is unreliable at best. Even when data are available, they argue that human data usually are from studies of healthy, adult, white males with occupational exposures.

In 1996, Superfund and SDWA legislation were being considered by House Committees. Regulatory Reform legislation compromise language is being circulated for comment by moderate members of both parties. Comparative risk assessment (CRA) legislation is being reviewed by various Senate and Congressional staff. This legislation calls for the development of CRA so that federal agencies can use it to prioritize environmental issues and potentially their budgets. The draft legislation calls for studies on CRA methods and their application to health and environmental concerns. However, comparative risk assessment is not sufficiently developed to be used by agencies to prioritize their budgets. Agency heads are asked to prioritize risks to maximize risk reduction for the resources expended. While statutory constraints are recognized, the degree of impact on any CRA process is not addressed. Comparative risk assessment is somewhat technically defensible within media, i.e. rankings within drinking-water contaminants, air pollutants or hazardous waste sites. However, procedures are not available for mixed media, i.e. comparing human health issues to ecological issues, and death versus injury or illness.

Table 10.7 Key issues in regulatory reform legislation.

Prescriptive principles of risk assessment and risk characterization
Comparative risk assessment
Peer review
Scope—definition of major rule, which agencies covered
Judicial review
Decision criteria
Cost–benefit

10.4 APPLICATION OF RISK ASSESSMENT

10.4.1 Public policy and risk management

When establishing public policy in environmental protection the decision maker must consider the various risk management parameters. These include: cost–benefit, engineering, societal issues, political issues and risk assessment. The goal of risk management is to reduce the risk or hazard of an environmental problem to humans and the environment. Risk assessment estimates whether there is a risk and, if so, the uncertainty bounds. The product of a risk assessment is judged, by the risk manager, as to how serious the risk is, availability of mitigation measures, cost to mitigate and other management concerns. Examples of the management concerns are: Who or what is impacted? Are there technologies available to mitigate? Does the cost reduce the risk significantly? Does the public care about this specific issue? Are there political concerns?

A White House Office of Science and Technology Policy (OSTP) paper listed several pertinent questions (OSTP 1995).

1 How should a risk assessment be used and presented? How can risk assessment best be used to inform environmental policy and management decisions?
2 As a means to ensure accountability and transparency, what is the most effective means to characterize and communicate information about risks, uncertainty in assessments and limits to assessments?
3 How might uncertainty be weighed into policy decisions?

10.4.2 Risk assessment as a tool

Risk assessment is appropriate as an analytical tool to help identify problems, set regulatory priorities, compare effectiveness of risk management options, communicate to the public and identify research needs. Since the purpose of environmental regulations is to protect human health and/or the environment, risk assessment will estimate needed protection levels either quantitatively or qualitatively. Risk assessment is often involved in the generation of health or environmental criteria used in the regulations. Typically, risk assessment alone will not provide a hard and fast number for regulation. Other questions asked in the OSTP report were as follows.
1 When should a risk assessment be undertaken? When will it enhance policy decisions?
2 What is the appropriate level of effort and precision to dedicate to risk assessment?
3 What is the estimated incremental value of obtaining additional information through increased research, versus the incremental cost of delaying a decision?
4 How should a risk assessment be conducted? How should risk assessments be framed and bounded to inform policy decisions appropriately and most effectively?
5 How should environmental justice and other social, cultural and ecological concerns be integrated into framing a risk assessment and defining relevant data needs?

10.4.3 Implementation in environmental statutes

In the implementation of environmental statutes by the USEPA and other agencies, there are variations in how risk is addressed in the regulations. There is also variation between USEPA Offices in the parameters incorporated into risk assessment. Both of these variables affect the determination of the appropriate use of risk assessment in environmental statutes. A uniform approach would enhance the usefulness of risk assessment as a regulatory tool. The previous USEPA administration formed a Risk Assessment Council and initiated action toward a unified approach to risk assessment (Habicht 1992).

The environmental statutes are written as either narrative or numerical directives. The USEPA implements narrative directives primarily through three different approaches: health-based standards, technology-based standard and no unreasonable (balanced) risk standards (Table 10.8). The primary environmental laws use one or more of these approaches (Rosenthal *et al.* 1992).

Health-based

Health-based standards regulate on the protection of human health or the environment without regard to technology or cost factors. The primary health-based standard is the Delaney Clause. The Delaney Clause is a zero risk standard and prohibits the approval of any food additive that has been found to induce cancer. The FDA has made few rule-making decisions that might trigger Delany's prohibitions and largely avoided the strictness of the Delaney Clause covering food additives. However, the USEPA, which must consider the carcinogenicity of pesticide residues on foods when making registration and reregistration decisions about pesticide products, has had to address the clause. While the USEPA argued that the cancer risks posed by these pesticides were negligible, a federal appeals court ruled in July 1992 that the Delaney Clause prohibits food additives that are carcinogenic, regardless of whether or not the magnitude of the cancer risk is negligible. It has only been used twice by the EPA to refuse a new food use of a carcinogenic pesticide. The Endangered Species Act currently requires the protection of listed biota without regard to technology or cost factors. However, this Act is not a zero risk standard.

Table 10.8 Example of risk-related provisions in selected Federal Laws (OSTP 1995).

Statute	Regulatory authority	Effects of concern	Approach to risk
Consumer Product Safety Act	Consumer products	'An unreasonable risk of injury'	Balance risks against product utility, cost and availability
Occupational Safety and Health Act	Risks in the workplace	'Material impairment of health or functional capacity'; what is reasonably necessary or appropriate to provide safe and healthful employment?	Attain highest degree of health and safety protection; best available evidence; technical and economic feasibility
Federal Water Pollution Control Act (Clean Water Act)	Waste water discharges	'The toxicity of the pollutant, its persistence, degradability ...'	Best available technology that is economically achievable
Clean Air Act			
Section 109	National ambient air quality standards	Protect public health	Set standards to provide ample margin of safety
Section 112	Emissions standards for hazardous air pollutants	Adverse effects to health and the environment	Reduce emissions using maximum achievable control technology, and later address 'residual risk'
Section 202	Emissions standards for new motor vehicles	Unreasonable risk to health, welfare or safety	Greatest degree of emission reduction achievable through technology available, taking into consideration cost, energy and safety factors
Federal Insecticide, Fungicide and Rodenticide Act Section 3	Pesticides	Unreasonable risks to health and the environment	Balance risks against economic benefits to pesticide users and society
Toxic Substances Control Act Section 6	Existing chemicals in commerce	Unreasonable risks to health and the environment	Balance risks against economic benefits, considering alternative technologies
Comprehensive Environmental Response, Compensation and Liability Act			
Section 313	Toxic release inventory	Hazards to human health or the environment	Reporting is based largely on hazard and quantity used
Section 9621	Hazardous waste site remediation	Persistence, toxicity, mobility and propensity to bioaccumulate, short- and long-term health effects	Protect human health and the environment in cost-effective manner
Safe Drinking Water Act Section 300g-1(b)	Drinking water	Known or anticipated adverse effects on human health	Set a goal (maximum contaminant level goal, MCLG) with an adequate margin of safety, and define a maximum contaminant level (MCL) as close as feasible to the goal

Section 112 under the Clean Air Act (CAA) requires values that protect the public health allowing an ample margin of safety. This is defined, by the USEPA, as an increase in risk by no more than one in a million. Initially this Act was designed to regulate as a health-based standard. However, between 1970 and 1990 extensive litigation occurred over the role of risk assessment under section 112. The wording, 'ample margin of safety', was found to be too vague and it is a good example of where risk assessment cannot be used to establish a single number for regulation. In 20 years, only seven air toxics were listed and regulated. Congress turned to technology-based standards, in 1990, in the hope that the USEPA would be able to reduce the delay in standard setting. The technology-based standards are the initial phase of risk reduction. If, after 8 years, the residual risk to the maximum exposed individual (MEI) is greater than one in a million, then the USEPA must apply health-based standards. During the Congressional debate there was much dissension over risk assessment. This is reflected in the directive to NAS to evaluate USEPA methods. Also, Congress created the Commission on Risk Assessment and Management to study the proper use of risk assessment and propose legislation.

Technology-based

Technology-based standards require best practicable control technology, best available technology and other controls for pollution reduction or treatment. Parameters evaluated include the determination of the effectiveness of methodologies on concentration reduction and often costs, rather than reduction of risk. The SDWA sets drinking-water standards using the best technology, treatment technique and other methods. When maximum contaminant level goals (MCLG) are established, risk assessment provides the scientific basis for the value. The enforceable standards are called maximum contaminant levels (MCL). They are set as close to the MCLG as is 'feasible with the use of the best technology, treatment techniques, and other means ... which are available (taking cost into consideration)'. The enforceable standards are set using technology criteria such as the affordability analysis of analytical technology for detecting contaminants (Rosenthal *et al.* 1992). While MCLG have limited impact under SDWA, they are used in the Comprehensive Environmental Response, Compensation and Liability Act (CERCLA) to establish clean-up levels. The radon issue in the proposed SDWA reauthorization is addressed with a unique multimedia risk reduction approach. Radon exposure from drinking water does not have to be controlled if the locality can regulate the reduction of radon in indoor air. The Clean Water Act requires technology-based controls to treat water pollution. Two sets of water quality criteria are used to protect aquatic life and to protect human health. These water quality criteria are based on risk assessment input. More recently, criteria to protect aquatic life have been set based on effluent bioassays or instream criteria. The criteria are the goal that the technology-based standards seek to achieve. The use of risk assessment to support technology-based standards is an appropriate use of risk assessment.

Balanced risk-based

The use of 'no unreasonable risk' in several statutes requires the balancing of risk assessment results against cost–benefit to determine the risk management approach. Risk assessment is used under the Toxic Substances Control Act (TSCA) and it is balanced against the benefits of the substance, availability of substitutes and the economic cost of the action to limit or ban. The TSCA uses a balanced cost–benefit approach to limit unreasonable risk of injury to health and the environment. In the Fifth Circuit's ruling on the asbestos ban, '... the court did not forbid the agency from controlling toxic chemicals. Rather it ruled that the USEPA had to follow every step in the analysis of risks and costs required under the law's Section 6—which authorizes permanent controls up to bans—before issuing a rule' (Rosenberg & Wheeler 1993). The court refused to let the USEPA ban asbestos, under the Act, because the USEPA did not show that the ban on asbestos represented the least burdensome process for reducing risk. The data on risk reduction, in terms of cancer cases avoided, were not significant when compared with the costs. The court required the USEPA to use the balanced risk management that the Act required. Rosenberg & Wheeler (1993) stated that the TSCA is effective because it requires balancing, it has the voluntary cooperation by Premanufacture Notice submitters, and it has the unseen benefits of the chemicals that were never produced because of the scrutiny of the TSCA.

However, few regulatory decisions have been made and the TSCA is regarded by many as ineffective legislation. This is due to several factors. The term 'unreasonable risk' has an unclear meaning and is open to wide interpretation. The fear of judicial oversight has lead to inaction. There are few instances where the TSCA has primary authority. The discretionary nature of the narrative standard had made clear implementation difficult. Also, the focus on individual chemicals combined with the expensive and time-consuming regulatory process leads to a limited number of regulatory decisions (Shapiro 1990; Rosenthal *et al.* 1992).

The Federal Insecticide, Fungicide and Rodenticide Act (FIFRA) also balances the health and environmental impact of a chemical versus economic cost–benefit and societal concerns. This statute, instead of setting a level of acceptable risk, provides the USEPA with flexibility to make determinations on a case-by-case basis using a balance of diverse factors. However, FIFRA has been criticized for being a slow process, with industry using the risk assessment process as a delaying technique.

Under CERCLA, sites are initially evaluated for their hazard to humans or the environment by the hazard ranking system (HRS). This system uses a health assessment along with exposure and persistence data on the chemicals to rank sites for inclusion as a Superfund site. Those HRS values above 28.5 require the site to be listed on the National Priority List (NPL). At an NPL site, risk assessment is used in two ways. First, a baseline risk assessment of health and ecological concerns is conducted to establish whether the risk justifies mitigation. Second, in the remedial investigation/feasibility study (RI/FS), risk assessment is used to establish risk levels for different areas of the superfund site. In CERCLA, health and ecological risk are relatively equal in importance. Applicable, relevant and appropriate requirements (ARAR) are standards from other federal or state environmental statutes. These are used, when available, to set clean-up criteria. The CERCLA also stresses a preference toward permanence in remediation decisions. Risk assessment can identify areas of the site that have elevated risk. These risks can be related to chemical levels for remediation. An example of where risk assessment is used would be to support the Endangered Species Act. The risk assessment could establish the chemical concentration (with uncertainty bounds) to protect a given organism. The results from risk assessment are used with the other risk management tools to determine the mitigation of a site. Risk assessment is also used to evaluate alternative remedial actions.

Section 307 of the Clean Water Act requires the USEPA to set effluent limitations for discharges of toxic pollutants to surface waters achievable by applying the best available technology that is economically achievable and 'will result in reasonable further progress toward the national goal of eliminating the discharge of all pollutants' [paragraph 301(b)(2)(A)]. In addition, the Act requires effluent standards to provide an ample margin of safety, taking into account 'the toxicity of the pollutant, its persistence, degradability, the usual or potential presence of the affected organisms in any waters, the importance of the affected organisms and the nature and extent of the effect of the toxic pollutant on such organisms, and the extent to which effective control is being or may be achieved under other regulatory'. The Act does not instruct the agency in how it should balance these considerations relative to one another.

In summary, each environmental statute approaches the problem of controlling risk from a different vantage point and authorizes consideration of different factors by the USEPA. Some statutes authorize several different approaches for controlling different kind of risk. One statute, CERCLA, incorporates all of the other statutory approaches to risk, at least in effect. These diverse statutes, however, seem to conform to a few general rules: they generally allow consideration of the costs of regulation at some stage of risk management, either explicitly or by reference to feasible, practical or available technology (the Delaney Clause is an exception to this rule); they tend to exclude costs from consideration in the development of scientific documents (e.g. water quality criteria), safety goals (e.g. safe drinking-water goals), or health-based standards of ambient environmental quality (e.g. primary air quality standards), all of which clearly are meant to be protective of health and the environment; and they require consideration of costs when the USEPA directly regulates commerce, i.e. the production, distribution and use of commercial products.

10.4.4 Importance of flexibility in risk management

Flexibility in the management of environmental risk reduction is an important consideration in the establishment of public policy. Portney (1990) discussed how decisions on risk reduction are eventually made on the costs of protection versus the added benefits. At some point, society will decide that the reduction in risk is too small to justify the cost. The health-based approach is inflexible to make these trade-offs. Technology-based standards may not be flexible enough to recognize that densely populated or pristine areas may need special regulation. These trade-offs may be difficult to do under technology-based standards. The balanced risk-based approach has the potential to make these trade-offs for effective management of risk reduction. The trade-offs of the multimedia approach and pollution prevention will work best under a balanced-based regulatory framework. Ruckelshaus (1985) also supported flexibility in environmental regulations. He called for agencies to be given greater flexibility to address environmental problems at the local level and to confront economic and social trade-offs required for effective risk management. A well-known example where Ruckelshaus had the affected public decide the issue was in Tacoma, Washington. Arsenic was released into the air from a copper smelter. After controls were in place, it appeared impossible to eliminate the carcinogenic risk from arsenic. To eliminate the risk would mean shutting down the plant. Ruckelshaus turned the decision over to the Tacoma citizens. The public considered the risk management trade-offs and came up with ideas to minimize the risk. Their rational approach to risk and the use of trade-offs enabled the plant to stay open.

Hoskins *et al.* (1994) calculated the occupational safety and health risks associated with hazardous waste site remediation. The two occupations with the highest death rates, truck driver and labourer, contributed most to total exposure hours for the alternatives considered. The calculated expected number of fatalities was converted, using the Poisson distribution, to the risk of experiencing at least one fatality, as follows: 0.149 for excavation and landfill, 0.012 for capping, and 0.014 for capping plus slurry wall. They stressed the need for a more scientific, quantitative approach to remediation decisions involving risks to workers.

Breyer (1993) (recently appointed to the US Supreme Court) describes a 'vicious circle' that inhibits proper prioritization of environmental risks. The circle is caused by the public's irrational reaction to risk, Congress' attempt to regulate risk through detailed legislation, and uncertain and irrational methods in technical regulation. He proposes to address the problem through restructuring risk regulation in the executive branch agencies. Through his various experiences, Breyer has observed three main problems in the regulation of risk: tunnel vision, random agenda selection and inconsistency. An example of tunnel vision ('getting the last ten per cent') is where regulation of a single medium requires strict treatment of an environmental problem. The strict treatment imposes high cost without significant additional environmental benefits. The random agenda problem refers to the lack of prioritization of environmental problems. This can be minimized through the ranking of risks. The inconsistency problem refers to agencies and offices using different methods to estimate risk and cost–benefit. This results in varied and confusing information on the impact of an environmental problem.

10.4.5 Canada

Lind (1995) reported on the development and use of policy goals for health and safety in Canada. In 1990, the Royal Society of Canada and the Canadian Academy of Engineering formed a joint Committee on Health and Safety. The committee proposed three guiding principles for economic scrutiny of health management and safety regulation. The first was that accountability, which was defined as decisions for the public, must be open and apply across the complete range of hazards. Balancing all important benefits, hazards and costs must be demonstrated and made available openly for comparison with other potential or realized alternative uses of the resources. Next was net benefit, which states that risks shall be managed to maximize the total expected net benefit to society. The net benefit of an option is the excess of the total of all benefits over the total of all costs and hazards. Since benefits and detriments are usually different in kind and accrue to different degrees to individuals, it is necessary to arrive at a common basis of measurement. The last principle was the measurement value. The committee recommended the safety benefit to be promoted as quality-adjusted life-expectancy (QALE). The committee proposed to maximize the net benefit to all members of society at all ages, and QALE aggregates net benefit over all individuals, health states and ages in a society or group. They discussed and acknowledged that the net benefit of any particular intervention is not evenly distributed. Lind (1995) suggested that two other issues be considered: procedure (processes for public health and safety must be quantified and follow accepted protocol) and sovereignty (management of risk to the public should be subject to public consent).

The Government of Canada is introducing a Toxic Substances Management Policy (TSMP), which is based on science (Environment Canada 1995; see Chapter 2). The policy calls for virtual elimination from the environment of toxic substances that result from human activity, which are persistent and bioaccumulative. The policy also calls for cradle-to-grave management of all other substances of concern that are released to the environment. It establishes a preventive and precautionary approach to deal with substances that enter the environment that could cause harm to human health or the environment.

The key objectives are as follows.

1 Virtual elimination from the environment of toxic substances that result predominantly from human activity and are persistent and bioaccumulative (i.e. Track 1 substances).

2 The management of other toxic substances and substances of concern, throughout their entire lifecycles, to prevent or minimize their release into the environment (i.e. Track 2 substances).

A substance that meets all four criteria outlined in Table 10.9, i.e. persistent, bioaccumulative, toxic and primarily the result of human activity, will be targeted for virtual elimination from the environment (Track 1 substance). This objective will be achieved by addressing sources of release to the environment or by removing or managing the substance if it is already in the environment. Pollution prevention strategies will be used to prevent the measurable release of a Track 1 substance from domestic sources. A Track 1 substance that cannot be managed successfully throughout its lifecycle will be targeted for phase-out of generation and uses. Through bilateral or multilateral agreements, the federal government will work to eliminate Track 1 substances that originate from sources outside the country. Remediation may be undertaken when a Track 1 substance is already in the environment. Where

Table 10.9 Criteria for the selection of substances for Track 1 for Canadian toxic substances (Environment Canada 1995).

Persistence				
Medium	Half-life	Bioaccumulation*	Toxicity	Predominantly anthropogenic
Air Water Sediment Soil	≥2 days ≥182 days ≥365 days ≥182 days	BAF ≥ 5000 or BCF ≥ 5000 or $\log K_{ow} \geq 5.0$	CEPA-toxic or CEPA-toxic equivalent	Concentration in environment largely resulting from human activity

* BAF, bioaccumulation factor; BCF, bioconcentration factor; $\log K_{ow}$, octanol/water coefficient.

the benefits to the ecosystem or to human health of removing the substance outweigh clean-up costs, remediation will be considered. Otherwise, management strategies will focus on minimizing exposure and the site's potential risks.

The federal government will identify toxic substances and other substances of concern subject to management under Track 2 through a variety of existing programmes. Substances that do not satisfy all four criteria are candidates for lifecycle management to prevent or minimize their release into the environment. Management strategies, including pollution prevention, pollution control and remediation, will be based on a lifecycle approach. Risk assessment and risk management approaches will be used to identify Track 2 substances and management options. Socio-economic factors will be considered when determining long-term environmental goals, targets, strategies and time lines for Track 2 substances.

Criteria to determine Track 1 or 2 have been documented. A substance is considered toxic if, after rigorous scientific assessment and based on decisions taken under federal programmes, it either conforms or is equivalent to 'toxic' as defined in Section 11 of the Canadian Environmental Protection Act (CEPA). The document entitled 'Toxic Substances Management Policy—Persistence and Bioaccumulation Criteria' provides details about these criteria, including their numerical values, the process and rationale used in establishing them, and information about how they are applied. Many Track 1 and Track 2 substances are already subject to federal statutory management strategies that are consistent with this policy—the Canadian Environmental Protection Act, the Pest Control Products Act, and the Food and Drugs Act. No new action will be required for a substance that is adequately managed under existing programmes.

10.5 SUMMARY

In summary, risk assessment is used in three basic ways by environmental regulations: health-based, technology-based and a balanced risk-based approach. In health-based standards, where risk is used without regard to cost or technology considerations, risk assessment alone has been found to be too restrictive, inflexible and has resulted in reduced enforcement. The use of risk assessment in technology-based standards is primarily to set a treatment goal. Technology treatments are enacted to approach the goal with the best or most appropriate technology available. This use of risk assessment has worked well. Standards can be implemented and enforced. However, technology-based standards may not be flexible enough to consider multimedia approaches. A balanced risk-based approach, where risk assessment is used along with economic, technological and societal concerns, provides an encompassing risk management approach to environmental problems. The balanced approach may take more time for implementation. The TSCA, FIFRA and CERCLA each have had problems with the implementation of this approach. Enforcement will not be as clear. There are case-to-case variations depending on the regulator or site-specific information. However, the inclusion of the public in the decision-making process is important. The flexibility of the risk management can provide an avenue for novel mitigation of environmental issues.

10.6 APPENDIX (CCRP 1994)

10.6.1 Human health caveats

When reviewing the risk-rankings of the Human Health Committee, the following caveats must be considered.

1 The human health risks estimated by the Human Health Committee are only a portion of the total risk, because comprehensive quantitative data are not available for all the compounds released into the environment.

2 Risk assessments are presented as numerical results. This gives an appearance of accuracy which can be misleading. Due to methodological limitations (e.g. the quantity and quality of data vary considerably between topic areas), results should be interpreted as order of magnitude indications of potential health impacts, not actual predictions of disease incidence.

3 Risk-ranking results are never determined by quantitative analyses alone. Selecting the data used, adopting risk assessment methods and extrapolating from analysed risks involves making major assumptions based on scientific judgement. The results of the Human Health Committee reflect the expertise and values of the scientists participating in the analysis. No single risk-ranking is based only on scientific data.

4 The technical approach of the Human Health Committee was not designed to evaluate emerging environmental problems. The focus on current risks, for example, cannot be used to identify problems that could be prevented by making proactive management decisions.

10.6.2 Ecological health caveats

When referring to the risk-rankings of the Ecological Health Committee, the following caveats must be considered.

1 The ranking of aggregate threats by the Ecological Health Committee incorporated evaluation of available data together with decisions based on scientific judgement. The aggregate threats and the rankings assigned were influenced by the experience and knowledge of Ecological Health Committee members.

2 The Ecological Health Committee believes that the aggregate threats, as presented, reflect the major potential threats to California ecological systems and that there is a major difference in the magnitude of the threat between high, medium and low groups.

3 The Ecological Health Committee was generally dissatisfied with the results of the translation from their 'Aggregate Threat List' to the CCRP's general environmental topic lists, although the translations would need substantial revision to conform with Ecological Health Committee members' perceptions of threats to California ecological systems.

4 Members of the Ecological Health Committee maintain that an aggregate threat-ranking provides the best means to evaluate risks to ecological health in California.

10.6.3 Social welfare caveats

When referring to the risk-rankings of the Social Welfare Committee, the following caveats must be considered.

1 The nature of social welfare impacts precludes a systematic weighing and comparison of topic areas. However, in order to 'rank', artificial separations and groupings had to be made in terms of the topic areas and lists, the impacts on health, ecology, and social welfare, and the aspects of social welfare impacts themselves (in the form of criteria and measures).

Currently, there is no effective way to validate whether such an approach can adequately capture the social welfare impacts present in the complex California system.

2 No well-established methodology or conceptual framework exists for assessing the social welfare impacts of environmental problems. Nor are there any systematic measures or databases available to use in these assessments.

3 The Social Welfare Committee was a group of diverse professionals, but did not comprise a representative cross-section of the State's population.

4 The rankings do not include a consideration of social welfare benefits, mitigation or regulation costs.

5 The Social Welfare Committee was constrained by insufficient time, data and resources.

6 Linking environmental problems to traditional social welfare impacts like economics and aesthetics is often tenuous, as any activity can have both beneficial and detrimental effects. Linking environmental problems to less-traditional social welfare impacts like anxiety or community fragmentation can be even more difficult, as people's definitions of what is pleasing or desirable depend greatly on their background, circumstances and personal taste.

10.7 REFERENCES

Breyer, S. (1993) *Breaking the Vicious Circle Toward Effective Risk Regulation*. Harvard University Press, Boston, MA.

CCRP (1994) *Toward the 21st Century: Planning for the Protection of California's Environment*. California Comparative Risk Project, Sacramento, CA.

CENR (1995) *Draft Research Strategy and Implementation Plan: Risk Assessment Research in the Federal Government*. Committee on Environmental and Natural Resources, National Science and Technology Council, Washington, DC.

Environment Canada (1995) *Toxic Substance Management*

Policy. Message from Minister of the Environment, June, 1995. Environment Canada, Ottawa.

Executive Order 12866 of September 30 (1993) Federal Register. 58(190) 51735–51744. Monday, October 4, 1993.

Gaudet, C. (1994) *A framework for ecological risk assessment at contaminated sites in Canada: review and recommendations*. Environment Canada, Ottawa, Canada.

Habicht, F. (1992) *Guidance on Risk Characterization for risk managers*. US Environmental Protection Agency Memorandum, 26 February 1992, Washington, DC.

Hoskins, A.F., Leigh, J.P. & Planek, T.W. (1994) Estimated risk of occupational fatalities hazardous waste site remediation. *Risk Analysis*, **14**(6), 1011–1017.

Lind, N.C. (1995) Policy goals for health and safety. *Risk Analysis*, **15**(6), 639–644.

NRC (1983) *Risk Assessment in the Federal Government: Managing the Process*. National Research Council, National Academy Press, Washington, DC.

NRC (1993) *Issues in Risk Assessment*. National Research Council, Committee on Risk Assessment Methodology, National Academy Press, Washington, DC.

NRC (1994) *Science and Judgment in Risk Assessment*. National Research Council, Committee on Risk Assessment of Hazardous Pollutants, National Academy Press, Washington, DC.

NRC (1995) *Science and the Endangered Species Act*. National Research Council, National Academy Press, Washington, DC.

OSTP (1995) *Science, Risk, and Public Policy*. Office of Science and Technology Policy, Executive Office of the President, Washington, DC.

Portney, P. (1990) The evolution of federal regulation. In: *Public Policies for Environmental Protection* (ed. P. Portney), pp. 7–26. Resources for the Future, Washington, DC.

RPR (1995) *Lessons for the national risk debate?* Risk Policy Report, California's Prop 65, 20 January 1995, 40–41, Washington, DC.

RPR (1996) *White House Issues Benefit–Cost Guidance, Ending Months of Debate*. Risk Policy Report, 31 January 1996, Special Report, Washington, DC.

Rosenberg, E. & Wheeler, J. (1993) Unreasonably at risk. *The Environmental Forum*, **10**, 18–22.

Rosenthal, A., Gray, G. & Graham, J. (1992) Legislating acceptable cancer risk from exposure to toxic chemicals. *Ecology Law Quarterly*, **19**, 269–362.

Ruckelshaus, W. (1985) Risk, science, and democracy. *Issues in Science and Technology*, **1**, 19–38.

Schierow, L. (1993) *Environmental Risk and Public Policy, Part I: Cost–Benefit–Risk Analysis of Environmental Regulations*. Congressional Research Service, The Library of Congress, Washington, DC.

Schierow, L. (1994) *Risk Analysis and Cost–Benefit–Risk Analysis of Environmental Regulations*. 94–961 ENR, Congressional Research Service, The Library of Congress, Washington, DC.

Schierow, L. (1995) *The Role of Risk Analysis and Risk Management in Environmental Protection*. IB94036. Congressional Research Service, The Library of Congress, Washington, DC.

Shapiro, M. (1990) Toxic substances policy. In: *Public Policies for Environmental Protection* (ed. P. Portney), pp. 195–242. Resources for the Future, Washington, DC.

Slovic, P., Malmfors, T., Krewski, D., Mertz, C.K., Neil, N. & Bartlett, S. (1995) Intuitive Toxicology. II. Expert and Lay Judgments of Chemical Risks in Canada. *Risk Analysis*, **15**(6), 661–675.

Suter, G.W., II (1993) *Ecological Risk Assessment*, 538 pp. Lewis Publishers, Chelsea, MI.

USDOE (1994) *Incorporating Ecological Risk Assessment into Remedial Investigation/Feasibility Study Work Plans*. US Department of Energy, DOE/EH-0391, Office of Environmental Guidance, Washington, DC.

USEPA (1987) *Unfinished business: a comparative assessment of environmental problems*. US Environmental Protection Agency, Office of Policy Analysis, Washington, DC.

USEPA (1990) *Reducing risk: setting priorities and strategies for environmental protection*. US Environmental Protection Agency. Science Advisory Board, SAB-EC-90-021, Washington, DC.

USEPA (1992) *Framework for ecological risk assessment*. US Environmental Protection Agency, EPA/630/R-92/001, Risk Assessment Forum, Washington, DC.

USEPA (1993) *A Guidebook to Comparing Risks and Setting Environmental Priorities*. Office of Policy, Planning, and Evaluation. US Environmental Protection Agency, EPA 230-B-93-003, Washington, DC.

USEPA (1995a) *Draft Proposed and Interim Guidelines for Carcinogen Risk Assessment*. US Environmental Protection Agency, National Center for Environmental Assessment, Office of Research and Development, Washington, DC.

USEPA (1995b) *Draft Proposed Guidelines for Neurotoxicity Risk Assessment*. Risk Assessment Forum, US Environmental Protection Agency, Washington, DC.

USEPA (1995c) *Draft Proposed Guidelines for Ecological Risk Assessment*. Risk Assessment Forum, EPA/630/R-95/002, US Environmental Protection Agency, Washington, DC.

Van Houtven, G. & Cropper, M. (1993) *When is a life too*

costly to save? Resources for the Future, CRM93-02, Washington, DC.

Wentsel, R.S. (1994) Environmental Policy and Risk Assessment. *Environmental Toxicology and Chemistry*, **13**(9), 1381.

Wentsel, R.S., LaPoint, T.W., Simini, M., Ludwig, D., Brewer, L.W. & Checkai, R.T. (1994) *Procedural Guidelines for Ecological Risk Assessments at U.S. Army Sites*, Vol. 1. ADA-297968, ERDEC TR-221. US Army, Aberdeen Proving Ground, MD.

Part 3
Balancing Risks with Other Considerations

European legislation on the assessment of risks associated with existing commercial chemicals requires that where problems are identified that might lead to proposals for restrictions on marketing and use of a substance, the advantages and drawbacks and the availability of substitutes have to be taken into account. Similar requirements are written into US legislation under the Toxic Substances Control Act. This is explicit recognition that factors, other than the likelihood of causing harm, have to be taken into account in developing environmental protection policy, here with regard to chemicals. For example, what benefit (advantages) to human health, food supply and quality of life might society be denied by controls designed to reduce the adverse effects (drawbacks) of chemicals to human health and ecological systems? This balance depends importantly upon how we perceive problems and value one aspect over another and, of course, applies to environmental protection issues in general, not just those involving hazardous substances. Chapters in this part of the Handbook deal with the principles of these 'balancing' issues as a prelude to putting principles both from risk assessment and risk perception and valuation into practice in various management scenarios in the next part of the Handbook. Clearly, recognizing the complex interactions between economic development, for example as represented by the production and use of commercial chemicals as pharmaceuticals, fertilizers, fabrics, fire-retardants, fuels, etc., and environmental degradation, making them explicit, and seeking to optimize them for long-term benefit is an integral part of the sustainable development approach.

Chapter 11
The Psychology of Risk and Uncertainty

NICK F. PIDGEON AND JANE BEATTIE*

11.1 INTRODUCTION

In this chapter we review recent theory, empirical research, and literature addressing the question of how psychological and wider social factors influence people's evaluations of uncertainty and risk. More generally, and when considering people's evaluations of, and behaviour towards, hazards involving actual environmental or technological threats, the term *Risk Perception Research* is often used as the generic label for this field of social science. It involves not only psychologists, but includes input now from a range of disciplines including sociology, anthropology, geography, decision theory, economics (Chapters 12 and 13) and policy studies.

The aims and agenda of researchers working on risk perception are, accordingly, as diverse as in any other research community, making any single characterization of the field difficult. Seen from a psychological perspective, however, some of the key research concerns are: (i) the ways in which hazards are perceived, along with the cognitive processes through which they are interpreted and mentally represented; (ii) the ways in which particular classes of hazards come to be viewed as risky (or not), both in terms of how accurately risk quantity can be subjectively expressed by people (where that is possible with respect to some prior normative standard of risk assessment) and also with respect to the more qualitative dimensions of hazards which influence the comparability of different risks; and leading on from both of the above (iii) the factors that influence the perceived acceptability (or tolerability) of particular hazards to experts and the public.

Beyond the psychological literature on individual risk perception and evaluation, however, other social science perspectives have sought to determine the influence of wider group, societal and historical factors upon risk perception and acceptability. For example, whether all groups in society tend to represent specific risks similarly, how the historical and other local contexts within which a risk arises frame collective attitudes towards riskiness and acceptability, and the ways in which social processes (such as the influence of the mass media) serve to amplify certain risks while seeming to neglect others.

The combined literature is enormous, and continues to expand rapidly. This chapter is therefore not intended as an exhaustive account, but rather points to the principal lines of argument and to accessible sources for further more detailed reading. In particular, regarding risk perception the interested reader is referred to recent reviews by Pidgeon *et al.* (1992), Slovic (1992) and Brun (1994), along with writings on the related issue of risk communication (e.g. NRC 1989; Kasperson & Stallen 1991; Fischhoff 1995; see also Chapter 20). We pay particular attention to how individuals evaluate environmental risks. This includes both harm to the quality of the environment from ongoing human activities, such as large-scale technological systems and industrial processes, as well as threats to human well-being with environmental origins. Note, however, that the latter, as in the case of ozone depletion, in turn often derive themselves from human activities rather than purely 'natural' events.

*After a short period of severe illness, Jane Beattie died at the end of March 1997. She is sadly missed by many of us in the UK and the US social science community.

Environmental risks are particularly interesting from the point of view of risk perception research, as they tend to have a number of characteristics which have been shown to be both particularly salient or difficult to deal with when people make risk evaluations or judgements of acceptability. It is also not without significance that a number of these characteristics ensure that environmental risks are very difficult to deal with in formal risk assessment terms. For example, consider the following.

1 Many environmental threats stem from very low probability but high consequence events (e.g. major industrial hazards such as process plant accidents), which pose particular difficulties of evaluation and perception.

2 Conversely, the possible health impacts of low levels of toxic contamination (as with 'routine' low-level chemical emissions to the environment) may be difficult to identify, let alone quantify.

3 There are often considerable uncertainties associated with the assessment of environmental hazards, and experts themselves may disagree (e.g. about the likelihood that containers of a hazardous waste will leak at some time in the future).

4 Assessment of environmental hazards often requires combining many low probability events (e.g. the chance that various chemicals in combination, if accidentally consumed in tiny amounts, might harm humans or other species).

5 Risk invariably presents both hazard and opportunity. It is now well known that people evaluate outcomes very differently depending upon whether they personally view them as 'losses' (relative to that person's current position) as opposed to 'gains'.

6 Many environmental hazards involve making difficult trade-offs over time (e.g. genetic modification of plants offers higher crop yield now, but with possible risks later), and the consequences may be very far into the future, such that long-term effects must be anticipated on people not yet born.

7 Some activities require inequitable trade-offs to be made across different groups (e.g. using pesticides on crops will profit business and farmers, but may lead to cancers in children who may never eat the crops produced), or are ones where those at risk believe they have been exposed involuntarily or have little personal control over the outcomes.

8 Difficult trade-offs may even be required between people and other species, or between people and the physical environment (e.g. the probable impact on wildlife and the countryside versus the likely benefits to the economy of siting a new motorway).

9 Many hazards will never have been directly experienced by those directly involved and who must be persuaded to act to mitigate their consequences (e.g. as in the case of naturally occurring radon gas).

10 The causes of environmental hazards vary (e.g. man-made versus 'natural') in ways that many people judge is relevant to the acceptability of the risk.

11 Finally, certain environmental hazards (e.g. the location of hazardous waste facilities) are associated with such extreme societal conflicts about risk acceptability that they appear to be a proxy for disagreements less over scientific facts than over different groups' values, politics or ways of life.

Many of the above issues pose basic and often intractable societal decision problems, both for laypeople and for environmental risk assessment, regulation and management. In seeking to address at least some of them from the perspective of the social sciences, we shall review a range of studies of the judgements that people, both experts and lay, make regarding the formal properties of negative events, such as the likelihood of harm and the potential costs and benefits attached to an activity. Also important are people's intuitive evaluations of a much wider range of characteristics of hazards, including such things as an activity's perceived voluntariness and controllability, or more widely the trustworthiness of the political and institutional arrangements for managing risk in society. The latter, of course, are not easily modelled in formal risk assessment terms, and provide the first clue as to why levels of public acceptance (or not) of a hazardous activity may at times diverge from formal expert assessments of risk. Research on the psychology of risk and uncertainty also shows us why, in the arena of environmental risk assessment, policy decisions may be particularly prone to conflict and mis-communication between the many varied stakeholders involved.

The chapter is organized into five sections. We start in section 11.2 by discussing a number of issues surrounding the definition of risk in both the formal risk assessment and the social science literatures. In section 11.3 experimental research on the psychology of uncertainty and probability judgements is reviewed, followed in section 11.4 by the considerable body of empirical findings concerning the psychology of risk perception. In section 11.5 we broaden the discussion to incorporate the more recent social theories of risk acceptance and perception. Finally, in section 11.6 we consider the complex question of whether (and how) risk perceptions should be incorporated into policy decisions for environmental programmes and regulation.

11.2 THE MEANING OF 'RISK'

Before examining the research evidence, we briefly consider various meanings of 'risk' in different literatures. This is important both because there are many different ways of defining and operationalizing risk in the formal assessment literature (see Vlek & Stallen 1980; Pidgeon *et al.* 1992; Gray 1996), and because lay usage ascribes very different meanings to the term too.

In the formal environmental regulation and risk assessment literature, there are typically three generic classes of definition.

1 Risk as the *uncertainty* attached to possible states of the world. Using this approach, risk is typically interpreted as the probability (*P*) of an undesired state or harm over some defined sampling space. Hence, a woman's risk of death during pregnancy and childbirth is, by this definition, far greater in any one year than that of being struck by lightning. A conceptual distinction can also be drawn here, as Blockley (1980) points out, between: (i) uncertainties in the values of known parameters—or *parametric uncertainties*—which can, as in the chance of developing asthma as a result of a given exposure to respiratory sensitizers, in principle be measured in terms of statistical probability; and (ii) non-statistical uncertainties such as those associated with the fundamental modelling of the risk generating system—so-called *systemic uncertainties*. For example, the contribution of human or organizational error to industrial process plant failures, while now acknowledged to be a very real and serious problem (Perrow 1984; Pidgeon 1988, 1991; Turner & Pidgeon 1997), is inherently difficult to model, and in many cases not interpretable, in terms of frequentist probability.

2 Risk as the magnitude of the negative *consequence* (*C*) that might flow from an action, decision or possible state of the world. Although some uncertainty is also typically implicit in such a definition, the measure of risk here depends primarily upon the maximum size of anticipated loss or detriment (hence, to many people, the potential for a Chernobyl-type accident is judged more risky than for drowning in the sea). A primary difficulty with this definition is, as noted earlier, that consequences (human deaths, morbidity, effects on wildlife or degraded quality of the environment) may have no ready common scale when decisions concerning trade-offs between different outcomes are required.

3 Risk as a *weighted combination of uncertainty and severity* of loss. Typically, although not always, this is interpreted as the product of probability and loss ($P \times C$). Deriving primarily from the intellectual tradition of decision analysis (see, for example, Raiffa 1968), this is the most common definition found in formal quantitative risk assessments. Once again, this definition does not of itself resolve the problems of dealing with non-statistical uncertainties, or the measurement issues surrounding what is meant by a 'loss'.

Interestingly, in a direct study of the meaning of the term 'risk' to laypeople, Drottz-Sjöberg (1991) finds that all three of the above definitions will be readily accepted by respondents, along with a fourth possibility which is typically absent from the more formal definitions—that of the *nature* of the risk event. And in much of the social science work on risk perception, the motivation is less to define a workable and conceptually consistent risk *measure* than to establish what people's own *understandings* of risk are. Here, two conceptually distinct approaches to psychological risk can be identified, which encompass to some extent all four of Drottz-Sjöberg's possibilities.

The first approach, which can be labelled the 'economics' perspective, views risk in terms of a judgement about uncertainty. This might be a probability or likelihood (possibly 'objective', although more often thought of as a subjective representation or degree-of-belief; Wright & Ayton 1994) or as an evaluation of higher order uncertainties associated with an adverse consequence, including such things as vagueness, ambiguity and second-order likelihood (Smithson 1989).

For example, in Chapter 12 on the economics of risk, a risky decision is defined as one between two or more prospects—involving a risky option in which one of two or more outcomes might occur, each having an *ex ante* probability of occurrence of more than 0 and less than 1. Thus in economics and finance, the term 'risk' has been tied closely to probability of occurrence. Note, also, that this definition of risk is strictly independent of whether the outcomes are undesirable or not, which is at variance with most formal risk assessment definitions (as well as lay understandings) of the concept.

What is more, in economic treatments the term 'decision-making under risk' is often used when the uncertainty present in a situation can be interpreted unambiguously in terms of statistical probability, and 'decision-making under uncertainty' when it cannot. Historically, in psychology there has been a long tradition of work which adopts this first 'economics' approach to risk, known as Behavioural Decision Research (Edwards 1961; Slovic *et al.* 1977; Payne *et al.* 1992). Here researchers have sought to identify and describe in the experimental laboratory how people make decisions in the face of statistical and other forms of uncertainty, together with the ways in which actual behaviour departs (or does not) from the formal predictions of normative economic theories, such as Expected Utility Theory (outlined in Chapter 12).

The second approach to risk in the social sciences has been adopted particularly by those working within the more recent tradition of risk perception research. It reflects the theoretical need to adopt a much wider concept than merely some subjective analogue of uncertainty, probability, or expected loss if lay conceptions of risk and hazard are to be adequately described and investigated. This has also led risk perception researchers to design studies with greater levels of realism (or external validity)—preferring to investigate people's judgements regarding realistic risk management problems. Hence, risk here is taken to refer to the full range of beliefs and feelings that people have about the nature of hazardous events (cf. Drottz-Sjöberg's (1991) fourth lay definition noted above), their qualitative characteristics and benefits, and most crucially their acceptability.

Accordingly, in the UK Royal Society report on *Risk: Analysis, Perception and Management* (Royal Society 1992), risk perception was defined as 'people's beliefs, attitudes, judgements and feelings, as well as the wider cultural and social dispositions they adopt towards hazards and their benefits' (p. 89). This definition was drawn up to be deliberately broad* since it is a wide range of multidimensional characteristics of hazards, rather than just an abstract expression of uncertainty and loss, which people appear to evaluate in forming perceptions—such that the risks, for example, from toxic chemical contamination and *Salmonella* poisoning, are seen as fundamentally and conceptually distinct. Furthermore, such evaluation may vary with the social or cultural group to which a person belongs (e.g. depending upon their different value commitments), the historical context in which a particular hazard arises, and may also reflect aspects of both the physical and human or organizational factors contributing to hazard, such as the trustworthiness of existing or proposed risk management.

Our discussion of the social science literature on risk and uncertainty in the following two sections (11.3 and 11.4, respectively) is structured around these two rather different views of the psychological risk concept.

11.3 THE PSYCHOLOGY OF PROBABILITY AND UNCERTAINTY JUDGEMENT

We begin with a discussion of some of the commonly held findings from experimental studies of people's judgements of objective and subjective probabilities. Typically, such studies have had

* The first author is indebted to Professor Richard J. Eiser of Exeter University for originally pointing out a definition along these lines.

the objective of demonstrating *biases* of judgement (availability bias, hindsight bias, ambiguity effects), expressed in terms of departures from some standard, such as normative probability theory. In recent years the focus of this work has been less upon errors *per se* than on the conditions which might prompt 'poor' and 'good' judgements, particularly in relation to the biases that expert judges might be prone to when estimating probabilities and uncertainties, and hence which should be explicitly considered in certain forms of environmental risk assessment.

11.3.1 Calibration of experts and novices

Research into formal risk assessment raises the question of whether people (experts and novices) can provide accurate evaluations of probabilities, or whether there are systematic biases in their judgements. For high-frequency cases where the outcome is well defined, the accuracy of an individual's probability judgements can be explored by relating the expressed subjective probabilities for a set of events to the actual outcome frequencies. The subjective probability predicted by the individual is plotted against the actual outcome probability in a so-called calibration curve. Normatively (for large enough samples of events) all points should lie on the identity line if an individual is personally well calibrated. There is now a large body of research on this issue (see Wright & Ayton 1994), and the standard finding is that novices in a particular prediction area tend to be overconfident, except for small probabilities, where they tend to be underconfident (Fischhoff & MacGregor 1982). Poor calibration also tends to increase with the difficulty of the judgement task (the 'hard–easy' effect).

Experts might be expected to be uniformly well calibrated, but this depends very much upon the particular judgement domain. While this is the case for professionals who formally collect and analyse feedback on their predictions, such as weather forecasters (Murphy & Winkler 1977), the picture is much bleaker from professions where feedback may be more difficult to obtain. For example, Christensen-Szalanski & Bushyhead (1981) found that physicians' judgements that a patient had pneumonia were almost uncorrelated with the outcome probability, and were generally extremely overconfident. These results are important as they make the divide between 'objective' experts and the 'subjective' public much less clear. They also suggest particular problems in attaining adequate calibration for judgements of many kinds of environmental risks. Such risks may not be realized for many years, and may be of such small probability that collecting sufficient feedback data to assess the numerical accuracy of predictions may be virtually impossible.

11.3.2 Estimating fatality frequency

We turn now to the body of evidence directly relevant to estimating fatalities from common causes of death. These kinds of risks are directly relevant to how well people are able to assess the dangers to which the environment exposes them (e.g. radon gas, tornadoes). This literature shows that when people are asked about their understanding of how many fatalities arise as a result of an activity X, their rank-ordered responses correspond by and large with those of available statistical estimates. This ranking is also consistent across different question formats (e.g. Fischhoff & MacGregor 1983; Eiser & Hoepfner 1991). Furthermore, there is some evidence (Brown & Green 1981) that evaluations of personal safety are also correlated well with statistical frequency. This pattern of results seems to support the idea that public evaluations of the *relative ordering* of frequency of death have a reasonably accurate foundation. The above are important findings because they suggest that: (i) where people then rank the *risks* of the same activities in a different order, this cannot be attributed to a lack of information or awareness concerning the actual frequencies of death; and therefore (ii) the notion of risk means something more to people than just expected fatalities.

Regarding *absolute* or *quantitative* judgements of frequency, the standard findings here, following the work of Lichtenstein *et al.* (1978), are that when asked for judgements in comparison to a specified anchor (such as the number of deaths from vehicle accidents across the USA), people tend to overestimate the prevalence of low-frequency causes of death (botulism) and underestimate those of high frequency (heart disease). This finding suggests a regression effect in which people's judgements are insufficiently discriminating, perhaps because of random error.

11.3.3 Heuristics and biases in probability judgements

The issue of systematic bias was raised by the research programme of Kahneman and Tversky (see, for example, Kahneman *et al.* 1982), in which they demonstrated a number of 'heuristics' (cognitive shortcuts) which people use in estimating probabilities (among other quantitative tasks). Kahneman & Tversky have proposed three main heuristics that would be expected to lead to systematic bias in risk estimation: availability, representativeness and anchoring and adjustment. While claimed to be generally adaptive, these heuristics can lead to large, systematic errors in probability assessment. These results damage the standard assumption from economics that individual errors in the public's risk evaluations will be random or unbiased, and hence will tend to cancel each other out, given large enough samples.

The availability heuristic (Tversky & Kahneman 1973) refers to the way in which the probability of an event is sometimes indexed by the ease with which it is accessed in memory. Slovic *et al.* (1979) find that vivid imaginable causes of death (tornado) receive similar likelihood estimates to non-vivid ones (asthma) which occur with a much higher frequency (in this case, by a factor of 20). Other studies of availability, including Fischhoff *et al.* (1978a), suggest that experts who are evaluating probabilities of safety failure will pay insufficient attention to causes that are not elaborated in their model. These results have implications for the accurate assessment of fault-tree probabilities in any large-scale mechanical or socio-technical system whose complete failure would cause damage to the environment or humans.

The representativeness heuristic (Kahneman & Tversky 1972) is implicated in a number of commonly researched biases of probabilistic reasoning (Kahneman *et al.* 1982). These include such things as insensitivity to the impact of base rate probabilities in Bayesian inference, and misunderstandings of chance phenomena. As an example of the latter, the sequence HTTHTH is regarded by many people as the more likely outcome of a sequence of six tosses of an unbiased coin than HHHTTT, on the grounds that the former is more representative of a 'typical' random sequence.

The anchoring and adjustment heuristic results in absolute estimates being influenced up or down by the provision of different anchors (starting points), even if the subject can readily see that the anchor provides no information about the correct answer. In the case of risk estimation, final estimates will be correlated with the starting anchor, and are thus open to manipulation by self-interested parties or the media. In discussing judgemental biases, however, it is important to note that the interpretation to be placed upon the findings from this research now occasions considerable disagreement amongst psychologists who study judgement and decision-making. A good overview of some of the arguments here is given by Jungermann (1986), and one issue concerns the precise conditions under which judgemental 'biases' investigated under artificial psychology laboratory conditions will generalize to real-world situations (see also Gigerenzer 1991).

11.3.4 Framing effects

Much of the work on heuristics and biases showed how the way in which a question is posed can influence the answer given. Tversky & Kahneman (1981) have described many such 'framing' effects, in which irrelevant changes in wording of the question produce substantial shifts in response. The most famous example is the 'Asian Diseases problem' (see Chapter 12), in which subjects must choose between a risky and a safe programme to combat the disease's effects on the population in the USA. The irrelevant wording change was to describe the effects either in terms of losses (lives lost) or gains (lives saved). Although the objective outcomes were the same in each case, people in the losses version made many more risk-taking choices than those in the gains version. This effect of gains/losses on risk attitude has proved robust in a variety of risk decision problems (Lopes 1987), and has potentially serious consequences for attempts to elicit preferences for environmental expenditure. The question is also raised of whether we can definitively state that any one framing of a problem is a better resolution than any other (see Bell *et al.* 1988).

Another kind of framing effect with particular relevance for environmental risk assessment is reported by Kahneman & Ritov (1994). They find that people are more willing to contribute money to prevent damage to the environment caused by humans than that caused by nature. This suggests that they view the human-caused risk as more serious than acts of nature (see also Brun 1992). These kinds of results suggest that people not only value the outcomes of actions, but also care about how these outcomes were obtained. Such findings violate the standard economic assumptions (Chapters 12 and 13) that people only care about ends, and ignore means, in making choices.

11.3.5 Motivational bias

A different kind of systematic distortion of probability estimates comes from possible motivational biases. The literature on unrealistic optimism (e.g. Weinstein 1980) appears to demonstrate that people believe that their personal chances of experiencing negative events are lower than the group mean. Svenson (1981) showed that 90% of drivers believe that they are safer than the average driver. While it might be comforting for an individual to believe that he or she is uniquely invulnerable to misfortune, such biased perceptions may be expected to reduce the use of protective measures (such as modifications to reduce radon in the home). Weinstein (1980) has proposed that unrealistic optimism is caused by individuals overestimating their personal control over events. If true, this has interesting links to the psychometric study of risk perception, reviewed below.

11.3.6 Group biases

It is tempting to imagine that the effect of individual biases in probability estimation might be reduced by group decision-making. Important expert decisions, for example, are often made by groups. However, the literature on group decision-making suggests a number of group biases. Janis' (1982) work on retrospective analysis of foreign policy decision-making disasters shows that insulated groups with insufficiently divergent views can become overconfident in courses of action that outsiders would regard as extreme (see also Vaughan (1996) for a contemporary organizational analysis of the flawed decision to launch the space shuttle *Challenger*).

Other studies by Stoner (1968) and Einhorn *et al.* (1977) all provide evidence that group processes appear to move consensus group decisions to a more extreme position (either more risky or more conservative, depending upon the initial average opinion) than members collectively hold initially. Janis' suggested solution is to improve collective problem-solving by ensuring that the group is comprised of individuals with differing orientations to the problem, operating in an atmosphere in which divergent opinions are valued (but also see 'T Hart *et al.* 1997). In these cases, group estimations may be more successful than the average, or even the best, of the individual estimations.

11.3.7 Appraisal of objectively given probabilities

The literature reviewed above largely examined how people assess probabilities from their knowledge and experience. A different question is how they deal with probabilities supplied to them, for example, by expert assessors. A variety of studies have shown that people do not always weigh objectively given probabilities adequately in decision-making. One problem lies in how people deal with very small probabilities. As we have seen from the literature on fatality frequency estimation, such probabilities are sometimes overestimated or overweighted. Other more artificial laboratory tasks confirm this, but also show how sometimes small probabilities get discounted all together (Kahneman & Tversky 1979). This issue raises particular problems because environmental risks (especially of human fatality, but also of consequences to the environment) are often, by their nature, very infrequent events. The reverse of this problem is that people have a tendency to overweight probabilities of 1.0 (Kahneman and Tversky call this the 'certainty effect').

It has been shown repeatedly that people have an aversion to dealing with second-order uncertainty (ambiguity) in probabilities. For example, Ellsberg (1961) showed that people are less willing to bet on the basis of ambiguous probabilities than on point estimates of the same mean value (see also Frisch & Baron 1988). In many areas of risk assessment for environmental damage there is uncertainty about the point estimate (e.g. the probability of a loss of coolant accident in a nuclear power plant). This may be due to disagreement between different experts' analyses of the problem, or, for example, because of uncertainties about the way that systems in particular states will operate or the unpredictable effects of human error (Pidgeon 1988). Again, some of the perceived differences in risk (see below) may be due to differences in the ambiguity of the probabilities. It may also be that a high degree of ambiguity correlates with particularly feared or 'dreaded' hazards (see 11.4.1 below).

11.4 THE PSYCHOLOGY OF RISK PERCEPTION

Here we move away from considering behaviour and judgements in relation to 'baseline' economic and statistical models of uncertainty, to consider the somewhat wider approaches adopted by risk perception researchers. To illustrate why this is necessary, consider one of the risk comparisons made earlier, between a Chernobyl-type accident and drowning while swimming in the sea. Although these hazards can be compared merely in terms of probabilities and consequences, intuition tells us, and research confirms the fact, that there are other important reasons why these are seen by people as very different types of event (Covello *et al.* 1988; Covello 1991; Freudenberg & Rursch 1994; see also Chapter 21). To take just a few aspects: the first risk is unfamiliar and affects both present and future generations, other species and the wider environment, while the second is everyday and only affects me; the former is much less under individual control as well as imposed involuntarily; and allied to this, the primary responsibility for risk management rests with society for major industrial hazards and with the individual in personal leisure activities. Psychological approaches to risk perception, and in particular the influential work of the so-called 'psychometric' research tradition, which we review first, have attempted to identify the dimensions along which such differences arise, as well as to predict where such differences will matter in people's judgements of risk magnitude and acceptability.

11.4.1 The psychometric paradigm and the 'quality' of hazard

As outlined above, when asked to give estimates of the annual expected *fatalities* from an activity, people's ordinal judgements agree quite well with those of actuarial statistics. However, an early finding of research on risk perception was that while people (experts and lay) might well agree on fatalities, they would disagree about the *risk* associated with a hazard. This implied to the researchers that the concept of risk means more to people than just expected fatalities (Slovic 1987). Many of the early empirical studies of individuals' risk perceptions were therefore designed to investigate this question, and revealed a pattern of complex qualitative understandings of hazards.

The possibility that qualitative differences might underlie different evaluations of hazards had first been raised by Starr (1969) in a revealed preference analysis of then current 'accepted' accident rates in the USA. The results suggested that the level of risk tolerated by society differed significantly for hazards to which people were exposed involuntarily (e.g. energy generation) compared to those undertaken voluntarily (e.g. leisure activities). Subsequent psychological research, designed in part to criticize the revealed preference methodology adopted by Starr, involved eliciting people's expressed preferences about risk (Otway & Cohen 1975; Fischhoff *et al.* 1978b). That is, people were given the opportunity in questionnaire surveys to say what risk means to them, what activities are risky, and which might be acceptable or not.*

* In particular, as Fischhoff (1990) notes, the results of the early psychometric work illustrated quite clearly that what is 'accepted' within society (in terms of current accident rates) might not in fact be *acceptable* to people

These early investigations, the findings of which are discussed in greater detail below, provided the foundation for the influential psychometric approach to risk perception research. The term *psychometric* refers to the methodology used. McDaniels *et al.* (1995) summarize the typical methodology adopted in the following four steps.

1 Develop a list of hazard items or risky events, technologies and practices that span a broad domain of potential hazards of interest.

2 Develop a number of psychometric scales that reflect characteristics of risks that are important in shaping human perceptions of, and response to, different hazards (controllability, catastrophic potential, etc.).

3 Ask people to evaluate the list of items on each of the scales.

4 Use multivariate statistical methods (such as exploratory factor analysis) to identify and interpret a set of underlying factors that capture the variation in the individual and group responses.

In principle, the method maps aspects of respondents' psychological understandings of the presented hazards on to a set of metrics, in order to explore the underlying structure of perceived similarities and differences, and through this their relationship to the initial judgements of riskiness and acceptability. As Slovic (1992) notes, in such investigations, risk is typically left deliberately undefined in order to elicit people's own understandings of the concept.

Results of the first large psychometric study on attitudes to nuclear power during the 1970s (Otway & von Winterfeldt 1982) indicated that people's perception and acceptance of technological hazards were based in part upon certain characteristics of the nature of harm shown in Table 11.1. According to this work, a hazard will be seen as less acceptable if it is imposed involuntarily, one does not have control over the risk, it is difficult to imagine risk exposure, and if the benefits are inequitably distributed, etc.

A further important distinction identified in early work was that between perception of so-called 'individual' and 'societal' risks (Green 1979), more recently described by Sjöberg (1995) as 'personal' and 'general' risks. The former are seen to hold consequences for, and are managed primarily by, individuals (e.g. radon in the home) while the latter impact upon a whole sector of society or the environment more widely (as in a major oil pollution accident at sea, or the release of toxic chemicals such as occurred at Seveso and Bhopal). Different responses are obtained from respondents depending upon whether they are asked to evaluate general or personal risks. It is also clear that several of the characteristics shown in Table 11.1 might differentiate between the perception and acceptance of these two notions of perceived risk. For example, lack of personal experience and involuntary exposure is likely to be a particular feature of societal/general rather than purely individual risks, as would be an unequal distribution of the risks and benefits.

Perhaps the most influential psychometric work was conducted in the USA at Decision Research in Oregon (Fischhoff *et al.* 1978b; Slovic *et al.* 1980; for good overviews, Slovic 1987, 1992). The results indicate that ratings of the qualitative risk characteristics exhibit a systematic pattern, with three important factors emerging. The first two principal factors, the qualitative dimensions that they represent and location of the 90 rated hazards within the factor space are illustrated in Fig. 11.1.

Table 11.1 General (negative) attributes of hazards that influence risk perception and acceptance. (From Otway & von Winterfeldt, 1982.)

Involuntary exposure to a risk
Lack of personal control over outcomes
Uncertainty about probabilities and consequences of exposure
Lack of personal experience
Difficulty in imagining risk exposure
Effects of exposure delayed in time
Threats to future generations
Infrequent but catastrophic accidents
Benefits not highly visible
Benefits go to others (or inequity)
Accidents caused by human failure rather than natural causes

(*Footnote continued*) when they are asked directly, thus undermining the basis for the revealed preference methodology adopted by Starr (1969).

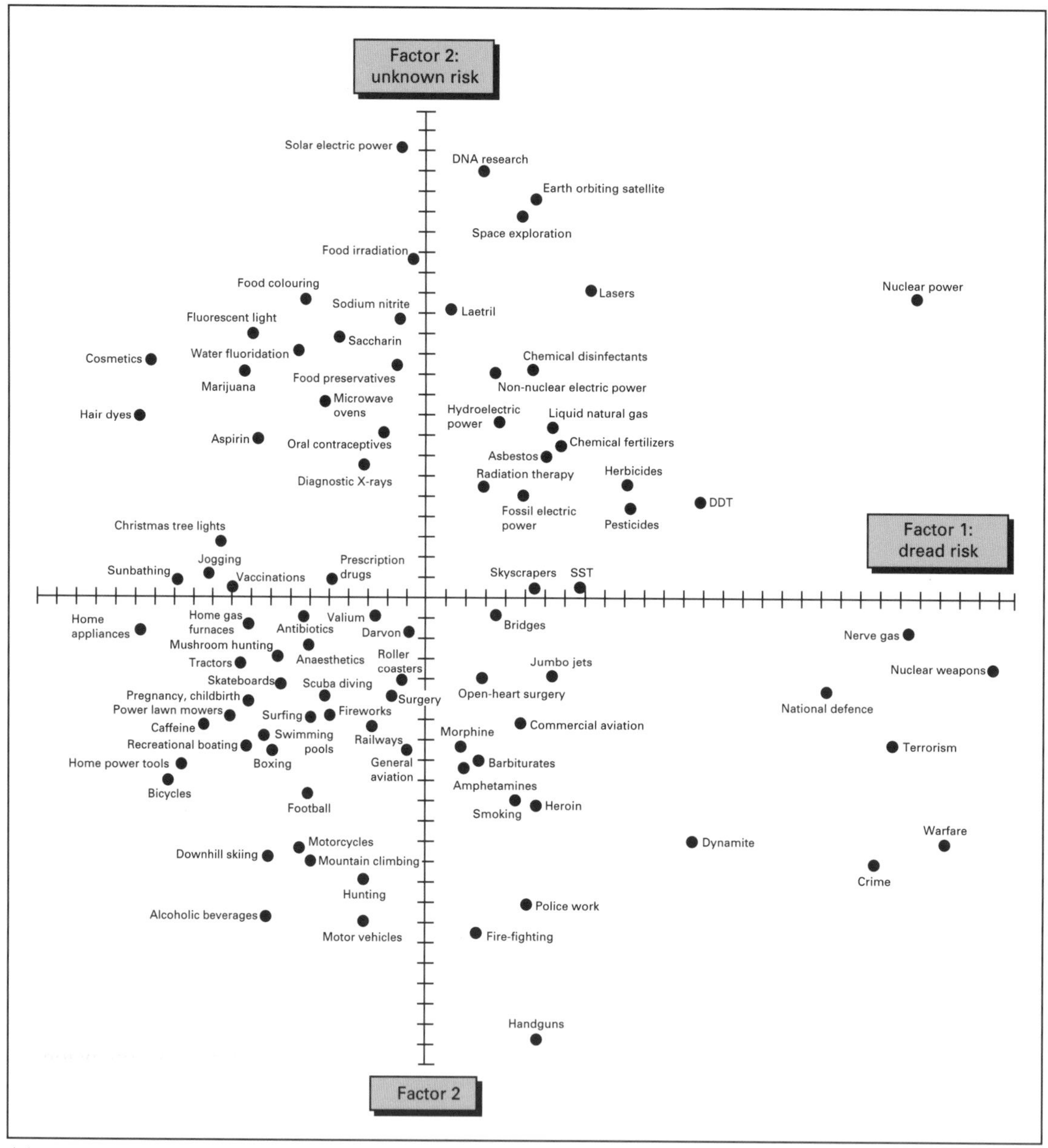

Fig. 11.1 Location of 90 hazards on factor 1 (dread risk) and factor 2 (unknown risk) of the three-dimensional factor space derived from the inter-relationships among 18 risk characteristics. (After Slovic *et al.* 1980.)

The first factor was labelled by Slovic *et al.* (1980) as 'Dread Risk'. This related judgements of scales such as uncontrollability, dread (or fear), catastrophic potential, involuntariness of exposure and inequitable distribution of risks. Hazards which rate high on this factor include nuclear power, pesticides and DDT, in contrast to those which rate low, such as home appliances and bicycles. A second factor, labelled 'Unknown

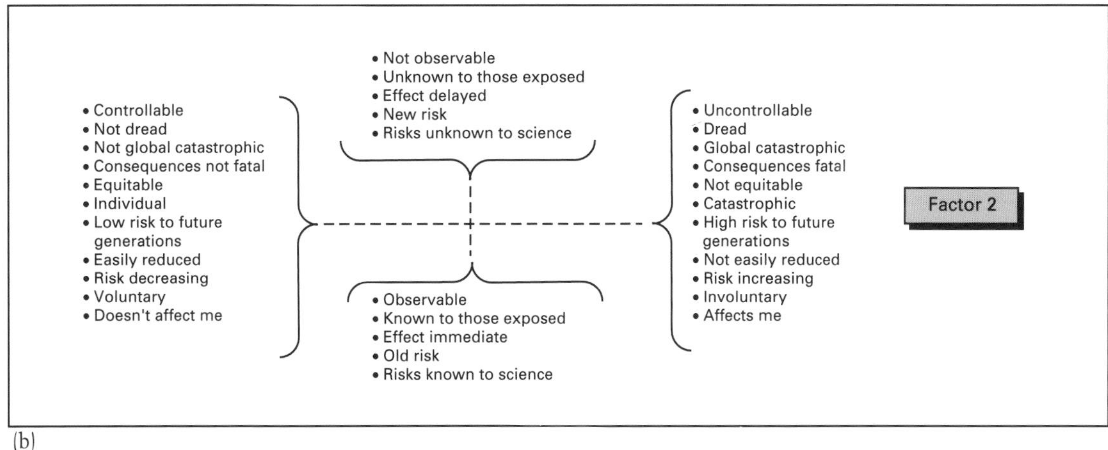

(b)

Fig. 11.1 (*continued*)

Risk', relates judgements of the observability of risks, whether the effects are delayed in time or not, the familiarity of the risk, and whether the risks are seen as known to science or not. Hazards that rate high on this dimension include solar electric power, DNA research and satellites, and those that rate low include motor vehicles, fire-fighting and mountain climbing. The analysis also identified a third factor, not shown in the figure, primarily related to the 'Number of People Exposed'.

The authors conclude that perceptions of risk are closely related to the position of an activity in the factor space. Most important here is the Dread Risk factor, according to Slovic, since 'the higher a hazard's score on this factor, the higher its perceived risk, the more people want to see its current risks reduced, and the more they want to see strict regulation employed to achieve the desired reduction in risk' (1987, p. 283). The findings of these studies indicated that the respondents were *not* satisfied with existing trade-offs between risks and benefits (the assumption underpinning the revealed preference approach). Fischhoff (1990) points out that one important subsidiary finding of the research programme was that judgements from a group of risk experts produced a *similar* factor structure to that of the lay subject groups. However, according to Slovic (1987) the expert judgements of risk were found to be more synonymous with their assessments of expected fatalities, and to be less influenced by the perceived qualities of the hazards. This does illustrate that the differences (and similarities) between expert and lay perceptions might be more subtle than a first reading suggests.

A related and key early study conducted in Europe by Vlek & Stallen (1981) used a large representative sample drawn from the general population in The Netherlands. Respondents completed a range of rating and similarity judgement tasks involving 26 separate hazards. They found risk to be characterized by two dimensions: 'size of a potential accident' (with some similarities to the 'Dread' factor) and 'degree of organized safety' (which might also be interpreted as lack of personal control over an activity). The 'size of accident' dimension accounts for the largest amount of variation in the riskiness ratings; perhaps not surprisingly, most respondents judge risk to increase for activities seen as higher on this dimension.

The early psychometric studies have provided a model for a growing body of research and literature (reviewed in Rohrmann 1995). In a study by Gardner & Gould (1989) a large representative sample of respondents was drawn from two states in the USA. They asked respondents to rate six hazards (motor-vehicle travel, air travel, nuclear power, nuclear weapons, handguns and industrial chemicals) on a number of dimensions of risk, benefit and acceptability. They also attempted to investigate the influence of socio-demographic variables such as age, gender, education level and social class on these judgements. They found that the relationships between 'overall risk' judgements and the other qualitative dimensions are different for specific hazards; for example 'catastrophic potential' is important for the nuclear and chemical technologies but less so for handguns, air and motor-vehicle travel. They conclude that *overall risk* is related mostly to judgements of number of deaths and a risk's dread, as well as (more weakly) to the degree of scientific disagreement about the risk and its catastrophic potential.

Like Fischhoff *et al.* (1978b), Gardner and Gould conclude that their findings cast doubt upon the revealed preference approach to risk acceptability, although they also report that the qualitative risk and benefit dimensions only explain a relatively modest amount (one-third) of the variance associated with respondents' judgements of the 'need for regulation'. Nor was the remainder of the variance explained by the influence of socio-demographic variables, which had only weak influence compared to the perceived qualitative properties of the technologies themselves. The fact that, when viewed singly, different hazards can have highly varied qualitative profiles for their particular perceived risks and benefits implies that the impact upon risk evaluations of any particular hazard's profile may have to be judged on a case-by-case basis, and related to the particular context within which that hazard arises (see also Brun 1994).

Numerous subsequent studies have varied the hazard sets rated by respondents. Some find broadly comparable factor structures; for example, with transport (Kraus & Slovic 1988), medical (Slovic *et al.* 1989) and food hazards (Sparks & Shepherd 1994). Others demonstrate striking differences. For example, Brun (1992) demonstrates a different factor structure for a set of 'natural' hazards as compared to a separate set of man-made hazards. And McDaniels *et al.* (1995), in a study of perceptions of ecological hazards, used a set of 65 items including such things as natural disasters (e.g. floods), effects of technologies and their applications (incineration), human practices with potential negative impact (outdoor recreation), and human beliefs and socio-political systems (big business). The researchers identify five different factors, in the following order of importance: impacts upon species; human benefits; impacts upon humans; avoidability of risk; knowledge of risk.

In methodological terms it has been demonstrated that different multivariate techniques can be used to highlight different features of the same data set (Arabie & Maschmeyer 1988; Johnson & Tversky 1984). One drawback to all of the psychometric studies is that where questionnaire scales are defined in advance, respondents are not allowed complete freedom to say what really matters to *them* about the risk question under investigation. This methodological issue remains critical if risk perception is to be properly investigated in terms of the *personal experiences* of respondents. Studies that have used more open-ended response methods, and which permit participants' flexibility of response, include Earle & Lindell (1984), MacGregor (1991) and Fischer *et al.* (1991). Henwood & Pidgeon (1992, 1994; see also Jasanoff 1993) argue that the value of such open-ended qualitative methods for generating genuinely new insights into respondents' own psychologies and understandings is undeniable, and often outweighs the extra effort needed to analyse and interpret the data obtained. Such approaches are increasingly important because, as some authors point out (NRC 1989; Pidgeon *et al.* 1992; Bostrom *et al.* 1994), a close understanding of how the audience themselves understand risk is critical for any attempts at risk communication.

In theoretical terms it is tempting to interpret the two principal psychometric dimensions of *Dread* (or sometimes risk *Potency*; Brun 1992) and (lack of) *Knowledge* as a direct mapping on to the more formal risk assessment concepts of 'consequence' and 'uncertainty'. However, a complexity here concerns the loading and interpretation of the critical dimension control (or lack of it) over risk. Some authors have argued (Vlek & Cvetkovich 1989; Teigen 1994) that control is a fundamental aspect of the *uncertainty* associated with an activity. And while certain European studies have indeed found that control loads on to the second 'Knowledge' dimension (Teigen *et al.* 1988; Brun 1992; Puy & Aragones 1994), in the most widely cited Slovic *et al.* (1980) study, as well as others, lack of control loads positively on to the primary 'Dread' dimension—presumably because what we cannot control we often fear, and hence is seen as risky. Conversely, in their study of the perception of a group of ecological hazards, McDaniels *et al.* (1995) find that control explains a relatively small amount of variance in overall judgements, and is unrelated to ratings of riskiness. These anomalies suggest that perceived control is a particularly complex and multifaceted concept, with its relationship to risk and uncertainty dependent upon the precise context in which the risk judgement is made. Hence, while in the Slovic *et al.* (1980) studies the hazards were primarily threats to human health and safety, which were to some extent amenable to direct control strategies, McDaniels *et al.* (1995) chose to study threats to the environment—such as pollution—where often the issue of direct control is both far more politicized and also difficult to achieve.

A key empirical question arising from the psychometric work is precisely how the qualitative factors relate to evaluations of riskiness and its acceptability or tolerability. This is a complex issue, which again must be viewed to some extent upon a case-by-case basis. Brun (1994) points out that the qualitative dimensions may impact upon several different evaluations, including: the perceived magnitude of a risk; judged benefits; the significance/importance of a hazard; who has responsibility for safety; willingness to pay for benefits (McDaniels *et al.* 1992; Savage 1993; Jones-Lee & Loomes 1995); and the signal potential of an event (that is, what an accident portends; Slovic 1987).

The original and subsequent research (see Rohrmann 1995) confirms that it is often the primary dimension (Dread/Potency or some variant) which is most consistently and positively correlated with directly expressed ratings of perceived risk, behavioural intentions (although, as some commentators have pointed out, intentions may not always predict behaviour: see, for example, Metz 1994; Erikson 1994), and attitudes towards tolerability or further reduction in risk. Other things being equal, the higher an activity's score on this factor the more risky it is seen to be, and again the greater is the desire for control and/or risk regulation. However, it may be that Dread by itself, despite its predictive power over a set of general technological hazards, holds the status of an intermediate variable only. This is because a hazard which engenders feelings of anxiety and unease will indeed be seen as (and be almost synonymous with) high risk, but what makes us uneasy will in turn differ from hazard to hazard and context to context (as Gardner & Gould (1989) demonstrate).

The relationship between the second psychometric dimension (Knowledge) and perceived risk appears to be very much weaker. And perceived benefits impact upon judgements of tolerability in a broadly predictable way (more benefit makes an equivalent risk seem more tolerable, and vice versa). Interestingly, however, Alhakami & Slovic (1994) report perceived risk and perceived benefit as *inversely* related, rather than independent dimensions as might be expected from a normative perspective, possibly because once an evaluative attitude towards an activity is formed (as either 'good' or 'bad'), this then drives both risk and benefit judgements in opposite directions.

In summary, the psychometric tradition has undoubtedly generated an impressive body of empirical data on individual risk perceptions. The evidence shows that human judgements of hazards and their benefits involve multiple qualitative dimensions related in quite subtle and complex ways. However, it can be argued that while this research tradition has provided extensive empirical descriptions, it has not, as yet, yielded substantive *theoretical* progress towards explaining risk perceptions in terms of underlying psychological processes or social factors. Recently it has been argued that such a theoretical understanding might be developed by closer integration of the psychometric findings with the more recent social theories of risk perception (see, for example, Pidgeon *et al.* 1992; Kasperson 1992).

11.5 SOCIAL AND CULTURAL INFLUENCES ON RISK PERCEPTION

There has been a growing consensus within the social sciences regarding the role played by social, cultural and institutional processes in the perception of risk (see Krimsky & Golding 1992; Holzheu & Wiedemann 1993). Social and cultural factors are important because the perceiver of risk is rarely an isolated individual but, rather, is a 'social being' defined through a range of group (e.g. family, friends and gender), institutional (place of work), and cultural (ethnic, national) relationships. Indeed, some contemporary European social theorists (Giddens 1990; Beck 1992) go so far as to claim that issues of risk and trust in institutions have replaced those of labour and capital as the organizing principles of the late-twentieth century industrial society. They comment that it is a paradox of the contemporary 'risk society' that, in an era of apparently greater safety than ever before, this has been achieved by ever-decreasing personal control over, and more novelty and complexity inherent to, the risks that we do face (see also Wildavsky 1988).

In empirical terms, the evidence suggests that there are indeed significant individual and group differences in risk perceptions; for example, in the differential importance placed upon particular hazards or their qualitative characteristics. In the human sciences, *individual* differences are generally held to be the result of relatively stable psychological predispositions of any particular person, which vary systematically across members of a given population; for example, the notion of 'personality' is held to be an individual difference, and in behavioural decision research a long-standing distinction is between risk-seeking and risk-avoiding individuals (see, for example, Lopes 1987). *Group* differences, on the other hand, are believed to reflect attitudes, beliefs and behaviours that result from an individual's identification with, and membership of, a particular social category, group or culture. Identification forms the basis for conformity with the norms, beliefs and behaviours of that group or culture, and an individual may identify with more than one group at any one time, as well as change affiliations in different social contexts or over time.

By going beyond purely psychological analyses, the social approaches to risk perception highlight first that the notion of a 'lay' public as an undifferentiated risk-perceiving entity is a misnomer: a society is always composed of many groups with very different attitudes towards, and appraisals of, what risk is, which risks should be run, and what values are relevant to making acceptability decisions. And, second, that risk perception and acceptance may be fundamentally related to social judgements of things such as responsibility, blame and trust in risk management and managers (with evidence to support this view coming from recent work on the controversies surrounding the threat to the environment from the storage of long-lived hazardous wastes: see, for example, Bord & O'Connor 1992; Flynn, *et al.* 1992, 1993).

11.5.1 Attitudes and world-views towards risk

If risk perceptions are closely related to individual differences in beliefs, then we should expect to find significant correlations between risk perceptions and measures of more general attitudes (such things as fundamental ethical principles, as well as views on lifestyle, technology or the environment), or between perceptions and socio-demographic variables (age, socio-economic status, gender, occupational and political affiliations; see section 11.5.2 below). The evidence to date, however, suggests relatively weak relationships, and is therefore far from clear cut.

In studies of belief salience in anti- and pro-nuclear groups (Eiser & van der Pligt 1979; van der Pligt *et al.* 1982), prior orientation on the issue of nuclear power led respondents to see different issues as salient or most important to them; pro-respondents saw the economic aspects of nuclear energy as most salient, and anti-respondents saw the risk of accidents and consequences for the environment. And, as noted earlier, the work by Alhakami & Slovic (1994) suggests that prior attitude towards a hazard (as good or bad) may drive evaluations of risk and benefits in opposite directions.

In the large psychometric study by Vlek & Stallen (1981) there was evidence of differences in the way participants viewed the (existing) degree of organized safety associated with a hazard. Half of their sample viewed the existence of a high degree of organized safety (as with a regulated industrial hazard process plant) as implying higher risk than the more individual harms (smoking). For the other half of the sample the reverse relationship was found. Vlek and Stallen point out that the dimensions on which their respondents disagree tend to involve socially controversial issues, and it can be argued that such dimensions are precisely those with contested institutional or political implications. Hence, the finding for one group of respondents that 'degree of organized safety' implies higher risk might indicate that such individuals do not trust existing institutional arrangements for managing risk.

Studies of the relationship between beliefs about technology and its impact upon the environment show that prior orientation is indeed often related to the aspects of environmental risks most salient to people. This research draws in particular upon the notion of a *world-view*: a constellation of attitudes and beliefs surrounding issues of industrial, technological and population growth that are related to people's perceptions of risk and preferences for risk management.

An extensive empirical survey in the UK (Cotgrove 1982) demonstrates that in relation to issues such as environmental pollution and nuclear power, two broad cultural paradigms of thought can be identified in people's discourse (Table 11.2). The dominant industrial (or 'cornucopian') paradigm values economic growth above all, and the environment is seen as a resource to achieve this through the control of nature. The alternative environmental ('catastrophist') paradigm stresses self-actualization and harmony with nature and the environment as intrinsically valued in and of itself. With regard to the latter, comparable results in the USA are provided by Dunlap van Liere (1978).

Research by Buss *et al.* (1986) indicates that those who favour a high-growth, high-technology society tend to see the benefits of technology as more important than its risks. The converse is found for those whose world view emphasizes such matters as concern about social and environmental impacts of growth, and equity in risk

Table 11.2 Competing social paradigms. (From Cotgrove, 1982, table 2.1, p. 27.)

	Dominant (cornucopian)	Environmental (catastrophist)
Core values	Economic growth Environment as resource Domination over nature	Self-actualization Environment intrinsically valued Harmony with nature
Economy	Market forces Risk and reward Rewards for achievement Differentials Individual self-help	Public interest Safety Income related to need Egalitarian Collective provision
Polity	Authoritative (Use expertise) Hierarchical Law and order	Participative (Public involvement) Non-hierarchical Liberation
Society	Centralized Large-scale Ordered	Decentralized Small-scale Flexible
Nature	Ample reserves Nature hostile/neutral Controllable	Earth's resources limited Nature benign Delicately balanced
Knowledge	Confidence in science Rationality of means Separation (of fact/value)	Limits to science and technology Rationality of ends Integration (of fact/value)

distribution. These findings are important because the different positions that stakeholders take up in environmental risk debates may derive from such fundamentally incompatible sets of attitudes and value commitments, and hence it is no surprise often to find them leading to considerable conflict (von Winterfeldt & Edwards 1984).

The picture concerning value orientation is, however, probably more complex than any simple categorization of people can capture. For example, consider pro- or anti-technology attitudes; here Gardner & Gould (1989) found that attitudes for stricter risk regulation often go hand in hand with the desire to see a technology more widely developed (suggesting in turn that public calls for stricter regulation cannot be lightly brushed aside as solely representing an anti-technology bias).

11.5.2 Sociodemographic variables

Of the general sociodemographic variables that have been studied, there is clear and consistent evidence that gender may influence evaluation of risk; women tend to give higher risk ratings of general or societal risk than men and see more threat to the environment (e.g. Pilisuk *et al.* 1987; Flynn *et al.* 1994; Barke *et al.* 1995; Greenberg & Schneider 1995). Single studies suggest also that age (Fischer *et al.* 1991), occupational affiliation (Sjöberg & Sjöberg 1991; Jenkins-Smith & Bassett 1994), and ethnic and socio-economic status (Vaughan & Nordenstam 1991; Vaughan 1995) may also be important in some contexts.

On the other hand, Gardner & Gould (1989) in their large US study found few correlational relationships between sociodemographic variables and measures of risk perceptions. Reflecting this, in a general review and empirical investigation of this question, Sjöberg (1995) concludes that while significant differences can sometimes be found (e.g. between fundamental value orientation and attitudes to technology, or between gender and risk appraisal as noted above), in many studies correlations are typically very weak (most $r \leq 0.3$) and therefore explain only a very small amount of the variation (typically less than 10% explained) in perceived risk scores. An obvious implication here is that the use of average scores to reflect a population's attitudes towards risk may in some circumstances be less problematic than once first thought.

11.5.3 Cross-national research

A number of cross-national comparisons of risk perception have been conducted, reflecting a growing interest in exploring risk issues from the perspective of cross-cultural psychology (see Cvetkovich & Earle 1991). In general terms, cross-cultural social science aims to investigate and explain differences and similarities in beliefs and behaviours between different cultural (such as national, ethnic or institutional) groups. Such studies have typically compared the psychometric findings from samples in the USA (some using the original Oregon data sets) to equivalent data obtained in other countries, including Hungary (Englander *et al.* 1986), Norway (Teigen *et al.* 1988; Brun 1992), Hong Kong (Keown 1989), Portugal (Lima 1993), Spain (Puy & Aragones 1994), China (Jianguang 1994), the former Soviet Union (Mechitov & Rebrik 1990), Japan (Kleinhesselink & Rosa 1991), and Poland (Goszczynska *et al.* 1991).

The overall pattern of factors obtained from such analyses is often similar to the USA studies, although with some local differences too. Kleinhesselink & Rosa (1991) in their comparison of USA and Japanese samples closely replicate the factor dimensions of Slovic *et al.* (1980) in both groups, but also report differences with respect to the ordering of hazards on both the dread and the known–unknown dimensions (e.g. food safety and drug risks were more feared in the Japanese sample, while hazards such as nuclear power and acquired immunodeficiency syndrome (AIDS) risks were rated as 'unknown' in the USA but 'known' in the Japanese sample). In his review, Rohrmann (1995) concludes that it is far from clear on the combined evidence whether the original findings in the USA represent cross-cultural universals, despite the reported similarities across countries. Likewise, the between-country differences must be interpreted with caution (see Johnson 1991), as they might be due to cultural differences, personal knowledge, experience of, or direct exposure to, particular hazards, or different political and economic circumstances in the comparison countries.

11.5.4 The cultural theory of risk

Perhaps the most fully developed sociological approach to group differences in risk perception is the so-called 'cultural theory' of risk (Douglas & Wildavsky 1982). Developed initially in part as an antidote to the universalizing claims of the psychometric model, this theory posits that human attitudes towards risk and danger vary systematically according to four cultural 'biases'—individualist, fatalist, hierarchist and egalitarian*—which can be identified in diverse contexts and societies. Cultural biases are held here to serve the function of defending individuals' favoured institutional arrangements and ways of life. Risk is central to this process of institutional defence, with cultural biases orienting people's selection of which dangers to accept or to avoid, the fairness of distribution of risks across society (Rayner & Cantor 1987), and accordingly who to blame when things go wrong (Douglas 1992). Put simply, a hierarchist fears threats to the social order and believes that technological and environmental risks can be managed within set limits. For the individualist, risks present opportunity, save those that threaten freedom of choice and action within free markets. Egalitarians fear risks to the environment, the collective good and future generations, and will only accept such risks that are unavoidable for the common good. Finally, fatalists do not knowingly take risks, but accept what is in store for them.

Cultural theory has made a major and growing theoretical contribution to the risk debate in Europe and to a lesser extent the USA (Schwarz & Thompson 1990; Thompson *et al.* 1990; Wildavsky & Dake 1990; Dake 1991; Rayner 1992; Adams 1995; Peters & Slovic 1996). Difficulties with this school of thought, however, as Boholm (1996) comments, include a certain circularity of argument in the definitions of the four cultural biases, and the fact that the varied writings that have developed present a number of different, sometimes conflicting, interpretations of the theory and in particular its unit of analysis (i.e. whether cultural biases are inherent to individuals, groups, or institutional arrangements and the precise relationship between these three). A further criticism, and one which Cotgrove (1982) recognizes as applying equally to his two environmental paradigms or world views, is of the somewhat static nature of the categories. For example, environmental concern embraces a very broad spectrum of groups (Funtowicz & Ravetz 1985) from the relatively traditional, such as the National Trust and the Sierra Club, to the more radical, including Greenpeace and Friends of the Earth. Accordingly, categories such as 'individualist' or 'environmentalist' should be seen largely as stereotypes.

In a series of recent quantitative empirical tests, Dake (Wildavsky & Dake 1990; Dake 1991) has attempted to correlate measures of the specific cultural-theory world views with risk perceptions. However, in these studies it is not obvious from the evidence that individualists and hierarchists, as identified by his psychometric scales, are very clearly distinguished from one another in their attitudes to risk. Dake (1991) reports both hierarchy and individualism as positively correlated with a pro-risk stance towards societal risk-taking, and egalitarianism with an anti-risk stance. And other attempts to utilize Dake's measures of the four biases have met with mixed success (Sjöberg 1995; Peters & Slovic 1996). While it is now clear that *individuals* cannot easily be categorized in terms of the four world views using Dake's standardized quantitative techniques, this does not rule out empirical support for some variants of cultural theory, particularly where in-depth qualitative research of particular social groups is combined with the more quantitative measures (see the extensive discussion of this issue in Marris *et al.* 1996).†

The sum findings of research on cultural theory to date would be consistent with a view that people's value orientations are not so much stable traits but plastic and constructed from a set of available societal discourses about such things as danger, blame, trust and accountability (Kemp

* In some versions of the theory there is a fifth (theoretical) possibility—the hermit!

† We are indebted to Claire Marris for commenting upon the material in this section.

1990; Rayner 1992, 1993; Pidgeon 1995). In reality, then, positions concerning a specific risk will draw upon a complex admixture of arguments and discourses—some complementary, but others in conflict. The question then becomes less one of who holds what attitude, but of which discourses are available (and why) to be invoked in any particular context or environmental risk conflict, and which may contribute to social amplification effects when things do go badly wrong.

11.5.5 Social amplification of risk

To date there has been little direct interaction between the cultural and psychometric traditions of risk research within social science. A recent conceptual framework that makes a genuine attempt to unify psychological, social and cultural approaches to risk perception is the *social amplification of risk perceptions* (Kasperson *et al.* 1988; Kasperson 1992). The approach adopts a metaphor from communications theory, to explain why certain hazards and events (e.g. the controversy in Europe over the proposed offshore dumping of the oil-production platform *Brent Spar* in the Atlantic Ocean in 1995) are a particular focus of concern in society, while others (such as naturally occurring radon gas in homes) receive comparatively little attention.

The social amplification model suggests that hazards and their objective characteristics (e.g. deaths, injuries, damage and social disruption) interact with a wide range of psychological, social or cultural processes in ways that intensify or attenuate perceptions of risk. In this sense, hazardous events hold a 'signal value' (Slovic *et al.* 1984), which crucially may differ for different people or social groups. An individual or group can in this way be conceptualized as a receiver of such signals about hazards.

A key insight of Kasperson *et al.* (1988) is that such signals may be subject to predictable *transformations* as they are filtered through a variety of social 'amplification stations'. This results in intensification or attenuation of aspects of risk in ways predictable from social structure and circumstances (a point compatible with the claims of the cultural theory of risk too). Examples of such amplification stations would include groups of scientists, the mass media, government agencies and politicians, and activist groups within a community.

Kasperson *et al.* (1988) also argue that social amplification accounts for the observation that certain events lead to spreading ripples of secondary consequences, which may go far beyond the initial impact of the event, and may even impinge upon initially unrelated hazards. Such secondary impacts include loss of sales (perhaps through a consumer boycott of a product; e.g. the financial consequences to Exxon Corp. of the petrol sales boycott in outrage at the *Exxon Valdez* oil-spill off the Alaskan coast eventually ran into many millions of dollars), regulatory constraints, litigation, community opposition and investor flight.

A drawback of the amplification idea, despite its *prima facie* plausibility, is that it may be too general to subject to direct empirical test, although some progress has now been made here (see Burns *et al.* 1993; Freudenberg *et al.* 1996). Furthermore, the source-receiver framework relies too heavily upon a simple conceptualization of risk communication as a one-way process; i.e. from risk events, through transmitters, and then on to receivers. As discussed by Fitchen *et al.* (1987), the development of social risk perceptions is always likely to be the product of more interactive processes between source and receiver of a message, and both the National Research Council (NRC 1989) and the Royal Society (1992) reviews of risk communication discuss the importance of viewing this as a two-way process of dialogue.

A number of criticisms have also been raised of the amplification metaphor. First, that it might be taken to imply that there indeed exists a baseline or 'true' risk, which is then transformed, or distorted, by the social processes of amplification. However, it is clear that the proponents of the idea of risk amplification do not wish to imply that such a single true risk baseline always exists, particularly in the heavily politicized 'transcientific' settings (Funtowicz & Ravetz 1992) where amplification is most likely to occur. Their conceptualization of risk in terms of signs, symbols and images is compatible with the view that risk is a social construction (Johnson & Covello 1987; Holzheu & Wiedemann 1993).

A semantic point is that the term amplification is typically associated, in its more common usage, primarily with the intensification of signals (Rip 1988). However, the detailed framework, as formulated by the original authors, is intended to describe both the social processes that decrease (that is attenuate) signals about hazards, as well as those involved in intensification.

A possible extension of the model, which would stress the role of attenuating phenomena, is to link it to the considerable empirical evidence of organizational and micropolitical processes that prevent warning signs from being effectively utilized both prior to and after the onset of large-scale organizational failures and environmental disasters (see, e.g. Couch & Kroll-Smith 1985; Shrivastava 1992; Pidgeon 1996a; de Souza Porto & de Freitas 1996). Elucidating the conditions for such attenuation (and how it might be counteracted) would be a significant step in helping to prevent such events.

11.5.6 Risk and trust

A very recent set of arguments and studies focus upon the relationship between risk and trust. The question of trust and risk perception was first raised by Wynne (1980). He hypothesized that differences between expert and lay constructions of risk might depend upon the evaluation of the trustworthiness of risk management, and of the authorities to act both in the public interest and with regard to best possible technical standards and practice. One interpretation, then, to be placed upon several of the qualitative dimensions of risk identified in the psychometric studies (for example, Control, Equity, Voluntariness, Known to Science) is that they tap concerns about the *institutional* processes of hazard management (Bord and O'Connor 1990; Royal Society 1992, chapter 6).

Trust is accredited a number of definitions in the literature, but is typically discussed in terms of an implicit relationship between two or more parties. One of its functions is to reduce complexity in our social environment (Barber 1983), hence making life more predictable. In relation to risk communication, Renn & Levine (1991) list five attributes of trust.

1 *Competence* (do you have the appropriate technical expertise?)
2 *Objectivity* (are your messages free from bias?)
3 *Fairness* (are all points of view acknowledged?)
4 *Consistency* (of your statements and behaviour over time?)
5 *Faith* (a perception of your goodwill?)

Renn & Levine argue that trust underlies *confidence*, and where this is shared across a community one has *credibility*. A somewhat different view of social trust is provided by Earle & Cvetkovich (1995), who argue that it is similarity of our basic values, rather than attributes of technical competence, which will underlie who we trust or not.

Very recent empirical research in the USA concerned with the specific issue of hazardous facility siting appears to confirm the hypothesis that trust is related to community-level acceptance of hazard (Flynn *et al.* 1992; Slovic 1993), although the link may not be a direct one. Bord & O'Connor (1992) argue that trust in industry is linked to trust in government and in turn to a belief that specific hazards are amenable to control. A link with 'control' suggests that trust arguments may underlie part of this aspect of the qualitative profile of a hazard (which, as we have argued earlier, appears to be in a rather complex relationship with the other dimensions identified from the psychometric studies). Flynn *et al.* (1992) find that distrust in waste repository management is positively correlated with gross perceived risk ratings and opposition to siting (see also Kunreuther *et al.* 1993).

A key implication here is that we need to frame the principal goals of risk communication around building trust through participation (NRC 1989; Royal Society 1992). However, as Johnson & Slovic (1995) demonstrate, the presentation by an agency of risk uncertainties to people can signal honesty for some or invoke greater distrust in others. And given that conflict is endemic to many environmental risk management controversies, effective risk communication will only follow if a resolution of conflict is obtained first, perhaps by searching for decision options that address all of the stakeholder's principal values and concerns (Edwards & von Winterfeldt 1987; Renn *et al.* 1993; Webler *et al.* 1995), or by identifying superordinate goals to which all parties can agree. Despite these difficulties, broad participation is also increasingly seen as critical to the wider processes of risk assessment and management in and of themselves (see, in particular, NRC 1996).

Clearly there is much more fundamental work to be done on this question, and in particular to establish whether the trust finding is issue specific (e.g. to hazardous waste) or generalizes to other high-profile environmental issues, and whether it is as salient in the wider international context as it appears to be in the USA. Here, some authors posit that distrust is a particular manifestation of US socio-legal relationships (Slovic 1993; Earle & Cvetkovich 1995) although others demonstrate that it is equally important in other cultural contexts, such as Europe (Kemp 1990; Petts 1994).

11.6 THE SIGNIFICANCE OF RISK PERCEPTION RESEARCH FOR ENVIRONMENTAL RISK MANAGEMENT AND DECISION-MAKING

The chapter has so far dealt with the basic descriptive findings and social science theories regarding uncertainty judgements and risk perception. While such study is a valuable and interesting exercise in its own right, and of considerable importance to the practice of risk communication, it is less clear whether, and how, these findings should have an impact upon risk policy-making in the environmental arena. This is partly because risk analysis and management emerged initially as engineering–practitioner disciplines largely separate from the social sciences. More important, however, both complex philosophical and controversial ethical questions are raised by the suggestion that perceptions should influence environmental risk management decisions concerning the allocation of scarce, often public, resources for environmental and human health protection. In this section, therefore, we take a broader view on the issues already reviewed, posing the question of whether public risk perceptions should enter into policy decisions concerning expenditure and/or regulation.

11.6.1 Arguments against a role for risk perceptions

To allow the perceptions of the public to influence policy has seemed, to some, to be a great mistake. To state our own position at the outset (and it is one which we suspect many risk professionals would subscribe to as well), we are far less pessimistic about the potential role of the public than this (see also Pidgeon 1996b). We believe that public perceptions can be used to greatly enrich formal risk analyses and provide one form of guidance upon risk tolerability criteria to be adopted for policy. However, it has been argued by critics that perceptions do not rest upon a firm scientific understanding of risk issues *a priori*, and that therefore to use them as a guide to policy would, at best, introduce noise into the system and, at worst, systematic bias. Lives and scarce resources will be lost, so the argument goes (Weiner 1993; Okrent 1996), as priorities are systematically deflected from activities which do most harm to those which scare the most, all at the expense of the socially 'optimal' long-run objective of minimizing fatalities. In addition, rapidly changing public perceptions, manipulated by a hostile media only interested in the failures of technology, may lead risk management to lurch between hazards of the month. The following is a brief review of some of the arguments supporting this position.

Public perceptions are noise/bias

Probably the most frequently argued thesis is that public perceptions have no place in policy decisions because the individuals involved do not have the knowledge to evaluate accurately what will harm them, and thus the resultant judgements of risk and its acceptability will be subject to noise or bias. As a result, then, error will be introduced into the (appropriate) risk analyses performed by experts.

Certainly public perceptions can be in error, transient and biased. Perceptions of the likelihood of a major environmental oil spillage at sea taken one day versus one month after a high-profile accident would show their volatile nature. However, such effects are entirely predictable from the psychological theory of availability (see section 11.3.3), and thus can be controlled for by appropriate sampling of opinions. Public perceptions can also reflect a lack of direct information. It would be unreasonable to expect laypeople to deliver reasoned, informed judgements concerning many of today's risks, particularly where the available scientific knowledge is itself subject to considerable uncertainty (Funtowicz & Ravetz 1992); as, for example, in the accidental release to the environment of genetically modified organisms.

However, none of this means that public perceptions are *inevitably* ill-founded. For example, individuals may be in possession of much relevant information concerning their own exposure to risk from everyday hazards. And to some extent this sets the social science community a research task: to develop methods that will allow public risk judgements to be supported by information on (and reflection about) the basic science of the matter, perhaps using a process of interaction between experts and laypeople in a form of citizen's commission or public participation panel (see, for example, Renn *et al.* 1993; Morgan *et al.* 1994; Lynn & Busenberg 1995). Psychological theory would also predict that extended discussion of risk issues in groups would have other benefits, such as reducing framing effects (see section 11.3.4) by encouraging alternative framing of the question, improving reflection through public justification of arguments and providing additional information.

Public attitudes are prejudiced/discriminatory

A potentially serious charge that can be levelled against risk perceptions is that they may reflect a society's prejudices against particular social groups. This might lead to public policy instantiating undesirable, or even illegal, allocation procedures. While there is a plethora of psychological literature showing negative stereotyping of out-groups (for a review, see Sabini 1995), there has been little linking of this to decisions about risk.

A recent article by Irwin *et al.* (1996) extends the psychometric approach to look at the dimensions along which victims of risk (in this case of HIV infection) are represented. Perceived distastefulness and riskiness of the method of infection and the 'deservingness' of the victim largely explained the representation of the victims. The authors point out that these results can be related back to some of the findings of Slovic *et al.* (1980); for example, that illicit drugs promote *less* concern than pesticide risks. While this has been explained in terms of the voluntariness of exposure in each case, it is also possible that beliefs about the potential victims (e.g. drug abusers versus general public) invoke differential degrees of blame and concern for their condition, and hence would be valued differently by society. To date, little research has been conducted to establish the degree to which perceived differences of victims are likely to influence public values for risk reduction, but it would be surprising if this were not the case. How to deal with this issue in policy is both an under-researched and a currently unsolved problem.

Managing risk merely entails managing public perceptions

If perceptions were the only input into risk decisions, then rather than regulating and reducing risks, risk managers might choose to focus upon convincing the public that the threat is negligible (see Rimmington 1995). Under such circumstances the public would be open to abuse by the unscrupulous, and certainly would have been manipulated in the past by self-interested parties such as companies (Gephart 1984), and even governments have at times sought to cover up environmental hazards. For example, there is evidence that knowledge of the seriousness of environmental pollution was actively withheld in many of the former Eastern European states, such as the German Democratic Republic (see Hunnius & Kliemt 1993), and elsewhere in the developed world the byproducts of former industrial processes (e.g. toxic contamination of land from town gas manufacture; Pidgeon *et al.* 1988) have often remained, and continue to remain, concealed for many years. A corollary of this position is that use of public perceptions in risk allocation may lead to inequitable exposures: groups in the population who are more willing to bear (or unable to avoid bearing) risk, or who perceive the risk as being low, will then be asked to bear more absolute risk than others.

However, few would claim that public perceptions should be the *sole* concern of risk management. Acceptable risk decisions require judgements both about the scientific issues involved *and* the ethical or value principles to adopt in allocating resources and distributing the unavoidable harms across a society (Fischhoff *et al.* 1981). The question then becomes one of deciding how far expressed risk perceptions do tap value and ethical issues that should count in any appropriate decision process, and how then to incorporate these with the best available scientific information there is. Of course, the *status quo* is that governments and regulators do often aim to achieve this (by making their own political or ethical judgements about what their constituents will tolerate). For example, in UK regulatory philosophy (HSE 1992) certain risk levels are determined to be intolerable: i.e. too high to expose the public to, irrespective of possible benefits, and even if the public is sanguine about the risk.

The public is not homogeneous in its risk perceptions

It should now be clear, from our earlier discussion of the social and cultural theories of risk perception, that society should not be viewed as a homogeneous whole in terms of risk representation: i.e. different sectors of society select and represent risks in fundamentally different ways. Set against this, as we have noted earlier, the empirical evidence regarding approaches such as cultural theory is, at best, equivocal, and it is also clear that most of the obvious demographic predictors (age, gender, etc.) either show no measurable effect on risk perception or account for only a small percentage of the variance in judgements. Of course, social scientists may not yet have found the right set of predictors of risk attitudes, although the dilemma first raised by Otway & Thomas (1982) would still remain: of accommodating cultural plurality in risk representation into a single position for acceptability decisions. Here, again, public participation processes, aided by decision analytical techniques, may ultimately prove effective in reconciling groups with opposing views (e.g., Fischhoff *et al.* 1984; von Winterfeldt & Edwards 1984; Gregory *et al.* 1993; Webler *et al.* 1995).

The four arguments above present the main challenges to the view that allocation of resources in the risk domain should reflect, to some extent, public perceptions of risk. We believe that none is so damning that the effort to do this in a systematic way should be abandoned. Instead, this section leads us to conclude that we should take efforts to ensure that the public, from whom judgements of risk and acceptability are elicited, should be in groups of well-informed individuals, and that efforts should be made to detect prejudiced perceptions, with a view to later 'debiasing' of judgements. Care should also be taken to reflect the diversity of viewpoints in a culture, with the hope of finding common ground between different groups. We now turn to the arguments in favour of taking account of public risk perceptions, and draw more lessons concerning the 'best' conditions in which to elicit these perceptions.

11.6.2 Arguments in favour of allowing risk perceptions to influence risk assessment and management

Perceptions have consequences

This argument claims that policy ignores perceptions at its peril, because perceptions lead to actions with real consequences: either direct costs or new risks to the public or the viability of institutions. For example, cost–benefit analyses may determine that requiring oil tankers in British waters to be double-skinned is unwarranted, yet tanker spillages may lead people to withdraw custom from particular companies unless regulation is imposed. Failure to regulate thus might cause boycotts of non-compliers, or even politicians not to be re-elected!

Similarly, risk analyses might determine that evacuation of civilians following a particular low-level radiation leak at a nuclear power plant is unnecessary on health grounds. Yet the news of such a leak could produce widespread panic and flight, causing more direct harm than the leak itself would have been expected to produce. Thus, in hindsight, evacuation (however, 'irrational') might have been expected to produce the best outcome (see the literature on the social amplification of risk perceptions in section 11.5.5). This argument indicates that, at minimum, policy makers must show awareness of the public's likely perception of a given hazard, and a willingness to act on the perception if the risk produced by the perception will exceed that from not acting. Indeed, it would be quite possible to take account of such factors in a conventional risk assessment. However, to do so requires a good understanding of their operation, and a willingness to build them into the analysis.

People should have input into risk decisions that affect them

On strictly normative ground, to us at least, technocracy should always be subordinate to democracy (see also Stern 1991; Pidgeon 1996). Such a belief can be justified either by reference to political realism (the assessment that in socio-political systems such as the USA and certain countries of the European Union, the public will force its way into the process anyway), or to political idealism, such that more rather than less public input to policy is an end to be desired in and of itself.

And, as noted above, risk evaluation is a complex decision process involving not only accepted knowledge but also contested facts and values. Consequently, in his report following the long-running planning inquiry into the building of a pressurized water reactor (Sizewell B) on the Suffolk coast, the inquiry chairman Sir Frank Layfield stated that, as a matter of principle: 'The opinions of the public should underlie the evaluation of risk' (Layfield 1987, summary paragraph 2.101h). This has been incorporated to some extent in the UK in the regulatory philosophy on the tolerability of risk (HSE 1992), which states that people should only be asked to live with a risk if they can have confidence that it is under control. It is also recognized that, for certain hazards (such as major accidents at nuclear power plants; Rimmington 1995), the limit to tolerability may be set very low. Clearly, the psychometric work on risk perception sheds considerable light on this issue (see also Pidgeon *et al.* 1992).

Beyond setting tolerability limits, there are other ways in which the public's risk perceptions may be taken into account, and other reasons why this would be desirable. One motivation is ethical/philosophical: exposing an individual to risk without consultation may be viewed as contrary to the democratic process or an infringement of individual rights (see Hadden 1989). Such philosophical niceties may also have tangible consequences. First, participatory decision-making increases an individual's commitment to the course of action selected, particularly if the decision process is seen to be fair (see Rayner & Cantor 1987). Second, public participation may increase trust in the organizations who manage the risk (NRC 1989), and through this acceptance of hazards (e.g. Slovic 1993).

Experts can be biased too

Much of the risk literature in the 1970s and early 1980s drew a sharp distinction between assessment of objective (correct) risk and probabilities by knowledgeable impartial experts and subjective (possibly biased) evaluations by laypeople. It is now generally accepted that such a distinction is untenable in its strong form (Watson 1981; Fischhoff 1989). In part, this is because, as noted above, risk assessment itself requires some value judgements from the experts performing the analysis; for example, when the analyst decides which attributes of risk should be counted as important in the analysis, or the utility associated with particular outcomes (Pidgeon *et al.* 1992; Baron 1993).

Also there is now considerable literature suggesting that it is not only the general public who might exhibit systematic biases: experts may do so as well. Experts' overconfidence in their own judgements has been demonstrated repeatedly (Henrion & Fischhoff 1986), and they are also prone to availability and other biases (see 11.3.3). While it is tempting to believe that placing experts in groups will eliminate such biases, this is not always the case. Janis' (1982) work on Groupthink (section 11.3.6) shows through historical analysis that like-minded groups of experts can produce erroneous judgements and choices accompanied by extreme confidence. Hence, while scientific expertise is indispensable in providing relevant information concerning both facts and their associated degree of uncertainty, experts cannot be viewed as unbiased gold standards of judgement. The psychological literature provides prescriptions concerning debiasing of experts either individually (e.g. provide detailed feedback on the quality of an expert's predictions; see Yates 1990), and in groups (the antidotes to Groupthink noted in section 11.3.6).

Public risk perceptions should enrich expert risk analyses

The above analysis suggests that experts are in a privileged position in terms of information, even if their values and decision processes are not always employed optimally. On the other hand, while non-experts may not possess as much relevant factual information, they may be in a position to augment expert risk analyses with additional useful information, resulting in an overall superior analysis. There are many forms that this proposed enriching might take. For example, formal risk calculations often use expected number of fatalities as the outcome variable of interest. The public, on the other hand, may care about other outcomes as well, such as non-fatal injuries, the distribution of risks and benefits, or 'societal risk' (see section 11.4.1). A richer representation of the problem may therefore be arrived at by considering such issues in the decision process.

The public may also represent a plurality of perspectives unlikely to be found in a group of risk professionals. As noted above, the cultural view argues against a unitary public value system. Cotgrove (1982) makes the interesting point that risk analysts tend to be of similar cultural perspective (cornucopians), at least in their working lives. This suggests that other cultural viewpoints (catastrophists) may be under-represented in formal analysis.

The public has also shown itself to be concerned with such issues as the trustworthiness of the risk managers, the feasibility of evacuation plans and the plausibility of probability estimates—all variables which are influential upon risk but not readily or typically modelled in formal risk assessments. Of course, the philosophical problem remains of how multiple value systems should be integrated into an overall decision or valuation.

The public may also possess knowledge not readily available to experts, corporations or government, and through this be in a position to criticize hidden—possibly inadequate—assumptions underlying risk assessments. For example, knowledge of the actual conditions under which a hazardous product will be used in the real world (Wynne 1992). Here, special interest groups (SIGs) may have collected large amounts of data and expertise concerning the problem under investigation. And while data appear to reside in the realm of objective facts, SIGs are likely to have a different agenda from establishment institutions, leading their data collection and interpretation to have different emphases (e.g. suspected, but not formally documented, safety violations). The SIGs may also have access to different forms of information (e.g. from whistle-blowers) than that directly available to establishment bodies.

11.7 CONCLUSIONS

Research on the psychology of uncertainty and risk perception has moved away from a narrow conception of risk as merely the probability of an event's occurrence, toward a much richer framework for understanding how risks are represented and communicated, and how they are affected by social processes.

Risk managers (government, business, environmental interest groups) must understand the operation of these processes, in order to be able to predict the risk perceptions and activities of the various groups of which society as a whole is composed, as well as to promote effective dialogue with and between them. We also believe that a participatory decision process, involving both experts and the public, offers a richer and more complete understanding of the problems of environmental risk assessment and management than will be offered by conventional analysis on its own. The challenge for social science and policy in the late 1990s and beyond is to elucidate the conditions under which such processes can be best nurtured and sustained to achieve their objectives.

11.8 ACKNOWLEDGEMENTS

We wish to thank Lorraine Hopkins and Richard Cookson for researching sections of the bibliography. The preparation of the chapter was supported in part by a grant from the UK Health and Safety Executive (No. 3271/R73.04) to the University of Newcastle, and subcontracted to the Universities of Bangor and Sussex. However, the opinions expressed in this chapter remain those of the authors alone.

11.9 REFERENCES

Adams, J. (1995) *Risk*. University College Press, London.

Alhakami, A.S. & Slovic, P. (1994) A psychological study of the inverse relationship between perceived risk and perceived benefit. *Risk Analysis*, **14**, 1085–1096.

Arabie, P. & Maschmeyer, C. (1988) Some current models for the perception and judgment of risk. *Organizational Behavior and Human Decision Processes*, **41**, 300–329.

Barber, B.R. (1983) *The Logic and Limits of Trust*. Rutgers University Press, New Brunswick.

Barke, R., Jenkins-Smith, H. & Slovic, P. (1995) *Risk perceptions of men and women scientists*. Unpublished Ms, School of Public Policy, Georgia Institute of Technology.

Baron, J. (1993) *Morality and Rational Choice*. Kluwer, Dordrecht.

Beck, U. (1992) *Risk Society: Towards a New Modernity* (translated by M. Ritter). Sage, London.

Bell, D.E., Raiffa, H. & Tversky, A. (1988) *Decision Making: Descriptive, Normative and Prescriptive Interactions*. Cambridge University Press, Cambridge.

Blockely, D.I. (1980) *The Nature of Structural Design and Safety*. Ellis Horwood, Chichester.

Boholm, Å. (1996) Risk perception and cultural anthropology: critique of cultural theory. *Ethnos*, **61**(1–2), 64–84.

Bord, R.J. & O'Connor, R.E. (1990) Risk communication, knowledge, and attitudes: explaining reactions to a technology perceived as risky. *Risk Analysis*, **10**, 499–506.

Bord, R.J. & O'Connor, R.E. (1992) Determinants of risk perceptions of a hazardous waste site. *Risk Analysis*, **12**, 411–416.

Bostrom, A., Atman, C.J., Fischhoff, B. & Morgan, G.M. (1994) Evaluating risk communications: completing and correcting mental models of hazardous processes, Part II. *Risk Analysis*, **14**, 789–798.

Brown, R.A. & Green, C.H. (1981) Threats to health or safety: perceived risk and willingness to pay. *Social Science and Medicine*, **15C**, 67–75.

Brun, W. (1992) Cognitive components in risk perception: natural versus man-made risks. *Journal of Behavioural Decision Making*, **5**, 117–132.

Brun, W. (1994) Risk perception: main issues, approaches and findings. In: *Subjective Probability* (eds G. Wright & P. Ayton), pp. 295–320. Wiley, Chichester.

Burns, W.J., Slovic, P., Kasperson, R.E., Kasperson, J.X., Renn, O. & Emani, S. (1993) Incorporating structural models into research on the social amplification of risk: implications for theory construction and decision making. *Risk Analysis*, **13**, 611–623.

Buss, D.M., Craik, K.H. & Dake, K.M. (1986) Contemporary worldviews and the perception of the technological system. In: *Risk Evaluation and Management* (eds V.T. Covello, J. Menkes & J. Mumpower), pp. 93–130. Plenum, New York.

Christensen-Szalanski, J.J.J. & Bushyhead, J.B. (1981) Physician's use of probabilistic information in a real clinical setting. *Journal of Experimental Psychology: Human Perception and Performance*, **7**, 928–935.

Cotgrove, S. (1982) *Catastrophe or Cornucopia: The Environment, Politics and the Future*. Wiley, Chichester.

Couch, S.R. and Kroll-Smith, J.S. (1985) The chronic technical disaster: toward a social scientific perspective. *Social Science Quarterly*, **66**, 564–575.

Covello, V.T. (1991) Risk comparisons and risk communication: issues and problems in comparing health and environment risks. In: *Communicating Risks to the Public* (eds R.E. Kasperson & P.J.M. Stallen), pp. 79–124. Kluwer, Dordrecht.

Covello, V.T., Sandman, P. & Slovic, P. (1988) *Risk Communication, Risk Statistics and Risk Comparisons: A Manual for Plant Managers*. Chemical Manufacturers Association, Washington, DC.

Cvetkovich, G.T. & Earle, T.C. (eds) (1991) Special issue on 'Risk and Culture'. *Journal of Cross-Cultural Psychology*, **22**, 11–149.

Dake, K. (1991) Orienting dispositions in the perception of risk: an analysis of contemporary worldviews and cultural biases. *Journal of Cross-Cultural Psychology*, **22**, 61–82.

de Souza Porto, M.F. & de Freitas, C.M. (1996) Major chemical accidents in industrialising countries: the social–political amplification of risk. *Risk Analysis*, **16**, 19–30.

Douglas, M. (1992) *Risk and Blame*. Routledge, London.

Douglas, M. & Wildavsky, A. (1982) *Risk and Culture: an Analysis of the Selection of Technological Dangers*. University of California Press, Berkeley.

Drottz-Sjöberg, B.M. (1991) *Perceptions of risk: studies of risk attitudes, perceptions and definitions*. Centre for Risk Research, Stockholm School of Economics.

Dunlap, R.E. & van Liere, K.D. (1978) The 'new environmental paradigm'. *Journal of Environmental Education*, **9**, 10–19.

Earle, T.C. & Cvetkovich (1995) *Social Trust: Towards a Cosmopolitan Society*. Greenwood Press, New York.

Earle, T.C. & Lindell, M.K. (1984) Public perceptions of industrial risks: a free-response approach. In: *Low-probability-High-consequence Risk Analysis* (eds R.A. Waller & V.T. Covello), pp. 531–550. Plenum, New York.

Edwards, W. (1961) Behavioral decision theory. *Annual Review of Psychology*, **12**, 473–498.

Edwards, W. & von Winterfeldt, D. (1987) Public values in risk debates. *Risk Analysis*, **7**, 141–158.

Einhorn, H.J., Hogarth, R.M. & Klempner, E. (1977) Quality of group judgment. *Psychological Bulletin*, **84**, 158–172.

Eiser, J.R. & Hoepfner, F. (1991) Accidents, disease, and the greenhouse effect: effects of response categories on estimates of risk. *Basic and Applied Social Psychology*, **12**, 195–210.

Eiser, J.R. & van der Pligt, J. (1979) Beliefs and values in the nuclear debate. *Journal of Applied Social Psychology*, **9**, 524–536.

Ellsberg, D. (1961) Risk, ambiguity and the Savage axioms. *Quarterly Journal of Economics*, **75**, 643–649.

Englander, T., Farago, K., Slovic, P. & Fischhoff, B. (1986) A comparative analysis of risk perception in Hungary and the United States. *Journal of Social Behavior*, **1**, 55–66.

Erikson, K. (1994) Comment on William C. Metz' 'Potential impacts of nuclear activities on local economies: rethinking the issue'. *Risk Analysis*, **14**, 771–772.

Fischer, G.W., Morgan, M.G., Fischhoff, B., Nair, I. & Lave, L.B. (1991) What risks are people concerned about? *Risk Analysis*, **11**, 303–314.

Fischhoff, B. (1989) Risk: a guide to controversy. Appendix C. In: *Improving Risk Communication: National Research Council Committee*, pp. 211–319. National Academy Press, Washington, DC.

Fischhoff, B. (1990) Psychology and public policy: tool or toolmaker? *American Psychologist*, **45**, 647–653.

Fischhoff, B. (1995) Risk communication and perception unplugged: twenty years of process. *Risk Analysis*, **15**, 137–145.

Fischhoff, B. & MacGregor, D. (1982) Subjective confidence in forecasts. *Journal of Forecasting*, **1**, 155–172.

Fischhoff, B. & MacGregor, D. (1983) Judged lethality: how much people seem to know depends upon how they are asked. *Risk Analysis*, **3**, 229–236.

Fischhoff, B., Slovic, P. & Lichtenstein, S. (1978a) Fault trees: sensitivity of estimated failure probabilities to problem representation. *Journal of Experimental Psychology: Human Perception and Performance*, **4**, 330–344.

Fischhoff, B., Slovic, P., Lichtenstein, S., Read, S. & Combs, B. (1978b) How safe is safe enough? A psychometric study of attitudes towards technological risks and benefits. *Policy Sciences*, **9**, 127–152.

Fischhoff, B., Lichtenstein, S., Slovic, P., Derby, S.L. & Keeney, R. (1981) *Acceptable Risk*. Cambridge University Press, Cambridge.

Fischhoff, B., Watson, S. & Hope, C. (1984) Defining risk. *Policy Sciences*, **17**, 123–139.

Fitchen, J.M., Heath, J.S. & Fessenden-Raden, J. (1987) Risk perception in community context: a case study. In: *The Social and Cultural Construction of Risk* (eds B.B. Johnson & V.T. Covello), pp. 31–54. Reidel, Dordrecht.

Flynn, J., Burns, W., Mertz, C.K. & Slovic, P. (1992) Trust as a determinant of opposition to a high-level radioactive waste repository: analysis of a structural model. *Risk Analysis*, **12**, 417–429.

Flynn, J., Slovic, P. & Mertz, C.K. (1993) The Nevada initiative: a risk communication fiasco. *Risk Analysis*, **13**, 497–502.

Flynn, J., Slovic, P. & Mertz, C.K. (1994) Gender, race and perception of environmental health risks. *Risk Analysis*, **14**, 1101–1108.

Frisch, D. & Baron, J. (1988) Ambiguity and rationality. *Journal of Behavioral Decision Making*, **1**, 149–157.

Freudenberg, W.R. & Rursch, J.A. (1994) The risks of 'putting the numbers in context': a cautionary tale. *Risk Analysis*, **14**, 949–958.

Freudenberg, W.R., Coleman, C., Gonzales, J. & Helgeland, C. (1996) Media coverage of hazard events: analyzing the assumptions. *Risk Analysis*, **16**, 31–42.

Funtowicz, S.O. & Ravetz, J.R. (1985) Three types of risk assessment: a methodological analysis. In: *Risk Analysis in the Private Sector* (eds C. Whipple & V.T. Covello), pp. 217–232. Plenum, New York.

Funtowicz, S.O. & Ravetz, J.R. (1992) Three types of risk assessment and the emergence of post-normal science. In: *Social Theories of Risk* (eds S. Krimsky & D. Golding), pp. 251–274. Praeger, Westport, CT.

Gardner, G.T. & Gould, L.C. (1989) Public perceptions of the risks and benefits of technology. *Risk Analysis*, **9**, 225–242.

Gephart, R.P., Jr. (1984) Making sense of organizationally based environmental disasters. *Journal of Management*, **10**, 205–225.

Giddens, A. (1990) *The Consequences of Modernity*. Polity Press, Cambridge.

Gigerenzer, G. (1991) How to make cognitive illusions disappear: beyond 'heuristics and biases'. *European Review of Social Psychology*, **2**, 84–115.

Goszczynska, M., Tyszka, T. & Slovic, P. (1991) Risk perception in Poland: a comparison with three other countries. *Journal of Behavioral Decision Making*, **4**, 179–193.

Gray, P.C.R. (1996) *Risk Indicators: Types, Criteria, Effects*. (Programme Group: Men, Environment, Technology: Studies on Risk Communication, Vol. 56.) KFA Research Centre, Jülich.

Green, C.H. (1979) *Someone out there is trying to kill me: acceptable risk as a problem definition.* Unpublished paper presented at the *International Conference on Environmental Psychology*, University of Surrey, Guildford.

Greenberg, M.R. & Schneider, D.F. (1995) Gender differences in risk perception: effects differ in stressed vs. non-stressed environments. *Risk Analysis*, **15**, 503–511.

Gregory, R., Lichtenstein, S. & Slovic, P. (1993) Valuing environmental resources: a constructive approach. *Journal of Risk and Uncertainty*, **7**, 177–197.

Hadden, S. (1989) *A Citizen's Right to Know: Risk Communication and Public Policy*. Westview Press, Boulder, CA.

Henrion, M. & Fischhoff, B. (1986) Assessing uncertainty in physical constants. *American Journal of Physics*, **54**, 791–798.

Henwood, K.L. & Pidgeon, N.F. (1992) Qualitative research and psychological theorizing. *The British Journal of Psychology*, **83**, 97–111.

Henwood, K.L. & Pidgeon, N.F. (1994) Beyond the qualitative paradigm: a framework for introducing diversity in qualitative psychology. *Journal of Community and Applied Social Psychology*, **4**, 225–238.

Holzheu, F. & Wiedemann, P. (1993) Introduction: perspectives on risk perception. In: *Risk is a Construct* (ed. Bayerische Rück), pp. 9–20. Knesebeck, Munich.

HSE (1992) *The Tolerability of Risk From Nuclear Power Stations*, 2 edn. Health and Safety Executive, HMSO, London.

Hunnius, G. & Kliemt, J. (1993) Risk as a social construction: the perception and communication of risk in the Federal Republic of Germany and the German Democratic Republic—a comparison between systems. In: *Risk is a Construct* (ed. Bayerische Rück), pp. 221–236. Knesebeck, Munich.

Irwin, J.R., Jones, L.E. & Mundo, D. (1996) Risk perception and victim perception: the judgement of HIV cases. *Journal of Behavioral Decision Making*, **9**, 1–22.

Janis, I.L. (1982) *Victims of Groupthink*, 2 edn. Houghton-Mifflin, Boston.

Jasanoff, S. (1993) Bridging the two cultures of risk analysis. *Risk Analysis*, **13**, 123–129.

Jenkins-Smith, H. & Bassett, G.W., Jr. (1994) Perceived risk and uncertainty of nuclear waste: differences among science, business and environmental group members. *Risk Analysis*, **14**, 851–856.

Jianguang, Z. (1994) Environmental hazards in the Chinese public's eyes. *Risk Analysis*, **14**, 163–167.

Johnson, B.B. (1991) Risk and culture research: some cautions. *Journal of Cross-Cultural Psychology*, **22**(1), 141–149.

Johnson, B.B. & Covello, V.T. (eds) (1987) *The Social and Cultural Construction of Risk*. Reidel, Dordrecht.

Johnson, B.B. & Slovic, P. (1995) Presenting uncertainty in health risk assessment: initial studies of its effects on risk perception and trust. *Risk Analysis*, **15**, 485–494.

Johnson, E.J. & Tversky, A. (1984) Representations and perceptions of risk. *Journal of Experimental Psychology: General*, **113**, 55–70.

Jones-Lee, M. & Loomes, G. (1995) Scale and context effects in the valuation of transport safety. *Journal of Risk and Uncertainty*, **11**, 183–203.

Jungermann, H. (1986) The two camps on rationality. In: *Judgment and Decision Making* (eds H.R. Arkes & K.R. Hammond), pp. 627–641. Cambridge University Press, Cambridge.

Kahneman, D. & Ritov, I. (1994) Determinants of the stated willingness to pay for public goods; a study in the headline method. *Journal of Risk and Uncertainty*, **9**, 5–38.

Kahneman, D. & Tversky, A. (1972) Subjective probability: a judgment of representativeness. *Cognitive Psychology*, **3**, 430–454.

Kahneman, D. & Tversky, A. (1979) Prospect theory: an analysis of decision under risk. *Econometrica*, **47**, 263–291.

Kahneman, D., Slovic, P. & Tversky, A. (1982) *Judgment under Uncertainty: Heuristics and Biases*. Cambridge University Press, Cambridge.

Kasperson, R.E. (1992) The social amplification of risk: progress in developing an integrative framework. In: *Social Theories of Risk* (eds S. Krimsky & D. Golding), pp. 153–178. Praeger, Westport, CT.

Kasperson, R.E. & Stallen, P.J.M. (1991) *Communicating Risks to the Public*. Kluwer, Dordrecht.

Kasperson, R.E., Renn, O., Slovic, P., Brown, H.S., Emel, J., Goble, R., Kasperson, J.X. & Ratick, S. (1988) The social amplification of risk: a conceptual framework. *Risk Analysis*, **8**, 177–187.

Kemp, R. (1990) Why not in my backyard? A radical interpretation of public opposition to the deep disposal of radioactive waste in the United Kingdom. *Environment and Planning A*, **22**, 1239–1258.

Keown, C.F. (1989) Risk perception of Hong Kong versus Americans. *Risk Analysis*, **9**, 401–407.

Kleinhesselink, R.R. & Rosa, E.A. (1991) Cognitive representations of risk perceptions: a comparison of Japan and the United States. *Journal of Cross-Cultural Psychology*, **22**, 11–28.

Kraus, N. & Slovic, P. (1988) Taxonomic analysis of perceived risk: modeling individual and group perceptions within homogeneous hazard domains. *Risk Analysis*, **8**, 435–455.

Krimsky, S. & Golding, D. (1992) *Social Theories of Risk*. Praeger, Westport, CT.

Kunreuther, H., Fitzgerald, K. & Aarts, T.D. (1993) Siting

noxious facilities: a test of the facility siting Credo. *Risk Analysis*, **13**, 301–315.

Layfield, F. (1987) *Sizewell B Public Inquiry: Summary of Conclusions and Recommendations*. HMSO, London.

Lichtenstein, S., Slovic, P., Fischhoff, B., Layman, M. & Combs, B. (1978). Judged frequency of lethal events. *Journal of Experimental Psychology: Human Learning and Memory*, **4**, 551–578.

Lima, M.L. (1993) *Seismic risk perceptions: fear and illusions of control*. PhD thesis, Instituto Superior de Ciências do Trabalho e da Empresa, University of Lisbon.

Lopes, L.L. (1987) Between hope and fear: the psychology of risk. *Advances in Experimental Social Psychology*, **20**, 255–295.

Lynn, F.M. & Busenberg, G.J. (1995) Citizen advisory committees on environmental policy: what we know, what's left to discover. *Risk Analysis*, **15**, 147–162.

MacGregor, D. (1991) Worry over technological activities and life concerns. *Risk Analysis*, **11**, 315–325.

Marris, C., Langford, I. & O'Riordan, T. (1996) *Integrating sociological and psychological approaches to public perceptions of environmental risks: detailed results from a questionnaire study*. Centre for Social and Economic Research on the Global Environment (CSERGE Working Paper GEC 96-07), University of East Anglia, Norwich.

McDaniels, T.L., Kamlet, M.S. and Fischer, G.W. (1992) Risk perception and the value of safety. *Risk Analysis*, **12**, 495–503.

McDaniels, T.L., Axelrod, L.J. & Slovic, P. (1995) Characterizing perception of ecological risk. *Risk Analysis*, **15**, 575–588.

Mechitov, A.I. & Rebrik, S.B. (1990) Studies of risk and safety perception in the USSR. In: *Contemporary Issues in Decision Making* (eds K. Borcherding, O.J. Larichev & D.M. Messick), pp. 261–270. North-Holland, Amsterdam.

Metz, W.C. (1994) Potential negative impacts of nuclear activities on local economies: rethinking the issue. *Risk Analysis*, **14**, 763–770.

Morgan, G., Fischhoff, B., Lave, L. *et al.* (1994) *A Procedure for risk ranking for federal risk management agencies*. Unpublished manuscript, Department of Engineering and Public Policy, Carnegie Mellon University.

Murphy, A.H. & Winkler, R.L. (1977) Can weather forecasters formulate reliable probability forecasts of precipitation and temperature? *National Weather Digest*, **2**, 2–9.

NRC (1989) *Improving Risk Communication*. National Research Council Committee, National Research Council, National Academy Press, Washington, DC.

NRC (1996) *Understanding Risk: Informing Decisions in a Democratic Society* (eds P.C. Stern & V. Fineberg). National Research Council, National Academy Press, Washington, DC.

Okrent, D. (1996) Risk perception research programme and applications: have they received enough peer review? In: *Probabilistic Safety Assessment and Management '96*, Vol. 2 (eds P.C. Cacciabue & I.A. Papazoglou), pp. 1255–1260. Springer-Verlag, London.

Otway, H.J. & Cohen, J.J. (1975) *Revealed preferences: Comments on the Starr benefit–risk relationships*. Research Monograph 75-5, International Institute for Applied Systems Analysis, Laxenburg, Austria.

Otway, H.J. & Thomas, K. (1982) Reflections on risk perception and policy. *Risk Analysis*, **2**, 69–82.

Otway, H.J. & von Winterfeldt, D. (1982) Beyond acceptable risk: on the social acceptability of technologies. *Policy Sciences*, **14**, 247–256.

Payne, J., Bettman, J.R. & Johnson, E.J. (1992) Behavioral decision research: a constructive processing perspective. *Annual Review of Psychology*, **43**, 87–131.

Perrow, C. (1984) *Normal Accidents*. Basic Books, New York.

Peters, E. & Slovic, P. (1996) The role of affect and worldview as orienting dispositions in the perception and acceptance of nuclear power. *Journal of Applied Social Psychology*, **26**, 1427–1453.

Petts, J. (1994) Effective waste management: understanding and dealing with public concerns. *Waste Management and Research*, **12**, 207–222.

Pidgeon, N.F. (1988) Risk assessment and accident analysis. *Acta Psychologica*, **68**, 355–368.

Pidgeon, N.F. (1991) Safety culture and risk management in organizations. *The Journal of Cross-Cultural Psychology*, **22**, 129–140.

Pidgeon, N.F. (1995) Risk perception, trust and stakeholder values: framing the risk communication question. In: *Her Majesty's Pollution Inspectorate Seminar Proceedings: Risk Perception and Communication* (eds D. Galson, R. Kemp & R. Wilmott), pp. F1–F17. Report DOE/HMIP/RR/95.011, Department of Environment, London.

Pidgeon, N.F. (1996a) The limits to safety? Culture, politics, learning and man-made disasters. *Journal of Contingencies and Crisis Management*, **4**, 1–14.

Pidgeon, N.F. (1996b) Why does risk assessment need risk perception research? In: *Probabilistic Safety Assessment and Management '96*, Vol. 2 (eds P.C. Cacciabue & I.A. Papazoglou), pp. 1261–1266. Springer-Verlag, London.

Pidgeon, N.F. (1996) Technocracy, democracy, secrecy and error. In: *Accident and Design* (eds D. Jones & C. Hood), pp. 164–171. University College London Press, London.

Pidgeon, N.F., Blockley, D. & Turner, B.A. (1988). Site investigations: lessons from a late discovery of hazardous waste. *The Structural Engineer*, **66**, 311–315.

Pidgeon, N.F., Hood, C., Jones, D. & Turner, B.A. (1992) Risk perception. In: *Risk: Analysis, Perception and Management*, pp. 89–134. The Royal Society, London.
Pilisuk, M., Parks, S. & Hawkes, G. (1987) Public perception of technological risk. *The Social Science Journal*, **24**, 403–413.
Puy, A. & Aragones, J.I. (1994). *Risk perception and the management of emergencies*. Unpublished Manuscript, Department of Social Psychology, University of Complutense, Madrid.
Raiffa, H. (1968) *Decision Analysis: Introductory Lectures on Choice under Uncertainty*. Addison-Wesley, Reading, MA.
Rayner, S. (1992) Cultural theory and risk analysis. In: *Social Theories of Risk* (eds S. Krimsky & D. Golding), pp. 83–116. Praeger, Westport, CT.
Rayner, S. (1993) Risk perception, technology acceptance and institutional culture: case studies of some new definitions. In: *Risk is a Construct* (ed. Bayerische Rück), pp. 197–220. Knesebeck, Munich.
Rayner, S. & Cantor, R. (1987) How fair is safe enough? The cultural approach to social technology choice. *Risk Analysis*, 7, 3–9.
Renn, O. & Levine, D. (1991) Credibility & trust in risk communication. In: *Communicating Risks to the Public* (eds R.E. Kasperson & P.J.M. Stallen), pp. 175–210. Kluwer, Dordrecht.
Renn, O., Webler, T., Rakel, H. Dienel, P. & Johnson, B. (1993) Public participation in decision making: a three step procedure. *Policy Sciences*, **26**, 189–214.
Rimmington, J. (1995) Risks and the regulator: puzzles and predicaments. *Transactions of the Institution of Chemical Engineers*, **73**(B), 173–181.
Rip, A. (1988) Should social amplification of risk be counteracted? *Risk Analysis*, **8**, 193–197.
Rohrmann, B. (1995) *Risk perception research: review and documentation* (Programme Group: Men, Environment, Technology: Studies on Risk Communication, Vol. 48). KFA Research Centre, Jülich.
Royal Society (1992) *Risk: Analysis, Perception and Management*. The Royal Society, London.
Sabini, J. (1995) *Social Psychology*, 2 edn. Norton, London.
Savage, I. (1993) An empirical investigation into the effect of psychological perceptions on the willingness-to-pay to reduce risk. *Journal of Risk and Uncertainty*, **6**, 75–90.
Schwarz, M. & Thompson, M. (1990) *Divided We Stand: Redefining Politics, Technology and Social Choice*. Harvester Wheatsheaf, Hemel Hempstead.
Shrivastava, P. (1992) *Bhopal: Anatomy of a Crisis*, 2 edn. Paul Chapman Publishing, London.
Sjöberg, L. (1995) *Explaining risk perception: an empirical and quantitative evaluation of cultural theory*. Center for Risk Research, Report No. 22, Stockholm School of Economics.
Sjöberg, L. & Sjöberg, B.M. (1991) Knowledge and risk perception among nuclear power plant employees. *Risk Analysis*, **11**, 607–618.
Slovic, P. (1987) Perception of risk. *Science*, **36**, 280–285.
Slovic, P. (1992) Perception of risk: reflections on the psychometric paradigm. In: *Social Theories of Risk* (eds S. Krimsky & D. Golding), pp. 117–152. Praeger, Westport, CT.
Slovic, P. (1993) Perceived risk, trust and democracy. *Risk Analysis*, **13**, 675–682.
Slovic, P., Fischhoff, B. & Lichtenstein, S. (1977) Behavioral decision theory. *Annual Review of Psychology*, **28**, 1–39.
Slovic, P., Fischhoff, B. & Lichtenstein, S. (1979) Rating the risks. *Environment*, **21**(3), 14–20, 36–39.
Slovic, P., Fischhoff, B. & Lichtenstein, S. (1980) Facts and fears: understanding perceived risk. In: *Societal Risk Assessment* (eds R. Schwing & W.A. Albers), pp. 181–124. Plenum, New York.
Slovic, P., Lichtenstein, S. & Fischhoff, B. (1984) Modeling the societal impact of fatal accidents. *Management Science*, **30**, 464–474.
Slovic, P., Kraus, N.N., Lappe, H., Letzel, H. & Malmfors, T. (1989) Risk perception of prescription drugs: report on a survey in Sweden. *Pharmaceutical Medicine*, **4**, 43–65.
Smithson, M. (1989) *Ignorance and Uncertainty*. Springer-Verlag, Berlin.
Sparks, P. & Shepherd, R. (1994) Public perception of potential hazards associated with food production and consumption. *Risk Analysis*, **14**, 799–806.
Starr, C. (1969) Social benefit versus technological risk. *Science*, **165**, 1232–1238.
Stern, P.C. (1991) Learning through conflict: a realistic strategy for risk communication. *Policy Sciences*, **24**, 99–119.
Stoner, J.A.F. (1968) Risky and cautious shifts in group decisions: the influence of widely held values. *Journal of Experimental Social Psychology*, **4**, 442–459.
Svenson, O. (1981) Are we all less risky and more skilful than our fellow drivers? *Acta Psychologica*, **47**, 143–148.
Teigen, K.H. (1994) Variants of subjective probabilities: concepts norms and biases. In: *Subjective Probability* (ed. G. Wright & P. Ayton), pp. 211–238. Wiley, Chichester.
Teigen, K.H., Brun, W. & Slovic, P. (1988). Societal risks as seen by a Norwegian public. *Journal of Behavioral Decision Making*, **1**, 111–130.
'T Hart, P., Stern, E.K. and Sundelius, B. (1997) *Beyond Groupthink: Political Group Dynamics and Foreign Policy Making*. Michigan University Press, Ann Arbor.
Thompson, M., Ellis, R. & Wildavsky, A. (1990) *Cultural Theory*. Westview Press, Boulder, CO.
Turner, B.A. & Pidgeon, N.F. (1997) *Man-Made Disasters*, 2 edn. Butterworth-Heinemann, London.

Tversky, A. & Kahneman, D. (1973) Availability: a heuristic for judging frequency and probability. *Cognitive Psychology*, **5**, 207–232.

Tversky, A. & Kahneman, D. (1981) The framing of decisions and the psychology of choice. *Science*, **211**, 453–458.

van der Pligt, J., van der Linden, J. & Ester, P. (1982) Attitudes to nuclear energy: beliefs, values and false consensus. *Journal of Environmental Psychology*, **2**, 221–231.

Vaughan, D. (1996) *The Challenger Launch Decision: Risky Technology, Culture and Deviance at NASA*. University of Chicago Press, Chicago.

Vaughan, E. (1995) The significance of socio-economic and ethnic diversity for the risk communication process. *Risk Analysis*, **15**, 169–180.

Vaughan, E. & Nordenstam, B. (1991) The perception of environmental risks among ethnically diverse groups. *Journal of Cross-Cultural Psychology*, **22**(1), 29–60.

Vlek, C.J.H. & Cvetkovich, G. (1989) Social decision making on technological projects: review of key issues and a recommended procedure. In: *Social Decision Methodology for Technological Projects* (eds C. Vlek & G. Cvetkovich), pp. 297–322. Kluwer, Dordrecht.

Vlek, C.J.H. & Stallen, P.J.M. (1980) Rational and personal aspects of risk. *Acta Psychologica*, **45**, 273–300.

Vlek, C.J.H. & Stallen, P.J.M. (1981) Judging risks and benefits in the small and in the large. *Organizational Behavior and Human Performance*, **28**, 235–271.

von Winterfeldt, D. & Edwards, W. (1984) Patterns of conflict about risky technologies. *Risk Analysis*, **4**, 55–68.

Watson, S.R. (1981) On risks and acceptability. *Journal for the Society for Radiological Protection*, **1**, 21–25.

Webler, T., Rakel, H., Renn, O. & Johnson, B. (1995) Eliciting and classifying concerns: a methodological critique. *Risk Analysis*, **15**, 421–436.

Weiner, R.F. (1993) Comment on Sheila Jasanoff's guest editorial in Risk Analysis, Vol. 13, Number 2. *Risk Analysis*, **13**, 495–496.

Weinstein, N.D. (1980) Unrealistic optimism about future life events. *Journal of Personality and Social Psychology*, **39**, 806–820.

Wildavsky, A. (1988) *Searching for Safety*. Transaction Books, New Brunswick.

Wildavsky, A. & Dake, K. (1990) Theories of risk perception: who fears what and why? *Dædalus*, **119**, 41–60.

Wright, G. & Ayton, P. (1994) *Subjective Probability*. Wiley, Chichester.

Wynne, B. (1980) Technology, risk and participation: on the social treatment of uncertainty. In: *Society, Technology and Risk* (ed. J. Conrad), pp. 167–202. Academic Press, New York.

Wynne, B. (1992) Risk and social learning: reification to engagement. In: *Social Theories of Risk* (eds S. Krimsky & D. Golding), pp. 275–300. Praeger, Westport, CT.

Yates, J.F. (1990) *Judgment and Decision Making*. Prentice-Hall, Englewood-Cliffs, NJ.

Chapter 12
The Economics of Risk

CHRIS V. STARMER

12.1 INTRODUCTION

The purpose of this chapter is to introduce and explore certain aspects of the economic theory of risk. Readers unfamiliar with economics might wonder why economic theory should feature in this volume. It is common for non-economists to assume that economic theory is primarily, if not exclusively, concerned with matters financial. In the assessment of environmental problems like, say, the environmental impact of alternative routes for a new road, or the risks associated with competing programmes for national power generation, the financial implications of choosing between such alternatives may appear as only one dimension of concern to individuals (or social groups) alongside, say, the physical risks to human life, health and the ecosystem. As such, the role of economic theory in relation to such problems might appear quite limited. It would be misleading, however, to think of economic theory as concerned only with pounds and pence (or dollars and cents).

Economics has, in part, concerned itself with developing strategies for valuing (non-marketed) 'goods' in monetary terms, and techniques of this kind (e.g. the Contingent Valuation Method, discussed in Chapter 13) have been widely applied. Such techniques, however, do not set aside concerns with the values that humans place on life, safety or the intrinsic beauty of natural habitats and so on. Quite the reverse: they constitute an attempt to provide a method by which different dimensions of value can be compared on a common scale. Whatever the merits and limitations of such techniques, the general scope of economic theory is broader still. Economics has been described as the 'science of choice'. This neatly captures the essence of (microeconomic) theorizing; it is about understanding decision-making.

The economic theory of decision-making under risk constitutes one branch of this 'science' and during the last two decades this has been an area of intense theoretical discussion in economics. My intention in this chapter is to review some key theoretical developments in this area, to assess the empirical support for competing theories, and to point to two potentially distinct roles which theories of decision-making may serve. In developing theories of risky decision-making, economists have been motivated by two main goals. One goal has been to develop a satisfactory *descriptive* model of choice behaviour, i.e. a model capable of predicting observable regularities in human responses to risks. The other has been to develop models of 'rational choice'; such models are to be interpreted *normatively* as theories that identify optimal choice strategies in certain contexts. In principle, the two projects could coincide, but only if real people choose 'optimally' in all relevant situations.

Empirical evidence, some of which is discussed below and in Chapter 13, suggests at least some divide between optimal behaviour and actual behaviour, in some contexts, and so both lines of investigation are relevant to issues of environmental risk assessment. Indeed, the two approaches are quite complementary. For instance, in order to assess whether ordinary people are reacting 'sensibly' in relation to a particular hazard—say, for example, exposure to ultraviolet radiation—one must understand how people do, in fact, respond to such risks (the descriptive project) and formulate some notion of how people ought to behave (the normative project). The model of optimal behaviour can then serve as a benchmark against which actual behaviour can be judged.

These two projects have each raised questions which have yet to be fully resolved. The conventional economic account of rational choice (see section 12.3) remains controversial and economists have yet to develop a fully satisfactory descriptive model of behaviour in relation to risk. Indeed, it may well be unreasonable to expect that the 'economic perspective' can provide all the answers in this domain. The purpose of this chapter is to explore what the economics literature has to contribute to debates in this arena.

The theory and evidence reviewed in this chapter constitute fundamental research examining the foundations of economic theory. Much of the theory is relatively new; a large proportion of the available evidence derives from observations of behaviour in simple laboratory contexts, and the implications it has for policy in relation to environmental issues has yet to be fully explored. Yet its potential importance for matters of policy in relation to environmental issues should not be underestimated. Many of the standard techniques applied in the context of environmental policy—for example, cost–benefit analysis (see Johansson 1993), contingent valuation methods (Chapter 13; Mitchell & Carson 1989) and the standard gamble method (see below)—rest on a particular conception of individual behaviour deriving from economics. Much of the evidence reviewed in this chapter suggests that this model of behaviour, at least in its conventional formulation (expected utility theory), may be unsatisfactory.

My prime objective will be to evaluate the basic theoretical assumptions which underlie such techniques (although I shall offer illustrations and refer to applications along the way). In section 12.2 I introduce the conventional economic theory of decision-making under risk. Subsequently, I discuss a range of empirical anomalies which have given rise to doubts about the validity of this model as a general descriptive account of individual choice under risk. I discuss a range of alternative models which have been proposed in the economics literature. Finally, in section 12.7, attention turns towards normative issues where I outline the basis for the use of expected utility theory as a normative guide to choice.

12.2 THE ECONOMIC CONCEPTION OF RISK

In this chapter I shall be working with a particular conception of what constitutes a risk. This is not the only way of conceiving of what risk means, and alternative conceptions of risk have been discussed earlier. I shall, nevertheless, argue that the framework applied here has a reasonable degree of generality.

In economics, a situation of 'risky choice' is any choice that can be characterized in the following way. An individual (or collection of individuals) faces a choice between two or more risky options, one of which must be selected. The decision-maker can identify a well-defined set of 'outcomes' $(x_1, \ldots, x_n)$ which contains all the possible consequences of all relevant alternatives, considered together. For any available option, the decision-maker can attach a probability to each consequence such that each option can then be represented simply by a probability distribution $(p_1, \ldots, p_n)$ over the outcome set, where p_i is the probability that the option under consideration results in consequence x_i, with $p_i \geq 0$ (for all i) and $\Sigma p_i = 1$. I will follow the convention in the economics literature and refer to any risky option described purely by a probability distribution over the outcome set as a 'prospect'. Notice that an option which offers one outcome for sure can be represented as a degenerate prospect in which the probability of one consequence is unity (hence, $p_i = 1$ for some i, and $p_j = 0$ for all $j \neq i$). A situation of risky choice, then, is to be understood as any situation where the decision-maker faces a choice between two or more prospects.

While this abstract characterization of a risky choice setting may not capture all elements of everyday conceptions of risk—for instance, in many real situations of concern, the decision-makers may not be aware of all of the possible implications of their actions, or may have only partial knowledge of the relevant probabilities* —the framework is

* Economists following Knight (1921) refer to circumstances where at least some of the consequences of action, or their associated probabilities, are unknown as situations of 'uncertainty' as opposed to risk.

perhaps more general than it may first appear. For instance, it is not necessary to assume that individuals have objective knowledge of the relative frequencies of potential outcomes (cf. Chapter 1): we could think of the probabilities as subjective representations of the decision-maker's beliefs about the likelihood of different consequences (see Savage 1954). Second, there is very little restriction on what can count as an outcome*—in principle, sources of 'hazard' might include, for example, risks of financial loss (or gain), risks of physical harm (ranging, for example, from minor effects of noise or litter pollution to the potentially dramatic effects of global climatic change), or even risks of psychological or emotional trauma associated with, say, living in proximity to some real or perceived environmental hazard. Moreover, we could think of the outcome set as containing those consequences imagined by the decision-maker as opposed to those which objectively exist.

This still leaves open many interesting questions about how decision-makers arrive at their subjective characterizations of a choice situation. These are questions that I will not address explicitly in this chapter (but see discussion of risk perception in Chapter 20). The focus in this chapter will be upon how individuals choose, or ought to choose, *given* their perceptions of the risks that they face. With these preliminaries in mind, we are now in a position to begin consideration of the economic theory of decision-making under risk.

12.3 EXPECTED UTILITY THEORY

Economic theories of decision-making start from a common basic assumption: that individuals have *preferences*. Initially it may be helpful to think of an individual's preferences as their likes or dislikes, although I shall have more to say about this shortly. Descriptive theories then typically proceed by assuming that individuals make those choices which are expected to lead to the most preferred outcome, i.e. they attempt to maximize their 'utility', or well-being, given their preferences.

The theory of decision-making under risk assumes that individuals have preferences over prospects and, from the point of view of descriptive theorizing, the problem is to find a characterization, or model, of preferences which allows us to predict behaviour. One theory of preferences under risk has dominated the literature in economics: expected utility theory (or EUT for short). The core hypothesis of EUT is that individuals are *expected utility maximizers*, i.e. they will behave *as if* calculating an 'expected utility' for each prospect under consideration, choosing the one with the highest expected utility. The expected utility of any prospect, P, is given by:

$$EU(P) = \Sigma_i p_i U(x_i) \quad (1)$$

where $U(x_i)$ is the 'utility' of consequence x_i. An expected utility maximizer will choose some prospect P over some other available prospect Q if, and only if, $EU(P) \geq EU(Q)$.

In the original statement of EUT, due to Bernoulli (1954), the utility of a consequence was to be interpreted as a (subjective) measure of the pleasure or pain associated with the experience of that consequence. On this interpretation, the expected utility of a prospect is then simply a weighted average of the possible 'satisfactions' and we may think of individuals as choosing whichever prospect generates, *ex ante*, the highest expected satisfaction.

This interpretation is now rather unfashionable.† Contemporary economists (at least those concerned with the descriptive project) tend to stress that theories like EUT are intended as theories of *choice* as opposed to theories of the mental processes which give rise to choices. On this latter interpretation, a proponent of EUT need not assume that individuals actually try to calculate expected utilities in their minds, indeed it is not even necessary to imagine that anything remotely corresponding to a utility function exists in the mind of the

* See Sugden (1997) for a more detailed discussion of this point.

† This is partly a consequence of the development of axiomatic reformulations of EUT, which give rise to a different interpretation of the utility function. These developments are discussed in section 12.7.

chooser. If the model is understood as a theory about the choices people make, as opposed to a theory of how they come to make them, the relevant question to ask is whether the model is successful in predicting the behaviour of real people in situations of interest; or, more bluntly, we may ask, does the theory fit the facts? It is to this question that I now turn.

12.4 ASSESSMENT OF EUT AS A DESCRIPTIVE THEORY

The question of whether EUT fits the facts can be addressed at a number of levels. One approach is to ask whether certain basic properties of the theory accord with our own intuitions, or expectations, about how ordinary people behave. For instance, in situations of risk, it seems quite obvious that, in general, it will matter to individuals what consequences they may face and how likely those consequences are. Any sensible theory must make reference to both of these dimensions. The expected utility function (expression (1) above) is one very simple way of combining these two dimensions into a single 'value measure' which gives rise to at least some intuitively plausible implications with respect to the way in which individuals will respond to changes in probabilities and/or consequences. For example, in EUT, increasing the probability of more highly valued consequences in a prospect (i.e. those with relatively high utilities), at the expense of less highly valued consequences, will increase the value assigned to a prospect, as will replacing some consequence of a prospect with one that is more highly valued. In more technical parlance, these properties reflect the fact that EUT satisfies *monotonicity*.

Monotonicity can be defined as follows. Let $x_1, \ldots, x_n$ be consequences ordered from worst (x_1) to best (x_n). We may say that one prospect ($p_1, \ldots, p_n$) stochastically dominates another prospect ($q_1, \ldots, q_n$) if for all $i = 1, \ldots, n$:

$$\sum_{j=i}^{n} p_j \geq \sum_{j=i}^{n} q_j$$

with a strict inequality for at least one i. Monotonicity is the property that stochastically dominating prospects are preferred to prospects which they dominate.

Since it is widely held that individuals do not violate monotonicity, in most cases monotonicity is typically regarded as a property which any satisfactory descriptive theory should satisfy. But while many theorists take this to be an intuitively appealing property of the expected utility function, it must be said that a variety of other functional forms which combine probability and outcome are also monotonic (we shall encounter some of these alternatives later). This feature of EUT, then, is no more than one benchmark test of the theory.

A further, more challenging, test would be whether the theory is capable of accommodating certain stylized facts about real-world behaviour. Casual observation reveals that people differ in their willingness to take risks: their willingness to engage in gambling or dangerous sports, for example, or their propensities to insure against certain kinds of risk. It is also quite apparent that the majority of ordinary individuals actively choose to undertake some risky activities while simultaneously insuring against other risks (e.g. I take it that many players of the National Lottery also insure their household belongings against theft). A credible theory of decision-making under risk should be able to accommodate such general facts.

In EUT, the 'shape' of the utility function $U(.)$ can be interpreted as reflecting the individual's attitude to risk. Consider, for ease of exposition, the class of prospects whose outcomes can be measured along a single dimension, such as wealth, so that x represents level of wealth. Let us assume, as seems reasonable for most individuals, that at most wealth levels the utility of wealth function $U(x)$ is increasing (i.e. $dU(x)/dx > 0$) such that more wealth is preferred to less, other things being equal. The sign of the second derivative of the utility of wealth function (denoted U'') can then be interpreted as reflecting the individual's (local) attitude towards risk.

For instance, an individual with a strictly concave utility function, such that $U'' < 0$, can be said to be 'risk averse' in the following sense: any such individual, if offered a choice between a wealth level of x^* for sure, and any (non-degenerate) prospect which induces an expected wealth† of x^*, will choose x^* for sure. The reverse will be true for individuals with convex utility functions (i.e. where $U'' > 0$) and such individuals may be described as 'risk seekers'. Since EUT allows the utility function to vary from person to person, the theory can accommodate differences in attitudes to risk across individuals. The theory can also accommodate certain conjunctions of risk seeking and risk averse behaviour, for a given individual, by invoking utility functions with concave and convex segments. Indeed, a famous paper by Friedman & Savage (1948) argues that a relatively simple specification of the utility of wealth function is capable of accommodating certain typical patterns observed in insurance and gambling behaviour (see also a subsequent paper by Markowitz (1952)).

Expected utility theory thus provides a simple account of risky decision-making which is consistent with certain easily observable aspects of real-world behaviour. There is evidence to suggest, however, that EUT has limitations as a descriptive account of human decision-making under risk. Much of this evidence has emerged from experimental settings which allow direct tests of the implications of EUT: these tests have revealed a variety of empirical 'anomalies', relative to the predictions of EUT. I shall discuss a small number of these anomalies to illustrate some important aspects of this literature. Readers seeking a more comprehensive treatment are recommended to consult Camerer (1995).

The most famous empirical challenge to EUT came in the form of a simple pair of choice problems invented by Allais (1953). I shall discuss a version of these problems due to Kahneman & Tversky (1979). An individual is asked to make two (hypothetical) pairwise choices between four alternative prospects labelled *A*, *B*, *C* and *D*. The choices are presented here in a state/payoff matrix with three states of the world ($s1$, $s2$ and $s3$) whose associated probabilities are 0.33, 0.01 and 0.66. In choice 1, an individual who chooses option *A* 'gets' £2400 for sure [in the original example the units of payoffs are Israeli pounds.] Whereas if they choose *B* they enter a gamble which gives £2400 with probability 0.66 and £2500 with probability 0.33, otherwise nothing. Options *C* and *D* should be interpreted in a similar way. Before reading further, take a moment to consider which option you would select in each of these two choice problems.

	State:	$s1$	$s2$	$s3$
	Probability:	0.33	0.01	0.66
Choice 1				
Choose either	*A*:	2400	2400	2400
A or *B*	*B*:	2500	0	2400
Choice 2				
Choose either	*C*:	2400	2400	0
C or *D*	*D*:	2500	0	0

It is easy to show that EUT implies that an individual who prefers *A* to *B* must also prefer *C* to *D* (and, likewise, a preference of *B* over *A* entails a preference of *D* over *C*). To see why, let us apply EUT to choice 1 assuming $U(0) = 0$ (which is a legitimate normalization of the utility function involving no loss of generality). We can then derive the decision rule:

$$A \gtrless B \Leftrightarrow U(2400) \gtreqless 0.66U(2400) + 0.33U(2500) \quad (2)$$

where $\succ$ is the relation 'is preferred to' and $\sim$ is the relation 'is indifferent to' (i.e. just as good as). Rearranging gives:

$$A \gtrless B \Leftrightarrow 0.34U(2400) \gtreqless 0.33U(2500) \quad (3)$$

But notice that the right-hand side of rule (3) is the decision rule that results from application of EUT to choice 2. Hence:

$$A \gtrless B \Leftrightarrow C \gtrless D \quad (4)$$

However, Kahneman & Tversky (1979) find that the modal response of subjects is to choose *A* then *D*. Contrary to EUT, it appears that individuals are more willing to take the 'riskier' option in choice 2, where the overall chances of winning are smaller. This is the Allais paradox, which is

† Note that this is the mathematical expectation of wealth induced by the prospect as opposed to the expected utility of wealth.

one example of a more general pattern of EUT violation that has become known as the common consequence effect (notice that choices 1 and 2 are identical apart from the value of the common consequence assigned to state *s*3). Instances of the common consequence effect have been reported by, among others, Moskowitz (1974), Slovic & Tversky (1974), Kahneman & Tversky (1979) and MacCrimmon & Larsson (1979).

A closely related phenomenon is revealed in the so-called 'common ratio effect'. Consider choices 3 and 4 below.

Choice 3	
Choose either	*E*: £3000, 1
E or *F*	*F*: £4000, 0.8
Choice 4	
Choose either	*G*: £3000, 0.25
G or *H*	*H*: £4000, 0.20

Option *E* offers £3000 for sure, option *F* offers a 0.8 chance of winning £4000, otherwise nothing, and so on. Kahneman & Tversky (1979) report a marked tendency for individuals presented with problems like this to select *E* in choice 3 and *H* in choice 4. This pattern of choice violates EUT, which implies that if *E* (*F*) is chosen in choice 3, then *G* (*H*) should be chosen in choice 4. To see why, notice that these two problems each offer a choice between a pair of prospects of the form:

$$S: \ (p, x; 1-p, 0)$$
$$R: \ (\lambda p, y; 1-\lambda p, 0)$$

where *S* here denotes the 'safer' option and *R* the 'riskier' option. In choices 3 and 4, $x = £3000$, $y = £4000$ and $\lambda = 0.8$. The only difference between the two choices is the value of p, which is 1 in choice 3 and 0.25 in choice 4. If we apply EUT to the choice between *S* and *R*, we may derive the decision rule:

$$S \gtreqless R \Leftrightarrow pU(x) \gtreqless \lambda pU(y) \qquad (5)$$

Since p can be eliminated from this expression (by dividing through by p on each side of the inequality), it follows from EUT that choices between *S* and *R* should be independent of the value of p.

Numerous studies involving pairs of problems with this structure, including Tversky & Kahneman (1981), Loomes & Sugden (1987), Starmer & Sugden (1989) and Battalio *et al.* (1990), have revealed a tendency for individuals to switch their choice from *S* to *R* as p falls. This switch of preference is the common ratio effect, and this effect has been replicated in a wide range of studies, many of which have used significant financial incentives to motivate subjects to respond thoughtfully and truthfully.

There is now a substantial body of evidence revealing widespread violation of EUT in experimental settings. The common consequence and common ratio effects are just two examples of such violation (for a more general discussion of these and other violations of EUT, see Appleby & Starmer (1987), Camerer (1995) and Machina (1987)). Although the tests revealing these violations of EUT have, admittedly, tended to focus on a narrow class of risk (i.e. quite simple choices between prospects with monetary payoffs), this does not undermine the significance of the evidence for our purposes here. If a theory fails to explain behaviour in simple controlled settings, this must raise serious doubts about its applicability to more complex choices in more naturalistic settings.

If individuals' preferences in relation to risk do diverge in general from the received theory of choice, then serious problems are created for policy evaluation methods, such as cost–benefit analysis. Such methods, when applied to public decisions which increase or reduce risks, require data on individuals' preferences in relation to risk. Any method for eliciting such preferences must be based on a theory of preferences which imposes some structure on the data generated. Conventional elicitation methods (see the standard gamble method below) presuppose preferences which satisfy standard coherence properties, and so break down when applied to data which exhibit anomalies.

It would, of course, be surprising to find a theory capable of predicting behaviour 100% of the time. Perhaps the most that one could reasonably expect of a theory is that the departures from it be 'random'. These anomalies, however, appear quite systematic, i.e. they are predictable. Moreover, a number of theorists have suggested that anomalies like the common consequence and common ratio effects might be interpreted as revealing some underlying feature of preferences which, if properly understood, could form the basis of a *common* or unified account of such phenomena. What might that feature of preferences be?

A possible connection between these two phenomena can perhaps be identified using a diagrammatic device known as a unit probability triangle. Consider the class of prospects defined over three outcomes x_1, x_2 and x_3 such that $x_1 < x_2 < x_3$. Any such prospect can be described as a vector of probabilities of the form $(p_1, 1 - p_1 - p_3, p_3)$ and can be located in a unit probability triangle. Figure 12.1 is a unit probability triangle where the horizontal axis measures the probability of consequence x_1 (i.e. p_1), which increases from 0 to 1 as we move left to right. The vertical axis measures the probability of the best consequence, with p_3 increasing from 0 to 1 as we move from bottom to top.

The prospect (0,0,1), which gives x_3 for sure, would be located at the top left corner of the triangle. The prospect (0.5, 0.5, 0), which offers a 50/50 mix of x_1 and x_2, would lie exactly half-way along the base of the triangle and any prospect giving a non-zero probability for each consequence would lie in the interior of the triangle.

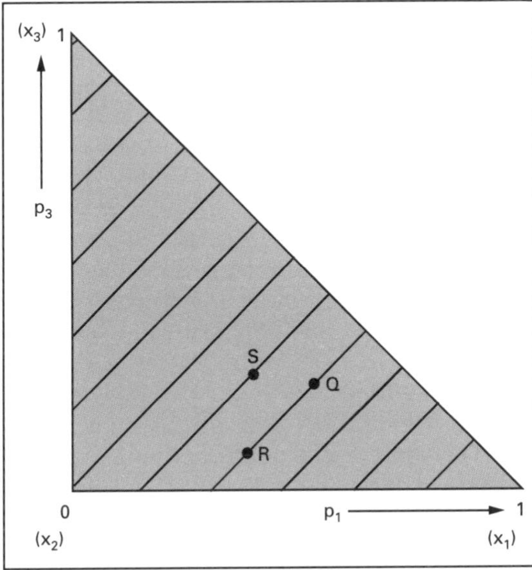

Fig. 12.1 The unit probability triangle with expected utility indifference curves.

Expected utility theory implies that preferences over prospects located in such a triangle can be represented by a set of *indifference curves*. The lines drawn inside the triangle represent indifference curves. An indifference curve is a locus of prospects which are equally valued (hence if two prospects R and Q lie on the same indifference curve, $R \sim Q$). Expected utility theory places quite strong restrictions on the form that indifference curves will take in the unit triangle: they will be upward sloping (left to right), linear and parallel. Finally, preferences are increasing as we move northwest so that, for example, $S \succ R$ and Q since S lies on a 'higher' indifference curve than R and Q. In fact only one feature of the indifference curves is left undetermined in EUT: their slope. This may vary from individual to individual, reflecting differing attitudes to risk; the steeper the slope of the indifference curves, the more risk-averse the individual. For a simple explanation of why EUT implies indifference curves of this form, see Sugden (1986).

Now consider Figs 12.2a and 12.2b. Figure 12.2a locates the prospects from choice problems 1 and 2 (the Allais paradox problems) in a unit probability triangle. Figure 12.2b does the same for choice problems 3 and 4. Notice that, in each of the triangles, the lines joining the prospects are parallel. Consider Fig. 12.2a. Suppose that the indifference curve through A is steeper than the line joining A to B. That implies that A is on a higher indifference curve than B and hence $A \succ B$. But if this is the case then, since EUT implies that indifference curves must be parallel, it follows that C must lie on a higher indifference curve than D. Likewise, if the slope of indifference curves is such that $B \succ A$, then $D > C$. Hence, albeit via different reasoning, we have arrived at the conclusion stated above in expression (4). Similar reasoning can quickly establish a corresponding result for the common ratio problems in Fig. 12.2b. I shall simply state this as follows:

$$A \gtreqless F \Leftrightarrow G \gtreqless H \qquad (6)$$

With the aid of this diagrammatic device, it is easy to see that the choice of A then D (the Allais

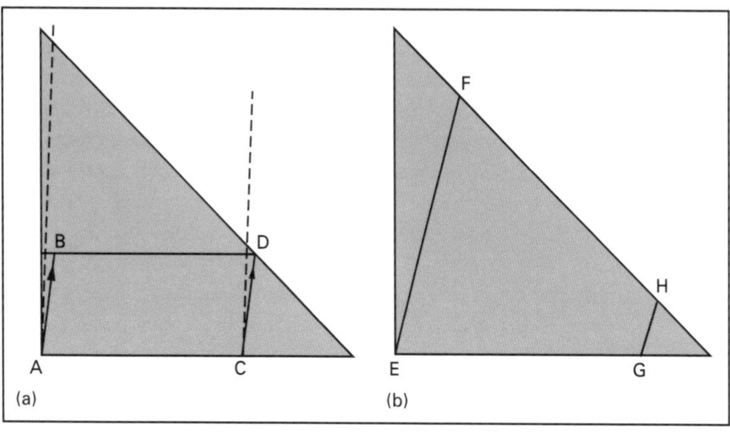

Fig. 12.2 (a) The Allais paradox problems. (b) The common ratio problems.

paradox in Fig. 12.2a) or the choice of E then H (the common ratio effect in Fig. 12.2b) violates EUT. Presenting the violations in this way also helps to suggest a possible connection between the two phenomena. In each case, the behaviour in the bottom right-hand corner of the triangle reflects less risk aversion than EUT would allow, given choices further to the left in the triangle.

If this is suggestive of a common pattern in at least some of the violations of EUT, the question then arises as to whether it is possible to develop an alternative model which retains certain desirable properties of EUT (like being able to explain the conjunction of insurance and gambling behaviour) but at the same time generates implications consistent with the anomalies. A number of theorists have addressed this issue and a substantial literature has emerged. In discussing this literature, my strategy will be to present a small number of models which are illustrative of different approaches to the problem of modelling behaviour under uncertainty. For a more detailed survey of alternatives to EUT, see Sugden (1997).

12.5 ALTERNATIVE MODELS OF CHOICE UNDER RISK

The models I will discuss are all, in one way or another, developments of EUT in the sense that each encompasses EUT as a special case. However, different theorists have followed discernibly different modelling strategies. I will identify three different types of model which I will refer to as the *conventional*, *exotic* and *procedural* models, respectively. I will present theories in each category, outlining the basics of each theory and indicating how the theories can explain anomalies relative to EUT. After presenting all of the theories, I shall discuss some more recent empirical evidence which reflects on the merits of these theories as descriptive models of choice behaviour.

12.5.1 The conventional approach

I will use the term 'conventional' here to refer to any model that assumes that choices can be explained purely in terms of utility maximization, using a preference function which assigns a unique value to each prospect. Expected utility theory is one model of this form and alternatives in this class simply assume different, though more complex, value functions. Most theorists in economics have followed this route (hence the label 'conventional').

Machina (1982) has sought to extend EUT by invoking a more sophisticated specification of the utility function. In Machina's generalized expected utility theory (GEUT) individuals are assumed to maximize expected utility, but the utilities are assigned to consequences by a *local utility function*. The local utility function allows the utility assigned to a consequence to vary from prospect to prospect, hence we may write the local utility function as $U(x; P)$, where P is a particular prospect. Machina assumes that the local utility function is increasing in x; this ensures monotonicity. If the local utility function is the same for all prospects (i.e. if $U(x,P) = U(x,Q)$ for all P, Q) then GEUT reduces to EUT. The distinctive features of Machina's theory derive from an assumption which he refers to as 'Hypothesis II'. This hypothesis assumes that the degree of (local) risk aversion increases (or at least does not decrease) as we move from stochastically dominated to stochastically dominating distributions.

This second assumption has a straightforward interpretation in terms of preferences in the unit probability triangle. As we move northwest in the triangle, we move from stochastically dominated to stochastically dominating prospects. Since the slopes of indifference curves reflect the degree of risk aversion, with steeper slopes reflecting higher degrees of risk aversion, Hypothesis II implies a tendency for indifference curves to become steeper, or to fan-out, as we move northwest in the triangle. The general pattern of indifference curves implied by Machina's theory is illustrated in Fig. 12.3. Notice that the indifference curves are drawn as wavy lines since GEUT does not imply that they must be linear (although they may be).

A number of other theories generate the implication that indifference curves fan-out. The GEUT can thus be thought of as one member of a class of fanning-out theories. Another theory in this class is Chew & MacCrimmon's (1979) 'ratio form', although this theory differs from Machina's by requiring indifference curves to be straight lines.

It is very easy to see how theories with this fanning-out property generate implications consistent with the common consequence and common ratio effects. Figure 12.4 imposes GEUT-type preferences on the Allais paradox problems. Since indifference curves are steeply sloped in the neighbourhood of prospect A, A lies on a higher indifference curve than B, and hence $A > B$. But the much flatter indifference curves in the bottom right-hand corner of the triangle are such that $D > C$.

An alternative (but still conventional) way to extend EUT is to assume that individuals attach subjective 'decision weights' to probabilities (as well as 'utilities' to consequences). A number of theorists have taken this route. Given this modification to EUT, the basic functional form, sometimes called

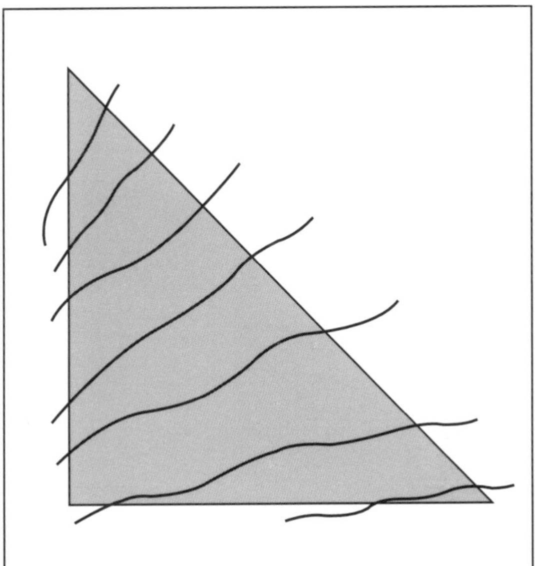

Fig. 12.3 Indifference curves in generalized expected utility theory.

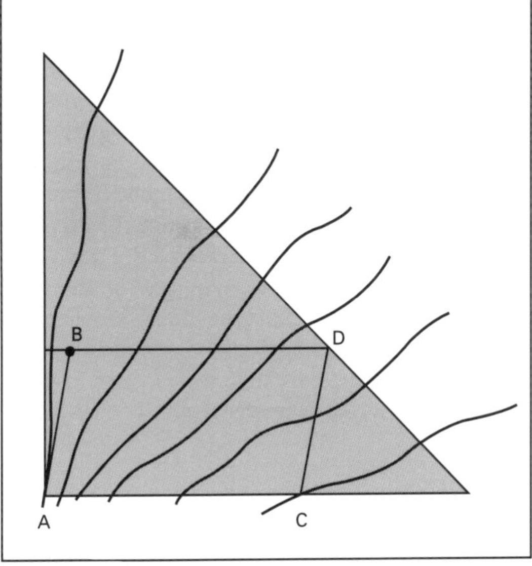

Fig. 12.4 Generalized expected utility and the Allais paradox.

the 'subjective expected utility function' (SEU for short), can be stated as follows:

$$SEU(P) = \sum_{i=1}^{n} w_i \cdot u(x_i) \qquad (7)$$

In expression (7), the utility of each outcome is combined with a decision weight, w_i. The weights are some function of the probabilities and, at the intuitive level, they allow for particular probabilities to have particular significance in a prospect. For instance, it is often argued that certainties loom particularly large in the minds of individuals; there is also evidence to suggest that many individuals are attracted to low-probability, high-payoff gambles (witness the popularity of the National Lottery for instance). Extending EUT by the incorporation of decision weights can thus be interpreted as a way of capturing certain features of individual dispositions towards objective probabilities apparently revealed in observed behaviour.

Some of the original models using variants of this functional form (e.g. Handa 1977) were the subject of criticism. In Handa's theory the decision weight attached to each outcome was a simple function of the probability of each outcome. This way of constructing decision weights, however, leads to the implication that preferences will not generally satisfy monotonicity (except in the special case where decision weights are equal to objective probabilities, in which case SEU reduces to EUT). Given the widely held belief that a satisfactory theory should satisfy monotonicity, this is generally regarded as a weakness in such theories.

More recent models using the SEU form have proposed more sophisticated probability transformation functions and, currently, one of the most popular models of this type is rank-dependent expected utility theory (RDEUT for short), which was first proposed by Quiggin (1982). In RDEUT, the decision weights are defined by:

$$W_i = \left[q\left(\sum_{j=1}^{i} p_j\right) - q\left(\sum_{j=1}^{i-1} p_j\right) \right] \quad \text{for } i = 2, \ldots, n$$

$$W_i = q(p_i) \quad \text{for } i = 1$$

where the utility function $U(.)$ is defined over a set of consequences numbered such that $U(x_i) < U(x_j)$ for all $i < j$.

Unlike earlier models based on the SEU form, in RDEUT the decision weight attached to any consequence of a prospect depends not only on the true probability of that consequence but on the entire probability distribution: more specifically, the decision weight attached to a consequence depends on its ranking relative to the other outcomes of the prospect. This construction of the weights ensures that the SEU function is generally monotonic.

The exact properties of RDEUT depend crucially on the shape of the function $q(.)$. Quiggin (1982) has proposed an S-shaped function, as in Fig. 12.5, although there has been a great deal of discussion about the most appropriate functional form (see Quiggin (1993) for a discussion of alternative specifications and their implications for indifference curves in the unit probability triangle). This S-shaped function has the property that small probabilities are 'overweighted' relative to EUT. Quiggin argues that this would be consistent with violations of EUT such as the Allais paradox and common ratio problems. To understand why this would be so, consider the Allais paradox problems: such a function would give very similar weights to the winning outcomes in options C and D.

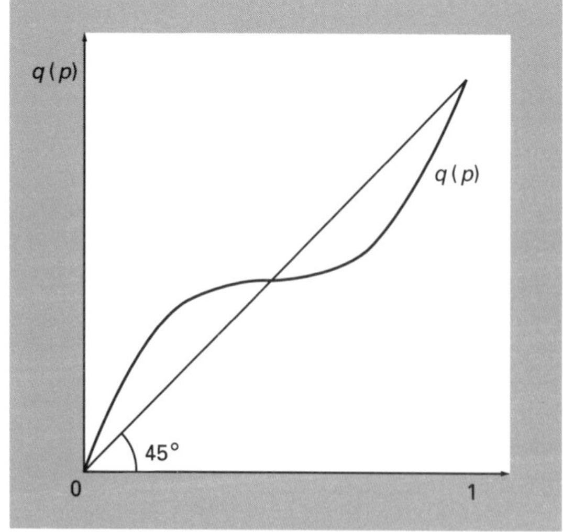

Fig. 12.5 A possible weighting function in RDEUT.

Consider now an individual who chooses D (in this case the extra utility of the higher outcome must outweigh the difference in weights). Such an individual may nevertheless prefer A to B since an S-shaped weighting function would give particular weight to the 1% chance of winning nothing in choice 1, as it is a low probability event.

The theories of Quiggin and Machina are perhaps two of the best known (conventional) extensions of EUT. However, a variety of other conventional models have appeared in the literature. These include the theory of 'disappointment' (Loomes & Sugden 1986) and the more recent formulation of Choquet expected utility theory (see Schmeidler 1989). This last theory is an extension of RDEUT, which allows subjective probabilities to be non-additive. Readers wishing to explore these alternative models will find Sugden (1997) and Machina (1987) useful references.

12.5.2 The exotic approach

It is possible to retain the assumption that individuals are utility maximizers but to part company with conventional theories by allowing the function to be maximized to refer to some feature of the choice problem beyond the information contained in individual prospects. I will refer to any theory of this type as an exotic theory. One well-known theory in economics fits this description: regret theory.

Regret theory is primarily a theory of pairwise choice in which preferences are defined over *actions* rather than prospects. Actions contain more information than prospects; in an action, the consequences of a risky alternative are assigned to particular states of the world (as in the Allais paradox problems above). This additional information is relevant in regret theory, which assumes that the utility derived from the outcome of a chosen action may be influenced by the outcomes of the alternative action foregone. Consequently, in regret theory values are not assigned independently to prospects: choices depend partly on comparisons between the risky alternatives, or actions.

Regret theory, at least the version presented by Loomes & Sugden (1982), is based on a psychological intuition about the preferences of ordinary people. The distinctive feature of the theory rests in the proposition that, if the outcome of a chosen action is worse than the outcome that would have resulted from the alternative action, the individual is assumed to experience a sense of regret which detracts from the utility of the consequence. Thus, as Loomes and Sugden have put it, the way you feel about what you have, is affected by 'what might have been'.

For example, consider an individual who bets on the National Lottery each week using the same numbers. If the numbers do not 'come up' in a given week, the person gets nothing. Now imagine how that same person would feel if those numbers did come up in a given week, but they had forgotten to buy their ticket. They get nothing, just as before, but would they feel the same way about the outcome of 'winning nothing' in each case? Regret theory assumes that the answer may be no because a person might well feel worse in the second case and experience a sense of regret at having missed out on the big win. The theory assumes that individuals will anticipate potential regrets and take this into account in the decisions that they make.

Loomes & Sugden (1987) model this formally as follows. Let A_i and A_j be any two actions which result in outcomes x_{is} and x_{js}, respectively, in state of the world s. The utility of x_{is} is described by a function $M(x_{is},x_{js})$ which is increasing in its first argument and decreasing in its second and represents the experience of having x_{is} supplemented by 'regret' when $x_{is} < x_{js}$, or 'rejoicing' when $x_{is} > x_{js}$. Loomes and Sugden then propose that individuals seek to maximize the mathematical expectation of this modified utility $M(.,.)$. For ease of exposition, define a further function $\psi(x_{is},x_{js}) \equiv M(x_{is},x_{js}) - M(x_{js},x_{is})$, where $\psi(.,.)$ is skew symmetric by definition, i.e. $\psi(x_{js},x_{is}) = -\psi(x_{is},x_{js})$, which implies that $\psi(x_{js},x_{is}) = 0$ for $x_{is} = x_{js}$.

The choice between any two actions i and j is then determined by the following condition:

$$A_i \gtreqless A_j \Leftrightarrow \Sigma_s P_s \psi\,(x_{is},x_{js}) \gtreqless 0 \qquad (8)$$

Regret theory contains EUT as the special case in which $M(x_{is},x_{js}) = U(x_{is})$. The assumption which gives regret theory its predictive novelty is 'convexity' (see Loomes & Sugden 1982, 1987), which entails that for any three consequences $x > y > z$, $\psi(x,z) > \psi(x,y) + \psi(y,z)$. This can be roughly interpreted as assuming that large regrets loom disproportionately large. To explain the evidence surrounding the Allais paradox and the common ratio effect, it is also necessary to assume that individuals treat actions as statistically independent prospects. Since this amounts to assuming that individuals effectively ignore the assignment of consequences to states of the world, this explanation for these anomalies perhaps begins to appear a little contrived.

But whatever the merits of regret theory as an account of these standard anomalies, the idea that regret is a real psychological phenomenon appears to accord with many people's intuitions. There is also at least some empirical support for the existence of 'regret effects' at work in the decisions of ordinary people (see below). If this is correct, and individuals are concerned to avoid 'large regrets', the theory serves to highlight a dimension to decision-making that is ignored in most theories that make no reference to comparisons between alternative actions. Such considerations may well be relevant in the area of environmental risk, where important decisions have consequences which could be a source of significant regret. The questions of whether considerations of regret can, or should, be built into formal decision analysis are discussed in Weinstein *et al.* (1988).

12.5.3 The procedural approach

The distinctive aspect of what I refer to as the procedural approach is to assume that choices are determined, at least in part, by certain decision heuristics or rules. Depending on the theory, these rules may either supplement or completely replace a maximizing element. One theory of this kind which has received considerable attention in the economics literature is prospect theory, developed by the psychologists Kahneman & Tversky (1979).

Prospect theory assumes a two-phase decision process. In the second phase, the individual is

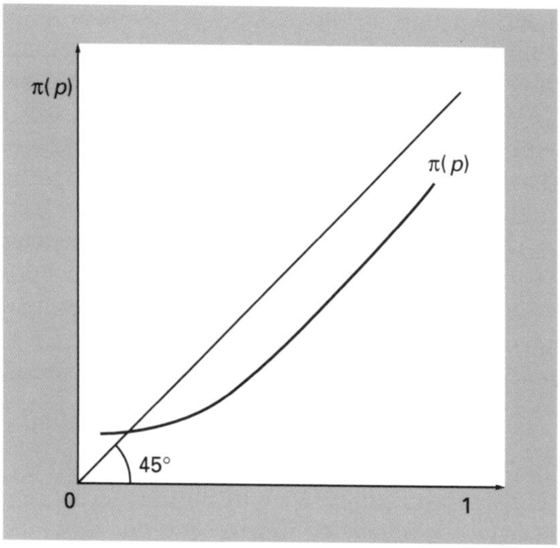

Fig. 12.6 Decision weights in prospect theory.

assumed to maximize an SEU function similar to expression (7) above.* The utility function has three essential properties: it is defined over changes in, as opposed to levels of, utility so that consequences are viewed as either gains or losses relative to the reference point of the *status quo*; it is concave for gains; and normally convex, and more steeply sloped, for losses. This implies risk aversion in the domain of gains and risk seeking in the domain of losses. Kahneman and Tversky propose a simple probability transformation function $\pi(p)$ of the form depicted in Fig. 12.6. This transforms objective probabilities into decision weights, hence, $w_i = \pi(p_i)$. This function overweights 'small' probabilities, underweights most 'larger' probabilities, and is discontinuous at both ends (Kahneman and Tversky impose the restrictions $\pi(1) = 1$ and $\pi(0) = 0$). Kahneman and Tversky demonstrate that, given these assumptions, many of the anomalies surrounding EUT can be explained in terms of SEU maximization.

This theory provides a possible account of certain anomalies in the context of risk valuation

* Prospect theory is intended to apply to prospects of the form $(x,p;\ y,q)$, which have at most two non-zero outcomes. The function assumed in prospect theory coincides with expression (7) for what Kahneman and Tversky call 'regular prospects', where $p + q < 1$, $x \geq 0 \geq y$, or $x \leq 0 \leq y$.

and assessment which apparently contravene EUT. For example, Pearce (Chapter 13) argues that the 'loss aversion' implied by the utility function in prospect theory might provide part of the explanation for the observed disparity between the willingness-to-pay and willingness-to-accept valuations elicited via the contingent valuation method. The probability transformation function assumed in prospect theory is also consistent with apparent biases observed in individual probability assessments. For instance, Pidgeon *et al.* (1992) report the finding that when individuals are asked to make judgements about the likelihood of death occurring from different causes, there is a tendency for individuals to underestimate deaths from relatively frequent causes, while underestimating the number of deaths from relatively infrequent causes. This is consistent with the transformation function in prospect theory for which high (low) probabilities are underweighted (overweighted).

Relative to the theories so far discussed, the most distinctive aspect of the original form of prospect theory is the editing phase.* It is this which I wish to focus on here. Kahneman and Tversky assume that, prior to the second stage of evaluation using the SEU form, individuals will 'edit' prospects using a variety of heuristics. These editing heuristics are to be understood as ways of simplifying decision-making, which transform prospects into a form that can be more easily handled in the second phase.

The editing routines proposed by Kahneman and Tversky include the following operations. A *combination* rule simplifies prospects by combining the probabilities associated with identical outcomes. For example, the combination rule would simplify the prospect (100, 0.2; 100, 0.3; 0, 0.5) to (100, 0.5; 0, 0.5). It is also assumed that decision-makers may simplify prospects by rounding of the probabilities and/or outcomes. So, for example, the prospect (101, 0.49; 0; 0.51) may be rounded to (100, 0.5; 0, 0.5). These first two routines are applied individually to each prospect.

Other routines are applied to sets of prospects. One such rule is *cancellation*, which involves the elimination of elements that are common to the prospects under consideration. Hence a choice between prospects $P = (150, 0.30; 50, 0.30; 0, 0.4)$ and $Q = (150, 0.3, 100, 0.25; -20, 0.45)$ may be evaluated as a choice between $P' = (50, 0.30; 0, 0.4)$ and $Q' = (100, 0.25; -20, 0.45)$. A final rule proposed by Kahneman and Tversky eliminates stochastically dominated options from the choice set prior to evaluation. In the absence of this final rule, the SEU function maximized in prospect theory would imply violations of monotonicity. In fact, prospect theory still admits violations of monotonicity by allowing for the possibility that individuals may fail to detect dominated prospects in some cases. This may occur in complex problems or even simple problems where the problem presentation obscures the relation of dominance, rendering it 'non-transparent'.

The editing phase of prospect theory makes explicit reference to the mental processes used in decision-making. Prospect theory is only one theory of this kind and numerous models of decision processes have been proposed (see Payne *et al.* 1993). Most of the literature relating to decision processes has emerged from the discipline of psychology and, while there are exceptions, most of the models proposed by economists make no reference to mental processes. Indeed, many economic theories would seem highly implausible if interpreted as models of thought processes. For instance, I suspect that few individuals would claim to assign utilities to outcomes using local utility functions, or to assign weights to utilities using a cumulative probability weighting function. It is patently unrealistic to assume that such entities feature in the conscious thought processes of ordinary people.

The standard economist's defence of such unrealism is to argue that theories intended as descriptive theories should be judged according to their predictive ability and not according to the perceived realism of their assumptions (see

* Conceptually, the second phase of prospect theory is very similar to RDEUT; both theories are based on the SEU form (although the detailed specification of the functions differ). Indeed, more recently Tversky & Kahneman (1992) proposed a revised version of prospect theory—cumulative prospect theory—which assigns rank-dependent probability weights. The key difference between cumulative prospect theory and RDEUT is that the former theory maintains a distinction between the evaluation of gains and losses.

Friedman (1953) for a classic statement of this point of view). The relevant question, then, is whether theories which make no reference to the actual mental processes of humans are likely to succeed in providing satisfactory descriptive accounts of actual behaviour? Kahneman & Tversky (1979) think not, and argue that:

> Many anomalies of preference result from the editing of prospects .... More generally, the preference order between prospects need not be invariant across contexts, because the same offered prospects could be edited in different ways depending on the context in which it appears. (Kahneman & Tversky 1979, p. 275.)

There is some evidence to support this claim. Perhaps the most compelling comes in the form of so-called 'framing effects'. A framing effect occurs when a change in the presentation of a choice problem, which does not affect the objective characteristics of the situation (i.e. the outcomes or probabilities), has a significant impact upon choice behaviour. Here is one example discussed by Tversky & Kahneman (1986):

> Consider the following pair of lotteries, described by the percentage of marbles of different colours in each box and the amount of money you win or lose depending on the colour of a randomly drawn marble. Which lottery do you prefer?

Option A

90% white	6% red	1% green	1% blue	2% yellow
$0	win $45	win $30	lose $15	lose $15

Option B

90% white	6% red	1% green	1% blue	2% yellow
$0	win $45	win $45	lose $10	lose $15

It is very easy to see that option B dominates option A since, for every colour, the prize for option B is always at least as good as the prize for option A and in some cases it is better. Kahneman and Tversky presented this problem to 88 subjects and found that all of them chose B. Now consider this slightly modified version of the above problems:

Option C

90% white	6% red	1% green	3% yellow
$0	win $45	win $30	lose $15

Option D

90% white	7% red	1% green	2% yellow
$0	win $45	lose $10	lose $15

Options C and D are stochastically equivalent to A and B, respectively, the only difference being a minor change in the presentation which 'simplifies' the options by assigning each prize to a single colour. This modification, however, also makes it more difficult to detect the fact that the first option is dominated by the second. Kahneman and Tversky presented this choice problem to 124 subjects and found that a majority of subjects (58%) chose the dominated option C. So, not only does the change in presentation bring about a very significant change in responses, but, given the relation of dominance between the two options, this second presentation leads people into making erroneous judgements.

If this example seems a little contrived (a mere trick of the presentation specifically designed to trip up the chooser), a further example provided by Tversky & Kahneman (1981) illustrates that framing effects may be highly relevant to issues of public decision-making. They presented the following cover story to two groups of subjects:

> Imagine that the US is preparing for the outbreak of an unusual Asian disease, which is expected to kill 600 people. Two alternative programmes to combat the disease have been proposed. Assume that the exact scientific estimate of the consequences of the programmes are as follows:

Each group was then faced with a choice between two policy options. For one group of subjects the policy options were described as:

> If programme A is adopted, 200 people will be saved.
>
> If programme B is adopted, there is a 1/3 probability that 600 people will be saved, and a 2/3 probability that no people will be saved.

For a second group of subjects, they were described as:

If programme C is adopted, 400 people will die.

If programme D is adopted, there is a 1/3 probability that nobody will die, and a 2/3 probability that 600 people will die.

Again, the two pairs of options are stochastically equivalent. The only difference is that the first description presents the information in terms of lives saved; the second in terms of lives lost. Tversky and Kahneman found a very striking difference in responses to these two presentations (72% of subjects preferred option A to option B, while only 22% of subjects preferred C to D) and claim to have found similar response patterns amongst groups of undergraduate students, university faculty and practising physicians.

Perhaps part of the explanation for this effect is that the initial framing makes option A look 'good' because it is presented as a gain (lives saved), whereas option C sounds 'bad' because it is framed in terms of lives lost. But whatever the precise explanation for this particular effect, one thing seems clear: framing effects cannot be explained by any theory of utility maximization in which choices depend purely on the objective characteristics of the problem setting. Some feature of the problem beyond the objective outcomes and probabilities must be at work.

Examples like this suggest that a full understanding of choice behaviour in the context of risk requires an account of how people conceptualize or 'make sense' of risky choice settings. Issues of this sort run beyond the scope of this chapter (see instead Chapter 20). The important message here is that we have apparently encountered a boundary to the domain of conventional and exotic theories: there are some features of real behaviour which such theories simply cannot explain. While this undoubtedly reveals a limitation of such theories as descriptive accounts of behaviour, they may nevertheless serve some useful purposes.

It may be that utility maximizing theories are able to account for a significant range of behaviour, even though some phenomena lie outside their scope. In section 12.6, I shall offer a brief review of the evidence which has emerged from the testing of alternatives to EUT. This evidence offers at least some insights into the usefulness of these theories as descriptive accounts of behaviour. But whatever the status of these theories as descriptive theories, the economic theory of decision-making may have a role as a prescriptive tool. Framing effects demonstrate that individuals do not consistently make utility maximizing choices, but perhaps they would be better off if they attempted to do so? In section 12.7 I shall discuss the interpretation of EUT as a normative guide to choice.

12.6 ASSESSING ALTERNATIVES TO EUT

Faced with an array of alternative theories, it seems natural to ask whether any of these theories may be regarded as offering a significant improvement over the predictive ability of EUT and, if so, which is the best of the available alternatives? In this section I shall provide a brief review of a range of evidence which has emerged from experimental tests designed to assess competing theories relative to each other, and to EUT. For a more extensive review, see Camerer (1995).

It goes without saying that each of the alternatives considered will be able to account for more observations than EUT. Since each of the new theories encompasses EUT as a special case, it is obvious that each new theory will be able to explain at least as much as EUT can. And since each theory was specifically designed to explain anomalies relative to EUT, it should be no surprise to find that each new theory is consistent with a wider body of data than EUT. But simply being able to account for a wider range of data is probably not enough to count as a 'significant improvement' in predictive power. For instance, there may be a case for accepting a trade-off between the simplicity of a theory and its overall predictive accuracy: a simple theory capable of explaining a substantial proportion of observed behaviour might well be preferred, for at least some purposes, to a much more complex theory which brings only a slight improvement in the range of phenomena explained.

Given a range of theories, all claiming to explain a similar range of anomalies, it is tempting to ask which, if any of them, offers the correct explanation. While this last question is tempting, it is also problematic. It is quite apparent that the models we have been discussing are all abstractions from reality. The objective of such models is to provide a relatively simple account of choice with 'satisfactory' predictive power. Such models do not aspire to be more than approximations to reality; as such, none of them can be regarded as 'true'. This said, there may be ways of assessing the fruitfulness of alternative modelling strategies.

Although the alternatives to EUT generate similar predictions in relation to the 'standard anomalies', since each theory offers a somewhat different account of them, the theories are not equivalent and, consequently, they give rise to at least some distinctive, or novel, implications for specific classes of choice problem. Testing novel implications of new theories provides one way to shed light on the fruitfulness of alternative approaches.

Numerous studies have investigated more general implications of theories for choice behaviour in the unit probability triangle, with a view to assessing the performance of alternatives to EUT. For example, GEUT explains the common ratio effect as part of a more general pattern of preferences—preferences which have the property that indifference curves generally fan out across the unit probability triangle. This theory organizes much of the data generated from the testing of EUT, but they are not usually rich enough to allow us to ask the question: do indifference curves generally fan out? Using suitably designed experiments, it is possible to investigate whether preferences do display patterns consistent with this general proposition about indifference curves. Examples of this kind of study include Camerer (1989), Chew & Waller (1986), Battalio *et al.* (1990) and Starmer (1992).

Camerer (1995) argues that comparisons across these and other studies suggest some stylized facts about the patterns of indifference curves which reflect on the merits of competing theories. Among these 'facts' are the following. There is evidence that indifference curves are non-linear and that the slopes of indifference curves (and hence the degree of risk aversion) vary across the triangle. The patterns in the data are not consistent with generalized fanning-out of indifference curves; there are systematic, but more complex, patterns consistent with fanning-out in some regions of the triangle, and fanning-in in others. Finally, indifference curves are more nearly parallel on the interior of the triangle compared to the edges.

The last of these three stylized facts probably has the most obvious interpretation in terms of its implications for environmental issues. To the extent that parallel indifference curves are a reasonable approximation for prospects on the interior of the triangle, EUT may be a reasonable characterization of preferences (and hence behaviour) in the relevant region. But there are cases where it now seems well established that EUT is a poor predictor of behaviour: this is particularly true on the 'edges' of the triangle. One interpretation of this is that EUT is relatively poor at accounting for behaviour in situations of risk where there is some 'outlying' consequence—i.e. a small probability of an extreme consequence. Since many environmental or ecological risks carry small probabilities of extreme consequences—for example, the most serious effects of global warming, nuclear accidents or ozone-layer depletion may be both low probability and potentially catastrophic—such outlying events may be highly relevant to environmental risk assessment. These therefore constitute cases where we should not expect EUT to be able to provide an adequate characterization of individual attitudes toward risks.

In the absence of a generally accepted theory of preference capable of organizing the data from studies of behaviour, it is hard to know what to make of such features of individual attitudes to risk. For example, if individuals place a high, and seemingly inordinate, weight (relative to EUT) on a very small probability of a nuclear accident, is this evidence of some bias which should be discounted for the purposes of public policy, or does it reflect some underlying feature of the ordinary individual's attitude to outlying events which should be respected, and hence reflected, in any appropriate policy appraisal? The answers to such questions require further theoretical investigation.

Camerer argues that there are clear signs to indicate which directions provide the most promising lines of investigation. The stylized facts presented by Camerer are not consistent with any model of generalized fanning-out but would be more supportive of a model based on the SEU form, because, 'the right kind of non-linear probability weighting function' (Camerer 1995, p. 637) can, broadly speaking, generate behaviour with the right properties. But although models based on the SEU form, like prospect theory or RDEUT, do appear to out-perform models of the fanning-out variety in relation to choices in the triangle, the general conclusion resulting from studies of this type is that, as yet, no single theory provides a fully satisfactory descriptive account of choice behaviour. The choice patterns observed have tended to be more complex than any of the theories predict. Moreover, the testing of new theories has, in its turn, given rise to the discovery of fresh anomalies which await explanation. A series of such anomalies has arisen out of attempts to test regret theory.

Regret theory predicts that in some circumstances individuals may exhibit systematic 'cyclical' or 'non-transitive' (see below) patterns of choice (see Loomes & Sugden 1983). That is, for some risky alternatives *A*, *B* and *C*, we may find that individuals exhibit cyclical preferences by choosing *A* over *B* (i.e. choosing *A* in a pairwise choice between *A* and *B*), *B* over *C*, and *C* over *A*. Regret theory also predicts systematic violations of monotonicity in certain circumstances. While most theories of choice developed by economists rule out such patterns of choice on *a priori* grounds, tests of regret theory reported by Loomes *et al.* (1989,1991a,b) have revealed both the violations of monotonicity and transitivity in exactly the circumstances predicted by regret theory.

The prediction of these phenomena, which were not anticipated prior to the development of regret theory, appeared to be a striking success for that theory and bad news for most of the other alternatives to expected utility theory that assume monotonicity and rule out cyclical choices. On the other hand, much of the data derived from tests of behaviour using problems in the triangle are inconsistent with regret theory. Once it was realized that regret may be an influence on choices, many of the later experiments in this genre 'controlled' for regret effects by designing choice problems in such a way that the predictions of regret theory coincide with EUT. These experiments have revealed widespread violation of EUT, hence regret theory fails against this evidence too. The implication seems to be that although regret may be a significant factor in choices in some contexts, regret theory, at least as currently formulated, does not provide a satisfactory general model of choice behaviour.

This conclusion has been reinforced by further experiments—for example Starmer & Sugden (1993) and Humphrey (1995)—which have raised doubts about whether some of the so-called regret effects have anything to do with regret after all. These two studies report the occurrence of a new phenomenon, an *event splitting effect* which explains at least some of the data that were previously attributed to regret. It appears that an outcome which occurs with a fixed objective probability in a given prospect (i.e. a p chance of x) will be given more weight when the probability of the outcome is described as two smaller probabilities, $(q,x; r,x)$, even though $q + r = p$. This is the event splitting effect. Starmer and Sugden, and Humphrey, demonstrate that some of the phenomena previously interpreted as 'regret effects' may be the product of event splitting effects.* As yet, there is no well-developed model of choice which explains event splitting effects, although Starmer and Sugden have argued that a suitably specified version of SEU theory could, in principle, accommodate such effects.

The overall picture emerging from experimental tests of new theories is that economists still have some way to go in developing a satisfactory descriptive account of choice behaviour, even in quite simple choice contexts. Nevertheless, there are reasons to be hopeful for future progress, particularly when we recognize that research in this area is still at a relatively early stage of development.

With hindsight, it seems quite apparent that alternatives to EUT were developed on the basis of a very 'thin' evidential base, driven primarily by evidence from a narrow class of anomalies. The testing of new theories has produced a much richer body of evidence, making it possible to address

* Not all of the effects attributed to regret can be explained in this way. For example, no other explanation of the cycles of choice predicted by regret theory is currently available.

some broader theoretical questions. This richer evidential base is now allowing theorists to identify some more promising avenues for theoretical development and to close off others. The accumulating evidence also forms a basis for at least some judgements about which of the available theories currently offers most predictive power.

In a recent, quite painstaking study, Harless & Camerer (1994) compared the predictive power of EUT and a variety of alternatives against a data set consisting of thousands of risky choices from 23 prior studies. The econometric technique that they employed also allowed them to make trade-offs between the goodness of fit and the complexity of theories. They draw a number of interesting and useful conclusions from their study.

Models of generalized fanning-out perform poorly against their data. They also note that the over-weighting of small probabilities seems to be an empirically robust phenomenon. Hence any satisfactory model should be able to accommodate this feature. They observe that every theory excludes at least some systematic patterns in choice which could be explained by a more refined theory. But they suggest that while some theories like EUT 'are too lean: They could explain the data better by allowing a few more common patterns', other theories such as RDEUT 'are too fat: They allow a lot of patterns which are rarely observed' (Harless & Camerer, 1994, p. 1285).

Harless and Camerer also offer some reflections on which of the available theories performs 'best'. They argue that it is not possible to identify a single 'best' theory because the choice depends upon 'one's trade-off between parsimony and fit'. However, they are able to offer this advice: for those who place most emphasis on predictive accuracy (and least on parsimony), the best theory is either prospect theory or a theory which allows some mixture of fanning-in and fanning-out behaviour; for those who place a premium on parsimony—for example, researchers wishing to build and examine the implications of models of aggregate behaviour—a restricted version of EUT may be the best choice: EUT with a linear utility function (this implies maximization of expected value for choices with monetary outcomes).

Whatever view is taken of the progress that has been made in modelling choice behaviour, the search for a descriptive theory has, without question, led to the discovery of many regularities in choice behaviour. Even in the absence of a well-specified general account of such phenomena, it is still possible to see their potential importance. The discovery of the event splitting effect provides one case in point.

Consider an individual, or governmental body, attempting to assess the risks associated with some environmental hazard. Take, for example, the risks of skin cancer from exposure to ultraviolet radiation (UVR). There are a variety of different cancers which may result from UVR. If event splitting effects generalize to this context, then the perception of the risks from UVR may be affected by whether the risks are described collectively as a single risk of 'skin cancer' with a given probability or, alternatively, described as a series of risks of different cancers, each occurring with a smaller probability. The evidence from the laboratory suggests that the risks would appear worse under the second, disaggregated, description.

On the basis of what we know about such affects, it is hard to draw any clear-cut conclusion about which description forms the basis of the best judgement in this case. Recognizing the existence of an event splitting effect, however, at least allows us to alert the decision maker to a possible source of bias in this sort of decision. But suppose a chooser recognizes that their own decision-making may be subject to biases of one kind or another. What should they do? It is to this question that we now turn.

12.7 EXPECTED UTILITY THEORY AS A NORMATIVE THEORY OF CHOICE

In developing descriptive theories, economists have sought models of preference which produce implications consistent with observed behaviour. For those concerned with the normative project, on the other hand, the key question is how people ought to choose, as opposed to how they actually choose. We may cast the problem as one about preferences by considering what properties the preferences of an ideal, or fully rational, chooser ought to have. One problem for the normative theorist is then to establish whether certain restrictions on preference are in some sense 'desirable', whether or not these are properties of the preferences of ordinary choosers.

The question of what exactly constitutes a desirable property raises some deep philosophical issues which I shall not discuss at any length here (see Hargreaves Heap 1989; Hargreaves Heap *et al.* 1992). Suffice it to say, for the present, that one possible test of whether a particular property is 'desirable' from a normative point of view is whether most 'reasonable' people would endorse it as a property that they would like their preferences to have.

It turns out that EUT can be derived from three assumptions about preference, or axioms as they are often called.* These are the axioms of ordering, continuity and independence. If an individual's preferences satisfy each of these axioms, then it follows as a matter of logic that there exists a utility function over outcomes $U(.)$ such that the individual chooses among prospects as if maximizing the expectation of the function $U(.)$. In this axiomatic version of the theory, the function $U(.)$ is a representation of preferences: it has no interpretation in terms of satisfaction. In fact, the utility function is only unique up to a linear transformation; i.e. if the function $U(.)$ is a valid preference representation, then the function $U^*(.) = a + bU(.)$, where a and b are two positive constants, is also a valid utility representation.

From what has been said about the descriptive failures of EUT, it is apparent that individuals' preferences do not, in general, satisfy these assumptions. But while the theory fails descriptively, a good number of economists would argue that each of the EUT axioms is desirable in the sense just described: each can be defended as a principle which any reasonable agent would wish their preferences to satisfy. If the axioms can be successfully defended in this way, then the reasonable agent would be bound to admit that any violations of EUT they commit must be reflections of errors or imperfections in their choices. If this is correct, then perhaps EUT has a role for helping people to make better decisions. This is a strong claim and one which merits some examination.

The axioms of ordering and continuity are relatively weak assumptions, or so most economists would argue. Taken together, they imply that preferences exist, are well-ordered and coherent, but they place very little restriction on the exact structure of preferences. It may be helpful to express this in terms of preferences in the unit probability triangle: ordering and continuity together imply the existence of indifference curves in the triangle while placing virtually no restriction on their exact form.

The ordering axiom requires that preferences be complete, in the sense that they specify a ranking of any pair of prospects. Roughly speaking, one might interpret this as assuming that the agent 'knows' their preferences (though allowing for the possibility that some alternatives may be deemed equally good). The ordering axiom also rules out non-transitive preferences: if P, Q and R are three prospects such that $P \succ Q$ and $Q \succ R$, then transitivity implies that $P > R$. Most, though not all, economists would view this as a transparently obvious principle of rational preference. (Notable dissenters to this view include Loomes & Sugden (1982), who present a model of rational, but non-transitive, preference.) Continuity implies that if $P \succ Q \succ R$, then there must be some probability p, such that the compound prospect $(p,P;\ 1-p\ R) \sim Q$. If P, Q and R are degenerate prospects, then con-tinuity implies that there must be some probability mix of a good and bad consequence, which is regarded as equivalent to a consequence in-between. Many theorists regard this as a relatively modest requirement.

While both of these axioms have been contested, I shall not attempt to justify them further. Instead, I shall focus the discussion on the most controversial of the axioms, namely, independence. For more extensive discussions of all of the axioms, their interpretation and justification, see Luce & Raiffa (1957) and Hargreaves Heap *et al.* (1992).

The independence axiom, or at least one version of it, can be stated as follows. If R, W and M are any three prospects such that $R \succ W$, and λ is some probability between 0 and 1, then independence requires that the compound prospect R', which is

* The original axiomatic derivation of EUT was due to Von Neumann & Morgenstern (1953), but numerous alternative axiomatizations have been developed.

a $\lambda{:}(1-\lambda)$ probability mix of R and M, will be preferred to the compound prospect W', which is a $\lambda{:}(1-\lambda)$ probability mix of W and M. This implies that preferences between R' and W' are determined solely by the preference ordering of R and W. In other words, the common element, M, of the compound prospects is deemed irrelevant. Some theorists, for example, Samuelson (1952), have attempted to justify this axiom by arguing that: (i) independence simply rules out *complementarity* between the elements of a prospect; and (ii) such a restriction is entirely reasonable. Let me briefly rehearse a version of this argument.

Imagine that M, R and W are three degenerate prospects where M is a meal of fish, R is a glass of red wine and W is a glass of white wine (each experienced with certainty). Consider a person who, when drinking wine on its own, prefers red to white. That same person may also have a strict preference for white wine over red, when eating fish. Hence, 'combining' wine with some other good (fish) reverses that person's ranking of red and white wine. In this case, such a reversal of preference seems entirely reasonable and merely indicates that white wine and fish are, as economists would say, complementary goods: for at least some people, drinking white wine in the presence of fish is a better experience than drinking white wine on its own.

The independence axiom rules out any analogous complementarity effect when wine and fish are 'combined' as elements in a prospect: according to independence, a person who prefers red to white wine (when wine is consumed on its own) must *not* prefer a prospect which 'combines' white wine and fish in a $\lambda, 1-\lambda$ mix to a prospect which 'combines' red wine and fish in a $\lambda, 1-\lambda$ mix. Why should complementarity be unreasonable in this latter case? The defender of the independence axiom will answer that the sense in which fish and wine are 'combined' here is very different from the former case. In the former case, the wine and the fish are to be consumed *together*. In a prospect, by comparison, the outcomes are mutually exclusive; i.e. they will never be consumed simultaneously. If the prospect results in wine, the consumer drinks wine on its own and does not eat fish with it. By ruling out complementarity, independence requires that the individual should let their preferences over wine (in the absence of fish) determine their ranking of prospects which offer different kinds of wine, each combined with identical components of fish.

If the case for independence seems convincing in this case, it may be less so in others. Consider the implications of independence when applied to choice 1 from the Allais paradox. Notice that state $s3$, which occurs with a fixed probability of 0.66, results in the same consequence, £2400, whichever option is chosen. Thus, independence implies that $s3$ is irrelevant to the choice between A and B. Similar reasoning applied to choice 2 implies that $s3$ is irrelevant to the choice between C and D. Hence, for both choice problems, independence implies that the only choice-relevant components are those which appear in states $s1$ and $s2$. Since the choice problems are identical in these states, it follows from application of the independence axiom that:

$$A \gtreqless B \Leftrightarrow C \gtreqless D$$

Notice that this rules out the commonly observed Allais paradox response of the form $A > B$ and $D > C$. Anyone who is convinced that the logic of the independence axiom is generally compelling must, if they are to be logically consistent, concede that this choice pattern reflects an error of some kind. But must any reasonable person be so convinced?

Consider the following possible justification for the 'anomalous' pattern of choice (A and D) in the Allais problems: 'In choice 1, I select option A because that gives me a good payoff (£2400) for sure and I do not want to risk even a small chance of getting nothing by choosing B; in choice 2, since the probability of winning is quite similar in each case, I choose the option with the higher prize (option D)'. To many people this seems like perfectly defensible logic. Indeed, studies by MacCrimmon (1968) and Slovic & Tversky (1974) have suggested that many people find the case for violating independence in Allais-type problems more appealing than arguments for conforming. This can be true even for experienced decision makers who have studied the case for and against conforming with independence. This suggests that the independence axiom does not have the status of a principle which any reasonable person would endorse. There may be cases where individuals violate EUT and are content to do so even when the logic of independence has been explained to them.

Given the controversial status of the independence axiom, it seems that no fully general case can be made for defending EUT as a normative guide to choice. Nevertheless, it is probably fair to say that the majority of decision theorists do endorse the axioms of EUT as sound principles of rational choice and, consequently, suggest that EUT may have a role as an aid to decision-making.

For example, Savage (1954), who presented his own axiomatization of EUT in *The Theory of Statistics*, recognizes that in general human choices, including his own, may not satisfy EUT. Indeed, in one part of the book, Savage explains that when first presented with the Allais problems (at lunch with Maurice Allais as it happens), his initial response was to violate EUT. Savage tells us that this incident led him to reconsider whether he was still convinced of the normative validity of his own axioms. On reflection, he concluded that axioms were indeed compelling and his resolution, therefore, was to admit that he must have made an error when making his initial choices.

Savage draws the following moral from this incident. Real people may well be tempted to make choices which on closer reflection they would come to view as mistaken, because they violate general principles of choice which they endorse as sound. Anyone similarly convinced by the normative appeal of EUT may wish to use the axioms to, as Savage puts it, 'police' their own decision-making. Suppose that we do endorse the axioms of EUT as normatively compelling, how can we use them to help us make better choices, or at least guard against making bad ones?

The axioms, taken individually, provide limited guidance with respect to how to choose in *isolated* choice settings: they are principles of consistency which typically rule out certain patterns of choice. For instance, application of the independence axiom to the Allais problems allows us to simplify the choice (by eliminating the third state *s*3 from consideration) and it rules out two patterns of preference ($A > B$ and $D > C$) and ($B > A$ and $C > D$), but it does not tell us which of the two remaining patterns is correct. Hence, we can test, *ex post*, whether a series of choices satisfy independence, but the principle itself cannot determine the choices initially, nor can it tell us how to change our choices if some inconsistency is revealed.

Although the individual axioms themselves are of limited applicability in practical decision-making, it may be possible to use the advice 'try to maximize expected utility' in at least some situations. In a well-defined situation of risky choice where the outcomes and probabilities are specified, if the decision maker were able to assign utilities to the outcomes of prospects, they may then proceed to maximize expected utility directly. The obvious question to ask then is how to assign subjective values to outcomes?

In some circumstances, it may be possible to find a suitable proxy for utility which is regarded as acceptable for the case in hand. For instance, when the outcomes of a prospect can all be measured on a common objective scale, such as 'lives saved' from alternative interventions to reduce an environmental hazard, it may be acceptable to use the objective levels of lives saved as a utility measure. Doing so would be to assume that the marginal utility of lives saved is constant, i.e. each extra life saved is equally valued. But even in cases where no obvious proxy is available, perhaps because the outcomes of prospects are multidimensional, it may still be possible to assign utilities to outcomes. A variety of methods for eliciting utilities has been proposed and one method for doing this—the so-called *standard gamble technique*—is based on EUT.

I shall briefly outline the logic of the standard gamble technique, but those interested in a full discussion of this and other such methods should consult Torrance (1986). For the purpose of illustration, consider the consequences of being exposed to some dangerous chemical substance which has a variety of potential effects, depending on the level of exposure. For example, low levels of exposure might cause minor problems such as skin or eye irritation; moderate levels of exposure might result in, say, respiratory problems which persist over a number of weeks; high levels of exposure might result in permanent physical damage, such as loss of sight or death. We can think of each of these outcomes as a 'health state' which could arise as a consequence of exposure. The standard gamble technique provides one method for eliciting utilities for consequences, such as health states.

The technique involves questioning a 'representative' subject in the following way. We begin by selecting two 'extreme' health states which will serve as end-points for a utility scale to be constructed. In health state evaluation, it is common to use perfect health and death as the end-points. We now wish to assign utilities to health states which, in utility terms, we expect to lie in-between perfect health and sudden death. The standard gamble technique works by asking subjects to imagine a choice between two alternatives as represented by Fig. 12.7. Alternative 2 is to experience health state H_i for sure, where H_i is the health state to be valued. Alternative 1 is a prospect which offers a p chance of perfect health and a $1 - p$ chance of death.

The individual is then questioned to find the value of p which makes them indifferent between the two options. Typically, this may involve an iterative process of adjusting the value of p and then asking 'which alternative do you prefer', using smaller and smaller adjustments in the value of p to focus in on a final value. If EUT holds, the value of p elicited by this method can then be used as a measure of the utility of health state H_i. To see why, note that if the individual is indifferent between the two alternatives, they must be of equal expected utility. That is:

$$EU(H_i) = p\ U(\text{Perfect Health}) + (1 - p)\ U(\text{Death})$$

Since a utility function in EUT, rather like a temperature scale, is only unique up to a linear transformation, we may arbitrarily assign two utility points on the scale, as long as the better alternative is assigned a higher utility. In the light of this, we have freedom to assign utilities such that $U(\text{Perfect Health}) = 1$, $U(\text{Death}) = 0$. Given this normalization, it follows that $EU(H_i) = p$. By repeated application of this method, we may construct a complete scale of utilities for all the relevant health states. The utilities elicited from this procedure may then be used to calculate expected utilities for risky alternatives which involve the health states as possible consequences.

The use of this technique is not limited to the evaluation of health-state utilities. Norton (1984, pp. 146–149) provides an illustration of how the standard gamble method could be used to construct a value function attaching utility numbers to levels of sulphur dioxide emissions from a power station. Norton argues that the values so constructed could play a useful role as part of a more widely based environmental impact analysis. Lathrop & Watson (1982) discuss how a variant of this procedure can be used to construct a multidimensional utility function over the health effects of a nuclear waste management system. Like Norton, they recognize that the analysis is 'partial' in the sense that, in addressing just the human health effects of the management system, it confronts only part of a broader public policy problem. Nevertheless, they argue that 'such partial analyses are much more likely to be used and useful than any attempt to bring the whole decision-making process for an issue of public policy under the hammer of hard analytic methods' (Lathrop & Watson 1982, p. 96).

The validity of the standard gamble method used in these studies relies on the axioms of EUT, so the values elicited via this technique must be treated with some caution. However, there are alternative methods for assigning utilities to outcomes which do not rely on EUT. Returning to the example of

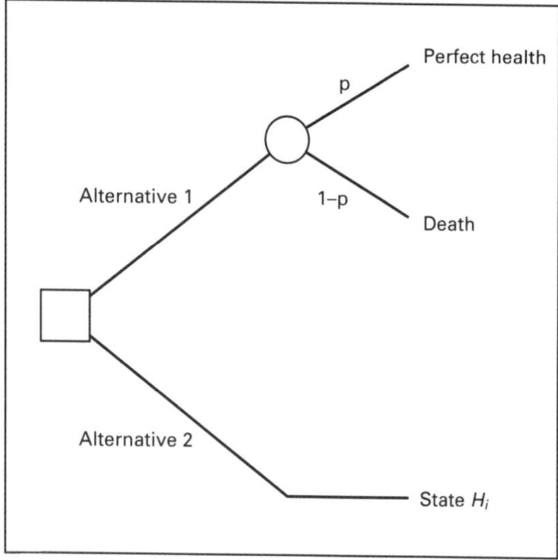

Fig. 12.7 The standard gamble method.

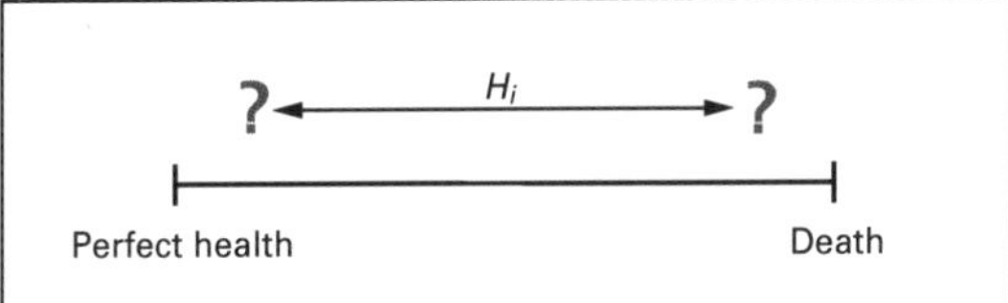

Fig. 12.8 A visual method for eliciting utilities.

health-state evaluation, one method uses a simple visual analogue of utility and asks subjects to locate alternative health states on a line, as in Fig. 12.8 below. The extreme points on the line represent the outcomes of perfect health and sudden death. For each health state, the 'representative' individual is asked to locate the state to be valued, such that the distance at which it is located along the line reflects how they value the health state relative to the two extreme states. If alternative procedures like this produce similar utility assignments, that may provide some cross-check as to whether the elicited values are meaningful.

When it is possible to construct an acceptable utility function over the relevant consequences for a decision problem, EUT can provide a powerful analytical device, particularly for tackling complex problems with a high informational content. For example, EUT can be used as a method for determining optimal choices in problems that can be represented in the form of decision trees (for a discussion of the mechanics of this approach, see Bunn (1982)). Kates (1978) discusses how such decision methods can be used to integrate and structure the multiple considerations, such as subjective judgements, scientific research, assessment studies and so on, which are typically relevant for issues of environmental risk assessment. As an illustration, Kates refers to a widely cited analysis by Howard *et al.* (1973) of the decision of whether (or not) to seed a hurricane.

It would be misleading to suggest that such methods provide 'optimal' solutions to complex decision problems in the sense of unequivocally correct answers. For instance, such analysis provides no means of escape from the many difficult value judgements involved in environmental assessment. The role of such analysis is rather to provide a logical framework within which the implications of alternative judgements can be clearly understood and communicated. While there are good grounds to be cautious about breeding overconfidence through an overly technocratic approach to decision-making procedures,* Howard *et al.* (1973, p. 52) offer the following defence of decision analysis:

> 'Decision analysis separates roles of the executive decision-maker the expert, and the analyst. The analyst's role is to structure a complex problem in a tractable manner so that the uncertain consequences of alternative actions may be assessed. Various experts provide technical information from which the analysis is fashioned, but it is the decision-maker who acts for society in providing the basis for choosing among the alternatives. The analysis provides a mechanism for integration and communication so that the technical judgements of the experts and the value judgements of the decision-maker may be seen in relation to each other, examined and debated. Decision analysis makes not only the decision but the decision process a matter of formal record. For any complex decision that may affect the lives of millions, a decision analysis showing explicitly the uncertainties and decision criteria can and should be carried out.'

There seems some force in this argument and it provides reason to think that reservations about the normative acceptability of the EUT axioms must be set against the potential pitfalls of attempting to make complex decisions without the use of systematic decision analysis methods.

12.8 CONCLUSIONS

Economists have yet to develop a fully satisfactory descriptive model of choice behaviour in relation to risk. The variety observed in real human behaviour is much richer than any available theory

* For example, Fischoff *et al.* (1978) found evidence that the use of 'fault trees' (i.e. diagnostic checklists presented in tree-like form) can 'blind' decision makers to possible sources of system malfunction not included in the analysis.

can currently accommodate, but there are indications that our understanding of human choices is evolving and there are reasons to be optimistic that further progress can be made.

Research in this area is still at an early stage of development and, in evaluating our state of knowledge, it is perhaps important to bear in mind that the theories that I have been discussing are only the first generation of offspring from EUT. These theories were developed with a view to explaining some quite specific phenomenon. And while none of the theories appears entirely satisfactory in the light of subsequent testing, there are clear indications that several of the new theories may each be identifying potentially important influences in decision-making.

For instance, it now seems clear the numerous 'anomalies' arise from the way in which human beings process probability information: in evaluating risks, objective probabilities appear 'distorted' with some probabilities, such as the probability of outlying consequences, or certainties receiving particular weight in decisions; and phenomena such as event splitting effects demonstrate that, in general, human beings do not evaluate probabilities according to the conventional rules of probability calculus. The SEU theories, which propose a subjective probability weighting function, seem to provide a simple account of facts like this. Other theories point to quite different influences in risky choice: for example, regret theory emphasizes the potential importance of cross-act comparisons, while prospect theory emphasizes the role that decision heuristics may play in determining choices. There seems to be at least some empirical support for the claim that each of these factors may be important. So, perhaps the time is right for cross-fertilization to develop a second generation of new theories, in the light of the much richer data base now available, which may integrate some of the intuitions offered by first-generation theories.

The normative project has arguably advanced more swiftly to maturity. While there are some notable dissenters, there is widespread agreement amongst decision theorists that the axioms of EUT are sound principles of rational choice. For those who accept this judgement, EUT provides a model for how people ought to choose and one which can be used to aid practical decision-making in at least some contexts. Indeed, even those who have reservations about the normative validity of EUT may find some value in its application.

While some may claim that violations of EUT, such as the Allais paradox, are normatively defensible, other anomalies, for example framing effects, event splitting effects, or violations of monotonicity, provide a clear warning that the judgements of ordinary human beings are prone to error and may be unduly influenced by contextual aspects of the decision setting, even in very simple choice problems. By contrast, real-world decision problems are typically quite complex and may have to be based on a large body of information. Consequently, those charged with making important decisions on the behalf of other people may be particularly troubled by the extent to which small changes in the contextualization of information can favour different judgements in even the simplest of choice settings. Under these circumstances, if there are decision analysis techniques available which can be endorsed by at least a majority of practitioners, there seems to be a strong case for using them.

12.9 REFERENCES

Allais, M. (1953) Le Comportement de l'homme Rationnel Devant de Risque: Critique des Postulats et Axiomes de l'École Americaine. *Econometrica*, **21**, 503–546.

Appleby, L. & Starmer, C. (1987) Individual choice under uncertainty: a review of experimental evidence, past and present. In: *Surveys in the Economics of Uncertainty* (eds J.D. Hey & P.J. Lambert). Basil Blackwell, Oxford.

Battalio, R.C., Kagel, J.H. & Jiranyakul, R. (1990) Testing between alternative models of choice under uncertainty: some initial results. *Journal of Risk and Uncertainty*, **3**, 25–50.

Bernoulli, D. (1954) Exposition of a new theory on the measurement of risk. *Econometrica*, **22**, 23–26.

Bunn, D.W. (1982) *Analysis for Optimal Decisions*. Wiley, New York.

Camerer, C. (1989) An experimental test of several generalised utility theories. *Journal of Risk and Uncertainty*, **2**, 61–104.

Camerer, C. (1995) Individual decision making. In: *Handbook of Experimental Economics* (eds J. Kagel & A.E. Roth). Princeton University Press, Princeton, NJ.

Chew, S.H. & MacCrimmon, K. (1979) *Alpha-nu choice theory: a generalisation of expected utility theory*, Working Paper No. 669, University of British Columbia.

Chew, S.H. & Waller, W.S. (1986) Empirical tests of weighted utility theory. *Journal of Mathematical Psychology*, **30**, 55–72.

Fischoff, B., Slovic, P. & Lichtenstein, S. (1978) Fault trees: sensitivity of estimated failure probabilities to problem representation. *Journal of Experimental Psychology—Human Perception and Performance*, **3**, 552–564.

French, S. (1989) *Readings in Decision Analysis*. Chapman and Hall, London.

Friedman, M. (1953) *Essays in Positive Economics*. Chicago University Press, Chicago, IL.

Friedman, M. & Savage, L.J. (1948) The utility analysis of choices involving risk. *Journal of Political Economy*, **56**, 279–304.

Handa, J. (1977) Risk, probability, and a new theory of cardinal utility. *Journal of Political Economy*, **85**, 97–122.

Hargreaves Heap, S. (1989) *Rationality in Economics*. Basil Blackwell, Oxford.

Hargreaves Heap, S., Hollis, M., Lyons, B., Sugden, R. & Weale, A. (1992) *The Theory of Choice*. Basil Blackwell, Oxford.

Harless, D.W. & Camerer, C. (1994) The predictive utility of generalised expected utility theories. *Econometrica*, **62**, 1251–1289.

Howard, R., Matheson, J. & North, D. (1973) *The Decision to Seed Hurricanes*. Stanford Research Institute, Stanford, CA.

Humphrey, S.J. (1995) Regret aversion or event-splitting effects? More evidence under risk and uncertainty. *Journal of Risk and Uncertainty*, **11**, 263–274.

Johansson, P. (1993) *Cost–Benefit Analysis of Environmental Change* Cambridge University Press, Cambridge.

Kahneman, D. & Tversky, A. (1979) Prospect theory: an analysis of decision under risk. *Econometrica*, **47**, 263–291.

Kates, R. (1978) *Risk Assessment of Environmental Hazard*. John Wiley & Sons, New York.

Knight, F.K. (1921) *Risk Uncertainty and Profit*. Sentry, New York.

Lathrop, J.W. & Watson, S.R. (1982) Decision analysis for the evaluation of risk in nuclear waste management. In: *Readings in Decision Analysis* (ed. S. French). Chapman and Hall, London.

Loomes, G., Starmer, C. & Sugden, R. (1989) Preference reversal: Information-processing effect or rational non-transitive choice?. *Economic Journal*, **99**, 140–151.

Loomes, G., Starmer, C. & Sugden, R. (1991a) Observing violations of transitivity by experimental methods. *Econometrica*, **59**, 425–441.

Loomes, G., Starmer, C. & Sugden, R. (1991b) Are preferences monotonic? Testing some predictions of regret theory. *Economica*, **59**, 17–33.

Loomes, G. & Sugden, R. (1982) Regret theory: an alternative theory of rational choice under uncertainty. *Economic Journal*, **92**, 805–824.

Loomes, G. & Sugden, R. (1983) A rationale for preference reversal. *American Economic Review*, **73**, 428–432.

Loomes, G. & Sugden, R. (1986) Disappointment and dynamic consistency in choice under uncertainty. *Review of Economic Studies*, **53**, 272–282.

Loomes, G. & Sugden, R. (1987) Some implications of a more general form of regret theory. *Journal of Economic Theory*, **41**, 270–287.

Luce, D. & Raiffa, H. (1957) *Games and Decisions.*, Wiley, New York.

MacCrimmon, K. (1968) *An experimental study of the decision making behaviour of business executives*. Unpublished dissertation, University of California, Los Angeles.

MacCrimmon, K. & Larsson, S. (1979) Utility theory: axioms versus paradoxes. In: *Expected Utility Hypotheses and the Allais Paradox* (eds Allais & Hagen). D. Reidel, Dordrecht.

Machina, M. (1982) 'Expected utility' theory without the independence axiom *Econometrica*, **50**, 277–323.

Machina, M. (1987) Choice under uncertainty: problems solved and unsolved. *Journal of Economic Perspectives*, **1**, 121–154.

Markowitz, H.M. (1952) The utility of wealth. *Journal of Political Economy*, **60**, 151–158.

Mitchell, R.C. & Carson, R.T. (1989) *Using Surveys to Value Public Goods: the Contingent Valuation Method*. Resources for the Future, Washington, DC.

Moskowitz, H. (1974) Effects of problem presentation and feedback on rational behaviour in Allais and Morlat-type problems. *Decision Sciences*, **5**, 225–242.

Norton, G. (1984) *Resource Economics*, Edward Arnold, London.

Payne, J.W., Betman, J.R. & Johnson, E.J. (1993) *The Adaptive Decision Maker*, Cambridge University Press, Cambridge.

Pidgeon, N., Hood, C., Jones, D., Turner, B. & Gibson, R. (1992) Risk Perception. In: *Risk: Analysis, Perceptions and Management*. Report of a Royal Society Study Group, The Royal Society, London.

Quiggin, J. (1982) A theory of anticipated utility. *Journal of Economic Behavior and Organization*, **3**, 323–343.

Quiggin, J. (1993) *Generalized Expected Utility Theory*, Kluwer, Dordrecht.

Samuelson, P.A. (1952) Probability, utility and the independence axiom. *Econometrica*, **20**, 670–678.

Savage, L. (1954) *The Foundations of Statistics*, Wiley, New York.

Schmeidler, D. (1989) Subjective probability and expected utility without additivity. *Econometrica*, **60**, 1255–1272.

Slovic, P. & Tversky, A. (1974) Who accepts Savage's axiom? *Behavioral Science*, **19**, 368–373.

Starmer, C. (1992) Testing new theories of choice under uncertainty using the common consequence effect. *Review of Economic Studies*, **59**, 813–830.

Starmer, C. & Sugden, R. (1989) Violations of the independence axiom in common ratio problems: an experimental test of some competing hypotheses. *Annals of Operational Research*, **19**, 79–102.

Starmer, C. & Sugden, R. (1993) Testing for juxtaposition and event splitting effects. *Journal of Risk and Uncertainty*, **6**, 235–254.

Sugden, R. (1986) New developments in the theory of choice under uncertainty. *Bulletin of Economic Research*, **38**, 1–24.

Sugden, R. (1997) Alternatives to expected utility theory. In: *Handbook of Utility Theory* (eds P.J. Hammond & C. Seidl). Kluwer, Dordrecht (in press).

Torrance, G.W. (1986) Measurement of health state utilities for economic appraisal. *Journal of Health Economics*, **5**, 1–30.

Tversky, A. & Kahneman, D. (1981) The framing of decisions and the psychology of choice. *Science*, **211**, 453–458.

Tversky, A. & Kahneman, D. (1986) Rational choice and the framing of decisions. *Journal of Business*, **59**, S251–S278.

Tversky, A. Kahneman, D. (1992) Advances in prospect theory: cumulative representation of uncertainty. *Journal of Risk and Uncertainty*, **5**, 297–323.

Von Neumann, J. & Morgenstern, O. (1953) *The Theory of Games and Economic Behavior*, 2 edn. Princeton University Press, Princeton, NJ.

Weinstein, M., Fineberg, H., McNeil, B. & Pauker, S. (1988) Minutes of a group discussion on clinical decision making. In: *Decision Making* (eds. D. Bell, H. Raiffa & A. Tversky). Cambridge University Press, Cambridge.

Chapter 13
Valuing Risks

DAVID PEARCE

13.1 INTRODUCTION: THE DOSE–RESPONSE MODEL

Changes in environmental quality and in the status of environmental assets give rise to changes in the risks that humans and ecosystems are exposed to. A deterioration in air quality, for example, alters the 'life chances' of individuals, particularly those who are susceptible to respiratory illnesses that can be exacerbated by air pollution. Life chances may be threatened by acute episodes, i.e. sudden peaks in ambient concentrations, and by continued exposure to pollution, chronic effects. Morbidity risks arise in a similar fashion and may range from eye irritation to higher than normal hospital admissions for asthmatics and bronchitics. Pollution changes also affect ecosystems: increases in the deposition of acidic pollutants (nitrogen, sulphur) will affect the quality of surface waters, groundwater, forests and soils generally. Changes in environmental assets also alter risk profiles: loss of forest cover, for example, might increase the risk of downstream sedimentation from forest soil runoff, putting fish stocks at risk and hence livelihoods at risk. Some outcomes may be localized and catastrophic, as with landslides due to the effects of monsoons on deforested hillsides.

Basic models for assessing such risks have a common form (Chapter 1). A functional relationship between the effect, for example mortality or ecosystem function, and ambient pollution concentrations is described as a 'dose–response function' (DRF). Such DRFs are in turn derived from broader functional relationships between, say, the health impact and the various factors thought to give rise to that impact: age, behavioural factors such as smoking, income levels, and so on. Thus, the broad relationship takes a general form:

$$H = f(A,S,Y,Q,E) \tag{1}$$

where H is health (mortality or morbidity), A is age, S is smoking, Y is income, Q is a vector of other influencing factors, and E is environmental quality. Since it is changes in health status that are of concern in risk assessment, Eqn. (1) will more typically be written:

$$\Delta H = f(\Delta A,\Delta S,\Delta Y,\Delta Q,\Delta E) \tag{2}$$

where 'Δ' simply means 'change in'.

Equation (2) can be estimated empirically using multiple regression techniques, and data may be in time series, cross-sectional, or time series plus cross-sectional ('panel') form. From such an empirical estimation process there will be a coefficient linking ΔH with ΔE. Let us call this coefficient the 'dose–response coefficient', d, i.e.

$$\Delta H = d\Delta E \tag{3}$$

To get an estimate of ΔH across a relevant population, the next stage requires that Eqn. (3) be multiplied by the 'stock at risk' (SAR). For human risk calculations, the SAR is the population at risk. This may range from localized and highly specific groups, for example workers in chemical factories, to entire national or regional populations. Thus:

$$\Delta H_{agg} = d\Delta E.SAR \tag{4}$$

where ΔH_{agg} serves to remind us that policy is usually concerned with risks across aggregate populations.

Equation (4) is the simplest form of a dose–response model linking human health risks to environmental quality. In principle, the same logic can be applied to ecological risks, i.e. risks of impairment of entire ecosystems, or to changes in environmental assets. In such cases, however, the DRF tends to consist of often complex sequences of interlinkages: rates of deforestation have to be linked to rates of soil loss, soil loss has to be linked to risks of landslide, landslides have to be linked to the population at risk. The basic contrast between human health risk and ecological risk is that there are no widely agreed measures of risk for ecological impacts.

13.2 THE VALUATION ISSUE: HEALTH

Other chapters in this Handbook (especially the preceding ones) deal with the problems of estimating risks. This chapter is concerned with the stage that follows Eqn. (4), the valuation stage. We consider ecosystem risks shortly.

Essentially, some value, call it V, needs to be attached to the health effect, completing the model as follows:

$$V.\Delta H_{agg} = d\Delta E.SAR.V \qquad (5)$$

The question that valuation seeks to answer is: how important is the health impact? A basic principle is that the more important the impact, the higher the value should be. The problem is how to measure value in such a way that this general relationship is preserved.

Few issues turn out to be more controversial than the valuation issue. To understand some of the reasons for controversy it is useful to distinguish between what we call the 'constituency' issue—i.e. whose values count?—and the 'measuring rod' issue—i.e. in what units are values to be expressed?

The constituency issue is whether the values in question are those of individuals (and, then, is it all individuals or just the individuals at risk?), experts or policy makers. If it is individuals who 'matter', should it be individuals now, or all individuals likely to have an interest in the risk decision whether those individuals exist now or not? And if all individuals are relevant, do they each have equal 'votes', or are some to have more say than others? What if an individual is not well informed about risks: should those valuations be downgraded?

The measuring rod issue overlaps to some extent with the constituency issue. Should values be expressed as single votes in 'yes/no' form? Or should some other measuring rod be used, such as money?

These questions already hint at the reasons for the extensive controversy surrounding risk valuation, for the questions strike at the very heart of the design of a political constitution. Basically, the issue is who decides and on behalf of whom? We first investigate these questions in a little more detail before elaborating risk valuation from an economist's standpoint.

13.3 VALUATION AND RESOURCE ALLOCATION

Valuation is intricately involved with the issue of how to allocate scarce resources. Risk reduction is not a costless activity and hence any resources used up in the reduction of one set of risks could have been used to reduce another set of risks. Taking a wider view, resources allocated to risk reduction might equally be allocated to some entirely different purpose: education, restoring national heritage, improving landscapes, and so on. Valuation attempts to provide the answer to the problem of choosing between alternative uses of resources. If risk reduction has a high value relative to other uses of resources, then risk reduction should have priority.

At the risk of some simplification, we categorize two schools of thought. The first is the 'rights approach'. On this approach, individuals have rights to human health and a clean environment, and such rights would have similar status to rights against discrimination (Bullard 1994). The implication of the rights approach is that all environmental risks should be reduced to zero, since any positive level of risk infringes individuals' rights. The second approach is based on 'trade-offs' between cost and risk reduction. Risk reduction is pursued up to some point where the costs of conservation are thought to be 'too high'. There are divergent views as to how this trade-off is to be made. In particular, there are those who favour a balancing of economically valued costs and benefits, and there are those who favour leaving the trade-off to the political system. This categorization is not meant to be all-encompassing. More detail of the considerable variation of views within these categories can be found in Turner (1993).

The issue, then, is one of values. Environmental values either transcend costs, or they are such that they have to be traded against costs. This is the environmental values debate.

One of the problems with the values debate is that much of the discussion takes place quite independently of the real-world context of environmental change. Whilst abstract discussion and 'philosophizing' must always be central to scientific advance, its divorce from the context of actual choice risks making it of limited relevance in the policy arena. To see why, we have only to remind ourselves of the existence of finite resources. If resources were infinite there would be no problem of trade-off, and hence no problem of determining priorities. Everything deemed to be 'good' or 'right' could be done. But the real world is not like this and it is necessary to choose. Adopting a rights-based approach implies that the choices surrendered by pursuing risk reduction as a matter of right are of a lower 'moral order' than risks to human health. As noted above, the problem then is that risk reduction has to be pursued regardless of the foregone values sacrificed. Moreover, all risk reduction has to be pursued: it cannot be correct to reduce some risks but not others unless the rights are attenuated in some way. Risk reduction may therefore conflict with other rights, for example rights to a decent livelihood, rights to education and, especially, rights to freedom of choice. The last is particularly important because risk reduction may involve actions which infringe individual rights to choose. In other words, rights approaches can easily be elitist and in conflict with other, equally 'reasonable', human rights. Much of the popularity of the rights approach arises from an understandable sense of frustration with the fact that trade-off approaches do involve 'acceptance' of some positive levels of risk. But it also has its foundations in a lack of appreciation of what 'cost' actually means, a perception fostered by the view that cost is 'just money', as if money is unrepresentative of human well-being. This is why the pursuit of zero risk has been described as the biggest risk of all.

The unreality of the rights approach does not mean that rights are irrelevant. Environmental rights are clearly not absolute. They are relative to other rights and their superiority would require some meta-ethical principle which, of course, many 'deep ecologists' espouse. The argument here is that the deep ecological view fails to acknowledge the reality of rights conflicts because it fails to identify just what moral absolutism entails in the real world. A further way in which rights remain relevant is in the way they influence individuals' own valuations of risk reduction. Even if rights are not formalized, it is clear that many individuals perceive that they have entitlements to certain levels of risk, such that an invitation to consider a higher level of risk compared to that currently experienced is regarded as an infringement on those entitlements (see section 13.7 on the difference between 'willingness to pay' and 'willingness to accept').

One response from the absolutists is to deny the trade-off by pointing to the small sums of money actually spent on risk reduction, i.e. arguing about priorities for risk reduction diverts attention from the real issue, which is to get a larger overall budget for risk reduction. The implication is that increasing risk reduction budgets has no true 'opportunity cost'—i.e. nothing of comparable value is being surrendered by relaxing environmental budgets. There are indeed good arguments for increasing risk reduction expenditures, but they do not obviate the trade-off problem. Resources simply are not unlimited, so that relaxing one budget necessarily entails tightening another. Something is always sacrificed.

A real-world view acknowledges trade-offs. But how are such trade-offs to be made?

13.4 THE TRADE-OFF VIEW: ECONOMICS

The economic approach to the trade-off issue operates through the aggregation of human preferences. The set of persons affected by a decision defines the set of people whose preferences count, where 'affected by' means that their well-being is, in one way or the other, partly dependent on the environment in question. This preference-base is inherently 'democratic'—it requires that policies be responsive to preferences however they are formed. Preferences are revealed in the marketplace through demand behaviour—i.e. as 'willingness to pay' (WTP). Indeed, the demand curve in textbook economics is a (marginal) WTP curve. This simple observation is important, for if WTP is rejected as a criterion for allocating resources to risk reduction, then some explanation has to be provided as to why environmental goods and services are different to other goods and services which are allocated on a WTP basis. Put another way, if supply and demand are allowed to determine the allocation of conventional goods, why should risk reduction not be allocated on the same basis? One does not have to be a free market economist to embrace the general attractions of markets: giving people what they say they want and minimizing paternalistic judgements and any overriding of those preferences.

But risk reduction often is different in one sense at least. It often has no market, i.e. the issue giving rise to risk is not bought and sold on the open market. Clean air would be an example. Thus the economic approach requires that preferences for risk reduction be *inferred* from human behaviour in other contexts.

13.5 THE THEORY OF ECONOMIC VALUATION

The theory of economic valuation has developed substantially in the last two decades. This section reviews, briefly, those techniques that relate to human health risks only. Other techniques are relevant to the valuation of other environmental changes. For a detailed review, see Freeman (1993).

13.5.1 A general statement

For a change in risk that threatens life and health generally we can say that the relevant valuation is the value that the individual at risk attaches to their own health and life chances, plus what others would be willing to pay to avoid the risk to that individual, plus any costs that society at large bears and that would not otherwise occur if the individual did not suffer the effects of the risk in question. These components are:

$VOR_{i,i}$ where 'VOR' refers to individual *i*'s valuation of risk to themselves, i.e. 'own risk'. The way in which these individual VORs are aggregated is dealt with shortly. Essentially, we will require the summation of such own valuations for all individuals at risk to give $\Sigma_i VOR_{i,i}$, more commonly known as the 'value of a statistical life'—see below.

\+

$VOR_{i,j}$ where the *i,j* notation now refers to *j*'s valuation of risks to *i*. Again, this will need to be summed for all *j*, i.e. for all people expressing some concern about risks to *i*, to give $\Sigma_j VOR_{i,j}$.

\+

COI_i where 'COI' refers to the 'cost of illness' suffered by *i* but whose costs are borne by the rest of society. An example would be hospital costs. The COI could be regarded as part of $VOR_{i,j}$.

13.5.2 Valuing statistical lives

Clearly, one form of health risk is the risk of premature mortality arising from some risk context, say, increased air pollution. What value should be attached to such risks of mortality? The sum of individuals' own valuations of risks to their own lives is known as the value of a statistical life, VOSL. The shorthand often used for the VOSL is 'value of life', which is unfortunate. Since the idea of 'valuing life' appears odd to some and morally offensive to others, it is important to understand what a value of a statistical life (VOSL) actually is.

The way in which VOSL is obtained is by aggregating up from a value (willingness to pay, WTP) of risk reduction. Imagine that the probability of dying next year is 0.004 for each person and suppose we have 1000 people in the population. Assume that there is some risk reduction policy that reduces the risk to 0.003: a change of 0.001. Each person is asked to express their WTP for this change in risk, and suppose the answer is £1000. The risk reduction policy is a public good: it affects everyone equally. Thus 1000 people say they are each willing to pay £1000 for the policy, i.e. their aggregate willingness to pay is £1 million. The change in risk will result in one statistical person being saved each year (1000 × 0.001). Thus the value of a statistical life is £1 million in this example. It is important to understand that no-one is being asked their WTP to avoid themselves dying at a specified time: they are being asked to express a WTP for a change in risk. As Freeman (1993) notes:

> '... the economic question being dealt with here is not about how much an individual would be willing to pay to avoid his or her certain death or how much compensation that individual would require to accept that death. In this respect, the term 'value of life' is an unfortunate phrase that does not reflect the true nature of the question at hand. Most people would be willing to pay their total wealth to avoid certain death; and there is probably no finite sum of money that could compensate an individual for the sure loss of life. Rather, the economic question is about how much the individual would be willing to pay to achieve a small reduction in the probability of death during a given period or how much compensation that individual would require to accept a small increase in that probability.' (p. 320)

These values can be expected to vary across different individuals. The two main reasons for this will be as follows.

1 People have differing attitudes to risk: some may even be 'risk lovers', i.e. positively enjoying risky contexts. Most people are risk avoiders, i.e. they will tend to reveal a positive willingness to pay for risk reduction. But there is no particular reason why their valuations of risk should be the same.

2 Incomes vary and hence WTP is likely to vary in such a way that those with higher incomes have higher WTPs. This is not a necessary result since attitudes to risk may vary in such a way as to offset an income effect. None the less, it raises an important equity issue about fairness between people, an issue that is not in fact confined to risk valuations but to the use of WTP measures in general.

A VOSL can also be measured by a 'willingness to accept' compensation for increased risk. It is well known that many people do make this trade-off between risk and money, for example, by accepting premia on wages to tolerate risk. It is tempting to think that the WTA approach will produce very much higher values for a VOSL than the WTP approach, simply because WTA is not constrained by income; WTP and WTA can, indeed, be different and WTA for environmental losses may exceed WTP for environmental gains by factors of 2–5 (Gregory 1986). Various explanations exist for this disparity, including the fact that individuals may feel they are losing an 'entitlement' if the issue is one of loss of an entitlement (WTA) rather than an increment to an existing entitlement (WTP). Another explanation, which is wholly consistent with economic theory, suggests that WTA > WTP arises mainly in contexts where there is no ready substitute for the environmental good in question (Hanemann 1991).

How, then, can WTP and WTA be estimated?

13.6 TECHNIQUES FOR ESTIMATING VOSL

A number of techniques have been developed to estimate VOSLs. The main ones are rooted in the general economic theory of valuation, i.e. they have a theoretical basis on the measurement of human well-being based on individual preferences. One widely used technique, however, has only a tenuous link to the theory.

13.6.1 Valuing mortality risks: wage risk models

The wage risk, or 'hedonic wage' model estimates a WTA measure of risk. Essentially, it looks at wages in risky occupations and seeks to determine the factors that determine wages. One of these factors is hypothesized to be the risk level. Other things being equal, workers will prefer jobs with less risk to jobs with high risk. This will result in a relative shortage of workers for risky jobs and hence wages in those jobs should be higher. This 'wage premium' then becomes a measure of risk valuation. It can be estimated by multiple regression techniques in which the wage is the dependent variable and the various factors influencing the wage are the independent variables. An example might be:

Wage = *f* (*Educ, Exp, Union, Risk, Occ*)

where *Educ* is education, *Exp* is years of experience, *Union* is an indicator of the degree of unionization of the labour force, *Risk* is the objective (or perceived) probability of fatal injury and *Occ* is some indicator of the desirability of the occupation. The coefficient linking *Wage* and *Risk* is then the WTA measure of risk.

One obvious problem with such approaches is that workers have to know about the differences in risks and, if they do, whether those perceptions coincide with 'objective' measures of risk, such as the probability of a fatality in that industry. If there is no perception of risk, but risk exists, then the 'hedonic wage' (i.e. the wage premium) may be zero, seriously understating risk values. If there is a perception of risk but it is exaggerated compared to objective risk, then risk may be overvalued. What limited evidence there is suggests that workers actually overstate the risk of their jobs. But as Freeman (1993) points out, what matters for the hedonic wage model is the perception of differences in risks between jobs, not the absolute level of risk in a given job.

Most hedonic wage studies have been carried out in the USA and suggest that VOSLs range from $2 million (1994 prices) to $3.5 million.

13.6.2 Valuing mortality risks: avertive behaviour

Individuals spend money in trying to reduce risks, so-called 'averting behaviour'. Under certain circumstances these expenditures approximate the economist's concept of WTP to reduce risk. The kinds of averting expenditures in question might be on smoke alarms, safety harnesses, tamper-proof drug storage containers and so on. These kinds of expenditures can be regarded as part of what is called a 'health production function' in which the state of good health is 'produced' by various factors, including expenditure on averting ill-health. Note that some apparent averting expenditures are not valid measures of risk reduction. Thus, it is quite widely assumed that life insurance expenditures are measures of WTP to avoid risk. But insurance expenditures do not have the effect of reducing risks. Indeed, they may actually increase risks by encouraging less-careful behaviour—the issue of 'moral hazard'. As Freeman (1993) notes, life insurance essentially values the earning capacity of the insured individual to the dependents, who are the ones who will gain from any insurance policy. This is not at all the same thing as the individual's willingness to pay to reduce risks to his or her own life, which is what is required.

The health production function can be written:

H = *f* (*Poll, Med, Avert, Other*)

where *Poll* is the level of pollution, *Med* is the level of medical treatment, *Avert* is the level of averting activity, and *Other* refers to all the other factors affecting health status: age, income, smoking behaviour and so on. Reducing pollution will reduce the time spent being unwell, say, from 4 days to 3 days. If by spending £*X* the same reduction in ill-health can be achieved, then £*X* should be the value of the reduced pollution to the individual. More formally:

WTP (Pollution reduction) = (Reduced time in ill-health) × (Extra cost of 'producing' health by mitigating activity)

In this way, expenditure on risk reduction can be interpreted as WTP for risk reduction. In practice, finding examples of avertive expenditures that are 'purely' health-producing has proved difficult. Studies include seat-belt use and smoke detectors, and suggest VOSLs of about $0.7 million to $2.2 million.

13.6.3 Valuing mortality risks: contingent valuation

The contingent valuation method (CVM) requires that individuals express their preferences in response to a questionnaire. It is therefore very much akin to market research in which the researcher seeks to find out how a respondent would behave, in terms of WTP, for a modified or new good. Questionnaires take two forms: (i) open-ended or continuous approaches simply ask what amount someone is WTP (or WTA), and no prompting of likely values is permitted; and (ii) discrete or dichotomous choice in which the value is posed and the respondent is then asked whether he or she is willing to pay that sum, yes or no. Yes/no questions that use the cost of providing some project or benefit as the sum to which the yes/no answer is sought are also known as 'referendum' approaches. The CVM is now the subject of a huge literature and, without question, has improved dramatically in light of many criticisms. The kinds of biases that may occur in the CVM include the following.

1 Starting point bias in the dichotomous choice format, i.e. respondents tend to produce WTP answers that tend towards the first 'price' put forward by the questioner. Such a bias is easily tested by seeing if the difference between the average stated WTP is statistically different to the starting point sum.

2 Strategic bias whereby the respondent understates the true value of their preference (they 'free-ride') in the expectation that others will state more and thus secure the goods in question for everyone. This phenomenon was long thought to be inherent with 'public goods', such as clean air, since if the clean air is provided for any one individual then it is provided for everyone (clean air is said to be 'jointly consumed'). Of course, the misstatement of preferences may be biased the other way: someone may be so keen to see the goods provided that they overstate their preferences, fearing that others will free-ride. Tests for strategic bias suggest that, contrary to expectation, it may not be of major significance. Such tests may involve stratifying the sample deliberately to give some people strong incentives to free-ride and others less of an incentive, and then seeing if their average WTPs diverge. Others involve indicating that unless more than a certain percentage of respondents vote for the goods, they will not be supplied.

3 Hypothetical bias: the respondent may produce answers that are purely hypothetical, i.e. if the good or policy in question is actually provided, their WTP will be less than stated in response to the questionnaire. Careful design of questionnaires can reduce the hypothetical bias problem to very low levels in WTP questionnaires. It may be more of a problem with WTA questions. Measuring the bias usually involves comparing stated preferences with actual preferences when real sums of money are involved in the CVM. Since respondents are less familiar with compensation contexts (the relevant context for WTA), their stated WTA is likely to exceed their actual WTA.

4 Part-whole or 'mental account' bias: here the problem is that, while the questionnaire may focus on a specific environmental benefit, the respondent may act as if he or she is valuing environmental improvement in general. Tests for this kind of bias involve varying the quantity of the goods in question, for example a 1% reduction in risk, a 10% reduction in risk, and so on, and seeing whether WTP varies significantly as the benefit is increased. A number of studies suggest that WTP is the same and that what individuals are 'purchasing' is not the benefit in question but the 'warm glow of giving' or 'moral satisfaction'. One response has been to ensure that questions are not asked until the respondent has been reminded that he or she has a specific budget to be allocated. The empirical evidence of part-whole bias remains mixed.

5 Information bias: the quality of information supplied to the respondent may affect the stated WTPs. Usually, the more information that is supplied about the risks in question, the higher is the WTP. Why this is regarded by some critics of the CVM as a flaw in the method is unclear. Information should influence WTP in exactly the same way as information influences WTP in the everyday marketplace for ordinary goods. While there is an interesting issue of how much information should be provided, what matters most is that the same level of information be provided to all respondents.

There are other problems with the CVM but modern CVM design is capable of minimizing the extent of error in stated responses. Good surveys of the CVM are to be found in Pearce *et al.* (1994) and Bateman & Turner (1993).

13.6.4 Valuing mortality: the human capital approach

Before the formalization of the hedonic wage, the CVM and the avertive behaviour approaches, the most commonly used technique to value risk to human life was the human capital approach. The idea is simple: an individual is 'worth' to society what he or she would have produced in the remainder of their lifetime, gross of taxes since the interest is in society's valuation of the individual. An argument did exist as to whether the earnings that are relevant should be net or gross of the individual's own consumption. If the individual's consumption is excluded, then the value concept is simply that of how the rest of society values the individual, and that is inconsistent with the WTP approach. If the individual's consumption is included, then at least some gesture is made towards including a value from the individual's own standpoint. But there is in fact nothing to suggest that an individual's WTP need be equal to the remaining lifetime income of that individual. The link to the WTP approach is clearly tenuous at least. The WTP approach is based on how individuals value risks to their own lives, whereas the human capital approach makes no obvious reference to that concept. Indeed, the human capital approach says little about individuals' attitudes to risk. Its one virtue is that it is thought to be very easy to calculate.

Even if the human capital approach is accepted as a rule of thumb, however, there are problems in its estimation. First, if the individual at risk has retired from work, the human capital valuation would suggest that their VOSL is zero or even negative (if consumption is deducted). This perhaps underlines the failure of the concept in terms of its theoretical underpinnings. If human capital was a WTP concept, those out of work should have positive WTPs. Second, someone may be of median age but still not be producing marketed 'output'. A houseperson, for example, produces non-market output and this would have to be valued. Third, future earnings cannot simply be added up year by year to get a total, since the individual will discount the future. Hence, a discount rate is needed.

While it would be better if the human capital approach was avoided altogether, it is still widely used, no doubt because of the relative ease of computation. If used, the computation in question would be:

$$\text{'VOSL'} = \Sigma_{i=1,\,T-t}(p_{t+i} \cdot Y_{t+i})/(1+r)^i$$

where $\Sigma_{i=1,\,T-t}$ denotes the sum over time from time t (the current age of the individual at risk) to time T (the age at which the individual ceases to work), p_{t+i} is the probability of the individual surviving from age t to age $t+i$, Y is income, and r is the discount rate.

The human capital approach does not produce a value of statistical life in the sense of $\Sigma_i\, VOR_{i,i}$ above. But can it be used to estimate the value that others put on the life at risk? For close relatives, friends, etc., the answer must be 'no': such people do not value risks in terms of the income forgone by the individual at risk. But what of society generally? There is a sense in which society loses the output of the individual less the consumption of that individual. But had the individual survived, the rest of society might have to produce transfer payments in the form of welfare payments, and illness costs arising from the ill-health that the individual would have suffered had they survived. Arguably, then, what the rest of society loses is i's income minus i's consumption minus i's claims on the rest of society over the expected lifetime of i without the risky event. This is perhaps the most that can be said for the human capital approach.

13.7 WILLINGNESS TO PAY VERSUS WILLINGNESS TO ACCEPT

The valuation of statistical lives rests on the WTP or WTA principle. It is easy to see that WTP is constrained by income and/or wealth. Willingness to accept appears not to be so constrained since it is an amount in compensation for accepting a risk. Both the contingent valuation and wage risk models have the capability to elicit WTA estimates, and practice has found that WTA figures are not infinite, i.e. people do not expect extremely large payments in compensation for losses of environmental quality. But WTA does tend to exceed WTP, as noted earlier, and these differences cannot generally be explained by issues of questionnaire design. There are genuine differences between WTP and WTA. This raises the issue of which measure is correct.

The answer depends on the context, and especially on property rights, although other factors

Table 13.1 Matrix explaining the property rights issue.

Valuation of a GAIN **Willingness to pay** = > Property rights **do not** rest with the individual	Valuation of a LOSS **Willingness to pay to avoid the loss** = > Property rights **do not** rest with the individual
Valuation of a GAIN **Willingness to accept compensation to forego the gain** = > Property rights **do** rest with the individual	Valuation of a LOSS **Willingness to accept compensation** = > Property rights **do** rest with the individual

help to explain the size of the discrepancy between WTP and WTA. The matrix in Table 13.1 explains the property rights issue. Generally, WTP is the right concept when the individual whose valuation is being sought does not have a right to the improvement being valued, whereas WTA is the correct concept when the individual does have a right to the *status quo* and is being asked to forego a benefit or accept a loss.

The relevance of property rights arises in two contexts. The first is where there are clearly defined legal rights, and the second relates to the individual's perception of rights. Thus, we might expect individuals to value a unit loss much more highly than a unit gain if he or she believes they have some right to the existing amount of environmental quality or asset. There is, indeed, evidence that individuals have 'loss aversion', i.e. they regard the *status quo* as some kind of reference point from which gains and losses are evaluated. This view is stressed by advocates of 'prospect theory'—see Kahneman & Tversky (1979).

A second factor explaining the wide divergence sometimes found between WTP and WTA is the degree of substitutability of the item being valued with other goods. Suppose, for example, that what is being valued is a unique environmental or material asset—the Grand Canyon or the Taj Mahal, say. Then, as there are no ready substitutes, one might expect WTA to be very much higher than WTP. And this turns out to be the case: the fewer the substitutes, the larger the discrepancy between WTA and WTP, as theory would predict (Hanemann 1991). Hanemann's explanation is not comprehensive because the same WTA/WTP discrepancy exists for commonplace goods, in which case the insights from prospect theory appear to be relevant.

The relevance to life risks is, of course, significant if individuals feel that risks to life or health constitute an invasion of their 'rights' not to have to tolerate those risks. There is some evidence to suggest that those rights will be especially pronounced when the risks are not voluntary, in contrast with, say, occupational risks. If so, we might expect wage risk models, which are WTA estimates for voluntary risk, to reveal risk valuations that are above, but not substantially above, WTP valuations. The CVM models, on the other hand, might reveal substantial WTA/WTP discrepancies if the risk in question is involuntary. The problem with the evidence is that most of the VOSL studies are either wage risk studies or CVM studies of transport risk. Transport risks may or may not be seen as voluntary compared to, say, radiation risks from a nuclear power plant, although most risk studies appear to treat transport-related risk as involuntary. Early analysis of the voluntary/involuntary risk valuation issue was fairly inconclusive. Starr (1972) attempted a comparison of risk levels and the associated benefits, and concluded that involuntary risk might be valued by a factor of 10 more than voluntary risk. Reworked by Otway & Cohen (1975), Starr's ratio appears far too high and a factor of two appears more appropriate. But further analysis by Fischoff *et al.* (1979) reinstates the large tenfold differential between involuntary and voluntary risk values. Substantial question marks hang over these studies, however, not because of the risk data but because of the use of measures of benefit based on actual expenditures on the activity in question or the contribution that the activity makes to an individual's income. There is some affinity here with the required benefit measure—WTP—but it is far from precise. Overall, then, the conceptual contexts in which WTP and WTA should be used are fairly clear and relate to the presumption about property rights. In practice, determining the assignment of property rights is far less straightforward.

13.8 THE EVIDENCE ON VOSL

Pearce *et al.* (1992) review the various estimates of VOSL and find the mean estimates across studies shown in Table 13.2 (updated to 1994 values). The estimates also show the ratio of WTA to WTP because of the presumption that WTA studies tend to find higher values than WTP studies. The wage risk studies are fairly consistent between the UK and USA, with a suggestion that a higher (average) value exists for WTA in the USA than in the UK. On the other hand, WTP studies appear to produce higher values in the UK. These data also suggests that the WTA > WTP inequality holds for the USA but not for the UK, but there is no ready explanation for these disparate results.

Other reviews for the USA suggest ranges of recommended VOSLs. Fisher *et al.* (1989) recommend a range of $2–10 million; Cropper & Freeman (1991) recommend $2–6 million; Viscusi (1992) recommends $3–7 million and Miller (1989) recommends $1–4 million.

Environmental risks are especially likely to affect the health of those already predisposed to illness, for example, the elderly. Hence, it is important to know if the VOSL is likely to vary with age. Are there any conceptual reasons for supposing that WTP will vary with age? At one extreme we could legitimately argue that we have no reason to suppose that this WTP will vary negatively with age. Indeed, older people may be all the more risk averse. But there are plausible reasons to suppose that WTP will fall as age increases (Freeman 1993, chapter 10; Cropper & Simon 1994). Freeman (1993) reviews lifecycle models and shows that, in general, one might expect WTP to decline with age. This is because lifetime utility is dependent on lifetime consumption in such models, and older people simply have fewer consumption years left. However, there are several reasons why such lifecycle WTP models understate 'true' WTP:

1 They tend to omit others' valuations of the life at risk (e.g. relatives, friends) (see section 13.9).

2 Lifecycle models assume that expected lifetime utility depends on expected lifetime consumption only, whereas individuals surely value survival as well. Note that this value of survival need not vary inversely with age at all, and could actually increase.

3 There is evidence to suggest that WTP for 'contemporaneous risk' is less than WTP for 'latent risk', i.e. WTP for avoiding accidents is less than WTP for avoiding risks of cancer (Jones-Lee *et al.* 1985). Yet the empirical VOSL literature is almost entirely based on accident risks. For pollution issues, then, transferring VOSLs from accident risk contexts to pollution contexts is likely to understate the 'true' degree of risk aversion.

Table 13.2 Values of statistical life (£m in 1994).

Value*	USA	UK
WTA (wage risk)	2.7–4.2	2.2–2.7
WTP (CVM, CRM)	1.1–2.0	3.1–4.9
(market)	0.8–0.9	0.5–2.6
WTA/WTP (WR/CVM)	1.5–4.2	0.4–1.0
WTA/WTP (WR/market)	3.4–6.1	0.9–5.4

* WTA, willingness to accept; WTP, willingness to pay; CVM, contingent valuation method; CRM, contingent ranking method; WR, wage risk.

13.8.1 Value of a statistical life and age

What evidence exists on WTP and age? Jones-Lee *et al.* (1985) produced some estimates of accident risk, which Cropper & Sussman (1990) have normalized to obtain the data in Table 13.3 (in 1994 UK£ million).

One could use this as evidence that older people have a lower WTP of around 70% of the WTP of a median age (40) person. Note that the curve of the ratios of WTP is actually bell-shaped, i.e. younger people also have a lower VOSL than the median-aged person: the overall WTP–age function becomes an 'inverted U'. It is not clear if

Table 13.3 Age and VOSL (£m).

Age	WTP	WTP/WTP(40)
18	2.03	0.67
20	2.34	0.77
30	2.72	0.89
40	3.04	1.00
50	2.41	0.79
60	2.13	0.70

this is consistent with the lifetime consumption model.

In more recent work, Jones-Lee *et al.* (1993, 1995) secure a best-estimate VOSL from a contingent valuation study of £4.25 million, which is considerably above the figures used in this report. Interestingly, they find no relationship between the VOSL and age. Age simply exerts no influence on WTP. Overall, and in the absence of further research, it appears to be a reasonable assumption that the VOSL is invariant with age.

13.8.2 Health states and VOSL

Finally, if those at risk are already ill, does this affect valuation? It probably does, but which way the valuation goes is not clear. To quote a recent study for the US Environmental Protection Agency (USEPA):

> '... it is possible that the reduced life expectancy and reduced enjoyment of life associated with many chronic illnesses may result in lower WTP to reduce risks of death. On the other hand, facing serious illness and reduced life expectancy may result in higher value [being] placed on protecting the remaining time.' (Chestnut & Patterson 1994)

13.9 OTHERS' VALUATION OF RISKS TO INDIVIDUAL *i*

So far, the discussion suggests that wage risk, contingent valuation and avertive behaviour techniques can be used to value 'own' life risks, but that the human capital approach is best avoided. The human capital approach might have some role to play in assessing the rest of society's losses, but even here there are problems.

The second component of the basic valuation equation was the value placed on risks to i by others who are close to i, i.e. relatives and friends. The literature that seeks to estimate such valuations is very much smaller, but suggestive of some results. Viscusi *et al.* (1988) surveyed consumers to elicit risk valuations for injury risks from the use of insecticides in the USA. Consumers were asked their WTP to reduce risks from 15/10 000 to 10/10 000 for two pairs of risk: inhalation and skin poisoning, and inhalation and child poisoning. The WTP figures of \$1.04 and \$1.84, respectively, therefore imply values of risk of \$2080 and \$3680 (1.04/0.0005 and 1.84/0.0005). Individuals were then asked their WTP for an advertising campaign to reduce risks by the same amount generally, i.e. to other people. The results implied valuations of the first risk pair of \$10 000 for North Carolina State—where the survey was conducted—and \$3070 for risks outside the state. For the second risk pair, the values were \$18 100 and \$4260. The state/non-state comparisons suggest that valuations decline as the individuals at risk become more 'anonymous' to the valuer, as one might expect.

An early study by Needleman (1976) sought the valuation of close relatives for reductions in risks. The study looked at kidney donors. Donors tended, at that time, to be close relatives to secure greater chances of acceptance of the transplanted organ. The kidney donor suffered a slight increase in risk while the recipient had dramatically improved chances of survival. By looking at data on actual kidney donations and at refusal rates—i.e. situations in which the relatives refused to make the donation—Needleman estimated a 'coefficient of concern'. An average coefficient of 0.46 implies that close relatives' valuations may be 46% of the value of risk of the individual at risk, i.e. one might write $VOR_{i,j} = 0.46VOR_{i,i}$, where j is now close relatives. Recall that $VOR_{i,i}$ is summed across all individuals at risk and expressing a positive WTP to obtain a VOSL. It follows that $VOR_{i,j}$ should be summed across all close relatives of those at risk. The effect could be substantial. For example, if each individual at risk has four close relatives, the effect would be to multiply

VOSL by $4 \times 0.46 = 1.64$ to obtain the summed valuations of close relatives. Schwab Christe & Soguel (1995) conducted a contingent valuation analysis of WTP to avoid the consequences of a road accident; WTP was estimated in two contexts: where the respondent was the hypothetical victim and where the respondent was a relative of the hypothetical victim. In each case, the pain and suffering of others is relevant. In the former case, WTP ($VOR_{i,i}$ in our notation) may already account for the pain and suffering of relatives and others, i.e. WTP is influenced by the concern the victim has for the effects of an accident to him/herself on others. In the second case, where the victim is a relative, WTP ($VOR_{i,j}$ in our notation) may reflect both the relative's own bereavement and also some judgement of the pain and suffering of the victim. Schwab Christe & Soguel try to distinguish these effects. The results are as follows.

1 $VOR_{i,i}$ for a death is 1.7 million Swiss francs, or around US $1.2 million.

2 $VOR_{i,i}$ for an accident involving severe and permanent disability is slightly higher than $VOR_{i,i}$ for death, at some 1.75 million Swiss francs.

3 $VOR_{i,j}$ for relatives (j) is *higher* than $VOR_{i,i}$ at around 2 million Swiss francs, and higher still for permanent and severe disablement. In general $VOR_{i,j}$ would appear to be equal to 1.25 $VOR_{i,i}$, which is about three times the effect found by Needleman's study.

Cropper & Sussman (1988) suggest that citizens in the USA have a WTP for children's statistical lives equal to 70–110% of their own values ($VOR_{i,i}$). This is consistent with a New Zealand study by Miller & Guria (1992), with a $VOR_{i,j}$ of 119% for family members. Blomquist *et al.* (1996) estimate a $VOR_{i,i}$ of $2 million and a $VOR_{i,j}$ for children by parents of $3–5 million, i.e. 1.5–2.5 times the $VOR_{i,i}$. Blomquist *et al.* (1996) also review other studies of $VOR_{i,j}$, finding a fairly consistent range of values between 23 and 50% of $VOR_{i,i}$ when the person at risk is not a family member.

The studies suggest that $VOR_{i,j}$ may be of the order of 100% for own family members and perhaps 20% for non-family members. Jones-Lee (1992) cautions against assuming that $VOR_{i,i}$ and $VOR_{i,j}$ can be added, but suggests a social value of a statistical life of 1.1–1.4 times the $VOR_{i,i}$. The implications are fairly significant. Not only would a typical valuation of $2 million be quadrupled because of close family valuations, but a further $0.4 million (20% of $VOR_{i,i}$) might need to be added for each person thought to exhibit a degree of concern for the individual at risk. The VOSL, then, could be seriously understated by focusing on $VOR_{i,i}$ alone.

13.10 WILLINGNESS TO PAY TO AVOID MORBIDITY AND INJURY

Comparatively few WTP studies exist for morbidity effects and for non-fatal injuries. Table 13.4 shows the results from the available studies. Note that some studies use an adjusted 'cost of illness' approach: this is explained in section 13.11.

Jones-Lee *et al.* (1995) report contingent valuation and 'standard gamble' results for non-fatal road injuries. Standard gambles involve offering a respondent the choice between a certain health state and a gamble in which a health state could ensue, which is better than the one offered with certainty but which also has the chance of being much worse, so much so that the injured person dies. While the two approaches—the CVM and standard gamble—should give similar results, Jones-Lee *et al.* (1995) found that the CVM produced valuations systematically higher than those implied by the standard gamble. They favoured the standard gamble. To convert the standard gamble into a monetary figure, the authors found the probability level at which respondents are, on average, indifferent between the certain outcome and the gamble. After adjustments, this was found to be 0.096. This was then 'anchored' to the VOSL as measured by WTP. The anchoring was done with respect to the UK 'official' VOSL used in transport planning, which is around £620 000 in 1990 values, so that the value of a non-fatal injury came to some £59 000, to which was added a COI element of £11 000 to give £70 000. This turned out to be a factor of three more than the prevailing non-fatal injury estimate used by the Department of Transport, a figure based on the human capital approach. Since the publication of the Jones-Lee work, the UK Department of Transport has revised its value for non-fatal injuries to £84 260 in 1993 prices. If the rough

Table 13.4 Values for morbidity effects, USA. (From Chestnut & Patterson 1994.)

Effect	Central estimate 1990$ per incident	Source	Nature of estimate*
Adult chronic bronchitis	$210 000	Viscusi *et al.* (1991) Krupnick & Cropper (1992)	WTP
Respiratory hospital admission	$12 000	Krupnick & Cropper (1992)	COI × 2
Emergency room visit	$400	Rowe *et al.* (1986)	COI × 2
Child bronchitis	$240	Krupnick & Cropper (1992)	COI × 2
Restricted activity day	$50	Loehman *et al.* (1979)	WTP and COI × 2
Asthma attack day	$33	Krupnick (1986) Rowe & Chestnut (1986)	WTP
Acute respiratory symptom day	$10	Loehman *et al.* (1979) Tolley *et al.* (1986)	WTP

* WTP, willingness to pay; COI, cost of illness.

guideline of 10% of VOSL is used for serious injury, and the VOSL is closer to £1.5–2 million (see above), then serious injuries would be valued at £150 000–200 000, which is twice the revised figure.

13.11 COST OF ILLNESS

Cost of illness relates mainly to morbidity, although, clearly, there are some medical and other costs associated with premature mortality that may not be present in the event of death from other causes. Cost of illness, in the form of medical costs and lost income can, of course, be part of the individual's own valuation of risks in contexts where the individual bears these costs. This will be the case where only private medical care exists—either in the form of direct payments for care when it is provided, or in the form of insurance premia—and where no system of sick pay exists. In other cases where a public health-care system exists, the cost of illness is borne by the rest of society through tax payments, including business through the payment of sick pay. Cost of illness measures typically encompass such medical costs and foregone income and productivity. It follows that COI, whether borne by the individual or society generally, is not a comprehensive valuation of morbidity. It omits the individual's own WTP for avoiding the ill-health over and above COI. Individual and social COIs also differ. Individuals bear some costs but society usually bears larger costs through wage compensation, and social COI will exceed individual COI even more in contexts where there is a public health-care system and limited private medicine (as in the UK, for example).

Krupnick & Cropper (1989) suggest that average lifetime medical costs per chronic respiratory disease patient may amount to around $100 000. This is small relative to the suggested estimates for VOSL, but across many such patients it implies a significant cost to society. Medical costs tend to be high in the USA relative to other countries, but there appear to be no COI studies for respiratory

Table 13.5 Relationship between WTP and COI.

Study	Health effect	WTP/COI (individual)	WTP/COI (social)
Rowe *et al.* (1984) Rowe & Chestnut (1986)	Asthma	1.6–2.3	1.3–1.7
Rowe & Nethercut (1987) (as perceived by respondent)	Cataracts	4.2 5.3	2.4 2.1
Chestnut *et al.* (1988) (as perceived by respondent)	Angina	2.5–4	NA

diseases with which to compare the Krupnick and Cropper estimates.

It has been suggested that COI bears a relationship to WTP measures of morbidity, such that WTP = 2COI (Chestnut & Patterson 1994). This tentative conclusion is based on three studies in which WTP and COI (individual and social) are estimated (see Rowe *et al.* 1984; Rowe & Chestnut, 1986; Rowe & Nethercut, 1987; Chestnut *et al.* 1988), and can only be regarded as holding for the USA. The results were as shown in Table 13.5. If such a ratio could be established, then the valuation of morbidity effects could be made much more simple, avoiding often costly WTP studies. It remains the case, however, that even COI studies are rare in Europe and are far more common in the USA.

13.12 VALUING FUTURE LIVES

Given that 'sustainable development' is a widely embraced goal of economic and environmental policy, and given that 'sustainability' raises the importance of impacts on future generations, one issue of some importance in risk valuation is that of how to value 'future lives'. Essentially, should a life at risk in, say, 50 years time be valued in the same way as a life today? This is an 'intergenerational equity' issue. Jones-Lee & Loomes (1993) have shown that, on balance, future lives should be valued at the current VOSL and should *not* be discounted. Or, put another way, the effective discount rate applied to future lives should be zero provided that the valuations being applied are the current VOSLs. In benefit–cost analysis a similar result would be obtained by valuing future lives at a future VOSL, i.e. one allowing for the expected growth of incomes which will therefore make future generations more willing to pay for risk reductions, and the discounting that value to get back to a current value. So, for a life risk 50 years hence we would have two alternative rules for valuation at the current period: $VOSL_{t=50} = VOSL_0$, the 'equal values no discounting' rule; or $VOSL_{t=50} = VOSL_0.e^{50g}.e^{-50r}$, the 'discounted future values' rule, where g = expected rate of income growth and r = the discount rate. So long as $r = g$, the two rules are the same. The rules become more complex when: we allow for the degree of aversion to inequality that might be displayed by the current generation; a distinction is made between the discount factor for future risks and the discount factor for future income; and survival probabilities vary between generations.

In general:

1 The greater the degree of aversion to inequality, the closer one gets to the equal values and no discounting case.

2 The greater the survival probabilities of future generations relative to current generations, the more justified is discounting of future risk reduction benefits.

3 Only if future well-being (as opposed to income) is discounted can discount rates greater than zero be justified in the context where the current VOSL is used to value future risks.

More generally, either future risks are valued at future WTP levels and then discounted in the same way as income, or future risks are valued at current VOSLs and no discounting is allowed, provided there is impartiality between current and future generations.

13.13 VALUING STATISTICAL LIVES WHEN INCOMES ARE UNEQUAL

Willingness to pay and, less obviously, WTA estimates of VOSL are constrained by income. Both WTP and WTA estimates are also averages, i.e. there is a frequency distribution from which the mean is taken, so that some people have much higher valuations of risk than the mean and some have much lower valuations than the mean. One of the reasons for these different valuations will be income differences within the nation. This procedure for deriving a VOSL has given rise to extensive misunderstanding. Imagine two countries, one rich and one poor, such that the rich country imposes a risk on the poor country through pollution. Global warming, which results from the emission of greenhouse gases, is often regarded as an example of such 'imposed' pollution costs. (Although the rich world (countries in the Organization of Economic Cooperation and Development) actually emits just under 40% of total greenhouse gases, with 60% coming from the developing world, oil-rich nations, and the ex-Soviet Union (World Resources Institute 1994).) Estimates of VOSLs determined by WTP estimates in the rich and poor countries will produce higher values for the rich country than the poor country, WTP being (partly) determined by income levels. Suppose that the rich country's pollution gives rise to an estimated 100 premature mortalities in the poor country. Assume the rich country faces the choice of spending resources on international pollution control to the benefit of the poor countries, or spending the same level of resources on a domestic issue which also saves 100 lives, i.e. the marginal cost of saving lives is the same in the two countries. A cost–benefit test will result in the resources being spent domestically because the 100 'domestic' lives will be 'worth' more than the 100 overseas lives due to the higher risk valuations. Yet if 'all lives are equal' in some sense, such an outcome seems very unfair, especially if the rich country can be said to impose the pollution on the poor country. Some have argued that if a VOSL is to be used at all, it should be the same VOSL for everyone and that VOSL should be the higher of the two figures, i.e. the VOSL for the rich country.

Is the cost–benefit test then invalid in some moral sense? There are several issues to be distinguished.

First, the VOSL within a country is an average, as noted above. In principle, then, the same procedure should be used where VOSLs differ across countries. The resulting VOSL will be an average of the two VOSLs, but it will not be the highest VOSL that is used. If the highest figure was chosen then, logically, it must also be chosen within a country, i.e. the average should not be used. Such an outcome is not logically tenable since the individual with the highest aversion to risk would then determine everyone's valuations. The idea of averaging valuations to reflect concern about the inequality of WTP is a longstanding idea in cost–benefit analysis. Pearce (1986, original edition 1971) discusses a rule in which WTP is weighted by a ratio of average income to actual income, i.e. an adjusted WTP for any country i becomes:

$$WTP_i^* = WTP_i . \hat{Y}/Y_i$$

and WTP_j is

$$WTP_j^* = WTP_j . \hat{Y}/Y_j$$

where $\hat{Y}$ is the average of Y_i and Y_j. The ratio of the two WTPs is then

$$WTP_i^*/WTP_j^* = \frac{WTP_i . Y_j}{WTP_j . Y_i}$$

This procedure will produce the same 'common VOSL value' if the value of risk as a proportion of income is the same in both i and j. Only if WTP_j as a proportion of Y_j is higher than WTP_i/Y_i will the resulting VOSL be higher in j than in i, and if the proportion is higher in i than in j, then the weighted VOSL will be higher in i than in j.

Second, regardless of the equity weighting procedure discussed above, the cost–benefit test need not produce the unfair outcome discussed in the example above. This is because the example assumes a common marginal cost of reducing risks in both countries. In practice, risk reduction is likely to be less costly in the poorer country than in the rich country. Even with 'unequal lives' then, nothing follows about the outcome of a cost–benefit test.

13.14 ECOSYSTEM RISK

The principles of valuation derived above are equally applicable to ecosystem risk. The wage risk model ceases to be relevant, but contingent valuation and, to some extent, avertive behaviour remain valid approach for ecosystem risks. Other valuation techniques also come into play, however. For example, ecosystems may be recreational sites visited by individuals. In this case the 'travel cost' method might be employed to capture some of the 'use values' that individuals have for the site. The travel cost model looks at the expenditure incurred by individuals travelling to the site. Typically, fewer and fewer people will visit the site the further away they are from it. This produces a 'distance decay' relationship which, with suitable modification, can be used to obtain a relationship between demand (visits) and price (cost of travel). The area under this resulting demand curve is then a measure of economic value of the site.

Contingent valuation of ecosystems has the advantage that it can capture 'non-use' values, i.e. expressions of willingness to pay to conserve an ecosystem regardless of whether that ecosystem is 'used' directly by the individual in question. This 'existence' value is intuitively understandable if one thinks of donations to conservation societies specializing in the protection of ecosystems and species that individual donors are never likely to see for themselves. The importance of existence value is best illustrated in the case of the *Exxon Valdez* incident in 1989. This tanker ran aground in Prince William Sound in Alaska, causing extensive damage to wildlife and amenities. The contingent valuation technique was used by a number of the parties at the subsequent damage hearings. A major exercise, for the State of Alaska, was conducted by Carson and others (Carson *et al.* 1995). The lowest estimate obtained by the questionnaire approach on WTP (rather than WTA compensation, which would have been the correct approach since the public is assumed to have the property rights) was $2.75 billion and differing assumptions produced significantly higher estimates. The major part of this damage estimate was non-use, or existence value (also called passive use value).

How are thresholds and economic values related in ecosystem damage estimation? An example from the acid rain context is instructive. While disputed by some experts, there is a widespread use in Europe of the concept of a 'critical load', i.e. a level of pollution deposition above which some identifiable form of ecosystem damage occurs and below which it does not. The analogous concept for ambient concentrations is a 'critical level'. Put another way, critical loads and levels are zero risk but positive deposition/concentration levels. In terms of international negotiations on acid rain control in Europe, critical loads have been adopted as a long-run goal of policy. But it is easy to see that, unless controlling acid rain is costless (in which case the problem would not exist), or unless damage from exceeding critical loads is infinite, achieving critical loads is not consistent with the way risk assessment and cost–benefit analysis proceed. Very simply, if it costs any finite sum of money to go from the existing position towards critical loads, then there must be some point before critical loads are reached at which the extra costs of control exceed the extra benefits of further control. This 'marginal condition' is fundamental to economics and requires that environmental standards are set where the marginal costs of control just equal the marginal benefits of control. Benefits are then measured in monetary terms using the dose–response approach indicated above, together with unit values of damage taken from the various valuation techniques.

But if the marginal equivalence rule is sensible, why bother to estimate critical loads? It would surely be more sensible to estimate damages and costs of control. The issue characterizes a classic debate between the 'zero risk' and 'cost–benefit' approach to risk assessment and decision-making. However, there must be some suspicion that both sides are right. Zero risk will be a long-run goal if there is a suspicion that tolerance of positive risks entails some process of irreversible environmental damage, for example total destruction of ecosystems. This is the public perception of the link between acid rain and forest loss in Germany, for example. While the perception is not correct (forest loss is related to acid rain in a complex and still not widely understood manner), it illustrates a valid idea, namely that failure to honour 'ecological limits' is an invitation to potential total loss of ecosystem function. Cost–benefit remains valid because there are very few contexts in which zero risk is affordable. It seems rational to set interim goals on a cost–benefit basis and some long-run goals on a critical-loads basis. Valuation remains relevant, but is subservient to the longer term goal of ecosystem health. Of course, even the irreversibility concept needs careful analysis. Costs and benefits still need to enter the picture because the costs of long-run conservation of ecosystems may well be prohibitive in the sense of demanding resources that would be better spent elsewhere. As always, money simply disguises the fact of opportunity cost: there is always some other benefit that is gone without.

The principles of ecosystem risk assessment are therefore essentially the same as those applying to health risk assessment, although the relevant valuation techniques are wider. Probably the greatest challenge to economic valuation in this context is the concern that conservation of a single ecosystem can be subjected to cost–benefit tests that might favour its destruction. Applied to all ecosystems, the cost–benefit logic might dictate total destruction on a sequential basis. A holistic approach, asking for the economic value of all ecosystems together, would surely produce extremely large figures. Pimentel *et al.* (1996) suggest just this with their estimate that world-wide, biological diversity produces economic functions that have a value of at least \$3 trillion (10^{12}). The economist's defence is that decision-making is sequential and incremental: the issue is not one of saving or destroying the whole set of ecosystems. None the less, the source of concern over the use of 'micro' risk assessment procedures to deal with 'big' ecological systems is easy to understand.

13.15 CRITICISMS OF THE ECONOMIC APPROACH TO RISK VALUATION

Numerous criticisms have been advanced against the economic approach to risk valuation (see also Chapters 11 and 12).

13.15.1 Relying on human preferences

The economic approach operates with human preferences. Humans are often poorly informed about the environment and the result of this poor information is that they will make choices which they may later regret, or which impose costs on others. There is indeed a problem here but it is far from being as serious as the critics make out.

First, no-one would argue that valuation is appropriate in all cases and all circumstances. If that were true, then society would not override human preferences for drugs, alcohol and tobacco, or even for education (or the lack of it). Rather, the presumption should be that economic valuation is the relevant approach unless there are good reasons for supposing otherwise. One of those reasons, as is widely recognized, is that the benefit in question is significantly 'distant' from individuals' familiarity and experience (Freeman 1986; Vatn & Bromley 1994). While this qualification applies to stated preference techniques, such as contingent valuation, it is far from clear that it applies to other procedures involving dose–response functions. Information is also important for valuing market goods as well. Where information is deficient, for example, with infrequent purchases such as housing, information markets emerge. To some extent, contingent valuation functions in this way—it supplies information to the respondent. In that respect it could be held to be superior to inferred valuations from behaviour in other markets, such as travel expenditures. This suggests that economic valuation will function more successfully the better the information. But lack of information is not an argument for rejecting WTP approaches. It is an argument for better information.

Second, information is often available on risks and is no worse than it is for other goods and services for which preferences are allowed to determine resource allocation.

Third, where information is sparse, it can be supplied, as it often is in contingent valuation studies for example.

Fourth, some economic valuation techniques quite explicitly adopt 'expert' assessments in determining the quantity of the environmental attribute to be valued. The most widely used valuation technique, for example, is that based on estimating dose–response functions and then using market or quasi-market prices to value the effect. The dose–response function itself is not determined by human assessment: it is determined by scientific evidence linking, say, doses of pollution to health effects, etc. But the basic distinction between a preference-based approach and an 'expert assessment' approach is that the actual values of the outcomes, whether they are states of human health or improved landscapes, reflect the values of those affected by the change in the state of nature. The role of experts is confined to conveying information. The 'lack of information' critique arises in part from over-concentration on techniques that appear to exclude the role of expert assessment.

A more basic criticism is that human well-being as interpreted in economics is not relevant at all—i.e. social decisions should not be based on economic well-being at all:

> 'The question we must ask is whether "welfare" has anything to do with happiness, well-being or any goal that appeals to common sense, morality, law or culture.' (Sagoff 1993)

On this view the 'quality' of desires is more important than 'desires' *per se*. The danger in such a view is that it quickly becomes elitist; A declares B's preferences to be of 'low quality' and therefore inadmissible, yet the appropriate strategy in a world where quality of desires matters is education and persuasion. It should not be the disenfranchisement of the individual because he or she does not qualify as having the 'right' desires. Vatn & Bromley (1994) also come close to this kind of preference imposition: 'the challenge is one of specifying the conditions for discourse over what is worth valuing by individuals—and why that is so' (p. 139).

13.15.2 The environment as 'public good'

Risk reduction could well be different to other marketed goods because in many manifestations it is not a private good but a public good. This means that its provision for any one person entails its provision for everyone else ('joint supply') and that it will generally be difficult to exclude others by some pricing mechanism (non-exclusion). In this respect, clean air has similarities with defence and public information. But this difference is not a reason for ignoring public preferences for the provision of environmental assets: it is an argument for being very careful to account for the pitfalls in finding out what those preferences are. This is exactly what economic valuation techniques are concerned with, and notably so in respect of the so-called 'free rider' problem in which people allegedly understate their preferences for such goods. As noted above, there is in fact not much evidence for such 'strategic bias' in economic valuation, but even if there were, it would be an argument for improving techniques to avoid it, not for ignoring preferences altogether.

13.15.3 Standard setting as social agreement

Another criticism of valuation is that it simply is not needed as a separate exercise. Society 'agrees' a set of risk standards. The cost of achieving those standards is then society's valuation of the standard. Thus:

> 'Where there are prior and firm commitments to maintaining environmental standards, policies which impact on environmental attributes are valued at the costs they entail for maintaining these standards. It is only when there is no commitment to such standards that the issue of valuing the environmental consequences of policies becomes a matter of debate.' (Bowers 1993)

It is, of course, possible to devise a social decision-making model in which the decisions made do reflect some sort of consensus of preferences of those affected. In such cases, the marginal cost of meeting the resulting standard is indeed an implicit social valuation of the standard. But how that implicit valuation is interpreted is important. The cost of achieving the standard is not an explicit social valuation of the standard. If it was, then *all* standards would have benefit : cost ratios of one or above. This is because the benefit of achieving the standard must be at least equal to the cost of achieving it, since the latter is used to measure the former. From this it follows that all standards must be for the best. There can be no mistakes and hence no need ever to revisit a standard to see if it is justified or not. Whatever is, is for the best. Worse, standards cannot be informed by preferences—they simply emerge. The existence of implicit valuations in no way impairs the case for explicit valuation.

13.15.4 Errors in willingness-to-pay estimates

Bowers (1993) offers various suggestions as to why WTP estimates are prone to error. The first is that sufferers of pollution may not correctly perceive the effects of pollution. But, as noted above, this criticism assumes all kinds of falsehoods: (i) that the aim is to 'value everything'; (ii) that one cannot inform people before seeking their WTP; and (iii) that the valuation method itself cannot elicit the 'hidden' effects, whereas, as noted above, dose–response approaches explicitly do this. More problematic is the implication that since people do not know the full effects of something, they cannot be allowed to express preferences for or against that outcome. Decision-making by experts is one alternative and 'political' decision-making another. Yet expert decision-making accounts for a fair number of modern environmental hazards, from nuclear waste to chlorofluorocarbons (CFCs), whilst decision-making by politicians assumes a model of information and decision-making that is at least no better than that for 'uninformed' citizens.

13.15.5 The problem of future generations

Critics of economic valuation are on stronger ground when they argue that the preferences elicited are those of one generation—here and now—and not of future generations. However, even this criticism is far from fatal. It assumes, for example, that generations do not overlap, whereas in practice they obviously do. Overlapping generations produce preference functions in which the interests of future generations are included. One can argue about the weights to be attached to those preferences, but it is wrong to suppose that future preferences count for nothing in WTP measures. Moreover, the issue of weighting reduces to an issue of discounting—see section 13.12. Otherwise, if all future generations' preferences were to count equally, generations for thousands of years to come would have to be accommodated. But that is neither feasible nor necessary. There are legitimate disagreements on the 'proper' discount rate, but they do not invalidate using WTP as a measure of economic value.

13.15.6 Citizen versus individual preferences

Another widely countenanced criticism of WTP is that it records individualistic preferences, whereas what matters for social policy are preferences expressed by the individual as a citizen: 'citizen's preferences' (Sagoff 1993). Arguably, citizen's preferences are best elicited in different contexts to those in which WTP is measured. But there are several problems with this view. When asked for my WTP, or when my WTP is inferred from surrogate market behaviour, the motivations for WTP are not easily uncovered. My WTP for an environmental asset may reflect my concern about the asset because I use it, because I want my children to use it, because I want it to exist and have no intention of using it, because I think it is 'right' that it exists, and so on. There is nothing in WTP approaches *per se* that requires the expressed valuation to reflect one particular motive—any or all motives may be present. As it happens, some attempts have been made to uncover motivation for preferences for environmental assets. This is how the differentiation between use and non-use values has come about, the latter relating to WTP unassociated with any use of the asset or intention to use it. Willingness to pay for non-use may well capture so-called 'ethical' preferences (Pearce & Turner 1992; Kopp 1993). The essential point is that the contrast between citizen and individualistic preferences may be valid, but it has no implications in terms of the validity of WTP measures. The point is the same as that for future generations. Vatn & Bromley (1994) repeat the presumption that moral judgements and 'commitment' somehow render WTP illicit, the 'commodity fiction' as they put it. But that can only be relevant if commitment and morality cannot enter into expressions of preference. They are tautologically counterpreferential in the sense of being inconsistent with the preferences that would be expressed if commitment did not exist. But that is quite different to saying they cannot be indicated by WTP, as 'commitment payments' to conservation organizations testify.*

However, this defence is open to one further criticism. If any kind of motive is consistent with WTP, and if revealed choice is the manifestation of preference, how can I tell if someone really prefers what they say they prefer? This is a further criticism by Sagoff (1993), but it again seems to rest on an illogicality. If we knew what people's 'true' preferences were, it would not be necessary to infer those preferences by looking at choices as the revelation of those preferences. In other words, the criticism implies some world of 'true' preferences, perhaps even distinguished by motive, against which recorded, inferred, or stated preferences can be measured. But the inference or statement is all that we have. There is nothing else, and to criticize observations of choice as a means of inferring value is to set up not just an arbitrary benchmark, but a non-observable one.

13.15.7 Economic valuation and the political process

Some attempts have been made to suggest alternatives to economic valuation, the main one being a resort to the political and legal system. On this argument, the political process is the 'best' reflector of citizens' preferences (rather than individualistic preferences), and the law treats environmental damage as a legal nuisance. The property rights of the polluter should therefore be attenuated. This approach does, of course, describe much environmental regulation in the rich world and, to a lesser extent, in the poor world.

But as a coherent alternative to a balancing of values, the political–legal approach suffers a wealth of problems:

1 The approach is indifferent to the opportunity cost of risk reduction. Sagoff (1993) suggests some trade-off through 'knee-of-the-curve' thinking—pursuing risk reduction until costs begin to rise significantly. This is certainly akin to 'best available technology not entailing excessive cost' thinking (BATNEEC), the problems of which are well known (Pearce & Brisson 1993). However, the intuition behind knee-of-the-curve thinking is weak. It assumes that opportunity cost curves all have 'knees', which is far from the case. It also assumes that benefits, measurable or otherwise, necessarily exceed costs up to the 'knee'. There is no evidence for such a proposition. However, there is some similarity between 'knee of the curve' thinking and 'safe minimum standards' (SMS) (see section 13.16).

2 It assumes that political systems are benevolent and at least as well informed, if not better informed, than citizens.

3 It assumes that political systems are representative, simulating referenda. But many of them are very distant from referenda, as with local land-use planning, for example, where participation is the watchword but is poorly practised. Many land-use planning decisions are made in response to interests expressed by elite groups of lobbyists who have time to lobby and to be 'represented' at planning inquiries (O'Doherty 1994).

13.15.8 Does value revelation capture everything?

The final criticism of economic valuation is that, while it can be done, it produces only partial results. In the biodiversity context, for example, it is argued that it excludes 'system functions' such as life support functions which, if impaired, seriously increases life and health risks. These values are variously described as 'primary value' (Gren *et al.* 1994) and even the 'glue' of the system. Here it is possible to have more sympathy with the criticism. It has close similarities to the informational issue, but it differs in that the information issue rests on individuals being ignorant but others having knowledge, whereas the 'primary value' argument says that neither individuals nor the experts really know what the true functions of some or even all environmental assets are. It is then more an argument about uncertainty than information. The 'glue' is then a

* Vatn & Bromley suggest, for example, that the 'rights' of wildlife to exist cannot be expressed through WTP. If so, it makes the existence of extremely large payments to wildlife conservation organizations difficult to understand.

condition for the existence of other values, and it is in that sense prior to other values.

The problems with the primary value critique are several, but there is an important issue that limits the comprehensiveness of human risk valuations.

First, if it is an argument about uncertainty of values, then the proper approach is to incorporate the uncertainty in the analysis. It need not be an argument for rejecting WTP measures as such. None the less, there must be some sympathy for a view that says, while such extensive ignorance prevails, some kind of precautionary approach is required. This issue is addressed in section 13.16.

Second, if the 'glue' value is present across all environmental assets, it offers little assistance in the choice context. Rather, it comes close to saying that since all environmental assets have unmeasured value, all assets should be protected. This runs into the budget constraint issue discussed in section 13.3.

Third, it would be relevant for non-marginal changes in assets but not so relevant for marginal changes. Many environmental choices are evidently marginal, but some modern environmental problems clearly are of a non-marginal nature, including the rapid rate of biodiversity loss. Even for marginal changes, the sum of those changes amounts to discrete change. In these contexts the value of avoiding irreversible changes can be very high and 'option value' becomes significant. In this sense, primary value concerns should make 'economic valuers' uncomfortable.

Of all the criticisms, the 'uncertainty of function' problem is the most important for economic valuation. Brief reflection will show that it will also be problematic for *any* choice approach. It is not, therefore, a criticism of valuation *per se* but a weakness of all decision-making procedures. It is here that the greatest challenge to research and pragmatism lies.

13.16 AVOIDING VALUATION: THE PRECAUTIONARY PRINCIPLE

If economic valuation does capture only 'part' of the 'true' economic value of environmental assets, if there is extensive uncertainty about the true social costs of environmental degradation, and if the negative pay-off to being wrong about environmental choices is potentially high, perhaps because of irreversibilities, then conventional economic theory would dictate a risk averse strategy. Such strategies have been propounded in several guises—'safe minimum standards' (SMS) (Bishop 1978), the 'precautionary principle' (PP), and 'sustainability constraints' (SC) (Pearce *et al.* 1989). In these cases there is a presumption to conserve unless the social costs of conservation are 'very high'. In a risk context this would be translated as a presumption to reduce risks to the highest level achievable, unless the social costs of doing so are very high. Unfortunately, neither SMS nor SC approaches have defined what is meant by 'very high' costs. The precautionary principle becomes a luxury because it implies that the cost of caution is not very high. But in many real-world contexts it is high. The trade-off becomes one between 'big' costs, either way.

13.17 VALUATION OF LIFE AND HEALTH RISKS: A CASE STUDY

The previous discussion has centred on concepts and issues in the valuation of risks. This section looks at a case study to show how valuation can be put into practice. The example focuses on particulate matter which comes from both stationary sources and vehicle transport. Small-sized particulate matter (PM) of less than 10 μm in diameter, so-called 'PM_{10}', has been implicated in health damage in a number of epidemiological studies (Schwartz & Marcus 1990; Schwartz, 1991a,b, 1993a,b, 1994; Schwartz *et al.* 1991, 1993; Dockery *et al.* 1992, 1993; Pope & Dockery 1992; Pope *et al.* 1992, 1995; Schwartz & Dockery 1992a,b; Xu *et al.* 1994). We first set out a general model of air pollution and health, following Ostro (summarized in Ostro 1994), and then focus on particulate matter. We estimate the monetized social costs of exposure to total PM concentrations and the benefits to be obtained by making marginal reductions in PM_{10} emissions. The results suggest that the social costs of PM emissions in the UK are substantial and that substantial benefits could be secured from their reduction, even disregarding the associated benefits that tend to accrue from the kinds of policy measures that would be needed (e.g. reductions in noise and congestion from traffic restraint). In particular, PM_{10} emissions are closely linked to road transport, and the control of the transport sector therefore becomes a matter of priority for environmental and health policy. A more complete picture of the links between transport and social cost is provided in Maddison *et al.* (1996). Our findings are similar, but somewhat larger in scale, to those of Small & Kazimi (1995) for California.

13.17.1 The underlying model

Health effects (H) are related to ambient air quality by:

$$dH_i = b.POP_i.dA_j \qquad (6)$$

where d = change in, i = health effect i, b = slope of dose–response function, POP_i = population at risk from health effect i, A = ambient air quality and j = jth pollutant.

Ambient air quality is related to emissions either through some diffusion model or through an approximation involving a relationship such as:

$$dA_j/A_j = dE_j/E_j \qquad (7)$$

where E = emissions. Equation (7) says that an X% change in emissions is associated with an X% change in ambient concentrations.

Each health effect, H_i, has a unit economic value P_i so that:

$$P_i.dH_i = P_i.b.POP_i.dA_j \qquad (8)$$

and the sum of damages, D, from pollutant j is:

$$D_j = \Sigma_i P_i.dH_i \qquad (9)$$

Factor b in Eqn. (6) is the slope of the dose–response function (DRF). The DRF could begin at some threshold value below which no damage is done. No firm evidence for thresholds for air pollutants appears to be available. For this reason, a working assumption is that the DRF begins at the origin. Health damage appears to be related to PM of less than 10 μm in diameter, i.e. PM_{10}. Particulates are associated with a wide range of respiratory symptoms. Long-term exposure may be linked to heart and lung disease. Particulates may also carry carcinogens into the lungs.

Concentrations of PM are measured in various ways: as total suspended particulates (TSP), sulphates and 'British Smoke' (BS) (or just 'smoke'). One problem is that the various measures need to be converted into units of PM_{10} since it is PM_{10} that is implicated in health damage. Ostro (1994) suggests the following conversions (all in $\mu g\,m^{-3}$):

$$PM_{10} = 0.55\,TSP \qquad (10)$$

$$\text{Sulphates} = 0.14\,TSP \qquad (11)$$

$$\text{hence Sulphates} = 0.25\,PM_{10} \qquad (12)$$

$$BS \approx PM_{10} \qquad (13)$$

Relation (10) is confirmed by Desvousges *et al.* (1993). The equivalence in relation (13) appears to be disputed, with some authorities claiming that there is no direct link between BS and PM_{10} and others claiming that there is a link, but it is more complex than is suggested by the simple equivalence in relation (10). The conversion problem arises because of different ways of measuring particulate concentrations. The British approach has been to use the black smoke method whereby air is drawn through a filter. The darkness of the stain is then measured and calibrated to indicate concentration. The darkness of the stain varies by the type of particle, and what is in dispute is the link between this measure and those obtained by gravimetric techniques which essentially measure the weight of the PM on the filter paper. Gravimetric approaches have been used in the UK only recently.

Based on a meta-analysis of DRFs, Ostro (1994) concludes that the consensus DRF for *mortality* from particulate matter is:

$$\%dH_{MT} = 0.096.dPM_{10} \qquad (14)$$

where dH_{MT} is the (central estimate of the) change in *mortality*. The coefficient makes due allowance for other compounding factors, such as smoking.

Equation (14) says that a $1\,\mu g\,m^{-3}$ change in PM_{10} concentration is associated with a 0.1% change in mortality, or a $10\,\mu g\,m^{-3}$ change in PM_{10} concentration is associated with a 1% change in mortality. The DRFs surveyed by Ostro include those in the work of Dockery, Schwartz and others. Upper and lower bounds for Eqn. (14) are given as:

$$\%dH_{MT} = 0.130.dPM_{10} \quad (15)$$

$$\%dH_{MT} = 0.062.dPM_{10} \quad (16)$$

To get the absolute change in mortality then, we require:

$$dH_{MT} = b.dPM_{10}.CMR.POP.1/100 \quad (17)$$

where $b = 0.062$, 0.096 and 0.130, respectively, for lower, central and upper bounds; and *CMR* is the crude mortality rate. The factor of 100 converts from percentages to absolute numbers.

The measurement problems relate to: what exactly the concentrations of PM_{10} are in the UK, since it is a measure of particulate concentration only recently introduced; what the population at risk is; and what the CMR is for that population. Somewhat surprisingly, all three issues cause difficulties for estimating the health–PM link. This may explain the variety of views concerning the importance of PM pollution in the UK.

Not all PM is 'anthropogenic'—some arises as a natural background level (dust, etc.). It is assumed here that 70% of PM_{10} in urban areas has an anthropogenic origin. In terms of assigning damage to anthropogenic sources, then, dPM_{10} in the above equations is taken to be 0.7 of the average PM_{10} concentrations.

13.17.2 Estimating deaths attributable to PM_{10}

Annual (arithmetic) average concentrations of PM_{10} in England and Wales in 1993 are available for Birmingham, Bristol, Cardiff, Leeds, Liverpool, London and Newcastle. The average figure for these seven cities in 1993, applying weights to measures of PM_{10}, is 28.5 $\mu g\,m^{-3}$. Using the figure of 28.5 $\mu g\,m^{-3}$ for average urban PM_{10} concentration, along with total urban deaths in 1992 of 243 707 (based on Key Population and Vital Statistics (KPVS), area aggregates data), and a dose–response function of 0.096% increase in all-cause/all-age deaths for every 1 $\mu g\,m^{-3}$ of PM_{10} (see Eqn. (14) above) total deaths in 1993 in England and Wales attributable to PM_{10} exposure are: $28.5 \times 243\,707 \times 0.00096 = 6668$ deaths in 1993. The upper and lower bounds, using the coefficient of 0.00062 and 0.00130, are 4306 and 9029 deaths, respectively. Taking the 70% anthropogenic source estimate has the effect of lowering the central mortality estimates to 4667 for England and Wales.

In summary, the 'best estimates' for urban PM_{10} mortality due to anthropogenic sources in England and Wales are shown in Table 13.6. For the UK, these estimates can be multiplied by 1.11 to allow for the larger population.

Table 13.6

	Low	Mid	High
Low urban population	3014	4667	6320
High urban population	4663	7220	9777

Brunekreef (1995) reports acute mortality associated with PM_{10} of some 5000 and serious health effects in a 'few tens of thousands' of people per year in the Netherlands, which appears to be consistent with the estimates here and may suggest that the mortality figures are understatements.

13.17.3 Problems of the dose–response function

The DRF is mainly derived from data linking changes in PM_{10} concentrations to health effects in USA cities. Transferring the use of such a DRF to calculate deaths attributable to total PM_{10} levels in urban areas of England and Wales is reliant upon a number of assumptions relating both the validity of the epidemiology underlying the DRFs and the validity of transferring the results. The main critical factors appear to be the following.

1 The argument that there is no threshold below which PM_{10} concentrations are harmless and are not a cause of mortality.

2 It is a linear DRF, so the relationship at one level of PM_{10} is the same at all levels.

3 Conditions in USA cities and measurements taken there are comparable to British cities.

4 There is some 'biological pathway' by which PM affects health.

5 The possible presence of confounding factors does not alter the relationship.

6 Statistical artefact.

For factor 1 there has as yet been no evidence produced of threshold levels. Thus, Schwartz (1994) asks 'is there a threshold or other non-linearities in the concentration–response relationship?' (with regard to the historical link between smog and daily mortality in London). 'No evidence is seen for a threshold.' Schwartz (1994, p. 49) (with regard to data from Philadelphia). Similarly, '... studies ... identify no threshold between quintiles of range of observed exposures. In general, the linear and exponential functional forms adopted by most researchers do not suggest a strong attenuation of effects at low levels, and we adopt the conservative assumption of no threshold.' (Desvousges *et al.* 1993, pp. 3–5).

Ostro (1995a) points out that the absence of an observed threshold may be the result of study design or of the fact that there is no threshold, but that the few studies that test for low-level effects suggest that effects continue to very low concentrations.

For factor 2, if the concentration to mortality function is not linear, then the dose–response calculated for a particular level will not be applicable at other levels of PM_{10}. However, graphs in Dockery *et al.* (1993, p. 1757) would seem to indicate that this relationship probably is approximately linear.

One implication of factor 3 would be the need to check age and health profiles of populations in USA and UK cities to see if there is a higher predisposition to ill-health in one country. This has not been done in this case study. The biological pathway by which PM affects human health appears to be very uncertain. Seaton *et al.* (1995) suggest that particulate clouds containing small acidic particles may both travel and persist in time, also penetrating indoors. Small particles may provoke alveolar inflammation, which exacerbates lung disease and may also affect the coagulability of blood. The latter may affect cardiovascular disease, a health effect not accounted for here. Utell & Samet (1993) review the various epidemiological studies and argue that the biological mechanism is unknown. Ostro (1995a) reviews the evidence and suggests that pollution exposure is an additional burden of inflammation which may make worse the conditions of those already affected by respiratory infections. He further suggests that the available evidence is consistent with the requirements of tests for causality: the relationship is consistent, specific, produces a dose–response relationship and fits other known facts.

One potentially significant issue is the extent to which the DRFs adequately control for all variables affecting health status. It is usual to control for factors such as smoking and diet, but recent work has suggested that 'social status' may be an important factor in determining illness. One possibility, then, is that exposure to PM_{10} is correlated with social status and with illness. This needs careful testing with panel data covering data that include income, other measures of social status, and indoor and outdoor pollution concentrations. Ostro (1995a) notes that time-series studies avoid such confounding factors because the population characteristics are fairly constant over time.

The importance of Ostro's consensus DRFs for premature mortality is signified by the fact that is has been used in a number of World Bank appraisals for air pollution control in various cities in developing countries (see Ostro 1994, 1995b). As such, significant expenditures could be affected by the results of the approach. The World Bank subjected Ostro's work to peer review—see McMichael *et al.* (1995) and Flanders *et al.* (1995). A large number of highly specific comments were made but the essential main points appear to be: (i) that the work neglects long-term chronic effects, especially permanent impairment of lung function and the development of diseases such as asthma and chronic obstructive pulmonary disease, and studies based on essentially 'acute' data will probably not capture long-term health risks; (ii) that it may not be appropriate to 'transfer' mainly North American DRFs to developing countries—the focus of the World Bank's interest—without further analysis of other factors, such as nutrition; and (iii) that the statistical procedure used will not be accurate if the DRFs are non-linear and/or there is a threshold. Criticism (ii) is not relevant to the current study, but clearly is relevant to extensions of the model to developing countries. Criticism (i) suggests that the procedures used may actually *understate* the health damage from air pollution generally and PM_{10} in particular. Criticism (iii) is discussed above. Clearly, there are uncertainties, but the peer review process does not add to the uncertainties already identified.

13.17.4 Mortality values

Applying a figure of £1.5 million per statistical life to premature deaths from anthropogenic particulate exposure, then, produces the following results for England and Wales (adopting the low urban population estimate):

Lower bound: 3014 deaths × £1.50 m = £4.52 billion
Central estimate: 4667 deaths × £1.50 m = £7.00 billion
Upper bound: 6320 deaths × £1.50 m = £9.48 billion

The previous analysis suggests that PM just in England and Wales produces social costs of mortality of £4–10 billion p.a., or £4.4–11.1 billion for the UK as a whole. No account has been taken of the value of a statistical life to other people. As noted above, this could increase the VOSL from a social standpoint significantly.

13.17.5 Particulate matter and morbidity

The damage estimates so far relate to excess *mortalities* only. Rowe *et al.* (1995) have assembled 'best estimates' of the relevant DRFs for morbidity. These are set out in Table 13.7. The assumption again is that such DRFs can be 'transferred' to the UK context. Certain simple adjustments could be undertaken, for example for variations in the age structure of populations.

Table 13.8 gives the resulting estimates of health incidence for England and Wales and the UK, assuming that 44% of the population is 'urbanized' in the sense defined above.

Table 13.9 provides unit prices taken from Rowe *et al.* (1995). These unit prices cannot simply be transferred since they relate to various studies in the USA. Table 13.9 therefore makes various kinds of adjustments to them to reflect the differences in incomes between the two countries. The results of these adjustments are to reduce the USA unit values by 10–25%. Improved adjustments could be made by looking more closely at medical costs in the USA and the UK.

Table 13.10 assembles the resulting total valuations and suggests that a few categories of morbidity effects dominate the overall picture, notably restricted activity days (RADs) (which are taken here to subsume the estimates for respiratory symptoms) and chronic bronchitis. The effect of computing morbidity effects is roughly to double the mortality costs.

Combining mortality and morbidity effects produces the result that anthropogenic PM_{10} may cost the UK something of the order of £14 billion p.a. (see Table 13.11).

Note, however, that these estimates are based on individuals' WTP to avoid the illness. They say nothing about the costs to the health services, i.e. COI. If they should, then the estimates here understate the true social costs of health damage.

Table 13.7 Morbidity dose–response functions for PM_{10} (effect per 10 $\mu g\, m^{-3}$ change in PM_{10}). (From Rowe *et al.* 1995.)

Morbidity effect*	Low estimate	Central estimate	High estimate
RHA/100 000	6.6	12.0	17.3
ERV/100 000	116.0	237.0	354.0
RAD/person	0.29	0.58	0.78
LRI/child/per asthmatic	0.010	0.017	0.024
Asthma attacks per person	0.33	0.58	1.96
Respiratory symptoms/person	0.80	1.68	2.56
Chronic bronchitis/ 100 000 (risk of new case)	30.0	61.2	93.0

* RHA, respiratory hospital admissions; ERV, emergency room visits; RAD, restricted activity days; LRI, lower respiratory illness (in children).

Table 13.8 Estimates of morbidity effects of anthropogenic PM_{10} for England and Wales and the UK using Table 13.6 meta-coefficients* (number of events or cases).

Morbidity effect	Lower bound	Central estimate	Upper bound
RHAs			
England and Wales	3007	5467	7882
UK	3345	6082	8768
ERVs			
England and Wales	52850	107977	161282
UK	58789	120112	179407
RADs			
England and Wales	13.21 m	26.42 m	35.53 m
UK	14.70 m	29.39 m	39.53 m
Child LRIs			
England and Wales	No estimate	No estimate	No estimate
UK			
Asthma attacks			
England and Wales	15.03 m	25.42 m	89.30 m
UK	16.72 m	29.39 m	99.33 m
Respiratory symptoms			
England and Wales	36.45 m	76.54 m	116.63 m
UK	40.54 m	85.14 m	129.74 m
Chronic bronchitis			
England and Wales	13668	27833	42371
UK	15204	31016	47132

* Source: Table 13.4 and assuming overall urban populations of 22.78 million and 25.34 million for England and Wales and UK population, respectively. The general formula for any estimate N is $N = b.POP.(28.5–8.5)\ 10 = 2.b.POP$, i.e. 45.56$b$ and 50.69b, respectively, where 8.5 is the non-anthropogenic concentration, and the division by 10 allows for the coefficients in Table 13.4 being expressed per 10 $\mu g\ m^{-3}$. Ratio of lower and upper bounds to central estimate from Rowe *et al.* (1995).

To illustrate, Action Asthma (1990) estimate that the effects of asthma as a whole resulted in a cost to the UK National Health Service in 1988 of some £344 million in terms of general practice costs and costs to the hospital sector.

13.18 CONCLUSIONS

Economic valuation of risks to human life and ecosystems remains controversial. But the reality is that life and health risks are traded off by individuals every day, as are the risks of ecosystem damage. A 'risk-free' world is simply not a realistic option for policy makers or for individuals. The valuation of risk is therefore unavoidable. What matters is that it should be done, as far as possible, in a rational way. The economic approach to risk assessment offers one coherent approach which is well founded in the basic economics of human well-being. It is not clear what the competing paradigm would be. None the less, there are problems. Lay perceptions of risk as revealed in human behaviour need not coincide with expert assessment of risks, nor with the policy-makers' perceptions. Lay perceptions tend to be favoured by strict adherents of the economic approach, regardless of whether they are under- or overstatements of 'objective' risk. The constituency of risk valuation is also debatable: is it all those exposed to risk, inclusive of those not yet

Table 13.9 Adjusting unit values* for morbidity effects.

Morbidity effect	Unit value (UK/USA) 1994£			Adjusted unit value = unit value $\times Y_{uk}/Y_{usa}$‡ (central value only)	Adjusted unit value = unit value $\times (Y_{uK}/Y_{uSa})^{e}$§ (central value only)
	Low	Central	High		
RHAs† (adjusted COI)	4900	9800	14 700	7450	8918
ERVs (adjusted COI)	185	370	555	280	337
RADs (WTP + adjusted COI)	25	50	75	38	46
Asthma (WTP)	9	25	40	19	23
Respiratory symptoms (WTP)	4	7	10	5	6
Chronic bronchitis (WTP)		147 000		111 720	133 770

COI, based on 'cost of illness' approach; ERVs, 1 emergency room visits; RADs, 1 restricted activity days; WTP, willingness-to-pay approach.
* Values in Rowe (1995) *et al.* adjusted from 1992 to 1994 prices by multiplying by 1.05 and from \$ to £ by dividing by 1.5.
† RHAs, respiratory hospital admissions;
‡ Ratio of UK to USA GNP per capita in 1992 = 0.76. This adjustment is the equivalent of setting e in the next column equal to unity.
§ e = 0.35; see Krupnick *et al.* (1993). The effect is to multiply the unit values in the first column by 0.91.

born? Risks to human life also excite considerable controversy, much of it due to a basic misunderstanding of what it means to speak of values of 'statistical lives'. Some of the controversy is, however, rooted in an ethical issue of how risks are to be valued when they accrue to different people with apparently different valuations of risk that are contingent on their incomes. Ethical weighting systems can be devised to deal with such issues. Finally, risk valuation in practice is fraught with difficulty, given the limited data that actually exist on many environmental risks. The case study of air pollution presented here illustrates many of the problems, but it happens to be one of the more tractable problems. In other cases, DRFs are barely known. Probably the largest single area of difficulty for economic valuation is in valuing risks to biological diversity. Here, only limited progress has been made. Despite the difficulties, risk valuation faces a vital issue head-on: we do not live in a world of infinite resources and choosing between risks is an unavoidable part of human existence.

Table 13.10 Total damage estimates for PM_{10} morbidity effects.*

Morbidity effect	Lower bound (£ million)	Central estimate (£ million)	Upper bound (£ million)
RHAs†			
England and Wales	26.8	48.8	70.3
UK	29.8	54.2	78.2
ERVs			
England and Wales	17.8	36.4	54.4
UK	19.8	40.5	60.5
RADs			
England and Wales	607.7	1215.3	1634.4
UK	676.2	1351.9	1818.4
Child LRIs			
England and Wales	No estimate	No estimate	No estimate
UK			
Asthma attacks			
England and Wales	345.7	584.7	2054.0
UK	384.6	676.0	2284.6
Respiratory symptoms			
England and Wales	218.7	459.2	699.8
UK	243.2	510.8	778.4
Chronic bronchitis			
England and Wales	1828.4	3723.2	5668.0
UK	2033.8	4149.0	6304.8
Totals‡			
England and Wales	2826.4	5608.4	9481.1
UK	3144.2	6271.6	10546.5

* Source: Tables 13.5 and 13.6; e, 0.35, i.e. the final column of Table 13.6 has been used.
† RHAs, respiratory hospital admissions; ERVs, emergency room visits; RADs, restricted activity days; LRIs, lower respiratory illness.
‡ Respiratory symptom costs are assumed to be subsumed in RAD costs, so that respiratory symptom costs are excluded from the totals.

Table 13.11 Total costs of PM_{10} health damage in the UK.

	Lower bound 1994 (£ billion)	Central estimate 1994 (£ billion)	Upper bound 1994 (£ billion)
England and Wales			
Morbidity	2.83	5.61	9.48
Mortality	4.52	7.00	9.48
Total	7.35	12.61	18.96
United Kingdom			
Morbidity	3.14	6.27	10.55
Mortality	5.02	7.77	10.52
Total	8.16	14.04	21.07

13.19 REFERENCES

Action Asthma (1990) *The Occurrence and Cost of Asthma*. Cambridge Medical Publications, Worthing.

Bateman, I. & Turner, R.K. (1993) The contingent valuation method. In: *Sustainable Environmental Economics and Management* (ed. R.K. Turner), pp.120–191. Belhaven Press, London.

Bishop, R. (1978) Endangered species and uncertainty: the economics of a safe minimum standard. *American Journal of Agricultural Economics*, **60**, 10–18.

Blomquist, G., Miller, T. & Levy, D. (1996) Values of risk reduction implied by motorist use of protection equipment. *Journal of Transport Economics and Policy*, **January**, 55–66.

Bowers, J. (1993) A conspectus on valuing the environment. *Journal of Environmental Planning and Management*, **36**, 91–100.

Brunekreef, B. (1995) *Quantifying the health effects for The Netherlands of exposure to PM_{10}*. Report 623710002, National Institute of Public Health and Environmental Protection, Bilthoven. (In Dutch.)

Bullard, R.D. (1994) Unequal environmental protection: incorporating environmental justice in decision making. In: *Worst Things First: the Debate over Risk Based National Environmental Priorities* (eds A. Finkel & D. Golding), pp. 237–266. Resources for the Future Inc, Washington DC.

Carson, R., Mitchell, R., Hanemann, M., Kopp, R., Presser, S. & Ruud, P. (1995) *Contingent Valuation and Lost Passive Use Damages from The Exxon Valdez*. Discussion Paper 95–02, Department of Economics, University of California, San Diego, CA.

Chestnut, L., Colome, L., Keller, L. *et al.* (1988) *Heart Disease Patients' Averting Behavior, Costs of Illness, and Willingness to Pay to Avoid Angina Episodes*. Report to Office of Policy Analysis, US Environmental Protection Agency, Washington, DC.

Chestnut L. & Patterson, A. (1994) *Human Health Benefits Assessment of the Acid Rain Provisions of the 1990 Clean Air Act Amendments*. US Environmental Protection Agency, Washington, DC.

Cropper, M. & Freeman, A. (1991) Environmental health effects. In: *Measuring the Demand for Environmental Quality* (eds J. Braden & C. Kolstad), pp. 165–226. North Holland, New York.

Cropper, M. & Simon, N.B. (1994) *Are we valuing risks to life correctly*? World Bank, Washington, DC.

Cropper, M. & Sussman, F. (1988) Families and the economics of risk to life. *American Economic Review*, **78**, 255–260.

Cropper, M. & Sussman, F. (1990) Valuing future risks to life. *Journal of Environmental Economics and Management*, **19**, 160–174.

Desvousges, W., Johnson, F.R. & Banzhaf, H. (1993) *Review of Health Effects Resulting from Exposure to Air Pollution*. Research Triangle Institute, Task Force on Externality Costing, Working Paper 1 (revised), November. Triangle Economic Research, Research Triangle Park, NC.

Dockery, D.W., Schwartz, J. & Spengler, J. (1992) Air pollution and daily mortality: association with particulates and acid aerosols. *Environmental Research*, **59**, 362–373.

Dockery, D.W., Pope, C.A., III, Xu, X. *et al.* (1993) An association between air pollution and mortality in six U.S. cities. *New England Journal of Medicine*, **329**(4), 1753–1808.

Fischoff, B., Slovic, P. & Lichtenstein, S. (1979) Weighing the risks. *Environment*, **21**(4), 17–20, 32–38.

Fisher, A., Chestnut, L. & Violette, D. (1989) The value of reducing risks of death: a note of new evidence. *Journal of Policy Analysis and Management*, **8**(1), 88–100.

Flanders, D., Haddix, A., Olson, D., Romieu, I., White, M. & Williamson, G.D. (1995) *Review of Urban Air Pollution Health Impact Methodology*. Center for Disease Control, Atlanta, GA.

Freeman, A.M., III (1986) On assessing the state of the art of the contingent valuation method of valuing environmental changes. In: *Valuing Environmental Goods: an Assessment of the Contingent Valuation Method* (eds R.G. Cummings, D. Brookshire and W. Schulze), pp. 148–161. Rowman and Allenheld, Totowa, NJ.

Freeman, A.M. III (1993) *The Measurement of Environmental and Resource Values*. Resources for the Future, Washington, DC.

Gregory, R. (1986) Interpreting measures of economic loss: evidence from contingent valuation and experimental studies. *Journal of Environmental Economics and Management*, **13**(4), 325–337.

Gren, A.-M., Folke, C., Turner, K. & Bateman, I. (1994) Primary and secondary values of wetland ecosystems. *Environmental and Resource Economics*, **4**(1), 55– 74.

Hanemann, M. (1991) Willingness to pay and willingness to accept: how much can they differ? *American Economic Review*, **81**, 635–647.

Jones-Lee, M. (1992) Paternalistic altruism and the value of statistical life. *Economic Journal*, **102**(410), 80–90.

Jones-Lee, M., Hammerton, M. & Philips, P. (1985) The value of safety: results of a national sample survey. *Economic Journal*, **95**, 49–72.

Jones-Lee, M., Loomes, G., O'Reilly, D. & Philips, P. (1993) *The Value of Preventing Non-fatal Road Injuries: Findings of a Willingness to Pay National Sample Survey*. Working Paper WP/SRC/2, Transport and Road Research Laboratory, Crowthorn.

Jones-Lee, M., Loomes, G. & Philips, P. (1995) Valuing the prevention of non-fatal road injuries: contingent valuation vs standard gambles. *Oxford Economic Papers*, **47**, 1–19.

Kahneman, D. & Tversky, A. (1979) Prospect theory: an analysis of decisions under risk. *Econometrica*, **43**, 263–291.

Kopp, R. (1993) Environmental economics: not dead but thriving. **Resources, Spring**(111), 7–12.

Krupnick, A. (1986) *A Preliminary Benefits Analysis of the Control of Photochemical Oxidants.* US Environmental Protection Agency, Washington, DC.

Krupnick, A. & Cropper, M. (1989) *Valuing chronic morbidity damages: medical costs, labor market effects and individual valuations.* Report to Office of Policy Analysis, US Environmental Protection Agency, Washington, DC.

Krupnick, A. & Cropper, M. (1992) The effect of information on health risk valuations. *Journal of Risk and Uncertainty*, **5**, 29–48.

Krupnick, A., Harrison, K., Nickell, E. & Toman, M. (1993) *The benefits of ambient air quality improvements in central and eastern Europe: a preliminary assessment.* Discussion Paper ENR93-19, Resources for the Future, Washington, DC.

Loehman, E.T., Berg, S., Arroyo, A. *et al.* (1979) Distributional analysis of regional benefits and costs of air quality control. *Journal of Environmental Economics and Management*, **6**, 222–243.

Maddison, D., Johansson, O., Littman, T., Verhoef, E. & Pearce, D.W. (1996) *Blueprint 5: the Social Costs of Transport.* Earthscan, London.

McMichael, A., Anderson, R., Elliott, P., Wilkinson, P., Ponce de Leon, A. & Simon Soria, F. (the 'London Review Group') (1995) *Review of Methods Proposed, and Used, for Estimating the Population Health Risks of Exposure to Urban Air Pollution.* London School of Hygiene and Tropical Medicine, London.

Miller, T. (1989) Willingness to pay comes of age: will the system survive? *Northwestern University Law Review*, **83**, 876–907.

Miller, T. & Guria, J. (1991) *The Value of Statistical Life in New Zealand.* New Zealand Ministry of Transport, Wellington.

Needleman, L. (1976) Valuing other people's lives. *The Manchester School*, **44**, 309–342.

O'Doherty, R. (1994) *Contingent valuation as a participatory process.* Department of Economics, University of West England, Bristol.

Ostro, B. (1994) *Estimating Health Effects of Air Pollution: a Method with an Application to Jakarta.* Policy Research Department, Working Paper 1301, World Bank, Washington, DC.

Ostro, B. (1995a) Addressing uncertainties in the quantitative estimation of air pollution health effects. Paper presented to EC/IEA/OECD *Workshop on the External Costs of Energy*, Brussels, January 1995.

Ostro, B. (1995b) *Air Pollution and Mortality: Results from Santiago, Chile.* Policy Research Department, Working Paper 1453, World Bank, Washington, DC.

Otway, H. & Cohen, J. (1975) *Revealed preferences: comments on the Starr benefit–risk relationships.* Research Memorandum 75-5, International Institute for Applied Systems Analysis, Laxenburg, Vienna.

Pearce, D.W. (1986) *Cost Benefit Analysis*, 2 edn. Macmillan, Basingstoke.

Pearce, D.W., Bann, C. & Georgiou, S. (1992) *The Social Cost of Fuel Cycles.* HMSO, London.

Pearce, D.W. & Brisson, I. (1993) BATNEEC: the economics of technology-based environmental standards. *Oxford Review of Economic Policy*, **9**(4), 24–40.

Pearce, D.W., Markandya, A. & Barbier, E. (1989) *Blueprint for a Green Economy.* Earthscan, London.

Pearce, D.W. & Turner, R.K. (1992) The ethical foundations of sustainable economic development. *Advances in Human Ecology*, **1**, 177–195.

Pearce, D.W., Whittington, D. & Georgiou, S. (1994) *Project and Policy Appraisal: Integrating Economics and Environment.* Organization for Economic Cooperation and Development, Paris.

Pimentel, D. Wilson, C., McCullum, C. *et al.* (1996) *Environmental and economic benefits of biodiversity.* College of Agriculture and Life Sciences, Cornell University, Ithaca, NY.

Pope, C.A., III & Dockery, D. (1992) Acute health effects of PM_{10} pollution on symptomatic and asymptomatic children. *American Review of Respiratory Disease*, **145**, 1123–1128.

Pope, C.A., III, Daily mortality and PM_{10} pollution, in Utah Valley. Schwartz, J. & Ransom, M. (1992) *Archives of Environmental Health*, **47**(3), 211–217.

Pope, C.A., III, Thun, M.J., Namboordi, M. *et al.* (1995) Particulate air pollution as a predictor of mortality in a prospective study of US adults. *American Journal of Respiratory and Critical Care Medicine*, **151**, 669–674.

Rowe, R. & Chestnut, L. (1986) *Oxidants and asthmatics in Los Angeles: a benefits analysis.* Report to Office of Policy Analysis, US Environmental Protection Agency, EPA-230-09-86-018, Washington, D.C.

Rowe, R. & Nethercut, T. (1987) *Economic assessment of the impact of cataracts.* RCH Hagler Bailly, Boulder, CO.

Rowe, R., Chestnut, L. & Shaw, W. (1984) Oxidants and asthmatics in Los Angeles: a benefits analysis. In: *Evaluation of the Ozone/Oxidants Standard*, (ed. S. Duk Lee). Air Pollution Control Association, Pittsburgh, PA.

Rowe, R., Chestnut, L., Peterson, D. & Miller, C. (1986) *The Benefits of Air Pollution Control in California*, Vols 1 and 2. Energy and Resource Consultants, Boulder, CO (for California Air Resources Board).

Rowe, R., Chestnut, L., Lang, C., Bernow, S. & White, D. (1995) The New York environmental externalities cost study: summary of approach and results. Paper presented to *European Commission, International Energy Agency and Organisation for Economic Co-operation and Development Workshop on External Costs of Energy*, Brussels, 30–31 January 1995.

Sagoff, M. (1993) Environmental economics: an epitaph. *Resources*, **Spring**(111), 2–7.

Schwab Christe, N. & Soguel, N. (1995) *The pain of victims and the bereavement of their relatives: a contingent valuation experiment*. Institute for Public Sector Management Studies, University of Lausanne.

Schwartz, J. (1991a) Particulate air pollution and daily mortality in Detroit. *Environmental Research*, **56**, 204–213.

Schwartz, J. (1991b) Particulate air pollution and daily mortality: a synthesis. *Public Health Reviews*, **19**, 39–60.

Schwartz, J. (1993a) Air pollution and daily mortality in Birmingham, Alabama. *American Journal of Epidemiology*, **137**, 1136–1147.

Schwartz, J. (1993b) Particulate air pollution and chronic respiratory disease. *Environmental Research*, **62**, 7–13.

Schwartz, J. (1994) Air pollution and daily mortality: a review and meta analysis. *Environmental Research*, **64**, 36–52.

Schwartz, J. & Dockery, D. (1992a) Particulate air pollution and daily mortality in Steubenville, Ohio. *American Journal of Epidemiology*, **135**, 12–19.

Schwartz, J. & Dockery, D. (1992b) Increased mortality in Philadelphia associated with daily air pollution concentrations. *American Review of Respiratory Disease*, **145**, 600–604.

Schwartz, J. & Marcus, A. (1990) Mortality and air pollution in London: a time series analysis. *American Journal of Epidemiology*, **131**(1), 185–194.

Schwartz, J., Slater, D., Larson, T., Pierson, W. & Koenig, J. (1993) Particulate air pollution and hospital emergency room visits for asthma in Seattle. *American Review of Respiratory Disease*, **147**, 826–831.

Schwartz, J., Spix, C., Wichman, H. & Malin, E. (1991) Air pollution and acute respiratory illness in five German communities. *Environmental Resources*, **56**, 1–14.

Seaton, A., Macnee, W., Donaldson, K. & Godden, D. (1995) Particulate air pollution and acute health effects. *Lancet*, **345**(January 21), 176–178.

Small, K. & Kazimi, C. (1995) On the cost of air pollution from motor vehicles. *Journal of Transport Economics and Policy*, **January**, 7–32.

Starr, C. (1972) Benefit cost studies in sociotechnical systems. In: *Committee on Public Engineering Policy, Perspective on Benefit–Cost Decision Making*. National Academy of Sciences, Washington, DC.

Tolley, G.S., Babcock, L., Berger, M. *et al.* (1986) *Valuation of reductions in human health symptoms and risks*. University of Chicago Report to US Environmental Protection Agency, Washington, DC.

Turner, R.K. (1993) Sustainability: principles and practice. In: *Sustainable Environmental Economics and Management* (ed. R.K. Turner), pp. 3–36. Belhaven, London.

Utell, S. & Samet, J. (1993) Particulate air pollution and health: new evidence of an old problem. *American Review of Respiratory Disease*, **147**, 1334–1335.

Vatn, A. & Bromley, D. (1994) Choices without prices without apologies. *Journal of Environmental Economics and Management*, **26**, 129–148.

Viscusi, W.K., Magat, W. & Forrest, A. (1988) Altruistic and private valuations of risk reduction. *Journal of Policy Analysis and Management*, **7**(2), 227–245.

Viscusi, W.K., Magat, W. & Huber, J. (1991) Pricing environmental health risks: survey assessments of risk–risk and risk–dollar trade offs for chronic bronchitis. *Journal of Environmental Economics and Management*, **21**(1), 32–51.

Viscusi, W.K. (1992) *Fatal Trade-offs; Public and Private Responsibilities for Risk*. Oxford University Press, Oxford.

World Resources Institute (1994) *World Resources 1994–1995*. Oxford University Press, Oxford.

Xu, X., Gao, J., Dockery, D. & Chen, Y. (1994) Air pollution and daily mortality in residential areas of Beijing, China. *Archives of Environmental Health*, **49**(4), 216–222.

Part 4
Risk Management

The dilemma is that we gain great benefits from economic developments involving, for example, the production of commercial chemicals, power generation, the use of complex and sophisticated 'life aids' such as vehicles, refrigerators, washing machines, etc., and all the chemicals that go with them. Yet the production of these items, the way we use them, and ultimately the way we dispose of them, cause potential harm to human health and ecological systems. And the relationship is not straightforward, for positives on one side do not necessarily lead to proportional negatives on the other. For example, economic development has positive effects on human health, welfare and education—all of which are likely to mean that society takes environmental protection more seriously; hungry and sick people are less likely to take environment seriously than well-fed and healthy people.

Taking all these complex issues into account in risk management is illustrated in this part by reference to a 'pot pourri' of topic areas, that are not meant to be comprehensive in the sense of covering all aspects of our activities that might be subject to environment risk management, but which do, nevertheless, address all the major issues. They range from risk management in business, especially the high-profile chemical industry (Chapters 14 and 15) and the disposal of waste (Chapter 16), its use in exploitation of the seas (Chapter 17) and inland waters (Chapter 18), to its use in international development projects (Chapter 19).

A number of high-profile recent events, involving both ecological and human health risks, have illustrated how the public is not necessarily persuaded by the objective outcome of assessment of risks in its demand for action. There is a complex interaction here between the public, pressure groups, media, risk-managers, regulators and policy makers. What matters is not just the logic of an argument, but the way that it is presented, something that is illustrated in a number of the chapters that follow but which is addressed explicitly, as an integral part of the risk management activity, in Chapter 20 on risk communication.

Finally, we return to the complication of balancing the risks that derive from economic development with the benefits that we obtain from this. The sustainable development philosophy presumes that an optimized balance can be struck that ensures long-term well-being for us and our environment. Managing potential environmental problems, be it at an international, national, company or personal level, in terms of an assessment of risks, making explicit the balance between risks, costs and benefits, and ensuring that these are made explicit in a way that can be appreciated and understood by all major players has to make a contribution to sustainable development. Considering how these principles might be applied, especially at company level in terms not only of process but also the design of new products, therefore fittingly draws the volume to a close.

Chapter 14
Corporate Chemical Management: a Risk-Based Approach

CHARLES A. PITTINGER, CHRISTINA E. COWAN, PETER HINDLE AND TOM C.J. FEIJTEL

14.1 INTRODUCTION

The great majority of corporations involved in chemical technology and product development today recognize their responsibilities to ensure that their products, services and operations are safe for humans and the environment. To satisfy responsibilities for environmental safety, effective environmental risk assessment mechanisms and a strong corporate commitment to sound risk management policies are required. This is a part of a corporation's commitment to environmental quality, which is required for several reasons: to prevent the degradation of ecological systems; to comply with the myriad of national and international regulatory requirements; to maintain long-term economic viability; and at the same time to serve the needs of consumers. In essence, sound environmental policies equate to good business practice.

This chapter describes how many corporations have managed to integrate business needs and environmental considerations into an everyday decision-making framework. Fundamental, risk-based assessment approaches are used routinely by many corporations in assessing product and process safety, in evaluating industrial sites and emissions, and in managing natural resources. Distinct differences exist in conducting risk assessments and making risk management decisions within the corporate culture, as compared with regulatory applications. Here we will attempt to characterize the applications, needs and management perspectives unique to the private sector.

14.1.1 Scope and terminology

The focus of this chapter is primarily on assessing ecological risks associated with the production of commodity chemicals and the manufacture of consumer products. We believe that the principles and approaches described here are broadly applicable across the private sector; however, we also recognize that individual companies will tailor the approach to meet specific organizational structures and needs. It is recognized that corporate applications of environmental risk assessment also include two other broad areas: site evaluations of industrial production and waste disposal facilities; and natural resource evaluations, principally involving agricultural, forestry, fishery or mining activities. Other chapters address the use of environmental risk assessment within these latter two applications.

Environmental risk assessment as used in this chapter is considered to be synonymous with ecological risk assessment (USEPA 1992; cf. Chapter 1). As mentioned in the previous chapter environmental/ecological risk assessment is defined as a process that evaluates the likelihood that adverse ecological effects might occur, are occurring or have occurred as a result of exposure to one or more physical, chemical or biological stressors. In simplest terms, it is a comparison of a concentration capable of producing an adverse effect with an actual or predicted exposure concentration in the environment. Fundamental to this concept is the recognition that environmental risk requires two elements: (i) a stressor with the inherent ability to cause adverse ecological effects; and

(ii) the stressor interacts with an ecological component (i.e. organisms, populations, ecological communities or ecosystems) at sufficient intensity and duration to elicit the adverse effect(s) (USEPA 1992). In this chapter our focus will be upon chemical (rather than physical and biological) stressors. Needs and priorities in environmental risk assessment have been reviewed in Chapter 1 and Part 1 and 2 (see also Fava *et al.* 1987).

Risk assessment is distinguished from risk management, the process by which risk assessment results are integrated with other information (e.g. technical feasibility, cost and offsetting benefits) to make decisions about the need for, method of, and extent of risk reduction (NRC 1994). Risk assessment is further distinguished from the process of hazard assessment, through which the intrinsic ability of stressors to elicit adverse effects is characterized (USEPA 1979). Hazard assessment does not involve consideration of exposure, yet has at times been proposed as a principal basis for chemical regulation in Japan (Ministry of International Trade and Industry (MITI) 1973 Chemicals Act), Western Europe (Council Directive No. 67/58/EEC on the classification and labelling of dangerous substances and its 6th Amendment, 79/831/EEC) and North America (International Joint Commission's Strategy for the Virtual Elimination of Persistent Toxic Substances in the Great Lakes Basin). Some of the more serious ramifications of excluding exposure considerations from chemical management will be discussed.

In contrast, risk assessment fundamentally involves consideration of *exposure*, the spatial and temporal distribution of a chemical stressor and its real or potential co-occurrence with an ecological component (USEPA 1992). We believe that risk-based approaches are an essential component of sound chemical management. Risk assessments of commodity chemicals and consumer products may be prospective in nature (e.g. predicting risks for new commercial chemicals), retrospective (e.g. confirming predictions for chemicals in commerce, through monitoring or field testing) or sometimes both (e.g. expanding commercialized products or ingredients to new applications, other geographies or products).

14.2 AN OVERVIEW OF CORPORATE ENVIRONMENTAL MANAGEMENT

Modern industrial society is extremely complex. As such, effective business management requires a clearly defined framework within which to compile and process the information critical to making sound business decisions. Such a management framework includes the separate business entities responsible for different functions: finance and purchasing, personnel administration, manufacturing and engineering, research and development, market research, marketing and sales. Each of these must participate in daily decisions using any and all of the appropriate management tools that are available. Successively higher levels of corporate management must ensure that these cross-functional decisions are consistent with corporate policies and can maximize business opportunities while minimizing vulnerabilities and trade-offs. Sound corporate management frameworks include environmental in addition to conventional business concerns.

Innovative companies with an eye on long-term success long ago recognized that environmental quality must be considered in conjunction with conventional business concerns. Pro-active companies have also implemented environmental policies to identify and manage potential problems associated with new products and new technologies. Environmental risk assessment represents an integral and indispensable tool to implement corporate environmental quality policies. Many companies routinely conduct environmental risk assessments for all significant product development and processing initiatives (AIHC 1996). These often require a substantial investment in a central corporate environmental research organization as well as professional environmental staff strategically located in key business sectors. Trained environmental risk assessors in specific business sectors can better understand and regularly input into highly dynamic product and process technology development activities.

14.2.1 Integrating environmental criteria and responsibilities into corporate culture

Corporate environmental management is no

different from other aspects of sound business management. An effective organizational and data-processing framework is imperative; daily decisions are the norm. These decisions must adhere to corporate environmental quality policies as well as satisfy the rigours of a dynamic business environment. To accomplish this, environmental criteria to assess for potential ecological risks must be incorporated early in the technology development process.

Environmental, human safety and regulatory criteria are integrated in the decision-making framework with a vast array of business and economic criteria. These include: consumer habits and practices; product performance and value; marketing; process feasibility and safety; efficacy and stability of ingredients in the product matrix; ingredient and/or manufacturing costs fluctuating on world markets; political stability in key manufacturing locations; labour costs and currency fluctuations; specifications of suppliers and contract manufacturers; national and international transportation and distribution networks. Balancing all of these considerations within the technology development process is complex. Close coordination and communication is required among a corporation's business, risk assessment and regulatory professionals as they evaluate these considerations across all aspects of a product's lifecycle (P.R. White *et al.* 1995).

Environmental management is also a responsibility shared between the corporation and its many external partners—suppliers, customers, consumers and the government. This is because corporations must operate within the context and constraints of society, and because many large-scale environmental issues cannot be solved by a single corporation acting alone. Ensuring the environmental safety of a new commercial technology often requires close coordination among the different corporations involved in the technology. Coordination is also needed between corporations and government institutions to ensure that proposed solutions to one environmental problem do not create or exacerbate other environmental problems.

Across the sequence or 'lifecycle' of products—those industrial activities required to source raw materials, produce commodity chemicals and formulate them into consumer products (Fig. 14.1)—each corporation is responsible for the safety of those activities and operations under its own purview. Few, if any, companies are 'vertically integrated' to manage all aspects of raw material sourcing, processing, formulation and marketing of consumer products. Supplier, processor and manufacturer are uniquely responsible for managing the environmental impacts of production operations within their own plants, as well as assessing impacts arising from the use and disposal of their products.

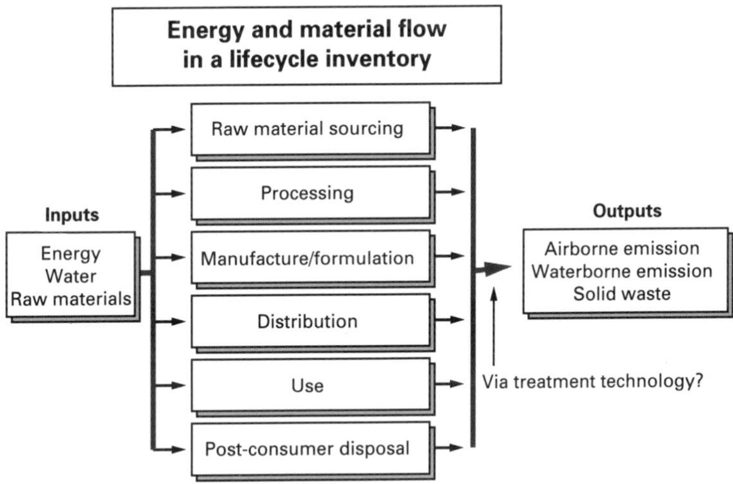

Fig. 14.1 Environmental management concerns extend across all stages of a product's lifecycle, including natural resource use and risks associated with industrial and post-consumer wastes.

Corporations exert influence upon other phases of the product's lifecycle, particularly upon 'upstream' suppliers which produce chemicals according to strict technical specifications. Technical specifications for commodity chemicals can, and often do, include human and environmental safety criteria set by the corporate buyer and/or the supplier. Environmental criteria include purity (e.g. the absence of potentially toxic precursors, intermediates or byproducts), consistency in quality, fate (biodegradability, bioaccumulation, sorptivity) and ecological effects (e.g. aquatic toxicity). Through these specifications, environmental quality policies of customer–supplier corporations are transmitted to other companies and become an inducement for them to adopt similar policies.

Because technical specifications are closely linked to production costs, criteria that do not add to performance or safety can result in unnecessary expenses. Thus, only valid and scientifically sound environmental criteria and technical specifications can be justified to Chief Executive Officers (CEOs), Boards of Directors and stockholders, and are included in the management framework. This results in a set of environmental criteria that are both scientifically sound and shown to contribute to the development of improved technologies.

In addition to corporate interests and responsibilities for developing safe chemicals and products, environmental regulations play a large role in shaping corporate environmental policies. European Union chemicals legislation and risk assessment legislation is in place for both new and existing substances (Article 3(2) of Directive 92/32/EEC; the 7th Amendment of Directive 67/548/EEC; Article 10(4) of Existing Substances Regulation EEC/793/93) (Chapters 3 and 9). A plethora of environmental statutes exist in the USA. Chief amongst these are the Toxic Substances Control Act (TSCA); the Comprehensive Environmental Response, Compensation, and Liability Act (CERCLA, or 'Superfund'); the Clean Water Act; the Resource, Conservation and Recovery Act (RCRA). At the present time, the USEPA is developing ecological risk assessment guidelines (Chapters 3 and 10).

The Canadian Environmental Protection Act governs Canada's approach to chemical management. Canada has developed a Toxic Substances Management Policy strategy and ecological risk assessment guidelines. Environmental regulatory requirements imposed by these and other statutes are addressed in other chapters and will not be detailed here, but some major ramifications of regulatory policies upon corporate environmental management will be discussed. One important factor in the management framework is a consideration of the existing regulatory systems and whether they support a 'level playing field' in a market category. Such systems help to create a business environment in which responsible corporations are rewarded rather than penalized for environmental quality criteria and measures.

14.2.2 The role of corporate and public partnerships in environmental research

A wide range of partnerships in environmental research exist among corporations, and between corporations and public institutions. The sharing of technical resources, expertise and costs often provides valuable technical and cost benefits, as well as promoting consensus and alignment on environmental goals and needs. Corporations involved in customer–supplier relationships frequently exchange safety information and cooperate in the conduct of environmental risk assessments. Companies supplying ingredients to consumer product manufacturers often work in tandem to generate ecological fate and effects data and to conduct joint risk assessments. 'Premanufacture notifications' required under the US Toxic Substances Control Act for registration of new chemicals sometimes contain environmental data contributed by both supplier and manufacturing corporations. Trade associations such as the European Centre for Ecotoxicology and Toxicology of Chemicals (ECETOC), the European detergent industry (Association Internationale de la Savonnerie, AIS), the American Industrial Health Council (AIHC) and the Chemical Manufacturers Association (CMA) regularly sponsor or participate in joint research programmes to support the common environmental needs and interests of their corporate members.

Partnerships among corporations and public institutions in generating environmental data are also common. Under the US Federal Technology Transfer Act, many 'Cooperative Research and Development Agreements' among corporations and US federal agencies have been initiated to address specific environmental research questions (USEPA 1993). Similar programmes have been developed by the European Union's fourth and fifth Action Programme Environment and Climate. In some cases, federal research agencies work closely with corporations or industrial consortia in advancing environmental risk assessment through technical dialogues on risk assessment (SETAC 1994) and research (Feijtel & van de Plassche 1995). Such cooperation is common around the world. However, there is still a need for further development of sound testing protocols and consistent data interpretation procedures to reduce redundant testing and registration procedures.

Partnerships between governments and industry, sometimes called 'voluntary agreements' or 'action plans', are also common in Canada and in some member states of the European Union. These are jointly prepared by the partners and lay down the principles and framework under which joint research or risk management actions will be undertaken. The contents of these documents vary depending on their objectives but commonly contain agreements on meetings, goals, methodologies, responsibilities, time schedules, etc. The Dutch covenants or 'Plan van Aapak' are initially non-published and non-publicized agreements (Dutch Ministry for Environment 1992, 1995). These covenants provide an opportunity for the different parties to reach concensus on the research project and the results before these are shared publically. The recently completed study of surfactants in the environment was the result of such an agreement between the Dutch Government and the Dutch Soap Association (Feijtel & van de Plassche 1995). In Canada, under the Accelerated Reduction/Elimination of Toxics (ARET) Programme, a voluntary agreement was used to reach an agreement between industry and the Canadian Government on the goals and time schedules for reducing the amount of certain chemicals released to the environment (ARET Secretariat 1995).

The consuming public also plays an environmental stewardship role when they use products in a manner compatible with their intended use, and dispose of products according to instructions that are compatible with local waste management infrastructures (P.W. White *et al.* 1995). Corporations encourage the safe use and disposal of products by consumers by providing environmental information, usage instructions and guidance for disposal on product labels or in supplemental information which is provided to the consumer upon request. Product and package design also contributes to safe use and proper disposal by the consumer. Municipalities participate in environmental stewardship through such activities as providing recycling services, ensuring effective waste treatment and promoting integrated waste management.

14.2.3 Business realities governing sound environmental management

Business realities shape the foundation of corporate environmental management. These reflect the constraints and fundamental necessities of doing business and achieving economic success in today's challenging international markets. The major business realities which have an impact on environmental management are presented below.

1 Only products that offer competitive performance and value, and that simultaneously offer an improved environmental profile, can survive in the marketplace long enough to deliver meaningful benefits to society (Hindle *et al.* 1993). A product with significantly reduced environmental burdens can only deliver its environmental benefit if it is successful in the same market niche as a less beneficial alternative, or if it competitively replaces that niche altogether. The superior environmental benefits of a product which is perceived by the consumer to provide lower performance or value will not be realized if the consumer does not purchase and/or use it.

2 All industrial and commercial operations—like all activities of society in general—have an impact on the environment in some way. Impacts range from natural resource use and energy consumption to the environmental risks of production and waste generation. Minimizing the potential environmental impacts while providing for economic growth is the goal of sound corporate environmental management. Corporations also recognize that achieving 'zero impact' is unrealistic. Therefore, the environmental attributes of chemicals and products are not judged in terms of absolute 'burden' or 'impact', but rather evaluated by comparison to the potential impacts of alternative products or services. Real improvements to products, chemicals and services can only be accurately evaluated in the context of both their value to society and their associated enviromental burdens at all stages of their lifecycles (Fig. 14.1).

3 Accurate information and data-based decisions are fundamental to sound management practices. It is imperative that the data upon which decisions are based can be traced back to their origins, along with an understanding of their quality, limitations and uncertainties. Decision-making requires explicit differentiation between, and consideration of, data-based and value-laden elements. This enables corporations to consistently adhere to successful business practices, ensures reproducibility and promotes organizational accountability. It is important to recognize, however, that the sharing of data by a corporation must be balanced against the needs of companies to restrict access to competitively sensitive business information.

4 Technological innovation and speed-to-market are fundamental to business success and environmental quality improvement. Few products or technologies will continue to grow in market share using obsolete technology. A four-stage cycle of technological innovation as applied to environmental quality improvements is presented in Fig. 14.2. Even highly recognized brand names with long histories of success in the market are continuously upgraded to satisfy changing consumer expectations and to adjust to changing international economic conditions.

Innovation is essential to allow the corporations to provide improved environmental quality as well as performance to the consumer. Innovation in chemical technology development provides such benefits as more readily biodegradable ingredients,

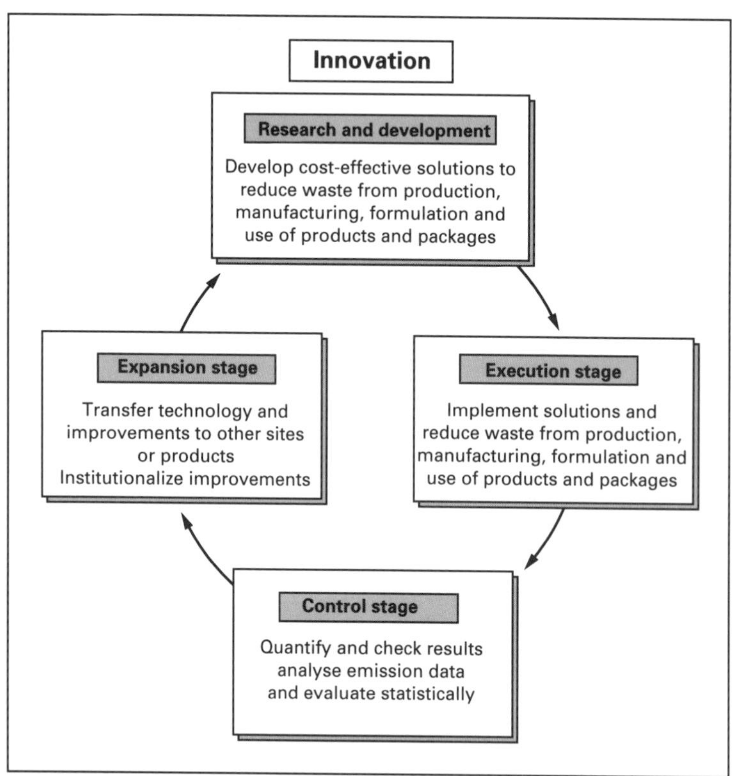

Fig. 14.2 A four-stage cycle illustrating innovative technology development within a corporation management framework.

greater purity, less production waste and greater recyclability. To be successful, a product offering an environmental benefit must gain an early foothold in the market. A social or regulatory environment which impedes responsible technology innovation or the speed at which these new technologies can be introduced to the marketplace can limit a corporation's ability to improve environmental quality.

14.2.4 A holistic environmental management framework

A variety of information and management tools are needed to ensure that all environmental aspects of the business are adequately addressed. Just as with business tools, no single tool is able to cover all dimensions of environmental management. Some of these tools, such as risk assessment, are based directly on the environmental and health sciences or are oriented toward regulatory compliance. Others are based on more traditional business tools which provide essential information on consumer behaviour. Decision-making in environmental and business management alike involves careful integration of the outputs from different tools, plus an interpretation based upon professional judgement and experience. Professional judgement is required to balance the trade-offs inherent in most decisions, to compensate for

Table 14.1 A needs-based environmental management framework, adapted from P.R. White *et al.* (1995). Reprinted from *Resources, Conservation and Recycling*, **14**, P.R. White *et al.*, 'Environmental management in an international consumer goods company', pp. 171–184, 1995, with kind permission of Elsevier Science NL, Sara Burgerhartstraat 25, 1055 KV Amsterdam, The Netherlands.

Corporate management needs	Assessment and management tools
Human and environmental safety	Human health risk assessment (occupational and domestic exposure) Ecological risk assessment (plant-site and consumer releases)
Regulatory compliance	Manufacturing site management system auditing Manufacturing site waste reporting Material consumption reporting (e.g. Dutch packaging covenant) New chemicals testing and registration Product and packaging classification and labelling
Efficient resource use and waste management	Manufacturing consumption monitoring and reduction Manufacturing site management system auditing Manufacturing site environmental auditing Auditing major and new suppliers Disposal company auditing Product lifecycle inventory 'Eco-design' for the environment Economic analysis
Addressing societal concerns	Understand Opinion surveys Consumer habits and practices Market research Interact Information through presentations and publications to key audiences Academic, policy and industry work groups (e.g. regulatory authorities, professional societies) Lobbying to influence future policy and regulations Corporate reporting Specific problem solving with others

uncertainties due to incomplete data sets or data of limited quality, and to factor in variabilities in the marketplace or, in this case, in environmental conditions.

Consistent with the business realities noted above, an environmental management framework has been developed which encompasses all corporate and governmental responsibilities for safety, regulatory compliance and economic viability (P.R. White *et al.* 1995). The four fundamental needs which must be addressed by environmental management are: (i) human and environmental safety; (ii) regulatory compliance; (iii) efficient resource use and waste management; and (iv) consideration of the concerns of society (Table 14.1). Each of these is supported by a range of environmental management tools, some of which may address two or more of these elements. These tools provide the data which the risk manager(s) and the business manager(s)—sometimes the same individual(s)—will use in making a decision. The tools have largely evolved independently as needs have arisen to answer specific questions. The abundance of such tools at times leads to confusion as to their precise role. The management framework demonstrates how the tools interrelate to meet the environmental management needs of a corporation.

There is a hierarchy to the four needs listed in the management framework. The first two elements—safety and regulatory compliance—are obligatory prerequisites for corporations to remain in operation. Companies must ensure safety of their operations and products under conditions of intended use and foreseeable misuse. They clearly must have a commitment to comply with regulations imposed at all levels of government, yet human and environmental risk assessment cannot, in themselves, accomplish the overall goal of an integrated corporate environmental management policy. An integrated policy must incorporate and enable both environmental improvement and economic success.

The latter two elements—efficient resource use and the satisfaction of social concerns—are less tangible but are increasingly important business needs essential to the long-term success of a corporation. These elements are addressed through less well-developed and validated tools than risk assessment and regulations. Emerging tools such as lifecycle analysis, 'eco-design' (e.g. designing waste out of production processes; designing durable goods for disassembly and recycling), and material consumption monitoring can aid in the management of resources and wastes. Efficient use of resources and minimization of waste is mutually beneficial to corporate economics as well as environmental protection. However, process changes can require significant investment and must be implemented within the constraints of business and governmental realities.

Successful corporations cater to the needs and wants of societies. Today, society as a whole increasingly values environmental quality in products and services, yet most consumers tolerate little compromise in performance or value to achieve it. Anticipating the myriad of social concerns related to environmental quality requires 'staying in touch' through opinion surveys and consumer and marketing research (see Chapter 20). Some of these concerns are scientifically supportable; others are pre-formed by myths and misperceptions. Corporations must respond to both, either by providing products with meaningful environmental benefits or by engaging in dialogue and educational initiatives to dispel invalid perceptions. Effective initiatives are those built upon trust between the corporation and its partners in society. This is a relatively new departure from the public's historical suspicion of industry. Hence, many corporations are responding by being more open and informative about their products and processes within the limits of legitimate competitiveness concerns. However, it is recognized that sometimes consumer perception will result in decisions being made by corporations even when scientific data and environmental risk assessments indicate that their perception is invalid.

14.3 A CORPORATE APPROACH TO ENVIRONMENTAL RISK ASSESSMENT

Although environmental risk assessment is sometimes viewed as predominantly a function of regulatory agencies, the same assessment approach is often applied by corporations as an internal technology development tool (Cowan *et al.* 1995a).

This section presents the fundamental process of environmental risk assessment as often applied with special focus on the considerations unique to industry. The regulatory overview of environmental risk assessment and additional details on the risk assessment process, especially testing, are provided in Part 1.

14.3.1 Corporate applications of environmental risk assessment

Corporations use environmental risk assessments in a variety of ways (AIHC 1996). First, corporate risk assessors use environmental risk assessment to evaluate the likelihood that the intended or unintended release of a product ingredient or industrial emission will result in particular impact(s) upon particular environmental compartments (e.g. aquatic, marine and terrestrial) (Gledhill & Feijtel 1992; Black *et al.* 1993; Larson & Woltering 1993; Feijtel & van de Plassche 1995). Corporate risk managers use the assessment to determine whether control measures can and should be taken (Chapter 1). These control measures range from constricted usages and special containment measures to options for alternative materials. Second, environmental risk assessments assist corporations in identifying additional data needs to reduce uncertainties or to account for variabilities. Third, risk managers use assessments to prioritize the allocation of corporate resources toward research, engineering, production and marketing initiatives to improve the long-term viability of the product.

As recommended by the National Research Council (NRC 1994), risk assessments must consider all segments of exposed populations, both human and ecological. Recent advances in probabilistic risk assessment have added valuable perspective on how exposure and associated risks are distributed across the range of exposed populations or species. The corporate risk manager must consider the full range, including the most highly exposed or most highly vulnerable segments of a population or species within an ecological community (e.g. above the 95th percentile). Yet these segments are more difficult to assess than 'average' or 'median' segments of an exposed population or species, and risk characterizations pertaining to these segments will contain greater uncertainty.

14.3.2 The corporate imperative for tiered testing

The same basic elements of environmental risk assessment as practised by government institutions (see Part 1) are employed by private corporations (Cowan *et al.* 1995a). This can generally be described as a tiered or iterative approach which refines the steps of problem formulation, analysis of exposure and effects, and risk characterization until the needs of the risk manager are satisfied. This approach reflects that advocated by the USEPA (1992) and by the European Commission (1994, 1995), and has been refined by the AIHC (Fig. 14.3). The refinements offered by the AIHC (1995a) are several: (i) the planning and evaluation discussions between the risk assessor(s) and risk manager(s) are emphasized; (ii) the need for conducting higher tiered testing is determined in the evaluation discussion and implemented before (as opposed to after) the risk management decision is made; and (iii) a tiered, iterative process leading from the evaluation discussion to a reformulation of the original problem is clearly identified.

Predictive and efficient methods for screening ecological fate and effects of chemicals is indispensable in technology development, as early product development efforts often include a large number of candidate substances. Early recognition of environmental pathways and partitioning is necessary to plan an appropriate risk assessment strategy before major commercial decisions and investments are made. Indiscriminate testing of all candidate substances early in technology development is neither economically feasible nor necessary in most cases to ensure safety.

An iterative approach is essential to efficient product testing, as it allows systematic collection of data, maximizes the efficient allocation of personnel and resources, minimizes costs to both the corporation and the regulatory body responsible for product registration or approval, and focuses the major expenditure of resources on those chemicals of greatest concern. The need to proceed in a testing programme from screening

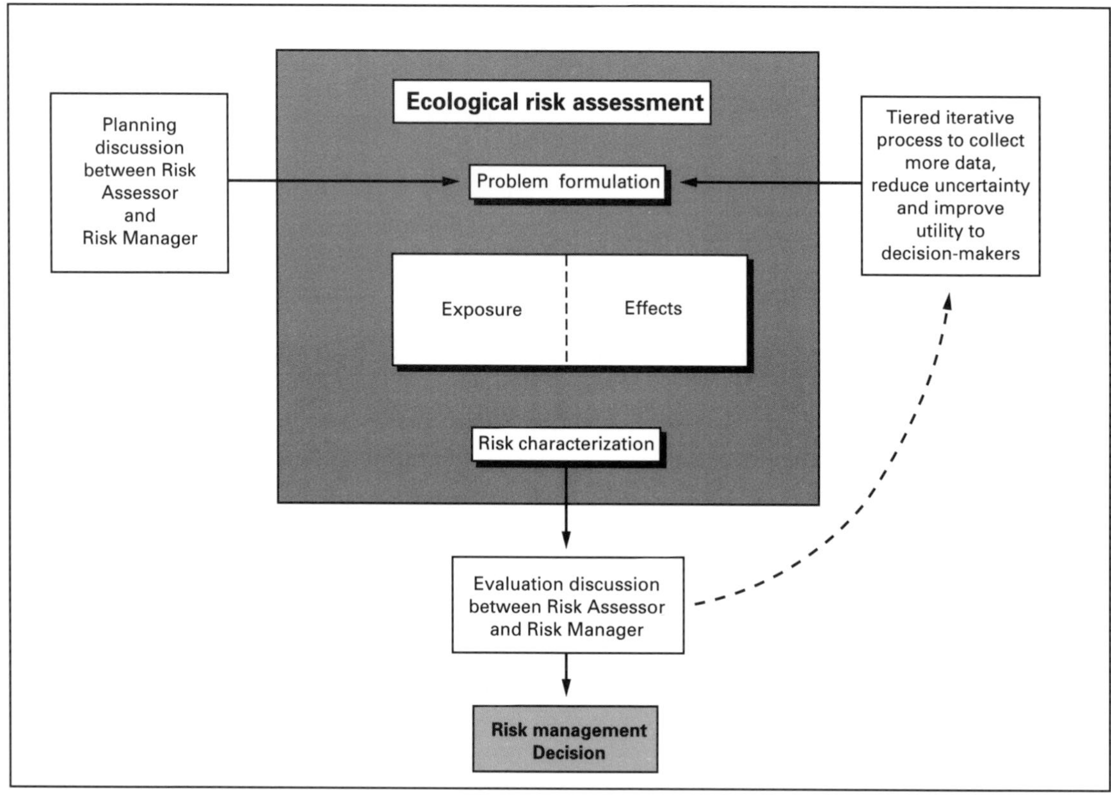

Fig. 14.3 Framework for ecological risk assessment (USEPA 1992) as refined by recommendations of the American Industrial Health Council (AIHC 1995a).

to more investigative research studies relies on a logical risk-based hierarchy for decision-making. Conservative assumptions are applied in initial screening to estimate 'worst-case' environmental concentration and effect levels. Successive tiers may be necessary to refine estimates through the use of more realistic and representative assumptions of test conditions. Often, more sophisticated methods (i.e. higher tiers of analysis) may be indicated for only one portion of the risk assessment (e.g. reducing uncertainty in the exposure estimate). The risk manager should be consulted at the completion of each tier to assess needs for higher-level testing.

Screening-level assessments using physico-chemical parameters, fundamental fate and toxicity trends extrapolated from acute testing, and the use of conservative assumptions and assessment factors can effectively be used to evaluate ecological safety early in the technology development process. For example, a substance discharged to surface water in an effluent may be assumed to enter the system undiluted and non-degraded, even though dilution and degradation in the environment are likely to occur based on available environmental and material characterization data. Thus, any judgement of safety made at this tier will result in a high level of assurance that there will be no impact on the environment.

More definitive exposure estimation techniques and direct testing of organism sensitivity may be necessary when a judgement of safety cannot be made upon initial screening or when regulatory requirements specify testing. In higher tiers, exposure and effects assessments are more sophisticated and expensive, and are specifically designed to

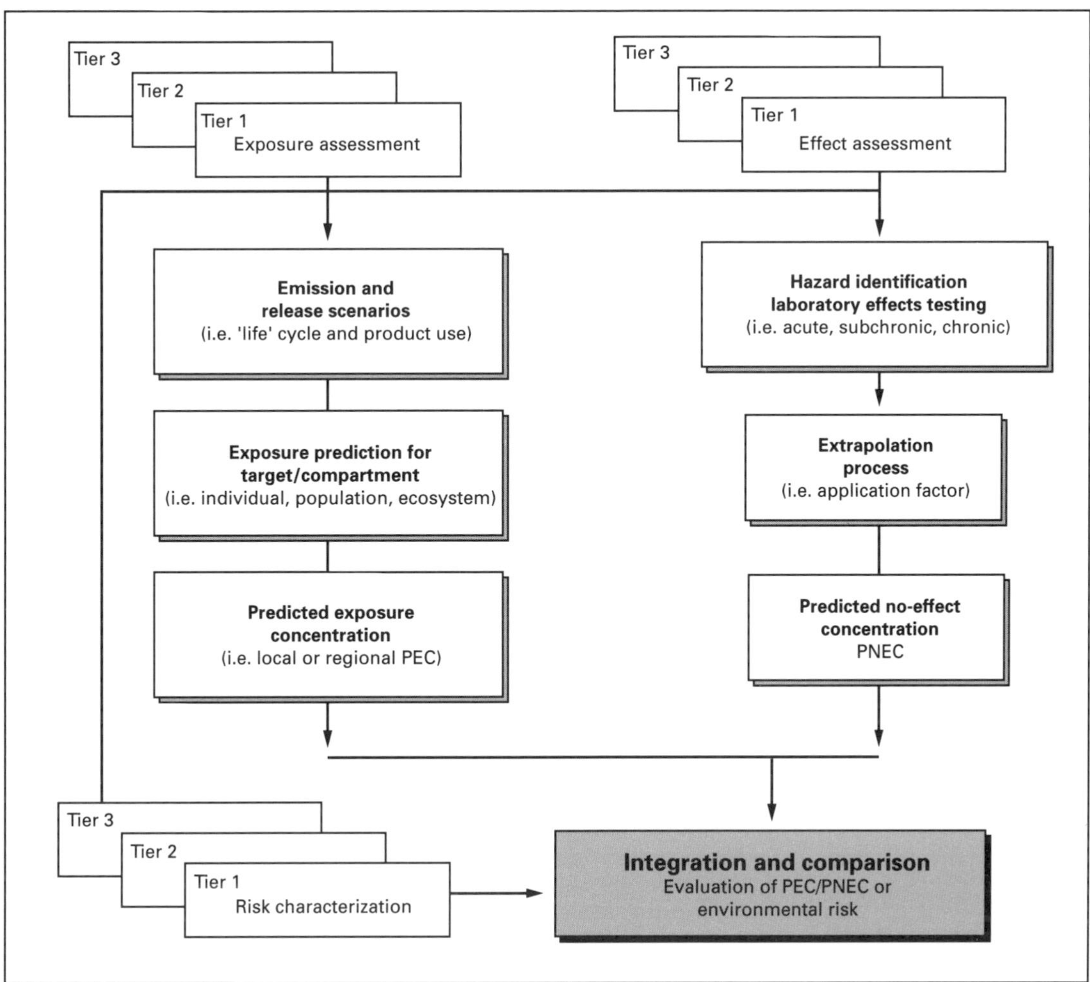

Fig. 14.4 A tiered process for exposure and effects analysis in environmental risk assessment, culminating in risk characterization.

reduce uncertainty in the assessment (Fig. 14.4). In the example above, the incorporation of realistic stream dilution factors and measured rates of biodegradability to project losses during conveyance to the discharge point may be included. The more sophisticated (highest tiered) risk assessments are conducted for product chemicals already in commerce, those intended for high-volume or direct environmental exposure (e.g.agrochemicals). These assessments frequently involve field testing and/or monitoring. Field monitoring often provides the most realistic estimate of direct exposure to organisms, but typically are cost- and time-intensive.

14.3.3 Problem formulation

Prior to initiation of virtually any environmental risk assessment, close communication and coordination between the corporate risk manager(s) and risk assessor(s) is required to ensure that the assessment satisfies management needs. The corporate risk manager typically represents one or a group of individuals directly responsible for technical development of a product. The risk assessors may represent an internal corporate environmental safety organization, a private contract laboratory, or both. If an assessment will be used to support chemical or product registration, the needs of the regulatory risk manager(s) responsible for product registration must also be considered at an early stage. Ideally, initial discussions can also take place between corporate and regulatory risk managers to identify the assessment needs of both the corporation and the regulatory institution.

Initial planning discussions between manager and assessor focus on framing the needs and scope of a formal risk assessment, the resources available for conducting the assessment, critical time-lines contingent upon other product development needs, and possible regulatory submissions involving or requiring the assessment. The problem formulation which follows this discussion begins with an evaluation of available product or chemical usage data: physical/chemical properties, intended uses and disposal scenarios, potential for spills and unintended environmental emissions, use and disposal of the same or similar materials by other corporations or public institutions, etc. All available environmental data are also compiled at this point. These data sets are used collectively to: set goals for the risk assessment; identify needs for appropriate exposure and effects end-points and models; determine decision-making criteria, including the amount of data required, its acceptable uncertainty and variability; and recognize and incorporate regulatory, societal or corporate policy issues which must be considered.

The selection of exposure and effects end-points for the risk assessment must broadly consider and incorporate scientific, social and regulatory goals and values (Suter 1993), at a scale which encompasses the geographies for which the product or chemical will be marketed. This is a particular challenge for corporations which conduct risk assessments for products aimed at international markets, where the values of society, habits and practices of consumers and waste management infrastructures may differ greatly across cultures and regions. Better harmonization of assessment end-points and agreement on sound testing protocols are needed to facilitate broader international acceptance of risk assessment results (OECD 1995).

14.3.4 Exposure and effects analysis

The exposure and effects analysis phase of environmental risk assessment assembles and generates the data needed to estimate the predicted environmental concentration (PEC) and the predicted no-effect concentration (PNEC) used in the risk characterization process (Fig. 14.4). These analyses should be closely linked to the needs of the risk manager as identified during problem formulation. Characterization of both parameters is a complex subject which can only be briefly summarized here.

In general, exposure analysis must include consideration of the route(s) of potential discharge to the environment and entry into specific environmental compartments. The spatial and temporal distribution of the product or chemical, including the magnitude, intensity, duration and route(s) of exposure to species and ecological communities, are evaluated across the geographical range of the product's market. Sophisticated exposure models are often adapted or developed by corporations to accurately simulate realistic conditions of product usage and disposal relevant to a particular market area (Cowan *et al.* 1995b). The development of the Geographical-Referenced Exposure Assessment Tool for European Rivers (GREAT-ER) is expected to provide very valuable information on both water quality and chemical exposure (ECETOC 1995). The model will link existing data bases on rainfall, river flow and location of sewage treatment plants to allow a mathematical analysis of the source and quality of the water at any point in the major European river systems. The emerging development of probabilistic exposure models to describe the distribution of exposures across broad geographical ranges or environmental conditions provides valuable insight across the full range of environmental conditions.

The analysis of environmental effects is typically conducted concurrently with the exposure analysis, and must be commensurate in scope with the geographical range for which the assessment is intended. Complexity is chiefly driven by the diversity of species and types of ecological communities which must be considered, the broad range of acute and chronic effects or phenomena (e.g. bioaccumulation and biomagnification) which must be considered, multiple routes of potential exposure to an organism, and the potential for indirect, community-level effects (e.g. effects on predation, competition among species). Quantitative structure–activity relationships (QSARs) are commonly used in the initial tiers of effects analysis to estimate effects within classes of similar chemicals. Laboratory toxicity testing using sensitive, representative species, microcosms and mesocosms affords greater realism and may be designed to test particular exposure conditions. Microcosm and mesocosm toxicity testing is commonly performed for higher tiers of risk assessment. Biological field surveys and *in situ* exposure of test organisms are generally reserved for corporate risk assessments involving high-volume chemicals which require maximum predictive capability and validation, or to address particular regulatory requirements.

Descriptions of some of the common types of effects testing and the considerations in using QSARs and effects data are discussed in Chapter 3. Because most effects data are generated in single-species laboratory tests, a number of assumptions or 'assessment factors' related to data variability and uncertainty are commonly applied to estimate a PNEC protective of ecological communities and ecosystems (ECETOC 1993; Zeeman 1995; Chapter 9). Assessment factors (AF), sometimes called uncertainty factors (Chapters 1 and 3), are subjective, informed estimates of the variability among species and the uncertainty in extrapolating from the available data to the desired assessment end-point (ECETOC 1993). Different regulatory organizations world-wide have adopted different approaches to using assessment factors, complicating corporate risk assessments intended for world markets.

14.3.5 Risk characterization

Risk characterization is widely accepted as the final element of risk assessment for both human health (NRC 1983; European Commission 1994, 1995) and ecological applications (USEPA 1992; European Commission 1994, 1995). Environmental risk characterization is synonymous with the UK Department of the Environment (DOE 1995a) definition of 'risk assessment', combining their concepts of 'risk estimation' (concerned with the outcome or consequences of a potential hazard) and 'risk evaluation' (concerned with determining the significance of estimated risks for those affected). The risk characterization integrates the exposure and effects analysis to characterize the likelihood, significance and conditions of environmental risk in terms of the assessment end-points chosen. Ideally, the risk assessor communicates the risk characterization to the risk manager, who then makes decisions on the basis of this information as well as other relevant criteria.

The risk characterization process for environmental risks has been described in detail by other authors (Habicht 1992; Suter 1993) and will not be presented here. Recent recommendations (NRC 1994) suggest that a consensus is emerging that risk characterizations should be expanded to convey, to the fullest extent possible, the full range or distribution of risks across the spectrum of exposure and effects. This requires that risk characterizations clearly communicate the nature of the risk both quantitatively and qualitatively, including: the scope of the assessment, default assumptions used in the effects and exposure analysis; uncertainties and variabilities identified in the data sets used; limitations to extrapolating the results; lack of fundamental biological understanding; and the logic or rationale for using specific assessment factors (AIHC 1995b).

For the corporate risk manager considering the environmental safety of a new chemical or product, the risk characterization should provide a coherent and comprehensive understanding of the environmental risk(s). The manager must integrate this understanding with other social, economic, policy, legal and communication issues in making a decision. It is essential that risk characterizations be understandable and transparent, and that they articulate risks in simple and consistent language. The process of comparative risk analysis, using other products or chemicals in commercial use as points of reference, is valuable to risk managers in providing benchmarks with which to gauge the significance of environmental risks of new technologies.

14.4 ENVIRONMENTAL RISK MANAGEMENT IN CORPORATE AND SOCIETAL CONTEXT

Environmental risk management includes concerns well beyond the technical domain of the environmental scientist and risk assessor. Corporate environmental risk managers are faced with the unenviable task of simultaneously balancing diverse and sometimes conflicting value systems: the corporation's financial success; the norms of federal, state and municipal regulations; the criticisms of public interest groups; and, not the least, the needs and wants of its customers and consumers. In an inclusive environmental management framework, scientific and non-scientific value systems must be integrated within an overall decision-making process.

14.4.1 Translating environmental quality objectives into business decisions

Effective risk management requires clear transmission of environmental quality objectives. These objectives are the collective policies, goals and interpretations that each corporation uniquely uses to address questions like: 'How safe is safe (enough)?', and 'How clean is clean (enough)?'. Often the objectives are stated in broad terms in corporate mission statements, developed through a flexible and subjective process within which corporate managers interpret legal, regulatory, economic and social (e.g. risk perception) requirements or information. The objectives must be dynamic and evolve with the changing emphases that society—usually recognized as changes in the habits and practices of its customers—place on environmental quality.

Translating a corporation's environmental quality objectives into specific criteria and decisions is rarely a simple or straightforward task for the corporate risk manager, even in a corporation unequivocally committed to 'doing the right thing' for the environment. The risk manager still faces the obligation of identifying what is right to do in a given situation and context. All corporate actions and decisions (or inactions and postponed decisions) result in consequences which are difficult to precisely quantify, verify or validate, and even more difficult to weigh against others.

Few corporate environmental risk managers are called upon to make decisions between clearly good and bad alternatives. Rather, they face choices among wide-ranging sets of options that are superior in some environmental or economic dimensions (e.g. reducing an emission's exposure to aquatic fauna; boosting production capability) but inferior in others (e.g. generating solid waste; requiring more energy; resulting in impaired product performance). Only comparative (and integrated) consideration of costs (i.e. risks) and benefits will ultimately define the 'right thing to do' in technology development and production. The same approach applies in regulatory decision-making, where costs to society in setting and meeting standards must be justifiable in terms of 'measurable' risk reduction and benefits to both environment and human health.

Finally, consensus must be reached by all entities within a corporation that a given decision reflects the best balance of perceived opportunities and liabilities to the business. Cross-functional involvement and alignment ensures that human health, environmental and fiscal management decisions are integrated vertically and horizontally within a corporation. Strong parallels exist in the regulatory arena, where open communications and cooperation between regulators and the regulated community are also essential. 'Command and control' policies in single corporations or in public/private regulatory relationships typically address only a narrow set of criteria and overlook creative and unique solutions to problems. Effective environmental management and business management encourages innovation and seeks to identify 'win–win' solutions for all parties involved.

14.4.2 Environmental management strategies

There are a variety of instruments (command and control, economic, voluntary) that can be used to achieve good environmental practice. A clear statement of the objectives of the instrument, as well as of the non-scientific value systems that are commonly held by society at large, are prerequisites to choosing the appropriate instrument.

One important consideration is how and if the instrument will promote technological innovation. For standards, this can be achieved by basing them on risk, rather than on hazards which may never be expressed. Ideally, environmental regulatory standards can both restrict unacceptable operating and discharging activities and establish long-term quality objectives.

The complexity of decision-making is why 'command and control' management philosophies sometimes have appeal to corporate management organizations as well as the regulatory community. Rigid rules and narrow compliance frameworks (i.e. 'the box-checking mentality') can greatly simplify the corporate regulatory manager's decisions, though often by compromising technological innovation or meaningful environmental protection on a broader, more lasting scale. Many in society falsely equate regulatory compliance with environmental safety, and assign public institutions responsibility for both. Yet as depicted in Table 14.1, regulatory compliance is just one of several fundamental elements of environmental management. A corporation's commitment to all elements is required to achieve real environmental quality that exceeds the standards of regulatory compliance.

Because unanticipated circumstances can and do arise, strict adherence to a set of rules for making environmental decisions will not necessarily ensure that 'the right choice' will be made in all instances and for all contexts. There are sound scientific reasons why it may be appropriate, for example, to recycle a waste material in one locality but not in another. Similarly, risk scientists recognize that a concentration of a chemical may cause significant adverse effects in one situation (e.g. a particular body of water) but not in another. For these reasons, a more flexible, risk-based approach is preferable to strict 'command and control' management.

14.4.3 The corporate decision-making process

The unique roles of the risk assessor and risk manager in a public regulatory context as distinguished by the National Research Council (NRC 1983) are entirely consistent with corporate environmental management in technology development. Early communication by the corporate risk manager(s)—often one responsible for a new technology in a particular business sector—with those involved in assessing potential ecological risks is essential. These interactions help to ensure that the resulting risk characterization is relevant to the manager's needs, timely and cost-effective. The corporate risk manager needs to convey the intended use of the technology, what decisions are pending that may be relevant to the assessment, specific requirements imposed by any potential regulatory submissions, the resources available and deemed appropriate for the assessment, and critical timelines contingent upon other aspects of the technology development.

Often the process of assessing environmental risks of a new technology serves to identify additional questions or concerns which may modify the scope of the assessment and require reformulation of the original problem. An iterative process within which the risk assessment proceeds through successive tiers of testing is most effective when the risk manager is involved in the risk characterization of each iteration. This provides the risk manager with the opportunity to input into the design and direction of subsequent analyses. In cases where environmental goals and criteria are not clearly defined, however, there is the possibility that an iterative risk assessment process may continue indefinitely and exhaust valuable time and resources. For this reason, decision points and criteria should be clarified to the fullest extent possible, and resources for the assessment should be commensurate with the benefits of the technology and the magnitude of potential risks involved.

The decision criteria which a corporate risk manager uses to determine whether or not a chemical technology requires management to preclude or minimize risks depends on several factors. Major considerations are those criteria which will be used by regulatory authorities. Quantitative guidelines for the use of assessment factors to ensure an 'acceptable' margin of safety have been published by a number of authorities, including the European Union (EU Commission Technical Guidance Documents) and the USEPA (Smrchek *et al.* 1995). The corporate risk manager must take into account whether and how regulatory authorities will apply such criteria in reviewing a chemical registration. Corporations may develop or refine their own criteria for acceptability (Cowan *et al.* 1995a), but by necessity these must largely be consistent with regulatory norms. Often, the concerns of society which may result in public relations concerns may play a major role in defining the acceptability of a new chemical technology or consumer product. These are captured in the fourth element of the environmental management framework (Table 14.1).

Hard-and-fast rules for distinguishing what is safe and unsafe tend to be case-specific, and numerical estimates of 'safe' risk quotients should be interpreted as guidelines only. Sound risk management decisions require 'best professional judgement' in which the risk manager can consider by a 'weight of evidence' approach any circumstances and variables which may modify (either increase or decrease) risk.

More often than is commonly appreciated, corporate risk assessments do demonstrate that some prospective chemical technologies or applications carry unacceptable environmental risks or risks that need to be managed in some way. Relatively few of these cases can be publicly communicated, which has, to some degree, limited corporate credibility in using risk assessment in the eyes of the public. There are several practical obstacles to corporate communications of potential risks. First, technologies which corporations choose not to pursue due to environmental safety concerns remain purely hypothetical and highly conjectural. Assessments of 'risks that never were' have little practical interest to science editors and the scientific community. Second, the rapid pace of technology development can often resolve safety concerns by improvements in chemistry, process design, formulation or waste treatment. Encouraging a public perception of a particular technology as inherently unsafe may render that technology practically unusable regardless of any improvements in its safety profile. Third, risk perceptions can easily be transferred from one technology to another, whether or not the basis of the risk is the same. Negative perceptions of a particular chemical technology may incriminate an entire industry or market, even if the technology has limited application. Finally, there is the practical consideration that risk assessors and managers in large corporations often are involved in a number of concurrent technology initiatives. Corporations value such individuals primarily for their contributions to building the business rather than an impressive publication record. Technical publications are extremely time-intensive. Once an internal safety question is addressed, for better or worse, the impetus is to move quickly to the next project.

What options are available to the corporate risk manager faced with an unacceptable environmental risk? Often there are many. The corporate risk manager usually has a wider and more flexible array of risk mitigation options available than his or her regulatory counterpart. Risk mitigation options have been characterized in detail for some industries, such as the agrochemical industry where environmental risks are highly scrutinized (SETAC 1994). In general, these may include limiting the use of the technology in some way, such as: reduced usage in the formula composition; narrowing the technical specifications of commodity chemicals to address particular environmental concerns; limiting production output; or restricting application rates or the geographical range of the market. Consumer education of preferred use and disposal practices may be a vital aspect of managing post-consumer environmental risks. In some cases, corporations can gradually stage product introductions. Beginning with smaller markets or applications, they can monitor actual environmental fate and exposure concentrations to reduce uncertainties in the risk estimates, then increase applications as the refined risk characterization allows. Opting not to use a particular technology is another option of the corporate risk manager. In this event, it is necessary for the risk manager to clearly spell out concerns at an early stage of technology development in order to minimize unnecessary costs and resources usage.

Risk–benefit analysis, or cost–benefit analysis, is an important tool that the risk manager must use in sorting through risk reduction alternatives to arrive at a decision. According to EU legislation (Article 10 of the Council Regulation EEC 793/93), risk–benefit analysis of a compound has to be conducted whenever the results of the risk assessment clearly indicate the need for risk reduction measures. Risk–benefit analysis presents different options in comparable terms and facilitates decision-making. The analytical methodology is not based on a standardized procedure, however, and the results are highly influenced by how costs and benefits of a technology are valued. Clearly, ecological costs and benefits are difficult to quantify monetarily. This means that alternative valuation 'currencies' or considerations must be developed and incorporated in the analysis.

14.4.4 A regulatory climate supporting technology innovation

Technological innovation was previously identified as necessary to improve environmental quality and to deliver business success. Business development and hence economic growth are necessary to deliver the financial resources that are in turn invested in scientific developments and technology to deliver environmental improvements. This cycle was identified by the World Commission on Environment and Development (WCED) as an essential component toward implementing sustainable development (WCED 1987).

Many standards today—chiefly those based solely on hazard assessment, inaccurate or misinterpreted risk assessments, or inconsistent environmental objectives—are incompatible with economic growth and are stifling to technological innovation. Standard-setting should be seen as one part of the overall process of environmental management, and standards should only be established where there is consensus that they are the best instrument for achieving an environmental objective. Standards based on a sound, scientific and risk-based approach can lead to meaningful and cost-effective environmental improvement.

The difficulties with some existing standards are usually found in two particular areas: (i) the difficulty in defining acceptable, absolute risk levels, leading to hazard-based (as opposed to risk-based) standards with restrictive safety factors; and (ii) the definition of rigid test methods or procedures which can preclude innovation. There is a need for regulators and regulations to be flexible to account for developments in test methods, scientific understanding and technologies. Regulators should be encouraged as well to be flexible in their use of detailed methods and procedures. Most importantly, specific test protocols and methods should not be prescribed in long-term legal standards (e.g. legislation).

In order to set sound scientifically based standards, valid scientific data and quality control mechanisms are needed. This implies the use and application of EU Chemicals Legislation Risk Assessments of new and existing substances as found in Article 3(2) of Directive 92/32/EEC (7th Amendment of Directive 67/548/EEC), Article 10(4) of Existing Substances Regulation (Regulation EEC/793/93), the US Toxic Substances Control Act and the Canadian Environmental Protection Act. The HEDSET (Harmonized Electronic Data Sets) in the IUCLID data base authorized by the EU legislation is emerging as a primary source of (eco) toxicological data in Europe. Similar environmental data inventories exist in other geographies. These too require mechanisms for periodic updating as science and technology advance.

The key to effective and sound scientific risk assessment and management is the participation of all stakeholders in the selection and peer review of data used in standard setting. Scientists from academia, commerce, public interest groups and government should be involved in the standard-setting process. Their role should be as advisers to the regulator who maintains responsibility for the final choice and implementation of the standard. Multistakeholder involvement and peer review of proposed standards by not only the expert scientific community but also a representative cross-section of the public is required to set risk management goals.

14.4.5 Environmental risk management complementing the precautionary principle

The precautionary principle has recently emerged as a guiding principle in international environmental law. As embodied in the European Union's Maastricht Treaty and the United Nations Declaration from the Rio Conference, the principle urges that responsible institutions use precaution in technology applications (i.e. 'where there are threats of serious or irreversible damage, lack of full scientific certainty shall not be used as a reason for postponing cost-effective measures to prevent environmental degradation'—Rio Declaration Principle 15). It is not an enforceable standard *per se*, but some proponents have interpreted it as a universal requirement that takes precedence over existing environmental laws.

The basis upon which the precautionary principle should be invoked has not been clearly defined (but cf. Chapter 1), and there is a wide range of possible interpretations. Clearly there is a need for clarity on interpretation. The International Council of Chemical Associations (ICCA) and the Chemical Manufacturers' Association (CMA 1994) advocate that a science-based, precautionary approach to risk management be taken to prevent health or environmental harm when the weight of plausible scientific evidence establishes that serious or irreversible damage to health or the environment is likely to be caused by the activity or product in question. Most corporations, including our own, would readily agree that cost-effective, risk-based environmental management is entirely consistent with this interpretation. The ICCA, through the CMA's Responsible Care initiative (see Table 14.2), encourages corporations to work toward characterizing products with respect to their hazards and their risks and, in concert with their customers, take appropriate risk reduction actions.

Other groups have adopted more extreme interpretations of the principle. It has been argued that, in the absence of specific data, society must always presume the worst when evaluating environmental consequences of technology. Under this extreme interpretation, a chemical should be presumed to cause significant adverse environmental effects unforeseen at the time of usage. Thus, the use or discharge of a chemical may be deemed unacceptable at any level based upon its intrinsic properties or even upon uncertainties in the absence of data. Such an interpretation would require the outright banning of products and processes based on a potential to cause harm, regardless of whether realistic environmental exposure scenarios exist.

We believe that extreme interpretations of the precautionary principle can impede technological innovation. Innovative technologies can be an essential element in resolving serious environmental concerns as well as improving the quality of life for consumers. Implementing extreme interpretations of the principle could do the following: (i) place a 'reverse onus' on companies to demonstrate the absence of adverse effects as a condition for use in commerce; (ii) mandate restrictions based on uncertainty, whether or not

Table 14.2 Guiding principles of the CMA's Responsible Care® initiative. These statements of the philosophy of Responsible Care® outline each member company's commitment to environmental, health and safety responsibility in managing chemicals. Member companies of the CMA pledge to manage their business according to the following principles.

To recognize and respond to community concerns about chemicals and our operations
To develop and produce chemicals that can be manufactured, transported, used and disposed of safely
To make health, safety and environmental considerations a priority in our planning for all existing and new products and processes
To report promptly to officials, employees, customers and the public, information on chemical-related health or environmental hazards and to recommend protective measures
To counsel customers on the safe use, transportation and disposal of chemical products
To operate our plants and facilities in a manner that protects the environment and the health and safety of our employees and the public
To extend knowledge by conducting or supporting research on the health, safety and environmental effects of our products, processes and waste materials
To work with others to resolve problems created by past handling and disposal of hazardous substances
To participate with government and others in creating responsible laws, regulations and standards to safeguard the community, workplace and environment
To promote the principles and practices of Responsible Care® by sharing experiences and offering assistance to others who produce, handle, use, transport or dispose of chemicals

evidence of harm exists; and (iii) invoke unrealistic worst-case assumptions of effects and exposure variables in risk assessments.

How, then, is the precautionary principle factored into corporate risk management? There are two major ways. First, in conducting environmental risk assessment, precaution is needed in estimating the magnitude of the risk from the relationship between PECs and PNECs. Uncertainty and variability in data sets, modelling approaches, species and community sensitivity to stress, and the moderating or exacerbating effects of environmental variables requires that a precautionary safety margin be applied to any risk estimate or management decision based on that estimate.

Second, in evaluating cause and effect relationships, there is a need for precaution. On the one hand, precaution is needed not to jump to conclusions. This argues for greater emphasis on improving our mechanistic understanding of the effects of chemical stressors on ecological systems, providing better understanding of the mechanisms by which a hypothesized cause could lead to an effect. On the other hand, a severe or irreversible effect must be dealt with even if the cause is not firmly established. Here, there are useful legal parallels: in most situations preventive action should be taken when matters 'are beyond reasonable doubt'. For a severe effect (i.e. one with clearly unacceptable consequences), action should be taken when the weight of the evidence is clear. Even in such a case it is important to take the cost of action into account. Depending on the specific situation, it may be appropriate to implement responsible actions to ensure 'no regrets' outcome.

14.4.6 Environmental management using cost–benefit analysis

A risk-based approach to management raises an issue as to the criteria that should be used to determine which steps to take to reduce risk when the analysis indicates that there is a risk that should be managed. As discussed in the previous section, in a hazard-based approach, it is possible to invoke the precautionary principle in its simplistic form and argue that when there is any uncertainty regarding safety, the material or action under evaluation should be avoided. This approach discourages the inquiry into which specific actions result in risk, since management of that risk is precluded. Likewise, this approach discourages the evaluation of any alternative or action which may have risks associated with it that may have not yet been uncovered. It is worth noting that this simplistic approach to the precautionary principle takes it out of the context of the German legal system within which it was developed. Within the German legal system it is necessary to apply simultaneously a number of principles, of which the precautionary principle is only one. Other principles that must be considered include size of action (i.e. including cost) and regulatory proportionality (RCEP 1988). This broader approach to precaution within the context of other principles is also included in the Maastricht Treaty for the European Union.

In both European and USA regulatory procedures, there are requirements to consider both risks and benefits when deciding on controls of hazardous materials (European Commission 1993; US Library of Congress 1994). These risk–benefit approaches are based on cost–benefit methodology taken from welfare economics (also known as micro-economics or Pareto economics), a branch of economics derived from the Principle of Optimality developed by Pareto (1848–1923) (Beardshaw 1992). The principle seeks to provide guidance on how to identify opportunities to gain higher utility out of existing resources. Pareto Optimality states that the optimum allocation of resources results when any changes from this allocation would be detrimental to some party. This optimality can be achieved if, in all markets, there is perfect competition, no externalities and no market failure connected with uncertainty. Of course, such conditions never arise in the real world, but welfare economists argue that only by knowing the ideal conditions for optimality is it possible to recognize deviations and seek to correct or control them. A report by the Royal Society for the Prevention of Accidents (RSPA 1992) states: 'The optimal level of safety will be when risks have been reduced up to the point where the extra cost of any extra reduction just equals its benefits'. The challenge is to determine the balance of risks and benefits.

Mathematically, this balancing requires the use

of common units; money is generally that unit. There is considerable debate over the appropriateness and feasibility of such approaches, especially regarding the monetary evaluation of aesthetic or ecological resources (Chapter 13). Thus, there is heated debate regarding the monetary value of risks and benefits and the lack of reality assumed in the basic approach (Adams 1995). Attempting to deal with the factors involved in such debates while carrying out a risk–benefit analysis can lead to immense complexity and, hence, cost (US Library of Congress 1994).

A recent document (DOE 1995b) has stressed that there is a spectrum of possible approaches to be used in balancing the advantages and drawbacks of substances. Three particular options and their appropriate use are as follows.

Qualitative risk–benefit analysis. To be used when the risk characterization data are limited and/or the risk is obviously high. Such an analysis will provide a descriptive, but systematic, listing of risks and benefits to check that the proposed control measures do not lead to any enormous disadvantage.

Quantitative risk–benefit analysis. To be used when the risk characterization data are reasonable in quantity and quality and the level of risk is at the margin (for materials, the quotient of the 'predicted no-effect concentration' and the 'predicted environmental concentration' is close to the safety factor appropriate for the particular tier of testing (Cowan *et al.* 1995a). In this case the cost of reducing the risk to humans and the environment should be balanced against the cost of reduced consumer and producer benefits.

Cost–benefit analysis. To be used when judgements are both extremely fine and of great economic importance. Such a study will need to define more precisely the probabilities of risk and appropriately value these. Complexities such as differing discounting techniques will need to be considered, as well as sensitivities, in the analysis.

It cannot be emphasized too strongly that such studies do not provide *the* answer. They seek to provide objective and transparent guidance to the decision maker. Such objectivity and transparency is vitally important in the public policy arena where equal treatment of disparate cases is so important and where precedent and judicial review play such an important role. In corporations, decision-making can be (and usually is) rather more pragmatic. Nevertheless, there are increasing pressures on corporations to justify their management decisions to the public (P.R. White *et al.* 1995; Hindle *et al.* 1996).

14.5 CONCLUSIONS

We have described how corporations regularly apply environmental risk assessment and management to integrate business needs and environmental considerations in an everyday decision-making framework. Innovative companies frequently conduct environmental risk assessments for product development and processing initiatives, even when not compelled to do so by law. In the risk management process, environmental, human safety and regulatory criteria must be integrated with an array of commercial criteria for decision-making.

Environmental management is recognized as a shared responsibility between the corporation and its many external partners — suppliers, customers, consumers and the government. Coordination is needed to ensure that proposed solutions to one environmental problem do not create or exacerbate other problems elsewhere. Partnerships in environmental research among corporations and between corporations and public institutions can promote cost-effective and innovative risk management solutions.

Business realities shape the foundation of corporate environmental management, and several have been presented: (i) only products that offer competitive performance and value as well as an improved environmental profile survive in the marketplace to deliver meaningful environmental benefits; (ii) industrial and commercial operations inevitably have an impact on the environment through resource use, energy consumption and waste generation (minimizing such impacts in a manner which allows economic growth must be the goal of sound corporate environmental management); (iii) accurate information and data-based decisions are fundamental to sound management practices; and (iv) technological innovation is essential to environmental quality improvement, as well as business success.

A hierarchical, needs-based environmental management framework has been described which encompasses all corporate environmental management responsibilities. Four fundamental needs are: (i) human and environmental safety; (ii) regulatory compliance; (iii) efficient resource use and waste management; and (iv) consideration of society's concerns (Table 14.1). Each of these is supported by a range of environmental management tools and is consistent with sound business practices.

Predictive and efficient screening of chemicals for ecological fate and effects is indispensable early in the technology development process. Early recognition of environmental risks and potential needs for mitigation are most cost-effective and resolvable before major commercial decisions are made and invested. Indiscriminate testing of all candidate substances early in technology development is neither economically feasible or necessary to ensure safety.

An iterative approach to environmental risk assessment has been described which enables close communication and coordination between the corporate risk manager(s) and risk assessor(s). The environmental risk characterization should present a coherent and comprehensive understanding of the environmental risk(s) to the risk manager, who must integrate this information with other social and economical considerations in making a decision. Effective risk management was linked to both a strong commitment by the highest levels of management and a clear perception of the corporation's environmental quality objectives by the risk manager—objectives which evolve with societal needs and expectations for environmental quality.

It was recognized that few corporate decisions present clear, dichotomous environmental alternatives. Rather, difficult choices among options superior in some environmental or economic dimensions but inferior in others is the norm. Consensus must be reached by all entities within a corporation that a given decision reflects the best balance of perceived opportunities and liabilities to the business, as well as of corporate and societal expectations for environmental quality. Environmental management and business management operating in tandem can encourage innovation and help to identify 'win–win' solutions for all stakeholders. Stakeholders from all sectors should be involved in the debate to establish and continually refine environmental quality standards.

Flexibility is needed to account for circumstances and variables which influence the magnitude and nature of an environmental risk. Sound risk management decisions require best professional judgement and a weight of evidence approach. The corporate risk manager has a wide array of risk mitigation options available when product and process development organizations collaborate. Overly rigid prescription of what options a corporation may choose to achieve environmental improvement can inhibit innovation and cost-effective solutions.

Finally, the precautionary principle has been endorsed as being complementary with a science-based corporate approach to risk management. Cost-effective, risk-based environmental management is entirely consistent with the precautionary principle. Two ways in which precaution is factored into corporate risk management are in interpreting the uncertainties of the risk characterization and in evaluating environmental cause and effect relationships. Responsible corporate management of environmental risks requires cost-effective mitigation or avoidance when the weight of the evidence is clear. A number of options and considerations for evaluating costs (risks) and benefits have been discussed.

14.6 REFERENCES

Adams, J. (1995) *Risk*. UCL Press, London.

AIHC (1995a) AIHC Perspective on ecological risk assessment framework and guidelines. An invited presentation by R.A. Kimerle at *Colloquium on Developing an EPA Ecological Risk Assessment Guideline*, Old Town Holiday Inn, Alexandria, Virginia, May 3, 1995. Sponsored by the Risk Assessment Forum, Office of Research and Development, US Environmental Protection Agency, Washington, DC. Available from American Industrial Health Council, Suite 760, 2001 Pennsylvania Ave., N.W., Washington, DC, 20006.

AIHC (1995b) *Advances in Risk Characterization*, 11 pp. American Industrial Health Council, Washington, DC.

AIHC (1996) *Ecological Risk Assessment: Sound Science*

Makes Good Business Sense, 13 pp. Ecological Risk Assessment Committee, American Industrial Health Council, Washington, DC.

ARET Secretariat (1995) *Environmental Leader 1. Voluntary commitment to toxics through ARET*. Accelerated Reduction/Elimination of Toxics Programme, Environment Canada, Hull, Quebec.

Beardshaw, J. (1992) *Economics, a Student's Guide*, 3 edn. Pitman Press, London.

Black, J.A., Barnum, J.B. & Birge, W.J. (1993) An integrated assessment of the biological effects of boron on the rainbow trout. *Chemosphere*, **26**(7), 1383–1413.

CMA 1994. *Responsible Care Overview Brochure*. The Chemical Manufacturers Association, Arlington, VA.

Cowan, C.E., Versteeg, D.J., Larson, R.J. & Kloepper-Sams, P.J. (1995a) Integrated approach for environmental assessment of new and existing substances. *Regulatory Toxicology and Pharmacology*, **21**, 3–31.

Cowan, C.E., Mackay, D., Feijtel, T.C.J. *et al.* (eds) (1995b) *The Multi-Media Fate Model: a Vital Tool for Predicting the Fate of Chemicals*, 78 pp. SETAC Press, Pensacola, FL.

DOE (1995a) *A Guide to Risk Assessment and Risk Management for Environmental Protection*, Department of the Environment, 92 pp. HMSO, London. SW8 5DT.

DOE (1995b) *Risk–Benefit Analysis of Existing Substances. A Guidance Document Produced by a UK Government/Industry Working Group*. Department of the Environment, HMSO, London.

Dutch Ministry for Environment (1992) Letter to Parliament December 2. Ministry of the Environment, den Haag, The Netherlands.

Dutch Ministry for Environment (1995) Letter to Parliament October 30. Ministry of the Environment, den Haag, The Netherlands.

ECETOC (1993) *Aquatic Toxicity Data Evaluation* (ed. D.A. Stringer). Technical Report No. 56, European Centre for Ecotoxicology and Toxicology of Chemicals, Brussels.

ECETOC (1995) *GREAT-ER—Geography-referenced Regional Exposure Assessment Tool for European Rivers*. ERASM Project 27 November 1995, European Centre for Ecotoxicology and Toxicology of Chemicals, Brussels.

European Commission (1993) Council Regulation (EEC) No. 793/93 of March 23, 1993 on the evaluation and control of risks of existing substances. Official Journal of the European Communities, European Commission, Brussels.

European Commission (1994) *Risk Assessment of Existing Substances*. Technical Guidance Documents (XI/919/94-EN) in support of the Commission Regulation (EC) No. 1488/94 on Risk Assessment for Existing Substances in accordance with Council Regulation (EEC) No. 793/93, European Commission, Directorate-Generale XI, Brussels.

European Commission (1995) *Environmental Risk Assessment of New and Existing Substances*. Draft Technical Guidance Document in accordance with Council Regulation (EEC) No. 793/93, European Commission, Directorate-Generale XI, Brussels.

Fava, J.A., Adams, W.J., Larson, R.J., Dickson, G.W., Dickson, K.L., Bishop, W.E. (eds) (1987) *Research Priorities in Environmental Risk Assessment*, Society of Environmental Toxicology and Chemistry, Pensacola, FL.

Feijtel, T.C.J. & van de Plassche, E.J. (1995) *Environmental Risk Characterization of 4 Major Surfactants Used in The Netherlands*, 81 pp. National Institute of Public Health and Environmental Protection, Report No. 679101 025, Bilthoven.

Gledhill, W.E. & Feijtel, T.C.J. (1992) Environmental properties and safety assessment of organic phosphonates used for detergent and water treatment applications. In: *The Handbook for Environmental Chemistry*, (eds O. Hutzinger & N.T. de Oude), Vol. 3E, pp. 261–285. Springer-Verlag, Heidelberg.

Habicht, F.H. (1992) *Guidance on risk characterization for risk managers and risk assessors*. Memorandum from F.H. Habicht, USEPA Deputy Administrator, US Environmental Protection Agency, Office of Research and Development, Washington, DC.

Hindle, P., White, P.R. & Minion, K. (1993) Achieving real environmental improvements using value : impact assessment. *Long Range Planning*, **26**, 36–48.

Hindle, P., De Smet, B., White, P.R., Owens, J.W. (1996) Managing the Environmental Aspects of a Business: A Framework of Available Tools. The Geneva Papers on Risk and Insurance, 21 (No 80, July 1996), 341–359.

Larson, R.J. & Woltering, D.M. (1993) A risk assessment case study for linear Alkylbenzene sulfonate (LAS). In: *Fundamentals of Aquatic Toxicology* (ed. G.M. Rand), 2 edn, pp. 859–882. Taylor & Francis, London.

NRC (1983) *Risk Assessment in the Federal Government: Managing the Process*. National Research Council, National Academy Press, Washington, DC.

NRC (1994) *Science and Judgment in Risk Assessment*. National Research Council, National Academy Press, Washington, DC.

OECD (1995) *Report of the OECD Workshop on Environmental Hazard/Risk Assessment*. Environment Monographs No. 105, Organization for Economic Cooperation and Development, Environment Directorate, Paris.

RCEP (1988) *Twelfth Report. Best Practicable Environmental Option*. Royal Commission on Environmental Pollution, HMSO, London.

RSPA (1992) *Risk: Analysis, Perception and Management*. Royal Society for the Prevention of Accidents, London.

SETAC (1994) *Aquatic Dialogue Group: Pesticide Risk*

Assessment & Mitigation. Society of Environmental Toxicology and Chemistry and the SETAC Foundation for Environmental Education, SETAC Press, Pensacola, FL.

Smrchek, J., Zeeman, M. & Clements, R. (1995) Ecotoxicology and the assessment of chemicals at the U.S. EPA's Office of Pollution Prevention and Toxics: current activities and future needs. In: *Making Environment Science* (eds J.R. Pratt, N. Bowers & J.R. Stauffer), 271 pp. Ecoprint, Portland, OR.

Suter, G.W. (1993) *Ecological Risk Assessment*, 538 pp. Lewis Publishers, Chelsea, MI.

USEPA (1993) *Environmental Technology Partner-ships: Opportunities for Commercialization*. US Environmental Protection Agency. EPA/600/F-93/004, Office of Research and Development, Washington, DC.

USEPA (1992) *Framework for Ecological Risk Assessment*, US Environmental Protection Agency, EPA/630/R/001, Risk Assessment Forum, Office of Research and Development, Washington, DC.

USEPA (1979) Toxic Substances Control Act. Discussion of premanufacture testing policies and technical issues; request for comment. *Federal Register* 44, 16240–16292.

US Library of Congress (1994) *Risk Analysis and Cost–Benefit Analysis of Environmental Regulations*, 55 pp. Congressional Record Service, CRS Report 94–961 ENR, Washington, DC.

WCED (1987) *Our Common Future*. Report of the World Commission of Environment and Development (the 'Brundtland Commission'). Oxford University Press, Oxford.

White, P.R., De Smet, B., Owens J.W. & Hindle, P. (1995) Environmental management in an international consumer goods company. *Resources, Conservation and Recycling*, **14**, 171–184.

White, P.W., Franke, M., Hindle, P. (eds) (1995) *Integrated Solid Waste Management: a Lifecycle Inventory*, 362 pp. Blackie Academic & Professional, Glasgow.

Zeeman, M. (1995) EPA's framework for ecological effects assessment. In: *Screening and Testing Chemicals in Commerce*. (OTA-BP-ENV-166), 69–78, US Congress, Office of Technology Assessment, Washington, DC.

Chapter 15
Environmental Risk Assessment in Business

MICHAEL QUINT

15.1 INTRODUCTION

This chapter reviews the use of environmental risk assessment in business. It focuses on the ways in which risk assessment is being applied by environmental regulators, the regulated community and financial institutions across a broad range of industry sectors. The chapter is primarily concerned with chemical hazards, as opposed to others such as radiation, explosions or physical dangers, since this reflects the predominant use of risk assessment in business.

The intent of this chapter is to provide a review of how risk assessment is being used within the daily running of industry, as part of 'environmental management systems', rather than an appraisal of the techniques involved. This is achieved partly by descriptions of current practice and partly by reference to case studies. The theoretical aspects of environmental risk assessment are covered in other sections of this book; it is the purpose here to look at how they are being applied in what is perhaps the ultimate test of their usefulness.

It is important to highlight a dichotomy with regard to the present understanding of environmental risk assessment in the business and non-business communities. Currently, the bulk of the business community, especially financial institutions, regard environmental risk assessment as an assessment of financial (and legal) liabilities arising from environmental issues (e.g. Simmons & Simmons 1995). This contrasts with the view from regulators and scientists, who generally regard environmental risk assessment as an evaluation of whether there is a potential for adverse environmental effects to occur, based on factual knowledge, scientific theories, assumptions and predetermined acceptability criteria. While the latter view is more relevant in the context of this book, and will be used as a working definition in this chapter, an understanding of the former is critical to addressing the chapter's title. In fact, the chapter aims to look at both kinds of assessment, focusing on how they interact currently and how they may do so in the future.

There are several references within this chapter to the current situation in the UK, although the basic principles that it highlights apply in many other countries. The UK's current drive towards risk-based environmental regulation has spawned a number of initiatives and debates, and consequently provides a fertile source of ideas relating to the applicability of environmental risk assessment to business (e.g. POST 1996).

15.2 HISTORICAL BACKGROUND

Environmental risk assessment has been an essential part of environmental regulation in the USA for over a decade (Chapters 3 and 10). Its first wide-scale application outside of chemical products or food and drug regulation (where it was initially used) was in the assessment of contaminated land under the Comprehensive Environmental Response, Compensation and Liability Act of 1980 (CERCLA), or 'Superfund' (USEPA 1986). This brought about the rapid development and dissipation of risk assessment techniques, which were then applied in situations such as the permitting of waste management facilities (especially landfills and incinerators) and the regulation of atmospheric emissions from industrial plants. In fact, environmental risk assessment quickly became the method of choice for implementing policies connected with the 'protection of public health and the environment' from chemical pollution, such that it now forms a central role in regulatory decision-making in the USA (US CRARM 1996).

In contrast to the USA, most countries in Europe have only recently started applying risk assessment to environmental issues (although The Netherlands has been doing so in connection with contaminated land policy since the late 1980s). Food and drug safety has again been a source of technical expertise, as have, more recently, chemical product registration and labelling. Areas such as contaminated land and industrial emissions are currently the subject of primary legislation in many countries and this will undoubtedly bring about the application and importance of risk assessment techniques in these areas. Associated Europe-wide initiatives are underway, including the European Commission-funded 'Concerted Action on the Risk Assessment of Contaminated Sites' programme, the primary objective of which is the elaboration of recommendations and guidelines for assessing the risk of contaminated land (CARACAS 1996). Similar activities are also being sponsored by industry, including the 'Network for Industrially Contaminated Land in Europe' (NICOLE 1996) and the 'risk-based corrective action' (RBCA) guideline for contaminated European oil industry sites (CONCAWE 1996).

Within Europe, the UK is currently implementing a great deal of environmental legislation that will require the use of environmental risk assessment (Chapter 9). Although explicit official acknowledgement of the techniques and usefulness of environmental risk assessment has occurred only in the last few years, its application across a range of government activities appears to have quickly gathered momentum (DOE 1995a; HSE 1995; ILGRA 1996; POST 1996). The importance of environmental risk assessment to UK environmental regulation has been further enhanced by the wording and intent of both primary legislation (e.g. the Environment Act 1995) and associated statutory and non-statutory guidance (HMIP 1995; DOE 1996).

15.3 ENVIRONMENTAL RISK ASSESSMENT AND THE FINANCIAL LIABILITY

15.3.1 Introduction

Environmental risk assessment in relation to financial liabilities is becoming increasingly important to funding institutions (banks, venture capitalists, pension funds, etc.). This is partly due to the huge insurance claims that have arisen from pollution (and the resulting exclusions on cover imposed by insurers) and partly from a growing tendency among funders to steer clear of ventures that have the potential to come into conflict with regulators and, perhaps more importantly, public opinion (Chapters 19 and 20).

The relationship of environmental risk to the assessment of financial liabilities has been complicated by issues of terminology. It has been further affected by a paradox whereby many environmental risks that are 'significant' from a scientific standpoint are of limited relevance commercially, while others that are scientifically 'insignificant' can have a great deal of commercial relevance. For example, acid mine drainage from abandoned mines is a chronic and pernicious form of chemical pollution in the UK, whereas, according to Section 89 of the Water Resources Act 1991, in certain circumstances there is no liability for it (although this will shortly change under the 1995 Environment Act). Conversely, the planned dumping of *Brent Spar* in the North Atlantic was arguably without environmental risk from a scientific viewpoint yet, due to public opinion, it brought about financial implications for Shell of the highest order (Anon. 1995).

A possible cause of the above paradox is that legislation and its enforcement often lag behind scientific knowledge. In addition, the balance of costs and benefits recognized at the time when legislation is drafted may change with public opinion or scientific fact. Clearly, for environmental risk assessment to be fully effective in assessing financial liabilities, both legislation and public perceptions of risk must match scientific 'fact'.

15.3.2 Environmental audit

An increasingly important tool in the assessment and management of financial liabilities arising from environmental issues is the 'environmental audit'. The use of the term 'environmental audit' has received some criticism from other audit practitioners, given the methodological rigour applied in other fields. However, this rigour is being applied in the environmental area, particularly with reference to Environmental Management System (EMS) standards or specifications, including: the British Standard specification on Environmental Management Systems (BS 7750); the International Standard specification (ISO 14001); and the Eco-management and Audit Scheme Regulation (EMAS).

There are several types of audit, although the most important are 'systems audits', 'compliance audits', and 'due diligence audits': systems audits are designed primarily to test the effectiveness of system controls (generally within a standard EMS or for the purposes of a specification); compliance audits may be undertaken on an annual basis and may involve the checking of current practices against either legal requirements or corporate/adopted 'best practice'; and due diligence audits may be undertaken on a one-off basis, as part of a merger or acquisition, for example, in order to help assess possible financial liabilities arising from: (i) infringement of current or foreseeable environmental regulations; (ii) presence of land or water contamination that could give rise to a need for remediation; and (iii) poor environmental image, leading to public concern and regulatory pressure for ever-increasing levels of environmental control.

While the principles of risk assessment are of use to an environmental audit, they form only a part of a much wider appreciation of the legal and other pressures under which a business operates. As indicated above, these are as much about public/regulatory perceptions of risk as they are about the scientific realities of actual or potential environmental damage.

In many cases, environmental audits are used as a first-stage tool to highlight existing or potential sources of contamination or pollution that, if uncontrolled, can result in an issue for an organization, whether it is a strict liability issue or not. In terms of environmental affairs, such audits can assist in the implementation of controls in advance of the source manifesting itself within a pathway or at a receptor. The importance of this was originally recognized by the insurance industry in the USA, where the concept of environmental auditing arose. Consequently, although such audits do not seek to be quantitative, they can be used to identify or exclude potential or actual sources, pathways for contamination escape, and receptors prior to more costly investigations. Such information can also be useful in future activities including formal quantitative environmental risk assessment.

The distinction between hazard and risk assessment (see Chapter 1) is especially relevant to environmental audits because they are often only able to highlight the potential hazards associated with a given business. This may be due to time constraints (e.g. imposed so that a stock market bid can take place) or pressures from others to 'provide an answer'. In both cases, an auditor is likely to err on the side of caution in the face of uncertainty, in order to protect his/her professional indemnity. This can have the undesirable result of overemphasizing environmental liabilities (by focusing on hazards rather than risks) and, consequently, the unnecessary prevention or suppression of market forces in certain business sectors. The undoubted value of 'due diligence' assessments has, however, been recognized by vendors and purchasers of property in applying market measures by identifying potential costs that are of material concern within the deal, allowing suitable precautions to be taken in contracts.

A consequence of the tendency of environmental audits to focus on environmental hazards rather than risks is that it is sometimes the case that an industry that presents minimal financial risk can appear to be a worse investment than one that presents a higher financial risk. For example, waste incineration is well-regulated to the extent that risk (both technical and scientific) is generally minimized to extremely low levels (Chapter 16), although it is perceived to be highly dangerous. Scrap-metal dealers, on the other hand, have historically operated in a much less regulated manner and can present arguably greater threats to the environment and possibly large financial liabilities due to uncontrolled chemical spillages. A standard auditing protocol is likely to steer potential investors away from the former, since the hazards of hazardous waste incineration are perceived to be much greater than those of scrap-metal dealing, although the risks they actually pose may be much less. While this may make sense from a commercial perspective, where public perceptions are important (Chapter 20) it acts to compound the problem identified above of perceived risk not following 'real' risk.

The EMS specifications outlined above can be broadly divided into two types: those relying on system-based improvements (ISO 14001) and those requiring performance-based improvements (BS 7750 and EMAS). All rely on some means to: (i) identify current and potential 'aspects' or 'effects' (called issues generically below); (ii) rank or classify issues; and (iii) establish means to demonstrate controls relating to such issues. System-based improvements are internal improvements of the elements of control, whereas performance-based improvements require reduction of adverse effects (or impacts) or improvement of positive effects (or impacts).

The area where broader concepts of risk have been applied has included the classification of issues. The BS 7750 suggests that 'risk assessments or other techniques may be used in order to compare the effects identified'. This often takes the form of a strengths–weaknesses–opportunities–threats (SWOT) analysis, which at best can be described as qualitative. The application of risk methodology tends to be simplistic and may give rise to substantial subjectivity. However, the basic assumption that a risk is the product of the probability and consequences of an undesired event can be applied. Thus, for example, the likelihood of a catastrophic fuel tank rupture and subsequent escape to a river can be at least broadly defined for the purposes of a basic classification of multiple issues requiring further action, improvement or control. Quantitative assessment may be applied subsequently (see also Chapter 4).

Similar activities are required or given as a way forward for classification of environmental issues. For example, the guidance given to accompany ISO 14001 in ISO 14004 suggests that under the planning activity, identification of environmental aspects (causes or sources) and evaluation of environmental impacts should be undertaken. It suggests a four-step process. The suggested final step includes an evaluation of significance that may be quantitative and covers: (i) the scale of impact; (ii) the severity of impact; (iii) the probability of occurrence; and (iv) the duration of impact. Again, business concerns may be included, such as the potential regulatory and legal exposure, cost of changing the impact or concerns relating to public image or interested parties.

One criticism levelled at such activities in relation to standard specifications is that the business objectives may override such assessments. This relates to the business position and viability of taking action as perceived by the organization concerned, and is, by necessity, in line with a market-orientated position.

In terms of the implementation of an EMS, risk assessment can be used to quantify environmental effects and targets as required by the guidelines. Full risk-based quantification exercises can facilitate the exclusion of potential effects that could only ever be insignificant or trivial. The result is a system tailored to the most important aspects of an organization's activities, which can lead to RBCA procedures, should these be appropriate (see above).

Lifecycle analysis (LCA) can also come under the broader heading of environmental audit (Chapters 2 and 9), since it involves the in-depth appraisal of the impact of a consumer product (e.g. washing machine) on the environment from 'cradle to grave'. This is required under the EU's voluntary ecolabelling scheme. Risk assessment can allow the quantification of impacts to soil, water, air and ecosystems in a consistent manner, allowing the aggregation of potential impacts into a limited number of criteria. Comparison can then be made between particular product types and brands, at all stages of production, distribution, use and disposal.

15.3.3 Case study

An investment house (Company A) was seeking to finance the management buy-out of a company

(Company B) that manufactured resins and plastics at a small number of sites world-wide. Environmental issues were played down by the management, but Company A insisted on a full compliance audit. This was to be carried out prior to final negotiations on the loan.

The audit found that the UK sites were old and had routinely used hazardous materials. They had a mixed regulatory compliance history and soil beneath the sites was suspected of being contaminated. Recent UK legislation concerning industrial emissions was having an increasing impact on their operations and substantial investment would be required to upgrade a number of processes. The sites were also found to have significant quantities of asbestos, some of which required removal.

Upgrade plans and associated costs were assessed, along with the likely expenditure for asbestos removal. Intrusive site investigations were carried out and risk-based approaches were used to interpret the results in terms of liabilities due to soil and groundwater remediation. The total likely cost to the business from these environmental issues was estimated and factored into the financial deal.

The study illustrates how an environmental audit performed for due diligence purposes identified several environmental liabilities associated with an industrial operation. Had the audit not taken place, the financial institution funding the operation could have been exposed to significant future costs associated with environmental assessment and remediation. These could have arisen indirectly through defaults on its loans, or directly in the event that Company B was liquidated.

15.4 REGULATORY COMPLIANCE

15.4.1 Introduction

Risk assessment is rapidly becoming the method of choice for addressing regulatory compliance issues associated with the environmental impacts of historic and operational industrial facilities. It allows compliance to be judged in an objective and scientific way and has been advocated by many governments via primary legislation and official guidance. Two of the main areas of environmental regulatory compliance in which risk assessment is used are contaminated land and atmospheric emissions. These form the focus of the remainder of this section.

15.4.2 Contaminated land

Contaminated land exists in all countries with an industrial heritage and has received regulatory attention in many due to any or all of the following.

1 A desire to correct the 'sins of the past'.
2 A need to build on disused 'polluted' urban industrial sites due to restrictions on green-field development.
3 The potential hazard to water resources.
4 A small number of high-profile cases in which polluted soil has caused actual adverse health effects.

In several countries, nationwide strategies are underway to identify contaminated land and undertake a systematic process of remediation, where required. This is leading to the widespread development and use of risk assessment techniques.

In the UK, the Environment Act 1995 will bring about an important development in the way in which contaminated land is dealt with. Specifically, Section 57 of the Act, which will be implemented in 1997, will require the identification of contaminated land by local authorities and the instigation of clean-up (if necessary), paid for by 'appropriate' parties. Contaminated land has a distinct definition in the Act, being land which:

> 'appears to the local authority in whose area it is situated to be in such a condition, by reason of substances in, on or under the land, that: (a) significant harm is being caused or there is a significant possibility of such harm being caused; or (b) pollution of controlled waters is being, or is likely to be, caused'.

Each of these terms is legally defined in the legislation.

The identification of contaminated land under the Act will take place according to the Secretary of State's guidance, which is currently being worked on. A working draft of this guidance suggests that site-specific risk assessment will play a prominent role, with the scientific appraisal of pollutant 'linkages' (source–pathway–target relationships) being emphasized in favour of the previous use of comparisons of analytical data with tables of numbers (DOE 1996). A great deal of work funded by the DOE is currently underway to develop a methodology for risk assessment and this is gradually being published in the form of non-statutory 'research reports' (DOE 1995b).

Included in the research funded by the DOE is the development of individual chemical soil *guideline values* for specific land-uses, based on advanced quantitative human health risk assessment procedures. These values are likely to be available for use as part of an initial screen for identifying contaminated sites, within an overall risk assessment framework (Ferguson & Denner 1995). Chemical concentrations exceeding these values will not necessarily trigger a need for soil clean-up/removal, since more refined assessment procedures, including the collection of additional data, may show the risks to be insignificant. This approach recognizes that worst-case assumptions (as are likely to be used in setting the *guideline values*) are sufficient to indicate when there is no problem, but should be used with care in establishing the need for soil clean-up/removal, which may cost large amounts of money to carry out. In relation to the pollution of 'controlled waters' (ground and surface waters), a methodology is being developed to allow the site-specific evaluation of chemical concentrations in soil and the degree of soil clean-up/removal required for their protection, based on a similar tiered risk assessment approach (Environment Agency 1996).

The tiered approach to risk assessment being adopted with regard to contaminated land in the UK reflects a wider move towards the cost-effective use of such techniques world-wide. These approaches recognize the need for 'balanced conservatism' in environmental risk assessment and a restriction of the level of detail required by the analysis to that required to make an appropriate risk-based decision. In other words, if a simplistic model based on worst-case assumptions shows that there is not a problem, then no amount of model refinement or site-specific data will show that there is.

The tiered approach is perhaps best illustrated by the RBCA framework that has been developed in the USA by the American Society for Testing and Materials (ASTM 1995). This framework links the principles of risk assessment with guidelines on data collection and remediation strategies at contaminated sites in a tiered manner, consistent with the above. It has received widespread support from industry, regulators and financiers and is seen as a pragmatic approach to dealing with the large number of contaminated sites that exist in the USA. The ASTM approach is forming the basis for similar initiatives being conducted elsewhere, including those by CONCAWE (1996) and a New Zealand government/industry initiative.

15.4.3 Air emissions

Air emissions from industrial activities are increasingly coming under the spotlight, especially where there is evidence of, or the potential for, nuisance, public health damage or the infringement of air quality standards and guidelines. This is being driven not only by regulatory requirements but also by a desire within certain industry sectors to be 'good neighbours'. A further driver is the possibility of lawsuits alleging damage from planned or unplanned releases.

Several countries have adopted risk assessment methods as important tools in regulating chemical emissions to the atmosphere. This is because it can help decision makers reach conclusions on the following issues.

1 The necessity for, and scope of, retro-fitting emissions control measures.

2 The potential for air quality impacts from planned industrial facilities or changes to existing operations.

3 The selection of the 'Best Available Technology', 'Best Available Technology Not Entailing Excessive Cost' (BATNEEC) or 'Best Practicable Environmental Option' (BPEO) for ameliorating air quality hazards.

4 The level of compliance with existing statutes and guidelines.

5 The possibility of environmental harm.

6 The prioritization of region-wide emission-control programmes.

The risk assessment of air emissions generally involves the following.

1 Characterization of emission sources (stacks, fugitive releases, etc.).

2 Computer simulation of atmospheric dispersion.

3 Assessment of dose–response (by reference to published guidelines or toxicological literature).

4 Prediction of exposure via direct (e.g. inhalation) and indirect (e.g. crop ingestion) pathways.

5 Characterization of risk.

Both long-term routine emissions and short-term accidental releases should be considered in the assessment. It is also necessary to undertake a sensitivity or uncertainty analysis and to explore fully the range of likely risks that could be posed by the emissions.

The risk assessment approach to dealing with air emissions has been adopted in the USA under various regulatory programmes, including the Resource Conservation and Recovery Act, which provides the national framework for the permitting of waste facilities, including incinerators, and California's Assembly Bill 2588, which regulates operational industrial facilities (USEPA 1990; CAPCOA 1992). In the UK, the Integrated Pollution Control (IPC) requirements of the Environmental Protection Act 1990 are starting to be addressed using environmental risk assessment, and a risk-based methodology is currently being developed for this purpose (HMIP 1995; Edulgee & Jackson 1996; Chapter 9).

15.4.4 Case study

As a result of a firm-wide rationalization of its distribution network, a petrochemical firm decided to close a fuel products storage facility in the commuter belt of a large European city. The facility consisted of a number of large above-ground tanks that, at one time, had held petrol, diesel and heating oil, along with associated loading/unloading apparatus (railway sidings, gantries, etc.).

Previous site investigations had found contamination of the soil and groundwater at the site, primarily by lighter-end petroleum constituents (benzene, toluene, xylene, etc.). The regulatory agency responsible for the protection of water resources had been informed and the company had agreed to address their concerns.

On economic grounds, the company clearly favoured selling the site for residential development. To do so would require 'planning permission', however, which depended on satisfying local environmental health officers of the site's safety for human habitation and satisfying the appropriate regulator that no threat was posed to water resources.

A site-specific risk assessment was carried out and risk-based clean-up levels (RBCLs) were defined in terms of chemical concentrations in soil. The latter were necessary since, although the assessment showed an insignificant risk based on the available data, these data related to exposed areas of the site, whereas a significant portion of it was occupied by storage tanks and other site infrastructure. It was considered likely that following demolition of the infrastructure, further contamination would be discovered that would need evaluation and, it was thought, treatment.

The exposure pathways considered in the risk assessment were: (i) soil ingestion; (ii) dermal contact; (iii) indoor vapour inhalation; and (iv) home-grown vegetable ingestion. Computer simulation models were used to estimate human exposure based on conservative assumptions and, where appropriate, sophisticated Monte Carlo techniques. In addition to the above pathways, groundwater modelling was performed to assess the potential for adverse impacts on water resources.

The exposure estimates were combined with numerical dose–response criteria from the World Health Organization and other bodies in order to provide quantitative risk estimates. For humans, these were in the form of increased lifetime risks of cancer and non-carcinogenic hazard indices. Where necessary, the models were run in 'reverse mode' to produce the RBCLs.

Following the demolition of the site infrastructure, 'hot spots' of contamination were remediated using the RBCLs. On the basis of the risk assessment, the environmental health officer accepted that remediation of the site to the calculated RBCLs would make it safe for residential development. Further, the groundwater modelling demonstrated that the natural attenuation (i.e. dispersion, dilution and degradation) of contaminants in the aquifer would be sufficient for the negation of any risk to water resources, to the satisfaction of the regulator. A programme of off-site monitoring of the plume would be established, however, to ensure that the model's predictions were accurate. Agreement with the owners of the adjacent property was also necessary.

The case study illustrates how environmental risk assessment can be used to allow the cost-effective redevelopment of disused contaminated industrial land for residential purposes, and hence the realization of property assets. It also demonstrates the practical application of risk assessment techniques in deciding the degree of clean-up required and the effectiveness of natural attenuation as an appropriate remedial technology in certain circumstances.

15.5 LITIGATION

While government regulation, such as that mentioned elsewhere in this chapter, goes some way to providing environmental protection, civil law also plays an important role. Civil law protects the rights of individuals (and other legal entities, such as companies) to go about their business uninterrupted by others and it includes tort (harm) and nuisance considerations.

Tort and nuisance laws have become increasingly important influences on the environmental consequences of business activities in the UK, partly as a result of the willingness of the Legal Aid Board to fund cases in these areas. The Legal Aid Board issues public funds in support of an individual's legal expenses provided that certain criteria are met, and as a result facilitates access to professional legal advice for all. Such a system is judged to be necessary in the UK since, unlike the USA, contingency fee-based work has historically not been allowed. (It should be noted, however, that the structure of Legal Aid and the ability of lawyers to act on a contingency basis is currently under review.)

Legal cases involving alleged exposure to hazardous substances increasingly feature detailed environmental risk assessment arguments and expert witnesses as an essential part of one or both parties' cases. The techniques of risk assessment can be used to help both plaintiffs and defendants in connection with civil and criminal liability cases involving claims of environmental damage from hazardous chemicals.

Of critical importance to the use of environmental risk assessment in litigation is the difference between the burden of proof required for criminal cases and that required for civil cases. In the former, the test is 'beyond reasonable doubt', while in the latter it is based on the 'balance of probabilities'. Since much of risk assessment is based on simulation modelling and prediction, it is unlikely to be effective in a criminal case because doubts will always exist with regard to the conclusions of the analysis, in the form of model uncertainty, for example. It is, however, of use in civil cases because a consensus regarding the applicability of risk assessment techniques is emerging in the scientific community that makes it suitable for showing, on balance, that 'something gave rise to something else'. A discussion of the use of risk-related techniques in the legal system of the USA is provided in Brannigan *et al.* (1992).

Typically, in a case involving alleged exposure to a hazardous substance, environmental risk assessment will be used by the plaintiff to demonstrate how, based on the available information and any assumptions necessary, the characteristics of the release were such that exposure would have been at some 'significant' level, based on known toxicological attributes of the substance. It thus forms a vital link between the source of the substance and the ill effects actually suffered, demonstrating cause and effect to the degree necessary (see above).

Conversely, environmental risk assessment can be used by the defendant to disprove that a particular release gave rise to an alleged adverse effect. In this circumstance, the risk assessment may adopt 'worst-case' assumptions and show that, in spite of these, no impact is predicted to have occurred. Clearly, the application of environmental risk assessment to legal disputes makes the use of conservative simplifying assumptions and approaches less appropriate for the plaintiff than the defendant. As a result of this and other factors (e.g. cross-examination), environmental risk assessment produced for litigation purposes is often highly sophisticated, placing a great deal of pressure on the assessors and resulting in important advances in the 'state-of-the-art'.

15.5.1 Case study

Company C had operated an industrial process for many years close to a residential part of a suburban community. The process involved the use of large amounts of volatile solvents, which evaporated inside the factory prior to being vented into the surrounding atmosphere.

Production had increased over the years, and odour complaints with it, although local environmental health officials had been reluctant to confirm the existence of a statutory nuisance. A local family attributed specific health problems to the fumes and there was general agreement in the community that the smell was unpleasant and caused headaches and minor irritation.

An application to the Legal Aid Board from a local family to pursue a private nuisance claim was accepted and a solicitor retained. As part of the evidence for the case, an environmental risk assessment was commissioned which, based on publicly available data and information supplied by the Company, showed that the solvent use was such that there was a significant risk that airborne chemical concentrations at the plaintiff's house would have been above odour thresholds and health-related guidelines for prolonged periods in the past. This information, along with witness statements regarding the nature of the odours, was sufficient to convince the court that, on the balance of probabilities, a nuisance had been caused. This resulted in the Company being obliged to pay damages.

This case study shows how environmental risk assessment can be important for legal claims under nuisance law, because it forms the framework within which scientific information can be provided to the court to link 'cause and effect'.

15.6 GOOD CORPORATE CITIZENSHIP

Many large multinational industrial organizations are adopting corporate-wide environmental policies which go beyond the requirements of the individual countries in which they operate. This is partly due to a desire to be perceived as 'good citizens' and partly as a pre-emptive measure taken in the face of large liabilities elsewhere (e.g. the USA) and a recognition that regulations in other countries may follow suit. It has further been recognized that good environmental practices can reap direct rewards, for example through cost savings due to the more efficient use of resources.

One aspect of a corporate environmental policy may be to carry out 'compliance audits' (see above) at all facilities according to a corporate environmental policy. It may also involve using risk assessment to determine remediation needs, if any, where contamination has been found, regardless of whether any environmental regulations require it or whether such regulations are in place at all (as is the case in some parts of the world). Such a consistent approach enables the review and understanding of all environmental issues and expenditure by staff at corporate headquarters. It also allows environmental liabilities to be kept under control.

15.6.1 Case study

As a multinational involved in the chemical industry in the USA, Company D had suffered at the hands of 'Superfund' and was concerned that it should not suffer similar experiences elsewhere. Furthermore, in response to the ever-increasing public scrutiny and, at times, pillorying of its activities, it had implemented a world-wide system of environmental controls that were audited yearly, on every site, by independent consultants.

A recent audit identified poor hazardous material management practices at a facility in South America resulting in the potential for significant soil and groundwater contamination. This was confirmed by exploratory sampling and analysis undertaken to Western standards and protocols.

While there were as yet no regulations concerning contaminated land and waters in the country concerned, a comprehensive programme of environmental risk assessment and monitoring was undertaken to establish the need for and scope of, any remedial action. This focused on the possibility that contaminants could move off-site and impact local wells used for domestic and agricultural purposes, along with adjacent fields.

The risk assessment concluded that a potential risk was posed by the contamination to local people ingesting crayfish from a stream and farmers growing certain kinds of crops. Measures were suggested that would contain the contamination on-site, including the installation of groundwater pumping wells. The abstracted water was treated following removal, prior to its return into the aquifer. The risk assessment had allowed Western standards of environmental compliance to be followed in a developing country, thereby maintaining the company's positive public relations record and its control on potential environmental liabilities.

15.7 PUBLIC RELATIONS

As the general awareness of environmental issues increases in Western countries, so too does the public scrutiny of industrial operations. This has led to a greater emphasis being placed on public opinion than ever before, and the importance of public relations issues whenever problems related to chemical pollution arise.

The growing importance of public opinion in environmental matters is a reflection of the competitive marketplace in goods and services (i.e. it is always possible to buy from somewhere else), greater access to environmental information (often through legal provisions) and a free press. It has given rise to a situation whereby historical lines of communication between regulators and the regulated community are often irrelevant in the face of exchanges between a far more important audience, the consumer and the mass media.

An environmental risk assessment can prove useful in helping to manage the public relations implications of an environmental incident. A well-reasoned, scientific appraisal that pinpoints the precise issues, if any, of concern can help to defuse tense situations in which positions regarding the presence or absence of a risk are often rapidly arrived at more for political expediency than anything else. Further, it can help to allay public concerns and fears, which are often at their greatest in the face of uncertainty.

15.7.1 Case study

A manufacturing plant involved in metal-plating activities had several episodic, unplanned releases of a substance due to the repeated failure of an emissions control system. The substance was carried by the wind over a large area, including a nearby residential estate, and traces of it were identifiable in the form of brown specks on exposed surfaces.

Newspaper reports, along with public complaints, led to local public health officials issuing a warning over the radio that the specks should not be touched and that any affected fruit and vegetables should not be eaten. Public concern consequently grew over the possibility of poisoned gardens and adverse effects from direct inhalation if the releases were to continue.

The company responsible was quickly overwhelmed with enquiries from members of the public and demands from local media groups to make a statement and offer redress. On the basis of an environmental risk assessment, it was able to assert that while a short-term hazard had existed during the releases and had to be prevented from reoccurring, there was no long-term risk from the residues. This position was backed up by a detailed modelling study, which showed that the substance converted rapidly to its benign form in the ambient environment and that consequently there were no long-term exposure scenarios. The risk assessment was reviewed by local officials who, following their concurrence with its findings, prepared and broadcast a second radio bulletin that retracted the first.

The environmental risk assessment provided the framework within which the company could admit responsibility for the accident, declare its intentions to prevent a recurrence, and demonstrate the absence of any ongoing risk to health. This resulted in its locally positive public image being maintained.

15.8 CORPORATE PLANNING

Larger companies often incorporate environmental risk assessment into corporate planning activities, either in response to regulatory requirements relating to development projects or as a useful tool in specific circumstances. With regard to the former, large projects must often undergo a detailed environmental assessment (EA) that identifies direct and indirect environmental effects and appropriate mitigation measures necessary for their prevention. In Europe, potential impacts on the following must be addressed in an EA (Council Directive 85/337/EEC):

1 Human beings, fauna and flora.
2 Soil, water, air, climate and the landscape.
3 The interaction between factors 1 and 2.
4 Material assets and the cultural heritage.

The use of risk assessment as a part of EA is generally restricted to more complex projects, such as incinerators (DOE 1995c). In such situations it can demonstrate the magnitude of any risks to human beings, fauna and flora and allows suitable ameliorative measures to be included in the engineering design. As an example, EAs undertaken in support of planning applications for incinerators typically include detailed risk assessments involving computer simulations of routine and accidental releases from stacks and fugitive sources. These can demonstrate the effectiveness of pollution control measures and the adequacy of emergency response measures.

The use of environmental risk assessment in determining the possibility of disruption to a corporate activity is of primary importance to businesses that rely on a steady supply of an environmental resource. For example, the food processing industry requires large amounts of clean air to dry products, some of which are susceptible to 'tainting' by unwanted industrial chemicals. An environmental risk assessment can establish the likelihood of this occurring from nearby sources and, by using predictive modelling techniques, establish suitable protective measures. Alternatively, a supply of clean groundwater may be required by a brewer or a water supply company (e.g. via abstraction boreholes). Where there is a possibility of water quality being adversely affected by chemical spillages within the well catchment, this must be assessed and suitable measures implemented. If necessary, this may involve the moving of the borehole to a less sensitive location.

15.8.1 Case study

As part of a planning application for a waste treatment and transfer station in the UK, a company undertook a detailed EA. The assessment indicated that emissions would be minimal and that there would be no adverse effects if the development was to proceed.

The application was met with fierce opposition from the local community, including an adjacent food products factory. The latter argued that the proposed facility would present a significant risk of tainting of its products and threaten the long-term livelihood of its business and workforce. Partly in response to these pressures, the application was turned down by the planning authority.

An appeal against the decision was lodged with the Secretary of State for the Environment and a Public Inquiry was held at which evidence was presented regarding the presence/absence of a risk. Detailed environmental risk assessment arguments were submitted by both the food products factory and the appellants, including complex analysis and atmospheric modelling studies. The Inquiry Inspector concluded that the evidence suggested that the risk was significant and the appeal was consequently rejected.

A second appeal was made to the High Court for judicial review of the Inspector's decision. A judge reviewed the evidence presented at the Inquiry and concluded that the Inspector had been incorrect in reaching his conclusions. The appeal was upheld and the Inspector's decision was quashed (Queen's Bench 1995).

This case study illustrates the use of environmental risk assessment in corporate planning, both for the purposes of environmental assessment of development projects and for the assessment of raw material supplies. A third facet of the study is that the holding of a Public Inquiry resulted in a highly sophisticated risk assessment being carried out in response to the needs of the legal setting (see above).

15.9 CHEMICAL PRODUCTS REGISTRATION

Since 1967, when the Dangerous Substances Directive was promulgated, Member States in the European Union (EU) have had a framework under which the hazardous properties of chemicals marketed throughout the EU can be examined, evaluated and appraised (Directive 67/548/EEC; see Chapters 3 and 9). Numerous amendments and 'adaptations' to the original Directive have produced schemes for Member States that cover the following:

1 Classifying the toxicological, environmental and physical hazards of major commodity chemicals.

2 Testing the toxicological, environmental and physical properties of such chemicals.

3 Testing new substances prior to marketing.

4 Restricting or banning the use, sale or transportation of certain hazardous substances.

In addition, to ensure that all 110 000 chemical substances already marketed throughout the EU are tested and evaluated in an analogous manner, the European Commission has introduced the Existing Substances Regulations, which require the provision of hazard information (toxicological and environmental) by manufacturers (Directive 93/793/EEC).

In terms of risk assessment, major developments to both items of legislation were made in the 1990s, with the introduction of a risk assessment requirement. This resulted in a holistic approach covering the entire lifecycle of a substance from manufacture through use to disposal. It considers effects on all environmental compartments and in people (workers and consumers).

For new substances, the risk assessment requires the predicted environmental concentration and estimated human exposure level to be compared with levels anticipated to be of concern to environmental receptors or human health. Where a concern is demonstrated, further information or risk reduction measures may be required. These may delay or restrict the marketing and use of new substances and may add significant costs and other burdens to industry. While it is the duty of the competent authority in each Member State to provide assessments for appraisal, notifiers (or their consultants) can carry out their own risk assessments by following technical guidance published by the Commission of the EU.

For the 110 000 existing substances already marketed in the EU, a priority process for hazard data submission has been defined. For priority substances environmental risk assessment is required to define whether further information or testing is necessary or whether risk reduction measures are required. Again, the outcome of the risk assessment could impose significant costs and other burdens upon industry, particularly where data for risk evaluation purposes are lacking.

15.9.1 Case study

A case study of the use (or abuse) of risk assessment methods in relation to chemical products is provided by the 'Alar scare' of 1989 in the USA (Ames & Gold 1989). Alar is a trade name for a preparation of daminozide, which was used to extend shelf-life, improve appearance and expedite harvests of certain fruits, especially apples. A breakdown product of daminozide had been found to be carcinogenic in animals by the Environmental Protection Agency (EPA) of the USA and formed the focus of a campaign by the Natural Resources Defense Council (NRDC 1989). The campaign called for a tightening up of pesticide residue limits and, in some cases, the outright banning of certain chemicals.

The basis of the NRDCs claim was that the amount of pesticide residues ingested by the average child in the USA was sufficient to warrant a significantly increased lifetime risk of cancer above that allowed under most regulatory programmes. The increased risk was calculated using standard environmental risk methods that included low-dose carcinogenic potency extrapolation procedures. Media attention was galvanized as nursery school population risks in the USA as high as 11 000–15 000 were postulated. This resulted in the dramatic boycott of apple juice and other similar products and consequent wide-scale disruption to fruit farmers in the USA.

Following the initial scare, debates raged over whether the risks were 'significant' and the 'risk–risk trade-offs' likely to arise from banning Alar (i.e. was a diet containing fruit more healthy in spite of Alar than one without?). These were eventually resolved and the USEPA restricted its use to flowers only.

This case study illustrates how environmental risk techniques have been used by non-governmental lobbying organizations to influence regulators in the area of chemical product safety. The application of the techniques was not without criticism, however, and whether the outcome of the 'Alar scare' was positive or negative for public health is perhaps open to debate.

15.10 FUTURE

The previous sections of this chapter have focused on the various applications of environmental risk assessment to business. Many of these are clearly only applicable to businesses of a certain size and nature, but it is hoped that they illustrate how the fundamental principles can be applied across a range of sectors and issues.

In order to assess the likely longevity of environmental risk assessment as a service to business in its own right, a shortened 'strengths–weaknesses–opportunities–threats' (SWOT) analysis has been carried out. This has revealed the following.

Strengths. The increasing acceptance of environmental risk assessment by business is a reflection of its usefulness as a framework within which reasoned debate about the environmental impacts of industrial activities can take place. It provides a mechanism for addressing environmental concerns with regulators and other stakeholders in a manner that focuses debates on scientific knowledge and uncertainty, rather than political opinions (although inevitably these are sometimes linked). It also facilitates the comparison of environmental risk with other, more familiar risks. Above all, it focuses on what matters to most people with regard to environmental pollution (i.e. the prevention of adverse health or ecological effects).

Weaknesses. A current weakness of environmental risk assessment is that it suffers from some level of public distrust, possibly because it is relatively new. Further weaknesses are more inherent to the approach. For example, by focusing on individual sites and incidents, it can obscure the wider implications of prolonged low-level contamination (this is especially true of site-specific risk assessments of persistent xenobiotics, such as chlorinated compounds). It can also neglect breakdown products of contaminants, which in some cases are either of unknown or greater toxicity than the parent compounds. For a thought-provoking review of these and other weaknesses, the reader is referred to Johnston *et al.* (1996).

Opportunities. There are currently many opportunities for environmental risk assessment to fill an ever-increasing role in environmental protection. This stems from the initiatives described above and the desire of many governments to allow businesses to justify their own actions, rather than forcing the adoption of particular technologies or control measures through deregulation initiatives (POST 1996). Environmental risk assessment is the perfect framework for demonstrating compliance with particular regulatory criteria, especially those based on 'protection of public health and the environment'.

Threats. The biggest threat to environmental risk assessment is probably from assessors themselves. This arises because, by necessity, a large amount of public trust is vested in the person carrying out a risk assessment, due to the complexity of the subject matter. This trust can potentially be abused for financial gain and such abuse, if detected, could rapidly result in the widespread distrust of environmental risk assessment. A possible solution to this is to ensure that adequate peer review takes place, although this could lead to disagreement among experts (which has its own implications in terms of public credibility). Another option is for regulators to provide greater guidance on appropriate methodology, although this could reduce any incentive to search for and use innovative assessment approaches.

In general, the future of environmental risk assessment in business is positive and it will act as an important interface between scientists, the public, regulators and politicians. These stakeholders all have differing views on environmental matters, and risk assessment can play a vital part in reconciling them as we move into the next millennium.

15.11 CONCLUSIONS

Environmental risk assessment is applied widely by businesses to aid in issues as diverse as regulatory compliance, litigation and corporate planning. Its application is likely to increase, especially as it becomes a mandatory aspect of environmental regulation in many countries.

A shortcoming of environmental risk assessment is that it can fail to predict accurately the financial liabilities faced by businesses due to environmental matters. This is because it focuses on scientific aspects only, whereas financial exposure is just as likely to arise from legal and public relations issues, which are often unrelated to science.

Over time, the level of financial risk arising from environmental matters is likely to reflect more closely the level of 'assessed' risk. This is because regulatory enforcement actions (fines, etc.) will gradually become based on environmental risk assessment, financial externalities related to environmental impacts will be better quantified, and the public and/or media will become better informed as to 'real' versus 'perceived' risks. Environmental risk will therefore develop a more direct role in the assessment of financial liabilities associated with environmental matters.

It is critically important to the future of environmental risk assessment that its practitioners do not abuse the level of trust that is often vested in them. To the majority of people, risk assessment will remain a 'black art' and they will be obliged to accept the assessor's judgements at face value. Public scepticism of 'experts' is relatively high and this must serve as an incentive for diligence and truthfulness on the part of assessors. One option for removing this scepticism is for governments to provide greater guidance on appropriate methodology. This presents its own dangers by reducing the incentive to search for and use innovative approaches.

15.12 REFERENCES

Ames, B.N. & Gold, L.S. (1989) Pesticides, risk and applesauce. *Science*, **244**, 755–757.

Anon. (1995) Brent Spar, broken spur. *Nature*, **375**, 708.

ASTM (1995) *Standard Guide for Risk-Based Corrective Action Applied at Petroleum Release Sites*. American Society for Testing and Materials, Philadelphia.

Brannigan, V.M., Bier, V.M. & Berg, C. (1992) Risk, statistical inference, and the law of evidence: the use of epidemiological data in toxic tort cases. *Risk Analysis*, **12**, 343–351.

CAPCOA (1992) *Air Toxics 'Hot Spots' Program Risk Assessment Guidelines*. California Air Pollution Control Officers' Association, Sacramento.

CARACAS (1996) *Concerted Action on Risk Assessment for Contaminated Sites*. CARACAS Office, Vienna.

CONCAWE (1996) *CONCAWE Review*. The Oil Companies' European Organization for Environment, Health and Safety, Brussels.

DOE (1995a) *A Guide to Risk Assessment and Risk Management for Environmental Protection*. Department of the Environment, HMSO, London.

DOE (1995b) *Contaminated Land Research Programme. Reports and Current Research Projects*. Department of the Environment, Contaminated Land and Liabilities Division, London.

DOE (1995c) *Preparation of Environmental Statements for Planning Projects that Require Environmental Assessment. A Good Practice Guide*. Department of the Environment, Planning Research Programme, HMSO, London.

DOE (1996) *Part IIA of the Environmental Protection Act 1990. Contaminated Land*. Draft Statutory Guidance (Final Working Draft), Department of the Environment, Marine, Land and Liabilities Division, London.

Edulgee, G.H. & Jackson, A.P. (1996) Overview of risk assessment and its application. In: *Environmental Impact of Chemicals: Assessment and Control* (eds M.D. Quint, R. Purchase & D. Taylor). Royal Society of Chemistry, London.

Environment Agency (1996) *Methodology to Determine the Degree of Soils Clean-Up Required to Protect Water Resources*. R&D Technical Report P13, Environment Agency, Foundation for Water Research, Solihull.

Ferguson, C.C. & Denner, J.M. (1995) UK action (or intervention) values for contaminants in soil for protection of human health. In: *Contaminated Soil '95* (eds W.J. van den Brink, R. Bosman & F. Arendt), pp. 1199–1200. Kluwer Academic Publishers, Dordrecht.

HMIP (1995) *Environmental, Economic and BPEO Assessment Principles for Integrated Pollution Control*. Her Majesty's Inspectorate of Pollution, London.

HSE (1995) *Generic Terms and Concepts in the Assessment and Regulation of Industrial Risk*. Health and Safety Executive, London.

ILGRA (1996) *Use of Risk Assessment Within Government Departments*. Interdepartmental Liaison Group on Risk Assessment, Health and Safety Executive, London.

Johnston, P., Santillo, D. & Stringer, R. (1996) Risk assessment and reality: recognizing the limitations. In: *Environmental Impact of Chemicals: Assessment and Control* (eds M.D. Quint, R. Purchase & D. Taylor). Royal Society of Chemistry, London.

NICOLE (1996) *A Network for Industrially Contaminated Land in Europe*. Apeldoorn, The Netherlands.

NRDC (1989) *Intolerable Risk: Pesticides in Our Children's Food*. Natural Resources Defense Council, New York, USA.

POST (1996) *Safety in Numbers?—Risk Assessment in Environmental Protection*. Parliamentary Office of Science and Technology, London.

Queen's Bench (1995) *Envirocor Waste Holdings Limited and The Secretary of State for the Environment, Humberside County Council and British Cocoa Mills*

(Hull) Limited. High Court of Justice, Queen's Bench Division, Crown Office List, CO/3288/93, London.

Simmons & Simmons (1995) *Companies and Environmental Risk*. A Special Report Produced by Simmons and Simmons London.

US CRARM (1996) *Risk Assessment and Risk Management in Regulatory Decision-Making*. US Commission on Risk Assessment and Risk Management, Washington, DC.

USEPA (1986) *Superficial Public Health Evaluation Manual*. US Environmental Protection Agency, EPA 540/1-86/060, Office of Emergency and Remedial Response, Washington, DC.

USEPA (1990) *Methodology for Assessing Health Risks Associated with Indirect Exposure to Combustor Emissions*. US Environmental Protection Agency, EPA 600/6-90/03, Environmental Criteria and Assessment Office, Cincinnati, OH.

Chapter 16
Risk Assessment and Management for Waste Treatment and Disposal

JUDITH PETTS

16.1 INTRODUCTION

The management of solid waste from the point of production through collection and transport to the point of final disposal has emerged as a political and social issue in developed countries only since the late 1960s. The first half of the century can be viewed as a period of enforced expert and public ignorance as to the potential risks arising from waste in the face of rapid urban and industrial development. The health dangers posed by sewage pollution reduced the amount of attention paid to solid waste (Tarr & Jacobsen 1987). Only with the control of infectious diseases and the rise in awareness of chronic diseases such as cancer did environmental health begin to replace public health as a concern.

The growth of the environmental movement in the 1960s saw the hazards of waste equated with a chemical and industrial problem. 'Out of sight, out of mind' waste disposal practices (Coleman 1985) were the result of the low level of management status afforded to non-income-generating materials. The misuse of resources, a lack of control over industrial operations and an increasing list of pollution incidents involving industrial waste became a public, and hence political, issue which finally brought waste management to the statute books: for example through the Deposit of Poisonous Waste Act 1972 and Control of Pollution Act 1974 in the UK and the Resource Conservation and Recovery Act 1976 (RCRA) in the USA.

Early regulatory action in the UK focused on the safety, environmental and amenity aspects of waste disposal, the definition of acceptable standards and the imposition of effective penalties for irresponsible practice. The definition of acceptable standards for facility design and operation was primarily based upon the selection of features from a range of alternatives which could achieve a definable goal. There can be no doubt that the management of risks provided the spur to regulatory action. However, understanding and assessment of those risks was largely *ad hoc*. For example, while landfill gas was recognized as a problem as early as the 1950s, it was not until the explosion incident at Loscoe, Derbyshire, in 1986 resulting in injury to two members of the public and demolition of their bungalow (Anon. 1986) that public and hence regulatory attention to monitoring and control measures of both operational and closed landfill sites was forthcoming.

Risk assessment as part of an objective and structured decision-making process for environmental risks has become a political priority in the UK in the 1990s (DOE 1995a). While UK risk assessment practice has strong roots in the assessment of technological and major accident hazards and also the assessment of the radiological safety of the deep disposal of radioactive wastes (Chapter 9), it is only comparatively recently that it has become an attractive decision tool for non-accident risks. The basic four-part assessment process familiar to engineering and safety assessments underpins the guidance on environmental risk assessment: i.e. (i) hazard identification; (ii) hazard assessment; (iii) risk estimation; and (iv) risk evaluation.

In relation to waste management risk assessment, UK practice preceded regulatory attention as public concerns about the siting of waste incinerators in particular, and the health risks of dioxins, placed pressure upon proponents of new facilities to undertake risk assessments of emissions, with the first assessments incorporated into Environmental (Impact) Assessments (Petts & Eduljee 1994).

Political and regulatory encouragement of the use of risk assessment is now based on a view that it provides for a non-prescriptive regulatory approach which also meets public demands for evidence of control through a focus on site-specific risk decisions. With policy now focused on sustainable waste management (DOE 1995b) and the choice of the best practicable environmental option (BPEO) (RCEP 1988), risk assessment is seen as a relatively sophisticated analytical tool for assisting decisions as to alternative options to minimize waste creation at source, to prevent releases wherever possible and to render harmless any unavoidable releases to the environment. Importantly, risk assessment should contribute to publicly accountable, sustainable waste-management decisions. The latter presents one of the greatest challenges for the next decade, with continuing evidence of a strong divergence between the priorities of society to achieve effective waste management and the willingness of individual communities to accept facilities to handle waste 'in their own back yards' (Petts & Eduljee 1994; Petts 1995).

This chapter discusses the use of risk assessment in the management of waste treatment and disposal. Practice in the UK in relation to facility design, siting and control provides the focus of the discussion, with landfills and incineration used as examples to illustrate risk assessment practice and technical issues. The uncertainties and data limitations within the risk assessment process are considered, although these problems are not specific to the issue of waste treatment and disposal.

First, however, it is necessary to consider the nature of the risks presented by waste treatment and disposal activities and the pressures which have led to the use of risk assessment.

16.2 THE RISKS OF WASTE TREATMENT AND DISPOSAL

16.2.1 The waste management cycle

The management of waste arising from households, industry and commercial activities (some 245 million tonnes of 'controlled' waste per annum in the UK) is determined after consideration of a hierarchy of options (DOE 1995b):

1 Reduction at source.
2 Reuse.
3 Recovery—including recycling, composting and the use of waste as a source of energy.
4 Disposal.

The interaction between the use of these different options can be considered as a management cycle—see Fig. 16.1 (Petts & Eduljee 1994, pp. 21–22). The primary objective is for waste producers (including households) to minimize the amount of waste which they generate. From the point at which waste is generated, segregation of materials for recycling and reuse and of hazardous wastes for specialized handling becomes the next priority. Unavoidable waste is then packaged, collected and transported either: (i) to interim storage facilities, such as transfer stations, civic amenity sites or materials recycling facilities where further sorting of materials for recycling and recovery takes place; or (ii) directly to treatment and disposal sites.

The hierarchy of waste management presented above does not stress the value of waste treatment *per se* as a means of volume reduction, achieving a reduction in toxicity or changing the physical form of the waste. Treatment of waste can be achieved by physical (e.g. screening, sedimentation, filtration), chemical (e.g. precipitation, solidification, neutralization), thermal (e.g. incineration) or biological (e.g. composting or anaerobic digestion) processes. The hierarchy of options emphasizes recovery of materials and/or energy; thus, any treatment process which does not do this is relegated to the bottom tier, equivalent to direct disposal. However, at the point of final disposal of either unavoidable waste or the residue from treatment processes, such as ash from incineration, the management objective should be to continue to utilize any inherent characteristics of the waste to optimize pollution potential reduction and to extract latent byproducts. Final disposal of waste is predominantly to controlled landfill or landraise facilities (approximately 70% of all controlled waste generated in the UK).

Within the waste management cycle, waste treatment or disposal options can be characterized in terms of the following lifecycle components:

1 Design and planning.

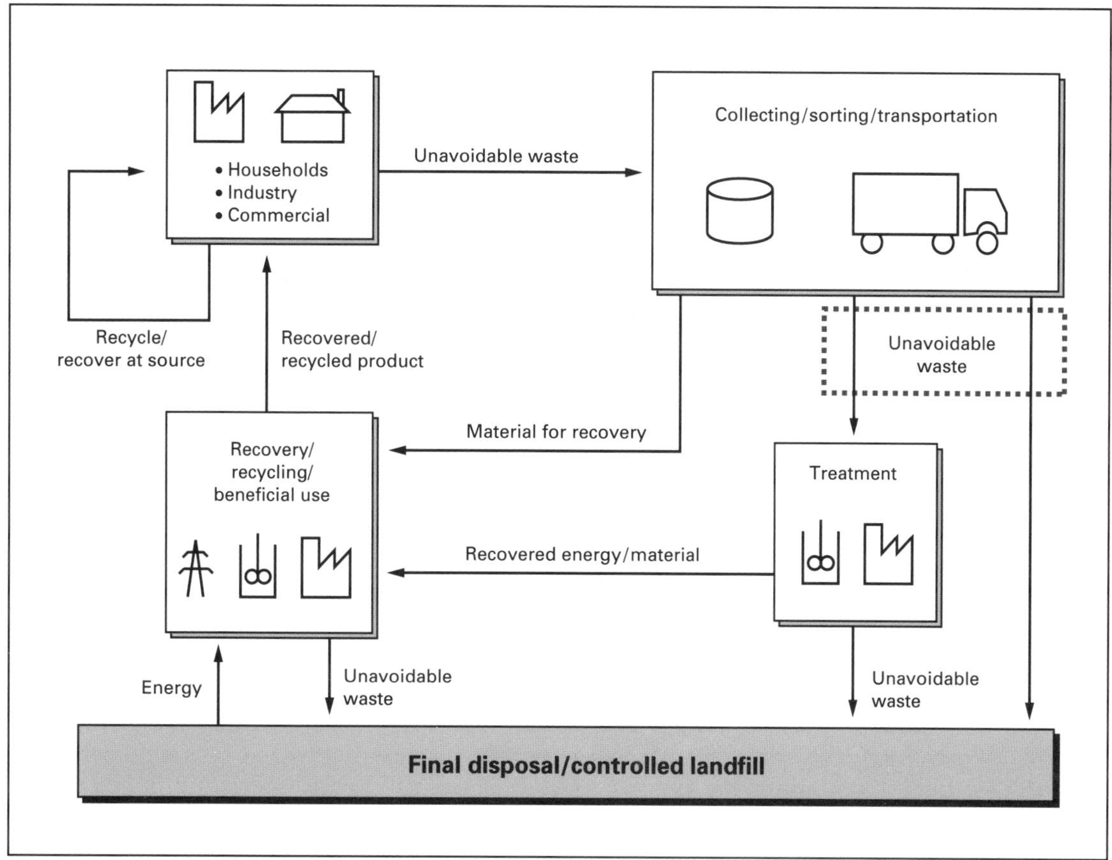

Fig. 16.1 The Waste Management Cycle—a generalized waste cycle indicating the interaction between various options in the waste hierarchy from reduction at source through recycling and recovery to final disposal of residual and unavoidable waste. (From Petts & Eduljee 1994.)

2 Permitting.
3 Construction.
4 Commissioning and operation.
5 Closure or decommissioning.
6 Post-closure monitoring or aftercare.

The assessment and management of risks should be an important consideration in each of these phases.

16.2.2 Waste management risks

At each stage of implementation of the waste management, and facility, lifecycles there is the potential for risk. Wynne (1987) identified two types of environmental hazards associated with waste management: an intrinsic hazard which is defined by the waste's physical, chemical or toxicological properties; and a situational hazard defined by the efficacy of containment, handling, transport, treatment and disposal. It is the latter which translates an intrinsic hazard into harm or a risk. Large parts of this Handbook are concerned with observable and measurable risks—being a combined consideration of the probability and magnitude of a hazard or adverse outcome being realized, and a degree of uncertainty over the timing or magnitude of that outcome (Covello & Merkhofer 1993). The concept of 'harm' which underpins UK environmental protection legislation is defined within the Environmental Protection Act 1990 as including 'harm to man, living organisms, offences to any of man's senses and harm to property' and in this context risk can be equated with harm. However, the inclusion of 'offences to man's senses and harm to property' in the definition of harm extends what might traditionally be considered to constitute targets in the environment at risk.

Common to all risk assessments discussed in this Handbook, the management of risks from waste treatment and disposal is achieved by a detailed understanding of the source–pathway–target chain. Figure 16.2 illustrates the source–pathway–target chain in relation to a landfill. Hazards are presented by the intrinsic properties of the waste, by the emissions or residues (leachate and gases) generated by the degradation of the waste, and by the operational activities which may either lead to an uncontrolled release of the latter or which in themselves may generate noise and dust, attract vermin or produce visual effects.

The potential for the hazards to be realized is enhanced by the disposal of waste that is inappropriate for landfilling (e.g. infectious waste, flammable materials, highly toxic waste) or for the particular site, given its environmental setting, poor site engineering, poor operational practices and the proximity of potentially sensitive targets such as potable drinking water or residential properties. This combination of intrinsic waste hazard, siting and proximity to sensitive targets, engineering design and operational practice determines the risk potential of any waste treatment or disposal activity.

Figure 16.2 presents what might be termed the risk assessor's view of the potential risks from

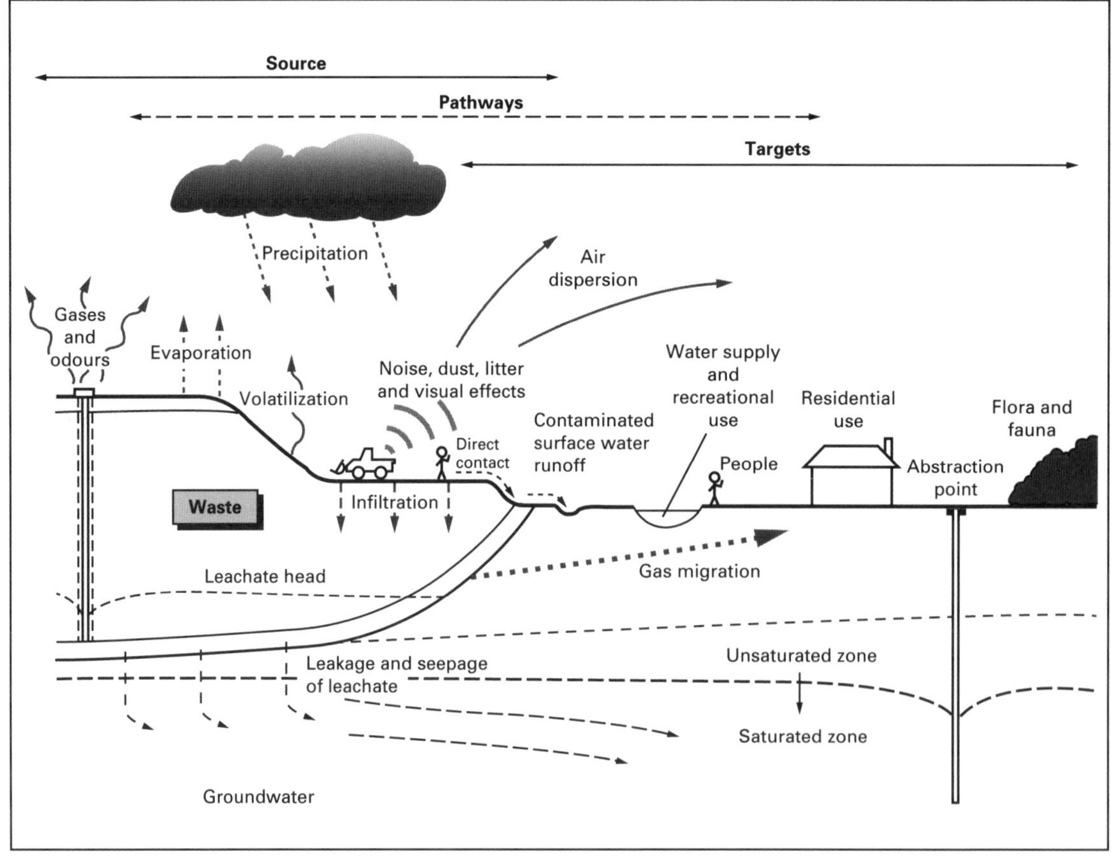

Fig. 16.2 Conceptual model of linkages between landfill exposure sources, environmental pathways and on- and off-site targets. (After DOE 1995c.)

landfilling. However, management of risks requires understanding of those risks or harms of importance to different stakeholders — regulators, operators, financial institutions, local residents, the general public, etc. Each of these stakeholders will not necessarily have the same priorities in terms of the identification of risks and of risk management. Just as waste is a socially defined concept (Thompson 1979), so is risk. Even the intrinsic hazards (toxicity, flammability, corrosivity, reactivity, ecotoxicity, etc.) presented by different wastes are subject to political, economic and social selection in terms of management priorities (Wynne 1987).

Most studies of public perception of waste facilities suggest that where waste has been defined as 'hazardous' and a treatment or disposal facility is or will be licensed to handle 'hazardous' waste, perceived risks to public health or to the environment are enhanced, with consequent demands upon the management process. Hazardous waste is usually equated with industrial operations linking to the view of industry as the primary polluter. However, current observation of public response to waste facilities indicates that the risks presented by facility operation can be viewed as adversely as the risks presented by the intrinsic properties of the waste. Table 16.1 lists some of the potential risks illustrated in Fig. 16.2, identifying the adverse consequence or harm and the source. Table 16.2 presents a similar listing in relation to a municipal incinerator. Both Tables 16.1 and 16.2 present risks which may be important to different stakeholders. The definition of risk inherent in Tables 16.1 and 16.2 is broad and undoubtedly presents a challenge to the traditional view of risk as a quantifiable physical impact. Included are risks to the following.

1 Human and ecological health.
2 Environmental assets.
3 Material assets.
4 Economic activity.
5 Amenity.
6 Social and individual well-being.

They are all risks which have been observed to be of public concern in relation to waste treatment and disposal activities (Petts 1992, 1994a,b).

Importantly, the risks have different spatial and temporal implications. For example, methane is approximately 30 times more 'effective' as a greenhouse gas than carbon dioxide. Landfills in the UK have been estimated to contribute a significant proportion (approximately 39%) of the total UK methane atmospheric load (Anon 1993). By contrast, the risk arising from potential accidents involving vehicles delivering waste to a facility will have a spatial scale limited to the transport route, and a temporal scale limited primarily to the operational period of the site. Landfilling presents a potential long-term liability if the operational period (e.g. 25 years) plus the post-closure period to the point of stabilization of the deposited material (possibly 30+ years) is considered. For landfills which are monodisposal sites that have only received one type of waste (usually hazardous), this post-closure period could extend over hundreds of years.

16.3 RISK ASSESSMENT FOR WASTE MANAGEMENT: DEVELOPMENT AND ISSUES

16.3.1 Risk assessment definition

The term 'risk assessment' is often loosely applied by both experts and non-experts, and can be used to describe a wide range of procedures from a very simple statement about possible hazards and risks, to highly formalized, quantitative procedures (Chapter 1). In the 'expert' community, risk assessment is most often discussed as a predominantly scientific process concerned with adverse outcomes that are measurable, whereas risk management is viewed as a primarily, legal, political and administrative process (Royal Society 1992). In the technical literature, and not least in the US National Academy of Sciences conceptual scheme (NAS-NRC 1983), risk assessment is usually considered as a systematic means of developing a scientific basis for regulatory decision-making: i.e. it is separate from risk management.

An alternative view sees risk assessment as a component of, not separate from, risk management (Chapter 1). This may seem something of a semantic difference. Such a definition seeks to achieve a more explicit recognition that all risk assessments are directly influenced by social, economic and political objectives and priorities and are subject

Table 16.1 Examples of source–pathway–target linkages and resultant risks arising from landfill engineering and operation.

Source	Pathway	Target	Risk
Liner failure; leachate leakage	Hydrogeological	Groundwater; potable water; people; rivers and associated flora and fauna	Pollution; loss of supply; health risks; damage or loss of flora and fauna
Leachate discharge	Sewer	Sewage treatment works	Effects upon biological processes
Contaminated surface water	Run-off	Soils; flora; fauna; watercourses; public via ingestion	Pollution; damage or loss; health risk
Landfill gas	Geological; soils; landfill cap to air	Buildings and people; flora; climate	Explosion and fire; asphyxiation; damage to crops; health risk from trace components; greenhouse gas
Noise—vehicles, alarms, machinery, etc.	Air	People and fauna; buildings; cultural heritage	Nuisance; vibration damage; loss of amenity
Dust; fumes—waste emplacement vehicles	Air	People and flora	Health risk; damage
Odour	Air	People	Loss of amenity and nuisance
Exposed waste	Direct contact	People and fauna	Health risk
General disposal activities and facilities	Visual	People/community; properties; landscape	Fear; loss of amenity; lowering of value; loss of investment in area; accidents
Litter	Air	Flora and fauna; people	Loss of visual amenity; odour; health risk
Vermin; birds	Air and contact; visual	People (potentially aircraft)	Health risk; loss of amenity; accidents

to the inherent values, biases and uncertainties of the technical expert and knowledge.

Literature in the USA in relation to risk assessment and waste (e.g. Asante-Duah 1993) stresses the quantified process for producing numerical estimates of the probabilities of possible adverse outcomes (particularly relating to public health risks). Covello & Merkhofer (1993, p. vii) boldly suggest that the emphasis on quantitative methods reflects their great advantage: that they express risk in numbers which are 'precise and leave less room for misinterpretation than words do'.

Clarity of definition is important if the benefits of risk assessment are to be realized. However, there is a danger that in stressing quantitative output and advocating this as 'precise', the inevitable uncertainties that then become apparent when the risk assessment output is discussed will amplify the lack of faith and trust in the assessment itself and in the risk assessor. Such an outcome has been evident in the USA (Silbergeld 1992). This Handbook supports the adoption of risk assessment as a means of improving decision-making. However, while expert support in this context is often founded upon a view that technological and environmental conflicts are predicated upon risk, the political science and social literature reminds us that conflicts are

Table 16.2 Examples of source–pathway–target linkages and resultant risks arising from incineration plant and operation.

Source	Pathway	Target	Risk
Stack emissions	Air; foodchain	People; flora and fauna; soil; crops	Health risk; damage; pollution
Effluent discharge	Sewer; water	Treatment works; water	Effects upon biological processes; pollution
Heat recovery; cooling water	Water	Water	Temperature change
Waste deliveries	Air; soil/water	People; flora; water; soil	Accidents; pollution/damage; loss of amenity
Waste storage and handling	Air; soil; water	People; soil; water; flora and fauna	Odour; fire, explosion or toxic release
Noise—vehicles, machinery, etc.	Air	People and fauna; buildings; cultural heritage	Nuisance; vibration damage; loss of amenity
Ash removal	Air; soil/water	People; water resources; soil	Health risks; dust pollution
General disposal activities	Visual	People/community; properties; landscape	Fear; loss of amenity; lowering of value; loss of investment; accidents
Stack and plant	Visual	People; cultural heritage	Fear; loss of amenity

actually based on more substantial differences between experts and the public: i.e. risk conflicts are fundamentally political (Stern 1991; Earle & Cvetkovich 1995; Vlek 1995).

A more pragmatic approach to the definition of risk assessment is to stress it simply as a structured process for understanding uncertain adverse outcomes. This process may result in a qualitative output, or in a semi-quantitative output (illustrated by a hazard ranking scheme for landfills in section 16.4.4), or in a probabilistic output discussed in relation to public health risks in sections 16.4.3 and 16.5.2. Such a pragmatic definition, providing for a simple conceptual framework which can be applied with increasing complexity according to the potential degree of risk, is more likely to find support from potential users who are not risk assessment specialists (e.g. operators of waste treatment and disposal facilities), so encouraging the adoption of the tool.

16.3.2 Development of use of risk assessment

The drive to the use of risk assessment for waste management appears to have a number of roots, including the following.

1 Advances in scientific and technical knowledge of the risks presented by waste treatment and disposal.

2 Public demands for information on potential risks.

3 Demands for operators and regulators to be publicly accountable for their decisions.

4 The opportunity afforded to implement a non-prescriptive approach to regulation and control which allows decisions to be made that are appropriate to the environmental setting of each facility (contrasting with the technology-based or generic standard regulatory approach).

5 An increasing recognition of the need to assess alternative options for managing waste, particularly to promote a movement up the hierarchy of management options, where appropriate.

As indicated in the introduction, risk assessment has only formed an important component of waste management in the UK since the late 1980s, and at the time of writing might still be described as an 'emerging art'. This contrasts with the relatively early development and application of risk assessment to radioactive waste disposal. In the latter case, the defined and separate regulatory controls, the need to find strategic disposal sites, and the public demands for fail-safe controls and proof of low (zero) risk to health combined to make this a high-profile issue. The siting policy for the disposal of such wastes in the UK was defined in probabilistic terms: i.e., a one in a million (1×10^{-6}) individual risk of a fatal cancer in any one year (DOE 1984). The risk assessment procedure requires the assessment to provide for long-term (tens to hundreds of thousands of years) safety (OECD 1990).

In contrast, solid waste regulation has focused on setting 'end-of-pipe' release limits and control standards to achieve these, rather than the consideration of alternatives for waste management, and ensuring that when releases did occur the impact on the environment was acceptable. On the one hand, there appears to have been a degree of 'expert faith' in the effectiveness of the regulatory regime and inherent control standards engendered by a relatively low level of proof that waste treatment and disposal facilities were presenting risks to the environment. On the other hand, a simplistic approach which applied specific standards to a relatively immature industry still on a 'learning curve' in terms of scientific understanding of waste processes and residue impacts (particularly for landfill) appeared to offer the potential for greatest control.

In the USA, a legacy of abandoned waste disposal sites began to provide early evidence of the potential risks (Levine 1981; Tarr & Jacobsen 1987). The resulting regulatory regime provided for the development of risk assessment within the RCRA, including in the selection of appropriate sites for hazardous waste facilities, and particularly within the Comprehensive Environmental Response and Compensation and Liability Act (CERCLA) 1980 to establish clean-up standards for contaminated sites. Risk assessment for waste in the USA developed within a strong culture of the need to understand and protect against toxic and, particularly, carcinogenic risks.

The application of risk assessment to the different waste treatment and disposal problems has varied in development terms. In the UK, the first significant development was in relation to incineration. The Second Report of the (then) UK Hazardous Waste Inspectorate (1986) recognized that risk assessment methodologies common to major hazard assessments could be relevant to chemical treatment plant and incinerators. However, this consideration of the potential role of risk assessment was largely exploratory. It was primarily public pressures which led to the first risk assessment relating to incineration emissions, linked to a 1989 proposal for new hazardous waste incineration capacity at Seals Sand, Billingham (northeast England) (Petts & Eduljee 1994, chapter 24). The specific proposal, combined with growing public concerns about the importation of hazardous waste into the UK for incineration and with continuing local opposition to the operation of a hazardous waste incinerator in South Wales, provided for the spotlight of public information demands to be placed upon incineration (Petts 1992, 1994b).

The role of risk assessment in providing for consideration of the acceptability of potential risks and optimization of engineering design in the site-specific context has now been addressed in UK guidance relating to landfill design, construction, and operational practice (DOE 1995c) (see section 16.4). It was in 1990 that the UK Department of the Environment (DOE) initiated a research programme involving the development of risk assessment for landfills. At that point in time the reliance upon landfill as a disposal route, the lack of apparent major pollution incidents (other than gas explosions), and the relatively low dependence upon groundwater resources for drinking water supply had helped to engender an acceptance of landfill based upon experience and faith in engineered systems, rather than upon an objective consideration of the risks. However, the UK's reliance on landfill as a waste management option was beginning to be challenged by European discussions about the banning of co-disposal landfill (the joint landfilling of industrial and household waste) and support for the containment landfill, which was seen to be in line with the requirements of the European Commission Groundwater Directive (CEC 1980). Public concerns about landfill have probably had less of an impact upon its development than in the case of incineration, as these concerns, until relatively recently, have largely focused on amenity issues such as noise, dust, etc.

To date, UK experience of risk assessment for waste treatment and disposal is still limited in terms of the hazards addressed, primarily to: waste storage and process stream failure; stack gas emissions; fugitive emissions from process plant; transport accidents; landfill liner failure and leachate migration. The risks addressed have primarily been public health effects from continuous releases and from accidents, and from surface and groundwater pollution. Ecological risk assessment is still in its infancy. Qualitative assessments of levels of effects on protected ecosystems have been included in proposals for the siting of facilities in potentially sensitive ecological locations, and some incineration risk assessments have addressed incremental concentrations of pollutants in soils and plants as a result of deposition of stack emissions (see section 16.5).

Separate from the development of risk assessment for the management of operational facilities, the UK has seen a rapid development of guidance on its use in relation to contaminated sites, including old waste disposal sites (e.g. CIRIA 1995). Here, the move to risk-based decisions has followed practice in other countries, including the USA, The Netherlands, Denmark and Germany, and is seen as a means of promoting consistent and robust decisions about site remediation requirements across the primary responsible authorities—the local authorities. Contaminated land management in the UK is within the policy context of the 'suitable for use approach': i.e. land must be able to support the current or intended use. In 1998 consolidated local authority powers will provide for the identification of sites which are, or have the potential to be, a 'significant risk' or which 'pollute controlled waters'. A set of procedural guidance is being published by the DOE to explain how decisions about significant risks are to be made.

Common to other risk assessment areas and to developments in the USA in particular, the UK is witnessing an increasing move to probabilistic methods for risk assessment as opposed to deterministic point estimates of risk. The degree of uncertainty and potential conservatism inherent in the latter has been a source of expert concern (Cullen 1994), although conservatism or caution has often been demanded by the public, not least in siting decisions where risk assessments have been discussed. The use of Monte Carlo analysis to produce distributions of risk reflecting uncertainty and/or variability has traditionally formed an important component of the risk assessment of radioactive waste disposal. Computationally more intensive than point estimate approaches, there is potential for mistakes and abuse of Monte Carlo techniques (Morgan & Henrion 1990; Burmaster & Anderson 1994) and, as yet, in the UK there is relatively little experience of its discussion in the public arena and also relatively little experience among regulatory authorities and operators of its application.

16.3.3 Risk assessment within the waste management framework

From the above discussion it can be seen that risk assessment has a role to play in the different components of the waste management framework.

1 Waste management policy and strategy development.
2 Setting of generic emission standards and regional and local environmental capacity indicators (not discussed further in this chapter).
3 Consideration of the best practicable option for managing specific waste streams.
4 Site selection for new facilities.
5 Environmental assessment linked to planning applications for new facilities.
6 Licence and authorization applications for waste treatment and disposal facilities.
7 Consideration of post-closure risks from sites.

Risk assessment and the BPEO

Support for the hierarchy of waste options defined earlier can be viewed as a risk reduction as well as resource conservation policy, with a move up the hierarchy away from landfill seen as the most sustainable waste management approach. The UK approach is to view the hierarchy as a guide rather than as a rigid policy tool (DOE 1995b). This view is in contrast to some European Union Member States, such as Germany, which are using regulation to restrict the use of landfill (e.g. through restrictions on the percentage of putrescible waste disposed).

The UK approach requires that the relative risks of the different options for waste management be considered in the locational, environmental and waste generation context. The relative risks of the different options when compared with the benefits and cost should provide for the BPEO (RCEP 1988) to be chosen. It is in such a situation that the adoption of structured assessment tools, such as risk assessment, becomes very important.

Sewage sludge provides an example of a waste stream for which alternative disposal options have had to be sought following the UK Government's signing of the North Sea Conference Declaration requiring the cessation of sea dumping by 1998. Thames Water Utilities Limited's assessment of alternative disposal options for sludge arising within its catchment area included an assessment of the relative risks of the short-listed options in its BPEO assessment (Thames Water Utilities 1991a,b; Petts & Eduljee 1994). Step 5 of a BPEO methodology applying to sewage sludge (Powlesland & Frost 1990) provides for the evaluation of short-listed options and includes a consideration of environmental impacts of each option, including public health, safety, nuisance, land contamination, water quality and greenhouse gas emissions. The Thames Water assessment included a qualitative and semi-quantitative ranking of eight variants of the three main disposal routes—recycling to agricultural land, landfill and dedicated incineration.

Regulatory requirement for consideration of the BPEO in the management of waste is currently restricted in the UK to the management of specific industrial processes at the operational stage. This is, in essence, too late in the waste management process to ensure consideration of appropriate alternatives for treatment and disposal (Turner *et al.* 1992; Petts & Eduljee 1994). Such policies should be incorporated into the waste management and waste local plans which have formed the basis of long-term waste planning in the UK, but these plans have not been required to be subject to a BPEO assessment.

The formation of the Environment Agency in England and Wales (the Environmental Protection Agency in Scotland) in 1996 and adoption of a National Waste Strategy based on the concept of the BPEO could provide an important opportunity to integrate strategic waste management planning with the plans drawn up at the local level, which identify potential sites for new facilities. There is a role for risk assessment as part of a system of environmental appraisal within such a strategic planning system.

A few local authorities have attempted to consider the relative risks of different waste management options when drawing up strategic plans. No formal study of their experience has as yet been possible, but informal discussions with officers have noted difficulties in obtaining relevant information and data to assess the relative risks (occupational and environmental) of different options—for example, comparing anaerobic digestion of household waste with landfilling. In the international literature, Asante-Duah (1993, pp. 218–249) presents an interesting case study of a risk comparison of disposal alternatives for dry-cell batteries arising in household waste in the Regional Municipality of Waterloo, Ontario, Canada. Limited only to a public health risk comparison between landfilling and incineration, the assessment followed the standard health risk assessment methodology in the USA (see section 16.5). The results suggest that neither disposal route presents a risk but that nickel–cadmium batteries would be better landfilled. Interestingly, the work suggests that recycling may present more significant risks from health-related problems associated with separate collection, storage and disposal.

Site-specific planning applications

The choice of an appropriate site is the key to reducing environmental risks. The standard of management at any site is subject to transient factors (ownership, skill levels, regulatory monitoring) and the possibility of an on-site incident leading to a potentially adverse environmental impact can never be reduced to zero. Therefore, the first line of defence is the choice of a site which provides protection for sensitive targets, either through appropriate separation or the absence of a vulnerable pathway.

The UK has witnessed the application, and primary development, of risk assessment for waste treatment and disposal at the siting and land-use planning stage (Petts & Eduljee 1994). Not only providing for a strong preventative regime in relation to environmental risks, the openness of the land-use planning system in terms of a statutory requirement for public participation has provided for the direct public questioning of proposals and proposed risk reduction measures, for the public discussion of social priorities and the value that people place on environmental quality.

Article 4 of the EC Waste Framework Directive (75/442/EEC amended by 91/156/EEC and 91/689/EEC) requires that planning decisions ensure that waste is recovered or disposed of without harming the environment, and in particular without endangering human health or causing a nuisance through noise. The assessment of environmental effects as required by the Town and Country Planning (Assessment of Environmental Effects) Regulations 1988 provides an important opportunity for planning decisions to be informed in this regard and, within environmental assessment, risk assessment can play an important part, including the consideration of alternatives, site selection, prediction of impacts of the proposed facility, and identification of appropriate risk mitigation measures.

Environmental assessment guidance has not recommended the use of risk assessments, and planning guidance has suggested a rather limited view of risk assessment as being relevant only to accident hazards and that it is not the role of planning authorities to undertake detailed assessment of releases to the environment, which is the responsibility of the pollution control authorities (DOE 1988, 1994c, 1995d). Such guidance has been in line with the limited view of risk assessment as only being relevant to measurable physical impacts, as discussed earlier. However, as discussed in section 16.5, despite the lack of regulatory support, risk assessment has been used where the siting process, and more particularly public questioning, has required its adoption.

Licensing and authorization

The design and operation of waste facilities in the UK are by means of a system of authorization and licences. Within the authorization process under Part 1 of the 1990 Environmental Protection Act, incineration plants burning more than 1 tonne per hour of waste are required to evaluate the impact of their releases on the environment, specifically in justification of the BPEO where their operation will result in releases of substances into more than one environmental medium. Following the earlier discussion, the BPEO consideration can be viewed as requiring a consideration of risks/harm and, in recognition of this, Her Majesty's Inspectorate of Pollution (HMIP) (now Agency) has been developing an approach to assessment which can be used by operators to demonstrate that no harmful environmental effects will occur (Environment Agency 1997). The approach provides an example of the development of tools for industry to use to provide proof to regulators that the environmental impacts (including risks to the environment) have been predicted and the relevant mitigation adopted.

It is interesting that the term 'risk assessment' is not used in relation to this 'environmental assessment' tool except in the discussion of the assessment of harm from releases to air and water, which will be by reference to environmental assessment levels (EALs) for each environmental medium. The latter are broadly based on the ALARP (as low as is reasonably practicable) principle, which underpins health and safety regulations in the UK and requires a balancing of the costs of making further reduction in risks against the actual reduction achieved and requires that further provisions are made until their costs (in time, trouble and expense) are grossly disproportionate to the improvement in risk. The ALARP principle requires the setting of upper and lower boundaries of what is an acceptable risk, with the tolerable region in-between these boundaries requiring judgements about the extent to which releases need to be prevented, minimized and rendered harmless to meet the requirements of the BPEO and the choice of the 'best available techniques not entailing excessive cost' (BATNEEC) for preventing releases to the environment. At the time of writing it has been suggested that releases should be considered a priority for control if the predicted environmental concentration resulting from the release is greater than or equal to 0.8 of the EAL. The latter may either be the relevant environmental quality standard for the substance or a level derived from other standards.

In the development of the approach, the Environment Agency has drawn an analogy with the approach to risks contained within European legislation relating to the control of new and existing chemical substances (CEC 1994) and the risk assessment approach developed in The Netherlands to regulate the risks from chemical substances (RIVM 1994).

Other waste disposal and treatment facilities are subject to licensing requirements under Part II of the 1990 Act. Regulation 15 of the Waste Management Licensing Regulations 1994 provides for implementation of the Groundwater Directive through control over waste management licences to ensure that discharge of List I substances is prevented and List II substances is controlled. Regulation 15 is the first UK waste-specific statute which requires an assessment of risk.

Specifically, Regulation 15 provides for an investigation to be conducted (by the operator) to provide for consideration of the possibility of the risk of pollution of groundwater. Where the groundwater is permanently unsuitable for other uses (especially domestic and agricultural), then where substances could reach groundwater this risk may be acceptable as long as it does not impede exploitation of the resources or lead to harm to other aquatic resources and ecosystems.

Post-closure

The Environmental Protection Act 1990 also provided for extended controls over the post-closure phase of licensed facilities (all waste disposal and treatment activities except incinerators). Licences will not be able to be surrendered until the regulatory authority is satisfied that pollution of the environment or harm to human health could not arise. Although not specifically requiring a 'risk assessment', consideration of surrender will require the holder of the licence to provide the Environment Agency with sufficient information about wastes disposed, site activities, types of contaminants likely to be present and information on actual contamination from sampling data. The Agency will have to make an assessment of 'risk' from this information (DOE 1994a) and we might expect this requirement to become increasingly more structured.

16.4 LANDFILL RISK ASSESSMENT

16.4.1 The sustainable landfill

Landfill in the UK has gone through a developing design philosophy (Westlake 1995a,b).

1 Dilute and disperse/attenuate—i.e. the principle of disposal in unconfined sites with little or no engineering of the boundary in which leachate (the liquid formed within a site that is composed of all liquids entering the site, leached from the waste by percolating liquids, and from rainfall) is allowed to migrate into the surrounding environment where natural attenuation mechanisms are relied upon to reduce its polluting potential.

2 Containment—i.e. a site where the rate of release of leachate into the environment is extremely low, with polluting components retained within the site by engineered barriers for a period of time sufficient to allow biodegradation and attenuation mechanisms within the site to reduce the pollution potential of leachate.

3 Sustainable landfill—'returning the contents of a landfill site to the environment in a controlled manner at a rate which the environment can accept without harm, generally using proactive measures over a limited timescale to diminish polluting capability, in a way which does not leave a long-term legacy of active monitoring and management' (DOE 1995c).

Until the mid-1980s the relatively low-cost dilute and attenuate landfill dominated, supported by research which confirmed that attenuation mechanisms could be used to 'treat' migrating leachate (DOE 1978). The UK's favourable geology and relatively low reliance on groundwater resources appeared to support the philosophy.

Reassessment was only sparked by the requirement to implement the EC Groundwater Directive. From the mid-1980s onwards, regulators began to require new sites to be contained, specifying permeability coefficients for clay barriers (1×10^{-9} m s^{-1}) and maximum leachate heads and increasingly requiring composite liner systems: i.e. utilizing high-density polyethylene (HDPE) liners and natural materials such as clay or bentonite with engineered drainage blankets and incorporating leachate collection systems. In 1992 the National Rivers Authority published its groundwater protection policy (NRA 1992), which is based on the vulnerability of water resources and the proximity of polluting activities to the source. The policy states, for example, that landfills accepting domestic, commercial and industrial waste within the inner zone (defined by a travel time of 50 days from any point below the water-table to the source) would not be acceptable to the NRA as the statutory consultee on applications for new facilities.

Publication of the NRA's policy immediately led to speculation about the development of site-specific predictive techniques for assessing potential groundwater pollution, as it would be for any proponent of a new site to convince the NRA of the acceptability of design. At this point there was still a reliance on the installation of systems to current best practice design. There was a strong belief that installation of a composite liner system would provide containment and that failure to do so was only due to poor materials or installation. The containment landfill has frequently been discussed as presenting 'no risk'. The difference of approach to landfill design between the USA and UK is demonstrated by the fact that in 1985 the US Environmental Protection Agency (EPA) had published a study of liner failure mechanisms using fault-tree techniques common to risk assessment (USEPA 1985) and work on the reliability of flexible membrane liners had developed from this time (Hartley 1988).

A few UK Environmental Statements for planning applications for new landfills began to incorporate some simple risk assessments for the prediction of liner leakage (Petts & Eduljee 1994). However, these were the exception rather than the rule. The sustainable landfill became the official UK landfill design philosophy in 1995 (DOE 1995c). The sustainability debate following the 1992 Earth Summit focused attention on the longevity of waste degradation processes. Experts were talking in terms of a leachate generation period in excess of 100 years (Mather 1992), and the need for leachate management over several centuries (e.g. Belevi & Baccini 1989; Knox 1990). A dichotomy was evident with, at one extreme, a danger that containment landfill could lead to a false sense of security as to the degree of environmental protection offered, and, on the other, a potential for over-design to encourage a reliance upon engineering rather than management (Petts 1993). Some other European Member States, while favouring the containment philosophy, saw this more in terms of entombment of waste with the intention being to return at a later date when treatment methods for the waste became available.

The fail-safe principle (Loxham 1993), inherent in technology assessment, came to the fore. This principle argues that whatever the containment system utilized, it will eventually fail and/or institutional control will cease. At that point it is necessary for the releases to the environment to present an acceptable risk. There are a number of potential risk management approaches which could achieve this.

1 Landfilling of only inert wastes.

2 Pretreatment of waste to produce a stabilized waste for landfilling.

3 Operational control through rapid flushing of the landfill to optimize the bioreaction process.

4 Design (including siting and engineering) to provide for risk mitigation in the event of failure and/or a lack of confidence in the ability to achieve stabilization.

Option 1 would require a significant and fundamental change to waste management practices in most countries. Option 2 is untested on the scale necessary to handle the biodegradable waste stream currently landfilled. Options 3 and 4 present the most robust and pragmatic approaches currently available.

Landfill risk assessment is founded upon a performance assessment: i.e. it is possible to quantify how the landfill may perform in the future, allowing for consideration of uncertainties in knowledge about the processes which will take place in the site and in the management of these processes over time. Landfill risk assessment currently stops primarily at the point at which a hazard can be realized—for example infringement of water quality criteria such as drinking-water standards. There are limitations in this assessment to our understanding of risk, for example:

1 It tells us nothing about whether the water already fails to meet standards.

2 It presupposes that the standards provide for protection of public health under all situations of use of the water and by the most sensitive targets.

3 It assumes that standards which are acceptable now will be acceptable in the future.

4 It does not adequately recognize other pathways of exposure.

16.4.2 Risk assessment for containment failure and groundwater pollution

The process for prediction of leaching to groundwater and estimation of the risk to water resources is relevant both to landfill leachate and to the leaching of surface spillages or leaks from tanks. The primary difference is in the characterization and estimation of the primary event: i.e. failure of containment. In relation to landfill, four phases are important.

1 Calculation of leachate generation and estimation of quality.

2 Estimation of the amount and rate of leakage.

3 Estimation of the concentration of specific pollutants in the groundwater.

4 Comparison of estimated concentrations with acceptable standards.

In the UK a new risk assessment decision tool (LandSim) has been developed (DOE 1995c; Hall *et al.* 1995) as a user-friendly, probabilistic risk assessment tool for use by regulators and developers. The tool has a number of potential uses.

1 Selection of initial liner and leachate management systems to achieve performance objectives within the site-specific geological and hydrogeological setting.

2 Fine-tuning of designs to achieve required mitigation of potential impacts.

3 Evaluation of the relative risks from different potential designs.

4 Review of the appropriateness of site investigation data and identification of outstanding information requirements.

5 Assessment of old landfills which may pose a future risk.

The model is user-friendly in as far as it has been programmed as a Microsoft Windows package with values entered typically as the minimum credible values, most likely values and maximum credible values. However, the user of such a decision tool does require understanding of, and familiarity with, landfill processes, the specification of liner systems and the potential impact of leachate.

Following the four-phase framework presented above, the LandSim model consists of the following elements, with the output of one forming the input to the next.

1 Source term.

2 Engineered barriers.

3 Geosphere.

4 Biosphere.

Table 16.3 provides a summary of some of the input data and information for each of these elements.

Hazard identification

The 'source term' represents the hazard identification, i.e. the nature of the waste, the biodegradation process and management strategies. The output of this hazard identification is, for assumed water infiltration rates, a list of contaminants together with their concentrations and volumes. Both of the latter will vary with time. The LandSim model holds leachate data from a UK inventory derived from a large number of sites. This allows for the user to default to minimum, average or maximum concentrations for a range of chemicals based on experience. However, local variations in waste composition, temperature, infiltration rate/compaction and landfill residence times will all affect digestion of waste within the site, which determines leachate quality (DOE 1995c).

The water balance of a landfill is normally determined by equating the rate of change of water storage within the site to the difference between water inputs (precipitation, surface run-on, groundwater discharge, liquid wastes, leachate recycle, irrigation water) and water outputs (evaporation, transpiration, surface runoff, surface seepage, groundwater recharge, leachate abstraction). Water balance equations with spreadsheet methods offer a convenient means of computing a range of leachate generation rates dependent upon alternative site designs, operational phasing, and different waste input rates. It is always difficult to make precise predictions and, traditionally, in landfill design undue sophistication in the model may not be warranted. For example, experimental figures for the absorptive capacity of waste are available (e.g. Blakey 1982); however, relatively few data are available for actual sites.

Leaching tests provide a laboratory-based test for estimating leachate quality under different waste conditions. Bagchi (1990) identifies a number of predictive models which have been developed to estimate leachate quality but notes that the techniques are not developed sufficiently to make a reliable prediction of 'field' leachate quality.

Given these difficulties, it is common to input data relating to a conservative non-retarded chemical species such as chloride, which will provide an indication of the extent of the leachate plume. A species that is strongly retarded, such as ammonia, can be used to indicate the longest expected travel time to the target, with the majority of other species falling between these two 'extremes'. A representative immobile metal or organic can be used in the same way.

Hazard assessment

The hazard assessment stage provides for a performance assessment of the engineered system, which can be defined as including the geomem-

Table 16.3 Examples of data/information requirements for the components of a risk assessment of landfill liner failure, based on UK LandSim model.

	Output	Input data/information considerations
Source term	Concentrations and volumes of chemicals in leachate	Waste inputs Leachate chemical species and concentrations Water balance/water infiltration rates Landfill cell size Decay processes
Barrier system	Leakage rate of leachate from base of site	Type of engineered barrier Hydraulic properties of clay Liner thickness Cation exchange capacity Defect distribution in high-density polyethylene Presence of quality assurance system Drainage system—design, hydraulic conductivity of drainage system
Geosphere	Time delay and attenuation of chemicals reaching aquifer	For the unsaturated and saturated zone—linear distance to aquifer; porosity; hydraulic conductivity; longitudinal and transverse dispersion; cation exchange capacity; contaminant retardation
Biosphere*	Health risk to target	Target characteristics Activity patterns Uptake rates Dose–response data

* At time of writing, not developed within LandSim model.

brane and/or clay layers together with the drainage system, and also the capping system and soil cover (see Fig. 16.3). The performance of such a barrier system relies on its integrity. This integrity can be breached for a number of reasons:

1 Design faults and choice of inappropriate materials and equipment.
2 Construction faults—e.g. holes in HPDE liner.
3 Installation faults—e.g. poor welding of liner seams; puncturing by vehicle or equipment.
4 Poor maintenance—e.g. of cover system leading to cracks and infiltration.
5 Equipment failure—e.g. failure of pumps on leachate pumping system.
6 Human intrusion—e.g. vehicle damage of liner or pipework.
7 Animal damage—e.g. burrowing animals.
8 Natural events—e.g. upward seepage of groundwater; uplift by gas beneath liner; extremes of weather.
9 Chemical attack by waste.

The extent of leakage through the base of a site is directly related to the head of leachate on the base. The latter is dependent upon the rate and distribution of infiltration and the drainage system installed to prevent build-up of excessive head. However, the parameters describing these two criteria are rarely known in detail. Simple assumptions can be made within a model: for example the infiltration rate is assumed to be consistent across the entire area and, rather than calculate the head at every position across a landfill cell, only the maximum expected head is calculated and the flow of leachate through the drainage system is not assumed to reduce the leachate head.

Leakage through clay barriers with no synthetic liner is normally calculated using Darcy's Law. The key inputs to this equation are the predicted leachate flow rate, the hydraulic conductivity of the unsaturated material directly below the waste, the hydraulic gradient across the barrier and the thickness of the barrier. Assumptions have to be made, for example about hydraulic conductivity, as this changes with time as the moisture content of the material increases.

Leakage through synthetic liners has been studied extensively and empirical equations are available

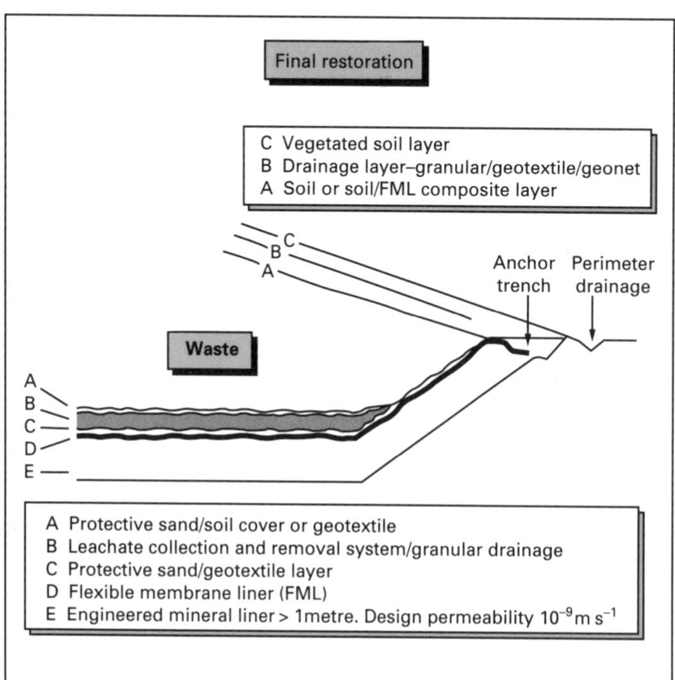

Fig. 16.3 Engineered landfill containment—typical composite basal liner utilizing synthetic and natural liners with engineered drainage blanket together with engineered capping system. (From Petts & Eduljee 1994.)

for the rate of leakage based on analytical studies and empirical work (Giroud & Bonaparte 1989; Giroud *et al.* 1992).

The output from this stage is simply a release rate of leachate through the base of the site. This rate, along with the individual concentrations of any particular defined chemicals within the leachate, is used to examine how the leachate will behave within the 'geosphere'—i.e. the unsaturated and saturated zones beneath the site.

For the leachate transport to groundwater, pathway factors which need to be considered include advection and dispersion of contaminants, retardation, and possible breakdown of contaminants to other chemical species. Typical reactions which attenuate the concentration of chemicals in solution and retard movement include ion exchange, degradation, sorption, precipitation and complexation. Retardation factors are site-specific and specific to particular contaminants.

Simple models use Darcy's Law to model groundwater flow conditions. Complex models incorporating both flow and solute transport models have been adopted by the USEPA, primarily for the purpose of assessing abandoned hazardous waste sites, and the USEPA (1988) provides a list of over 30 available computer-based models for modelling the fate of chemicals in the geosphere—MOC, MODFLOW, SESOIL and VHS (Vertical and Horizontal Spreading Model) are examples that have been used by UK landfill consultants.

Risk estimation

Within LandSim, the output from the geosphere component is the possible time delay and attenuation of contaminants reaching the 'biosphere' or targets. At the time of writing, the model stops at the contaminants reaching groundwater as the target, with protection of this target acting as a surrogate for protection of human health. However, there is scope to add a final model that would provide for the assessment of direct risks to humans, and development of this seems likely over the next few years. Potentially, fauna and flora could be added as targets, although currently deterministic methods remain predominant in ecological impact assessment. Currently, LandSim also does not model gas generation and transport.

The significance of the predicted risk to groundwater can be evaluated against a number of criteria which might include general objectives to be achieved, a comparison of predicted levels against background levels or direct comparison with statutory criteria relating to water use. An example of a general objective might be that groundwater quality should remain suitable for existing and likely future uses (as in the EU Groundwater Directive).

Uncertainty

A probabilistic model suits the problem of a system which is subject to considerable uncertainty or known variation. Issues relating to groundwater contamination have been viewed as more uncertain than those arising from other environmental media, because assessments need to address a long and technically complex set of causal relationships linking waste disposal to human exposure (Reichard *et al.* 1990). The performance of liner systems under different conditions and management regimes, water infiltration rates, leachate chemistry and production rates, leakage rates, drainage pipe blocking and the geological and hydrogeological characteristics for a site are examples where uncertainty can be significant. Performance of any landfill is time dependent, with the performance of a new facility expected to be very different to that decades after closure. At the latter point in time, degradation of the cap, leading to water infiltration, drainage systems blocked by fines or the leachate pumping system being no longer operational, could all affect performance. The LandSim model uses a Monte Carlo simulation to deal with such uncertainties and assesses the landfill at discrete points in time—for example new, immediately after closure and restoration, and 30 years after the latter. The Monte Carlo simulation technique randomly chooses inputs to the assessment which look like real-life possibilities (or realizations). These random possibilities are selected from the input distributions describing the system uncertainties. Thus, for example, in relation to the concentration of a chemical in leachate at a specific point in time, the input distributions might include the minimum, medium and maximum concentrations recorded across a selection of actual sites. For each 'realization' the output is calculated and repeat runs provide a distribution of calculations.

The use of a probabilistic risk assessment model allows for repeat runs to be made using different engineering design permutations and different default criteria (i.e. using worst-case, conservative assumptions or more realistic assumptions) until the results provide for the acceptability criteria to be met to an acceptable degree of confidence. Most importantly it should also provide for points of weakness within the design, which appear to have a significant affect upon the predicted 'risk' to be considered in terms of potential mitigation.

Inherent in the use of risk assessment is the possibility that unlikely events will occur. This can provide for consideration of possible responses to such events and whether contingency or emergency measures need to be built-in from the start—for example installation of a leachate leak detection system.

16.4.3 Landfill health risk assessment

Development of a health risk assessment methodology for landfills began in 1989, with initial work testing-out the use of USA protocols in relation to existing and closed landfills where hazardous wastes are or have been deposited (Eduljee 1992; DOE 1994b). In 1995 the DOE formed a new research contract to consider the health risks associated with household waste disposal and specifically to develop a methodology for use by regulators and operators for assessing site-specific applications at the development stage.

Figure 16.2 and Table 16.1 provided an introduction to landfill health risks. The health risk assessment framework includes: release (runoff, leachate, dust, volatilized chemicals and gas) analysis, inter-media transfer and fate and transport analysis through soil, air and water (surface and ground), and exposure assessment and risk estimation (Eduljee 1992).

Dependent on site-specific conditions such as operational stage of the site (i.e. operating or closed), types of waste disposed and presence and proximity of targets, and use of water resources, possible release mechanisms could include the following.

1 Leachate production—discussed in section 16.4.2.
2 Contaminated surface runoff—chemicals leaching from soil and waste being carried by surface runoff.
3 Gas production and migration due to pressure gradients carrying volatile gases away from the site.
4 Volatilization of chemicals directly from the waste mass to the atmosphere.
5 Dust emissions, with windborne particulates carrying adsorbed chemicals.
6 Direct contact—people coming into direct contact with the waste and then potentially carrying contaminated soil on their skin and clothes.

Fugitive dust emissions are usually considered to be negligible in relation to active landfills where the normal operational activities on landfills relate to waste emplacement. However, where large areas of soil may be left exposed and could be subject to wind erosion, then it may be considered a significant risk. The USEPA guidance (1989) deals with wind erosion emissions and presents empirical equations to predict wind-blown particulates from areas containing erodible and non-erodible surface materials. The assessment focuses on particles of $10\,\mu m$ or less as being the most significant in terms of inhalation and lung penetration. Assumptions have to be made about the concentration of chemical in the dust and, for simplicity, this has been usually assumed to be the same as the concentration of the chemical in the soil. Plume dispersion models can then be used to predict the downwind air concentration.

The volatilization of volatile organic chemicals from soil or from standing liquids is possible and is particularly relevant to operational landfills where liquids are deposited in pits or trenches (increasingly uncommon). It is also relevant where organics could be entrained in landfill gas emissions. The USEPA guidance (1988) uses an equation that derives an emission rate for each volatilized chemical using the chemical concentration in the landfill, a diffusion coefficient, a measure of soil porosity and the depth of clean soil cover. The predicted downwind concentrations can then be modelled as in relation to fugitive dusts, using meteorological data from a representative weather station. Both short-term and long-term impacts can be modelled.

The UK landfill research programme has been investigating the development of models to aid in the prediction and quantification of the hazards from gases migrating through surrounding strata. While progress has been made in modelling gas generation on sites, much fewer data have been available on gas migration in and around landfills (Ghabaee & Rodwell 1989). The ability to predict the pathway and distance over which gas migration could occur in the event of failure of the gas control system is currently limited due to problems in understanding the relative contribution of different mechanisms in the surrounding geology, including diffusion and advection and the effect of water.

The assessment of exposure and estimation risk once a concentration of chemical at the point of exposure has been calculated for each relevant pathway follows generic health risk protocols, which are discussed in detail in section 16.5.2.

The testing of the USA protocol for the assessment of public health risk from hazardous waste landfills on three UK sites revealed data availability deficiencies. For all of the pathways, the concentration of the contaminants in the emplaced waste provided the source term. For the USA sites this source term could be sufficiently well defined; for the UK sites it could not. As noted in section 16.4.2, the characterization of hydrogeology is essential for application of risk assessment models, however *in situ* data were sometimes insufficient. The study concluded that suitable default values for concentrations of contaminants in different waste types were needed, and where a detailed risk assessment was required an appropriate sampling programme must be instigated to supply suitable information. The availability of such information would allow for site screening using default values and then full site assessment based on the sampling programme.

In general, UK landfill health risk assessment has not followed the extended pathway and exposure scenarios of the USA protocols; not least because application and testing is focusing on new and current sites where these risk scenarios are considered to be manageable rather than upon abandoned hazardous waste sites. The current focus is upon the determination of exposure concentrations in the media of concern, particularly air and water, where health-related standards can be applied.

16.4.4 Site screening or ranking

For certain applications an alternative to a full risk assessment is a semi-quantitative risk assessment or hazard assessment by structured checklist. Primarily these approaches provide an indicator rather than an estimate of risk. Such approaches are practicable and efficient as a means of the following.

1 A regulatory authority determining priorities for further investigation and assessment across a number of landfills in their area.

2 Landowners determining priorities for remediating sites or taking relevant risk-reducing actions to reduce future liabilities.

3 Financial institutions and insurance companies determining the potential for future liabilities prior to investment or the provision of environmental liability insurance.

4 Screening out insignificant potential risks so as to help focus a full risk assessment.

The USEPA's Hazard Ranking Scheme (HRS) is perhaps the most well-known prioritization tool (USEPA 1982, 1990a,b,c). The HRS is the principal method used to rank the potential risks posed by sites reported to the USEPA under CERCLA. Once an identified site is placed on the CERCLA information system, it is subject to a preliminary assessment (primarily desk-top). This is followed by an initial site investigation and the data gathered are used as the source for applying HRS. The HRS is a typical ranking system in that it relies upon expert judgement to allocate scores related to defined criteria of high to low risk. Sites with high scores (see below) are then considered for placement on the National Priorities List for more detailed risk assessment and remediation.

The HRS assigns numerical scores to a variety of factors that affect the risk potential, including characteristics of waste toxicity, exposure potential and potentially exposed populations. The three original pathways were groundwater, surface water and direct exposure by air. Revisions have added to these, including the potential for contaminated groundwater to reach surface water and the addition of a soil exposure pathway. Each of the pathways consists of a set of factors that represent the potential to cause harm: (i) the manner in which the contaminants are contained, actual releases and the potential of a contaminant being released; (ii) the characteristics and amount of the materials, soils, etc.; and (iii) the nature of the likely targets.

A score for each of the pathways is obtained by allocating a numerical value or 'no score' (to denote not significant) for each of the characteristics or factors within each pathway. Each value is multiplied by a specified weighting factor and then the factor scores are multiplied together to give an overall pathway score:

$$\text{Pathway score} = \text{Likelihood of release} \times \text{Material character} \times \text{Exposed targets}$$

The pathway scores are normalized and combined using a root mean square approach rather than a simple average. This has the effect of emphasizing a facility recording a high score for one particular pathway, which can then be used to discriminate between sites with severe problems in one respect compared with sites with a combination of less-severe problems. Sites are ranked on the basis of the total score and the score for each pathway; a score of 28.5 (on a scale of 0–100) warrants inclusion on the National Priorities List. The score of 28.5 does not imply a specific level of risk, merely a screening level to allow priorities for further investigation and assessment to be set.

Table 16.4 provides, as examples, the factor categories and individual factors which are scored in relation to the groundwater, and air migration, pathways. Scores are assigned according to prescribed guidelines. For example, when considering the likelihood of release for a pathway, the maximum score possible is 550 for an observed release. Where there is no observed release, then a score not exceeding 500 is assigned to the potential for release. This score is composed of a series of pathway-specific factors. Taking as an example from Table 16.4 the groundwater pathway factor relating to the potential for release, the following maximum values can be assigned: factor 2a, containment — 10; factor 2b, precipitation — 10; factor 2c, depth to aquifer — 5; factor 2d, travel time — 35.

Reviews of the HRS have revealed characteristics that unintentionally bias its results. For example, waste type is scored on the basis of the most hazardous component of the waste, while waste quantity is scored on the basis of all components. Therefore, a large amount of a relatively low-hazard material could receive the same final score as a small quantity of a highly toxic material. However, like all such systems, the HRS is not designed to distinguish between the risks presented by two sites where the scores are similar, but rather to provide a meaningful indicator of the relative risks between sites with very different scores: i.e. it is a method for *prioritizing* sites so that those with the highest scores (and hence potentially the most significant risk) receive attention first.

Similar prioritization or ranking systems have been developed (but not used so widely) in Canada under the National Classification System (CCME 1992); the US Department of Defense has a waste-site prioritization model which considers both ecological effects and human health effects in relation to the surface and groundwater pathways (Smith & Barnthouse 1987), and New South Wales, Australia, developed a scheme for the ranking of sites by local authorities to enable priorities for detailed site investigation to be set (McFarland 1992). The latter contrasted with HRS in that it used a simple scoring system, where each of 10 factors were scored on the same basis of 1–5, indicating low to high risk, respectively. Simplicity, and hence potential assessor variance, was weighed against the expense of collecting site-specific data and the need for expert skills.

In the UK, work in the late 1980s was started on the Hazard Assessment of Landfill Operations (HALO) package. The HALO was designed to rank the operations and environmental impact of existing operational waste disposal sites. The methodology was loosely based on the USEPA's HRS, with scores developed for: material pollution potential; landfill operations; groundwater pathway; surface water pathway; landfill gas; and public perception. As with HRS, each component and its related factors are assigned a numerical value which increases as the degree of hazard increases. These scores are then normalized and subsequently weighted according to their relative

Table 16.4 Examples of two pathways considered by the Hazard Ranking System, including the factors and factor categories (numbered) scored in relation to these pathways. (From USEPA 1990b.)

(a) *Groundwater pathway*

Likelihood of release	×	Waste characteristics	×	Targets
1 Observed release or **2** Potential to release **2a** Containment **2b** Net precipitation **2c** Depth to aquifer **2d** Travel time		**4** Toxicity/mobility **5** Hazardous waste quantity **6** Waste characteristics		**7** Nearest well **8** Population **9** Resources **10** Wellhead protection

Groundwater pathway score = ⟨Likelihood of release {higher of 1 or [2a × (2b + 2c + 2d)]} × (6) × (7 + 8 + 9 + 10)⟩

(b) *Air migration pathway*

Likelihood of release	×	Waste characteristics	×	Targets
1 Observed release or **2** Potential to release; route characteristics **2a** Gas potential to release **2b** Particulate potential to release		**4** Toxicity/mobility **5** Hazardous waste quantity **6** Waste characteristics		**7** Nearest individual **8** Population **9** Resources **10** Sensitive environments

Air migration pathway score = {Likelihood of release [higher of 1 or (2a or 2b)] × (6) × (7 + 8 + 9 + 10)}

importance. The HALO has never been published for general use and considerable debate has focused around the scoring and weighting, its relevance to non-operational sites, and the extent to which the relatively complex evaluation system would be able to be applied by individuals in the field. An adapted HALO methodology has been developed for the hazard assessment of methane-generating sites (DOE 1990).

16.5 INCINERATION RISK ASSESSMENT

16.5.1 Issues and application

Incineration is a versatile waste treatment option which: reduces the volume and weight of bulky solids with a high combustible content; destroys and detoxifies some wastes to render them more suitable for final disposal (e.g. pathologically contaminated materials, toxic organic compounds or biologically active compounds); and destroys the organic component of biodegradable wastes which, when landfilled, generates landfill gas. Incineration can be used to recover energy from organic wastes and to replace fossil fuel for energy generation. There is more than 100 years of experience of use of the technology for municipal waste incineration with application to industrial waste since the 1940s. Despite the versatility of the treatment method, opposition, particularly to commercial or merchant sector facilities, has developed to such an extent over the last two decades that in many countries proposals for new plant have faced long delays and often refusal, existing facilities have closed, and even national waste management programmes have had to be delayed or modified following protest (e.g. that of Spain and also Australia's proposals for handling hazardous wastes).

The 1970s witnessed a rapid growth in concern over incineration as a public health risk, particularly with the identification of polychlorinated dibenzo-*p*-dioxins (PCDDs) and dibenzofurans (PCDFs) in municipal incinerator emissions (Olie *et al.* 1977), which coincided with the release of 2,3,7,8-TCDD (tetrachlorodibenzo-*p*-dioxin) and subsequent environmental contamination in a chemical accident at Seveso in Italy. Public perceptions of health risks are underpinned by the reaction of specific communities against existing and proposed facilities in their local area, including concerns about management and control capabilities and the management of waste generally (Petts 1992). National and international environmental groups have led significant campaigns against incineration proposals, yet governmental policies support incineration and in 1993 the UK's Royal Commission on Environmental Pollution (RCEP), after a detailed review, commended incineration as having an important place within a national waste management strategy (RCEP 1993).

While public perception appears to regard incineration as a major pollutant source, work in relation to the UK municipal incinerators operating in 1989–1990 (i.e. prior to the imposition of tighter emissions standards) suggested that they might contribute: less than 0.015% of total emissions to air of volatile organic compounds; less than 0.2% of SO_2 and NO_x; about 1% of CO_2; less than 2% of Cr, Cu, Ni and Pb; and less than 3% of HCl (Clayton *et al.* 1991). Cadmium and mercury contributions were estimated at approximately 24% and 13%, respectively, these figures since having been revised to 32% and 11%, respectively (RCEP 1993). It has been estimated that dioxins from municipal and clinical waste incinerators may contribute about 40% of the national atmospheric burden (Harrad & Jones 1992a), although all man-made sources of dioxin in the environment possibly represent only 12% of the total environmental load (Harrad & Jones 1992b). Updated estimates suggest that dioxins from municipal waste incinerators contribute between 460 and 580 g I-TEQ $year^{-1}$ of total dioxin releases, with total releases from all sources including traffic and the metals industry between 560 and 1100 g I-TEQ $year^{-1}$ (HMIP 1995). (*Note*: International Toxic Equivalents, abbreviated to I-TEQ, is the weighting of the summation of toxic PCDD and PCDF congeners relative to the toxicity of 2,3,7,8-TCDD, which is the most toxic of the family of 210 compounds or congeners referred to collectively as dioxins.)

The singling out of incineration as a polluting technology on the basis of emissions of PCDDs and PCDFs has been challenged in a recent piece of work compiling an inventory of releases of these substances to the environment from different components of the waste management hierarchy (Eduljee *et al.* 1995). The inventory suggests that waste management strategies with a greater level of recycling could be larger overall emitters of these chemicals to the atmosphere, the medium in which they have the greatest mobility, and also introduce PCDD/Fs to land in a form that is potentially more transferable to the foodchain.

Despite the detection of higher than background dioxin soil concentrations in the vicinity of some incinerators (e.g. Welsh Office 1995), there has been no proof of a UK incinerator being the major contributor to local contamination. Consumption of foodstuffs such as meat, milk, fish and eggs accounts for the primary (98%) background intake of dioxins, with direct inhalation constituting the rest (ECETOC 1992). In The Netherlands, high levels of dioxins in milk have been correlated with high emission rates from both municipal waste incinerators and other industrial sources (Liem *et al.* 1991). Several epidemiological studies have been undertaken following concerns about elevated levels of cancers, birth abnormalities and respiratory problems in areas surrounding incinerators (Lloyd *et al.* 1988; Shy *et al.* 1995; Elliot *et al.* 1996), but no correlation between source and reported health defects has been proven to date.

In 1994 the USEPA released a draft review of the health risks of dioxins (USEPA 1994), which was widely reported in the press but at the time of writing this chapter had not been approved for final publication. The review confirmed dioxins as potential carcinogens and suggested that they may affect the immune system and reproduction. It is against this background of health risk concern that risk assessment has become an important component of the consideration of incineration development and siting.

In the late 1970s and early 1980s in the USA and Canada, risk assessments in relation to municipal waste incineration tended to concentrate upon dioxin stack emissions and inhalation risks (e.g. USEPA 1981). One of the earliest assessments to consider multiple exposure pathways was in relation to a proposed facility in New York City (Hart Associates 1984). In 1987 the USEPA published a formal methodology for the health risk assessment of emissions from municipal waste incinerators (USEPA 1987a,b). A study by Levine *et al.* (1991) considered 21 assessments conducted in North America over the period 1986–1989, noting the significant differences in assumptions made, although the methodologies used were broadly similar. The late 1980s saw the development of methodologies primarily relating to indirect exposure via the terrestrial foodchain pathway (e.g. Travis *et al.* 1986; Hattemer-Frey & Travis 1991) and in 1994 a revised methodology with a validation of the terrestrial foodchain methodology was published for dioxin uptake via the air–soil/grass–beef–human pathway (Lorber *et al.* 1994).

The USEPA methodology has provided the basis of UK work, as countries in Europe have not developed a systematic risk assessment framework applicable to incinerators. An exception to this is the work undertaken by the National Institute of Public Health and Environmental Protection (RIVM) in The Netherlands (Slob *et al.* 1993).

Risk assessment for proposed incineration and treatment plant has been applied to consideration of accidents involving waste storage and handling (particularly for chemical waste), operational failure events due to component or human errors, fugitive emissions, and could be relevant to the consideration of the impact of effluent discharges and to transport risks. In relation to accident events (e.g. storage tank fire or leakage) and releases (e.g. of stack gases) under abnormal or accident conditions, the primary hazard identification and hazard assessment methodologies borrow heavily from the techniques used in the process industries, such as HAZOPs (Hazard and Operability Studies) (Petts & Eduljee 1994, chapter 18).

However, following North American practice it is in relation to the potential health risks arising from routine (normal operating conditions) stack emissions that most new development has taken place in the UK, significantly without the benefit of official guidance or even of explicit support from the regulatory agencies. Testing of the application of risk assessment has, to date, largely been in the adversarial arena of the public inquiry where the assumptions, methodologies and risk acceptability criteria suggested by proponents of new plant have been subject to challenge by opposition groups and also by local authorities. Table 16.5 provides examples of applications for new incineration capacity where a risk assessment has contributed to the consideration, often as part of the environmental assessment.

16.5.2 Incineration development and health risk assessment

Following the earlier discussion of public health risks from landfill, the general source–pathway–target factors that influence the scale of impact of an incinerator on public health are related to the following factors.

1 The properties of the wastes handled on the site.

2 The environmental medium into which the chemicals are released.

3 The location of sensitive targets relative to the site and their activity patterns.

Hazard identification

Hazard identification involves the identification of: (i) the chemicals to be accepted and handled on the site; (ii) the processes on site which may result in the releases; (iii) the sources and identification of those releases; and (iv) indicators to apply to the risk estimation. An inventory of chemicals can be achieved through an understanding of the waste to be incinerated and the processes producing the waste, and those chemicals which will be produced through the combustion process itself (such as dioxins). However, it is also necessary to consider those releases which are, or may be, regulated.

Table 16.5 Incineration risk assessment in the planning process in the UK.

Year	Proposal	Developer	Planning authority
1988	Seal Sands—industrial waste	Cory Environmental	Cleveland CC
1989	Calder Valley—sewage sludge	Yorkshire Water	Calderdale MDC
	Doncaster—hazardous waste	Leigh Environmental	Doncaster MBC
	Portrack—sewage sludge and industrial waste	NEM and ITC	Tyne & Wear Dev. Corp.
	Howdon—sewage sludge and industrial waste	NEM and ITC	Tyne & Wear Dev. Corp.
1991	Crossness—sewage sludge	Thames Water	LB of Bexley
	Beckton—sewage sludge	Thames Water	LB of Newham
	Knostrop—sewage sludge	Yorkshire Water	Leeds City Council
	Knostrop—clinical waste	Yorkshire Water	Leeds City Council
	Salt End—industrial waste	ITC	Humberside CC
	Belvedere—energy from waste	Cory Environmental	LB of Bexley
1992	Knostrop—energy from waste	Yorkshire Renewable Energy	Leeds City Council
1993	Belfast—energy from waste	NiGen	DoE Northern Ireland
	Shell Green—sewage sludge	North West Water	Cheshire CC
1994	Bolton—clinical waste	Whiterose	Bolton MBC
1995	Bournemouth—clinical waste	Whiterose	Dorset CC
	Belvedere—energy from waste	Powergen	LB of Bexley

CC, County Council; LB, London Borough; MBC/MDC, Metropolitan Borough (District) Council; Dev. Corp., Development Corporation; DoE, Department of the Environment; NEM, Northumbria Environmental Management; ITC, International Technology Corporation.

Herein lies one of the first problem areas for the risk assessor. Limiting the list of assessed chemicals to those directly currently regulated may fail to identify specifically those substances that may be of public concern (e.g. polychlorinated biphenyls in hazardous waste incineration) and/or substances of toxicological concern, or may not address future regulatory requirements, including existing tighter standards in other countries.

Generic regulatory standards may not provide for protection of the most sensitive target in specific locations. Regulators generally bulk emissions into one of two categories: for example in relation to organic micropollutants, either total hydrocarbons (THCs) or total organic carbon (TOC). Emission limits are then assigned to these groups—for example, 20 mg m^{-3} for TOC emissions under UK legislation (HMIP 1992). However, the bulking up of chemicals, which can then be more readily sampled and monitored, does not provide for the assessment of the health effects of individual chemicals, a fact which can be used by opposition groups as support for the argument that the risk cannot be determined and therefore the plant should not be allowed to operate. The USEPA has addressed this issue in assessments to provide for a worst-case generic composition for the organic fractions in emissions from hazardous waste and sewage sludge incinerators by adding an additional 75 compounds of toxicological concern (Eduljee 1995).

In order to focus upon the most important chemicals and reduce the work required for data gathering, it is necessary to screen the list of chemicals, selecting those that should be taken forward to the full health risk assessment while eliminating those that do not contribute significantly to the risk. This type of screening analysis as a form of risk assessment can also be used as a means of final site selection, where a short-list of sites which each appear to meet topographical, land-use and population criteria has been derived.

Screening analysis or 'indicator chemical selection' involves the calculation of exposure and risk, overlapping with the subsequent stages of the risk assessment but differing from the more detailed assessment in several ways, including the use of measurement detection limits rather than measured emission rates, and maximum as opposed to measured emission rates. At the screening stage, transport is modelled and intakes by the MEI (maximally exposed individual) estimated for both carcinogenic and non-carcinogenic risks. Screening criteria are then set: for example, all chemicals that contribute more than 5% of the total carcinogenic risk may be included, while those contributing a less than 1 in 10 million lifetime risk might be excluded.

The concept of the MEI has underpinned health risk assessments. This hypothetical individual depicts an extreme or 'worst-case' scenario with respect to the location, lifestyle and habits of the exposed population. Typical assumptions concerning exposure of the MEI are as follows.

1 They live at the location of maximum exposure to each pollutant (due to all pathways) and this is typically out-of-doors.

2 Exposure is over a lifetime.

3 Most of the food and water consumed is subjected to maximum exposure.

The setting of health risk criteria for an incineration risk assessment presents problems. For example, there is no widely agreed terminology for describing the air concentration of a substance below which health effects are very unlikely to occur (i.e. a safe ground-level concentration). The European Commission and USEPA air quality standards are incomplete in terms of substances of concern from incineration. The World Health Organization's Air Quality Guidelines (WHO 1987) are intended to be used for both risk management decisions on existing emissions sources and in the development process. The Guidelines indicate that compliance with the values will not have adverse human health effects, but that highly sensitive groups may be affected at or near the guideline concentrations. Where an incinerator may be proposed in a location where WHO guidelines are already exceeded during peak episodes, the risk assessment becomes the focus of attention as to the potential incremental air quality/health burden.

In the light of the lack of accepted standards, the approach in UK assessments for setting health criteria of long- and short-term air quality standards (LAQs and SAQs) has been to refer to occupational exposure standards, applying safety factors (see below) to these to provide for the different exposure duration of the general population as opposed to workers and for possible variability in human response. Thus, the LAQs equal 1/100th of the occupational exposure standard (OES). In the UK, maximum exposure limits (MELs) are also set for occupational exposure to chemicals where the nature of the toxicity is of particular concern, and this produces an additional safety factor of 5 over the OES. In incineration assessments, therefore, a conservative risk management approach is to set the LAQs at 1/500th of the MEL. The SAQs against which peak hourly average ground-level concentrations can be compared have been set at STEL/10 (where STEL is the short-term exposure limit).

Such an approach is common to the setting of thresholds below which non-carcinogens are considered to fail to induce adverse health effects. In the USA this threshold is termed the reference dose (RfD), and in the UK it is termed the acceptable daily intake (ADI) or tolerable daily intake (TDI). The reference dose is set for the most dose-sensitive of the various non-carcinogenic effects that a chemical may induce, by first determining a no-observed-effect level (NOEL) and then applying a safety factor. Generally, safety factors consist of a series of multiples of 10, each representing a specific area of uncertainty inherent in the available data. For example, a factor of 10 may be introduced to account for possible differences in response between humans and animals in prolonged exposure studies. A second factor of 10 may be used to account for variability in susceptibility among individuals in the exposed population. The resulting factor of 100 has commonly been used and dates back to the work by the US Food and Drug Administration (FDA) in the 1950s and used by convention since. Where chemicals are involved for which data bases are incomplete, an additional factor of 10 may be used (resulting in a factor of 1000) (Barnes & Dourson 1988).

The USA approach to the estimation of risk from carcinogens is to assume that there is no safe exposure level (i.e. no dose is thought to be absent of risk). Therefore, the USEPA applies the so-called linearized multistage model to estimate the largest possible linear slope (within the 95% confidence limit) in order to extrapolate from high doses typical of experimental studies to the much lower doses typical of environmental exposures. A carcinogenic slope (potency) factor (CSF) is computed from the data for each carcinogen which, when multiplied by the lifetime average dose of a carcinogen, yields the lifetime cancer risk resulting from exposure to that dose.

This approach has not found favour in Europe, being considered to overestimate risk. There is a lack of agreement over non-genotoxic chemicals in particular. These are carcinogens which do not act through an initial effect on DNA (i.e. do not act as initiators of cancer, but only as promoters). General European concern about treating chemicals such as PCDD/Fs as initiators has been driving work to derive threshold doses with the application of safety factors providing for TDIs to be set below which the likelihood of an adverse health effect is regarded as being very low (Maynard *et al.* 1995). This approach has been followed in a study by HMIP entitled 'A Risk Assessment of Dioxin Releases from Municipal Waste Incineration Processes' (1996).

Hazard assessment

There are potentially six exposure pathways of concern following release of substances to air (see Fig. 16.4).

1 Inhalation.
2 Ingestion of food.
3 Ingestion of drinking water.
4 Dermal contact with soil.
5 Ingestion of soil.
6 Contact with water.

Inhalation is the most direct route and would normally comprise one of the most important exposure pathways. The remaining pathways are indirect, although consideration of the chemical properties of individual substances, persistence in the environment, and the bioaccumulation potential provides for understanding of the primary exposure routes. Most health risk assessments of incinerator emissions have concluded that dermal adsorption and ingestion of soil are uptake routes that are typically at least one order of magnitude less efficient than lung absorption (Paustenbach 1989). In the case of ingestion of soil, it can be presumed that the risk is adequately covered when considering ingestion of food. The aquatic pathways of human exposure to contaminated water (e.g. through swimming, fishing and other recreational activities) may be considered to be of sporadic frequency.

Inputs to the hazard assessment are as follows.

1 Estimated emission rates—using regulated emission limits and the application of destruction and removal factors to the feed rates of the pollutants into the system to provide particularly for the emission rate of residual chemicals, such as polychlorinated biphenyls and lead.
2 Site characteristic information to provide input to transport and fate modelling—climate, meteorology, vegetation, soil type, etc.
3 Population information—location of potentially exposed populations, including: subgroups that may be at increased risk from exposure (e.g. in a rural setting farmers and young children (aged 1–6 years) whose diet consists of a high proportion of home-grown food products); activity patterns; and future land uses, which may result in either more sensitive people being located in the area or a change in activity patterns.

Notwithstanding factor 3 above, worst-case risk estimates (i.e. use of the MEI) have been the convention in virtually all incinerator risk assessment to date. The rationale for building-in a considerable degree of conservatism is to ensure full protection for all members of the public; i.e. it is presumed that if the MEI is adequately protected, all other members of the public will be at negligible risk. Although the approach may be considered to provide 'comfort' to the concerned lay-person, the approach can be misinterpreted by the latter as identifying the actual risk. Worst-case assessments have to be presented and interpreted with care. Maxim (1989) identified the following problems in their use.

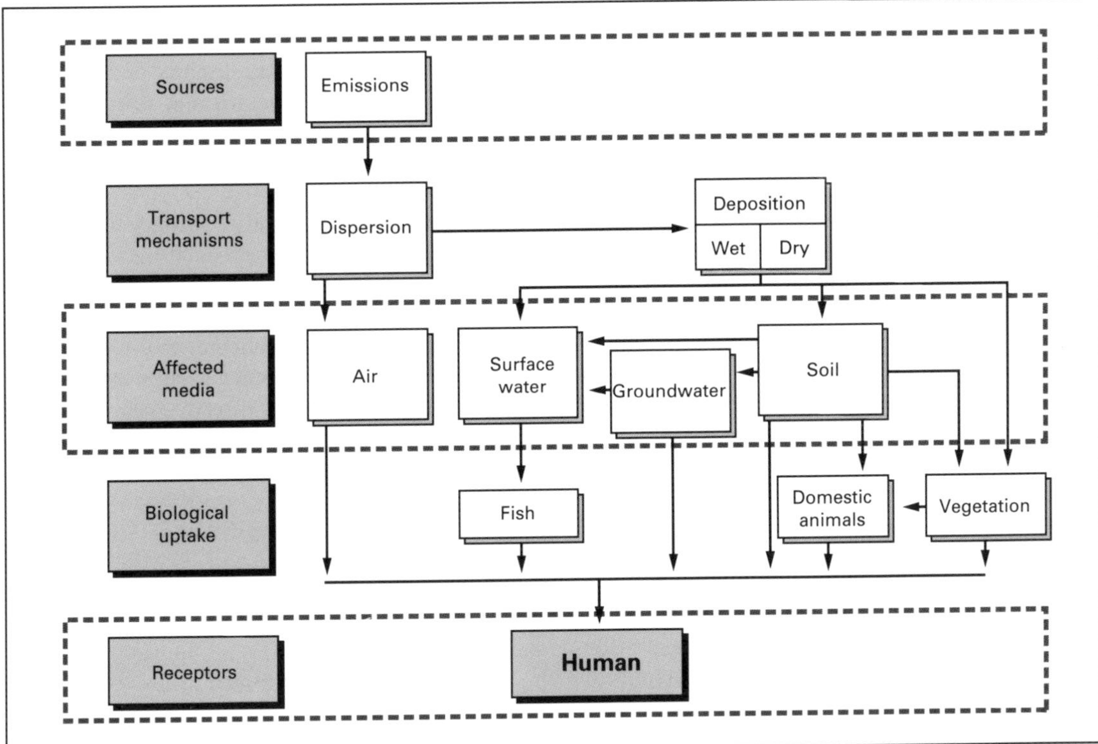

Fig. 16.4 Source–pathway–receptor (target) linkages for incinerator stack emissions, including direct (inhalation) and indirect (e.g. consumption of animal products where animals have grazed upon contaminated grass) pathways. (From Petts & Eduljee 1994.)

1 Risk estimates based on compounded conservative assumptions may lead to the regulation of insignificant risks.
2 Risk estimates with extremely conservative biases blur the distinction between risk assessment and risk management.

In the USA, criticism of the concentration upon the MEI, not least in Superfund assessments, and the misuse and misinterpretation of the most sensitive target and species contributed to new Guidelines for Exposure Assessment released in 1992 (USEPA 1992). Rather than referring to hypothetical maximum exposures, the guidelines were designed to promote more accurate characterization of the targets at risk at a particular site and the 1992 Guidelines suggest that the MEI should only be used as a screening tool to determine whether exposure may be insignificant, not as the basis for characterizing actual or plausible human health risks.

A large number of mathematical formulations of modelling atmospheric dispersion exist (Szepesi 1989) and Gaussian plume models provide the basis for dispersion up to 50 km. The USEPA has 'approved models'; among these the Industrial Source Complex (ISC) model, in its long- and short-term forms (USEPA 1987c), and the Rough Terrain Diffusion Model (RSDM) (USEPA 1987d), which provides short-term ambient ground-level concentrations, are commercially available; the ISC in particular has been extensively used in the UK. Although any model is only an approximation of the true behaviour of a pollutant in the environment for dispersion over the short distances common to incineration, annual average concentrations and maximum hourly concentrations can be predicted within a factor of 2 (Jones 1988).

In considering ingestion exposure, concentrations in food have to be calculated. Deposition from the air on to soil, plants, water bodies and other surfaces is effected through two processes: dry and wet deposition (see Travis (1991) for discussion). Any component of the foodchain that is in contact with the affected media can incorporate the chemical and either transfer it further up the foodchain and ultimately to humans, or directly ingest it. Foods grown in soil will be affected by aerial deposition and uptake through the soil system, while fish and shellfish will be in direct contact with surface water and sediments, which may themselves be affected by aerial deposition.

No models have been developed in the UK for simulating the uptake of non-radioactive chemicals into plants. The US 'Terrestrial Food Chain' (TFC) model (Travis & Cook 1989; Hattemer-Frey & Travis 1991; Lorber *et al.* 1994) considers the different routes of uptake in plants: for example deposition and uptake through leaves and stem; direct vapour absorption; deposition to soil and root uptake; wind mobilization of soil particles and adhesion on to plant surfaces; and volatilization from soil and absorption above ground. The TFC also models uptake by animals via consumption of plants and grain. Tables 16.6a & 16.6b present the results of calculations using the TFC model, showing the contribution of the various pathways to the total concentrations of 2,3,7,8-TCDD in vegetative and meat products, and in animal products from the ingestion of soil, forage and grain. Table 16.6a confirms, for example, the dominance of air-to-leaf transfer over root uptake pathway for above-ground vegetation.

The UK assessments have used the TFC modified, for example, to allow for UK food consumption statistics (e.g. MAFF 1995) to be used. Typically, the food groups included in the assessments are: vegetable produce and fruits; meats and other animal food products; milk and milk products; and fish where relevant to the site-specific foodchain route. Assumptions have to be made about food consumption patterns. The UK Ministry of Agriculture, Fisheries and Food (MAFF) (1994) provides data on the difference between the quantity of food consumed and the quantity purchased; for example, in relation to green vegetables, 24% more is consumed than is purchased. Such data can be used to assess potential consumption patterns relating to locally growth foods. In the case of the MEI, it might be assumed that 100% of food is locally produced.

The intake of chemicals by humans has been typically obtained by combining the following information.

1 The concentrations of the chemicals in the affected media.

2 The activity patterns of the potentially exposed populations, defining the types of exposure, their duration and their frequency.

3 The contact rate with the affected medium, for example rate of ingestion of water or foods.

A generic risk assessment equation can be used for estimating the intake (dose) resulting from different exposure scenarios, the starting point being the assessed concentration of the chemical of interest in the receiving medium at the point at which the target (MEI) is exposed. This generic equation is (USEPA 1989):

$$I = \frac{C \times CR \times EF \times ED}{BW \times AT}$$

where I, intake or dose (mg kg^{-1} body weight); C, average concentration of the chemical in the medium of interest (mg kg^{-1}, mg l^{-1} or mg m^{-3}); CR, contact rate (e.g. l day^{-1} of water or m^3 day^{-1} of air inhaled); EF, exposure frequency (days per year); ED, exposure duration (years); BW, body weight (the average body weight over the exposure period); and AT, average time or period over which exposure is averaged (days). The equation can be modified to take into account chronic daily intakes or shorter term subchronic or acute intakes by varying C or the various time factors. It may be important to assess exposure for the shortest time period that could result in an effect—for acute toxicants, this is usually one event or one day.

The equation is also modified to account for the variety of different routes of exposure, for example: chemical contact with soil; ingestion of soil; ingestion of water; dermal contact with water; inhalation of airborne chemicals and dust; ingestion of contaminated food. The USA guidance provides default values in relation to the components of the equations. For example, typical USA values for the exposure period are 70 years for lifetime exposures and 9 years at one residence. Typical body weight

Table 16.6 Results of calculations using the Terrestrial Foodchain Model showing the contribution of various pathways to the total concentrations of 2,3,7,8-tetrachlorodibenzo-*p*-dioxin (TCDD) in vegetative and meat products. (After Hattemer-Frey & Travis 1991.)

(a) *Percentage of vegetative contamination due to deposition, foliar uptake and root uptake of TCDD*

	Percentage due to deposition	Percentage due to foliar uptake	Percentage due to root uptake
Crops consumed by humans			
Potatoes	0	0	100
Leafy vegetables	60	39	0.3
Legumes	12	88	0.7
Root vegetables	0	0	100
Fruits	45	54	0.4
Grain	0	0	100
Weighted-average concentration crops consumed by animals			
Forage	71	29	0.2
Grain	0	0	100

(b) *Predicted TCDD concentration in animal products from ingesting contaminated soil, forage and grains*

Product group	Percentage due to ingesting forage	Percentage due to ingesting grain	Percentage due to ingesting soil
Beef/beef liver	76	0.2	24
Dairy products	85	0.1	15
Pork	0	1	99
Chicken	0	1	99
Lamb	82	0	18
Eggs	0	1	99

values are 70 kg for men, 60 kg for women and 20 kg for children over 6 years. However, these should always be replaced by site-specific data where available. Significant uncertainties can result from predicting population behaviours that may contribute to, or result in, exposures. Even when sufficient information is available for computing average exposure rates, the use of these might hide population variations. In the USA, for example, to estimate ingestion of drinking water, risk assessors have often assumed a lifetime average ingestion rate of 0.03 l of water kg^{-1} body weight. However, some sensitive population groups, such as milk formula-fed infants, may have much higher rates (Covello & Merkhofer 1993).

The generic equation is used to combine the exposure factors and the exposure point concentrations to estimate the received dose. Dose may be measured at any location between the point of contact and the location within the body where the contaminant takes effect (blood, specific target organs, etc.). Thus, the equation has been developed to estimate both the dose at the point of entry into the body and the dose that is absorbed through the body boundary and is available for biological action. For example, Paustenbach *et al.* (1987), in reviewing the bioavailability of 2,3,7,8-TCDD from soil ingestion, suggest a high end estimate of 30%. Where the extent of intake from a dose is unknown or cannot be estimated by a defensible argument, then usually dose and intake are taken to be the same (i.e. 100% absorption from contact).

Compounds such as PCDD/Fs can accumulate in the body, especially in adipose tissue. They therefore have the potential to be present in the breast milk of nursing mothers. Typically site-specific risk assessments have not considered exposure of nursing infants to contaminated breast milk. However, it has been addressed in the UK work on risk assessment for dioxins from municipal waste incineration (HMIP 1996). While the combination of low body weight and a diet comprising almost exclusively breast milk might result in a high daily intake relative to the TDI (MAFF 1995), the period of breast-feeding is short compared with the half-life of PCDD/Fs in the body and so, when averaged over a lifetime, the cumulative effect of an increased intake during breast-feeding is not considered to be significant (Department of Health 1995).

Risk estimation and evaluation

The commonly adopted approach in health risk assessments for waste facilities was discussed above in relation to the screening assessment for hazard identification.

1 For non-carcinogens and for the non-carcinogenic effects of carcinogens, the ratio of the calculated dose (chronic daily intake, CDI) to the relevant acceptable or tolerable daily intake (ADI or TDI) is computed (e.g. risk = CDI/ADI), and this results in a Hazard Index (HI). It is important to note that the magnitude of the HI does not indicate either a probability or the severity of the effect. Relevant reference doses for intake via ingestion are available in Technical Reports of the WHO and the Food and Agriculture Organization (FAO), and the USEPA has a number of relevant databases, including the on-line Integrated Risk Information System (IRIS), the Public Health Risk Evaluation Data Base (PHRED), the Agency for Toxic Substances and Disease Registry (ATSDR) and the Health Effects Assessment Summary Tables (HEAST).

2 For carcinogens, risk is estimated by multiplying the calculated intake of the carcinogen (the CDI) by the cancer potency factor for that carcinogen, and this produces an estimate of the additional or incremental cancer risk to the individual as a result of exposure to the waste (i.e. risk = CDI × CSF). The risk is expressed as an estimate of incremental annual or lifetime risk, or the chance of mortality from cancer (e.g. one in a million or 1×10^{-6}).

The USA approach to the handling of exposure to mixtures of chemicals is to assume additivity of risks from individual chemicals. For carcinogens, this means that the overall carcinogenic risk is the sum of carcinogenic risks from individual substances, while for exposures to mixtures of non-carcinogens, the individual HIs are summed. It has to be noted, however, that different toxicants affect different target organs and therefore this approach is conservative. For example, the principal targets of cadmium are the kidneys and the principal targets of polychlorinated biphenyls are the liver and skin. Furthermore, a target may not be exposed to multiple media during the same exposure period. Hazard quotients for different exposure durations (e.g. chronic, subchronic) are not additive. Therefore, a more appropriate use of additivity is as a screening mechanism to indicate whether or not a risk may be presented by mixtures of chemicals.

The significance of the HI in terms of considering the acceptability of the risk was referred to above as being based on whether or not HI is greater than 1. For cancer risks it is necessary to set numerical risk targets that define acceptability. The USEPA has set the range for acceptability as a $1/10\,000$–$1/10^6$ risk of death over a lifetime (i.e. 70 years). In 1984 the DOE set a target risk value of 1×10^{-6} per annum, equivalent to a lifetime risk of 7×10^{-5}. In guidance on accident risks, the Health and Safety Executive suggested an acceptability band of $1/10\,000$ (1×10^{-5}) per annum as the upper level of risk of receiving a 'dangerous dose' by a member of the public, and $1/10^6$ (1×10^{-6}) as a broadly acceptable level of risk (HSE 1989, 1992).

In its report on incineration, the UK RCEP (1993) stated that:

> 'There is a consensus amongst regulatory authorities in the UK and USA that incremental risks of death greater than 1 in 10 000 (10^{-5}) are too high to be acceptable; and that a risk of 1 in 1 000 000 (10^{-6}) represents a reasonable upper bound beyond which measures to achieve a further reduction in the risk would not be justified in terms of the benefits gained.'

It is important in the use of any such criteria to remember that the health risks are usually presented in terms of carcinogenesis rather than death. In incineration risk assessments (e.g. Powergen CHP Ltd 1995) it is assumed that there is a 50% probability of death following carcinogenesis, with the calculated risks divided by two prior to comparison with such criteria.

16.6 CONCLUSIONS—RISK MANAGEMENT FOR WASTE DISPOSAL AND TREATMENT

Guidance in the UK (DOE 1995a) defines risk management as 'the process of implementing decisions about tolerating or altering risks'. This author views the assessment of risk as a component of risk management in that it should provide for a structured understanding of the nature of potential adverse outcomes in the context of taking decisions. Ideally, the assessment should consider risks without accepted mitigation in place and then address the residual risk with appropriate mitigation. Only by this admittedly more time-consuming approach can the relative effectiveness of different mitigation measures be understood.

Risk management for waste treatment and disposal can include one or a combination of the following strategies.

1 Managing risk through the strategic consideration of the BPEO for specific waste streams at the national or regional level.

2 Preventing risk through the siting process.

3 Incorporating appropriate engineering, monitoring and management controls during the design, operational, decommissioning and post-closure stages of a project.

4 Interrupting a complete exposure pathway, thereby breaking the link between source and receptor (often only possible in extreme cases, as it may involve facility closure).

It is clear that risk management for waste treatment and disposal has become a high priority in the UK in the 1990s. This has resulted from: a combination of public pressures for control over facilities which can present inequitable risks (i.e. local communities having to bear risks arising from the management of waste not generated by themselves); expert and public recognition that waste management in the past has not always been conducted with a full appreciation of potential adverse impacts; and deregulatory pressures encouraging non-prescriptive approaches to design and operation. Risk assessment experience and practice in the USA has provided the basis of the development of methodologies, but with increasing application of risk assessment we can expect more UK specific guidance. Difficult areas for resolution remain in relation to the use of conservative assumptions and the understanding of the impact of different types of uncertainty and variability upon assessment outputs. Further work on probabilistic uncertainty analysis seems likely, including detailed development of distribution functions and ranges for inputs to model algorithms and exposure assumptions. It can be argued that a careful balance is required between encouraging pragmatic conceptual risk assessment frameworks, which are more likely to be readily adopted by the industry, and providing for an adequate degree of conservatism to deal with uncertainty, to address social concerns and to apply the precautionary principle to environmental protection.

There are whole areas where risk assessment methodologies are underdeveloped, largely due to a lack of data on some fundamental processes: for example on discontinuous releases during the start-up and shut-down periods of incineration; and in relation to the behaviour of gases in soils. Ecological risk assessments remain largely unconsidered in the context of proposed or operational waste treatment and disposal facilities. Ecological assessments, particularly within environmental assessments, largely rely on deterministic methods for considering impacts of waste disposal where sensitive species and/or habitats are identified. It may be that the industrial nature of waste disposal and treatment facilities has favoured siting of many facilities in locations remote from sensitive ecological environments. However, the lower attention to ecological risk assessment appears to be partly as a result of a decision to address human health risks first, for political reasons, and partly from a mistaken belief that human health protection based on conservative assumptions is universally protective (Suter 1993). The concept of ecological and human health risk assessments as parallel activities is currently missing in UK policy and practice. The greater complexity of ecological risk assessment in terms of potentially a larger number and variety of targets, pathways and responses, combined with less guidance and fewer consensus models and standard assumptions, does present data and knowledge problems.

For some, the use of risk assessment is an admission that an amount of risk is acceptable and that some individuals can be affected to a greater degree than others. This has been a particularly powerful argument amongst groups opposed to incineration and concerned about the health risk posed by emissions. There is no doubt that the presentation of risk assessment output requires far greater attention to explanation of assumptions, methodology, etc. However, the expert view that risk communication is about the transfer of information from the assessor to other interested parties and that information provision and the presentation of risk comparisons will help to overcome opposition to proposals for new facilities indicates a significant failure to understand the basis of social concerns.

The siting of waste treatment and disposal facilities provides a significant example of the failings of traditional processes for making risk management decisions (Petts 1994b). There is a need for far greater attention to the use of consensus-building approaches to waste management strategy development (Petts 1995) and recognition that involvement of different interested parties in the whole risk management process (including risk assessment) is going to become essential. Waste treatment and disposal provides a classic example of the loss of trust and credibility in operators, regulators and decision makers on the part of the public (Petts 1994c). Risk assessment and management decisions which are open to public input and scrutiny present a significant challenge to traditional institutional arrangements, however, if trust and credibility is to be regained this represents an essential requirement.

16.7 REFERENCES

Anon. (1986) *Report of the Non-Statutory Inquiry into the gas explosion at Loscoe, Derbyshire, 24 March 1986*. Derbyshire County Council, Matlock.

Anon. (1993) Landfill under pressure as methane emission data soar. *ENDS Report*, **217**(February), 7.

Asante-Duah, D. Kofi (1993) *Hazardous Waste Risk Assessment*. Lewis Publishers, Boca Raton, FL.

Bagchi, A. (1990) *Design, Construction and Monitoring of Sanitary Landfill*. John Wiley, New York.

Barnes, D.G. & Dourson, M. (1988) Reference dose (RfD): description and use in health risk assessments. *Regulatory Toxicology and Pharmacology*, **8**, 471–486.

Belevi, H. & Baccini, P. (1989) Long-term behaviour of municipal solid waste landfills. *Waste Management and Research*, **7**, 43–56.

Blakey, N.C. (1982) Absorptive capacity of refuse: WRC research. In: *Proceedings of the Harwell Symposium on Landfill Leachate*, May 1982, Environmental Safety Centre, Harwell Laboratory, Harwell.

Burmaster, D.E. & Anderson, P.D. (1994) Principles of good practice for the use of Monte Carlo techniques in human health and ecological risk assessments. *Risk Analysis*, **14**(4), 477–481.

CCME (1992) *National Classification System for Contaminated Sites*. Report CCME EPC-CS39E, Canadian Council of Ministers of the Environment, Winnipeg, Manitoba.

CEC (1980) Directive on the protection of groundwater against pollution caused by certain dangerous substances, 80/68/EEC. *Commission of the European Communities Official Journal*, **L20**, 43–47.

CEC (1994) *Technical Guidance Document on Risk Assessment of Existing Substances*. Document No. XI/919/94/EN, Commission of the European Communities, Brussels.

CIRIA (1995) *Remedial Treatment for Contaminated Land*, Vol. III. *Site Investigation and Assessment*. Construction Industry Research and Information Association, London.

Clayton, P., Coleman, P., Leonard, A. *et al.* (1991) *Review of Municipal Solid Waste Incineration in the UK*. Report LR 776 (PA), Warren Spring Laboratory, Stevenage.

Coleman, A.K. (1985) Alternatives to landfill. In: *Hazardous Waste Management Handbook* (ed. A. Porteous), pp. 209–255. Butterworths, London.

Covello, V.T. & Merkhofer, M.W. (1993) *Risk Assessment Methods: Approaches for Assessing Health and Environmental Risks*. Plenum Press, New York.

Cullen, A.C. (1994) Measures of compounding conservatism in probabilistic risk assessment. *Risk Analysis*, **14**(4), 389–393.

DOE (1978) *Cooperative Programme of Research on the Behaviour of Hazardous Wastes in Landfill Sites*. Department of the Environment, Final Report of the Policy Review Committee, HMSO, London.

DOE (1984) *Disposal Facilities on Land for Low and Intermediate-Level Radiocative Wastes: Principles for the Protection of the Human Environment*. Department of the Environment, HMSO, London.

DOE (1988) *Environmental Assessment: A Guide to the Procedures*. Department of the Environment, HMSO, London.

DOE (1990) *Appraisal of Hazards Related to Gas Producing Landfills*. Report No. CWM/016/90, Department of the Environment, London.

DOE (1994a) *Licensing of Waste Management Facilities*. Department of the Environment, Waste Management Paper no 4, HMSO, London.

DOE (1994b) *Health Effects from Hazardous Waste Sites*. Report no. CWM/057/92, Department of the Environment, Wastes Technical Division, London.

DOE (1994c) *Planning and Pollution Control. Planning Policy Guidance Note No. 23*. Department of the Environment, HMSO, London.

DOE (1995a) *A Guide to Risk Assessment and Risk Management for Environmental Protection*. Department of the Environment, HMSO, London.

DOE (1995b) *Making Waste Work: a Strategy for Sustainable Waste Management in England and Wales*. Department of the Environment, HMSO, London.

DOE (1995c) *Landfill Design, Construction and Operational Practice*. Department of the Environment, Waste Management Paper No. 26B, HMSO, London.

DOE (1995d) *Preparation of Environmental Statements for Planning Projects that Require Environmental Assessment: A Good Practice Guide*. Department of the Environment, HMSO, London.

Department of Health (1995) *COT statement on the USEPA Draft Health Assessment Document for 2,3,7,8-Tetrachlorodibenzo-p-dioxin and related compounds*. Department of Health, London.

Earle, T.C. & Cvetkovich, G.T. (1995) *Social Trust: Toward a Cosmopolitan Society*. Praeger, Westport, CT.

Eduljee, G.H. (1992) Assessing the risks of landfill activities. In: *Proceedings of the 1992 Harwell Waste Management Symposium, New Developments in Landfill*, May 1992, Environmental Safety Centre, Harwell.

Eduljee, G.H. (1995) Emissions to the atmosphere. In: *Issues in Environmental Science and Technology*, No. 3. *Waste Treatment and Disposal* (eds R.E. Hester & R.M. Harrison), pp. 69–89. The Royal Society of Chemistry, Cambridge.

Eduljee, G.H., Dyke, P. & Cains, P.W. (1995) PCDD/PCDF releases from various waste management strategies. *Warmer Bulletin*, **46**, 22–23.

ECETOC (1992) *Exposure of man to dioxins: a perspective on industrial waste incineration*. Technical Report no. 49, European Centre for Ecotoxicology and Toxicology of Chemicals, Brussels.

Elliot, P., Shaddick, G., Klienschmidt, I. *et al.* (1996) Cancer incidence near municipal solid waste incinerators in Great Britain. *British Journal of Cancer*, **73**, 702–710.

Environment Agency (1997) Best Practicable Environmental Option Assessments for Integrated Pollution Control, 2 volumes. HMSO, London.

Ghabaee, K. & Rodwell, W.R. (1989) *Landfill Gas Modelling: a Literature Survey of Landfill Gas Generation and Migration*. AEEW-R2567, Atomic Energy Establishment, Winfrith, Dorchester.

Giroud, J.P., Badu-Tweneboah, K. & Bonaparte, R. (1989) Leakage through liners constructed with geomembranes, Parts I and II. *Geotextiles and Geomembranes*, **8**(1–2), 27–67, 78–111.

Giroud, J.P. *et al.* (1992) Rate of leakage through a composite liner due to geomembrane defects. *Geotextiles and Geomembranes*, **11**, 1–28.

Hall, D.H., Jefferies, M., Gronow, J. & Harris, R.C. (1995) Probabilistic risk assessment for landfill regulation. In: *Contaminated Soil '95* (eds W.J. van den Brink, R. Bosman & F. Arendt), pp. 617–619. Kluwer Academic Publishers, Dordrecht.

Harrad, S. & Jones, K. (1992a) A source inventory and budget for chlorinated dioxins (PCDDs) and furans (PCDFs) in the UK environment. *Science of the Total Environment*, **126**, 89–107.

Harrad, S. & Jones, K. (1992b) Dioxins at large. *Chemistry in Britain*, **December**, 1110–1112.

Hart Associates (1984) *Assessment of potential public health impacts associated with predicted emissions of PCDDs and PCDFs from the Brooklyn Navy Yard Resource Recovery Facility*. Submitted to the New York Department of Sanitation.

Hartley, R.P. (1988) Assessing the reliability of flexible membrane liners. In: *Land Disposal of Hazardous Waste* (eds J. Gronow, A.N. Schofield & R.K. Jain), pp. 245–259. Ellis Horwood, Chichester.

Hattemer-Frey, H.A. & Travis, C.C. (eds) (1991) *Health Effects of Municipal Waste Incineration*. CRC Press, Boca Raton, FL.

Hazardous Waste Inspectorate (1986) *Second Report—Hazardous Waste Management: Ramshackle and Antediluvian?* HMSO, London.

HMIP (1992) *Waste Disposal and Recycling: Municipal Waste Incineration—Process Guidance Note IPR5/3*. Her Majesty's Inspectorate of Pollution, HMSO, London.

HMIP (1995) *Review of Dioxin Emissions in the UK*. DoE/HMIP/RR95/004, Her Majesty's Inspectorate of Pollution, London.

HMIP (1996) *A Risk Assessment of Dioxin Releases from Municipal Waste Incineration Processes*. Her Majesty's Inspectorate of Pollution, London.

HSE (1989) *Risk Criteria for Land-use Planning in the Vicinity of Major Industrial Hazards*. Health and Safety Executive, HMSO, London.

HSE (1992) *The Tolerability of Risks from Nuclear Power Stations*, 2 edn. Health and Safety Executive, HMSO, London.

Jones, J.A. (1988) What is required of dispersion models

and do they meet the requirements? *NATO-CCMS 17th International Technical Meeting on Air Pollution Modelling*, Cambridge, 19–22 September.

Knox, K. (1990) The relationship between leachate and gas. Paper presented at the *Energy and Environment Conference*, Bournemouth, Energy Technology Support Unit/Department of the Environment, ETSU, Harwell.

Levine, A. (1981) *Love Canal: Science, Politics and People*. Lexington, MA.

Levine, A., Fratt, D.B., Leonard, A., Bruins, R.J.F. & Fradkin, L. (1991) Comparative analysis of health risk assessments for municipal waste combustors. *Journal of the Air and Waste Management Association*, **41**, 20–31.

Liem, A.K.D., Hoogerbrugge, R., Koostra, P.R., van der Velde, E.G. & de Jong, A.P.J.M. (1991) Occurrence of dioxins in cows' milk in the vicinity of municipal waste incinerators and a metal reclamation plant in The Netherlands. *Chemosphere*, **23**(11–12), 1675–1684.

Lloyd, O.L., Lloyd, M.M., Williams, F.L.R. & Lawson, A. (1988) Twinning in human populations and in cattle exposed to air pollution from incinerators. *British Journal of Industrial Medicine*, **45**, 556–560.

Lorber, M., Cleverly, D., Schaum, J., Phillips, L., Schweer, G. & Leighton, T. (1994) Development and validation of an air-to-beef food chain model for dioxin-like compounds. *Science of the Total Environment*, **156**, 39–65.

Loxham, M. (1993) The design of landfill sites—some issues from a European perspective. In: *Landfill Tomorrow—Bioreactors or Storage, Proceedings of a Seminar held at Imperial College*, pp. 7–12. Centre for Waste Management, Imperial College, London.

MAFF (1994) *Household Food Consumption and Expenditure*. Ministry of Agriculture, Fisheries and Food, HMSO, London.

MAFF (1995) *Dioxins in Food—UK Dietary Intake, Food Surveillance Information Sheet, No. 71*. Ministry of Agriculture, Fisheries and Food, HMSO, London.

Mather, J.D. (1992) Current landfill design—a short term engineering solution with a long term environmental cost. In: *Planning and Engineering of Landfills, Proceedings of the Midland Geotechnical Society Conference*, July 1991. Midland Geotechnical Society, University of Birmingham.

Maynard, R.L., Cameron, K.M., Fielder, R., McDonald, A. & Wadge, A. (1995) Setting air quality standards for carcinogens: an alternative to mathematical quantitative risk assessment—discussion paper. *Human and Experimental Toxicology*, **14**, 175–186.

Maxim, L.D. (1989) Problems associated with the use of conservative assumptions in exposure and risk analysis. In: *The Risk Assessment of Environmental and Human Health Hazards: a Textbook of Case Studies* (ed. D.J. Paustenbach), pp. 526–560. John Wiley & Sons, New York.

McFarland, R. (1992) Simple quantitative risk assessment technique for prioritisation of chemically contaminated sites in New South Wales, Australia. In: *Risk Assessment, Proceedings of a Conference on Risk Assessment*, 6–9 October, Health and Safety Executive, London.

Morgan, M.G. & Henrion, M. (1990) *Uncertainty, a Guide to Dealing with Uncertainty in Quantitative Risk and Policy Analysis*. Cambridge University Press, New York.

NAS-NRC (1983) *Risk Assessment in the Federal Government: Managing the Process*. NAS–NRC Committee on the Institutional Means for the Assessment of Risks to Public Health, National Research Council. National Academy Press, Washington, DC.

NRA (1992) *Policy and Practice for the Protection of Groundwater*. National Rivers Authority, Bristol.

Olie, K., Vermeulen, P.L. & Hutzinger, O. (1977) PCDD & PCDF trace constituents of flyash and flue gas of some municipal incinerators in The Netherlands. *Chemosphere*, **6**, 455–456.

OECD (1990) *International Symposium on Safety Assessment of Radioactive Waste Repositories*. Organization of Economic Cooperation and Development, Paris.

Paustenbach, D.J. (1989) A comprehensive methodology for assessing the risks to humans and wildlife posed by contaminated soils: a case study involving dioxins. In: *The Risk Assessment of Environmental and Human Health Hazards: a Textbook of Case Studies* (ed. D.J. Paustenbach), pp. 296–330. John Wiley & Sons, New York.

Paustenbach, D.J., Shuh, H.P. & Murray, F.J. (1987) Assessing the potential human health hazards of dioxin-contaminated soil. American Chemical Society, Symposium Series, Vol. 338, Washington, DC.

Petts, J. (1992) Incineration risk perceptions and public concern: experience in the UK improving risk communication. *Waste Management and Research*, **10**, 169–182.

Petts, J. (1993) Risk assessment, risk management and containment landfill. In: *Options for Landfill Containment, Proceedings of the Harwell Waste Management Symposium*, 19 May, pp. 5–16, Environmental Safety Centre, Harwell.

Petts, J. (1994a) Stress and public concern over hazardous waste. In: *Human Stress and the Environment* (ed. J. Rose), pp. 181–208. Gordon Breach Science Publishers, Switzerland.

Petts, J. (1994b) Effective waste management: understanding and dealing with public concerns. *Waste Management and Research*, **12**, 207–222.

Petts, J. (1995) Waste management strategy development:

a case study of community involvement in consensus-building in Hampshire. *Journal of Environmental Planning and Management*, **38**(4), 519–536.

Petts, J. & Eduljee, G. (1994) *Environmental Impact Assessment for Waste Treatment and Disposal Facilities*. John Wiley & Sons, Chichester.

Powergen CHP Ltd (1995) *Thameside Energy from Waste Power Station. Environmental Statement*. Prepared by Environmental Resources Management, Powergen CHP Ltd, London.

Powlesland, C. & Frost, R. (1990) *A methodology for undertaking BPEO studies of sewage sludge treatment and disposal*. Report No. PRD 2305-M/1, Water Research Centre, Medmenham.

RCEP (1988) *Best Practicable Environmental Option*. Royal Commission on Environmental Pollution, Twelfth Report, HMSO, London.

RCEP (1993) *Incineration of Waste*. Royal Commission on Environmental Pollution, Seventeenth Report, HMSO, London.

Reichard, E., Cranor, C., Raucher, R. & Zapponi, G. (1990) *Groundwater Contamination Risk Assessment*. IAHS Publication No. 196, International Association of Hydrological Sciences, Wallingford, Oxford.

RIVM (1994) *Uniform System for the Evaluation of Substances (USES)*, Version 1.0. National Institute of Public Health and Environmental Protection (RIVM), The Hague.

Royal Society (1992) *Risk: analysis, perception and management*. Report of a Royal Society Study Group, Royal Society, London.

Shy, C.M., Degnan, D., Fox, D.L. *et al.* (1995) Do waste incinerators induce adverse respiratory effects? An air quality and epidemiological study of six communities. *Environmental Health Perspectives*, **103**, 714–724.

Silbergeld, E.K. (1992) Risk assessment—the perspective and experience of the environmentalists. In: *Risk Assessment, Proceedings of a Conference on Risk Assessment*, 6–9 October, Health and Safety Executive, London.

Slob, W., Trrost, L.M., Krijgsman, M., de Koning, J. & Sein, A.A. (1993) *Combustion of Municipal Solid Waste in the Netherlands*. Report No. 730501052, RIVM, Bilthoven.

Smith, E.D. & Barnthouse, L.W. (1987) *User's Manual for the Defense Priority Model*. US Defense Department, Washington, DC.

Stern, P.C. (1991) Learning through conflict: a realistic strategy for risk communication. *Policy Sciences*, **24**, 99–119.

Suter, G.W. (ed) (1993) *Ecological Risk Assessment*. Lewis Publishers, Chelsea, MI.

Szepesi, D.J. (1989) *Compendium of Air Quality Simulation Models*. Akademiai Kiado, Budapest.

Tarr, J.A. & Jacobsen, C. (1987) Environmental risk in historical perspective. In: *The Social and Cultural Construction of Risk* (eds B.B. Johnson & V.T. Covello), pp. 317–334. D. Reidel Publishing Company, Dordrecht.

Thames Water Utilities (1991a) *Environmental Statement—Crossness Sewage Sludge Incinerator*. Prepared by Ove Arup, Thames Water, Reading.

Thames Water Utilities (1991b) *Environmental Statement—Beckton Sewage Sludge Incinerator*. Prepared by Ove Arup, Thames Water, Reading.

Thompson, M. (1979) *Rubbish Theory: the Creation and Destruction of Value*. Oxford University Press, Oxford.

Travis, C.C. (ed.) (1991) *Municipal Waste Incineration Risk Assessment*. Plenum Press, New York.

Travis, C.C. & Cook, S.C. (1989) *Hazardous Waste Incineration and Human Health*. CRC Press, Boca Raton, FL.

Travis, C.C., Holton, G.A., Etnier, E.L. *et al.* (1986) Assessment of inhalation and ingestion population exposures from incinerated hazardous wastes. *Environment International*, **12**, 533–540.

Turner, K., Brown, D., Vickers, J. & Powell, J. (1992) An assessment of BPEO in practice. In: *Integrated Pollution Control—a Practical Guide for Managers* (eds T. O'Riordan & A. Blowers). IBC Technical Services Ltd, London.

USEPA (1981) *Interim Evaluation of Health Risks Associated with Emissions of Tetrachlorinated Dioxins from Municipal Waste Resource Recovery Facilities*. US Environmental Protection Agency, Washington, DC.

USEPA (1982) Appendix A—Uncontrolled Hazardous Waste Site Scoring System: a User's Manual. US Environmental Protection Agency, *Federal Register*, **37**(137), 31219–31243.

USEPA (1985) *Liner Location Risk and Cost Analysis Model*. Report No. PB87-157210/AS, US Environmental Protection Agency, Economic Analysis Branch, Office of Solid Wastes, Washington, DC.

USEPA (1987a) *Methodology for the Assessment of Health Risks Associated with Multiple Pathway Exposure to Municipal Waste Combustor Emissions*. US Environmental Protection Agency, Cincinnati, OH.

USEPA (1987b) *MWC Study: Assessment of Health Risks Associated with Municipal Waste Combustion Emissions*. US Environmental Protection Agency, EPA/530-SW-87-021g, Washington, DC.

USEPA (1987c) *Industrial Source Complex (ISC) Dispersion Model User's Guide*, 2 edn. US Environmental Protection Agency, EPA/450/4-88-002a, Office of Air Quality Research, Research Triangle Park, NC.

USEPA (1987d) *User's Guide to the Rough Terrain Diffusion Model (RTDM)*, Version 3.2. Report P-D535-585, US Environmental Protection Agency, Office of Air Quality Research, Research Triangle Park, NC.

USEPA (1988) *Superfund Exposure Assessment Manual*. US Environmental Protection Agency, EPA/540/1-88/001, Office of Remedial Response, Washington, DC.

USEPA (1989) *Risk Assessment Guidance for Superfund Volume 1: Human Health Evaluation Manual (Part A)*. US Environmental Protection Agency, EPA/540/1-89/002, Office of Remedial Response, Washington, DC.

USEPA (1990a) Hazard ranking system—final rule. US Environmental Protection Agency, *Federal Register*, **55**, 51532–51667.

USEPA (1990b) *The Revised Hazard Ranking System: an Improved Tool for Screening Superfund Sites*. 9320.7-01 FS Nov. 1990, US Environmental Protection Agency, Office of Solid Waste and Emergency Response, Washington, DC.

USEPA (1990c) *The Revised Hazard Ranking System: Background Information*. 9320.7-03 FS, Nov. 1990, US Environmental Protection Agency, Office of Solid Waste and Emergency Response, Washington, DC.

USEPA (1992) Final Guidelines for Exposure Assessment. US Environmental Protection Agency, *Federal Register*, **57**.

USEPA (1994) Estimating exposure to dioxin-like compounds. US Environmental Protection Agency, Review draft, EPA/600/6-88/005ca. Washington, DC.

Vlek, C.A.J. (1995) Understanding, accepting and controlling risks: a multistage framework for risk communication. *European Review of Applied Psychology*, **45**(1), 49–54.

Welsh Office (1995) *Polychlorinated biphenyls, dioxins and furans in the Pontypool environment*. Seventh Report to the Welsh Office by the Environmental Risk Assessment Unit, University of East Anglia. Welsh Office, Cardiff.

Westlake, K. (1995a) Landfill. In: *Issues in Environmental Science and Technology*, No. 3. *Waste Treatment and Disposal* (eds R.E. Hester & R.M. Harrison), pp. 43–67. Royal Society of Chemistry, Cambridge.

Westlake, K. (1995b) *Landfill Waste Pollution and Control*. Albion Publishing, Chichester.

WHO (1987) *Air Quality Guidelines for Europe*. World Health Organization, Regional Office for Europe, European Series No. 23, Copenhagen.

Wynne, B. (1987) *Risk Management and Hazardous Wastes: Implementation and the Dialectics of Credibility*. Springer-Verlag, Berlin.

Chapter 17
Risk Assessment and Management in the Exploitation of the Seas

JOHN S. GRAY

17.1 INTRODUCTION

Firstly, a disclaimer. This review is not concerned with the environmental risks associated with the exploitation of living resources, nor is it concerned with human health risks of eating contaminated fish and shellfish, and nor with risks of effects of radionuclides in the environment. These aspects relate to classical risk assessment and are well covered in other reviews, for example by the International Atomic Energy Agency (IAEA) for human health risks of radionuclides (Anon. 1995; and see other chapters in this volume). Here, more general aspects of the environmental risks caused by physical disturbances and contaminants will be covered.

Classical risk assessment has been relatively little used in a marine context. Lloyd (1980) was one of the first to try to devise a framework for hazard assessment in the marine environment. Recently there has been a spate of new publications on environmental toxicology in general (Moriarty 1983); in a marine context, Forbes & Forbes (1994) and Chapman (1995) provide good overviews.

One marine example of risk assessment is that of the risks of oil tanker accidents leading to damage to the marine environment. Whilst the general public probably view these accidents as a major threat to life in the sea, recent data show that, globally, approximately 50% of the oil entering the seas comes from land-based sources (GESAMP 1993). Data show that in 1989 tanker accidents accounted for 0.114 million tonnes discharged per year, whereas 0.253 came from bilge and fuel oil discharges. Thus, the actual risk of tanker accidents to the marine environment is small. However, tanker accidents can have large local impacts, but effects seldom last for longer than 10 years (GESAMP 1993). If effects of accidents are considered on regional scales, the impacts are relatively slight. Public conceptions of what are severe environmental risks are often grossly inaccurate (see Chapter 20).

A good example of the issues involved in risk assessment and risk management in the marine environment is provided by a recent, much publicized event: the Greenpeace campaign to stop the dumping of the disused oil rig *Brent Spar* in deep water off the west coast of Scotland. Greenpeace used, as the main elements of their campaign, the fact that there had not been a recent inventory of the chemicals contained onboard and thus it was not possible to give a risk assessment of the likely environmental risks posed to the environment. They claimed (letter to me, dated 5/7/95 from Dr Helen Wallace, Science Unit, Greenpeace UK) that 'there is a lack of understanding of the deep sea environment and it is currently impossible to predict the effects of the proposed dumping on deep sea ecosystems'. They also claimed that the dismantling of the *Brent Spar* ashore was technically feasible and was Greenpeace's preferred option, i.e. they recommended risk management without there being any form of risk assessment.

The Greenpeace campaign resulted in a massive boycott of Shell petrol stations in Germany and The Netherlands, support being given to their campaign by the Chancellor of Germany and Environment Ministers from The Netherlands and Sweden. Shell UK was finally forced to abandon dumping in the deep sea as an immediate option. At the time of writing, *Brent Spar* is anchored in a

Norwegian fjord awaiting a verdict on how it is to be disposed.

This incident is an excellent example of problems related to risk assessment and risk management in a marine context. Firstly, from the risk assessment point of view, when the decision had been taken by Shell and the British Government to dispose of the *Brent Spar* in deep water there was doubt concerning the amounts and types of chemicals contained in the tanks. Shell's own survey was done in 1991 and strictly an assessment of environmental risks cannot be stated without having a full and detailed inventory. Analyses done independently by Den Norske Veritas showed that the chemical content and quantities did not vary much from those found by Shell in 1991 (Masood 1995). The contents were then estimated: oil, 9.2 t; cadmium, 5.8 kg; chromium, 2.1 kg; copper, 42.9 kg; nickel, 3.9 kg; zinc, 87.4 kg; arsenic, 0.3 kg; and mercury, 0.2 kg (AURIS 1994). What are the environmental risks posed by these chemicals to the environment at the disposal site at 2000 m deep? Will the chemicals leak out slowly over time or will they be released on impact on the sea bed? Greenpeace's claim that it is not possible to estimate the environmental risks in the deep sea is false. There are many studies done in the area and the current regime, sediment types and bottom fauna are reasonably well known. Experienced, international deep-sea biologists have responded to Greenpeace stating that the environmental risks to the deep sea in the area concerned were minimal (see Angel 1995). Greenpeace has not produced an analysis of the environmental risks of their preferred option — disposal on shore — and yet was able to persuade important politicians that it was right and that Shell was wrong.

The important lesson is that in terms of environmental management there was not one single environmental option — dispose at sea or do not dispose at sea — with the accompanying risks. Rather, environmental risks of deep-sea disposal had to be weighed against those of disposal on shore. Neither Shell nor Greenpeace gave the public (or politicians) the data on which they could make a rational choice. Clearly, the politicians judged it politically expedient to be seen to be 'green' without being concerned about the long-term consequences of their actions. The economic costs of such short-sightedness will become apparent later when the full costs of dismantling all North Sea oil structures on shore, the politically preferred option, are better appreciated (calculated at £6 billion; Williams 1994). The key question that has not been answered in the Greenpeace campaign is what are the environmental risks of alternative options?

In this review I shall first discuss the traditional ways that have been used to protect the marine environment from adverse effects of human activities. Then I shall examine newer ideas such as the precautionary principle and reversal of the burden of proof. This will be followed by the development of a strategy for marine environmental risk exposure and effects, and finally a consideration of how this information can be used in a management context.

17.2 REVIEW OF TRADITIONAL WAYS OF PROTECTING THE SEAS

There are a number of approaches that have been taken to protecting seas from potentially damaging influences, both physical and chemical. I present these in a historical framework.

17.2.1 Assimilative capacity

The environment has a capacity to cope with discharges of waste without suffering deleterious effects on its biological systems. This ability to cope is known as the assimilative capacity (Cairns 1977) or as the environmental capacity (Pravdic 1985; Portmann & Lloyd 1986) of a given area. For the marine domain a definition of environmental capacity proposed by the UN's Joint Group of Experts on Scientific Aspects of Marine Environmental Protection (GESAMP) is 'a property of the environment, defined as its ability to accommodate a particular activity, or rate of activity, without unacceptable impact' (Pravdic 1985). The concept is based on the following assumptions.

1 That a certain level of some contaminants may not produce any undesirable effect on the marine environment and its various uses.

2 That each environment has a finite capacity to accommodate some wastes without unacceptable consequences.

3 That such capacity can be quantified, apportioned to certain activity, and utilized.

The problems with these assumptions are that there is no definition of what is an undesirable effect or how it will be measured, nor how the capacity of a given area can be predicted or measured. The practice usually adopted is that a concession for a given discharge is given and then monitoring is done to see that the assimilative capacity is not exceeded, with the result that deleterious effects are observed on biological systems. Too often in the past, damage has occurred and then it is too late to do anything about it. Earll (1992) gives mercury in Liverpool Bay as an example; the mercury has been shown to be having negative effects on biological systems and yet there is no hope of removing the mercury that was permitted to be discharged.

The key problem with the environmental capacity concept is that the management and legislative systems surrounding the environmental capacity have not used all the scientific data and knowledge available. Instead, decisions have been made, often with 'application factors and other arbitrary rules of thumb' (Stebbing 1992), that lead to pollution problems that are 'a legacy of a weak scientific basis for consenting procedures and insensitive techniques, or imprecise toxicological data, while the regulators have been obliged not to deny industry and society its traditional access to the environment to dispose of wastes'. The failing of management, and not science, in relation to environmental capacity is also acknowledged by Earll (1992): 'in practice, science and its processess have often been totally ignored when development decisions have been taken'. Yet science has been mistakenly attacked by Greenpeace as 'failing the environment' (Wynne & Mayer 1993). It is not science that has failed in the use of the assimilative capacity of the marine environment, but the application of science by managers and legislators.

17.2.2 Black and grey lists

One of the earliest attempts to regulate discharges of toxic chemicals was the production of lists of chemicals that were so toxic that they should not be discharged to the environment, i.e. black-listed chemicals. Chemicals that were not known to be so toxic or had unknown properties were put on the grey list. The problems, of course, were that many substances not listed have subsequently been found to be highly dangerous if discharged to the environment.

The listing idea has, however, continued unabated and in a recent survey a European Science Foundation report gives 74 such lists (Peterson 1995). Over 7000 chemicals have been listed under various categories of posing environmental risks. The United Nations Environmental Programme (UNEP) has developed a PC-based Register of Potentially Toxic Chemicals, which classifies the chemicals into 17 variables from toxicology, ecotoxicology and bioaccumulation.

In a marine context, the work of the Oslo and Paris Commission (OSPARCOM) has been influential in compiling such lists. The North Sea Task Force was set up (by OSPARCOM and North Sea states) in order to assess the state of the North Sea and to recommend measures to improve its environmental quality. The Task Force (OSPARCOM 1993) lists 36 chemicals that enter the sea from land-based sources that are of concern and a subset of 17 that enter the marine environment from atmospheric inputs. The Task Force recommended that discharges of dioxins, mercury, cadmium and lead be reduced by 70% or more by 1995, with 1985 as the baseline year. Special attention was focused on the need to phase out the use of polychlorinated biphenyls (PCBs) and for special regulations on uses of tributyl tin (TBT) and hexachlorocyclohexane (HCH). Tributyl tin was used to prevent fouling on ships, but led to unexpectedly severe effects on marine molluscs, particularly oysters and dog whelks (Bryan *et al.* 1986, 1988; Gibbs *et al.* 1987; Alzieu 1991; Horiguchi *et al.* 1995; Minchin *et al.* 1995).

Perhaps the main problem with listings is that they are based on hazard assessment rather than on a proper assessment of risks. Environmental characteristics vary and thereby the potential problems (risks) posed by the chemical also vary. For example, in terrestrial environments top priority is often given to lead, cadmium, sulphur dioxide and ozone, which as airborne contaminants are known to lead to significant pollution effects on plants. Priorities for those working on invertebrate animals might be polycyclic aromatic hydrocarbons (PAHs) or organophosphorus pesticides. In freshwater habitats, acidification is known to be a major problem over many areas in Europe. In the marine environment, although in the 1960s heavy metals were often regarded as a problem, new clean-technology analysis methods show that there are seldom problems with heavy metals in the marine environment (GESAMP 1990). Organic chemicals such as TBT, PAHs, PCBs and HCHs are of more general concern.

17.2.3 Montreal Guidelines

The UNEP developed the Montreal Guidelines (UNEP 1985), which are a set of recommendations to governments to assist them in the development of legislation for the protection of the marine environment against pollution from land-based sources. The guidelines have taken common elements and principles from existing agreements, such as the Oslo, Paris and Helsinki Conventions, the Athens protocol and the UN convention on the Law of the Sea. Strategies for protecting, preserving and enhancing the quality of the marine environment are presented in Annex 1 to the Guidelines. Firstly, there are *control* strategies such as those based on marine quality standards, emission standards and environmental planning. The marine quality guidelines consider standards based on water, sediment, fish or their tissues and health or community composition, but do not give a framework for deciding how the set of standards should be derived. It is suggested that no change over ambient concentration of a given substance should be used, but this does not take into account the fact that a given ambient concentration may not have any deleterious effects. Dilution and loads are also mentioned, but other important components of hazard assessment are not considered.

The Guidelines suggest that technology standards relating to emission standards should be applied on a sector by sector basis. A distinction is drawn between *best practicable technology*, which takes into account costs, and *best available technology*, which does not but it is suggested that it should apply to the most noxious substances.

In relation to planning strategies: certain activities are stated as being incompatible with values or uses of the environment; uses of the environment will have different quality standards and the standards should be defined for the specific uses; environmental impact assessments are necessary for any activity affecting the environment; and regional management is needed for such areas as the coastal zone, drainage basins and specially sensitive areas.

The Guidelines give a useful checklist of factors that have to be considered, but they do not provide a logical scientific framework for protection and management of the marine environment. It is this framework that will be developed in the following section.

17.2.4 Precautionary principle

At the Second International Conference on the North Sea in London, 1987, the Ministerial Declaration agreed to: 'accept the principle of safeguarding the marine ecosystem of the North Sea by reducing polluting emissions of substances that are persistent, toxic and liable to bioaccumulate at source by the use of the best available technology and other appropriate measures. This applies especially when there is reason to assume that certain damage or harmful effects on the living resources of the sea are likely to be caused by such substances, even when there is no scientific evidence to prove a causal link between emissions and effects (the principle of precautionary action)'. (See Freestone & Hay (1995a) for a discussion of the origins and development of the precautionary principle.) In a recent book (Freestone & Hay 1995b) the legal implications of the precautionary principle and the challenge of implementing it are discussed in detail.

Acceptance of this principle poses a number of fundamental problems (Gray 1990, 1995). Firstly, the terms persistent, toxic and bioaccumulatable are not defined. At its most extreme, all elements should be banned from discharge because they are persistent. Likewise, does toxic mean toxic to all marine species, and at what concentration or dose is a substance defined as 'toxic', because all chemicals will become toxic at sufficiently high levels? In fact, aspirin has the same toxicity to man as DDT, and yet DDT, unlike aspirin, is bioaccumulatable and lethal doses can accumulate (Clark 1992). Therefore, in this example it is the combination of toxicity and bioaccumulatability that need to be measured together. Some chemicals bioaccumulate naturally and yet pose no stress symptoms in the accumulating organism, nor do they present health risks for other species consuming them. It is obvious that there is a need for both qualitative and quantitative definitions of these terms that will allow proper management of the marine environment and yet maintain the spirit of the precautionary principle.

A second major criticism of the precautionary principle is the acceptance of suspicion of effects rather than scientific evidence as sufficient to introduce discharge controls. The precautionary principle can be invoked to prevent discharge by simply arguing that at some future date a given chemical is likely to have an effect and discharge to sea therefore should be banned. Since the introduction of most substances to the marine environment will cause at least local disturbances, and because effect is not defined, there is a tendency for the Green Movement to argue for a very strict interpretation of the precautionary principle that nothing should be discharged to the sea.

In cases where little is known about the chemical concerned or where the biogeochemical cycle and risks for the chemical in the environment are so poorly understood, the precautionary principle does have a major role to play. Likewise, it has a further role in environmental management in that it would argue against discharges where concentrations are approaching environmental quality standards and/or critical loads at all times, and particularly in the early stages prior to the accuracy and predictions being proven.

Whilst it may seem reasonable, as is stated in the precautionary principle, to employ the best available technology to minimize discharges of toxic chemicals in order to protect the environment, there are problems with adopting this approach. In some circumstances the use of the best available technology may prove to be overprotective (e.g. a small discharge of a substance with limited persistence and toxicity to a large receiving environment) or, alternatively, in other situations it may prove to be totally inadequate (e.g. discharge of a highly persistent and toxic substance to a small receiving environment or in smaller amounts but from several different sources to the same area).

The best available technology approach has been adapted to include economical considerations. Britain argues strongly for using the best available techniques that do not entail excessive economic costs (BATNEEC, see p. 138 of Anon. (1990), 'This Common Inheritance'). However, whether or not the adoption of the best available techniques should include an assessment of economic costs and/or the consequences for other environmental sectors is a controversial and much debated point.

The sea does have a capacity to cope with waste material and there is very much more sea than land or freshwater; thus, the sea must remain an option for the disposal of human waste. If disposal at sea is prevented, as the Green movement seems to be campaigning for, then the remaining options are on land or in freshwater. Risk, cost–benefit and other relevant analyses should be done for all the environmental options in order to consider all environmental compartments.

17.2.5 Reversal of the burden of proof

At the Ministerial Declaration of the Fourth International Conference on the Protection of the North Sea, Esbjerg, Denmark, June 1995, in relation to nutrient discharges, the Ministers agreed to a new idea termed the 'reversal of the burden of proof' (Oslo and Paris Commissions 1995). Most present legislation on waste discharge requires that scientists must prove that there is an effect of a harmful substance before regulation can occur. This is clearly highly unsatisfactory and not in tune with a precautionary approach (Earll 1992). The new idea is stated in the text of the declaration in relation to concerns over nutrient discharges as:

> 'Therefore the Ministers agree to remain committed to reach the reduction targets set by the previous Conferences and to strengthen the implementation of measures as soon as possible.' (The agreement was for a 50% reduction of nitrogen and phosphorus discharges between 1985 and 1995.)
> 'Fundamental elements in fulfilling these goals in the EU and European Economic Area are *inter alia*: i) to apply in the North Sea and its catchment the measures for sensitive areas under the Urban Waste

> Water Directive and to apply the measures for vulnerable zones under the conditions of Directive 91/676, including the criterion of contribution to pollution as mentioned in Article of the Directive. These measures will be implemented for the whole of the North Sea and its catchment except for those parts of the North Sea where comprehensive scientific studies, to be delivered by 1997, demonstrate to the satisfaction of the Committees set up under the respective Directives or the relevant European Economic Area body *that nutrient inputs do not cause eutrophication effects or contribute to such effects in other parts of the North Sea*' [my emphasis].

As Stebbing (1992) has pointed out, the notion is flawed in that, harmlessness cannot be disproved because no finite number of observations of harmlessness can eliminate the possibility that there are effects somewhere at some time. Yet the politicians and managers have accepted this entirely flawed notion as a way of protecting the environment! As mentioned above, any introduction of a chemical or material will lead to some effect on the *status quo* of the environment. Thus, compliance to discharge consents has to be assessed in terms of acceptable effects. It will be difficult to say what the costs of this are. For Norway, it will mean that local councils along most of the Oslofjord and south coast will have to install nitrogen treatment plants at a cost of £100–120 million in areas where there are likely to be no measurable effects of this investment. This is simply because environmental management is now out of phase with environmental risk assessment and management is taking place without there being a proper risk assessment.

What is needed is a strategy of environmental risk assessment that ensures that the best possible techniques are available to assess whether or not there are environmental risks to biological systems and then, and only then, to assess what the environmental management options are.

17.2.6 Vulnerable marine areas

Recent European Union (EU) and European Economic Area (EEA) environmental directives have targeted areas that are vulnerable to pollution as being of particular concern. In relation to eutrophication—one of the major environmental disturbances in the marine environment (GESAMP 1990)—vulnerable areas are defined as waters affected by pollution and waters that could be affected by pollution if remedial actions were not taken. This definition requires that surveillance is undertaken to detect areas that are subjected to pollution, and that environmental risk assessments are conducted to detect areas that could be affected. Yet the procedures to do this for marine areas are not described.

17.3 EFFECTS OF CHEMICALS ON THE MARINE ENVIRONMENT

Chemicals discharged into the environment may lead to alterations in background concentration levels. If these changes are statistically significant increases, then this is called *contamination*. Contamination may or may not lead to effects on biological systems. It is usually only when effects can be measured that are related to the contamination that *pollution* is invoked. This distinction between contamination of the environment and pollution is important.

Chemicals discharged into the marine environment will be transported by currents and, depending on their physicochemical properties, will be chemically transformed or modified by both abiotic and biotic processes (mostly by microorganisms). The resulting products may have different environmental behaviour and toxic effects than the chemical discharged. It is important that these processes are understood because the type of exposure of the chemical and the nature of the organisms that are exposed will determine whether or not there are detrimental effects. Figure 17.1 shows the sequence of events that lead to the effects of a chemical at different levels of biological organization.

Although effects should be predicted from the risk assessment, it is necessary to monitor whether the predictions are upheld. Effects on biological systems can now be monitored with high precision. New techniques—biomarkers—have been developed based on biochemical and physiological responses of individual organisms to

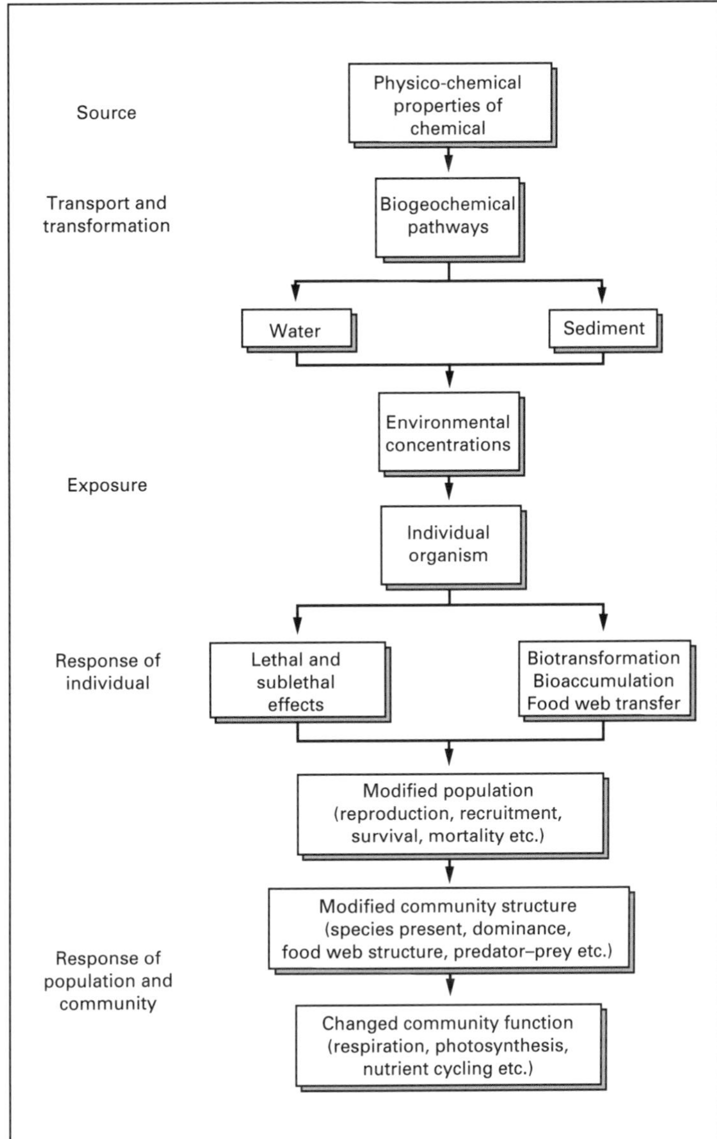

Fig. 17.1 The sequence of events leading to the effects of a chemical at different levels of biological organization. (After Rand 1994.)

stress (see McCarthy & Shugart 1990; Depledge & Fossi 1994; Peakall 1994; GESAMP 1995). Peterson (1995) divides biomarkers into three categories: biomarkers of exposure, effect and susceptibility. Some biomarkers are specific to particular contaminants, for example: metallothionein response with metal contaminants (see Hogstrand & Haux 1991; Chan 1995); acetylcholinesterase inhibition with organophosphorus pesticides (Galgani *et al.* 1992); and cytochrome P-450 system with PAHs (Stagg & Addison 1995). Such techniques indicate whether individual organisms have been exposed to contaminants or are showing symptoms of stress from the exposure.

Figure 17.2 shows the biomarker approach using ethoxyresorfin O-de-ethylase (EROD) and acetylcholinesterase applied in a decreasing gradient of contamination in the North Sea. Whereas the

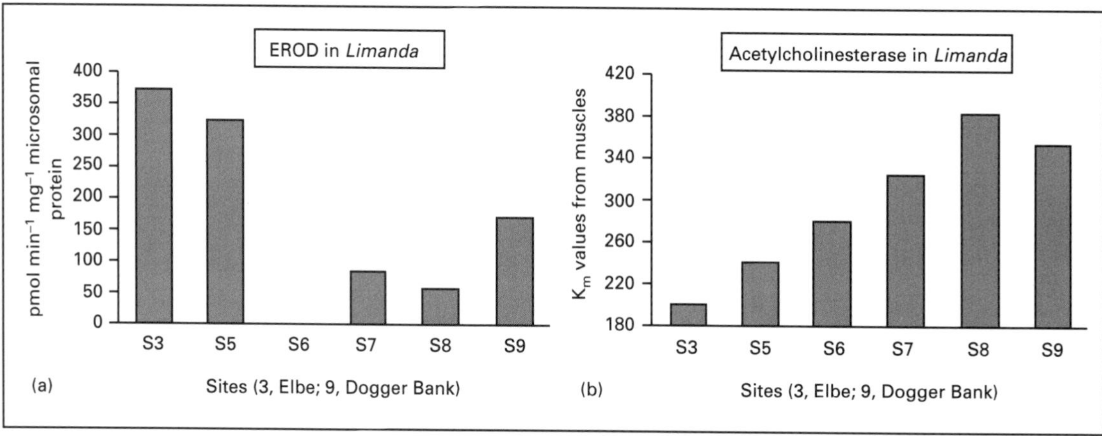

Fig. 17.2 The biomarker approach illustrated by two biomarkers. The sample sites are of a gradient from the highly contaminated River Elbe, Germany (site 3), out across the North Sea to the Dogger Bank (site 9). EROD concentrations are induced by polycyclic aromatic hydrocarbons and increase with increased stress and suggest a decreasing trend from the Elbe. Acetylcholinesterase concentrations decrease with increased organophosphate and suggest a decreasing trend from the Elbe. (After (a) Galgani *et al.* 1992, (b) Renton & Addison 1992.)

EROD data show a decrease in the induced biomarker along the gradient, indicating a decrease in effect of inducers such as PAHs and/or PCBs, the acetylcholinesterase biomarker shows an increase with increased stress. However, both data sets suggest that at the Dogger Bank there is unexpected evidence of effect of stress, so that this station is more affected than Station 8. Subsequent chemical analyses have shown that the Dogger Bank area is indeed impacted and studies are being done to establish likely causes.

Figure 17.3 shows an experimental and field approach using another biomarker—scope for growth in the mussel *Mytilus edulis* (Widdows & Johnson 1988). In this test the energy demand of the mussel under different conditions of stress is measured and that remaining, the scope for growth, is used as an indicator of stress. The data show

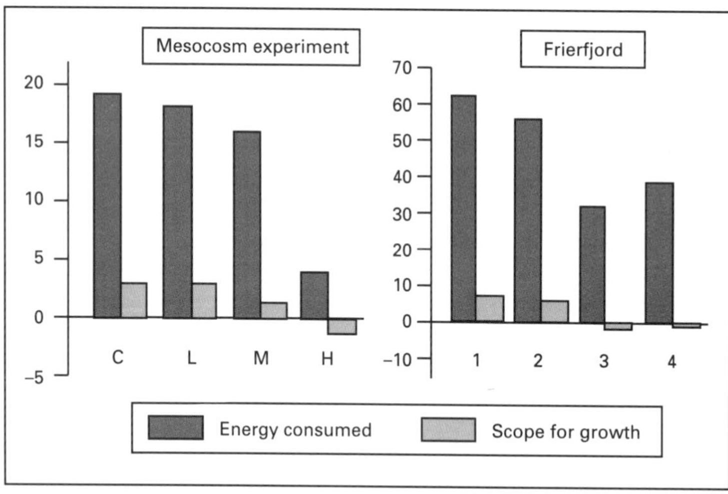

Fig. 17.3 The biomarker approach illustrated by changes in scope for growth in *Mytilus edulis* in a mesocosm experiment and in a contaminated fjord, Frierfjord, Norway. Contamination increases from C to L to M to H in the mesocosm and from 1 to 4 at the field site, as indicated in Table 17.1. The data show that more energy is used to cope with environmental contaminants under high doses. (After Widdows & Johnson 1988.)

Table 17.1 Contaminant levels for total hydrocarbons (THC), (CU) copper and total PAH and mesocosms used by Widdows & Johnson (1988), Fig. 17.3.

Mesocosm	THC ($\mu g l^{-1}$)	Cu ($\mu g l^{-1}$)	Field site	Total PAH ($mg kg^{-1}$)
C	3.0	0.5	1	2.24
L	6.4	0.8	2	5.87
M	31.5	5.0	3	11.43
H	124.5	20.0	4	15.45

both in mesocosm experiments and in the field that the scope for growth is reduced in response to the doses applied.

It is important also to assess whether or not the communities in the area subjected to discharges are affected. With the development of computer-based analyses it has been shown that analyses of changes in community patterns in space and/or time in the suspected impact area is an effective way of assessing impacts (Gray *et al.* 1990; Clarke 1993). Olsgard & Gray (1995) show clearly how, using a simple sampling design, these techniques can extract clear patterns of impact of oil activities. Recent methodologies establish correlations between the faunal data and chemical data and generate hypotheses of what variables might cause the effects on the fauna found.

Figure 17.4 shows the development of effects of pollution around the Gyda oil field, North Sea from 1987—the baseline survey when little drilling waste had been discharged—to 1993. The area of altered benthic community structure is clearly shown. Table 17.2 shows an analysis of the environmental data, showing that whereas Cu, Sr and Zn were significantly correlated with the fauna in the baseline survey, after 6 years total hydrocarbon concentration (THC) and Ba that were discharged show the highest correlations. Clearly, oil activities are related to the effects. These data suggest that one can assess with high accuracy the effects of disturbances on benthic communities, in order to test the risk assessments made. Such monitoring and analysis techniques should be part of modern environmental management practices.

Figure 17.4 and Table 17.2 show clearly that in 1987—the year of the baseline survey—there was no pattern in the faunal distribution. The CANOCO analysis showed that the correlation with environmental variables was low in 1987 but that in 1990 there were clear influences of both hydrocarbons (THC) and barium, both of which derive from oil industry discharges. By 1993 the patterns of effect cover large areas (ca. 30 km^2) and there are very clear correlations with discharges. When permission was given by the Norwegian State Pollution Board for oil activities, the companies did environmental impact assessments (EIAs) and predicted that they would affect the sea bed to a 1 km radius, i.e. 3 km^2. The effects on many fields within the Norwegian sector have been shown to cover much greater areas, up to 50 km^2 at some fields (Olsgard & Gray 1995). Since these effects were not predicted by the companies, the Norwegian State Pollution Board has acted to reduce discharges of oil-based drilling cuttings. Technological solutions were found within 2 years and now water-based drilling muds are used and effects of pollution have been reduced. The lessons are that the predictions of the EIA must be testable if one is to protect the marine environment.

17.4 RISK ASSESSMENT IN A MARINE CONTEXT

17.4.1 Environmental impact assessment

Environment impact assessments (EIAs) are now required almost world-wide before new activities that may affect the environment are approved. An EIA is a general form of risk assessment and predicts the likely impact of a planned activity. These activities are often physical constructions rather than the impact of a specific chemical and yet by disturbing the physical environment often biogeochemical processes are altered. For example, construction of a bridge (the new link between Copenhagen in Denmark and Malmö in Sweden) or an airport in a coastal area (e.g. Hong Kong's new airport; see Morton 1994) involves much excavation and creation of sediment and nutrient plumes, which are likely to have effects on the natural biota (seagrass beds, mussel beds, mangroves and coral reefs, etc.). Thus an EIA is a general form for risk assessment that involves much more than simply assessing the impact of one chemical.

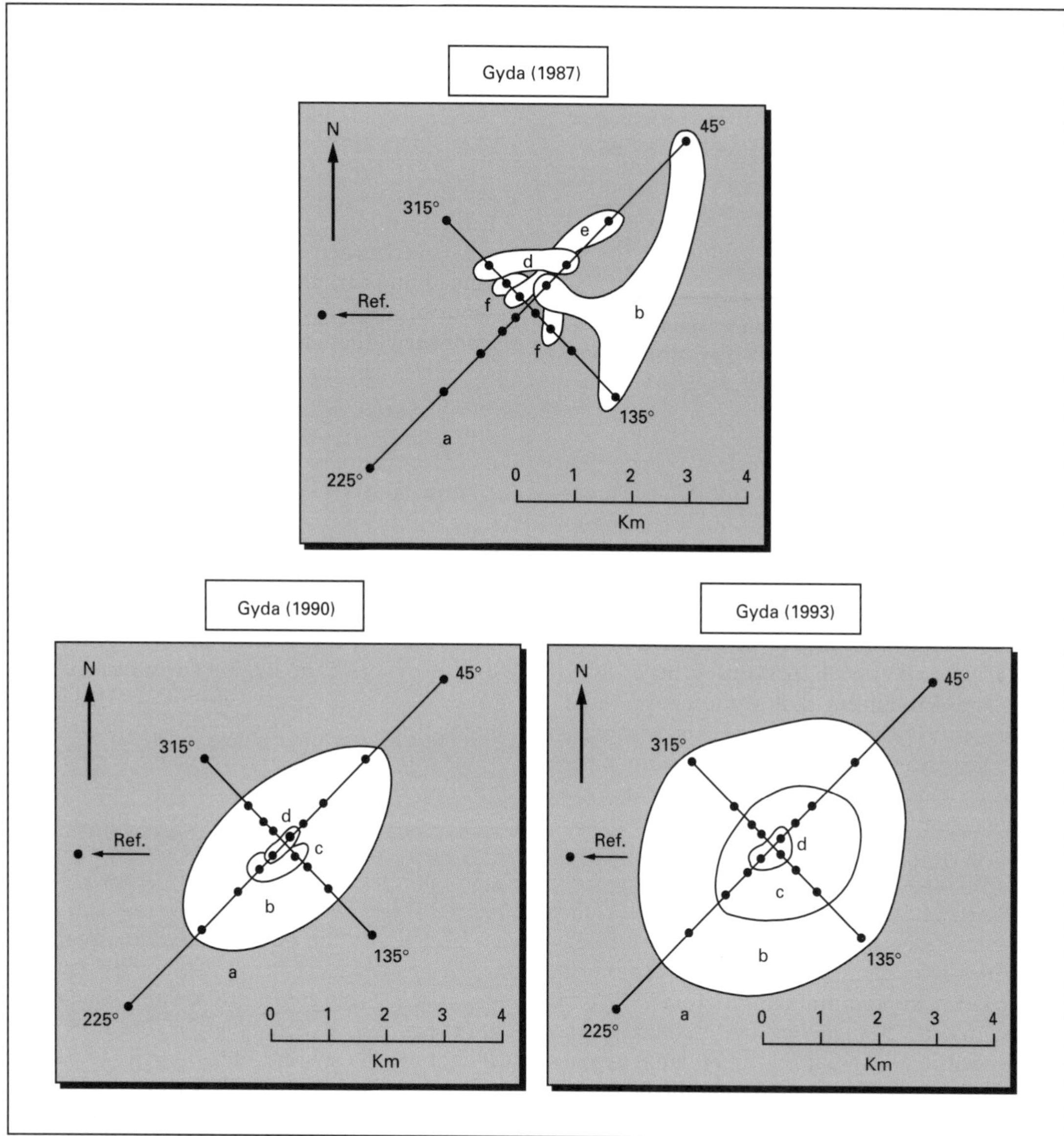

Fig. 17.4 Development of pollution around the Gyda oil field, North Sea. Species groupings from multivariate statistical analyses: in 1987 there is no clear effect and the groupings do not show the gradient; in 1990 and 1993 the groups are: (a) unpolluted; (b) slightly polluted; (c) moderately polluted; (d) grossly polluted. (From Olsgard & Gray 1995.)

The process usually involves a baseline survey conducted before the activity begins and then a detailed consideration of the scale of the likely impacts, based often on modelling techniques for assessing dispersion and other impacts. From these data, predictions are made of likely effects. Then environmental authorities make an evaluation of the predicted effects and give their consent, usually with conditions on how the activity will be conducted. In order to assess whether or not the predictions are correct, monitoring must be undertaken. Yet usually there is little follow-up

Table 17.2 Relationship between environmental variables and fauna using the forward selection procedure in CANOCO (ter Braak 1988, 1990).

1987		1990		1993	
Cu	14.7%**	Cu	38.1%**	THC	39.7%**
Sr	9.4%**	THC	37.7%**	Ba	39.2%**
Mdφ	7.3% n.s.	Zn	37.2%**	Sr	39.2%**
TOC	6.3% n.s.	Sorting	34.5%**	Zn	37.5%**
		Ba	33.6%*	Cu	37.0%**

The environmental variables best explaining the faunal patterns are given. The percentage explanation of each environmental variable to the total faunal variance is given together with the level of significance: $^{**}p = 0.01$; $^{*}p = 0.05$; Mdφ, median particle diameter (phi units); n.s., not significant. TOC, total organic carbon.

as to whether the predictions were borne out, nor are there specific tests that have a consequence on the constructor or pollutant discharger if the predictions are incorrect. Often *post hoc* remedial action or even litigation is necessary to stop damaging effects that were not predicted. Clearly, such procedures do not give adequate protection to the environment, a point emphasized by Earll (1992). Recently there have been new developments in EIAs where predictions are tested, and by feedback monitoring there are tests of both the predictions and the consequences for those making them (Gray & Jensen 1993).

An example of an EIA using the newly developed principles relates to construction of the largest bridge in Europe over the Great Belt in Denmark. The main concern with the environmental effects of the bridge was water flow, because approximately 70% of the water entering the Baltic Sea passes through the Great Belt. However, this will not be covered here because I want to emphasize another lesson.

The more general aspects were that the bridge construction would need excavation of bottom sediments and construction of artificial islands. These activities were expected to lead to increased sediment turbidity and nutrients would be released into the water column. From baseline surveys and hydrographical models the extent of the effects in space and time were predicted. One of the major concerns was that of the possible effects of the nutrients released from the sediments. As a result of eutrophication, the Great Belt is subject to low oxygen concentrations in late summer–early autumn each year. Increased nutrient supply could enhance plankton blooms and lead to even lower oxygen concentrations and consequent adverse effects on marine life.

Once predictions had been made by the scientists of the likely spatial and temporal scales of the planned impact, they were fully evaluated by a control authority. Here a Control Panel was established that involved the environmental authorities, the company doing the construction, scientific consultants who had made the predictions, local authorities and representatives from the Green movement. They decided, not just for the oxygen conditions but for all environmental impacts, just what was acceptable and what was unacceptable. The most important development in this EIA was that the predictions made by the consultants to the company were actually put to the test in a way that had a consequence for the company. A feedback loop agreed that if oxygen fell below 4 mg l^{-1} then, automatically, without having to ask the company, the consultants would undertake a wide-scale survey of a wide area in order to assess whether or not the low oxygen values were limited to the area of the construction. This feedback loop was activated in 1991; it was found that oxygen was low over the whole area, so it was concluded that the Great Belt company was not contributing to the low oxygen. However, had the low oxygen been limited to the near-field, then a feedback loop would automatically have come into operation that would have stopped excavation of sediments until oxygen conditions were again above 4 mg l^{-1}. This was by no means a trivial agreement because the company was using the world's largest cutter-dredger and the daily hire-rate was extremely high. The feedback loop is illustrated in Fig. 17.5.

This example further illustrates that the conditional hypotheses and feedback loop ideas derived a priori had direct consequences on the construction activities. Other feedback loops involved the risk of enhancing phytoplankton blooms where monitoring involved measuring fluorescence *in situ*. If the chlorophyll increased by a certain amount over a given time period, then again widespread

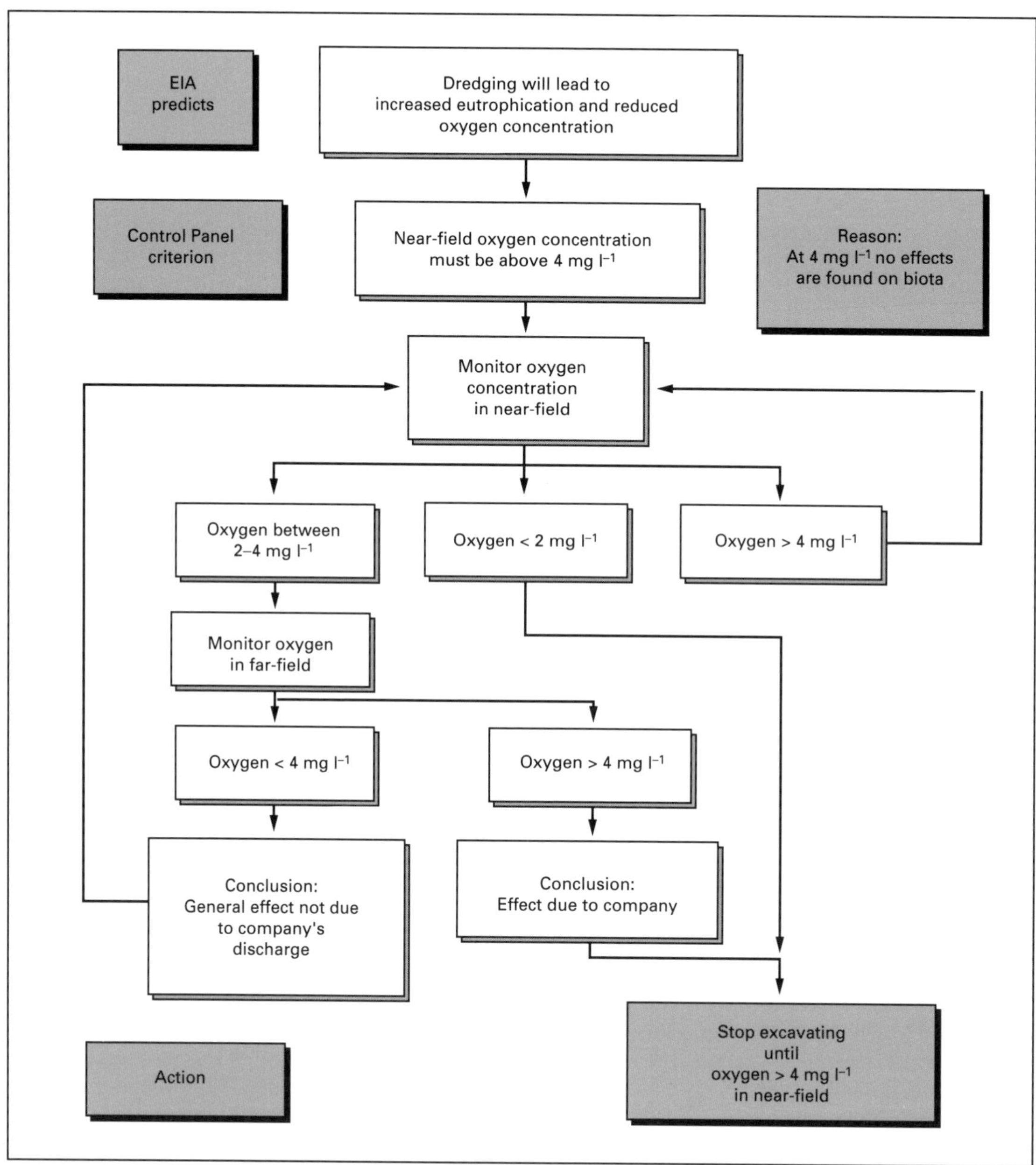

Fig. 17.5 An example of feedback monitoring applied in an environmental impact assessment. See text for further details. (From Gray & Jensen 1993.)

monitoring was started to check whether or not it was an effect that could be attributed to the construction activities. Again, sediment excavation would stop if a link to the construction activity was found.

This example shows that the most general form for risk assessment, EIA, has now been developed so that it is possible to make testable predictions. Environmental impact assessments no longer need to be paper exercises that do little to protect the environment. Instead, if properly designed, they can be a major influence in protecting the marine environment.

The classical form of risk assessment relates to human health issues, such as the risk of cancer from exposure to given chemicals (see Paustenbach (1994) for a recent and comprehensive review). In a marine context the classical approach can be used for estimating the environmental risks posed by discharging chemicals to the marine environment. I develop below a protocol for risk assessment in a marine context.

17.4.2 Classical risk assessment

Paustenbach (1994) has given a comprehensive review of environmental risk assessment. In this, based on a review by the US National Academy of Sciences (NAS 1983), he draws a distinction between risk assessment and risk management. Risk assessment involves the characterization of adverse environmental effects of exposure to hazards. Risk management, on the other hand, involves the process of evaluating alternative regulatory actions and selecting among them. This entails consideration of the political, social, economic and technological information with risk-related information to develop, analyse and compare regulatory options. The NAS report emphasize that risk assessment should be clearly separated from risk management, yet this is often not the case.

Firstly, attempts are made to assess the risk of exposure to the chemicals discharged into the environment. The precautionary principle lists three characteristics of chemicals that are considered as important: toxicity, persistance and bioaccumulatability. Gray *et al.* (1991) list seven characteristics.

1 *Quantities*: the production methods, loads and sources, and discharge patterns.
2 *Distribution*: physicochemical characteristics, affinity for environmental compartments and sinks.
3 *Persistence*: kinetics of hydrolysis, photolysis and biodegradation.
4 *Bioaccumulation*: *n*-Octanol/water partition coefficients and metabolic pathways in different organisms.
5 *Toxicity*: measures of biological activity of the substance (ideally from cells to ecosystems).
6 *Biological system typologies*: biotic and abiotic characteristics of the structure and function of the biological communities.
7 *Targets of exposure*: consideration of the vulnerability and stability of potential target communities and lifecycle stages of potential target species.

Up until recently there has been little attempt to quantify the loads and sources of chemicals reaching the marine environment. The North Sea Task Force (1993) has produced a valuable review of sources of chemicals entering the North Sea from the land via rivers and via the atmosphere. A study by GESAMP (1990) shows that on a global scale atmospheric inputs dominate riverine inputs for lead, cadmium, zinc and most high-molecular-weight hydrocarbons, whereas riverine inputs dominate for copper, nickel, arsenic, iron, nitrogen and phosphorus.

Figure 17.6 shows the outline of a comprehensive approach to risk assessment and risk management in a marine context (adapted from Gray *et al.* 1991).

Once released to the environment the chemicals are subjected to physical, chemical and biological processes that will determine where and in what concentration they will be found. It is often possible from knowledge of the chemicals and their physical forms, and from the environment in which they are to be discharged, to predict what environmental compartment they will be found (e.g. affinity for the water-bound or sediment-bound phase). Paustenbach (1994) gives a comprehensive review of the properties that can be measured, such as water solubility, photodegradation either directly or indirectly, biodegradation, the dissociation constant, sorption and desorption, and the partition coefficient (to predict bioaccumulation and bioconcentration). Fugacity models have been developed for terrestrial systems, where from physicochemical properties it is possible to predict the relative amount of a substance into each environmental compartment (Scheunert *et al.* 1994). This has not been done in the marine environment.

Since many chemicals adsorb on to particles and are sedimented out of the water column, it is usually marine sediments that are the sinks for

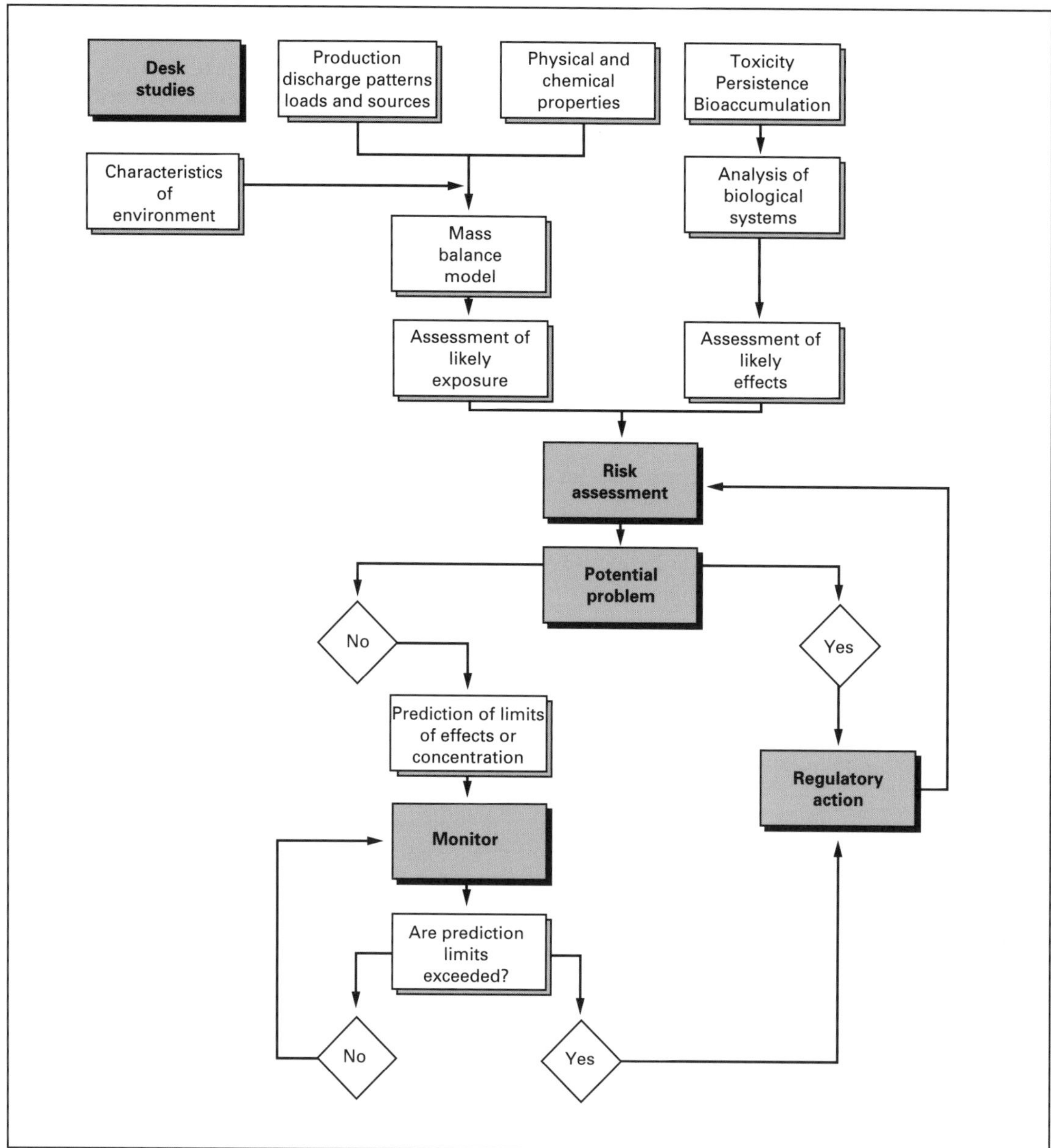

Fig. 17.6 Outline of a comprehensive approach to risk assessment and risk management in a marine context. (After Gray *et al.* 1991.)

contaminants. Should the area be one subjected to periodic erosion and resuspension, then sediments may also provide a source of contaminants to the water column. Harland *et al.* (1993) review the prediction of fate and of exposure of organic chemicals in aquatic sediments. They stress the importance of the sedimentation and resuspension processes and the fact that sorption/desorption is highly important to understanding the prediction of fate and exposure in aquatic sediments. In particular, the sorption of hydrophobic organic chemicals by partitioning in or on to organic matter is a key process (Lyman 1983; Elzerman & Coates 1987; USEPA 1989). Oxidation/reduction (the redox potential) of sediments also play a major role in the chemistry of sediments , particularly for iron and manganese. There has been much recent work on the fate of chemicals in the environment. Through knowledge of the kinetics of hydrolysis and photolysis Klöpffer (1993) and Altschuch & Brüggemann (1993) provide useful reviews), and on degradation (Vasseur *et al.* 1993), it is possible to predict their persistence.

These data are then combined with data on the environment itself (e.g. physical, chemical and biological characteristics of the environment, including hydrography and sedimentology and dispersion models) to produce a mass balance model that will help to confirm that all sources and sinks are known. The importance of the biological systems should not be underestimated. Failings of mass balance models to predict, for example eutrophication events, may be due to the fact that the significant roles played by biological organisms in biogeochemical cycling were not known. All these data, in principle, allow prediction of the concentration and physicochemical form of the chemical in a given environmental compartment (e.g. the sediment surface or in sediment pore water) at a point in space and time. This gives a prediction of the likely exposure of organisms to the chemical.

Then, from knowledge of the physical attributes of the chemicals, it is often possible to predict the effects on biological systems. Quantitative structure–activity relationships (QSARs) have been developed for several classes of chemicals and within these classes reliable predictions can be made of the toxicity (Connell 1995).

Assessment of the toxicity of chemicals in the marine environment has produced many new techniques for the measurement of acute and subacute effects. It is not the place here to review toxicity testing. Forbes & Forbes (1994) have reviewed this topic and should be consulted. It is usually assumed from such tests that a certain concentration can be established which is expected to have no adverse effects in nature: the no-observed-effect concentration (NOEC). Forbes & Forbes (1994) have a good discussion of the problems of measuring the NOEC and the interpretation of such data.

In addition to the toxicity data, information must be obtained on the biological systems in the areas that are likely to be affected by the chemical. In the USA, a National Research Council panel (NRC 1989) lists the ecological information that is necessary before oil and gas exploitation on the outer continental shelves can begin. This provides a useful checklist for any coastal area:

> '... a characterization of the environment, including its biological resources, and a basic knowledge of ecological relationships are needed. The desired information includes:
> **1** a characterization of major habitat types
> **2** a catalogue of representative species (or major species groups) present in the area
> **3** seasonal patterns of distribution and abundance
> **4** basic ecological information on feeding behavior and reproduction
> **5** basic information on factors determining the vulnerability of various species
> Where unique habitats or endangered or rare species exist, more extensive characterization of the biota ... may be required'.

Using these data to predict risks is not an easy task. Long-term studies on effects of heavy metals on bivalves in San Francisco Bay (Cain & Luoma 1990) show that there are strong seasonal variations in heavy metal content associated with seasonal variations in soft tissue weight. Cain & Luoma (1990) found that there was a strong correlation between the age of the organisms and heavy metal concentration, but a weaker relationship between shell length and metal concentration. More recently, Luoma (1996) shows that whereas tissue concentrations of organic chemicals can be used to predict toxic effects in bivalves, this does not hold for metals. Metal uptake is related to concentration of metals in the food source and not to concentrations in the water. These results emphasize that there is a need for extensive field studies if we are to be in a position to assess the risk of a given concentration of chemical in water or sediment to organisms living there.

Once toxicity and field data are available and the exposure concentrations have been predicted, a risk assessment can be made. It is possible to assess the likely risks posed by discharge of a given chemical in a given quantity to a given environment with its biological system. This is the risk assessment process. Quantifying the risk is, however, hardly ever done. The usual procedure is to accept a 100-fold safety factor, i.e. to divide the NOEC by 100 as a safe discharge. Paustenbach (1994) gives a clear and full account of how these factors are applied in relation to human health and the problems with assessment of risk factors. There is clearly a need for much more and better data on effects of chemicals on marine organisms, such as that of Luoma and co-workers, so that it will be possible to quantify the risks to populations and communities. This also implies the need for better sampling designs so that it is possible to detect changes in populations and communities in the variable environment. (See Underwood (1991, 1993, 1996) for a full discussion of sampling designs relevant to detect change.)

17.5 RISK MANAGEMENT

An analysis of the regulatory options available that follow from the risk analysis is the risk management process. Figure 17.6 shows that in the suggested strategy a prediction is made of the concentration and/or the level of effect that is deemed acceptable, and then monitors to check whether or not that level is actually found. Too often monitoring is done without there being any given criteria for what is being monitored over what area, nor are estimates given of likely effect levels and/or concentrations. Peterson (1993) has given an excellent review of a number of case studies of environmental impacts and the pitfalls of inadequate aims, sampling design, data collection analysis methods and management responses. In the strategy presented here monitoring is done to test predictions and thus has clear objectives.

17.5.1 Monitoring and surveillance

Gray *et al.* (1991) have suggested that in the marine environment there are two different basic strategies to environmental risk assessment, dependent on whether the discharge is diffuse and/or the effect of chronic disturbance, or whether the discharge is a point source and/or effect of local disturbance. With diffuse sources and/or chronic disturbance it is suggested that *surveillance* is undertaken, in which levels of contaminants or disturbances are compared with control sites that are not thought to be contaminated. If the level of one or more contaminants or the state of a physically impacted site are shown to be different to controls then a risk assessment should be carried out.

Both surveillance and monitoring to test predictions of the risk assessment should take fully into account the widely accepted before–after control–impact (BACI) design (Underwood 1991, 1993). This suggests that a series of control sites, not just one, are necessary, and that if effects are expected over time, sampling should be random with seasons. Accompanying this sampling design are complex statistical methods based on analysis of variance, which allow proper analysis of effects. Before–after control–impact is not the only sampling design that can detect effects. In monitoring the impact of oil and gas explorations on the benthic fauna of the North Sea, a design with a cross of transects centred at the site of the pollution is used, with sampling stations set at logarithmically increasing intervals (namely 250, 500, 1000, 2000 and 4000 m). The furthest stations are set so that they are beyond the areas affected, so that they remain as undisturbed controls (Gray *et al.* 1990; Olsgard & Gray, 1995).

Another important aspect related to interpretation of surveillance and monitoring data is the plea by Peterman & M'Gonigle (1992) and Fairweather (1991) for statements of power when presenting results. If the means of a variable from a suspected impacted site are compared with a control site and no statistically significant difference is found, this may mean that there is no difference or, alternatively, that the sampling methods were so poor that even with a large difference it would not be possible to detect it. Power analyses take into account variance and number of samples and allow a simulation of how many samples for a given variance would be needed to detect a significant difference (see Cohen (1988) for a thorough description of methods). Power analyses are slowly being incorporated into surveillance and monitoring programmes.

As an example, a baseline study was done in Denmark using various biological variables to measure the condition of an eelgrass bed. The environment authorities accepted that a 25% reduction in these variables was an acceptable impact. A power analysis was then done to see whether this could be detected using these variables (Fig. 17.7). The accepted criterion is that a change can be detected with 80% power. None of the measured variables met this criterion! Figure 17.7 shows that some only had a 20% power of detecting a change. Further analyses were done to try and improve power. Increasing the sampling effort fivefold after the baseline survey made little difference, but a fourfold increase in sampling effort before and after gave the necessary power to detect a change. This example illustrates that the application of these techniques will greatly improve the chances of detecting trends in monitoring programmes. However, the decision of whether to use such techniques rests not with scientists but with managers who fund monitoring programmes. A fourfold increase in sample number may not be economically acceptable and the consequence of not accepting this as a change will not be detected, even though one is present.

From this analysis it is clear that a strategy for risk assessment in the marine environment has been developed that is soundly based on science and can lead to testable predictions. This ensures that monitoring of the condition of an area is possible and that this can be efficient. Too often, today, monitoring is done without clear goals, with a lack of proper sampling designs and without using power analyses (see review by Peterson 1993). A good example of this poor approach is that of the North Sea Task Force. The Ministerial Declaration in 1987 agreed to reduce discharges of heavy metals and other contaminants discharged to the North Sea by 50% between 1985 and 1995. Analyses done in 1993 showed that this had in fact been achieved. The politicians expected that the Quality Status Report on the North Sea (OSPARCOM 1993) would show the results of these costly efforts, yet the results merely show no change in the trends of contaminants in sediments and fish. This is largely due to the poor sampling design and to the fact that no power analyses have been done to assess what magnitude of change can be detected with the present sampling design or, alternatively, the number of samples that will be needed to detect a change with the present design. Clearer and better use of scientific knowledge is needed if managers are to properly protect the marine environment from damage. However, there is a tendency today to accept more extremist views of environmental organizations than to use science in management. Bewers (1995) shows that public perceptions of the nature and severity of marine pollution frequently differ from scientific

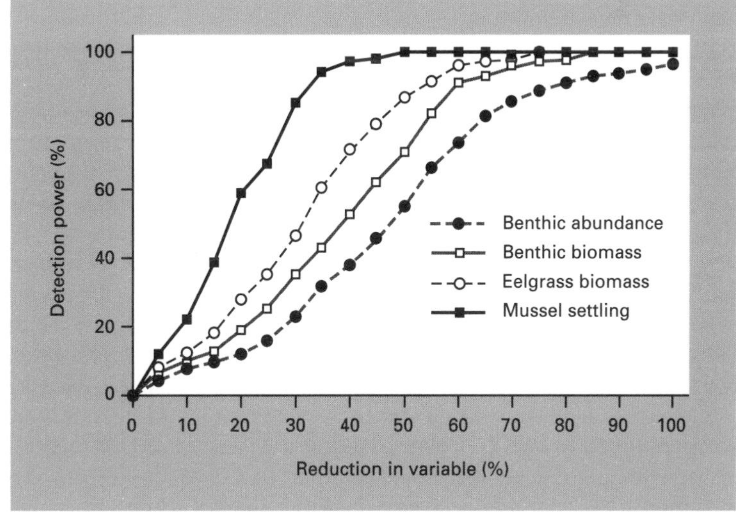

Fig. 17.7 Illustration of power analyses as applied to measurement of biological variables used to assess impact on benthic communities in the Øresund, Denmark (unpublished data). Power is the risk of wrongly assuming that there is no difference between means. A power of 80% is usually used as the accepted statistical criterion and can be envisaged as equivalent to the familiar use of 95% certainty in most statistical analyses.

assessments. This leads to a marine environmental protection policy that is simplistic and/or unnecessarily extreme. The *Brent Spar* case as outlined in the introduction highlights this increasing trend. Rational management decisions based on sound risk assessment is what is needed. How can we improve on the present system?

17.6 TOWARDS BETTER MANAGEMENT

I believe that what is needed is a way to marry the spirit of the precautionary principle with the latest scientific findings. If such a way forward can be found then it will not be necessary to take the extreme Green view that nothing must be discharged into the oceans until we know everything about all chemicals. A possible way forward is to devise a Control Panel which comprises scientific experts, the Greens, environmental authorities, industry and, if possible, the general public. This would evaluate the evidence provided within the framework of Fig. 17.6 and incorporate the latest scientific findings in their analysis. Figure 17.8 shows what the Control Panel would do. What will be achieved by adoption of such a framework, with a Control Panel formed from all concerned parties, is that all decisions will be transparent and the data will be openly available and by agreement a priori to remedial action or stopping of a damaging discharge. This should prevent the unnecessary and expensive adversarial type of environmental assessments common in the USA (see example of the *Exxon Valdez*, quoted by Peterson (1993)). With a sound monitoring programme and with agreed limits for the discharge/construction and their effects on biota, there should be little or no danger to the environment resulting from using only objective scientific criteria to judge whether or not limits have been exceeded. The precautionary principle will be applied, but in the context of using the best available scientific expertise and acknowledging that there is an assimilative capacity in the natural environment to cope with man's wastes.

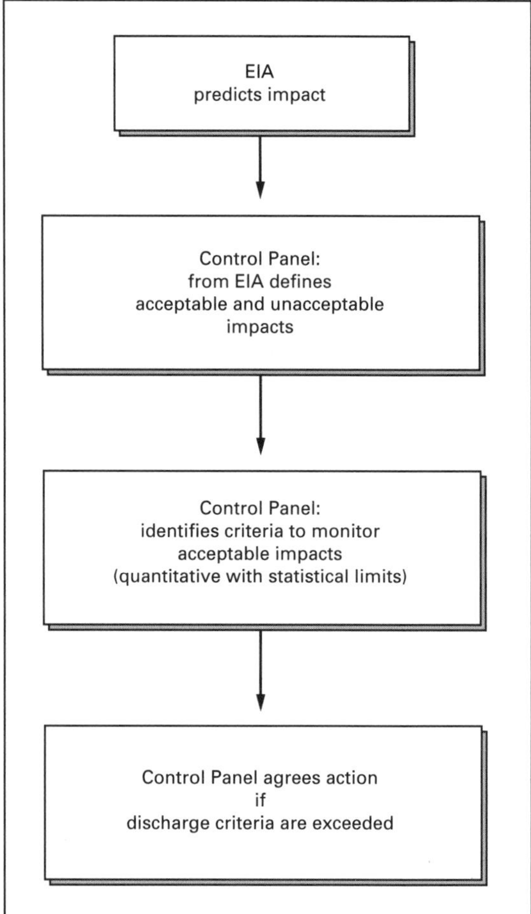

Fig. 17.8 Functions of a Control Panel formed from scientific experts, the Green movement, environmental authorities, industry and the general public.

A point not considered by Peterson (1993) or Bewers (1995) is that in environmental science there is almost a vested interest in so-called 'green science', which attempts to prove that some environmental effect is occurring. Yet the most generally accepted philosophy of science is that one should try to disprove rather than prove hypotheses (Popper 1983). One clear example of this is marine eutrophication. It is a widespread view of politicians and the general public that the whole of the North Sea suffers from eutrophication. The available data shows that most of the North Sea is nutrient-poor and runs out of nutrients during the summer; only the coastal strip is nutrient-enriched. Sewage outfalls are one of the ways that nutrients enter the environment. In a given area a full analysis of the sources and loads of nutrients is needed before management decisions are made. Whilst I believe that primary and secondary treatment should be compulsory everywhere, some governments have accepted data purporting to show nutrient enrichment is general and are investing vast amounts of money in unneeded tertiary sewage treatment to remove nitrogen. They do this in the naïve belief that they are saving the North Sea.

As an example of this dangerous trend in trying to prove hypotheses, in 1988 along the south coast of Norway an algal bloom of *Chrysochromulina polylepis* occurred that caused massive mortalities to fish and benthos (Draget & Holthe 1989). Data show that at the peak of the bloom the requirement for nitrogen could be met by normal seasonal levels, yet this bloom is cited (e.g. in the reports of the influential Oslo–Paris Commission) as an example of how eutrophication leads to algal blooms. The evidence, however, does not necessarily support this hypothesis.

One can conclude from this brief survey that environmental scientists are producing new methods and new data that give a much better possibility of risk assessment in the marine environment. In addition there are a range of new approaches to the risk assessment process, to EIAs and to monitoring that should give managers a better chance of properly managing the environment.

In addition, members of the general public should be involved, as indeed is the case with NRC panels in the USA that consider environmental issues (Peterson 1993). These Control Panels should agree that the predictions made are acceptable impacts and prior to the construction or discharge beginning they should agree what will be the consequences if the prediction limits are exceeded. This may take time and much debate, but it is the spirit of the precautionary principle and is a vitally important step. Panels such as this have been widely used in Denmark and Germany to arrive at a consensus opinion as to whether or not genetically modified organisms could be released into the natural environment.

17.7 REFERENCES

Altschuch, J. & Brüggemann, R. (1993) Limitations on the calculation of physicochemical properties. In: *Chemical Exposure Predictions* (ed. D. Calamari), pp. 1–11. Lewis, Boca Raton, FL.

Alzieu, C. (1991) Environmental problems caused by TBT in France: assessment, regulations, prospects. *Marine Environmental Research*, **32**, 7–17.

Angel, M.V. (1995) Brent Spar: no hiding place. *Biologist*, **42**, 192.

Anon. (1990) *This Common Inheritance*. Britain's Environmental Strategy, HMSO, London.

Anon. (1995) *Potential exposure in nuclear safety—a report by the International Nuclear Safety Advisory Group*. International Atomic Energy Agency, Vienna.

AURIS (1994) *Removal and disposal of the Brent Spar*. Aberdeen University Research and Industrial Services Ltd, Aberdeen.

Bewers, M. (1995) The declining influence of science on marine environmental policy. *Chemical Ecology*, **10**, 9–23.

Bryan, G.W., Gibbs, P.E., Hummerstone, L.G. & Burt, G.R. (1986) The decline of the gastropod *Nucella lapillus* around south-west England: evidence for the effect of tributyltin from antifouling paints. *Journal of the Marine Biological Association of the United Kingdom*, **66**, 611–640.

Bryan, G.W., Gibbs, P.E. & Burt, G.R. (1988) A comparison of the effectiveness of tri-*m*-butyltin chloride and five other organotin compounds in promoting the development of imposex in the dogwhelk, *Nucella lapillus*. *Journal of the Marine Biological Association of the United Kingdom*, **68**, 733–744.

Cain, D.J. & Luoma, S.N. (1990) Influence of seasonal growth, age, and environmental exposure on Cu and Ag in a bivalve indicator, *Macoma balthica*, in San Francisco Bay. *Marine Ecology Progress Series*, **60**, 45–55.

Cairns, J. (1977) Quantification of biological integrity. In: *The Integrity of Water* (eds R.K. Ballantine & L.J. Guarraia). Stock No. 055-001-01068-1, US Environmental Protection Agency, US Government Printing Office, Washington, DC.

Chan, K.M. (1995) Metallothionein: potential biomarker for monitoring heavy metal pollution in fish around Hong Kong. *Marine Pollution Bulletin*, **31**, 411–415.

Chapman, P.M. (1995) Ecotoxicology and pollution—key issues. *Marine Pollution Bulletin*, **31**, 167–177.

Clark, R.B. (1992) *Marine Pollution*, 3 edn. Oxford University Press, Oxford.

Clarke, K.R. (1993) Nonparametric multivariate analyses of changes in community structure. *Australian Journal of Ecology*, **18**, 117–143.

Cohen, J. (1988) *Statistical Power Analysis for the Behavioural Sciences*, 2 edn. L. Erlbaum Associates, Hillsdale, NJ.

Connell, D.W. (1995) Predictions of bioconcentrations and related lethal and sublethal effects with aquatic organisms. *Marine Pollution Bulletin*, **31**, 201–208.

Depledge, M.H. & Fossi, M.C. (1994) The role of biomarkers in environmental assessment (2). Invertebrates. *Ecotoxicology*, **3**(3): 173–179.

Draget, H.B.H. & Holthe, T. (1989) *Oppblomstring av Chrysochromulina polylepsis 1988*. DN-rapport 12, 53 pp. Directorate for Nature Management, Trondheim.

Earll, R.C. (1992) Commonsense and the precautionary principle—an environmentalist's perspective. *Marine Pollution Bulletin*, **24**, 182–185.

Elzerman, A.W. & Coates, J.T. (1987) Hydrophobic organic compounds on sediments. Equilibria and kinetics of sorption. In: *Sources and Fates of Aquatic Pollutants* (eds R.A. Hites & S.J. Eisenreich), pp. 263–317. American Chemical Society, Washington, DC.

Fairweather, G.P. (1991) Statistical power and design requirements for environmental monitoring. *Australian Journal of Marine and Freshwater Research*, **42**, 555–568.

Forbes, V.E. & Forbes, T.L. (1994) *Ecotoxicology in Theory and Practice*. Chapman & Hall, London.

Freestone, D & Hay, E. (1995a) Origins and development of the precautionary principle. In: *The Precautionary Principle and International Law: the Challenge of Implementation* (eds D. Freestone & E. Hay), pp. 3–15. Kluwer Law International, The Hague.

Freestone, D. & Hay, E. (1995b) *The Precautionary Principle and International Law: the Challenge of Implementation*. Kluwer Law International, The Hague.

Galgani, F., Bocquené, G. & Cadiou, Y. (1992) Evidence of variation in cholinesterase activity in fish along a pollution gradient in the North Sea. *Marine Ecology Progress Series*, **91**, 77–82.

GESAMP (1990) *The state of the marine environment*. Group of Experts on Scientific Aspects of Marine Environmental Protection, UNEP Regional Seas Reports and Studies, **115**, pp. 1–111. United Nations Environment Porgramme, Nairobi.

GESAMP (1993) *Impact of oil, individual hydrocarbons and related chemicals on the marine environment, including used lubricant oils, oil spill control agents and chemicals used offshore*. Group of Experts on Scientific Aspects of Marine Environmental Protection, GESAMP Regional Seas Reports and Studies, **50**, pp. 1–178. IMO, London.

GESAMP (1995) *Biological indicators and their use in the measurement of the condition of the marine environment*. Reports and Studies, Group of Experts on Scientific Aspects of Marine Environmental Protection, GESAMP Reports and Studies, **55**, pp. 1–56.

Gibbs, P.E., Bryan, G.W., Pascoe, P.L. & Burt, G.R. (1987) The use of the dogwhelk *Nucella lapillus*, as an indicator of tributyltin (TBT) contamination. *Journal of the Marine Biological Association of the United Kingdom*, **67**, 507–523.

Gray, J.S. (1990) Statistics and the precautionary principle. *Marine Pollution Bulletin*, **21**, 174–176.

Gray, J.S. (1995) Integrating precautionary scientific methods into decision-making. In: *The Precautionary Principle and International Law: The Challenge of Implementation* (eds D. Freestone & E. Hay), pp. 133–146. Kluwer Law International, The Hague.

Gray, J.S. & Jensen, K. (1993) Feedback monitoring: a new way of protecting the environment. *Trends in Ecology and Evolution*, **8**, 267–268.

Gray, J.S., Clarke, K.R., Warwick, R.M. & Hobbs, G. (1990) Detection of initial effects of pollution on marine benthos: an example from the Ekofisk and Eldfisk oilfields, N. Sea. *Marine Ecology Progress Series*, **66**, 285–299.

Gray, J.S., Calamari, D., Duce, R., Portmann, J.E. Wells, P.G. & Windom, H.L. (1991) Scientifically based strategies for marine environmental protection and management. *Marine Pollution Bulletin*, **22**, 432–440.

Harland, B.J., Matthiesen, P. & Murray-Smith, R. (1993) The prediction of fate and exposure of organic chemicals in aquatic sediments. In: *Chemical Exposure Predictions* (ed. D. Calamari), pp. 115–136. Lewis, Boca Raton, FL.

Hogstrand, C. & Haux, C. (1991) Binding and detoxification of heavy metals in lower vertebrates with special reference to metallothionein. *Comparative Biochemistry and Physiology*, **100C**, 137–141.

Horiguchi, T., Shiraishi, H., Shimizu, M., Yamazaki, S. & Morita, M. (1995) Imposex in Japanese gastropods (Neogastropoda and Mesogastropoda): effects of tributyltin and triphenyltin from antifouling paint. *Marine Pollution Bulletin*, **31**, 402–405.

Klöpffer, W. (1993) Photochemical processes contribution to abiotic degradation in the environment. In: *Chemical Exposure Predictions* (ed. D. Calamari), pp. 13–26. Lewis, Boca Raton, FL.

Lloyd, R. (1980) Toxicity testing with aquatic organisms: a framework for hazard assessment and pollution control. *Rapports et Procès–Verbaux des Réunion, Conseil International pour l'Explorations de la Mer*, **179**, 339–341.

Luoma, S.N. (1996) The developing frame work of marine ecotoxicology: Pollutants as a variable in marine ecosystems. *Journal of Experimental Marine Biology and Ecology*, **200**, 29–55.

Lyman, W. (1983) Adsorption coefficients for soils and sediments. In: (eds W.J. Lyman, W.F. Reehl & D.H. Rosenblatt) *Handbook of Chemical Property Estimation Methods*. McGraw-Hill, New York.

Masood, E. (1995) Expert panel to judge on deep sea disposal. *Nature*, **377**, 670.

McCarthy, J.F. & Shugart, L.R. (1990) *Biological Markers*

of Environmental Contamination. Lewis, Boca Raton, FL.

Minchin, D., Oehlmann, J., Duggan, C.B., Stroben, E. & Keating, M. (1995) Marine TBT antifouling contamination in Ireland, following legislation in 1987. *Marine Pollution Bulletin*, **30**, 633–639.

Moriarty, F. (1983) *Ecotoxicology: the Study of Pollutants in Ecosystems*. Academic Press, London.

Morton, B. (1994) Here comes summer. *Marine Pollution Bulletin*, **28**, 576–577.

NAS (1983) *Risk Assessment in the Federal Government: Managing the Process*. National Academy of Sciences, National Academy Press, Washington, DC.

North Sea Task Force (1993) *North Sea Quality Status Report 1993*. Oslo and Paris Commissions, London, and Olsen & Olsen, Fredensborg, 132 pp.

NRC (1989) *The Adequacy of Environmental Information for Outer Continental Shelf Environment Studies Program II. Ecology*. National Research Council, National Academy Press, Washington, DC.

Olsgard, F. & Gray, J.S. (1995) A comprehensive analysis of the effects of offshore oil and gas exploration and production on the benthic communities of the Norwegian continental shelf. *Marine Ecology Progress Series*, **122**, 277–306.

Oslo and Paris Commissions (1995) *North Sea Ministerial Declaration, Esbjerg, June*. Oslo and Paris Commissions, London.

OSPARCOM (1993) *North Sea Quality Status Report 1993*. Oslo and Paris Commissions, London.

Paustenbach, D.J. (1994) A survey of environmental risk assessment. In: *The Risk Assessment of Environmental and Human Health Hazards: a Textbook of Case Studies*. (ed. D.J. Paustenbach), pp. 27–124. Wiley, New York.

Peakall, D. (1994) The role of biomarkers in environmental assessment (1). Introduction. *Ecotoxicology*, **3**, 157–160.

Peterman, R.M. & M'Gonigle, M. (1992) Statistical power analysis and precautionary principle. *Marine Pollution Bulletin*, **24**, 231–234.

Peterson, C.H. (1993) Improvement of environmental impact analysis by application of principles derived from manipulative ecology: lessons from coastal marine case histories. *Australian Journal of Ecology*, **18**, 21–52.

Peterson, P. (1995) *Final report on the environmental toxicology programme and the environmental damage and its assessment programme of the European Science Foundation*. European Science Foundation, Strasbourg.

Popper, K. (1983) *Realism and the Aim of Science*. Rowman & Littlefield, Totowa, NJ.

Portmann, J.E. & Lloyd, R. (1986) Safe use of the assimilative capacity of the marine environment for waste disposal—is it feasible? *Water Science and Technology*, **18**, 233–244.

Pravdic, V. (1985) Environmental capacity—is a new scientific concept acceptable as a strategy to combat marine pollution? *Marine Pollution Bulletin*, **16**, 295–296.

Rand, G.M. (1994) An environmental risk assessment of a pesticide. In: *The Risk Assessment of Environmental and Human Health Hazards: a Textbook of Case Studies* (ed. D.J. Paustenbach), pp. 899–934. Wiley, New York.

Renton, K.W. & Addison, R.F. (1992) Hepatic microsomal mono-oxygenase activity and P450IA mRNA in North Sea dab *Limanda limanda* from contaminated areas. *Marine Ecology Progress Series*, **91**, 65–69.

Scheunert, I., Schroll, R. & Trapp, S. (1994) Prediction of plant uptake of organic xenobiotics by chemical substance and plant properties. In: *Chemical Exposure Predictions* (ed. D. Calamari), pp. 137–146. Lewis, Boca Raton, FL.

Stagg, R.M. & Addison, R. (1995) An inter-laboratory comparison of measurements of ethoxyresorfin O-deethylase activity in dab (*Limanda limanda*) liver. *Marine Environmental Research*, **40**, 93–108.

Stebbing, A.R.D. (1992) Environmental capacity and the precautionary principle. *Marine Pollution Bulletin*, **24**, 287–295.

ter Braak, C.J.F. (1988) *CANOCO—a FORTRAN program for canonical community ordination by (partial) (detrended) (canonical) correspondence analysis, principal components analysis and redundance analysis (version 3.10)*, pp. 1–95. TNO Institute of Applied Computing Science, Statistics Department.

ter Braak, C.J.F. (1990) *Update notes: CANOCO version 3.10*. Agricultural Mathematics Group, Wageningen.

UNEP (1985) *Protection of the marine environment from land-based sources*. Montreal Guidelines, Environmental Policy and Law, 14/2/3, pp. 77–83, United Nations Environmental Programme, Nairobi.

Underwood, A.J. (1991) Beyond BACI: experimental designs for detecting human environmental impacts on temporal variations in natural populations. *Australian Journal of Marine and Freshwater Research*, **42**, 569–587.

Underwood, A.J. (1993) The mechanics of spatially replicated sampling programmes to detect environmental impacts in a variable world. *Australian Journal of Ecology*, **18**, 99–117.

Underwood, A.J. (1996) *Ecological Experiments: their Logic, Design, Analysis and Interpretation Using Analysis of Variance*. Cambridge University Press, Cambridge.

USEPA (1989) *Briefing report to the EPA Science Advisory Board on the Equilibrium Partitioning Approach to Generating Sediment Quality Criteria*. USEPA Environmental Protection Agency, EPA 440/5-89-002, Washington, DC.

Vasseur, P., Kuenemann, P. & Deviliers, J. (1993) Quantitative structure–biodegradability relationships for predictive purposes. In: *Chemical Exposure Predictions* (ed. D. Calamari), pp. 47–61. Lewis, Boca Raton, FL.

Widdows, J. & Johnson, D. (1988) Physiological energetics of *Mytilus edulis*: scope for growth. *Marine Ecology Progress Series*, **46**, 113–121.

Williams, H. (1994) The economics of when to abandon North Sea structures. The technical, safety and cost questions. A marine contractor's view. *Institute of Petroleum, North Sea Abandonment Conference*, London, February 1995. Heerema UK Ltd.

Wynne, G. & Mayer, S. (1993) Science and the environment. *New Scientist*, **1876**, 33–35.

Chapter 18
Risk Assessment and Management for Inland Waters

STUART HEDGECOTT AND TONY J. DOBBS

18.1 INTRODUCTION

A great deal of technical mysticism has grown up around the use of risk in its various guises, i.e. risk assessment, risk management, risk evaluation, etc. Also, the distinction between risk and hazard has absorbed many hours of debate that must appear somewhat arcane to many practical managers, decision makers and other observers (Chapter 1). These conceptual difficulties are unfortunate because the application of risk assessment is really very familiar to everybody who makes a decision in the face of uncertainty, which is truly everybody. The jargon and mysticism have tended to serve as a barrier to the wider application of what is, in many respects, only applied common sense. The demystification of risk thinking is much too grand a task for this chapter. Rather, the tools and concepts described in this chapter should be viewed as aids to decision-making in the face of uncertainty. They cannot remove uncertainty or remove the need for expert judgement and experience to play a key role in the decision-making process, nor can they create knowledge that does not already exist about the issues involved: risk assessment is not a panacea. They can, however, help structure thinking and thus help to make decision-making more consistent, more transparent and more auditable.

Risk assessment has been undertaken by dischargers and pollution control authorities, in the course of exercising their duties to protect inland waters, for many years. However, in the past such assessments (regulatory or otherwise) have often not been formalized and, when considered overall, have sometimes been based on differing views and approaches. The demands of increasing economic and political harmonization mean that this situation is no longer tenable. Rather, authorities need now to employ consistent methodologies. These must be based on best practice and sound technical principles and the decisions made should be auditable. In a similar vein industry has to be seen to be managing its environmental impacts in a responsible way. In both cases risk assessment and management provide relevant tools.

The essence of risk assessment is the dual consideration of the probability that a particular event will occur and the consequences of such an event. Often the consequences of the event also have a probability associated with them. Many approaches to pollution control in the past, as we shall see, have been fairly 'static', concentrating, for example, on worst cases rather than estimating probabilities. The advantage of a risk-based approach is that it can '... direct finite resources to where they can be most usefully employed ...' (Pritchard 1995).

Most descriptions of risk assessment emphasize the importance of specifying the objective of the risk assessment activity (Chapter 1). Clarity of thought at this stage greatly facilitates the successful application of the risk assessment process. In describing the application of risk assessment and management to inland waters it is important to emphasize that inland waters have a wide variety of uses, and risks to these uses are similarly varied. This chapter focuses on risks posed by chemicals but there is a considerable body of knowledge and experience related to other issues, such as flood protection, water resources, and other management tasks (e.g. species introductions; Chapter 6). Some of the pertinent aspects of these issues are outlined below but the main emphasis of the chapter is risks from chemicals. Such a division is useful in a book such as this to keep within the scope of a single chapter, but in real life the regulators, industry and the community have to take an integrated view of all the risks to inland waters if they are to sustain these valuable resources (Jordaan *et al.* 1993).

18.1.1 Non-chemical risk assessments

Non-chemical risk management of inland waters encompasses a diverse range of topics, such as the following.

1 Hydrological change (e.g. flooding, structural damage, impoverished flow).

2 Resource availability (including seasonally fluctuating supply and increasing demand in the long term).

3 Physical pollution (such as fly-tipping and other litter).

4 Biological perturbation (such as ecologically aggressive competition of an introduced species and release of genetically modified organisms).

5 Microbiological contamination (threatening potable supplies for humans and livestock).

The use of risk assessment in water management probably has its longest history in association with flood protection, where for many years decisions on investment have been made on the basis of risks of adverse weather conditions in combination with the value of the land areas at risk. Procedures for making judgements about investment in flood defence in the UK have been published (MAFF 1993), and 'standards of service' for flood defences are defined as the return period of the flood events, typically in the range 1–10 years for poor agricultural land up to 100 years for urban areas.

A simplified risk assessment procedure has been developed by the Environment Agency in England and Wales to aid decision-making in relation to investment in flood defences. This was primarily motivated by the need to manage the financial liabilities involved in flood defence failures. The system is based on a combination of a 'risk factor' with an estimate of the value of the land and property at risk from flood defence failure. The risk factor is constructed from three main elements.

1 An inherent risk factor involved with flood defence design and construction.

2 A system factor involved with operation and maintenance.

3 An associated factor involved with the infrastructure for monitoring, flood prediction and warning.

The system has been trialled but is still being developed and refined.

A paper by the UK Health and Safety Executive (HSE 1994) concerned with quantified risk assessment compares the level of risk considered acceptable applied to the Thames Barrier, built to alleviate flooding in London, with the corresponding much lower level of risk considered acceptable in building nuclear power plants. However, the paper emphasizes that such direct comparisons should not be made in a simplistic way. A major difference between these two risks is that one, the Thames flooding, is an existing risk. All management action can achieve in this case is a reduction in the level of risk involved. The other, a new nuclear power plant, is an additional risk and one option is not to build the plant at all, and thus avoid the risks involved. This concept of additional versus existing risks can also be usefully applied to considerations of chemical releases to inland waters, particularly in relation to existing and additional sources.

18.2 CHEMICAL RISKS FOR INLAND WATERS

The interest and application of risk assessment and risk management for chemicals in the environment has increased greatly in recent years. Most regulatory authorities and industrial groups in developed nations are now using these tools in some form for the management of problems associated with chemical production, use and release. There are no hard-and-fast guidelines indicating what specific approach(es) to use under particular circumstances, although there are some applications of risk assessment that are prescribed for certain circumstances (such as the European Union's (EU) risk assessment Directive for New Chemicals and Regulation for Existing Chemicals) (COM 1993a; CEC 1994). Although this chapter concentrates on approaches for risk assessment and management for chemicals associated with inland waters, developments in environmental management and regulation are such that it is increasingly difficult to distinguish between protection of inland waters and other environmental sectors.

Comprehensive controls for chemicals in inland waters were first facilitated in the USA by the passing of the Clean Water Act (originally the 1972 Water Pollution Control Act Amendments (PL 92–500)), which included concepts such as a zero discharge goal, the National Pollutant Discharge Elimination System (NPDES) for permitting of point-source releases, minimum treatment standards for sewage discharges and certain industrial discharges, control of toxics and water quality-based standards. Significantly these requirements were enforceable in law.

Following this, management and regulatory options for controlling chemicals have developed a great deal and now cover a variety of approaches ranging from end-product controls such as product registration to process and industry controls such as the EU's Integrated Pollution Prevention and Control Directive (COM 1993b; HMSO 1993). The regulations are increasingly being supplemented by approaches that seek to use market forces to encourage environmentally friendly products and production like ecolabelling and the Eco-Management and Audit Scheme (EMAS) (Council Regulation (EEC) 1836/93). These regulations applied to toxic substances have been summarized elsewhere (Crathorne *et al.* 1996; Dobbs & Zabel 1996). The key to the effective implementation of such regulatory 'packages' is the application of the most appropriate controls to meet the circumstances, which must be based on an integration of knowledge (or prediction) of the potential effects of the chemicals or effluents under scrutiny.

18.2.1 Description of risk assessment and terminology

Two very useful guides have been published by UK Government Departments: the Health and Safety Executive (HSE) provides a clear description of the terms used in risk assessment and management in non-technical language (HSE 1995) and the Department of the Environment (DOE) provides a guide to risk assessment and management for use in environmental protection (DOE 1995a). The diagram in Fig. 18.1 is based on that given in the DOE guide (see also Chapter 9). It illustrates the procedures leading to risk management. Also in Fig. 18.1 is an example of the guidance, considering the steps needed to apply the framework to the risk management of a new sewage treatment works at a particular location on a river. The risk estimation stage in this example involves comparing the probabilities of a range of estimated exposure concentrations (low dissolved oxygen) and durations with the probabilities of a range of estimated effect concentrations.

A quantified risk assessment would involve consideration of the conditions that lead to extreme exposure concentrations along with extreme effect concentrations and appropriate allowance for the joint probabilities. For example, periods of high rainfall may lead to high storm tank overflows into the river with high biochemical oxygen demand (BOD) values, but such conditions may also lead to high river flows so there will be a high dilution capacity. Consideration of the probability distributions of dissolved oxygen in the effluent discharge and in the river, along with the likely duration of different events, is obviously complex, as is estimating probabilities of the environmental effects of different dissolved oxygen levels and durations. In view of the complexities involved, current practice is often to take 'worst-case' conditions as a basis for decision-making and to predict 'expected' exposure and effect concentrations based on, for example, extreme low river flows and effects on the most sensitive aquatic species. This is not a truly risk-based approach despite sometimes being referred to as such. In true risk assessment, probabilities should be taken into account in a variety of ways, for example, by taking a low flow statistic that is unlikely (less than 5% chance) to be exceeded more frequently than once in every 50 years as a worst case, but also considering seasonal and annual average flows as well as high flows that might be particularly associated with increased probability of pollution events.

The approach of comparing expected concen-

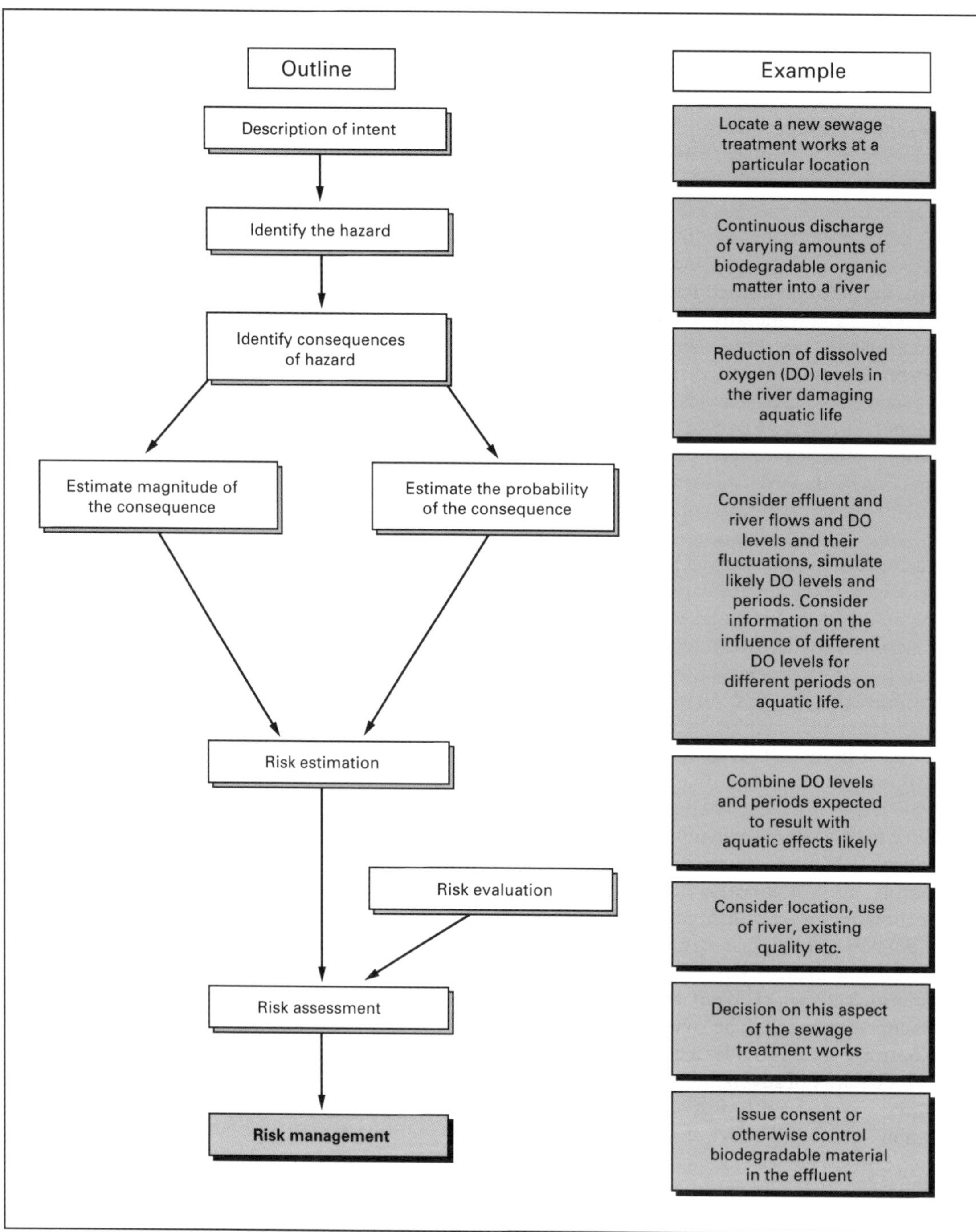

Fig. 18.1 Risk assessment and management outline with an example for inland waters.

trations with effect concentrations is widely used in a variety of regulatory activities involved in the safety and environmental aspects of chemical registration and control. Often, the risk estimation is facilitated by establishing concentrations of chemicals that have acceptably low probabilities of effect, such as environmental standards (Hedgecott & Fawell 1996) or predicted no-effect concentrations (PNECs). The risk estimation then involves creating the ratio of the predicted environmental concentration (PEC) to this PNEC, and generally if this value is $\gg 1$ the risk is considered unacceptable. There are, however, some deficiencies with this approach. The essential elements of Fig. 18.1 are those concerned with the identification of both the magnitude and the probability of each consequence. As indicated above, much of the current activity in chemical control associated with inland waters is not fully risk based because these probabilities are not considered in a rigorous way: predicted exposure levels are often considered as single values based on specific (often worst-case) scenarios with little consideration of the possible frequency of occurrence of these scenarios; similarly effect concentrations for ecotoxicological impacts are often established as no-effect levels (by the use of ecotoxicological data with safety factors), which provide no information on the probability of significant effects actually happening in the environments. To be of greatest use to risk managers, assessments must make some consideration of different scenarios.

Recent work in North America on the procedure and application of risk assessment to chemicals in water specifically consider the probability distributions of both exposure concentrations and effect concentrations. The overlap of these two distribution curves provides a quantified estimate of the probability of adverse effects. The technique has been applied with full rigour to atrazine (Solomon *et al.* 1996), vividly illustrating the sophistication that can be achieved if there are adequate data. Solomon and co-workers conducted a comprehensive aquatic ecological risk assessment for atrazine, a herbicide with extensive agricultural and non-agricultural uses. The environmental exposure assessment used extensive monitoring data, including background levels and short-term levels associated with storm runoff events, but also applied mathematical fate models to extrapolate to unmonitored sites. The effects of reductions in application rates were also evaluated using fate models. The effect assessment considered laboratory data for major groups of freshwater organisms, but laid greater emphasis on evidence from field studies that indicated at what levels atrazine had ecological impacts, the significance of these impacts and, importantly, at what levels ecosystems recovered. Based on an integration of these laboratory and field effects studies with the measured and predicted environmental levels, Solomon *et al.* were able to assess the risks associated with atrazine in terms of: (i) types of organisms that might be affected; (ii) the severity of these effects and probability of recovery; and (iii) sites where the risk of impact was highest. With the exception of certain reservoirs in areas with intensive atrazine usage—for which site-specific assessments were recommended—the herbicide did not appear to pose a significant risk to inland waters in North America.

As illustrated to some extent by Solomon's work, the outcome of risk assessments usually falls into one of three categories.

1 The risk arising from the hazard is so low that it can be regarded as 'negligible' in comparison with the benefits of the intention.

2 The risk is so high that it is 'intolerable' in comparison with the benefits.

3 The risk is 'tolerable' but might be reduced, and if action can be taken to further reduce it this should be done.

The terms 'negligible', 'intolerable' and 'tolerable' obviously require a great deal of technical and political consideration.

18.2.2 Chemical risks and inland waters

The EU's Dangerous Substances Directive (CEC 1976) defines pollution of waters as: '... the discharge by man, directly or indirectly, of substances or energy into the aquatic environment, the results of which are such as to cause hazards to human health, harm to living resources and to aquatic ecosystems, damage to amenities or interference with other legitimate uses of water'. Thus, in the context of inland waters, risk assessment should be applied to each of these possible environmental impacts. This is still a developing field and is an area where the uncertainty involved is particularly acute. For example, in assessing the environmental effects of chemicals we are frequently uncertain about how uncertain we are! The use of risk assessment does not remove uncertainty but does enable different aspects to be clearly identified, and thus the impact they make on decision-making can be assessed and audited. A wide variety of activities or intents (see Table 18.1) will lead to a correspondingly wide variety of hazards associated with contamination of the aquatic environment. The hazards can be grouped together into a number of different sources of chemical contamination, which are described in section 18.2.3. These sources can produce effects or consequences (see Table 18.1) that can be broadly grouped into four areas:

1 Direct toxicity to aquatic or other organisms (including humans), including effects manifest through food chains.

2 Non-toxic effects on organisms leading to disturbance of the natural community structure, such as excessive growth, and the consequential changes, for example eutrophication and deoxygenation.

3 Physicochemical impairment, for example foaming, smell and coloration.

4 Damage to human processes, activities or structures using or associated with water.

For risk management purposes each of the above needs to be considered in the context of the sources of the chemical contaminants that are potential pollutants. Whatever the source, a contaminant's effects will depend on its intrinsic hazard, defined by its physicochemical and toxicological properties. The most obvious impacts are those resulting from high concentrations of contaminants, which result in observable degradation of the immediate environment. However, a longer term release of a chemical at low levels may have no immediate discernible effects but could result in degradation of an ecosystem over time. Continuous releases might also result in accumulation of chemicals in certain environmental sectors, such as areas of slack water or, more commonly, settled sediments, and adverse effects could arise at localities well removed from the original point of entry. Thus any management plan must ensure that both acute and chronic effects of chemicals are considered, and that these are related to patterns of input and environmental distribution.

18.2.3 Sources of contamination of inland waters

Point-source and diffuse-source releases

The distinction between point and diffuse sources is not well defined, but the broad characteristics are summarized as follows. Point-source releases of chemicals are discrete discharges that might best be considered as those whose point of entry can be readily circumscribed (and potentially stopped). Industrial and sewage works outfalls are 'classic' point sources, but less obvious ones include road drainage culverts and storm overflow sewers. In all cases they may be continuous or intermittent. Diffuse sources may be impossible to define quantitatively, or even impossible to identify. They include rainfall, surface water runoff and recharge of rivers from adjoining aquifers. Clearly the assessment and, in particular, the management of contaminants arising via these routes need to be adapted to the very different circumstances associated with them.

Accidental and planned releases

The risks of planned chemical use must be balanced against the economic or social benefits arising from that use. In contrast, the management of unplanned or accidental releases is orientated toward preventing or minimizing the likelihood of events that would have intolerable or unacceptable consequences. Here, more 'traditional' risk assessment tools considering, for example, estimates of frequency of system failure and probabilistic event analysis are critical, in addition to the chemical risk characterization. Also key to this decision-making process are estimations of what constitute intolerable consequences of accidental release. In the UK the Control of Industrial Major Accident Hazards (CIMAH) Regulations 1984 require operators of certain industrial sites to identify potential 'major accident' hazards and put in place systems that,

Table 18.1 Effects on the aquatic environment that constitute a 'major accident'.

Freshwater and estuarine habitats	Effects on a significant part of any stream, river, canal, reservoir, lake, pond or estuary that: • lower the chemical quality by one class for > 1 month, or • lower the biological quality by one class for > 1 year, or • cause long-term damage to the habitat overall
Marine environment	Permanent or long-term damage to an area of ca. 2 hectares or more of the littoral or sub-littoral zone or the benthic community adjacent to the coast or the benthic community of any fish spawning ground or to an area of ca. 250 hectares or more of the benthic community of the open sea or A casualty count of ca. 100 seabirds, excluding commoner species of gulls, or 500 seabirds of any species, or five sea mammals of any species found dead or unable to reproduce as a result of the accident
Aquifers and groundwater	Damage to an aquifer leading to contamination or other effects that preclude its use for public, domestic or agricultural supply or have significant adverse impact on the surface waters and biotic systems it supports
Water sources and supplies	Contamination of a water supply such that the supply to 10 000 or more consumers has to be interrupted because the source has been rendered unfit for human consumption or the water treatment works or distribution system has been damaged or contaminated, in circumstances where an alternative piped system is not readily available
Sewerage and sewage treatment	Damage to a sewerage system or treatment works that results in significant risk to public health by pollution of a water source used for supply to 10 000 or more persons, or unacceptable or widespread hazard to public health or safety through flooding
Crops, domestic animals and other foodstuffs	Contamination of a significant area of any aquatic habitat that prevents fishing or aquaculture or renders it inaccessible to the public
Public access	
National Nature Reserve, Site of Special Scientific Interest and other designated areas	Permanent or long-term damage to: • >10% or 0.5 hectares (whichever is the lesser) of the area of the site, or • >10% of the area of a particular habitat, or • >10% of a particular species associated with the site
Particular species	Death or inability to reproduce in a significant percentage of known or estimated national population of a particular species
The wider environment	Permanent or long-term damage to: • 2 or more hectares of scarce habitat, or • 5 or more hectares of intermediate habitat, or • 10 or more hectares of more widespread or unclassified habitat
Recreational facilities	Damage to locally significant amenity such that it no longer possesses its aesthetic, cultural, amenity or public enjoyment value
Built heritage	Damage to an area of archaeological importance, to a conservation area, to a Grade 1 listed or Category A building, or to a scheduled ancient monument
Release of persistent toxic substances	Release of 10% or more of the 'top tier' threshold quantity of a persistent dangerous substance

as far as possible, prevent their occurrence and minimize their impact should they occur. To facilitate appropriate decision-making in this respect, the DOE (1991) provides guidance on what constitutes 'major accidents' on the basis of the resulting impacts, and all these impacts can be directly or indirectly associated with impacts on aquatic environment, as summarized in Table 18.1.

Increasingly, pollution prevention activity is focusing on the reduction of foreseeable but unplanned releases as a key part of pollution control activity. In the UK, Her Majesty's Inspectorate of Pollution (HMIP) proposed the implementation of a programme known as Operator and Pollution Risk Appraisal (OPRA), which identifies two separate elements contributing to the overall risk associated with accidental releases: operator performance and pollution hazard. Scores are assigned to these in order to establish an overall risk rating. This scheme is being considered as a basis for prioritizing site visits by Inspectors of the Environment Agency (into which the HMIP has been incorporated). Subsequently, scores can be used to monitor improvements in management from a site-specific level to a national level.

Existing and new sources

All new releases of chemicals arising as described above can be considered sources of input to water. In any particular case, the source concerned could be controlled or diverted elsewhere in order to minimize the amount released or the impact of any release. In other cases, chemicals might be released into waters from existing sources. Some of these, such as factory discharges, are similar to additional sources in that they can become subject to controls. However, others—like the contaminated land around an existing factory, or contaminated sediment in a river—cannot be stopped, but can only be ameliorated to a greater or a lesser extent. The risk estimation of existing sources usually follows an exactly equivalent path to that for additional sources, but the risk evaluation leads to a lower level in the risk assessment than would apply to an additional source, as indicated earlier.

18.3 THE BUSINESS NEED AND AVAILABLE FRAMEWORK FOR RISK ASSESSMENTS

18.3.1 The need for aquatic risk assessment

Protection and improvement of inland water quality need to encompass some form of control on all sources of potentially harmful chemicals described in section 18.2.3. The key advantages of using risk assessment approaches for the control of chemicals in inland waters are twofold:

1 They enable chemicals to be compared, processes to be compared and localities to be compared in a quantitative manner. Thus, pollution prevention and control measures can be prioritized and resources can be targeted to challenge releases of greatest concern based on both the intrinsic hazard(s) of the chemical(s) involved and the local conditions, resulting in increased overall effectiveness.

2 They allow for quantitative (or at least semi-quantitative) estimates of impact to be determined. Thus the benefits to be gained from alternative management options can be compared, and this can be done in relation to the effort required to achieve them. Only by doing this can the risk assessor identify the optimal strategy for chemical control, considering not only environmental risks but also the economic and social benefits associated with the chemical.

Consistency in approach is essential if the demands placed on those involved in the manufacture, handling, use or discharge of potentially hazardous chemicals are to be equitable. Risk assessments need to be based on sound principles and practice to ensure that any risk management demands resulting from them are reasonable, because these may have significant practical and economic implications. For example, one extreme outcome might be that risk management requires an existing process to be stopped (or a new process prevented), and if this is to be put into practice it must be clear beforehand that the environmental benefits outweigh any adverse economic and social impacts. Guidance on the application of risk–benefit analysis to existing chemical substances has been produced by a UK Government–Industry Working Group (DOE 1995b).

More generally there are currently a number of regulatory and industrial operational needs converging on the requirement to undertake environmental impact assessment and to incorporate estimates of costs and benefits. Examples include the development and implementation of Environmental Management Systems (Hunt & Johnson 1995), the use of lifecycle analysis (SETAC 1993; Chapters 9 and 29) for product development, the production of environmental impact statements and the need for best practicable environmental option (BPEO) assessments. There is a need for a protocol of risk assessment and cost–benefit analysis to be drawn up to provide a consistent approach to this important area. The cost–benefit protocol might usefully cover decisions about whether to adopt standard pollution control measures or undertake a cost–benefit analysis, and how the iterative process of considering precautionary measures might take place in a quantitative risk assessment. A consistent approach to such assessments might also reduce the replication of effort devoted to assessments for a particular chemical or scenario.

18.3.2 The legislative basis for aquatic risk assessment

Early regulations focused on procedures for the control of point sources, mainly through the issue of discharge consents. Alongside these were various actions to reduce the use of chemicals that were considered very hazardous to the aquatic environment, such as through the identification and creation of priority pollutant lists and the requirement to control and reduce emissions of these chemicals. More recently, attention has focused on diffuse sources and their control. Now the concern is to attempt to integrate environmental management, which involves the consideration of environmental impact at all stages of the lifecycle of products and services, i.e. production, use and disposal (HMSO 1994). Environmental regulations in the future will have to adopt different approaches to accommodate these more modern integrated procedures. The regulatory mechanisms used have been reviewed previously (Chapters 9 and 10). All that will be attempted here is to view the regulatory approach from a risk perspective.

The assessment of risks associated with chemicals is considered in legislation from separate bases: chemical-specific, site-specific (i.e. location-specific) and process- or activity-specific. These are considered below.

Chemical-specific

Chemical-specific approaches have until now mostly been based on simple approaches to exposure and effect assessments. The most common process is the derivation of one or a number of quality criteria for a chemical in water. Historically this has tended to take the form of an environmental quality standard (EQS), which might be considered as a PNEC for all freshwater organisms. Such standards have been established at various levels of regulatory organization, including international legislation (such as European Union Directives arising from the Dangerous Substances Directive (CEC 1976)), national legislation (such as mandatory EQS values proposed by the UK DOE (DOE 1989) and the US Environmental Protection Agency (USEPA 1986)), and national or regional advisory values. They are most often established as long-term values (such as annual average concentrations) and short-term values (such as maximum admissible concentrations or 95-percentile concentrations, the latter being more realistic for statistical determination). However, in some cases more sophisticated considerations are applied. For example, the USEPA has established several water quality criteria on the basis that aquatic ecosystems can recover from chemical perturbation. Thus values are established as concentrations (either 1-h average or 4-day average concentrations, depending on the chemical) that, if exceeded once, must not be exceeded again for a period of 3 years in order to allow the system to recover (USEPA 1986).

One of the more comprehensive schemes for chemical-specific risk assessments is the Toxic Substances Control Act (TSCA) in the USA. Enacted in 1976, this provides for the regulation of chemicals that pose an 'unreasonable' risk to human health or the environment (Chapter 10). Assessments to determine whether risk is 'unreasonable' integrate chemical hazard assessment and exposure assessment to provide a quotient value (e.g. Rodier

& Mauriello 1993), indicative of the margin of 'safety' between effects concentrations and likely exposure concentrations—a concept that will recur in this chapter. A risk assessment would normally comprise three steps: worst-case risk assessment, identification of the type of risk (e.g. acute versus long term), and quantification of the degree of risk of potential environmental impact (Nabholz 1991).

In the EU the situation is similar, with a number of regulations requiring risk assessment for existing and new substances, considering threats to all environmental compartments arising from the production and use of chemicals (see Chapters 3 and 10). These assessments have a number of tiers of complexity and detail, with progress up the tiers based on undesirable or uncertain results from the lower tiers. This strategy for risk assessment is quite common, as is discussed elsewhere in this volume. More recently the EU has been investigating strategies for the use of environmental risk criteria, suggesting that future legislation may focus on site or activity risk assessments rather than chemical-specific assessments.

Recently there have been some advances towards a more truly risk-based approach to effects assessment. In France, the Agences de l'Eau are trialling a scheme where priority chemicals are assessed to determine a number of quality criteria that are associated with incremental levels of impact (i.e. no impact, possibility of chronic impact, possibility of acute impact, acute impact). The intention is that the different action levels can be compared with data from receiving water-monitoring programmes in order to prioritize chemicals in terms of the biological significance of their impact, as well as to provide a system for early warning of increasing impact. Additionally, determination of levels of contamination of priority chemicals and comparison with the action levels can be used to aid in water quality classification and thus prioritization for chemical control at different localities, which leads to a site-specific consideration of risk.

Site-specific

In Europe, risk management of sites to protect inland waters is conducted in a number of ways. Under regulation there are three broad areas of control: environmental impact assessment at the planning stage; authorization of discharges of effluent to surface waters during normal operation; and, for the larger sites, consideration of the risks from major accidents. Control of the site by the operator usually uses some or all of the procedures associated with Environmental Management Systems. For larger, more hazardous, sites in the UK the principle of Integrated Pollution Control applies involving a requirement to use Best Available Technology Not Entailing Excessive Cost (BATNEEC) to minimize, prevent or render harmless any discharges. Where discharges occur to more than one medium the BPEO has also to be applied (see below). All these activities require similar risk estimation and assessment activity. For single releases to water, estimates can most readily be performed for a chemical for which an EQS or other form of PNEC has already been derived. When these are considered in conjunction with information on the amounts of chemical released and patterns of dilution and dissipation, the probability of approaching or exceeding acceptable levels can be determined. Thus the 'risk' assessment process is a relatively straightforward and unsophisticated comparison of a no-effect level with one or a number of PECs. A more thorough risk basis can be achieved by comparing no-effect concentrations for different durations of exposure with PECs for different scenarios of release and dissipation, and their associated probabilities, but the situation remains an oversimplification in that it does not consider a gradation of environmental consequences. A true risk-based approach would need to consider the probabilities of release, exposure and effects.

A BPEO assessment requires a comparative consideration of risks to identify which alternative process or activity would have the most acceptable impact (HMSO 1988). For example, scrubbing of gaseous waste streams before emission may be a possible 'treatment' option, but before it is employed it is necessary to know that the resulting aqueous effluent will not be more harmful than the original gaseous emission. This type of consideration recognizes that inland waters cannot be considered in isolation from other environmental compartments. However, the fundamental barrier to this type of comparative assessment is the lack of a common denominator to enable a meaningful comparison. Our lack of understanding of ecosystem functioning and resilience is partly to blame for this and undoubtedly needs to be improved. Another factor is the lack of a consistent basis for establishing environmental standards for different environmental media. Such standards are usually set in a conservative fashion with the laudable intention of being 'on the safe side'. The problem is that it is not possible to say how far on the safe side they are, i.e. what are the implied levels of risk? This is essential information if these standards are to be used for crossmedia comparisons.

The approach to BPEO assessment proposed by the Environment Agency in England and Wales was published in 1996 (HMIP 1996). In this approach the harm from a release is assessed by reference to an upper level that is considered unacceptable and a lower level that is considered insignificant. Between these two levels the harm caused by the release is considered to be tolerable and judgements are made about the appropriate balance between environmental harm and the cost (and other) implications of control options. The approach is based on consideration of the concentration of the substance in the environment as the measure of the potential risk.

1 When an environmental standard is available for a substance in a particular medium, this value is used to establish the boundary of the upper, unacceptable, limit. When no standard is available, an Environmental Assessment Level (EAL) is established (Douben 1995).

2 If (a) the PEC arising from a process is greater than 80% of the standard or EAL for that compartment, or (b) the process contribution to the PEC is equivalent to more than 2% of the standard or EAL, then further control of the process is required.

3 If neither (a) nor (b) apply, then risks are insignificant and no futher evaluation of alternative techniques is needed.

In order to compare different release options the resulting estimated environmental concentrations are expressed as environmental quotients by dividing by the appropriate standard or EAL. The quotients for each substance in each medium are then added together, and these are summed for all media to give an integrated environmental index. The process option with the lowest integrated environmental index is then the one expected to have the least overall environmental impact. Other potential environmental effects are also considered, such as global warning potential from 'greenhouse' gas releases and low-level ozone production.

In order to focus management and regulatory attention on areas where improvement in performance is necessary, HMIP proposed a risk-based approach to process operation assessment (HMIP 1995). The scheme, known as OPRA (Operator and Pollution Risk Appraisal), involves rating a process according to two attributes: operator performance, essentially the probability of an event; and pollution hazard, essentially the magnitude of the consequence of the event (see Fig. 18.1). Each attribute has seven items of assessment, as indicated below.

1 Pollution hazard:
- (a) hazardous substances present,
- (b) techniques for prevention and minimization,
- (c) scale of process,
- (d) location,
- (e) frequency of operation,
- (f) techniques for abatement,
- (g) offensive substances in process.

2 Operator performance:
- (a) Compliance with limits and adequacy of records,
- (b) knowledge of authorization and implementation,
- (c) plant maintenance and operation,
- (d) management and training,
- (e) procedures and instructions,
- (f) frequency of incidents and justified complaints,
- (g) auditable environmental management system.

Each of the items is given a score from 1 to 5 (1 = worst, 5 = best) and is multiplied by a factor that represents the relative importance of the item. Such values are expected to evolve as experience of implementing the scheme develops. This exercise will give a single-value estimate of the pollution hazard and the standard of the operator's risk management performance for a single process. The main benefits of the scheme are seen as follows.

1 Improved pollution prevention by prioritizing attention on those areas where improvements are necessary.

2 Improved targeting of inspections by pollution inspectors through the identification of higher risk processes.
3 Information on changes in time in the overall performance of process sites and industrial sectors and on the performance of the regulators.
4 A framework for the possible introduction of charging that links to pollution risk.

In the USA, site-specific risk assessments are perhaps best developed in the case of National Priority List sites (or 'superfund' sites). At these highly contaminated localities, baseline risk assessments identifying and quantifying human and ecological risks can be the subject of major scientific and fiscal investment. One approach used for the assessment of risks (relating to any environmental compartment, not just water) is the triad approach. This entails chemical analysis to establish information on patterns of contamination, ecological surveys to determine the level of impact, and toxicological investigation to determine the link(s) between the two. By thus establishing both the extent and the cause of ecological damage, the risk assessor is then in a position to identify appropriate control measures or remedial strategies. This 'problem formulation' methodology is an essential initial step in the development of risk management plans for contaminated sites, such as inland sediments (Pascoe & DalSoglio 1994).

Proces- or activity-specific

Process-specific risk assessments may be split broadly into those where the process is associated with known chemical contaminants (e.g. mercury from chlor-alkali industry, considered in Europe by EC Directive 82/176/EEC (CEC 1982)) and those where chemical releases are not so well characterized but are known to be potentially toxic (e.g. paper pulping and milling). In the former case, risk assessments are effectively based on the hazard of the known contaminants and are very similar to the chemical-specific assessments considered above. The latter, however, are less amenable to such considerations. The potential for interactions between chemical toxicants means that hazard assessment can only really be performed through practical investigation, i.e. by determining the toxicity of the whole effluent or waste stream. This can then to some extent be treated as a single chemical entity, with risk assessment proceeding on the basis of measured effects and predicted, modelled or observed dissipation of the whole effluent (or at least its toxic components). Additionally, steps towards risk management might be taken through comparison of toxic effects with and without additional treatment applied to the effluent.

The alternative is to identify the process as being one that poses high environmental risks and regulate it appropriately. This is the basis for the Integrated Pollution Control regulations in the UK and the Integrated Pollution and Prevention Directive of the EU. These regulations put into place specific pollution prevention and control techniques to prevent and minimize discharges from the industries concerned. A major part of the regulatory activity is concerned with the use of best available technology (BAT) and/or techniques. Techniques include technologies but also include site operating principles and procedures, for example, which are essential factors in pollution prevention.

18.4 THE FRAMEWORK FOR RISK MANAGEMENT

18.4.1 Water uses

However risks are assessed or managed, they must be considered in the context of what impacts might be considered as tolerable or acceptable. As indicated in section 18.2, levels of tolerability and acceptability can only be defined subjectively. Central to this are considerations of uses for which the water resource may be exploited. A consideration of the risk posed to a specific inland water might encompass assessment of risks for several different uses.

1 *Protection of aquatic life*. Levels necessary to protect aquatic organisms from adverse effects are usually determined by extrapolation from laboratory toxicity data. A number of models have been developed (e.g. Stephan *et al.* 1985; Kooijman 1987; Wagner & Lokke 1991) that allow a reasonable estimate to be made of the concentration at or below which aquatic ecosystems should not be adversely affected to a

significant degree. These models have a statistical basis designed to ensure that 95% or more of species will not be affected. The potential effects on the remaining 5% will be dictated by the type of data on which the extrapolation is based. Thus, for example, if the risk assessor extrapolates to a safe level from data on no-effects levels in laboratory tests, then the uncertainty in this extrapolation means that 5% of species may experience some toxic effects. If, however, the extrapolation is based on LC_{50} data (i.e. concentrations of chemicals that would kill 50% of an exposed population), then the uncertainty means that 5% of species may, potentially, experience up to 50% mortality. It is, therefore, vital that any such extrapolation is always considered in the context of the quantity, quality and type of data used to derive a protective level or 'standard'.

2 *Protection of key aquatic communities.* In certain cases the level of protection afforded by statistical extrapolation methods as described in use 1 above is insufficient, and more specific considerations may be necessary. Examples might include protection of commercial fishery interests, including fish farming activities, other aquaculture (e.g. watercress beds), Sites of Special Scientific Interest or similar designations, and rare species. Risk levels will be determined by information specific to the vulnerability of the ecosystem or species of concern.

3 *Abstraction for drinking water.* Levels to protect the health of human consumers are usually determined from mammalian toxicology information extrapolated to the human situation. When establishing risk levels for raw drinking water sources, some consideration of removal of the chemical during treatment and distribution might be made. Drinking-water levels are more stringent than those to protect aquatic life because the acceptability of adverse effects is lower. Certainly a level of protection of 95% could never be considered acceptable!

4 *Protection of food for human (and wildlife) consumers.* Consideration of the health of human and wildlife consumers of aquatic organisms is necessary for chemicals that are persistent and readily associate with biological tissues, and may thus be accumulated by aquatic organisms and passed up the food chain (biomagnified). Such chemicals are normally identified by physico-chemical features determining their stability and their partitioning into organic membranes, and also considering the extent to which they are transferred from one trophic level to the next.

5 *Agricultural use.* Livestock watering can be considered to demand the same quality as water for human consumption, although there are differences in terms of water treatment and, arguably, fewer concerns if livestock are adversely affected. Water used for crop irrigation will have different quality demands, based on any herbicidal effects of chemical contaminants as well as considerations of chemical accumulation in herbage destined for livestock or human consumption, and possible long-term impacts on soil quality from contaminants that accumulate in the soil.

6 *Amenity value and recreational use.* Recreational uses can broadly be categorized according to the extent of human contact. Thus, water-sports with significant contact (e.g. bathing, canoeing) may require similar considerations to those involving potable water abstraction, angling can be considered along with protection of food consumers, and non-contract uses such as sight-seeing, natural history and navigation are primarily concerned with aesthetic value (including odour and colour) and maintenance of flow and open channels rather than with chemical toxicity.

7 *Sensitive industries and structures.* In addition to the food and beverage industries, which require high-quality water for processing and production of consumed goods, certain other industries (e.g. microelectronics production) also require high-quality water. The risk assessor must also take into consideration any potential damage to relevant structures associated with water, for example hydroelectric installations, which might be vulnerable to certain chemical action.

18.4.2 Catchment management planning

Clearly, the risks posed to the different uses of inland waters are very variable and they require different levels of protection (e.g. protection of all individual human consumers but only 95% or more aquatic species). The sensitivity of the uses varies between chemicals, and the hazards posed by a chemical varies between uses. Some assessment of existing water quality may also be relevant, for example a discharge of a chemical at a level determined to be tolerable from the perspective of protecting aquatic life may be deemed more acceptable if the receiving water is already significantly impacted and thus has depauperate flora and fauna. Therefore, although the outcome of a chemical risk assessment for a particular use might be considered constant, the risk evaluation of that particular consequence may vary and the risk management strategy adopted would reflect this.

The many legitimate uses of inland waters may potentially be in conflict. The ability to satisfy each use depends on water quality, water quantity, and physical features (such as flood defences or fish passes), and in satisfying a particular need others may be affected detrimentally. Demands therefore need to be balanced in a sustainable and cost-effective manner so that sufficient water of sufficient quality is available to sustain life and satisfy priority water uses. This can best be achieved by planning for environmental sustainability and improvement through an integrated approach to river catchment management, i.e. the development and application of catchment management plans. Catchment management plans allow an integrated approach to balance the major uses of water, including abstraction and discharging, recreation, navigation use, etc., within a catchment. They enable a formal and transparent approach to assessing and managing risks in a consistent and co-ordinated way. Thus they integrate the interests of all the functions of regulators at the natural hydrological planning level of the river catchment. Elements of a catchment management plan would typically include the following.

1 Detailed description of catchment geology and hydrology.

2 Data on demands for water from all users, with short-, medium- and long-term forecasts of future demand.

3 Categorization of land adjacent to surface waters according to usage and consequent demands for flood defences and warning.

4 Assessment of effects of water abstraction and impoundment, as well as augmentation, on flow at all points downstream, and planning of such to protect the quality of the aquatic environment. Also assessment of effects of water abstraction on groundwater resources.

5 Identification of aquifers vulnerable to pollution and of necessary protection measures.

6 Predicted impacts on water quality of identified sewage and industrial discharges, and of chemical and other releases from agricultural activities. Particular attention is also paid to the location of solid waste disposal sites within the catchment.

7 Necessary measures to ensure appropriate habitat protection and nature conservation, and sustainability of all other relevant water uses (as discussed above).

To be effective, catchment management plans need to address all aspects of inland water use and management, in particular the following.

1 They must be based on a detailed understanding of all the catchment's resources, and the uses and activities to which these are put.

2 They must balance conflicting uses and identify actions where conflicts or potential conflicts exist.

3 They must be based on consultation with all water users and other interested parties, including regulatory authorities, industry, environmental groups and the public.

4 They must establish a long-term perspective based on projected changes in demand, usage and uses.

5 They must target key issues and prioritize actions and investments to improve the water environment in the context of both overall quality and desirable quality in other catchments locally, regionally and nationally.

Above all, the critical feature of catchment management plans is their need to be proactive, so enabling, if necessary, management of risks through intervention activities before problems arise (Water Quality (1990)).

Within such a planning framework, point sources of contamination are primarily controlled through the establishment of discharge consents with subsequent monitoring of discharges and receiving waters to ensure that these are being complied with. Discharge consents contain upper limits of specific chemical determinands that are deemed to pose a threat to the aquatic environment, with these limits being based on legislative environmental quality standards where these exist or on other published or calculated PNECs where they do not. This mechanism potentially allows for risks associated with particular chemicals to be considered consistently and equitably. For example, in England and Wales the National Rivers Authority (subsequently incorporated within the Environment Agency) developed a comprehensive 'Discharge Consents Manual' that provided detailed guidance to ensure national consistency in consenting policy.

More recently the Environment Agency in England and Wales, the Scottish Environmental Protection Agency (SEPA), and the Department of the Environment for Northern Ireland have begun to institute a programme for the control of environmental effects from effluents through direct toxicity assessment, following similar developments in the USA (the USEPA's Water Quality-based Toxics Control scheme developed under the Clean Water Act). Thus in the future toxicity-based consents may be issued for certain effluent discharges that are particularly complex—making control of individual chemical parameters difficult or inappropriate—and for which there is evidence of adverse environmental effects resulting from toxic impacts in the receiving water. Such consents would be established according to acceptable or tolerable levels of measured toxicity in the effluent itself, based on the dilution potential in the receiving water and the quality and sensitivity of the receiving water ecosystem. Such controls aim to specifically protect aquatic ecosystems, rather than other water uses (Environment Agency 1996).

The control of diffuse contamination is less well developed in legislation because its nature makes it more difficult to identify, to quantify impacts and to control. Some regulations exist that enable proactive control of activities that might lead to diffuse contamination, such as the UK Control of Pollution (Silage, Slurry and Agricultural Fuel Oil) Regulations (1991) (Statutory Instrument 1991/324 and 1991/346), which grant regulators the powers to influence the activity of farmers so as to reduce the risks of these materials entering water and to mitigate their effects if they do, whilst recovering costs from those responsible for the materials. However, most catchment management planning activities for diffuse risk management concentrate on pollution prevention and waste minimization through advising and influencing chemical users and water users. In some countries national guidelines have been developed to assist with this process through the provision of risk-based criteria for prioritizing and assessing different activities. Overall, this process is particularly dependent on the successful encouragement of industry and the public to take a responsible attitude toward the water environment. With industry in particular this should be greatly aided by the introduction of environmental management techniques such as BS7750 (BSI 1992) and EMAS (Council Regulation (EEC) 1836/93), which help organizations come to terms with the terminology, strategies and approaches for improved environmental management at a local scale. Such environmental management techniques will also have an impact on point sources, as more cost-effective and environmentally effective alternatives to end-of-pipe 'controls' are introduced based on waste minimization.

18.4.3 Contaminant transfer

As indicated above, inland waters cannot be considered in isolation from other environmental media, including other waters (aquifers, coastal waters and oceans), air and land. Chemicals released to a particular compartment rarely remain in that compartment; rather, they are subject to dynamic processes of distribution and dispersal that generally will not reach equilibrium (due to continuing additional inputs to the environment and removal from the environment by degradation). It is, therefore, desirable that risk assessment and management strategies take account of impacts in other compartments besides inland waters. For releases made directly into surface waters, the following significant transfer processes need particular consideration in order to assess possible risks to other environmental compartments.

1 Chemicals that are highly volatile may undergo significant transfer from surface waters into air. Flow disturbance (e.g. weirs) will increase volatilization losses and aerosol dispersal.

2 Chemicals that sorb to, and persist in, sediments

may subsequently be deposited on land during periods of flooding or through channel dredging if spoil is disposed of on land.

3 Irrigation of agricultural and amenity land can result in exposure of crops, livestock, terrestrial wildlife and humans to chemicals in irrigation water.

4 Abstraction for drinking-water treatment may result in exposure of human consumers to contaminants, as well as producing waste sludges requiring disposal.

Risk assessment in such cases might involve consideration of whether these transfers are significant locally and, if so, whether they would result in undesirable environmental effects other than those considered for the water body itself.

The wider issue of the relative importance assigned to chemical water problems alongside other environmental issues, and thus the consequential resource allocation, is one with which risk assessment can also assist. The USEPA's comparative risk analysis approach to priority setting was designed to address this issue. The approach involves extensive public consultation in order to improve the public acceptability of the priorities and to attempt to move policy-making away from a short-term reactive position. Such introduction of public involvement emphasizes the importance of risk evaluation alongside risk estimation in the risk assessment process. The comparative risk assessment process is two-stage. First, the various problem areas are ranked based on a combination of the best technical analysis and the public perceptions of interest groups involved. Second, the problem areas where the greatest risk reduction can be achieved within practical constraints are identified. The guidebook produced by the USEPA provides methods and approaches that can be used at the various stages of analysis (USEPA 1993). Many of these techniques would be applicable in the production of catchment management plans that, as discussed earlier, are central to risk management of inland waters. It is also clear from the stance taken in this document that risk reduction is considered to be the key driving force behind future environmental protection strategy.

A further transfer of chemicals entering inland waters is their movement across geographical, rather than environmental, boundaries. This situation may arise across administrative boundaries within a single nation, but is perhaps of greatest concern when rivers cross national borders, transferring a contaminant load into a country downstream of the discharges and limiting the options there for further use of the water as a resource. An upstream input may have no overt effect and be considered to have acceptable risk, whereas an identical downstream input into an already contaminated watercourse may be intolerable. This situation can only be managed if approached in a consistent way that considers the entire catchment. Therefore, the harmonization of risk assessment protocols is paramount, and the adoption of political agreements to the management of risks is essential to achieve this. The formation of the International Commission for the Protection of the Rhine and its subsequent work is a basic model being followed by other countries to face up to the problems of international rivers. Essentially, the Commission imposes an obligation on contracting parties to improve the quality of this international river through a consistent risk management strategy applied to the entire catchment. Further examples of this type of international agreement can be seen in the Action Plans produced for the Danube and for Lake Constanz. It is laudable that the significance of trans-boundary pollution transfer in Europe is increasingly being recognized in regulation, with the EU proposing that this becomes an essential component of risk assessment strategies (COM 1993c).

18.5 APPLICATIONS OF RISK ASSESSMENT AND MANAGEMENT

Chemicals might be considered on a scale relating to the likelihood and degree (i.e. the risk) of impact and the extent to which these risks are deemed 'allowable'. This is represented in Fig. 18.2, where increasing environmental levels are associated with increasing risk and decreasing 'allowability'.

Risk assessment for any one chemical, process or activity determines where it falls in such a determination of incremental risk. Risk management is then concerned with lowering the risk

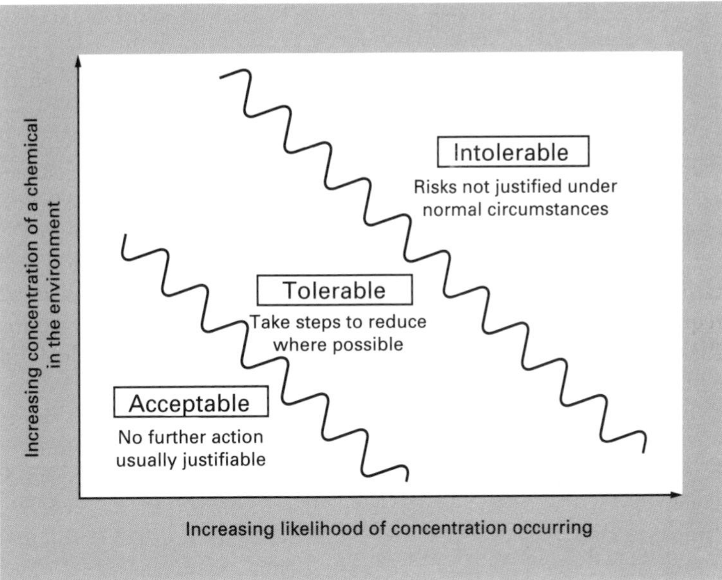

Fig. 18.2 Degree of impact (chemical concentration) and likelihood, and zones of acceptability, of risks.

below the maximum threshold of tolerability or acceptability, as appropriate. The classifications themselves (acceptable, tolerable, intolerable) may be defined on the basis of regulatory (guideline and mandatory) levels for a chemical, or levels determined by the risk assessment exercise to be associated with different degrees of effect on the defined water use.

Of course this representation is highly generalized and in particular we never really consider there to be discrete thresholds between what is acceptable, tolerable and intolerable, nor that the incremental scale is linear. Therefore, within the middle band in particular, judgements are needed to achieve the optimum balance between harm on the one hand and costs (and other impacts) of reducing the risk on the other. This can be achieved by use of tools identified earlier.

1 The environmental quotient approach, i.e. the ratio of the predicted environmental concentration of the chemical with an estimated quality standard (or a number of levels offering different levels of protection) for the specified water use, where the margin between the two is an indication of the margin of 'safety' between predicted levels and acceptable/tolerable levels.

2 Estimation of an integrated environmental index, i.e. the sum of the environmental quotients over all substances released by the process, as a ready indication of the overall environmental risks posed by a process.

3 Consideration of identified water uses (as discussed above) to determine which, and how many, of these are at risk.

Where the outcome of an assessment is uncertain—particularly where the apparent risk falls near to one of the conceptual risk boundaries—the risk manager should balance the reliability and certainty of the assessment against the implications of risk management activities needed to render the risk more acceptable. This may require moving on to a more sophisticated risk assessment.

It is also worth noting that this conceptual model of incremental risk can also be used to assist in catchment management activity after the planning stage. Thus the boundaries between risk categories can be used as triggers, such that increasing concentrations detected during monitoring programmes, and associated with increasing risk, stimulate investigation or intervention activities as appropriate. In this way, monitoring data indicating that the boundary between acceptable and tolerable risk has been reached, for example, may be used to trigger investigation and control to ensure that levels do not continue to rise towards the intolerable.

Section 18.5.1 considers in more detail a tiered assessment scheme with three levels of detail, based on approaches adopted by regulatory authorities in the UK, whilst section 18.5.2 considers specific examples of a proactive and a reactive risk management strategy.

18.5.1 Strategies for tiered risk assessment

Risk assessments for inland waters are designed to predict the impacts of planned or accidental activities on water resources, and the assessor must recognize the limitations of our ability to make such predictions. One way in which such limitations can be managed is to establish different 'tiers' of risk assessment with differing requirements for data and information which, therefore, make predictions with differing degrees of confidence. Thus, most strategies proposed for the assessment of risks in inland waters consider a framework with a number of levels of detail and sophistication to be appropriate. To demonstrate this, consideration is made here of a scheme with three tiers, referred to as 'Initial', 'Primary' and 'Secondary' risk assessment. At each level there is a gradation from qualitative to quantitative and from subjective to objective, which must be taken into account when interpreting the framework. Thus, at each level decisions about any risk management measures or proposed development must be considered in the context of the level of knowledge and confidence inherent in the assessment. As an example, Table 18.2 summarizes the types of information that might be sought when considering such tiered risk assessment for industrial sites that might threaten a vulnerable water course, such as a river used for potable supply.

Essentially, when applying the framework, a specific activity being assessed may progress to the next level of sophistication of risk assessment, may be considered to have a negligible risk, or may require some risk management or reduction actions to be taken (see Figs 18.3–18.5). In moving from initial to primary to secondary assessment, the effort required increases substantially, but fewer potentially polluting activities need the higher levels of assessment. Therefore, resources can be progressively targeted, with greater accuracy, as the process focuses on the high-risk

Table 18.2 Typical information requirements for tiered risk assessment of industrial sites.

Initial	Scale plan of site Inventory of chemicals on site Maximum amounts per vessel of each chemical in process operations, on-site storage and on-site transport
Primary	Scale plan of site detailing locations and quantities (maximum and normal) of each toxic chemical Location map of site and local area, including watercourses Inventory of chemicals on site and maximum amounts per vessel of each chemical in process operations, on-site storage and on-site transport Description of process and storage activities and holding conditions (temperatures and pressures), and the main controlling parameters of the process(es) Generic failure frequencies for significant components (tanks, pipework, fittings, tanker fill operations, etc.)
Secondary	Site-specific analysis of plant with detailed hazard identification and frequency evaluation procedures, along with an evaluation of pollution prevention, mitigation and control measures. This will require detailed information on: • all site geography, specifically including drainage, storage and protective measures • all site processes and support activities, including chemical transport to/from site • all on-site chemical storage and handling facilities and procedures • programmes of inspection for all site activities, and maintenance of plant and pollution control equipment • emergency plans in place, including lines of authority and responsibility, plus containment and clean-up procedures

sites and activities. A tiered system also affords the risk assessor greater flexibility, allowing for assessment of an activity to be carried out at a level of detail and confidence appropriate to the quality of the receiving water or the likely impact of the activity. For example, when establishing a catchment management plan, a large number of activities, processes or chemicals can be 'initially' assessed in order to identify priorities for further study. Conversely, a 'secondary' assessment might be performed if a high degree of confidence in the assessment's predictions is required, such as:

1 The local receiving water includes a commercial fishery or shellfishery, or fish spawning site.

2 The receiving water includes, or is part of, a site of particular biological interest, for example a site of special scientific interest or similar designation, or internationally important wildfowl feeding site.

3 The receiving water is abstracted for potable supply or has major recreational use.

4 An existing activity, chemical or process is suspected or known to have adverse environmental impacts and appropriate management measures need to be identified.

5 A new activity or process is likely to result in significant chemical release, for example chemical control of aquatic weeds.

6 The outcome of a primary assessment indicates unacceptable (or uncertain) predicted impact, and a secondary assessment is needed to confirm predictions and to identify the preferred management strategy.

Initial risk assessment

Initial (or 'scoping') assessments of risk (Fig. 18.3) will be performed most frequently and can perhaps best be thought of as a screening and prioritizing exercise. They should, therefore, not make intensive demands on staff, time and other resources, although it is difficult—and indeed counter-productive—to specify the amount of detail that is necessary for an initial assessment.

Initial risk assessments should be based on a combination of expert judgement and existing evidence of effects, considering the nature of the activity and the local environment and encompassing considerations of: the types and amounts of chemicals involved; the history of incidents for the activity and site; the likelihood of planned, accidental and malicious events; and the proximity and vulnerability of water resources. It is im-

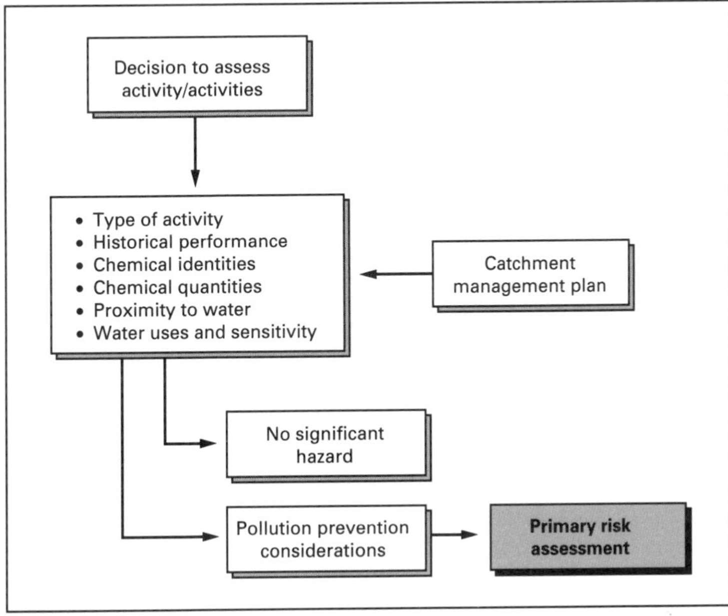

Fig. 18.3 Outline procedure for 'Initial' risk assessment.

portant that the assessment considers all aspects of a planned activity. For example, chemical storage facilities at a manufacturing plant may be distant from any watercourses and apparently pose minimal or no risk, but the access of delivery vehicles may be problematic. Thus the concept of 'lifecycle analysis' should always be applied at a local level to each activity and each chemical under consideration.

The answers to these questions should facilitate a decision as to whether the activity warrants further investigation and, if so, how high a priority this is. Alternatively there may be no significant perceived risk and no further consideration is required. It is unlikely that the outcome of an initial assessment would be to prevent the activity or require further risk management measures to be applied without further investigation. Instead, initial assessments might be applied in three ways.

1 Proactively to a proposed change in activity or a new activity, in order to identify whether further assessment or management is necessary before proceeding.

2 Comparatively to two or more options for achieving the same industrial, etc. goal, in order to determine which poses the least risk to inland waters.

3 Systematically to existing activities across a catchment or region, in order to screen out those with inconsequential risk and prioritize those needing additional consideration.

Primary risk assessment

A primary (or 'medium level') risk assessment may be required for a variety of regulatory reasons: it may be the result of significant perceived or actual environmental impacts, or it may be considered when a screening exercise using initial risk assessment suggests that the activity is a priority for further consideration (Fig. 18.4).

The requirements of a primary assessment are broadly similar to those of an initial assessment, but more detailed. Additionally, a primary assessment will assess both the probability of adverse effects occurring and the likely magnitude of their impact on each combination of water resource and water use. This assessment will normally only be qualitative or, at best, semi-quantitative, and should consider the following.

1 The type of activities involved, to determine whether there is potential for pollution.

2 Toxicity and other hazard information (e.g. ability to affect pH or dissolved oxygen levels) for all chemicals with potentially significant release, based on the quantities of each individual substance, as well as major interactions between chemicals where these might increase their toxicological hazard.

3 Route(s) by which substances might reach inland waters, and their persistence and fate in those waters.

4 Specific water uses under threat, in view of toxicity and other characteristics of each substance.

Primary assessments might be applied proactively to new risks or to existing activities that have been identified as priorities for further consideration. Thus a primary risk assessment may represent a reaction to one or more of the following scenarios.

1 Inclusion of the activity in a catchment management plan or other pollution prevention programme at a priority level that warrants action.

2 Identification of a potential hazard when considering some form of consenting of the relevant activity or release, including identification of releases of priority pollutants.

3 Concerns related to the adverse impacts of related activities.

4 Evidence for adverse effects in receiving waters identified by ongoing monitoring programmes.

5 A pollution incident associated with the activity.

The outcome of a primary assessment may be: (i) an indication that risks are acceptable and can be adequately managed by the use of existing (or, for a new activity, standard) control measures; or (ii) an indication that more stringent risk management controls are required. In the latter case, any costs associated with the introduction of new risk management measures need to be considered in the context of associated costs, or losses of the activity itself. Thus a more rigorous secondary risk assessment may be appropriate before such measures are made a requirement.

Secondary risk assessment

A secondary (or 'extended') risk assessment

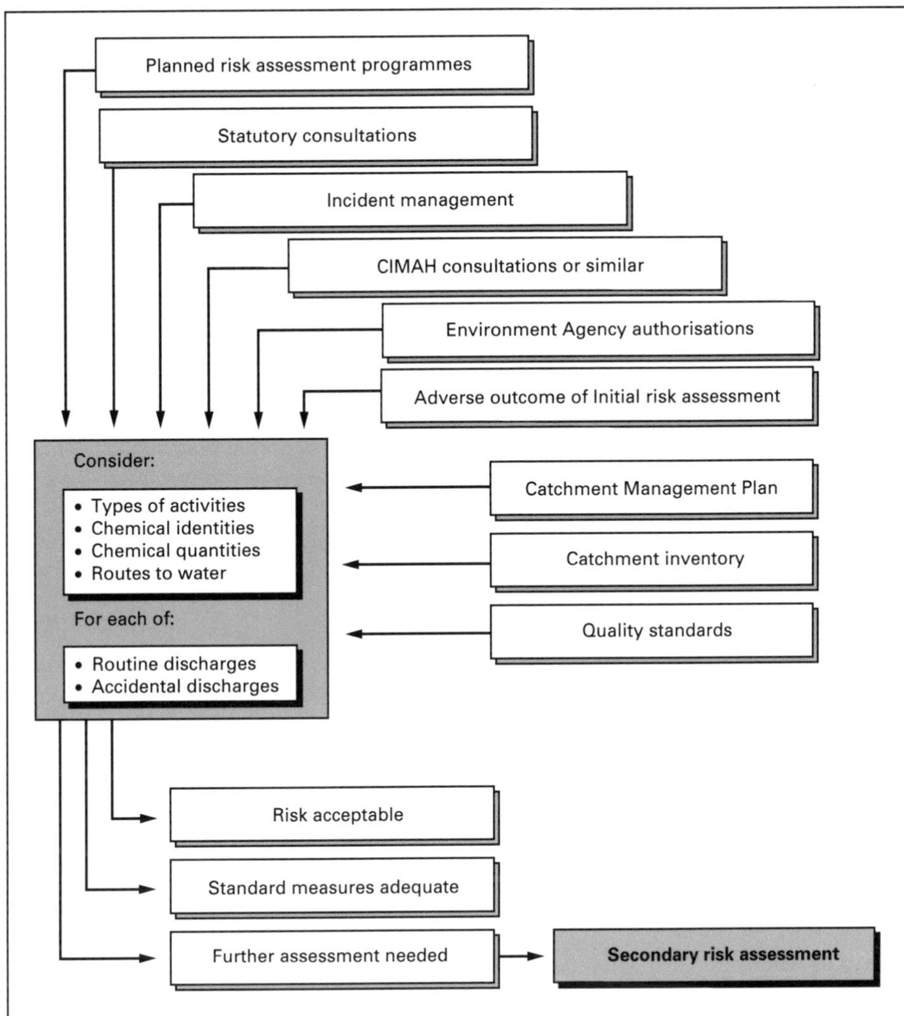

Fig. 18.4 Outline procedure for 'Primary' risk assessment.

(Fig. 18.5) is the corollary of an uncertain or unacceptable result from a primary risk assessment. Secondary risk assessments must make full consideration of all water uses that might be threatened by the activity, and in each case must encompass two major themes: event risk assessment and consequence risk assessment.

Event risk assessment is a prediction of how often incidents of differing magnitudes might occur. This can only be determined with a thorough knowledge of all elements of the activity and the specific implementation at the site under scrutiny and, therefore, can only be adequately obtained through discussion with those responsible for the activity. For example, details of any precautionary measures on site, such as bunding or interceptors in drains, are essential for proper event risk assessment. If additional precautions are being considered, these should be considered separately to determine how they might affect the outcome of the assessment. Thus there is some element of cost–benefit analysis built into the assessment even before risk management strategies are formally considered.

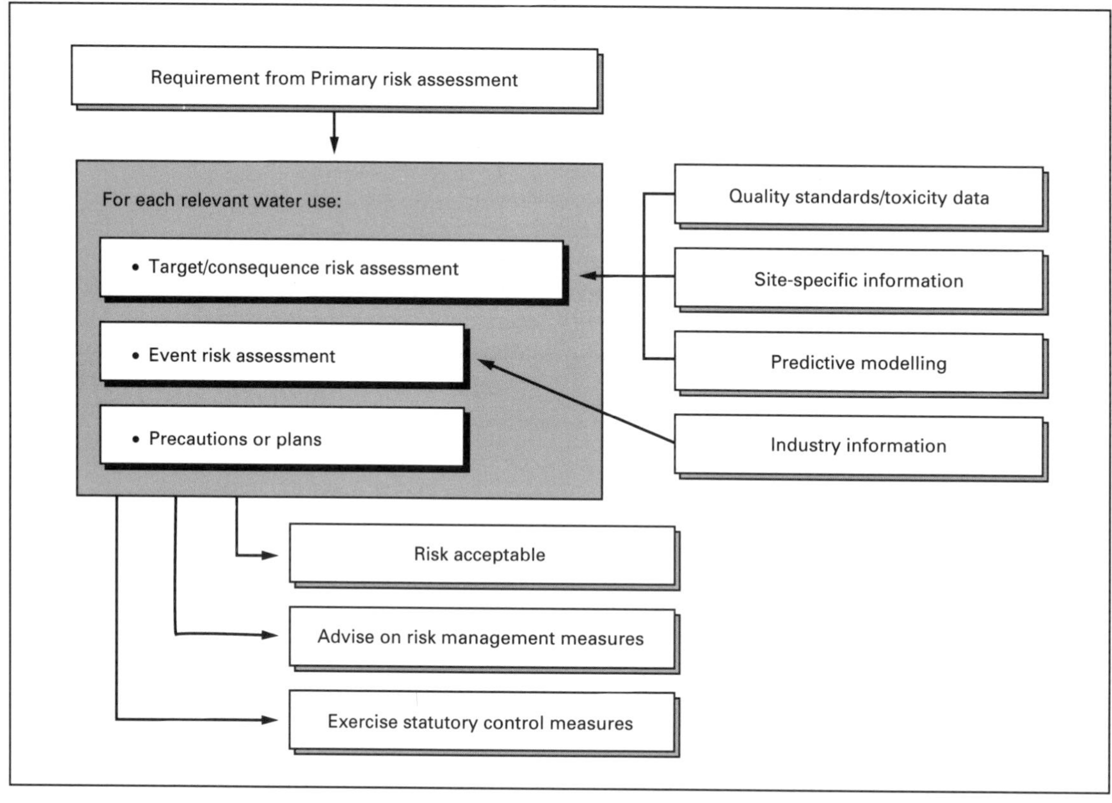

Fig. 18.5 Outline procedure for 'Secondary' risk assessment.

Consequence risk assessment is a prediction of the consequences, in terms of both magnitude and frequency, arising from the predicted events. In addition to the information required by a primary risk assessment, this requires detailed understanding of the local hydrology and hydrogeology of the receiving water, the environmental behaviour of the chemicals of concern, and their toxicological and other hazard characteristics. Whereas primary risk assessments concentrate on key chemicals (priority pollutants and those with potentially high volume release), secondary assessments should make consideration of all chemicals, and mixtures of chemicals, that might be released and their potential effects.

Secondary assessments should employ rigorous methodologies and should be based on the best data available. This will ensure that their conclusions do not falsely imply either the need for increased management or the safety of waterways without additional control measures. Consequently, these assessments can be time consuming and costly to perform and would normally only be initiated when there are clear indications that existing risk management activities are inadequate to ensure acceptable risk.

The outcome of the secondary risk assessment could be: (i) risk is fully acceptable under current management practices and no additional action necessary; or (ii) confirmation that additional pollution prevention measures are required. In the latter case, implementation requirements must be viewed against the costs of taking these measures. Thus, in such cases the secondary risk assessment could lead on to full cost–benefit analyses for the different risk management options available, to ensure an equitable balance between amelioration of environmental risk and continuing benefits accrued from the activity under scrutiny.

18.5.2 Management consequences

Institutional frameworks for risk management for inland waters

The roles of regulators and others involved in risk management strategies for inland waters vary from country to country according to the institutional mechanisms in place. Frequently the responsibilities are met by a variety of organizations, and as an example Table 18.3 summarizes those involved in the principal water management tasks in Germany, The Netherlands and the UK.

Given such complexities, strategies for risk management need to encompass features that delineate management responsibilities and ensure effective communication. The rest of this section considers two specific examples as an indication of the types of management strategies that might be involved in the protection of inland waters. The first considers the elements involved in risk management to protect potable water resources, and is thus an example of proactive risk management where the primary objective is to identify possible sources of unacceptable contamination and put in place measures to prevent these having

Table 18.3 Responsibilities for water management in a number of European states. (After Rees *et al.* 1995.)

Role in inland water management	Germany	The Netherlands	United Kingdom (E&W, England and Wales; S, Scotland; NI, Northern Ireland)
Assessment and monitoring	State Water Authority/ Regional and Local State Authorities	Regional Offices of Ministry of Transport, Public Works and Water Management (surface waters)/provinces (groundwaters)	Environment Agency (E&W)/ Scottish Environmental Protection Agency (S)/ Department of the Environment (NI)
Protection and control	State Water Authority/ Regional and Local State Authorities	State Water Office (surface waters)/provinces (groundwaters)	Environment Agency (E&W)/ Scottish Environmental Protection Agency (S)/ Department of the Environment (NI)
Water supply	Municipalities/private companies	Municipalities/private companies	Private companies (E&W)/ Regional and Island Councils (S)/Department of the Environment (NI)
Sewerage	Municipalities	Municipalities/Water Boards	Private companies (E&W)/ Regional and Island Councils (S)/Department of the Environment (NI)
Licensing of discharges	Regional and Local State Authorities	Regional offices of Ministry of Transport, Public Works and Water Management/ municipalities/provinces/ Water Boards	Environment Agency (E&W)/ Scottish Environmental Protection Agency (S)/ Department of the Environment (NI)
Licensing of abstractions	Regional and Local State Authorities	Regional Offices of Ministry of Transport, Public Works and Water Management/Water Boards (surface waters)/ provinces (groundwaters)	Environment Agency (E&W)/ Scottish Environmental Protection Agency (S)/ Department of the Environment (NI)

an impact. The second considers a scheme for the assessment and reduction of toxicity in a complex effluent in order to improve the ecological quality of receiving waters, as an example of reactive risk management for an existing pollution source.

Protection of potable water—an example of proactive risk management

Actions constituting part of a pollution risk management strategy to protect potable supplies drawn from surface and groundwater sources are summarized in Fig. 18.6. Essentially the strategy comprises four main elements.

1 Monitoring programmes and early warning networks, as a means of warning of increasing background contamination levels, which might indicate the need for improved chemical risk management practices in the catchment, and identifying a 'pollution event' that might require changes in abstraction or in the treatment applied.

2 Considerations of pollutant transport, to identify key sites where chemical spills or other releases could threaten raw waters, and to identify directions and time of travel in the event of a pollution incident.

3 Establishment of protection zones, providing legislative control over the use of chemicals within a designated area around a surface water or, more commonly, groundwater source.

4 Catchment auditing.

Each of these elements of risk management for raw waters constitutes a major topic in itself, and only catchment auditing is considered further here as an example of how risk management can be applied in practice.

A catchment audit is intended to reveal the potential for contamination of a source of raw water and, thus, to assist in specifying risk management measures needed to prevent unacceptable or intolerable consequences. All chemicals undesirable on grounds of toxicity or tainting potential, as well as those that might interfere with water treatment itself, should be considered in a catchment audit. In practice most attention will be paid to those chemicals that have been identified as priority concerns by inclusion in World Health Organization guidelines, national legislation on potable water, etc., as well as any additional high-volume chemicals in the catchment. As indicated in Table 18.3, it is generally national authorities who have the responsibility for maintaining the quality of inland waters, and in most instances it is they who will be responsible for the development of catchment plans and the implementation of catchment audits, although water suppliers may also be involved to a variable extent. Catchment audits can themselves be broken down into a number of elements.

1 Inspection of records (monitoring data, pollution incident records, survey results, hydrological information, etc.).

2 Definition of catchment boundaries (incorporating detailed hydrological and hydrogeological studies and patterns of flow).

3 Identification of potential diffuse and point sources (based on land-use data, records of diffuse chemical usage by farmers and local authorities, register of discharge consents, identification of non-consented discharges, identification of non-discharging sites using/storing chemicals, identification of transporting hazards such as accident blackspots near watercourses, etc.).

These processes, particularly step 3, will enable the auditor to identify those activities and sites that appear to pose a significant threat to water resources. In such cases the next step in risk management would be to conduct a site audit, which again can be broken down into a number of fundamental steps.

1 Audit planning.

2 Developing an understanding of internal management systems and procedures.

3 Assessing strengths and weaknesses.

4 Gathering evidence through site audit.

5 Evaluating audit findings.

6 Reporting and follow-up.

The actual planning and conduct of site audits, even if restricted to assessing only chemical risks, requires a thorough investigation of the site and its environs. There are, therefore, no standard requirements to support site-based risk auditing for risk management purposes, because these will vary according to local factors. However, the principal areas for consideration can be identified and associated with chemical handling and stor-

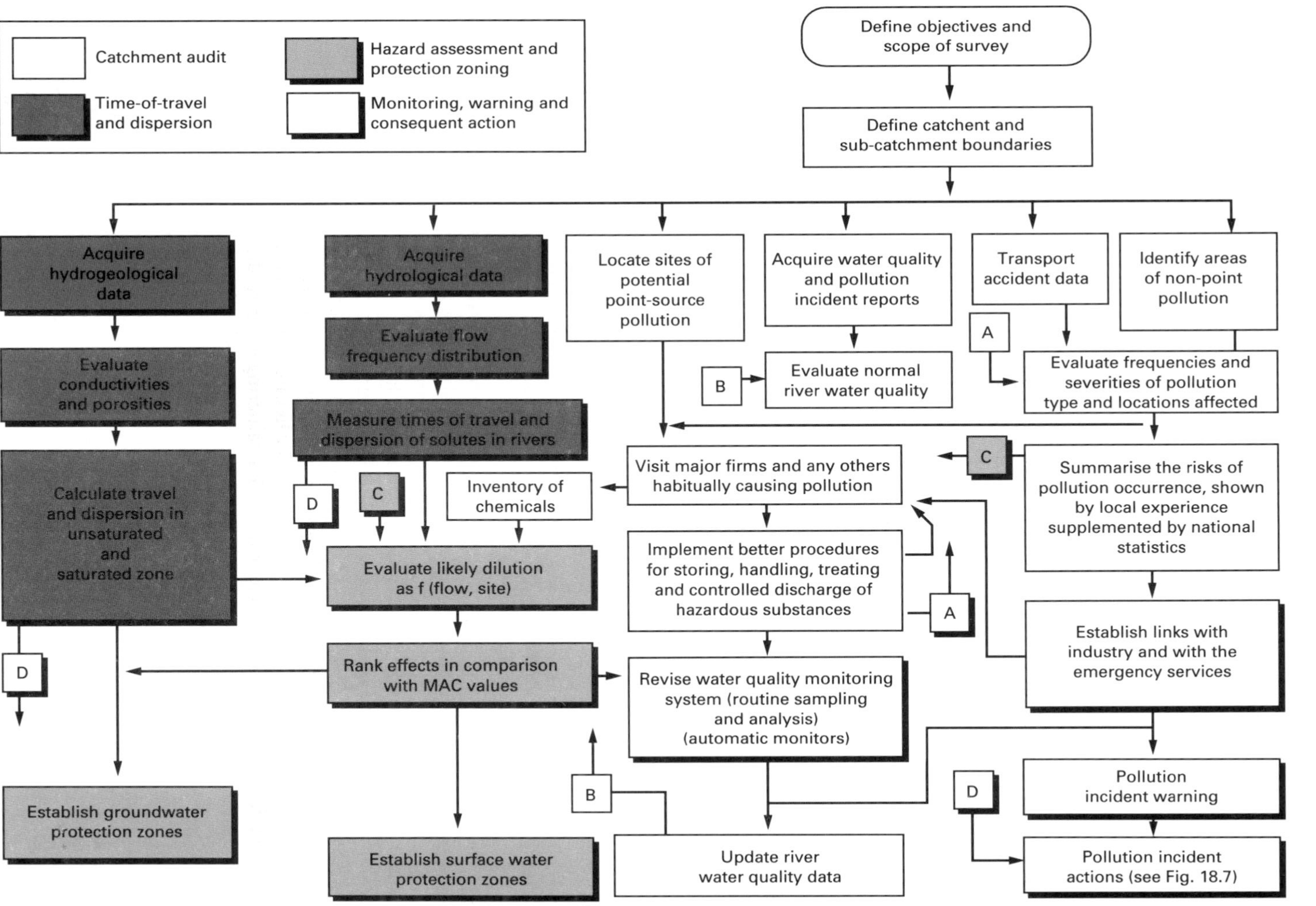

Fig. 18.6 Pollution risk management strategy to protect potable water supply.

Table 18.4 Outline checklist for site auditing associated with chemical risk management. (After Greeno *et al.* 1989; ICC 1991; Winter 1987.)

Chemical inventories
All chemicals on site identified and documented
Alternatives to hazardous chemicals investigated/used where possible
Chemical hazards (and associated precautions for handling, with procedures for dealing with spillage) identified and documented, and staff familiar with these

Chemical storage
All storage tanks appropriately designed and constructed, and labelled with nature and volume of contents
All storage tanks protected against corrosion (internal and external)
Bunding in place for above-ground storage
Bund walls and floor impermeable with no drains or valves, and walls secure and regularly checked
Rainwater regularly pumped from bunded areas
All fill pipes, valves, overflow pipes, etc. discharge to bunded area
Valves marked with open and shut positions, and locked shut when not in use
All drum stored in bunded areas (ideally roofed)
Appropriate containers used for all storage, and kept in good condition
Oil separators fitted to surface drains at risk from oils leakage
Contingency plans in place for incident management, and staff and contractors trained for action in event of spillage or other accident
Adequate security to prevent vandalism

Chemical deliveries
All deliveries authorized and supervized by receiving company
All loading/unloading areas and other chemical delivery points clearly identified, including chemical/product name, and well separated from surface water drain
Systems in place to prevent over-filling of storage facilities, including automatic cut-off valves on delivery pipes
Catch pits or sumps with isolating valves at delivery points
Pipelines sited above ground or, if underground, in protective sleeve or duct, with regular inspection and maintenance
Stocks of absorbent materials readily available for spill clean-up

Site drainage
Foul water drain and surface (clean water) drain clearly identified, all man-hole covers and drains colour-coded, and staff aware of distinction
Comprehensive drainage plans available, and key staff familiar with layout of drainage system
All drain connections correctly made (e.g., all sinks to foul water drain)
Washing operations carried out in area discharging to foul drain, and no detergents to be disposed of to surface drain
Oil interceptors fitted on surface drains in areas used for fuelling and vehicle parking
Adequate containment measures for fire-fighting water runoff

Waste storage/disposal
All wastes stored in designated, bunded areas completely isolated from surface water drains
Efforts made to minimize waste volume
All waste containers and equipment covered or roofed, and any rainwater entering pumped to foul sewer
Waste handled according to requirements of relevant legislation

Emergency plans
Emergency response plans in place, staff trained in these, and procedures periodically tested
Documentation maintained on all the above

age facilities and associated incident handling procedures (Table 18.4). These can be considered within the framework of OPRA (p. 485) such that pollution hazard and operator performance constitute the two key elements of an audit. Thus it is possible to demonstrate by working through this process how proactive risk management activities can be broken down into discrete but integrated activities and actions at an operational level, for which the benefits—in terms of reduced risk—can be readily determined and monitored.

Toxicity reduction evaluations—an example of reactive management of existing risk

In some countries, most notably the USA, a programme of toxicity-based (rather than chemical-based) discharge consents has been in operation for some time. This approach to the management of chemical risk is also now being adopted more widely in Europe. As mentioned in section 18.4.2, such consents are generally considered most suitable for discharges that are complex in terms of their chemical composition (and may also vary temporally) and for which not all important constituents can be identified, rendering ineffective control measures based on specific constituent chemicals. Generally a toxicity-based consent is most likely to be applied if such a discharge has also been demonstrated to have adverse effects on the receiving waters. Thus, this strategy is currently applied for the management of existing chemical risk with demonstrable environmental effects.

Of course, the detail of a toxicity-based consent will depend on conditions specific to the particular combination of discharge and receiving water. In general terms such consents will be designed to achieve, at worst, no further significant degradation in the biological quality of the receiving water, and at best, a significant improvement. Thus the consent will be set such that the predicted dilution of the effluent in the receiving water results in a 'concentration' of effluent (a PEC) below that which is predicted to have no adverse effects (a PNEC). Consequently, in cases where a consent is not met, this implies that the consent objectives will also not be met unless remedial action is taken—i.e. toxicity reduction measures (Fig. 18.7).

Some toxicity problems will be more difficult to resolve than others, and this needs to be recognized and reflected in the time allowed by the regulators for implementing toxicity reduction measures, as well as the extent to which toxicity reduction is required (e.g. elimination or minimization). Such decisions and the consequent actions necessary can best be facilitated through drawing up and implementing a toxicity reduction plan, which might include the following considerations.

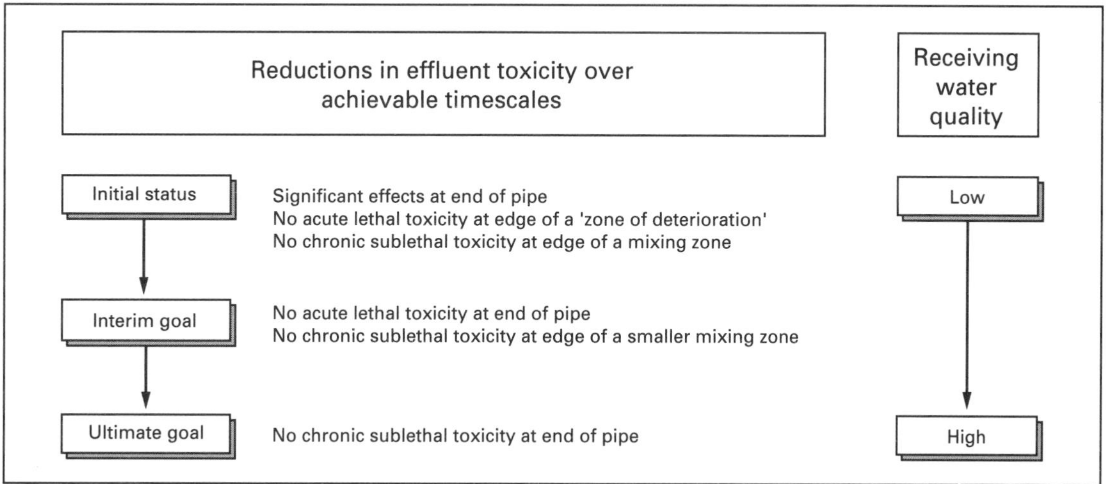

Fig. 18.7 Targets for reductions of effluent toxicity.

Information acquisition. The first step in a toxicity reduction programme would be to gather all available chemical, toxicological, plant process (design, operation and efficiency) and effluent treatment information. The combination of this information will aid in identifying processes or practices that are the source of the toxicity. This would then normally be followed by practical investigations to confirm or otherwise identify the main contributory source(s) of the toxicity ('toxicity tracking').

Toxicity tracking. Procedures for toxicity tracking may include:

1 Facility performance evaluation, which will determine whether a procedural, mechanical or other failure in the running of a plant is resulting in toxicant release that would not occur under 'normal' operating conditions.

2 Toxicant identification, which involves effluent fractionation using chemical and physical manipulation, combined with toxicity testing and chemical analysis of the fractions, to identify those chemicals or chemical groups responsible for observed toxicity. This identification need not be to the level of individual chemicals because classes of similar chemicals, or chemicals arising from the same process, might all be reduced using the same treatment or other remedial action.

3 Source identification, which is important where there are a number of effluent streams arising from different processes operating at the same time—or possibly at different times if batch processes are combined in an effluent stream—or even processes on different sites sharing a single discharge. In such cases toxicity testing can be used to trace toxicity though the waste production processes to identify its origins. Toxicity reduction measures might then be concentrated on the process or processes contributing most significantly to the final effluent toxicity.

Remedial action (toxicity reduction). Required actions arising from the desk-based and practical investigations could include:

1 Improved efficiency and operating practices to reduce quantities of toxic chemicals released.

2 Improved waste handling procedures, including strategies for chemical recovery and reuse.

3 Substitution of toxic chemicals with less-toxic or less-persistent chemicals that serve the same function.

4 Improved maintenance of plant, processes, storage facilities, waste separation facilities, etc.

5 Replacement of old process technology.

6 Introduction or upgrading of waste treatment facilities.

The final measures employed will be decided on a case-by-base basis, given the requirements of the regulators and recognizing the commercial and financial implications of alternative actions. The concept of 'best available technology not entailing excessive cost' (BATNEEC) is thus of particular relevance to such considerations. It should also be noted that there is evidence from those countries with longer standing toxicity-based consent protocols that remedial measures identified for purposes of toxicity reduction may also lead to cost–benefits associated with increased efficiency and improved operational management practices.

Post-reduction monitoring. Following a toxicity reduction programme the effluent should again be characterized to ensure that new measures have successfully reduced the potential for impact to tolerable or acceptable levels, based on the PEC/PNEC quotient assessment.

As in the example of proactive risk management for potable water protection considered above, this example of reactive management for an existing risk again demonstrates how the overall strategy is comprised of a number of tiered and progressive steps that facilitate the most cost-effective use of resources.

18.6 PRIORITIES FOR FUTURE DEVELOPMENT

This chapter can only make generalizations about risk management practices for inland waters, because of the diversity of strategies in operation. As has been shown, the current 'state of the art' in developed countries enables a number of approaches to be adopted, depending on the level of sophistication and the degree of confidence required. However, in general these approaches are arguably in the early stages of development and are definitely in the early stages of implementation. There is still much about which we have insufficient information or understanding to ensure adequate protection of inland waters and their various uses, and it is possible to identify a number of priorities for future developments.

18.6.1 Improve the risk manager's 'toolbox'

1 Improve and standardize as far as possible event risk assessment protocols relevant to inland water protection, perhaps by technology transfer of tools from sectors where numerical risk assessment tools have a history of usage, such as the nuclear industry. Further use of checklist and other semi-quantitative tools should help extend risk-based techniques to new users.
2 Further develop methodologies for the derivation of environmental quality standards and PNECs for water, to ensure that these are based on sound ecotoxicological principles and thus to enable meaningful quotient assessments. In particular there is a need to develop a better understanding of the requirements of extrapolating from test organisms to ecosystems and from simple, controlled systems to complex environments (collectively considered in laboratory-to-field extrapolations), with validations to verify any predictive models developed. If true risk-based pollution control is to be implemented, there is a need to develop approaches to the derivation of risk-based standard such that the probabilities of effects can be established in a quantitative fashion.
3 Further develop risk assessment tools for sediments and wetlands, based on procedures developed for the aquatic phase of inland waters.
4 Develop agreed environmental values that can be used to associate costs with environmental impacts, in order to allow risk-related cost–benefit analysis through the integration of fiscal, societal and environmental costs.
5 Further develop risk assessment and management strategies for chemicals in combination in the environment released as mixtures and resulting from multiple inputs, perhaps based on direct toxicity assessment combined with predictive modelling for chemical interactions.

18.6.2 Respond to the needs of risk assessment and management for the whole environment

1 Agree a standard approach to management planning that can also be applied to other environmental compartments besides inland waters.
2 Consider mechanisms for integration of direct toxicity assessment (or other tools for mixture assessment) to ensure equability across different environmental compartments, in support of BPEO assessments.
3 Validation and verification of predictive risk assessment tools through field-based research and environmental monitoring programmes.
4 Harmonize protocols for decision-making processes for environmental remediation, and develop remediation tools.

18.6.3 Ensure consistency in approach and use of appropriate methodologies

1 Develop an environmental risk culture, clarifying the importance of risk assessment and management, ensuring the financing of risk management, and integrating the role of risk managers into the overall corporate structure. Integrate environmental risk management with operational risk management and legal liability.
2 Enhance and standardize (as far as possible without sacrificing flexibility) methodologies for risk assessments, to facilitate information exchange between those conducting them.
3 Promote and formalize the concept of shared, rather than transferred, risks and associated costs.
4 Transfer of appropriate information, tools and strategies for risk assessment and management into developing and resource-limited countries.

18.6.4 Commitment to risk management planning for the future

1 Integration of environmental legislation with environmental effects of concern, to ensure that priorities are focused and consistent.
2 Address 'long-tail' environmental risks, such as release of persistent substances and substances contributing to long-term climate change.

18.7 CONCLUSIONS

The use of risk assessment and management is bound to grow as environmental problems increasingly involve fundamental social and economic issues and engage with complex and often long-term factors. The wider application of risk assessment holds no dangers for pollution control professionals and should be seen as an aid to decision-making rather than as a threat. However, a great deal needs to be done to make the terms, concepts and tools more widely available, better understood and easy to use. There are also many fundamental areas where further research and development activity is needed to enable the newly created disciplines of environmental risk assessment and management to develop.

18.8 REFERENCES

BSI (1992) *BS7750 1992—Specification for Environmental Management Systems.* British Standards Institute, Milton Keynes.

CEC (1976) Directive on pollution caused by certain dangerous substances discharged in to the aquatic environment of the Community (76/464/EEC). *Council of the European Communities Official Journal,* **L129/23**, 18 May 1976.

CEC (1982) Directive on limit values and quality objectives for mercury discharges by the chlor-alkali electrolysis industry (82/176/EEC). *Council of the European Communities Official Journal* **L81**, 27 March 1982.

CEC (1994) Council Regulation (EC) of 28 June 1994 laying down the principles for the assessment of risks to man and the environment of existing substances in accordance with Council Regulation (EEC) No. 793/93 (1488/94). *Council of the European Communities Official Journal,* **L161/3**, 29 June 1994.

COM (1993a) Commission Directive of 20 July 1993 laying down the principles for assessment of risks to man and the environment of substances notified in accordance with Council Directive 67/548/EEC (93/67/EEC). *Commission of the European Communities Official Journal,* **L227/9**, 8 September 1993.

COM (1993b) Proposal for a Council Directive on integrated pollution prevention and control. Commission of the EC (93/C 311/06) (COM(93) 433 final). *Commission of the European Communities Official Journal,* **C311/6**, 17 November 1993.

COM (1993c) Amended proposal for a Council Decision concerning the conclusion, on behalf of the Community, of the Convention on environmental impact assessment in a transboundary context (93/C 112/22). (COM(93) 131 final). *Commission of the European Communities Official Journal,* **C112**, 22 April 1993.

Crathorne, B., Dobbs, A.J. & Rees, Y. (1996) Chemical pollution of the aquatic environment by priority pollutants and its control. In: *Pollution: Causes, Effects and Control,* 3 edn (ed. R.M. Harrison), pp. 1–25. Royal Society of Chemistry, London.

Dobbs, A.J. & Zabel, T.F. (1996) Water quality control. In: *River Restoration* (eds G. Petts & P. Calow), pp. 44–59. Blackwell Science, Oxford.

DOE (1989) *Water and the Environment.* Department of the Environment Circular 7/89, HMSO, London.

DOE (1991) *Interpretation of Major Accidents to the Environment for the Purposes of the CIMAH Regulations: a Guidance Note.* Department of the Environment, HMSO, London.

DOE (1995a) *A Guide to Risk Assessment and Risk Management for Environmental Protection.* Department of the Environment, HMSO, London.

DOE (1995b) *Risk-benefit Analysis of Existing Substances.* Department of the Environment, London.

Douben, P.E.T. (1995) Derivation of environmental assessment levels for evaluating environmental impact. *Ecosystem Health,* **1**(4), 242–254.

Environment Agency (1996) *The Application of Toxicity-based Criteria for the Regulatory Control of Waste Discharges.* A consultation paper prepared by the DTA National Centre and issued by the Environment Agency on the proposed introduction of Toxicity-based Criteria in discharge consenting. Environment Agency, London.

Greeno, J.L., Hedstrom, G.S. & DiBerto, M. (1989) *Environmental Auditing. Fundamentals and Techniques,* 2 edn. Arthur D. Little Inc., Cambridge, MA. ISBN 0-915094-10-X.

Hedgecott, S. & Fawell, J.K. (1996) Derivation of acceptable concentrations for the protection of aquatic life. *Environmental Toxicology and Pharmacology,* **2**, 115–120.

HMIP (1995) *Operator and Pollution Risk Appraisal (OPRA).* April 1995. Her Majesty's Inspectorate of Pollution, London.

HMIP (1996) *Best Practicable Environmental Option Assessments for IPC. A Summary.* Guidance for Operators of IPC Processes, March 1996, Her Majesty's Inspectorate of Pollution, London.

HMSO (1988) *Best Practicable Environmental Option.* Royal Commission on Environmental Pollution Twelfth Report. Her Majesty's Stationery Office, London.

HMSO (1993) *Integrated Pollution Control: a Practical Guide.* Her Majesty's Stationery Office, HMSO, London. ISBN 0-11-752750-5.

HMSO (1994) *Life Cycle Management and Trends.* Her

Majesty's Stationery Office, London. ISBN 96-64-14148-0.

HSE (1994) *Quantified Risk Assessment: Its Input to Decision Making*. Health and Safety Executive, London.

HSE (1995) *Generic Terms and Concepts in the Assessment and Regulation of Industrial Risks, Discussion Document*. Health and Safety Executive, London.

Hunt, D.T.E. & Johnson, C. (1995) *Environmental Management Systems—Principles and Practice*. McGraw-Hill, London.

ICC (1991) *ICC Guide to Effective Environmental Auditing*. International Chamber of Commerce Publishing, Paris. ISBN 92-842-1115-8.

Jordaan, J., Plate, E.J., Prins, E. & Veltrop, J. (1993) *Water in Our Common Future*. UNESCO, Paris.

Kooijman, S.A.L. (1987) A safety factor for LC_{50} values allowing for differences in sensitivity among species. *Water Research*, **21**, 269–276.

MAFF (1993) *Strategy for flood and coastal defence in England and Wales*. Ministry of Agriculture, Fisheries and Food, London.

Nabholz, J.V. (1991) Environmental hazard and risk assessment under the United States Toxic Substances Control Act. *Science of the Total Environment*, **109/110**, 649–665.

Pascoe, G.A. & DalSoglio, J.A. (1994) Planning and implementation of a comprehensive ecological risk assessment at the Milltown Reservoir–Clark Fork River superfund site, Montana. *Environmental Toxicology and Chemistry*, **13**, 1943–1956.

Pritchard, P. (1995) *Managing Environmental Risks and Liabilities*. Stanley Thornes, Cheltenham.

Rodier, D.J. & Mauriello, D.A. (1993) The quotient method of ecological risk assessment and modeling under TSCA: a review. In: *Environmental Toxicology and Risk Assessment* (eds W.G. Landis, J.S. Hughes & M.A. Lewis). American Society for Testing and Materials, Philadelphia.

Rees, Y.J., Zabel, T.F. & Buckland, J. (1995) *EUROWATER: Institutional mechanisms for water management in Europe*. UK National Rivers Authority R&D Report 21, Bristol.

SETAC (1993) *Guideline for LCA: a Code of Practice*. Society of Environmental Toxicology and Chemistry. Pensacota, FL.

Solomon, K.R., Baker, D.B., Richards, R.P. *et al.* (1996) Ecological risk assessment of atrazine in North American surface waters. *Environmental Toxicology and Chemistry*, **15**(1), 31–76.

Stephan, C.E., Mount, D.I., Hansen, D.J., Gentile, J.H., Chapman, G.A. & Brungs, W.A. (1985) *Guidelines for deriving numerical national water quality criteria for the protection of aquatic organisms and their uses*. USEPA Office of Research and Development, Report no. PB85-227049, Washington, DC.

USEPA (1993) *A Guidebook to Comparing Risks and Setting Environ-mental Priorities*. US Environmental Protection Agency, EPA 230-B-93-003, Washington, DC.

USEPA (1986) *Quality criteria for water 1986*. US Environmental Protection Agency, EPA 440/5-86-001, Washington, DC.

Wagner, C. & Lokke, H. (1991) Estimation of ecotoxicological protection levels from NOEC toxicity data. *Water Research*, **25**, 1237–1242.

Water Quality 2000 (1990) *Phase II Report. Problem Identification*. Water Quality 2000, Alexandria, VA.

Winter, G. (1987) *Business and the Environment. A Handbook of Industrial Ecology with 22 Checklists for Practical Use*. McGraw-Hill Book Company, New York.

Chapter 19
Environmental Risk Assessment in Development Programmes: the Experience of the World Bank

OLAV KJØRVEN

19.1 INTRODUCTION

Over the last decade or so, multilateral development banks (MDBs) have increasingly embraced the principle of environmentally sustainable development and have come to view enhanced stewardship of the environment and natural resources as an integral and necessary part of the development process. At the World Bank, this shift has manifested itself in a number of ways.

1 The institution has adopted a host of environmental policies, procedures and guidelines, covering areas such as environmental assessment, forestry, natural habitats, cultural heritage, water resources management, agricultural pest management, pollution prevention and abatement, occupational health and safety, indigenous peoples and involuntary resettlement.

2 As a result, the Bank's lending portfolio has been transformed. First, projects with mainly environmental objectives—such as pollution abatement, biodiversity conservation, energy efficiency and environmental institution building—have reached a total value of US$12 billion. These projects make up the Bank's *environmental* project portfolio—no doubt the largest investment any organization has ever made to protect the environment in developing countries. Second, the more 'traditional' project portfolio has become 'greener'. By subjecting all proposed projects to screening for potential environmental impact (see Box 19.1) and requiring borrowers to undertake environmental assessment when needed, potentially significant impacts are identified and assessed, and measures designed to avoid, minimize, mitigate or compensate for adverse impacts. A growing number of projects also include specific environmental components, for example to strengthen environmental management capabilities of the project implementing agency.

3 Analytical approaches for planning at the national and sectoral levels intended to move policies and investments towards an environmentally more sustainable direction have been introduced.

4 Environmental staff has grown from some 20 in 1987 to over 300 today.

Throughout this process of change, *environmental risk* and *uncertainty* have become central concerns for MDBs and other institutions, as they have sought to improve their understanding of the environment and the interrelationship between society, societal change and natural systems—whether at local, regional or global levels. The goal has been to use this understanding to learn how to minimize and manage risks to people and natural systems, whether caused by humanly induced (usually gradual) environmental changes, natural disasters, or industrial and other accidents.

This chapter discusses how the World Bank is addressing environmental risk and uncertainty at three main levels: (i) as part of its efforts to ensure environmental sustainability at the project level, in normal lending operations; (ii) through strategic environmental planning to reduce risks to health and the environment, whether within the context of a project, or wider sectoral, national or even transboundary strategic work; and (iii) as part of its work to assist developing countries in preventing and limiting the effects of natural disasters and its efforts to protect the global environment.

Box 19.1 The Environmental Classification System of the World Bank.

At the time of project identification, the Bank screens all proposed investment projects for environmental impacts and classifies them into one of three environmental categories

Category A. A proposed project is classified as Category A if in the Bank's judgement it is likely to have significant adverse environmental impacts that may be sensitive (e.g. affect vulnerable ethnic minorities, lead to loss of a critical natural habitat), diverse or precedent-setting, or that affect an area broader than the sites or facilities subject to physical works. A full environmental assessment is required.

Category B. A proposed project is classified as Category B if its potential environmental impacts are site-specific in nature and do not significantly affect human populations or alter environmentally important areas, such as mangroves, wetlands and other critical natural habitats. Few if any of the impacts are irreversible, and mitigatory measures can more readily be designed. A partial environmental analysis is required, curtailed to the particular environmental issues of the project.

Category C. A proposed project is classified as Category C if it is unlikely to have adverse environmental impacts or if its impacts are likely to be negligible, insignificant, or minimal. Environmental assessment is not required for such projects.

19.2 ENVIRONMENTAL RISK ASSESSMENT TO 'DO NO HARM'

19.2.1 Historical background

The concepts of *uncertainty* and *risk* were first introduced to the World Bank within the context of economic and financial project analysis and concerned the risk of not achieving sufficient economic rates of return. In 1970, the Bank published two 'Staff Occasional Papers' called *Techniques for Project Appraisal Under Uncertainty* (Reutlinger 1970) and *Risk Analysis in Project Appraisal* (Pouliquen 1970).

The entry point for environment-related risk came with the growing appreciation of dangers associated with certain types of major industrial or infrastructure operations, particularly after the Bhopal disaster in India in 1984 (although the Bank was not involved in any way in that incident). In 1985 the Bank issued its first guidelines involving use of risk assessment techniques for prevention and management of industrial hazards: *World Bank Guidelines on Identifying, Analyzing and Controlling Major Hazard Installations in Developing Countries.* To support their implementation, the Bank issued a manual in 1988: *Techniques for Assessing Industrial Hazards: A Manual* (Technica Ltd 1988). The guidelines state that all energy and industry projects should include a formal plan to prevent and manage industrial hazards. The manual provides procedures for calculating the consequences of fires and explosions and methods for their avoidance and reduction. It also provides detailed guidelines on *Major Hazard Assessment*, which is required whenever there is a potential major hazard within a proposed Bank-financed project (see Box 19.2).

Subsequently, the Bank issued several environmental policies and guidelines related, in different ways, to managing environmental risk. A manual containing guidance and standards for control and prevention of industrial pollution (*Environmental Guidelines*) was issued in 1988. A policy on Dams and Reservoirs came out in 1989 (Operational Directive 4.00, Annex B), providing specific guidance for addressing environmental issues in the planning, implementation and operation of dam and reservoir projects, including guidelines for promoting dam safety (Le Moigne *et al.* 1990). Guidelines were also developed on: the safe manufacture, use, transport, storage and disposal of hazardous and toxic materials; selection and use of agrochemicals; occupational health and safety; and precautions against natural hazards. The Environmental Guidelines of 1988, which provided environmental guidance and norms for various industrial subsectors, are now being updated and

Box 19.2 Major hazard assessment.

As described by the Bank's guidelines on major hazard assessment, a major hazard would exist under the following circumstances: a release of toxic substances, highly reactive or explosive substances or flammable substances. The guidelines list the types of industrial installations, substances and quantities that could present a major hazard.

The objectives of the Major Hazard Assessment are as follows:

- to identify the nature and the scale of use of dangerous substances at the installation;
- to specify arrangements made for safe operation of the installation for control of serious deviation that could lead to a major accident and for emergency procedures at the site;
- to identify the type, relative likelihood and broad consequences of major accidents; and
- to demonstrate that the developer has appreciated the major hazard potential of the company's activities and has considered whether the controls are adequate.

revised into a handbook, which will include specific guidance on environmental risk assessment (World Bank 1996).

In 1989, with the issuance of the Operational Directive (OD) on environmental assessment (OD 4.00, Annex A, revised and amended (World Bank 1991) as *OD 4.01: Environmental Assessment*), the Bank established an overall framework and process for consideration of environmental impacts and risks of specific development proposals. This significantly broadened the Bank's use of risk assessment approaches within the area of environment, from being almost exclusively oriented towards industrial hazards to encompassing wider health and ecological risks associated with development proposals of different kinds. A forthcoming revised version of the environmental assessment policy (OP/BP/GP 4.01, where OP is Operational Policy, BP is Bank Procedure and GP is Good Practice) makes explicit reference to risk assessment as an instrument to be used in environmental assessment when confronted with uncertainty about impacts.

19.2.2 Environmental assessment: the main framework for considering environmental risk at the project level

The purpose of environmental assessment (EA) in the context of Bank operations is to ensure that the projects considered for Bank financing are environmentally sound and sustainable.* Environmental assessment is a *flexible process* that allows environmental considerations to be an integral part of project preparation (i.e. the process of prefeasibility and feasibility studies) and implementation. The process incorporates consultation with project-affected communities and local non-governmental organizations to ensure social acceptability and local ownership of projects. The mandate of EA is to:

1 Assess the environmental impacts of a given project proposal.

2 Compare the environmental impacts (or costs and benefits) of the proposal with those of feasible alternatives (taking into account design, technology and location variables), and on this basis recommend a preferred alternative.

3 Propose measures to prevent, minimize, mitigate or compensate for any adverse environmental (including social) impacts.

* By terming the process 'environmental assessment' rather than 'environmental impact assessment', the Bank signals that it is concerned about more than just expected impacts of a given proposal. Environmental assessment should proactively feed into project design by assessing not only a given proposal but also reasonable alternative sites, designs or technologies. It should also assess risk under conditions of uncertainty.

4 Ensure that such measures are integrated into the project.*

Following the issuance of the OD, the World Bank (1991) published the *Environmental Assessment Sourcebook* (three volumes), which provided more in-depth cross-sectoral and sector-specific EA guidance. The chapter on energy and industry, for example, provided detailed guidance on assessment, minimization and management of industrial hazards. Other chapters were focused on broader ecological and health impacts and risks. Several updates have appeared subsequently (World Bank 1993–1996). Figure 19.1 provides an illustration of the EA process in relation to the project preparation and implementation process.

The Bank's EA policy, then, provides the main context for consideration of environmental risk associated with specific projects. When there is uncertainty about an impact in terms of its likelihood of occurrence and its consequences should it occur, standard risk assessment techniques should be used. When uncertainty is not addressed explicitly in EA, realistic risk scenarios are not presented to decision makers, thereby weakening the basis on which to make decisions.

As is the case for the overall process (EA), environmental risk assessment should be viewed as a process rather than as a product such as a report. The process recommended by the Bank is consistent with internationally accepted norms: hazard identification, exposure assessment, dose–response assessment and risk characterization (Chapter 13).

Other MDBs have taken approaches to environmental risk and EA that are similar to that of the Bank. The Asian Development Bank (ADB 1990), for example, has developed specific guidance on environmental risk development in the form of a comprehensive training and reference document on environmental risk assessment: *Environmental Risk Assessment: Dealing with Uncertainty in Environmental Impact Assessment*. As the title suggests, the guidelines discuss environmental risk assessment as an extension of environmental impact assessment (EIA): as a method that 'kicks in' when an EIA team encounter uncertainty about the impacts of a proposed project. The foreword states that 'ERA [environmental risk assessment] builds upon environmental impact assessment in that risks are impacts where the likelihood of occurrence and magnitude of consequences are uncertain'.

The precise relationship between the normal EA process and environmental risk assessment will now be described.

Environmental risk assessment addresses three questions ADB 1990):

1 *What can go wrong with the project?* What impacts might occur to human health and the natural environment? What are some reasonable scenarios for the project that result in damage to health, the environment or productive capital?

2 *What is the range of magnitude of these adverse impacts?* What number of people could be affected? How much monetary damage? What geographical area would be affected? What is the maximum credible accident that could occur in the project over its lifetime? What are the risks of routine operations?

3 *How likely are these adverse consequences?* With what frequency might they occur? What historical or empirical evidence is available to judge their likelihood? What data are available on failure rates of processes or components in the project technology?

Any normal EIA should answer the first question and give at least a qualitative indication of the magnitude of the impacts. Environmental risk assessment, therefore, complements and extends EIA into addressing mainly the third question: consideration of frequency of occurrence of adverse impacts and how this relates to their magnitude or severity. The World Bank expects this added dimension to be part of any EA process whenever there is uncertainty surrounding existing hazards and potential impacts.

In practice, environmental risk assessment is often a follow-up of an EIA study (see Box 19.3). For example, the sequence may be that a standard EIA study is carried out as part of project

* The primary responsibility for the EA process lies with the borrowers; the Bank's role is to advise the borrowers throughout the process, make sure that practice and quality are consistent with EA requirements, and ensure that the process feeds effectively into project preparation and implementation.

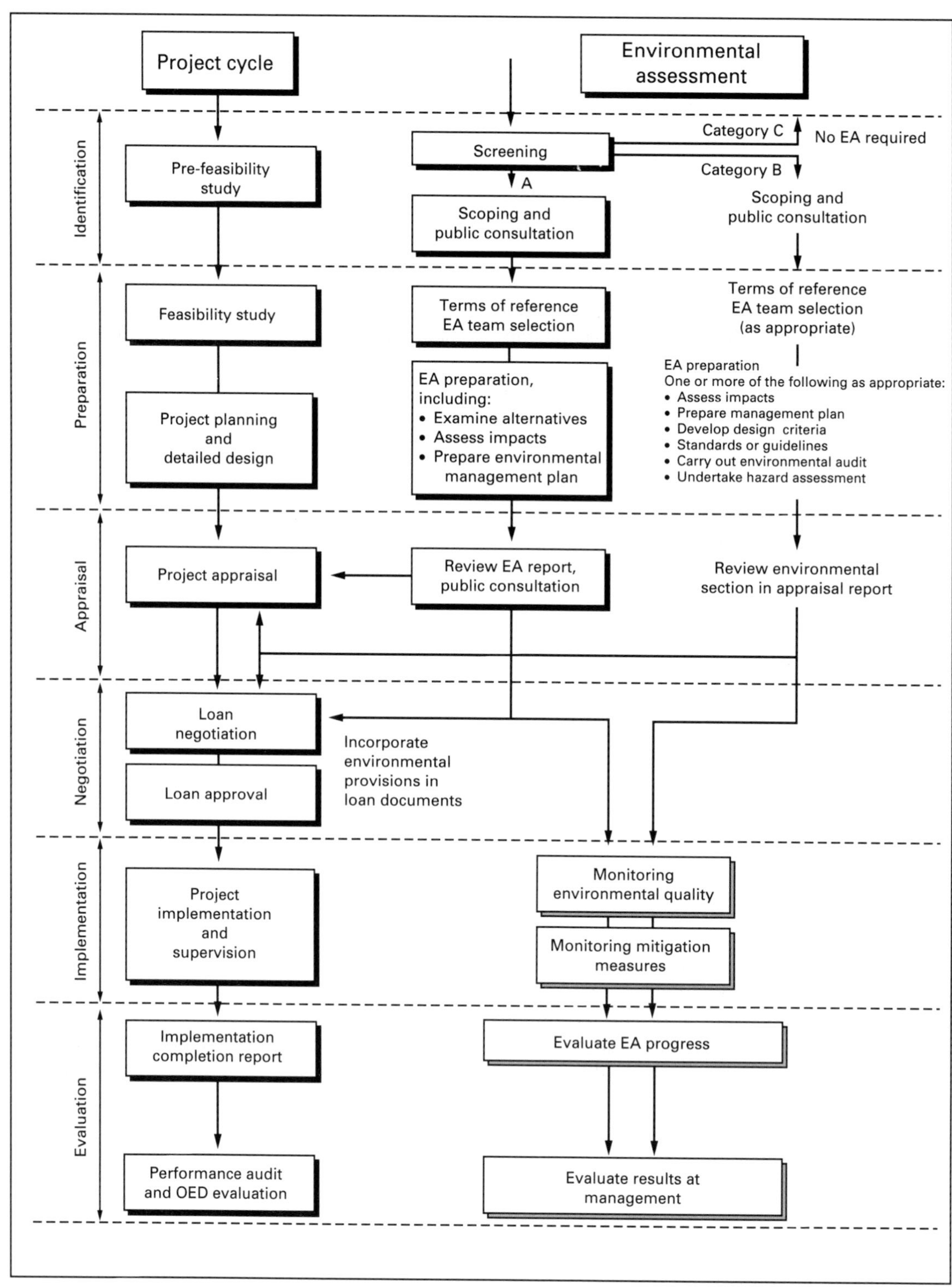

Fig. 19.1 Environmental assessment and the project cycle.

Box 19.3 Risk assessment as follow-up to EA: two energy projects in Pakistan.

One project supports: the development of oil and gas fields under a joint venture between the state-owned Oil and Gas Development Corporation (OGDC) and international petroleum companies; the construction of a condensate plant and gas pipelines; and instituting measures for environmental protection and institutional strengthening. Environmental risks were identified in the EA and played an important role in the selection of contractors. For example, one bid was rejected mainly on the basis of failing to meet an environmental requirement with regard to the siting of a gas processing facility, although it was the most financially competitive bid. However, the bidder wanted to build the facility on the Indus floodplain despite the fact that the EA had clearly recommended not doing so due to the risk of flooding. A formal safety and hazard assessment study to reduce the risk of accidents during construction and operation was subsequently carried out by independent consultants appointed by the contractor, as required by the bidding documents.

The second project involves the design, construction and financing of the hydrocracker complex by a joint venture company, National Crescent Petroleum, Ltd. (NCPL), incorporated in Pakistan. The refinery would process about 1.3 million tonnes of locally produced straight-run fuel oil. Following feasibility studies, which included a detailed EA, a set of job instructions was released to the turnkey contract bidders covering the necessary changes in the initial project design to conform with the EA mitigation plan. Among other things, the instructions included undertaking a risk and hazard assessment of the refinery's operating, handling and storage conditions before the refinery was commissioned.

preparation and identifying and assessing potential impacts, including risks. The EIA concludes that the project—with certain mitigation measures and design changes—is sound and can go forward. However, it recommends that a hazard and risk assessment be carried out for existing installations that could potentially represent a danger to the adjacent, planned facilities that would be financed under the project, unless they are properly managed. It also recommends that the risk assessment include consideration of potential seismic activity. Prior to initiation of construction works, the risk assessment is carried out, resulting in a clearer picture of the risks and a set of recommendations for avoidance and minimization during construction and operation of the facility, as well as for contingency plans.

Other times, the EA process is primarily focused on environmental risks from the very beginning, resulting in an EA report that is essentially an environmental risk assessment. A case in point is the Industrial Pollution Control Project in Algeria (see Box 19.4).

Lending to a country's banking sector, which in turn onlends for industrial and agricultural investments, provides a venue for a particular type of risk assessment that is rapidly emerging: environmental credit risk management for banks. The International Finance Corporation (IFC 1994), which is the private sector arm of the World Bank Group, has developed an environmental risk management approach targeted to commercial banks in developing countries (see Box 19.5).

19.3 STRATEGIC ENVIRONMENTAL RISK ASSESSMENT: USING RISK ASSESSMENT TO DESIGN PROJECTS AND STRATEGIES AIMED AT REDUCING RISKS

Risk-based techniques have been introduced for a variety of strategic environmental planning purposes, especially over the last 4–5 years. In India, risk assessment is being used in Bank-financed projects to improve industrial safety and avoid potential disasters. A human health risk-oriented methodology was developed to identify priority clean-up projects in the mining sector in Bolivia (see Box 19.6).

Health risk assessments have been central in the development of urban environmental strategies and action plans for large cities such as Bangkok,

Box 19.4 Environmental Assessment in the form of health risk assessment and environmental audits: an industrial pollution control project in Algeria.

Since its independence in 1962, Algeria has undergone rapid industrialization without due emphasis on protection of the country's environment. Industrial pollution problems exist in the major cities of Algiers, Oran, Constantine and Annaba. Pollution from public sector enterprises (PEs) of the northeastern coast and, in particular, those located in or around the cities of Annaba (population 465 000) and Skikda (population 167 000) is a growing problem that needs to be immediately addressed, especially given its impact on health.

The objective of this project is to assist the Government of Algeria to reduce pollution that causes health problems or serious ecological degradation. An institutional and legal component will, among other things: reinforce the technical and administrative capacity of the environmental institutions and selected ministries; enhance and update the legal and regulatory framework; and design a system for monitoring and enforcement. The investment component will, among other things, finance measures to reduce air emissions and treat and reuse industrial waste waters, and provide a hazardous waste disposal and storage facility that will serve the northeastern region.

The project falls within Category 'A' (see Box 19.1) of the Bank's classification procedure, mainly because of the construction of the storage and disposal site for hazardous wastes. The EA included the following activities:

- a health risk assessment;
- detailed environmental audits for the three beneficiary enterprises;
- mitigation plans for the respective investment subcomponents of the project.

The main benefits of this project are reduced environmental degradation and decreased negative public health impact as a result of reduced hazardous emissions to air and water, and leakages from wastes. The institutional and legal component will strengthen the national and local capabilities in the design, monitoring and enforcement of appropriate policies and regulations. The investment component will contribute substantially to the reduction of hazardous emissions and waste in the northeastern region, and improve industrial efficiency through the use of cleaner technology and waste minimization.

Manila, Bombay, Cairo, Mexico City and Quito.* In Bangkok, for example, risk assessment was used to rank environmental problems according to their effects on human health. Exposure to lead was found to be the top environmental health problem, resulting in up to 400 deaths, 500 000 cases of hypertension, 800 heart attacks and strokes, and 700 000 lost IQ points in children per year. The risk assessment helped identify cost-effective health risk reduction strategies (see Box 19.7). In Quito, food contamination from microorganisms and outdoor air pollution were determined to be the most serious environmental health risks. An innovative multimedia approach to health risk assessment from urban environmental degradation is discussed in Box 19.8.

At the level of national planning, national environmental action plans (NEAPs) have been developed with the Bank's support (World Bank 1995). These plans identify priority environmental problems for the country as a whole and propose ways to address these problems. Dealing with uncertainty is an intrinsic part of this process, as cause-and-effect relationships between environmental changes, human health and economic activities are not generally fully understood. For instance, the impact of toxic waste disposal on air or water can be long term, while the value of biodiversity, or the capacity of polluted ecosystems to purify themselves, is uncertain. Several NEAPs have addressed uncertainties surrounding pollution and impacts on human health by drawing on dose–response studies. In Santiago, Chile, for example,

* The Metropolitan Environmental Improvement Program (MEIP) for large Asian cities (sponsored by the United Nations Development Program (UNDP) and the World Bank) have been particularly active in promoting and using health risk assessment. For more information, see publications in the Urban Management Program Discussion Paper Series, World Bank, Washington, DC, particularly Bartone *et al.* (1994).

Box 19.5 Environmental risk management for the banking sector.

The World Bank and other multilateral financial institutions use lending to commercial banks and other *financial intermediaries* (FIs) to support private sector development and strengthen the financial sector in borrowing countries. The intermediaries in turn onlend the resources to private investors or provide equity investment. The projects or other activities financed by the FIs may cause environmental impact or, in other ways, entail some degree of environmental risk. Methods have therefore been developed to help FIs take environmental risk into account when considering investment opportunities, as opposed to relying solely on financial criteria. Such methods are broadly labelled as *environmental risk management*, and include steps to assess and manage environmental risk from the perspective of banks (or other financial institutions).

The major environmental issues affecting companies in which FIs typically invest include: site contamination, major hazards associated with construction or operation activities, and violations of regulations. The International Finance Corporation (IFC), the main private sector lending arm of the World Bank Group, has developed detailed environmental risk management guidance to help their FIs address these issues: *Environmental Risk Management for Financial Institutions*.

At the heart of IFC's guidance document is a corporate environmental checklist that poses a series of questions designed to:

- identify risks to the environment associated with a company's current operations, such as potential site contamination;
- identify risks to the environment associated with a company's planned investment, such as major hazards; and
- assess liabilities associated with current or future environmental risks.

Guidance is provided when more-detailed risk assessments are required, and on managing risks throughout the investment project cycle. A programme of training financial officers within IFC's FI lending partners in the application of environmental risk management techniques is ongoing.

studies of linkages between air and water pollution and health problems looked at changes in the incidence of premature death due to certain types of respiratory diseases. In Nigeria, uncertainty was addressed by having a group of experts assigning each identified environmental problem a subjective risk value, which was then used to rank a diverse set of problems. This type of assessment is implicit in many NEAPs. In general, however, NEAPs have not yet used risk (or vulnerability) assessment techniques to their fullest potential.

At the regional level, the Bank is involved in a number of initiatives that in different ways address environmental risk through international collaborative efforts. The Metropolitan Environmental Improvement Program (MEIP) in Asia is one example, and the Mediterranean Environmental Technical Assistance Program (METAP) is another. In Central and Eastern Europe, the Bank is supporting the implementation of a region-wide environmental strategy focusing especially on cost-effective measures to reduce human health risks associated with pollution (Hertzman 1995).

Of particular interest may be RAINS-Asia, an international programme (co-funded with the Asian Development Bank) to help assess environmental risks associated with a transboundary pollution problem: acid rain depositions in Asia. This programme is based partly on the framework of a pioneering acid rain prototype for Europe (RAINS-EUROPE) that is currently being used for policy analysis under the European Convention on Long-Range Transboundary Air Pollution. Phase 1 of RAINS-Asia (completed in 1995) developed the methodology and a PC-based software tool (RAINS-ASIA) to assess impacts of various acid rain emission reduction strategies in Asia. The model covers the countries of East, South and South-East Asia, with particular emphasis on Japan, India, China, Indonesia, Thailand and South Korea, and is used to show how different areas of Asia are likely to be affected by acid rain over the next 25 years, under different energy development and air pollution scenarios. The user can estimate current and future costs and impacts of alternative emission control strategies on a regional, country

Box 19.6 Risk-based methodology to plan environmental remediation in the mining sector in Bolivia.

In conjunction with a project in Bolivia aimed at improving environmental management in the mining sector and remediating existing pollution risks to the environment and human health, World Bank environmental economists developed a health risk-oriented methodology for setting priorities for remediation and for selecting the highest priority investments for environmental management. In other words, this methodology might be referred to as a sector-wide risk assessment.

The methodology follows the conceptual framework for risk assessment of hazard identification, dose–response assessment, exposure assessment and risk characterization, with the objective of targeting those investments and activities likely to provide the greatest benefits for the money spent (as opposed to a standards-based approach). However, the sequence of steps was different, because the initial emphasis was on the impacts of multiple sources rather than of any one pollutant. The methodology comprises four principal stages.

1 Screen all sites and produce an initial ranking of their hazard potential, based on minimal data.
2 Identify priority sites for remedial action from data obtained in short visits to the sites (and their surroundings) identified in Stage 1 as most likely to be hazardous. Define data needs for more complete assessments of the impacts and sources of damage.
3 Carry out audits on high-priority properties to ascertain the nature and extent of hazards. (For many of the mines, this had already been done as part of the privatization process currently under way in the Bolivian mining industry.)
4 Determine appropriate actions at each of the priority sites, based on estimates of the benefits of reducing environmental contamination and on the costs of different actions for reducing contamination. The environmental audits would provide much of the information needed to carry this out.

By going systematically through these steps, the outcome would be a list of possible remediation actions, ranked by their likelihood to provide significant benefits for the resources spent, thus providing input to the decision-making process of the Bolivian mining authorities.

(From Ayres & Hanrahan 1995.)

Box 19.7 Cost-effective risk reduction strategies for Bangkok.

The Bank commissioned a study in 1994, aimed at identifying cost-effective health risk reduction strategies for the Bangkok Metropolitan Region, Thailand. Of particular concern was the inevitable escalation of social costs associated with pollution from energy production and use, transportation and manufacturing as a result of continued economic expansion.

Prioritization of environmental problems was based on the efficiency criterion (maximizing net social benefits). The highest priority was assigned to air pollution by particulate matter, lead and traffic congestion, followed by microbiological contamination of water. The latter was not addressed in the study because it would not be exacerbated by economic growth. The next priority was water pollution due to organic and toxic wastes.

The study made recommendations for policy and institutional reforms, including:

- altering fuel taxes to remove price distortions that favoured lignite over less-polluting fuels;
- establishing emission standards for particulates and SO_2 from power generation facilities;
- implementing programmes to phase out lead in gasoline and reduce the sulphur content of diesel;
- establishing a presumptive charging system for waste waters to replace a standards-based system (to shift the onus for monitoring to firms);
- establishing an incentive framework to encourage better management of hazardous wastes; and
- streamlining and clarifying institutional responsibilities for pollution management.

Because Thailand is in the early stages of pollution control, it should be possible to reduce the impacts of the main pollutants in Bangkok at relatively low cost. The country's recent remarkable economic growth should also support the necessary investment.

Box 19.8 Human health and environmental degradation: a multimedia risk assessment approach.

The Manila metropolitan area (Metro Manila) shares the experience of many Asian cities over the last couple of decades. Rapid industrialization and migration from rural areas have dramatically increased the population both in terms of numbers and density. Coupled with growing motorization and generally inadequate urban infrastructure (roads, water supply and sanitation, etc.), this has lead to increasing problems with congestion and air and water pollution. As a result, many public health problems have become worse.

However, the links between environmental degradation and human health tend to be overlooked or poorly understood. Consequently, many municipalities in developing countries continue to treat the environment as a 'luxury problem' they cannot afford to address. It was partly in response to this that the World Bank took the initiative to develop an innovative tool that would do two things: (i) assess the links between environmental quality and health in terms of risks; and (ii) disseminate and explain this relationship in an understandable and persuasive way to policy makers as well as the general public in order to raise awareness and understanding and set premises for action. This tool, which is now fully developed as computer software and is being used by a new Health Risk Assessment Resource Centre within the Department of Health in Manila, is called the *Multimedia Health Risk Assessment Tool.* It contains a rich database on environmental conditions and indicators and public health data in the Metro Manila area. The information is georeferenced and a GIS capacity allows for detailed analysis and comparison across different communities or neighbourhoods within the metropolitan area. Many of the data are qualitative rather than quantitative in nature and the tool does not include the capacity to undertake quantifications of probability. Rather, the tool has the ability to organize imperfect information and correlate and present it. For example, it looks at occurrence of certain water-borne diseases in communities that use contaminated drinking water. It can then translate these relationships into potential risks for other communities whose drinking water quality might be threatened by certain developments. The Resource Centre serves as the analytical arm of an interagency task force on environment and health (which includes representation by non-governmental organizations), which will be responsible for developing policy responses.

The non-linear nature of the tool makes it possible to direct it to a wide variety of audiences: communities within Metro Manila; children and students through the schools; the media; health experts and officials; environmental experts and officials; industry; and political decision makers.

The risk assessment tool was established as part of a Bank-financed Urban Health and Nutrition Project and is likely to be used in forthcoming Bank-financed projects in the Metro Manila area. However, it will also be used in urban planning and development independent of any Bank financing. It cost approximately US$400 000 to develop and install the tool. A Japanese grant provided financing.

or subnational basis. This information can be utilized to: develop national and regional policies to address the rapidly emerging problems of atmospheric pollution in Asia; develop basic strategies for the World Bank, Asian Development Bank and other donor institutions concerning policy advice, institution building and investment initiatives in the borrowing countries to implement the findings; and strengthen Asian research capacity for the support of national and regional policies. Phase 2 of RAINS-Asia will verify methodology and data and strengthen institutional links and capacity in Asia to take full advantage of the programme.

19.4 RISK ASSESSMENT APPLIED TO NATURAL HAZARDS AND GLOBAL ENVIRONMENTAL CHALLENGES

The United Nations declared the 1990s the International Decade for Natural Disaster Reduction (IDNDR). The *ad hoc* group of experts assembled to support the initiative saw the decade as a moral imperative and urged the application of scientific and technical knowledge to alleviate human suffering and improve economic security. The World Bank has played a leading role in expanding the understanding of the relationship between poverty, economic development and disasters (Kreimer & Munasinghe 1991; Munasinghe & Clarke 1995).

The Bank's interest, and stake, in this topic stems from the fact that there is a close connection between poverty and the consequences of natural disasters. Poverty limits the resilience of developing countries—especially poor people—to natural disasters, both in terms of putting in place measures to protect communities *ex ante* and in terms of their impact when they occur. Economic development, on the other hand, enables *ex ante* precautionary measures to be taken and makes it possible to cushion the impact *ex post*. Put differently, development creates opportunity for risk reduction through growth and accumulation of wealth, the ability to save and invest, the creation of functioning institutions and investments in human capital. Perhaps the simplest expression of the benefits of economic development in the face of disaster is the contrast between the effects of similarly strong earthquakes in rich and poor countries. The 1989 earthquakes in San Francisco and Armenia were relatively similar in strength (with the one in San Francisco being the strongest). However, the Armenia earthquake killed tens of thousands of people whereas the one in San Francisco caused 62 deaths.

At the same time development is not enough and environmentally unsustainable development patterns can undermine key ecological functions and increase the frequency of natural disasters and the risks associated with them. The challenge is therefore to foster a development process that protects ecosystems that serve as buffers for human communities, such as coastal wetlands and forested, steep slopes along rivers, while at the same time investing in cost-effective measures to minimize the negative effects of unavoidable natural occurrences.* Box 19.9 describes current efforts to improve flood protection in three river basins in Argentina, using a *regional EA*, as a planning tool.

*While efforts to prevent and manage the effects of disasters are continuing and need to be strengthened, there will always be a need for emergency response. The Bank therefore frequently finances emergency projects to alleviate the effects of natural disasters.

Vulnerability assessment, a subset of risk assessment, has been developed to analyse the hazard and risk associated with a potential natural disaster—for example an earthquake, windstorm, flood or wildfire—in a given area (Anderson 1995). In this context, the term *vulnerability* has to do with future jeopardy and potential harm to people's health, life, property or resources on which health and life depend. The methodology follows the same basic steps as any risk assessment: identification of hazards; identifying exposure; identifying the complex sources of the hazard; and assessing changes in vulnerability over time and space. The approach takes into account development patterns (e.g. deforestation of watersheds or drainage of coastal wetlands) that may exacerbate the risk of disaster. To date, the methodology has had limited application in the context of development planning, although elements of the methodology have been used for dam construction (dam safety in relation to natural floods and earthquakes) and flood protection programmes. It is expected that the methodology will be applied increasingly to assess vulnerability associated with global environmental change, such as sea-level rise or violent storms caused or exacerbated by human-induced global climate change.

At the global level, the Bank's involvement in the Global Environmental Facility (GEF) is in itself a response to increasing environmental risks at the global level. The work of the Intergovernmental Panel on Climate Change is in many ways a global risk assessment of the consequences of emitting 'greenhouse gases', providing a rationale and basis for action by all parties to the Climate Convention, including the World Bank as a GEF Implementing Agency. Box 19.10 discusses risk and vulnerability assessment in the context of regional planning for adaptation to global climate change in the Caribbean.

19.5 THE ROAD AHEAD

Environmental risk assessment is likely to become increasingly important in development planning. As managing and utilizing increasingly scarce natural resources such as water and soil sustainably becomes more and more of a challenge, the need increases for improving our understanding

Box 19.9 A proactive regional EA: a flood protection project in Argentina.

The central objective of the Argentina Flood Protection Project is to improve flood protection for the human communities inhabiting the flood plains of the Parana, Paraguay, and Uruguay rivers in northern Argentina. It would finance implementation of a comprehensive programme of investments within the flood plains lying within the boundaries of seven provinces. This region has suffered enormous economic and human loss from floods, the last occurring in 1983 and 1992. At the same time, the periodic flooding sustains ecological systems and many forms of productive activities. The project has therefore adopted a 'living with floods' strategy. The investments include both structural and non-structural measures to protect important economic and social infrastructure and enhance the provincial capacity to deal with periodic flooding.

At the Bank's suggestion, a regional environmental assessment (REA) was initiated during the earliest stages of project preparation, to serve as an input to its design. The REA studied the interaction of natural and man-made systems within the flood plains, including the ecological functions of the periodic floods and the current state of critical ecosystems such as wetlands and gallery forests.

The study, undertaken by a local team led by a Colombian specialist, found that to a surprising extent many ecosystems and human activities depended on the floods. This had a direct impact on the way the project was designed. Criteria for the selection of investments were modified to ensure that flooding would continue but not threaten human well-being and economic infrastructure.

The study also documented the extent to which wetlands, gallery forests and aquatic ecosystems of the tributaries to the three rivers are threatened by ongoing human activities. The REA found that the most disruptive activities were road construction, followed by poorly planned urban expansion and effluent from the meat packing industry. Another significant finding was that poor urban sanitation services were directly undermining existing flood protection works. For example, many communities disposed of garbage along protective dykes, attracting rodents that weakened the dykes by digging tunnels, thus making them ineffective against floods.

The REA helped design four key project components to help improve the environmental and economic benefits of the project: (i) a component to strengthen EA procedures in key institutions within the seven provinces; (ii) technical assistance for urban environmental management; (iii) environmental education and awareness programmes in communities benefiting from protection works; and (iv) support to protection and management initiatives for wetlands and other ecosystems.

However, perhaps the most important contribution of the REA was its direct contribution to screening all potential investments under the project. From a total of some 150 possible investments, the REA helped reduce the number to 51 subprojects, all with a clear economic, social and environmental justification. Once these subprojects had been selected, project-specific EAs were prepared for all of them by the REA team. Upon completion of these EAs, the REA team went back and examined the likely cumulative impacts of all the 51 subprojects, ensuring that such impacts would be minimized. The total cost of the entire EA process did not exceed US$300 000—a small amount compared with the cost of the project and the expected economic, social and environmental benefits.

of environmental risk and its relationship to economic development and social welfare.

To date, risk assessment techniques have been applied mostly to potential industrial hazards and to the assessment of environmental health risk. The assessment of ecological risk is still not as common, although methods are increasingly being tested and applied, often in conjunction with environmental cost–benefit analyses (e.g. in forestry projects involving a combination of production and conservation activities, or in the assessment of impacts on natural habitats of water diversions). The main reason for this 'lag' when it comes to ecological risk is that the relationship between different human activities and ecological 'chain reactions' in different environmental settings

Box 19.10 Caribbean regional planning for adaptation to global climate change.

The members of the Caribbean Community (CARICOM) are primarily small island states with fragile coastal ecosystems. Agriculture and tourism are their principal sources of employment and foreign exchange earnings. Coastal areas are vital to the prosperity of these countries and are usually the most biologically productive areas, supporting a wealth of living marine resources. In recent years, these resources have come under increasing stress and the lack of a coordinated institutional structure prevents the integrated management of those resources.

Anticipated global warming, and consequent changes in sea-level, sea surface temperature and wind and ocean currents may seriously compound these problems. Sea-level rise, in particular, will likely affect freshwater supply, increase beach and coastal erosion, and aggravate the impact of tropical storms. It also threatens the disproportionate share of industrial, tourism, energy, transport and communications infrastructure concentrated in the coastal zone. The Intergovernmental Panel on Climate Change has calculated first-order costs for protection of Caribbean shorelines from future sea-level rises. For Caribbean island territories, the projected cost of new construction alone would be $11.1 billion, which is well beyond the combined investment capacity of their economies. Other more cost-effective adaptation measures are therefore called for.

The GEF is supporting Caribbean countries in preparing a project to cope with the adverse effects of global climate change—particularly sea-level rise—in coastal and marine areas through *vulnerability assessment*, adaptation planning and capacity building. The project will follow a regional approach; it will be executed through a cooperative effort of all participating countries and through a combination of national pilot/demonstration actions and regional training and technology transfer linked to adaptation planning.

The adaptation planning component is concerned with short-term planning of adaptation to global climate change in vulnerable areas. It will focus on regional sea/climate data collection and management, impact and vulnerability studies, and assessment of policy options. Its main activities will be to: (i) establish a comprehensive Sea Level/Climate Monitoring Network for the Caribbean; (ii) prepare a comprehensive inventory of physical and biological resources of coastal areas and their current uses and users for subsequent use in vulnerability assessment; (iii) develop a coral reef monitoring network seeking to obtain appropriate information for evaluating both the impacts of the large-scale temperature stress on existing coral reefs and the adaptive capacity of these reefs; (iv) prepare studies for assessing both the vulnerability of selected coastal areas to sea level and the impact of sea-level rise and storm surge risk in those areas—using the tropical storm hazard assessment model of the Organization of American States (OAS); (v) establish data bases and information systems on climate/sea-level data, impact and vulnerability; (vi) assess the economic value of coastal natural resources and the cost of their current loss; (vii) formulate a planning framework for integrated coastal and marine; and (viii) review existing regulations, legislation, and economic policies, and identify minimum-cost measures (including incentive and revenue instruments) that could be applied immediately to initiate implementation of such a strategy.

is still subject to great uncertainty. Consequently, there is some resistance towards applying the methodology on a systematic basis.

The World Bank will continue to encourage and expand the use of environmental risk assessment and seek to contribute to the development and refinement of risk assessment techniques and their application to new areas.

19.6 REFERENCES

Anderson, M.B. (1995) Vulnerability to disaster and sustainable development: a general framework for assessing vulnerability. In: *Disaster Prevention for Sustainable Development. Economic and Policy Issues* (eds M. Munasinghe & C. Clarke), pp. 41–59. The World Bank, Washington, DC.

ADB (1990) *Environmental risk assessment. Dealing with uncertainty in environmental impact assessment*. Environment Paper 7, Asian Development Bank, Manila.

Ayres, W.S. & Hanrahan, D. (1995) Setting priorities for

remediation of environmental problems due to mining in Bolivia. Environment Department, World Bank, Washington, DC. (Draft.)

Bartone, C., Bernstein, J., Leitmann, J. & Eigen, J. (1994) *Towards environmental strategies for cities: policy considerations for urban environmental management in developing countries.* Urban Management Programme Policy Paper 18, World Bank, Washington, DC.

Hertzman, C. (1995) *Environment and Health in Central and Eastern Europe.* The World Bank, Washington, DC.

IFC (1994) *Environmental Risk Management for Financial Institutions: a Handbook for Financial Officers.* International Finance Corporation, Washington, DC.

Kjørven, O. (1995) *Environmental assessment. Challenges and good practice.* Environment Department Paper 018, The World Bank, Washington, DC.

Kreimer, A. & Munasinghe, M. (eds) (1991) *Managing Natural Disasters and the Environment.* The World Bank, Washington, DC.

Le Moigne, G., Barghouti, S. & Plusquellec, S. (eds) (1990) *Dam safety and the environment.* World Bank Technical Paper 115, Washington, DC.

Munasinghe, M. & Clarke, C. (eds) (1995) *Disaster Prevention for Sustainable Development. Economic and Policy Issues.* The World Bank, Washington, DC.

Pouliquen, L.Y. (1970) *Risk analysis in project appraisal.* Staff Occasional Paper 11, The World Bank, Washington, DC.

Reutlinger, S. (1970) *Techniques for project appraisal under uncertainty.* Staff Occasional Paper 10, The World Bank, Washington, DC.

Technica Ltd (1988) *Techniques for Assessing Industrial Hazards. A Manual.* World Bank Technical Paper 55, Washington, DC.

World Bank (1991) *Environmental Assessment Sourcebook.* World Bank Technical Paper 154. Washington, DC.

World Bank (1993–1996) *Environmental Assessment Sourcebook Updates*: 1, *The World Bank and Environmental Assessment: An Overview.* 2, *Environmental Screening.* 3, *Geographic Information Systems for Environmental Assessment and Review.* 4, *Sectoral Environmental Assessment.* 5, *Public Involvement in Environmental Assessment: Requirements, Opportunities and Issues.* 6, *Privatization and Environmental Assessment: Issues and Approaches.* 7, *Coastal Zone Management and Environmental Assessment.* 8, *Cultural Heritage in Environmental Assessment.* 9, *Implementing Geographic Information Systems in Environmental Assessment.* 10, *International Agreements on Environment and Natural Resources: Relevance and Application in Environmental Assessment.* 11, *Environmental Auditing.* 12, *Elimination of Ozone Depleting Substances.* 13, *Guidelines for Marine Outfalls and Alternative Disposal and Reuse Options.* 14, *Environmental Performance Monitoring and Supervision.* 15, *Regional Environmental Assessment.* World Bank, Washington, DC.

World Bank (1996) *Handbook on Pollution Prevention and Abatement.* The Environment Department, World Bank, Washington, DC. (in press.)

World Bank (1991) *Operational Directive 4.01; Environmental Assessment.* World Bank, Washington, DC.

World Bank (1995) *National Environmental Strategies: Learning from Experience.* World Bank, Washington, DC.

Chapter 20
Risk Communication

VINCENT T. COVELLO

20.1 INTRODUCTION

Risk communication occurs whenever there is an exchange of information among interested parties about the nature, magnitude, significance or control of a risk. Information about risks can be communicated through a variety of channels, ranging from media reports and warning labels to public meetings or hearings. As government agencies, industry groups, unions, the media, scientists, professional organizations and special interest groups place a heavier emphasis on improved environmental communication in crisis and non-crisis situations, the value of research-based communication takes on an increasing and crucial measure of importance.

In recent years, the literature on risk communication has grown rapidly. Hundreds of articles and books have been published on the topic. Most of these focus on the challenges of communicating information during crisis and non-crisis situations about the risks of exposures to environmental and occupational risk agents—particularly the risks of exposures to chemicals, heavy metals, radiation in the air, water, land and food.

Why is there such a growing interest in risk communication? First, there is an increased number of hazard communication and right-to-know laws relating to exposures to environmental risk agents. Second, there is an increase in public fear about exposures to environmental and occupational risk agents from past, present and future industrial or related activities. Third, there is an expansion of media interest and reporting of environmental issues and concerns. Fourth, there is an increasingly strident nature to debates about environmental risks with advantages flowing to those with the most effective risk communication skills, A fifth explanation underlies the first four: the loss of faith and trust in established institutions as responsible environmental managers and as credible sources of risk information.

The purpose of this article is to review the research-based literature on risk communication. The specific aim is to provide the reader with a general outline of the field and to relate this work to the practical needs to those with risk communication responsibilities.

20.2 SCIENCE AND PERCEPTION

A significant part of the risk communication literature focuses on problems and difficulties in communicating risk information effectively. These problems and difficulties revolve around issues of science and perception and can be organized into four categories: (i) characteristics and limitations of scientific data about risks; (ii) characteristics and limitations of spokespersons in communicating information about risks; (iii) characteristics and limitations of the media in reporting information about risks; and (iv) characteristics and limitations of the public in evaluating and interpreting risk information. Each is described below.

20.2.1 Characteristics and limitations of scientific data about risks

One source of difficulty in communicating information about risks is the uncertainty and complexity of data on health, safety and environmental risks. Risk assessments, despite their strengths, seldom provide exact answers. Due to limitations in scientific understanding, data, models and methods, the results of most risk assessments are at best approximations. Moreover, the resources needed to resolve these uncertainties are seldom adequate to the task.

These uncertainties invariably affect communications with the public in the hostile climate that surrounds many environmental issues. For example, uncertainties in environmental risk assessments often lead to radically different estimates of risk. An important factor underlying many debates about risks is the different assessments of risk produced by government agencies, industry and public interest groups.

20.2.2 Characteristics and limitations of spokespersons in communicating information about risks

A central question addressed by the literature on risk communication is why some individuals and organizations are trusted as credible sources of risk information and others are not. For example, numerous studies have found that scientists and officials in industry and government — two of the most prominent sources of risk information — often lack trust and credibility. These same studies have found that public distrust of government and industry is grounded in several beliefs: that they have been insensitive to public concerns and fears about environmental and occupational risks; unwilling to acknowledge problems; unwilling to share information; unwilling to allow meaningful public participation; and negligent in fulfilling their responsibilities to protect the health and safety of workers and the public. Compounding the problem are beliefs that environmental laws are too weak and that government and industry have done a poor job protecting the environment.

Several factors compound these perceptions and problems. First, many scientists and officials have engaged in highly visible debates and disagreements about the reliability, validity and meaning of the results of environmental and occupational health risk assessments. In many cases, equally prominent experts have taken diametrically opposed positions on issues as diverse as the safety of nuclear power plants, hazardous waste sites, asbestos, electric and magnetic fields, lead, radon, polychlorinated biphenyls (PCBs), arsenic and dioxin. While such debates can be constructive for the development of scientific knowledge, they often undermine public trust and confidence.

Second, resources for risk assessment and management are seldom adequate to meet demands by citizens and public interest groups for definitive findings and rapid action. Explanations by scientists and officials that the generation of valid and reliable toxicological or epidemiological data is expensive and time consuming — or that risk assessment and management activities are constrained by resource, technical, statutory, legal or other limitations — are seldom perceived to be satisfactory. Individuals facing what they believe is a new and significant risk are especially reluctant to accept such claims.

Third, coordination among responsible authorities is seldom adequate. In many debates about risks, for example, lack of coordination has severely undermined public faith and confidence. Compounding such problems is the lack of consistency in approaches to risk assessment and management by authorities at the local regional, national and international levels. For example, few requirements exist for regulatory agencies to develop coherent, coordinated, consistent and interrelated plans, programmes and guidelines for managing risks. As a result, regulatory systems tend to be highly fragmented. This fragmentation often leads to jurisdictional conflicts about which agency and which level of government has the ultimate responsibility for assessing and managing a particular risk. Lack of coordination, different mandates and confusion about responsibility and authority also lead, in many cases, to the production of multiple and competing estimates of risk. A commonly observed result of such confusion is the erosion of public trust, confidence and acceptance.

Fourth, many scientists and officials lack the skills needed to communicate effectively information about risk. For example, many scientists and officials use complex and difficult technical language and jargon in communicating information about risks to the media and the public. The use of technical language or jargon is not only difficult to comprehend but can also create a perception that the official or expert is being unresponsive, dishonest or evasive.

Finally, scientists and officials have often been insensitive to the information needs of the public and the differences between expert and lay perceptions of risk. Scientists and officials often operate on the assumption that they and their audience share a common framework for evaluating and interpreting risk information. However, this is often not the case. One of the most important findings to emerge from risk perception and communication studies is that people take into consideration a complex array of qualitative and quantitative factors in defining, evaluating and acting on risk information.

One of the costs of this heritage of mistrust and loss of confidence is the public's reluctance to believe the risk information provided by government and industry. Programmes for overcoming this have focused on improvements in risk assessment, management and risk communication. In the risk communication arena, the two areas that have received the greatest attention are skills enhancement and credibility enhancement.

Skills enhancement is based on three fundamental risk communication principles: that perceptions are realities; that the primary goal of risk communication is to establish trust and credibility; and that risk communication is a skill (Table 20.1). The verbal and non-verbal skills listed in Tables 20.2 and 20.3 represent a sampling of the practical tools and techniques identified in the applied risk communication literature.

Table 20.1 Key formulae underlying effective risk communication.

$P = R$	Perceptions (P) equal realities (R): what is perceived as real will be real in its consequences
$G = T + C$	The primary goal (G) of risk communication is to establish high levels of trust (T) and credibility (C): the secondary goal is to convey facts and figures
$C = S$	Effective risk communication (C) is a complex skill (S): as with any complex skills, it requires significant amounts of knowledge, training and practice

Credibility enhancement is based on research indicating that trust and credibility are built on a foundation of: perceived caring and empathy; perceived competence and expertise; perceived honesty and openness; and perceived dedication and commitment (Table 20.4). Activities that enhance credibility include outreach efforts aimed at improving coordination and collaboration with

Table 20.2 Verbal risk communication skills.

Jargon	Avoid jargon; define all technical terms
Organizational identity	Use personal pronouns; avoid using the organization's name
Attacks	Attack issues; avoid attacking those with higher credibility
Humour	Avoid humour (jokes, cartoons, sarcasm, etc.)
Risk–benefit comparisons	Discuss risks and benefits in separate communications
Risk comparisons	Use risk comparisons for perspective and not acceptability
Negative allegations	Do not repeat negative allegations
Speculation	Do not speculate about worst cases or 'what if' situations
Key messages	Stress performance, trend lines and achievements
Technical details and debates	Avoid providing excessive amounts of technical detail
Quantitative health risk numbers	Avoid risk statements stated as quantitative probabilities
Guarantees	Talk about achievements and not about the lack of guarantees
Zero risk	Emphasize the value of zero risk as a goal
Comparisons with others	Avoid comparing your organization with other organizations
Length of answers to questions	Limit answers to 2 min or less
Length of presentations	Limit presentations to 15–20 min or less
Visuals	Use visuals and graphics as much as possible
Abstractions	Use examples, analogies and stories to establish meaning
Testing	Test and practise all answers and presentations

Table 20.3 Non-verbal risk communication skills.

Poor eye contact	Messages: dishonest, closed, unconcerned, nervous
Excellent eye contact	Messages: honest, open, competent, sincere, dedicated, confident, knowledgeable, interested
Frequent blinking	Messages: nervous, deceitful, inattentive
Sitting back in chair	Messages: uninterested, unenthusiastic, unconcerned, uncooperative
Sitting forward in chair	Messages: interested, enthusiastic, concerned, cooperative
Arms crossed on chest	Messages: uninterested, unconcerned; defiant, not listening, arrogant, impatient, defensive, stubborn
Frequent hand-to-face contact	Messages: dishonest, deceitful, nervous
Hidden hands	Messages: deceptive, guilty, insincere
Open hands	Messages: open, sincere
Speaking from behind barriers	Messages: dishonest, deceitful, formal, unconcerned, uninterested, superior
Speaking from an elevated position	Messages: superiority, dominance, judgemental
Clenched hands	Messages: anger, hostility, determination, uncooperative

Table 20.4 Indicators of trust and credibility.

Indicator	Example
Perceived caring/empathy (50%)	Perceived sincerity, ability to listen, ability to see issues from the perspective of the other
Perceived competence/expertise (15–20%)	Perceived intelligence, training, authoritativeness, experience, educational level, professional attainment, knowledge, command of information
Perceived openness/honesty (15–20%)	Perceived truthfulness, candidness, justness, objectivity, sincerity
Perceived dedication/commitment (15–20%)	Perceived altruism, diligence, self-identification, involvement, hard work

organizations and individuals perceived to be credible. Surveys indicate that organizations and individuals perceived to be relatively high to medium in credibility on environmental and occupational risk issues include health and safety professionals, educators, scientific organizations, professional organizations, the media, non-management employees, non-profit voluntary health organizations, environmental groups and citizen or employee advisory groups (Table 20.5).

20.2.3 Characteristics and limitations of the media in reporting information about risks

The mass media play a critical role in transmitting risk information. Given this importance, researchers have focused their attention on the role of the mass media and on characteristics and limitations of the media that contribute to problems in risk communication.

One of the major conclusions to emerge from risk communication research is that journalists are often biased toward stories that contain drama, conflict, expert disagreements and uncertainties. The media are especially biased towards stories that contain dramatic or sensational material, such as a minor or major accident at a chemical manufacturing facility or a nuclear power plant. Much less attention is given to daily occurrences that kill or injure far more people each year but take only one life at a time. In reporting about risks, journalists often focus on the same concerns as the public, for example, potentially catastrophic effects, lack of familiarity and understanding, involuntariness, scientific uncertainty, risks to future generations, unclear benefits, inequitable

Table 20.5 Trust and credibility of sources of environmental information based on 1996 sample survey of the population in the USA.

Top third	Local citizens that are perceived to be neutral, respected and well informed about the issue Health and safety professionals Educators (especially those from respected local schools) Non-profit voluntary health organizations Non-management employees Professional societies
Middle third	Media Environmental groups
Bottom third	Industry officials Federal government officials Environmental consultants from for-profit firms
Changes from previous years	Environmental groups: 10–15% loss of credibility Media: 5–10% gain in credibility Government and industry: 10% loss in credibility

distribution of risks and benefits and potentially irreversible effects.

Media coverage of risks is frequently deficient in that many stories contain oversimplifications, distortions and inaccuracies in reporting risk information. Media coverage is also deficient, not only in what is contained in the story but in what is left out. For example, analyses of media reports on cancer risks show that these reports are often deficient in: providing few statistics on general cancer rates for the purposes of comparison; providing little information on common forms of cancer; not addressing known sources of public ignorance about cancer; and providing little information about detection, treatments and other protective measures.

Many of these problems stem from characteristics of the media and the constraints under which reporters work. First, most reporters work to extremely tight deadlines that limit the amount of time for research and for the pursuit of valid and reliable information. Second, with few exceptions, reporters do not have adequate time or space to deal with the complexities and uncertainties surrounding many risk issues. Third, journalists achieve objectivity in a story by balancing opposing views. Truth in journalism is different from truth in science. In journalism, there are only different or conflicting views and claims, to be covered as fairly as possible. Fourth, journalists are source dependent. Under the pressure of deadlines and other constraints, reporters tend to rely heavily on sources that are easily accessible and willing to speak out. Sources that are difficult to contact, hard to draw out, or reluctant to provide interesting and non-qualified statements are often left out. Finally, few reporters have the scientific background or expertise needed to evaluate the complex scientific data and disagreements that surround many debates about risks. Given these limitations, effectiveness in communicating with the media about risks depends in part on understanding the constraints and needs of the media and adapting one's behaviour and information to meet these needs.

20.2.4 Characteristics and limitations of the public in evaluating and interpreting risk information

Much of the risk communication literature focuses on characteristics and limitations of the public in evaluating and interpreting risk information. These include: (i) inaccurate perceptions of levels of risk; (ii) difficulties in understanding probabilistic information related to unfamiliar activities or technologies; (iii) strong emotional responses to risk information; (iv) desires and demands for scientific certainty; (v) strong beliefs and opinions that are resistant to change; (vi) weak beliefs and opinions that are easily manipulated by the way the information is presented; (vii) ignoring or dismissing risk information because of its perceived lack of personal relevance; (viii) perceiving accidents and mishaps as signals; (ix) using health and environmental risks used as proxies or surrogates for other concerns; and (x) distinguishing perceptions of risk from judgements of risk acceptability. Each of these characteristics is described below.

Inaccurate perceptions of risk

People often overestimate some risks and underestimate others. For example, people tend to overestimate the risks of dramatic or sensational causes of death, such as accidents at manufacturing plants or waste disposal facilities, and underestimate the risks of undramatic causes, such as asthma, emphysema and diabetes. This bias is caused in part by the tendency for risk judgements to be influenced by the memorability of past events and by the imaginability of future events. A recent disaster, intense media coverage or a vivid film can heighten the perception of risk. Conversely, risks that are not memorable, obvious, palpable, tangible or immediate tend to be underestimated.

Difficulties understanding probabilistic information related to unfamiliar activities or technologies

A variety of cognitive biases and related factors hamper people's understanding of probabilities. These difficulties, in turn, hamper discussions of risk probabilities between experts and non-experts. For example, risk experts are often confused by the public's rejection of the argument that a cancer risk from a new activity is acceptable if it is smaller than one in a million. This is especially frustrating given that a one in a million risk is an extremely small number and that the background chance of cancer is approximately 1 in 4. In rejecting this argument, people respond with one or more objections. First, they personalize the risk (e.g.: What if that one is me or my child?). Second, they raise questions of trust (e.g.: Why should I believe you—didn't you and your colleagues do the calculations?). Third, they raise concerns about cumulative risks (e.g.: I am already exposed to enough risks in life—why do I need one more?). Fourth, they question whether the risks are worth the benefits (e.g.: Is the activity that generates the risk really worth losing one life?). Finally, they raise ethical questions (e.g.: Who gave government and industry the right to play God and choose who will live or die?). Exacerbating the problem is the difficulty people have in understanding, appreciating and interpreting small probabilities, such as the difference between one chance in a hundred thousand and one chance in a million.

Given these difficulties, the risk communication literature contains explicit cautions about the use of probabilistic information in explaining risk decisions. This is especially the case when there is little time available for discussion. A more effective strategy is to focus the risk communication on: (i) the degree to which the activity meets health, safety or environmental standards as set or reviewed by trusted and credible authorities; (ii) the relationship between the risk of the activity in question and other risks; and (iii) the degree to which the risk estimate is based on worst-case or pessimistic assumptions that are biased toward public health and safety.

These same problems hamper discussions between experts and non-experts on low-probability/high-consequence events and 'worst-case scenarios'. In many such cases, the imaginability of the worst case makes it difficult for people to distinguish between what is remotely possible and what is probable.

Strong emotional responses to risk information

Strong feelings of fear, hostility, anger, outrage, panic and helplessness are often evoked by exposure to unwanted or dreaded risks. These emotions often make it difficult to engage in rational discourse about risk in public settings. Emotions tend to be most intense when people perceive the risk to be involuntary, unfair, not under their personal control and low in benefits. More extreme emotional reactions often occur when the risk affects children, when the adverse consequences are particularly dreaded (e.g. cancer and birth defects) and when worst-case scenarios are presented.

Desires and demands for scientific certainty

People often display a marked aversion to uncertainty and use a variety of coping mechanisms to reduce the anxiety generated by uncertainty. This aversion often translates into a marked preference for statements of fact over statements of probability—the language of risk assessment. People often demand to be told exactly what will happen, and not what might happen.

Strong beliefs and opinions that are resistant to change

People tend to ignore evidence that contradicts their current beliefs. Strong beliefs about risks, once formed, change very slowly and are extraordinarily persistent in the face of contrary evidence. Initial beliefs about risks tend to structure the way that subsequent evidence is interpreted. New evidence, for example, data provided by a government or industry official, appears reliable and informative only if it is consistent with the initial belief; contrary evidence is dismissed as unreliable, erroneous, irrelevant or unrepresentative.

Weak beliefs and opinions that are easily manipulated by the way in which information is presented

When people lack strong prior beliefs or opinions, subtle changes in the way that risks are expressed can have a major impact. To test this hypothesis, one group of researchers asked two groups of physicians to choose between two therapies—surgery or radiation. Each group received the same information but with one major difference—probabilities were expressed either in terms of dying or in terms of surviving. Even though these two numbers are the same, the difference resulted in dramatic differences in the choice of therapy. Virtually the same results were observed for other test populations.

A variety of studies have demonstrated the powerful influence of such presentation or 'framing' effects. The experimental demonstration of these effects suggests that risk communicators can, under some circumstances, easily manipulate risk perceptions.

Ignoring or dismissing risk information because of its perceived lack of personal relevance

Most risk data relate to society as a whole. These data are usually of little interest or concern to individuals, who are more likely to be concerned about risks to themselves rather than about risks to society.

Perceiving accidents and mishaps as signals

The significance of an accident is determined only in part by its health, safety or environmental consequences (e.g. the number of deaths or injuries that occur). Of equal, if not greater, importance is what the accident or mishap signifies or portends. A major accident with many deaths and injuries, for example, may have only minor social significance (beyond that to the victims' families and friends) if it occurs as part of a familiar and well-understood system (e.g. a train wreck). However, a minor accident in an unfamiliar or poorly understood system—such as a leak at a radioactive waste disposal site—can have major social significance as a harbinger of future, possibly catastrophic events.

Using health, environmental or occupational risks used as proxies or surrogates for other concerns

The specific risks that people focus on reflect their beliefs about values, social institutions, nature and moral behaviour. Risks are exaggerated or minimized accordingly. Debates about risks often are a proxy or surrogate for debates about other, more general, social, economic or political concerns. The debate about nuclear power, for example, has often been interpreted as less a debate about the specific risks of nuclear power than about other concerns, including the proliferation of nuclear weapons, the adverse effects of nuclear waste disposal, the value of large-scale technological progress and growth and the centralization of political and economic power in the hands of a technological elite.

One conclusion that can be drawn from these observations is that risk is not an objective phenomenon perceived in the same way by all interested parties. Instead, it is a psychological and social construct with its roots deeply embedded in specific social contexts. A variety of scientific, psychological, social and cultural factors determine which risks will ultimately be selected for societal attention and concern. Scientific evidence about the magnitude of possible adverse consequences is only one of these factors.

Distinguishing between perceptions of risk and judgements of risk acceptability

Even though the level of risk is related to risk acceptability, it is not a perfect correlation. Two factors affect the way in which people assess risk and evaluate acceptability; these factors modify the correlation.

First, the level of risk is only one among several variables that determine acceptability (Table 20.6). Some of the most important variables that matter to people in evaluating and interpreting risk information are described below.

1 *Catastrophic potential.* People are more concerned about fatalities and injuries that are grouped in time and space (e.g. fatalities and injuries resulting from a major accidental release of toxic

Table 20.6 Factors important in risk perception and evaluation.

Factor	Conditions associated with increased public concern	Conditions associated with decreased public concern
Catastrophic potential	Fatalities and injuries grouped in time and space	Fatalities and injuries scattered and random
Familiarity	Unfamiliar	Familiar
Understanding	Mechanisms or process not understood	Mechanisms or process understood
Uncertainty	Risks scientifically unknown or uncertain	Risks known to science
Controllability (personal)	Uncontrollable	Controllable
Voluntariness of exposure	Involuntary	Voluntary
Effects on children	Children specifically at risk	Children not specifically at risk
Effects manifestation	Delayed effects	Immediate effects
Effects on future generations	Risk to future generations	No risk to future generations
Victim identity	Identifiable victims	Statistical victims
Dread	Effects dreaded	Effects not dreaded
Trust in institutions	Lack of trust in responsible institutions	Trust in responsible institutions
Media attention	Much media attention	Little media attention
Accident history	Major and sometimes minor accidents	No major or minor accidents
Equity	Inequitable distribution of risks and benefits	Equitable distribution of risks and benefits
Benefits	Unclear benefits	Clear benefits
Reversibility	Effects irreversible	Effects reversible
Personal stake	Individual personally at risk	Individual not personally at risk
Origin	Caused by human actions or failures	Caused by acts of nature or God

chemicals or radiation) than about fatalities and injuries that are scattered or random in time and space (e.g. automobile accidents).

2 *Familiarity.* People are more concerned about risks that are unfamiliar (e.g. leaks of chemicals or radiation from waste disposal facilities) than about risks that are familiar (e.g. household accidents).

3 *Understanding.* People are more concerned about activities characterized by poorly understood exposure mechanisms or processes (e.g. long-term exposure to low doses of toxic chemicals or radiation) than about activities characterized by apparently well-understood exposure mechanisms or processes (e.g. pedestrian accidents or slipping on ice).

4 *Uncertainty.* People are more concerned about risks that are scientifically unknown or uncertain (e.g. risks from a radioactive waste facility designed to last 20 000 years) than about risks that are relatively known to science (e.g. actuarial data on automobile accidents).

5 *Controllability.* People are more concerned about risks that they perceive to be not under their personal control (e.g. accidental releases of toxic chemicals or radiation from a waste disposal facility) than about risks that they perceive to be under their personal control (e.g. driving a car or riding a bicycle).

6 *Voluntariness.* People are more concerned about risks that they perceive to be involuntary (e.g. exposure to chemicals or radiation from a waste or industrial facility) than about risks that they perceive to be voluntary (e.g. smoking, sunbathing or mountain climbing).

7 *Effects on children.* People are more concerned about activities that put children specifically at risk (e.g. milk contaminated with radiation or toxic chemicals; pregnant women exposed to radiation or toxic chemicals) than about activities that do not put children specifically at risk (e.g. adult smoking).

8 *Effects manifestation.* People are more concerned about risks that have delayed effects (e.g. the development of cancer after exposure to low doses of chemicals or radiation) than about risks that have immediate effects (e.g. poisonings).

9 *Effects on future generations.* People are more concerned about activities that pose risks to future generations (e.g. genetic effects due to exposure to toxic chemicals or radiation) than to risks that pose no special risks to future generations (e.g. skiing accidents).

10 *Victim identity.* People are more concerned about risks to identifiable victims (e.g. a worker exposed to high levels of toxic chemicals or radiation) than about risks to statistical victims (e.g. statistical profiles of vehicle accident victims).

11 *Dread.* People are more concerned about risks that are dreaded and evoke a response of fear, terror or anxiety (e.g. exposure to radiation or carcinogens) than to risks that are not especially dreaded and do not evoke a special response of fear, terror or anxiety (e.g. common colds and household accidents).

12 *Trust in institutions.* People are more concerned about situations where the responsible risk management institution is perceived to lack trust and credibility (e.g. lack of trust in certain government agencies for their perceived close ties to industry) than they are about situations where the responsible risk management institution is perceived to be trustworthy and credible (e.g. trust in the management of recombinant deoxyribonucleic acid (DNA) risks by universities and by the National Institutes of Health).

13 *Media attention.* People are more concerned about risks that receive much media attention (e.g. accidents, leaks and other problems at waste disposal facilities) than about risks that receive little media attention (e.g. on-the-job accidents).

14 *Accident history.* People are more concerned about activities that have a history of major and sometimes minor accidents (e.g. leaks at waste disposal facilities) than about activities that have little or no history of major or minor accidents (e.g. recombinant DNA experimentation).

15 *Equity and fairness.* People are more concerned about activities that are characterized by a perceived inequitable or unfair distribution of risks and benefits (e.g. inequities related to the siting of waste disposal facilities) than about activities characterized by a perceived equitable or fair distribution of risks and benefits (e.g. vaccination).

16 *Benefits.* People are more concerned about hazardous activities that are perceived to have unclear, questionable or diffused benefits (e.g. waste disposal facilities) than about hazardous activities that are perceived to have clear benefits (car driving).

17 *Reversibility*. People are more concerned about activities characterized by potentially irreversible adverse effects (e.g. nuclear war) than about activities characterized by reversible adverse effects (e.g. injuries from sports or household accidents).

18 *Personal stake*. People are more concerned about activities that they believe place them (or their families) personally and directly at risk (e.g. living near a waste disposal site) than about activities that do not place them (or their families) personally and directly at risk (e.g. disposal of hazardous waste in remote sites or in other nations).

19 *Nature of evidence*. People are more concerned about risks that are based on evidence from human studies (e.g. risk assessments based on adequate epidemiological data) than about risks based on animal studies (e.g. laboratory studies of the effects of radiation using animals).

20 *Human versus natural origin*. People are more concerned about risks caused by human actions and failures (e.g. accidents at waste disposal sites caused by negligence, inadequate safeguards or operator error) than about risks caused by acts of nature or God (e.g. exposure to geological radon or cosmic rays).

These factors explain, in large part, the aversion of parts of the public toward activities and technologies such as nuclear power. They also help to explain phenomena such as the 'not in my back yard' (NIMBY) response to chemical, nuclear and related facilities. For example, many residents in communities where unwanted industrial facilities exist or are planned believe that government and industry officials: (i) have excluded them from meaningful participation in the decision-making process; (ii) have denied them the resources needed to evaluate or monitor independently the associated health, safety or environmental risks; (iii) have denied them the opportunity to give their 'informed consent' to management decisions that affect their lives and property; (iv) have imposed or want to impose upon them facilities that provide few local economic benefits; (v) have imposed or want to impose upon them facilities that entail high costs to the community (e.g. adverse impacts on health, safety, wildlife, recreation, tourism, property values, traffic, noise, visual aesthetics, community image and quality of life); (vi) have imposed or want to impose on them facilities that provide most of the benefits to other parties or to society as a whole; and (vii) have dismissed their opinions, fears and concerns as irrational and irrelevant.

Critical to resolving NIMBY and related risk controversies is recognition that a fairly distributed risk is more acceptable than an unfairly distributed risk. A risk entailing significant benefits to the parties at risk is more acceptable than a risk with no such benefits. A risk for which there are no alternatives is more acceptable than a risk that could be eliminated by using an alternative technology. A risk that the parties at risk have control over is more acceptable than a risk that is beyond their control. A risk that the parties at risk assess and decide voluntarily to accept is more acceptable than a risk that is imposed on them. These statements are true in exactly the same sense in which it is true that a small risk is more ac-ceptable than a large risk. Risk is multidimensional; size is only one of the relevant dimensions.

If the validity of these points is accepted, then a whole range of risk communication and management options present themselves. Because factors such as fairness, familiarity and voluntariness are as relevant as size in judging the acceptability of a risk, efforts to make a risk fairer, more familiar, and more voluntary are as appropriate as efforts to make the risk smaller. Similarly, because control is important in determining the acceptability of a risk, efforts to share power, such as establishing and assisting advisory committees or supporting third-party research, audits, inspections and monitoring, can be effective in making a risk more acceptable.

The second factor in deciding what level of risk ought to be acceptable is not a technical question but a value question. People vary in how they assess risk acceptability. They weigh the various factors according to their own values, sense of risk and stake in the outcome. Because acceptability is a matter of values and opinions, and because values and opinions differ, debates about risk are often debates about values, accountability and control.

20.3 COMMUNICATING UNCERTAINTY: THE ROLE OF RISK COMPARISONS

A significant part of the risk communication literature focuses on the difficulties of communicating uncertainty and the role of risk comparisons in accomplishing this task. Interest in risk comparisons derives specifically from perceived difficulties in communicating complex, quantitative, probabilitistic information to laypersons and the need to put risk information in perspective for informed decision-making to take place. Several authors have argued that risk comparisons provide this perspective.

In a typical risk comparison, the risk in question is compared with the risks of other substances or activities. Because comparisons are perceived to be more intuitively meaningful than absolute probabilities, it is widely believed that they can be used effectively for communicating risk information. A basic assumption of the approach is that risk comparisons provide a conceptual yardstick for measuring the relative size of a risk, especially when the risk is new and unfamiliar.

Risk comparisons have several strengths that address important facets of this problem. They present issues in a mode that appears compatible with intuitive, natural thought processes, such as the use of analogies to improve understanding; they avoid the difficult and controversial task of converting diverse risks into a common unit, such as dollars per life lost or per day of pain and suffering; and they avoid direct numerical reference to small probabilities, which can be difficult to comprehend and evaluate in the abstract.

Many risk comparisons are advanced not only for gaining perspective and understanding but also for setting priorities and determining which risks are acceptable. More specifically, risk comparisons have been advocated as a means for determining which risks to ignore, which risks to be concerned about, and how much risk reduction to seek. A common argument in many risk comparisons is that risks that are small or comparable to already accepted risks should themselves be accepted.

The risk comparison literature contains two basic types of risk comparisons: (i) comparisons of the risks of diverse activities; and (ii) comparisons of the risks of similar or related activities. Each type is described below.

20.3.1 Comparisons of the risks of diverse substances, activities and technologies

The basic strategy in this type of comparison is to compare—along a common scale or metric—the risk of a new or existing substance, activity or technology to the risks of a diverse set of substances, activities or technologies. For example, the health risks of a new pesticide might be compared to the risks of sunbathing, smoking and driving. An underlying but untested assumption is that the health risks of the new or existing substance, activity or technology can be more easily appreciated by people if placed in comparative perspective.

Approaches

A variety of different scales have been used by researchers for comparing risks, including scales based on the annual probability of death, the risk per hour of exposure and the overall loss in life expectancy. Data for constructing such scales are typically drawn from diverse sources, including public health and accident data collected by various government agencies.

One of the most commonly used scales for comparing risks is the annual death rate. Using such a measure, it can be shown, for example, that citizens in the USA face, on average, a 2 in 1000 risk of dying from cancer, a 5 in 10 000 risk of dying in an accident, a 2.5 in 10 000 risk of dying in a motor vehicle accident, a 1 in 10 000 risk of being murdered, and a 5 in 10 million risk of being killed by a lightning bolt. These risks can be contrasted with other risks. For example, the average smoker faces a 3 in 1000 risk of dying each year from smoking, the average mountain climber faces a 6 in 10 000 risk of being killed in a climbing accident and the average hang-glider faces a 4 in 10 000 annual risk of being killed in a hang-gliding accident. On an annual basis, the risk of smoking is: substantially greater than the risk of hang-gliding, mountaineering, boxing and working in a mine; somewhat greater than the risk of military service during the Vietnam era; and nearly as great as the risk of stunt flying.

One deficiency in these risk comparisons is their lack of sensitivity to age differences. For example, at age 5 years the risk of dying from all causes is less than 1 in 1000; at age 40 years it is about 2 in 1000; and at age 80 years it is about 83 in 1000. Given the large effect that age can have on risk estimates, an alternative procedure that takes this factor into account is to calculate the expected loss in life expectancy due to various causes. Several authors have taken this approach and have shown that the risk of dying from cigarette smoking is twice as great as the risk of being a coal miner; and that the risk of dying in a motor vehicle accident is twice as great as the risk of dying in an accident at home.

Other formats for comparing risks have also been developed. For example, one format is to compare a set of activities with approximately equal risks, such as the risk of activities estimated to increase a person's chance of death (during any year) by one in a million. Using this measure, researchers have found that each of the following activities presents the same risk: smoking 1.4 cigarettes, riding 10 miles by bicycle, eating 40 tablespoons of peanut butter, drinking 30 12-ounce cans of diet soda containing saccharin, and living within 5 miles of a nuclear reactor for 50 years.

Adopting a somewhat different approach, researchers have calculated the time needed to accumulate a one-in-a-million risk of death from a variety of activities. Using this measure, it can be shown that the risk of dying in a motor vehicle accident is approximately equivalent to the risk of dying from being on police duty; and that the risk of dying from employment in trade or manufacturing is approximately equivalent to the risk of dying from a fall.

Many of the risk comparisons described above have been advanced not only for gaining perspective and understanding but also for setting priorities and determining which risks are acceptable. More specifically, they have been advocated as a means for determining which risks to ignore, which risks to be concerned about and how much risk reduction to seek.

Based on such arguments, researchers have constructed scales ranking risks from acceptable to unacceptable. In one such study, activities falling in the upper zone, representing risks of death per year of exposure of less than one in a million, were deemed acceptable. The basic argument was that the risks of these activities were insignificant—insignificance being defined as the level of risk that individuals routinely accept in their personal and daily activities. For example, since individuals routinely accept the risk of being struck by lightning—which poses a risk of death of one in a million per year of exposure—risks of this size can be regarded as acceptable. Following the same logic, it was argued that activities representing risks of death that are greater than one in a thousand per year of exposure can be regarded as unacceptable. Activities falling in the middle zone of the scale were identified as the most problematic: the acceptability of these could not be determined a priori. Instead, they must be closely scrutinized and subjected to analysis and societal debate.

Numerous authors have criticized studies that use this type of approach for determining which risks are acceptable. The basic criticism is that such efforts fail to recognize the importance and legitimacy of basing decisions about the acceptability of risk on factors other than the size of the risk.

20.3.2 Comparisons of the risks of similar or related substances, activities or technologies

Some researchers have adopted a narrower approach to risk comparison—limiting their comparisons to risks that are similar or closely related. Several examples are described below.

Foods, food products and food additives

To gain perceptive and improved understanding, a large number of studies have compared the risks posed by different foods, food products and food additives. Some of the best-known comparative analyses of the risks of different foods and food products are the studies on food risks, diet and cancer by Professor Bruce Ames and his colleagues at the University of California, Berkeley. These studies compared the cancer risks of foods that contain synthetic chemicals (e.g. food additives and pesticide residues) with the risks of natural foods. An important conclusion is that synthetic chemicals represent only a very small fraction of the total carcinogens in foods. The basic argument underlying this conclusion is that natural foods are not benign. Large numbers of potent carcinogens (e.g. aflatoxin in peanuts) and other toxins are present in foods that contain no synthetic chemicals. Many of these natural carcinogens are produced by plants as part of their natural defence processes. Analysis shows that human dietary intake of these natural carcinogens in food is likely to be at least 10 000 times greater than the intake of potentially carcinogenic synthetic chemicals in food (although partial protection against the effects of natural carcinogens is provided by the many natural anti-carcinogens that also appear in food).

Some of Ames' critics have argued that his risk estimates are inflated. The same critics have argued against an implicit, and sometimes explicit, risk comparison argument that natural carcinogens in foods deserve greater societal and regulatory attention and concern than synthetic chemicals.

Energy production technologies

In the last two decades, a large number of studies have attempted to compare the risks of alternative energy production technologies. Perhaps the best-known comparison of risks from alternative energy production technologies was an analysis conducted by Dr Herbert Inhaber for the Atomic Energy Control Board of Canada. The study compared the total occupational and public health risks of different energy sources for the complete energy production cycle—from the extraction of raw materials to energy end-use. The study examined the risks of 11 methods of generating electricity—coal, oil, nuclear, natural gas, hydroelectricity, wind, methanol, solar space heating, solar thermal, solar photovoltaic and ocean thermal. Two types of risk data were analysed: data on public health risks from industrial sources or pollutant effects; and data on occupational risks derived from statistics on injury, death and disease rates for workers. Total risk for the energy source was calculated by summing the risks for the seven components of complete energy production cycle: (i) materials acquisition and construction; (ii) emissions from materials acquisition and energy production; (iii) operation and maintenance; (iv) energy back-up system; (v) energy storage system; (vi) transportation; and (vii) waste management.

The report concluded: (i) that most of the risk from coal and oil energy sources is due to toxic air emissions arising from energy production, operation and maintenance; (ii) that most of the risk from natural gas and ocean thermal energy sources is due to materials acquisition; (iii) that most of the risk from nuclear energy sources is due to materials acquisitions and waste disposal; and (iv) that most of the risk from wind, solar thermal and solar photovoltaic energy sources is due to the energy back-up system required (assumed to be coal). Alternative sources were compared on the basis of the calculated number of man-days that would be lost per megawatt-year of electricity produced.

The most controversial aspect of the report was the conclusion that nuclear power carries only slightly greater risk than natural gas and less risk than all other energy technologies considered. Inhaber reported, for example, that coal has a 50-fold larger mortality rate than nuclear power. The report also argued that, contrary to popular opinion, non-conventional energy sources such as solar power and wind pose substantial risks, and that the risks of nuclear power are significantly lower than those of non-conventional energy sources. The relatively high risk levels associated with non-conventional energy sources were traced by Inhaber, in part, to the large volume of construction materials required for these technologies and to the risks associated with energy back-up systems and energy storage systems.

Following publication of the report, its methodology was severely criticized. Critics claimed: (i) that the study mixed risks of different types; (ii) that it used risk estimators of dubious validity; (iii) that it made questionable assumptions to cover data gaps; (iv) that it failed to consider future technological developments; (v) that it made arithmetic errors; (vi) that it double-counted labour and back-up energy requirements; and (vii) that it introduced arbitrary correction factors. Perhaps the most damaging criticism was that the study was inconsistent in applying its methodology to the various energy technologies. For example, while the study considered materials acquisition, component fabrication and plant construction in the analysis of unconventional energy sources and of hydropower, critics have claimed that the study did not follow the same approach for coal, nuclear power, oil and gas. Furthermore, the labour figures for coal, oil, gas and nuclear power included only on-site construction, while those for the renewable energy sources included on-site construction, materials acquisition and component manufacture.

Despite these criticisms, Inhaber's study represented a landmark effort in the literature on risk comparisons. It made a significant conceptual contribution by attempting to compare, in a systematic and rigorous way, the risks of alternative technologies intended to serve the same purpose. Also important were Inhaber's observations that risks occur at each stage in product development (e.g. raw material extraction, manufacturing, use and disposal), and that risks from each stage need to be added together to obtain an accurate estimate of the total risk.

Cancer

A variety of studies have used risk comparisons to put cancer risks in perspective. In perhaps the best-known such study, Sir Richard Doll and his colleagues analysed data for a variety of causes of cancer, including industrial products, pollution, food additives, tobacco, alcohol and diet. Results of the study provided an important comparative perspective on cancer risks. The study found, for example, that the combined effect of food additives, occupational exposures to toxic agents, air and water pollution and industrial products account for only about 7% of cancer deaths in the USA. These results suggest that removing all pollutants and additives in the air, water, food and workplace would result in only a small decrease in cancer mortality (although even this small percentage represents a substantial number of lives). By contrast, the combined effects of alcohol, diet and smoking are related to 70% of cancer deaths in the USA. Consequently, even a modest change in personal habits would result in a significant decrease in cancer mortality.

Other authors have adopted a different approach for comparing the risks of cancer. For example, researchers have compared data on the annual risks of cancer from various common or everyday activities. Smoking clearly poses the largest risk of cancer, with an annual risk of 1.2 in 1000. Drinking one beer per day, receiving an average number of diagnostic X-rays, and background radiation at sea level all pose about the same annual risk of cancer—2 in 100 000.

20.3.3 Other types of comparisons

A large number of other studies have also compared the risks of similar or diverse activities and technologies. Because of their policy significance, two types of risk comparisons have received special attention in the risk communication literature: comparisons of different occupations or industries; and comparisons of different sources of radiation. In the occupational arena, one study found that the annual job-related risk of death experienced by timber workers, miners and pilots—the most risky occupations—is several orders of magnitude greater than for school teachers, dentists and librarians—among the least risky occupations. Occupational risks of death range from one chance in a thousand per year to one chance in a million per year. In the radiation field, one study compared different sources of radiation exposure (e.g. natural, medical and occupational) and found that the single largest source of radiation exposure is radon emanating from the ground or construction materials and accumulating in closed buildings. Such a finding is, of course, only as useful as it is accurate. Only a few years ago, diagnostic X-rays were identified as the largest source of radiation exposure. Radon was not even listed in most tables as a source of radiation exposure.

One of the most ambitious attempts to compare risks was a 1987 study by the US Environmental Protection Agency (USEPA) entitled *Unfinished Business: a Comparative Assessment of Environmental Problems*. A critical part of the study is a comparative ranking of 31 health and environmental risks. The ranking was performed by USEPA experts and covered nearly every environmental problem addressed by USEPA programmes, including air pollution, water pollution and hazardous waste. Risks were ranked in four major categories: cancer effects, non-cancer-causing human health effects, ecological effects and welfare effects (e.g. damage to materials). Factors not included in ranking environmental problems were: benefits to society, qualitative aspects such as whether the risk is voluntary or equitable and economic or technical controllability.

An important finding of the study was that the USEPA ranking of health and environmental risks did not correlate well with the public's ranking of health and environmental risks. For example, risks associated with hazardous waste sites were in the middle of the USEPA ranking for cancer effects and at the lower end of the USEPA ranking for non-cancer-causing health effects. By comparison, risks associated with hazardous waste sites were at the top of the public's rankings. Risks associated with radon were at the top of the USEPA ranking, while the same risks were at the bottom of the public's ranking. The report notes that the USEPA risk rankings also do not correspond well with the agency's programme priorities. Agency priorities appear to be more closely aligned with public opinion than with estimated risks.

Another type of comparison that attempts to put risk information in perspective is the 'concentration' comparison. In such a comparison, the concentration of a toxic substance in the environment is compared to other measurement units such as length or time (Table 20.7). For example, one part per billion of a toxic chemical in drinking water is equivalent to 1 inch in 16 000 miles and 1 s in 32 years. Although some insight is provided by such comparisons, the data often lack relevance and meaning unless coupled

Table 20.7 Concentration comparisons organized by unit categories.

Unit	1 part per million	1 part per billion	1 part per trillion
Length	1 in./16 miles	1 in./16 000 miles	1 in./16 000 000 miles (a 6-in. leap on a journey to the sun)
Time	1 min/2 years	1 s/32 years	1 s/320 centuries (or 0.06 s since the birth of Jesus Christ)
Money	1 cent/$10 000	1 cent/$10 000 000	1 cent/ $10 000 000 000
Weight	1 oz/31 tonnes	1 pinch salt/10 tonnes of potato chips	1 pinch salt/10 000 tonnes of potato chips
Volume	1 drop vermouth/80 'fifths' of gin	1 drop vermouth/500 barrels of gin	1 drop of vermouth in a pool of gin covering the area of a football field 43 ft deep
Area	1 square ft/23 acres	1 in./160-acre farm	1 square ft/the state of Indiana; or 1 large grain of sand on the surface of Dayton Beach
Action	1 lob/1200 tennis matches	1 lob/1 200 000 tennis matches	1 lob/ 200 000 000 tennis matches
Quality	1 bad apple/2000 barrels	1 bad apple/2 000 000 barrels	1 bad apple/ 2 000 000 000 barrels

with data on the toxic potency of a particular chemical.

In addition to concentration comparisons, several studies have attempted to compare the costs of different risk reduction programmes. For example, one study compared the cost-per-fatality-averted implied by various activities aimed at reducing risks. Researchers have also compared the cost per year of life saved of different various investments, and the cost per life saved of various risk-reduction regulations.

One general conclusion from these studies is that different health, safety and environmental programmes and investments vary enormously in their costs per lives saved. Disproportionate amounts of money are spent to reduce risks in some areas, while other areas are relatively neglected. One study found, for instance, that the cost-per-fatality-averted for different risk reduction programmes ranged between \$10 000 and \$1 billion; another study found that the cost-per-life-saved for various health investments ranged between \$540 and \$6.6 million per year; and a third study showed that the cost-per-life-saved of various proposed, rejected and final regulations ranged between \$100 000 and \$72 billion (for final regulations only, the range is \$100 000 to \$132 million, with an average and median of \$23 million and \$2 million, respectively).

20.3.4 Limitations of risk comparisons

As indicated in the examples cited above, critics have noted significant limitations of the risk comparison approach. The most important are: (i) failing to identify and emphasize uncertainties involved in the calculation of comparative risk estimates; (ii) failing to consider the broad set of quantitative dimensions that define and measure risk; and (iii) failing to consider the broad set of qualitative dimensions that underlie people's concerns about the acceptability of risks and technologies. Each is described below.

Failing to identify and emphasize uncertainties involved in the calculation of comparative risk estimates

A critical flaw in many risk comparisons is the failure to provide information on the assumptions underlying the calculation of comparative risk estimates. Since risk estimates are typically drawn from a variety of different data sources, tables of comparative risks may contain risk estimates based on one set of assumptions together with non-comparable estimates based on another set of assumptions. Similarly, tables of comparative risks may contain risk estimates based on actuarial statistics (e.g. deaths from motor vehicle accidents) together with estimates based on controversial models, assumptions and judgements (cancer deaths from chronic exposure to pesticides or air pollutants).

A related flaw in many risk comparisons is the failure to describe and characterize uncertainties. This flaw can seriously undermine the value and potential usefulness of risk comparisons for risk communication purposes. Tables of risks that report only single values for adverse health, safety or environmental consequences, for example, ignore the range of possibilities and may provide an inaccurate picture of the risk problem to the public. Given the various biases, errors and other sources of uncertainty that can undermine the validity and reliability of a risk assessment, it is critical that tables of comparative risks provide the fullest possible information on potential errors and inaccuracies in each computed risk value—including qualifiers, ranges of uncertainty, confidence intervals and standard errors. To date, it is more the exception than the rule for results of comparative risk studies to be presented with full disclosure of the strengths and limitations of the assessment and with full disclosure of the degree to which assessment results are based on controversial data and judgements. Risk comparisons that do not include such information can produce a false sense of certainty.

Failing to consider the broad set of quantitative dimensions that define and measure risk

Most lists of comparative risks are unidimensional. They present statistics for only one dimension of risk, such as expected annual mortality rates or reductions in life expectancy. The use of such narrow quantitative measures of risk can, however, obscure the importance of other significant quantitative dimensions, such as: expected annual probability of injury or disability; spatial extent; concentration, persistence, recurrence, population at risk; delay, maximum expected fatalities; transgenerational effects; expected environmental damage (e.g. ecological damage or adverse effects on endangered species) and maximum expected environmental damage.

Significant distortions and misunderstandings also result from comparative analyses that fail to provide the full range of relevant quantitative risk information. Consider some of the problems involved in comparing the risks of aeroplane travel to the risks of travelling by car or train. Using a measure of risk to an individual based on the number of deaths per hundred million passenger-miles, travelling as an aeroplane passenger appears to pose slightly less risk to an individual (0.38 deaths per hundred million passenger-miles) than being a car passenger (0.55 deaths per hundred million passenger-miles), and slightly more risk than travelling as a train passenger (0.23 deaths per hundred million passenger-miles). However, for aeroplane travel, the landing and take-off phases represent the periods of highest risk; thus it can be argued that a better estimate of individual risk is the number of passenger journeys rather than the number of miles travelled. Using this measure, travelling as an aeroplane passenger poses slightly greater risk (1.8 deaths per million passenger journeys) than travelling as a car passenger (0.027 deaths per million passenger journeys) or as a train passenger (0.59 deaths per million passenger journeys). As a result, if distance travelled is the selected measurement criterion, then aeroplane travel is marginally safer than car travel and marginally less safe than train travel; but if number of journeys is the selected measurement criterion, then aeroplane travel is marginally less safe than both car travel and train travel.

A related deficiency is the failure in most risk comparisons to estimate the total quantitative risk of technologies and activities included in the risk comparison. Technological activities encompass a variety of different components, stages of development (e.g. extraction of raw materials, production, consumption and disposal), and relationships (direct and indirect) with other technological and societal activities. Detailed examination of the risks of these different components, stages of development and relationships may significantly alter the overall ranking of a technology or activity. Consequently, any risk comparison that claims to be comprehensive must either present risk data for each of these aspects or explicitly acknowledge those aspects of the analysis that have been excluded.

Even when the analyst provides data on the total quantitative risk of an activity or technology, the comparison can, none the less, be misleading if it fails to provide risk data for sensitive, susceptible or high-risk groups. These include children, pregnant women, the elderly and individuals who are particularly vulnerable or susceptible because of illness or disease. Most lists of comparative risks present population averages. However, population averages often mask important subpopulation variations in susceptibility.

Important distinctions also can be masked in other ways. For example, it is not always clear from risk comparison tables what is included in the specific risk entries. For example, do deaths from smoking include cardiovascular disease and emphysema as well as lung cancers? Are the risk estimates based on the entire population or only the population that is exposed?

Even if the analyst carefully and accurately reports risk data, misunderstandings can develop if important situational qualifiers are left out. For example, the risk calculation for driving includes many different driving situations. Yet speeding home from a party just before dawn is two orders of magnitude more dangerous than driving to the supermarket. Similarly, the risk of being hit by lightning for people who remain on a golf course during a thunderstorm is much higher than the average risk for the population in the USA.

A related deficiency stems from the failure to recognize the importance of framing effects on risk comparisons. Different impressions are created by different presentation formats. Each format for presenting or expressing risk information, such as deaths per million people, deaths per unit of concentration or deaths per activity, is likely to have a different impact on the audience. Context can be equally important. For example, an individual lifetime risk of one in a million in the USA is mathematically equivalent to approximately 0.008 deaths per day, 3 deaths per year, or 200 deaths over a 70-year lifetime. Many people will view the first two numbers as small and insignificant, whereas the latter statistic is likely to be perceived as sufficiently large to warrant societal or regulatory attention.

A final deficiency is the failure in most risk comparisons to acknowledge deficiencies in the quality of the data. Most risk comparisons draw on diverse data sources that vary considerably in quality. Because of the high cost and difficulty of collecting original data, researchers seldom have access to data developed exclusively for the comparison. Instead, a variety of existing data sources are used, each varying in quality. As a result, comparative risks often contain data of high quality together with data of questionable scientific validity and reliability.

Failing to consider the broad set of qualitative dimensions that underlie people's concerns about the acceptability of risks and technologies

Risk comparison is often advocated as a means for setting priorities or for determining which risks are acceptable. A common argument is that risks that are small, or that are comparable to risks that are already being accepted, should themselves be accepted. A number of critics have argued, however, that such claims cannot be defended. Although carefully prepared lists of comparative risk statistics can provide insight and perspective, they provide only a small part of the information needed for setting priorities or for determining which risks are acceptable.

Judgements of acceptability are related not only to annual mortality rates—the focus of most risk comparisons—but also to a multiplicity of qualitative dimensions or factors. These factors were discussed earlier in the section on risk perceptions and include voluntariness, controllability, fairness, effects on children, familiarity and benefits.

Because of the importance of these factors, comparisons showing that the risk of a new or existing activity or technology is higher (or lower) than the risks of other activities or technologies may have no effect on public perceptions and attitudes. For example, comparing the risk of living near a nuclear power or chemical manufacturing plant with the risk of driving *x* number of hours, eating *x* tablespoons of peanut butter, smoking *x* number of cigarettes a day, or sunbathing *x* number of hours may provide perspective but may also be highly inappropriate. Because such risks differ on a variety of qualitative dimensions—for example, perceived benefits, extent of personal control, voluntariness, catastrophic potential, familiarity, fairness, origin and scientific uncertainty—it is likely that people will perceive the comparison to be meaningless.

For example, it is often tempting for risk communicators to use the following argument during a meeting: the risk of *A* (e.g. emissions from an incinerator or facility) is lower than the risk of *B* (driving to the meeting or smoking during breaks). Since you (the audience) find *B* acceptable, you are obliged to find *A* acceptable.

This argument has a basic flaw in logic and its use can severely damage trust and credibility. Some listeners will analyse the argument in this way: 'I do not have to accept the (small) added risk of living near an incinerator just because I accept the (perhaps larger, but voluntary and personally beneficial) risks of sunbathing, bicycling, smoking or driving my car. In deciding about the acceptability of risks, I consider many factors, only one of them being the size of the risk—and I prefer to do my own evaluation. Your job is not to tell me about what I should accept but to tell me about the size of the risk and what you are doing about it'.

The fundamental argument against the use of risk comparisons is that it is seldom relevant or appropriate to compare risks with different qualities for risk acceptability purposes, even if the comparison is technically accurate. Several reasons underlie the argument. First, there are important psychological and social differences among risks with different qualities. Risks that are involuntary and result from lifestyle choices, for example, are more likely to be accepted than involuntary and imposed risks.

Second, people recognize that risks are cumulative and that each additional risk adds to their overall risk burden. The fact that a person is exposed to risks resulting from voluntary lifestyle choices does not lessen the impact of risks that are perceived to be involuntary and imposed.

Finally, people perceive many types of risk in an absolute sense. An involuntary increased risk of cancer or birth defects is a physical and moral insult regardless of whether the increase is small or whether the increase is smaller than risks from other exposures.

Aggravating the problem is the lack of attention given in most risk comparisons to how people actually make decisions about the acceptability and tolerability of a risk. Judgements about risks are seldom separated from judgements about the risk decision process. Public responses to risk are shaped both by the characteristics of the activity or technology and by the perceived adequacy of the decision-making process. Risk comparisons play only a limited role in such determinations.

20.3.5 Guidelines for improving the effectiveness of risk comparisons

Despite the limitations reviewed above, researchers have found that a well-constructed and well-documented risk comparison can be useful in communicating risk information. It can, for example, provide: (i) a benchmark and yardstick against which the magnitude of new or unfamiliar risks can be calibrated and compared; (ii) a means for determining and communicating the relative numerical significance and seriousness of a new or existing risk; and (iii) a means for informing and educating people about the range and magnitude of risks to which they are exposed.

For a risk comparison to achieve these goals and purposes, however, the limitations and deficiencies of the approach must specifically be addressed. Results from experimental studies and case studies suggest the following guidelines: (i) the risks that are compared should be as similar as possible; (ii) dissimilarities between the compared risks should be identified; (iii) sources of data on risk levels should be credible and should be identified; (iv) limitations of the comparison should be described; (v) the comparison should have only one purpose — numerical perspective; and (vi) all comparisons should be pilot tested.

In summary, the risk comparison approach can be a powerful tool in risk communication. However, the simplicity and intuitive appeal of the method is often deceptive. Many factors play a role in determining the legitimacy and effectiveness of a risk comparison. The success of the comparison as a risk communication will depend on the degree to which these factors have been adequately recognized, considered and addressed.

20.4 CONCLUSIONS

Given the passage of increasing numbers of right-to-know laws, and given increasing demands by workers and the public for risk information, risk communication will be the focus of increasing attention in years to come. The findings reported in this article represent only a sampling of results from the emerging area of risk communication research. However, several general principles and guidelines for communicating information about risks can be extrapolated from this literature. Although many of these principles and guidelines may seem obvious, they are so often violated in practice that a useful question is why are they so frequently not followed?

Principle 1: accept and involve the public as a legitimate partner

Discussion. Two basic tenets of risk communication in a democracy are generally understood and accepted. First, people and communities have a right to participate in decisions that affect their lives, their property and the things they value. Second, the goal of risk communication should not be to diffuse public concerns or avoid action. The goal should be to produce an informed public that is involved, interested, reasonable, thoughtful, solution-oriented and collaborative.

Guidelines.

1 Demonstrate your respect for the public and your sincerity by involving the community early, before important decisions are made.

2 Make it clear that you understand the appropriateness of basing decisions about risks on factors other than the magnitude of the risk.

3 Involve all parties that have an interest or a stake in the particular risk in question.

Principle 2: plan carefully and evaluate performance

Discussion. Different goals, audiences and media require different risk communication strategies. Risk communication will be successful only if carefully planned.

Guidelines.
1 Begin with clear, explicit objectives—such as providing information to the public, motivating individuals to act, stimulating emergency response or contributing to conflict or dispute resolution.
2 Evaluate the information you have about risks and know its strengths and weaknesses.
3 Classify the different subgroups among your audience.
4 Aim your communications at specific subgroups in your audience.
5 Recruit spokespersons who are good at presentation and interaction.
6 Train your staff—including technical staff—in communication skills and reward outstanding performance.
7 Whenever possible, pretest your messages.
8 Carefully evaluate your efforts and learn from your mistakes.

Principle 3: listen to your audience

Discussion. People in the community are often more concerned about issues such as trust, credibility, control, competence, voluntariness, fairness, caring and compassion than about mortality statistics and the details of quantitative risk assessment. If you do not listen to people, you cannot expect them to listen to you. Communication is a two-way activity.

Guidelines.
1 Do not make assumptions about what people know, think or want done about risks.
2 Take the time to find out what people are thinking: use techniques such as interviews, focus groups and surveys.
3 Let all parties that have an interest or a stake in the issue be heard.
4 Recognize people's emotions.
5 Let people know that you understand what they said, addressing their concerns as well as yours.
6 Recognize the 'hidden agendas', symbolic meanings and broader economic or political considerations that often underlie and complicate the task of risk communication.

Principle 4: be honest, frank and open

Discussion. In communicating risk information, trust and credibility are your most precious assets. Trust and credibility are difficult to obtain, and once lost they are almost impossible to regain.

Guidelines.
1 State your credentials; but do not ask or expect to be trusted by the public.
2 Disclose risk information as soon as possible (emphasizing any appropriate reservations about reliability).
3 Do not minimize or exaggerate the level of risk.
4 Speculate only with great caution.
5 If in doubt, lean towards sharing more information, not less—or people may think you are hiding something.
6 Discuss data uncertainties, strengths and weaknesses—including the ones identified by other credible sources. Identify worst-case estimates as such, and cite ranges of risk estimates when appropriate.

Principle 5: coordinate and collaborate with other credible sources

Discussion. Allies can be effective in helping you communicate risk information. Few things make risk communication more difficult than conflicts or public disagreements with other credible sources.

Guidelines.
1 Take time to coordinate all inter-organizational and intra-organizational communications.
2 Devote effort and resources to the slow, hard work of building bridges with other organizations.
3 Use credible and authoritative intermediaries.
4 Consult with others to determine who is best able to answer questions about risk.
5 Try to issue communications jointly with other trustworthy sources, such as credible university

scientists, physicians, trusted local officials and opinion leaders.

Principle 6: meet the needs of the media

Discussion. The media are a prime transmitter of information on risks. They play a critical role in setting agendas and in determining outcomes. The media are generally more interested in politics than in risk; more interested in simplicity than in complexity; and more interested in danger than in safety.

Guidelines.
1 Be open with and accessible to reporters.
2 Respect their deadlines.
3 Provide information tailored to the needs of each type of media, such as graphics and other visual aids for television.
4 Prepare in advance and provide background material on complex risk issues.
5 Follow-up on stories with praise or criticism, as warranted.
6 Try to establish long-term relationships of trust with specific editors and reporters.

Principle 7: speak clearly and with compassion

Discussion. Technical language and jargon are useful as professional shorthand, but they are barriers to successful communication with the public.

Guidelines.
1 Use language appropriate to the audience.
2 Use vivid, concrete images that communicate on a personal level.
3 Use stories, examples and anecdotes that make technical risk data come alive.
4 Avoid distant, abstract, unfeeling language about deaths, injuries and illnesses.
5 Acknowledge and respond (both in words and with actions) to emotions that people express—anxiety, fear, anger, outrage and helplessness.
6 Acknowledge and respond to the distinctions that the public views as important in evaluating risks.
7 Use risk comparisons to help put risks in perspective, but avoid comparisons that ignore the distinctions that people consider important.
8 Always try to include a discussion of actions that are under way or can be taken.
9 Promise only what you can do, and be sure to do what you promise.
10 Never let your efforts to inform people about risks prevent you from acknowledging—and saying—that any illness, injury or death is a tragedy.

Analyses of case studies suggest that these principles and guidelines can form the basic building blocks for effective risk communication. Each principle recognizes, in a different way, that effective risk communication is an interactive process based on mutual trust, cooperation and respect among all parties. Each principle also recognizes that effective risk communication is a complex art and skill that requires knowledge, training and practice.

20.5 FURTHER READING

Covello, V. (1989) Issues and problems in using risk comparisons for communicating right-to-know information on chemical risks. *Environmental Science and Technology*, **23**(12), 1444–1449.

Covello, V. & Allen, F. (1988) *Seven Cardinal Rules of Risk Communication*, US Environmental Protection Agency, Office of Policy Analysis, Washington, DC.

Covello, V.T., Sandman, P. & Slovic, P. (1988) *Risk Communication, Risk Statistics, and Risk Comparisons*. Chemical Manufacturers Association, Washington, DC.

Covello, V., McCallum, D. & Pavlova, M. (eds) (1989) *Effective Risk Communication: the Role and Responsibility of Governmental and Non-Governmental Organizations*. Plenum Press, New York.

Davies, J.C., Covello, V.T. & Allen, F.W. (eds) (1987) *Risk Communication*. The Conservation Foundation, Washington, DC.

Fischhoff, B. (1985) Managing risk perception. *Issues in Science and Technology*, **2**, 83–96.

Fischhoff, B. (1987) Treating the public with risk communications: a public health perceptive. *Science, Technology, and Human Values*, **12**(3–4), 13–19.

Fischhoff, B., Lichtenstein, S., Slovic, P., Derby, S.L. & Keeney, R.L. (1981) *Acceptable Risk*. Cambridge University Press, New York.

Johnson, B. & Covello, V. (eds) (1987) *The Social and Cultural Construction of Risk: Essays on Risk Selection and Perception*. Reidel, Boston.

Kasperson, R. (1986) Six propositions on public participation and their relevance to risk communication. *Risk Analysis*, **6**, 275–282.

Kasperson, R. & Kasperson, J. (1983) Determining the acceptability of risk: ethical and policy issues. In: *Risk: a Symposium* (eds J. Rogers & D. Bates). The Royal Society of Canada, Ottawa.

Kasperson, R. & Stallen, P.J. (eds) (1991) *Communicating Risks to the Public.* Kluwer Academic Publishers, Boston.

Krimsky, S. & Plough, A. (1988) *Environmental Hazards: Communicating Risks as a Social Process.* Auburn House, Dover, MA.

Lowrance, W.W. (1976) *Of Acceptable Risk: Science and the Determination of Safety.* Kaufman, Los Altos, CA.

National Research Council (1989) *Improving Risk Communication,* National Academy Press, Washington, DC.

Nelkin, D. (1989) Communicating technological risk: the social construction of risk perception. *American Review of Public Health,* **10**, 95–113.

Sandman, P.M. (1986) *Explaining Environmental Risk.* US Environmental Protection Agency, Office of Toxic Substances, Washington, DC.

Sandman, P., Sachsman, D., Greenberg, M. & Gotchfeld, M. (1987) *Environmental Risk and the Press.* Transaction Books, New Brunswick.

Slovic, P. (1987) Perception of Risk. *Science,* **236**, 280–285.

Slovic, P., Krauss, N. & Covello, V. (1990) What should we know about making risk comparisons. *Risk Analysis,* **10**, 389–392.

Wilson, R. (1979) Analyzing the daily risks of life. *Technology Review,* **81**, 40–46.

Wilson, R. & Crouch, E. (1987) Risk assessment and comparisons: an introduction. *Science,* **236**, 267–270.

Chapter 21
A Framework for Sustainable Product Development

JAMES A. FAVA, FRANK J. CONSOLI, KEVIN H. REINERT AND SVEN-OLOF RYDING

21.1 OVERVIEW

There is ample evidence that enhancing the quality of the current environment as well as establishing policies to sustain the environment must take on greater priority within this generation and those that follow. While this belief is commonly held by governments, environmental groups, industry, academics and the general public, there is considerable debate on how to achieve improved environmental conditions without sacrificing current economic benefits. The potential risks of human activities on the global environment have been a concern particularly with issues associated with consumption of energy, materials and the release of hazardous and toxic materials.

To assist in these debates, environmental professionals from all sectors are developing tools (e.g. risk assessment, lifecycle assessment, environmental auditing, environmental impact assessment, design for the environment, etc.)* that are a major step forward in integrating environmental considerations across a spectrum of possible operations from 'cradle to grave'. These provide excellent mechanisms to understand and manage the risks associated with activities, products and services. For example, risk assessment/management is being increasingly used to evaluate the potential impact of new chemicals to the environment.

However, as powerful as these new developments appear to be, they do not balance the needs for environmental improvement with economic considerations and sound social responsibility. Thus, environmental improvement is usually in a classic struggle against current economic considerations (such as the capital cost to implement the improvement) or against current social responsibility (such as jobs). This usually leads to winners and losers on one of the important dimensions, such as the environment or economic growth, and ultimately impedes progress on all dimensions because the fear of losing is great. What is needed is a concept and ultimately a process that will help society make decisions regarding the best balance of environmental needs, economic realities and societal wishes—sustainable development may be this concept.

Sustainable development is a concept that attempts to integrate the importance of environmental protection along with economic health and social responsibility. Developed on the belief that environmental improvement can be achieved along with economic and social improvements, Sustainable development provides a powerful new way of thinking with regard to environmental health for everyone (Fava *et al.* 1995).

The purpose of this chapter is to define the fundamental elements of sustainable development and illustrate where tools such as risk assessment (USEPA 1992) and lifecycle assessment (Fava *et al.* 1991; Consoli *et al.* 1993) can be used to assist in improving our decision-making processes. Additionally, the chapter also suggests that tools like risk assessment and lifecycle assessment, when used alone, can only provide a limited amount of information. Additional tools in economic and

* The authors would like to acknowledge that the concepts and portions of this chapter were originally discussed during a Society of Environmental Toxicology and Chemistry and Ecological Society of America Workshop on Sustainable Environmental Management held in Pellston, Michigan, in 1993.

social welfare are needed to complement risk and lifecycle assessments.

21.2 DEFINITION OF SUSTAINABLE DEVELOPMENT

The concept of sustainable development has been defined a number of ways by different multinational authors and organizations over recent years. Generally, these definitions are based on the belief that current generations should be entitled to meet their needs without compromising the ability of future generations to meet their needs (Brundtland Commission 1989). For the purposes of this publication, we have defined sustainable development as:

> the capability of individuals and society to evolve and enhance their quality of life without compromising the needs and desires of future generations.

This definition for sustainable development recognizes the role of the individual in helping to obtain a sustainable future while establishing that improvements in the quality of life are an important consideration for all generations. In addition, this vision for sustainable development stresses that the choices of current generations should not just stop at the *needs* of future generations but establishes that the possible *desires* of future generations should also be considered.

While a well-defined strategy or vision is important to establish what it is that society would like to achieve, it is equally as important to set forth a plan to reach the goal as well as to identify the necessary tools needed to measure progress towards, and attainment of, the goal. From a strategic or planning perspective, the vision of sustainable development can be achieved by establishing a process to integrate environmental protection, economic health and social responsibility into decision-making through effective education programmes and through policy initiatives in both the public and private sectors. The integration of these three components of sustainable development is illustrated in Fig. 21.1.

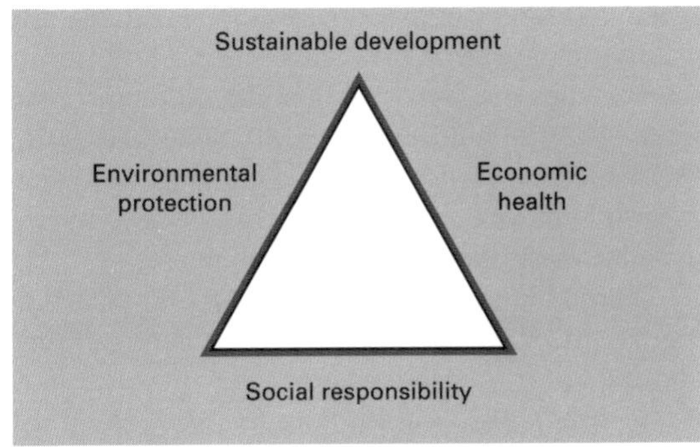

Fig. 21.1 Sustainable development component model.

The tools needed to deliver this strategy must be based on sound scientific principles that are achievable in a practical way and influenced by the values of all affected parties. Possible tools include, but are not limited to, risk assessment, design for the environment (DfE), total quality management (TQM), lifecycle assessment (LCA), full cost accounting (FCA) and other decision-making tools. These and other tools are discussed in section 21.9.2.

21.3 THE VALUE OF SUSTAINABLE DEVELOPMENT

The shift to implement sustainable development principles, processes and tools will require commitment and education at all levels of the organization. Not unlike the experience in implementing TQM systems or other major programmes, organizations will be faced with the need to explain and gain alignment to the value of the sustainable development approach versus some other approach or versus *status quo*. In undertaking this task, organizations may find some of the following points useful.

1 Unlike other methods of considering either single-issue environmental needs or even holistic environmental issues, the sustainable development approach integrates and balances the needs of the environment with the organization's economic and social responsibility needs. This approach brings together all parts of decision-making and improves overall efficiency in the process. It enables organizations to 'do it right the first time'.

2 The sustainable development approach helps an organization to gain control of its own destiny.

Often in the environmental area, organizations are reactive to governmental regulations and never seem to gain the ability to plan effectively for the future. Sustainable development encourages both short- and long-term action, enabling organizations to influence the direction of this issue and ultimately to get out of the reactive mode.

3 Sustainable development thinking is a natural evolution of the needs of society to balance an ever-increasing array of priorities. This approach is a trend that is unlikely to be reversed in the near term, so organizations will inevitably need to begin to think in this manner.

4 Embracing sustainable development does not necessitate economic trade-offs; to the contrary, organizations that embrace the sustainable development axiom and thought process should actually improve their economic situation by balancing these issues at the outset of debate versus other trade-offs. This thought process has the power to save and create jobs, influence the direction of technology development and ultimately to improve the competitive standing of organizations and even nations.

5 Sustainable development can become an organizing mechanism within organizations. It is a new, holistic way of thinking that promotes interdisciplinary problem solving and provides a useful framework to work in cooperation among different sectors (e.g. non-governmental organizations, industry, government and academics). This transformation within an organization helps to develop a new and more effective language across disciplines, helps to identify gaps in knowledge and information and facilitates innovative solutions to problems.

6 More importantly, sustainable development helps to identify the environmental, economic and societal issues associated with an organization's actions or plans, and helps to identify and implement lasting solutions.

7 Finally, embracing and implementing sustainable development generates a morale lift for organizations and their stakeholders. Under this vision, management, employees and the local community can share a common goal for economic, environmental and societal stability and enhancement.

21.4 COMPONENTS OF THE SUSTAINABLE DEVELOPMENT MODEL

The components of sustainable development, as illustrated in Fig. 21.1, include environmental protection, economic health and social responsibility. This section provides a broad overview of each of these important areas. Subsequent sections describe specific elements or subelements in more detail.

21.4.1 Environmental protection

Environmental protection occurs as a result of utilizing approaches, systems and tools to identify, implement and sustain environmental improvements associated with any activity. Environmental areas of concern include ecological health, human health and resource depletion. For example, concerns include the environmental or human health risks associated with the release of hazardous materials and/or the consumption of materials. Unlike current environmental protection solutions, which tend to focus on single-issue, point-source problems, environmental protection in the context of sustainable development requires that the approach, system or tool include the entire lifecycle of the product, process or activity, encompassing all steps from raw material acquisition to final disposition (i.e. 'cradle to grave'). Environmental protection requires the development and deployment of a number of tools, such as risk assessment techniques, understanding impact assessment techniques, uncertainty analysis, assessment of biodiversity, climate change assessments, valuation techniques and various environmental remediation technologies.

21.4.2 Economic health

Economic health is achieved by assessing and improving the organization's current economic conditions through processes that consider and implement programmes to effect competitive advantage from which a flow of short- and long-term benefits for the entity (such as profits for the private sector) are generated. Key current practices to be maintained include understanding and forecasting market needs, understanding resource availability, investing capital responsibly and managing overall employment and growth. Sustainable development requires the organization to modify accounting practices to include the full or true costs of any activity. In addition, sustainable development creates an organizational shift to include societal considerations in the full cost of this activity. Organizations practising these economic health considerations should be rewarded with stability and long-term stakeholder welfare.

21.4.3 Social responsibility

Achieving social responsibility requires the acknowledgement of the importance of the direct and indirect effect of an organization's actions on society. For example, an environmental concern that affects an organization's economic health, which in turn leads to a reduction in the labour force, would have profound effects (at least locally) on society and could result in further environmental damage. The interconnectedness of societal values and needs to environmental protection and economic health is well established. From a sustainable development perspective, societal considerations include demographic impacts, economic impacts, community impacts, social impacts, sociocultural impacts/way-of-life considerations, family impacts and others. A more complete description of the types of societal impacts to include was developed at a SETAC workshop on lifecycle impact assessment (Fava *et al.* 1993b). A list of these is included in Table 21.1.

Table 21.1 Social welfare impact categories. (After Fava *et al.* 1993b.)

• Demographic impacts	• Community impacts
• Economic impacts	• Family impacts
• Fiscal impacts	• Sociocultural impacts
• Sociopolitical impacts	• Psychosocial impacts
• Social impacts	

Perhaps the most difficult stage in integrating environmental protection, economic health and social responsibility will be the valuation step, which attempts to combine all factors into a decision framework. Here, organizations should be guided by valuation techniques and should consider the appropriate involvement of all affected stakeholders. Valuation techniques (see Fava *et al.* (1993b) for a description of valuation techniques in lifecycle studies) and specifically decision tools are an area of immediate interest to organizations embracing sustainable development and will be examined later in this chapter.

21.5 IMPLEMENTATION STRATEGIES AND TOOLS

In order to effectively achieve sustainable development, an organization will probably have to

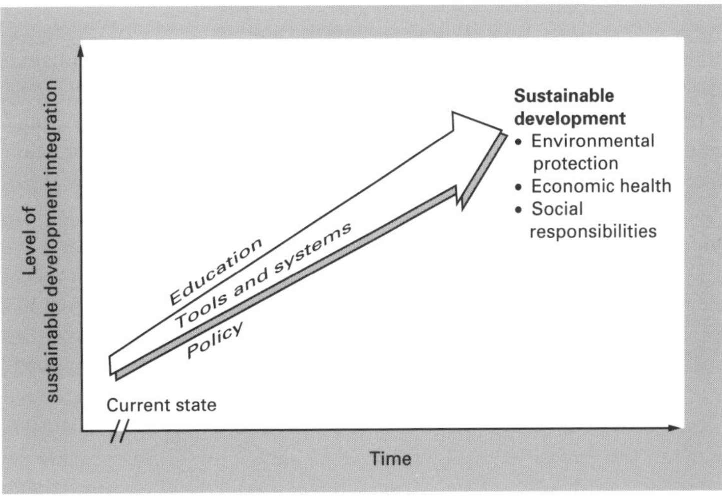

Fig. 21.2 Implementing sustainable development.

consider a host of strategies and develop/deploy an array of tools to achieve results. Figure 21.2 illustrates the transformation from current state to the future sustainable development vision, demonstrating the deployment of specific strategies and the utilization of tools and systems. Note that as an organization starts to implement parts of the vision of sustainable development, the number of tools and systems deployed increases.

Two strategies available to organizations to deploy sustainable development include education and policy.

21.5.1 Educational strategies

In the area of education, an organization first should decide who is the target of the educational programme. Possible focus areas include constituents/citizens, consumers/customers, employees, teachers, children, mass media, thought leaders and generally any practitioner of sustainable development (i.e. in a private setting, management, designers, engineers, finance, manufacturing, procurement, human resources, etc.). Once the target audience is identified, the message may have to be tailored to the individual audience. The educational message should build on previously successful initiatives. For example, consider positioning sustainable development as a natural extension of the current focus on recycling. Further suggestions include developing training programmes for the teachers, utilizing practical case examples and building sustainable development education into school curricula at all levels as well as furthering the advancement of sustainable development through professional conferences. Whatever the path for the educational programme, one should recognize the importance of this dimension in achieving sustainable development and provide the details in a manageable and digestible fashion.

21.5.2 Policies

Like education, specific development and deployment of policies are an effective way for an organization to implement sustainable development. Within the broad area of sustainable development, a number of organizations have already suggested policies and principles to utilize during the transformation to a sustainable development vision. Many of these programmes suggest that organizations consider the following.

1 Development of a statement of basic philosophy.

2 Development of guidelines for action to possibly include: general management policies, corporate organization, a statement of concern for the environment, commitment to technology development, commitment to technology transfer, emergency measures, public relations and education, community relations, positions on overseas operations, contribution to public policies and response to global issues.

3 Organizations should tailor the above suggestions to best fit their needs. It should be recognized that there likely will not be a 'one size fits all' set of policies to deploy sustainable development and, in fact, organizations should be encouraged to be creative in this area and thus continue to advance society as a whole.

4 Once the policies are established, a host of methods are available to any organization to implement them. These methods range from prescriptive to voluntary and include various incentives to possible use of disincentives. Once again, the choice of implementation options is dependent on the specific organization.

Finally, in order to deploy sustainable development, an organization will need an array of current and emerging tools. These tools include environmental management system programmes such as risk assessment, design for the environment (DfE), total quality environmental management (TQEM), product stewardship and pollution prevention. Additional environmental methodologies include lifecycle assessment (LCA), integrated pollution prevention and control (IPPC), industrial ecology, environmental auditing, environmental impact assessment (EIA) and site assessments. A tool necessary for the economic health dimension of sustainable development is full cost accounting (FCA). Within social responsibility, effective reward and recognition systems must be developed and employed to effect and reinforce the behaviour of the organization. Many of these methods and tools are described more fully in previous chapters. Sustainable development practitioners should not let the absence of specific tools or the state of development of a specific tool discourage their implementation of sustainable development. As with all new areas, organizations must utilize the best available technology and continue to help advance the field.

21.6 CURRENT STATUS—INTEGRATION OF SUSTAINABLE DEVELOPMENT

This section discusses how well sustainable development principles and policies are integrated into current approaches to product support, product and process design and redesign and development of governmental and academic sponsored or supported products and processes. At this time, products and processes are considered to be meeting the regulations (environmental, safety, human health and social) governing them; if regulations do not apply, then we can assume that no or limited stimulus exists for product or process redesign, unless directly requested from a customer. In some cases, integration of sustainable development is limited to a 'gate-to-gate' perspective, with product and process design based on traditional considerations of function and cost, without direct (and, in many cases, indirect) consideration or integration of environmental protection. Certainly other aspects of sustainable development, economic health and social responsibility are not addressed in any manner in the current state of affairs. Table 21.2 outlines the current driving issues for main positions responsible for decision-making within a manufacturing company. The majority of positions are driven primarily by economic health, with fewer driven by environmental protection. Social responsibility drives only two of these functions.

Some of the more progressive companies, agencies and academic departments have begun to integrate single-issue sustainable development concepts and tools such as product stewardship (e.g. chlorofluorocarbon reduction and handling of solid waste) into their product and process development schemes. For example, Rohm and Haas has applied a Health and Environmental Regulatory Index to prioritize their product portfolios (Reinert *et al.* 1996) in several business units. Such integration is primarily due to 'market' pressure such as federal procurement policy, ecolabelling and supplier selection based on environmental criteria. Many of these chemical companies have signed on to Responsible Care® programmes, but are still struggling with how to integrate and implement such far-reaching policies and 'way of life' into everyday thought processes and activities. Some progress has been seen towards a fuller integration of environmental issue consideration into the design of products, processes, policies, etc.

Table 21.2 Decision issues for various functions in a multiproduct manufacturing company—current status.

	Issue		
Function	Social responsibility	Economic health	Environmental protection
Marketing/sales		X	
R&D		X	X
Manufacturing		X	X
Compliance		X	X
Government affairs			X
Finance		X	
Procurement		X	
Human resource	X	X	
Legal		X	X
Public relations	X	X	X
Business management		X	X

Coordination among various groups interested in fostering and integrating sustainable development into everyday decision-making has generally been lacking or limited in application.

From the above discussion, the current state of integration of sustainable development and other high-level organizing policies into product, process, policy, etc., development and redesign is not developed. We see evidence that environmental protection aspects are under integration in some situations; however, these efforts are isolated, uncoordinated, potentially overlapping and potentially misdirected. Integration, both directly or indirectly, of economic health and social responsibility aspects into this process is currently essentially non-existent.

21.7 SUSTAINABLE PRODUCT DEVELOPMENT IS THE NEXT LOGICAL STEP

The chemical industry, through its Chemical Manufacturers' Association (CMA), established its programme that laid out a series of principles and management practices to manage the risk of chemicals in the environment. One of the six management codes is product stewardship, which identifies practices to be performed by a company to develop and implement proactive product stewardship programmes.

According to the CMA, the purpose of the product stewardship code of management practice is to make safety, health and environmental protection an integral part of designing, manufacturing, distributing, using, recycling and disposing of products. The code provides guidance as well as a means to measure continuous improvement of the practice of product stewardship.

When examined, this product stewardship code was assumed to be adequate for the sustainable product strategy. However, when compared with the three components in the sustainable development framework, it was found to be incomplete because it did not address social responsibility and nor did it adequately address the full cost accounting aspects of the economic health component. As such, the term 'sustainable product strategy' more accurately reflected the intent to fully integrate social responsibilities, economic health and environmental protection into product design, raw materials acquisition, manufacture, use and disposal.

Sustainable product development is available to any organization that provides a set of products or services, whether it be a public entity such as a state government or national laboratory, or a private, profit-seeking firm. The organization can address real and perceived problems related both to ecological issues, such as pollution, energy consumption and natural resource depletion, and to societal concerns, such as the social cohesion and well-being of the community in which it is located.

Sustainable development policies can be applied early on in the strategic and business planning processes. This strategy recognizes the interrelationships among the ecosystem and society at large with individuals and organizations who manufacture, buy, sell and use products, processes, services and technologies. The actions and activities associated with each stage of the product's lifecycle (i.e. materials acquisition, manufacture, use and disposal) consume energy and materials and release emissions to the air, water and land. These burdens have an effect on ecological systems, human health, resource use, social welfare and economic health. Thus, focusing on a product-oriented strategy allows the organization to fully understand and integrate sustainable development policies into product design, raw materials acquisition, manufacturing and intended use and disposition.

Lastly, it serves as a framework whereby organizations can have more complete control in assuming customer or client satisfaction. The organization can deepen its external relationship beyond its clients (constituents or citizens, in the case of a government), to include considerations of its natural and social environment within an integrated and holistic framework.

21.8 CURRENT PRACTICES VERSUS SUSTAINABLE DEVELOPMENT

One distinction that sheds light on the difference between the current, traditional practices and the new sustainable development way of thinking concerns the way in which it addresses the needs and problems presented by its clients.

Currently, organizations generally serve their clients by addressing single, separate issues in a disjointed fashion. Clients' needs are met sequentially, item by item, and solutions are sought from a narrow perspective with little regard to ecological and social implications. Often, environmental and social issues are addressed well after large investments in technology, organizational change or political capital have been made. The organization that practises sustainable development identifies and brings a wider set of stakeholders into the decision-making process, serving clients' needs and advancing the organization, while directly addressing the ecological and social implications of its actions at the outset. Furthermore, members of this enlarged set of stakeholders would have a broader mix of skills with which to approach problems in a more multidisciplinary fashion.

Several hypothetical cases are presented to illustrate these ideas (Boxes 21.1–21.3).

Box 21.1 Sustainable product development (SPD)—industry case study.

A multinational company with offices in the United States and Sweden producing numerous chemical- and paper-based products has decided to integrate sustainable development into the new and redesigned process and product development schemes. Concurrent with this high-level policy deployment, an urgent need to redesign a product has just been received by the business director for one of the company's eight divisions. A major customer of the division requires that a biodegradable component be developed to replace a more persistent component that eventually degrades but is not considered readily biodegradable under current testing schemes.

Under the old, presustainable product development process, the business director calls a meeting with the marketing director (who received the original request) and the research and development director for new or redesigned products (or the manufacturing director if an optimized product or process is required only). The company specifically addresses customer dimensions and cost (see table); other issues such as technical fit of the request into current business, economic health, social responsibility and health and safety issues for the new product are interpreted and eventually addressed by the 'team'.

Issues specifically addressed during product redesign.

Issue	Pre-SPD	Using SPD
Customer dimensions	X	X
Cost	X	X
Technical fit		X
Economic health		X
Environment protection		X
Social responsibility		X
Health and safety		X

Such an approach does not include all parties needed to provide integrated, sustainability-based decisions. This pre-SPD approach may have many total quality management (TQM) attributes; however, this does not imply that all of the aspects of SPD have been covered. Under the pre-SPD scenario, a series of options would result that would then need to be discussed with the manufacturing director and other personnel involved with economic health, social responsibility and environmental protection issues at a later date.

Under the SPD, the business director would need to utilize a multidisciplinary, expanded skill team consisting of the business director, marketing director, research and development director, manufacturing director, financial manager (for full cost accounting) and procurement officer. Additionally, environmental protection and social responsibility personnel would attend so that the three components of sustainable development are represented. This team approach is consistent with TQM and will directly address the issues during the product redesign featured in the table.

During the SPD phase, a single, well-thought-out and supportable option would be developed and implemented to solve this business opportunity. From this exercise, full cost accounting and an appropriate rewards and recognition programme would be identified as obstacles to the full implementation of SPD.

Box 21.2 Sustainable product development (SPD)—federal agency case study.

Consistent with the industrial case study, a non-industrial (e.g. government) case study is also described. The following illustrates how this process could work in this situation using the USEPA as the example governmental agency. Pre-SPD was initiated at the Assistant Administrator (AA) for the Office of Solid Waste (OSW) based on an executive order from the White House. The executive order involved redefining policy concerned with the management of non-hazardous solid waste. The AA invited the appropriate division head (decided by the AA) to discuss how to develop and implement the programme. Several options would be proposed that would then go to the branch chief responsible for non-hazardous solid waste (who owns the technical fit aspect) who, with respective section heads, would decide which of the options seem viable and could be developed within the allotted time frame and cost. Economic health, social responsibility and environmental protection issues are usually addressed after the options have been collected and assessed for their applicability to the programme under development.

Under SPD, the AA calls each of the division heads from the respective offices to the meeting. Other AAs from other USEPA offices and Executive Branches (e.g. Office of Management and Budget) could be asked to attend the meeting so that there is a consistency among the options discussed and holistic participation in the selection of the primary pathway for dealing with the programme development. This solution involves a multidisciplinary team with an expanded skill base to identify the best, most appropriate, solution to the problem posed by the Executive Branch. The option would include the appropriate customer dimensions (i.e. provide the desired result), it would place the solution in the appropriate branch or section and would directly assess the environmental protection, economic health and social responsibility of the solution.

Box 21.3 Sustainable product development (SPD)—municipal government case study.

Consider a municipal government. Its clients are inhabitants of the city and it is charged with meeting citizens' needs by offering a wide range of public goods, such as providing social services, maintaining the city's physical infrastructure and making citizens as safe as possible from fire and crime, to name but a few. Suppose the mayor of a city must respond to new waste-water treatment needs. Currently, the mayor might simply contact the city sanitation engineers and charge them with procuring a new sewerage system subject to a predetermined budget constraint. If any consideration needs to be made concerning ecological impact, the administration would be doing so only in compliance with existing regulations and laws concerning public health and the environment. Indeed, the mayor would most likely be responding to increased waste water demand under such laws. A narrow problem has presented itself and, from an equally narrow perspective, it is addressed. What if the new capacity causes unpleasant odours in a certain neighbourhood? What if the waste-water treatment plant's emergency contingency plans include options that are considered ecologically or politically unacceptable? Will increasing capacity encourage unwanted growth in the city? What should be the fee structure for waste-water services? For how long is it expected to meet the city's needs? It is likely that these and other questions will be somehow resolved, but usually at a point in time late in the process and thus likely to be at great expense in time and resources.

In the city whose government fully subscribes to sustainable development, the city mayor would convene with the budget director, the environmental health and safety director, the legal counsel, local citizens' groups and others, *in addition* to the chief sanitation engineer, and then in concert with these interests, the mayor would formulate a response to new waste-water treatment needs under an integrated strategy that addresses environmental and social issues. By relying on this multidisciplinary and multiparty team, the administration

Continued

Box 21.3 *Contd.*

resolves the issue of increased waste-water treatment demand and has the potential to proactively address any other related issues before they metamorphose into real problems that ultimately, for the administration, become political. Questions such as those concerning human health, ecological risk or population growth are addressed at the outset, before the first hole is dug, and not after the new facility goes on-line. Under such circumstances, citizens' needs are addressed in a timely fashion, and trust in the effectiveness and sensitivity of the city government is reinforced.

21.9 PROPOSED FRAMEWORK

A framework to implement sustainable product development (or design for sustainable development—DFSD) would carry out the organization's sustainable development policy and make use of available tools to develop the best option for action (Fig. 21.3).

The DFSD framework will consist of a variety of policies that describe the sustainable product development concept and a set of tools to implement them. Typically, during the design of a new product (generically speaking, this may include a public policy response), several stages are followed.

1 Objectives: does the product fulfil the client's requirements?

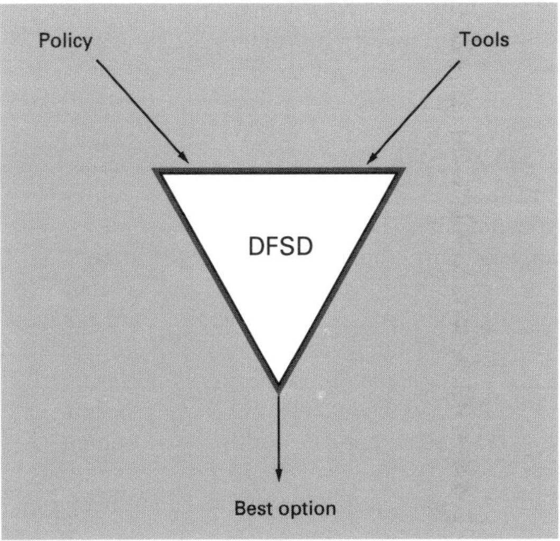

Fig. 21.3 Design for sustainable development (DFSD).

2 Internal capabilities: does the product fit with the organization's core competencies?

3 Other considerations, such as implementation, etc.

This process is followed iteratively until the final design is obtained. In the case of a DFSD framework, the three aspects concerning economic viability, environmental stability and societal issues are considered at every stage. A valuation of final options takes place and the final option chosen depends on an interactive process that relaxes or tightens the constraints of any of the three categories.

21.9.1 Policies

As part of the effort to implement design for sustainable development, key criteria will have to be developed for each of the major categories of environmental protection. While there are many major environmental protection areas, six major ones are discussed: energy and transboundary pollution, preservation of biodiversity, watershed management, environmental restoration, product stewardship and urban environment.

To guide the product or technology designers, a matrix has been established that lists examples of key environmental issues for one of the major topic areas (Table 21.3). These issues represent only those that should be considered in the design of a product or technology from raw material acquisition, manufacture, use and disposition. The matrix does not attempt to determine how and whether the issue can be resolved.

One of the policies for a design for a sustainable development designer might be to optimize the use of materials and energy and to minimize their loss to the system. For example, a sustainable

Table 21.3 Design for sustainable development environmental protection issues (examples).*

Environmental topic	Lifecycle stages Raw materials acquisition (energy, water, materials)	Manufacture	Use	Disposition
Biodiversity†				
• Forest products	• Habitat destruction[1] • Monoculture plantings[2]	• Sludge and water toxics[4]	–	• Landfill (habitat)[5] • Incinerators (air toxics)[3]
• Agriculture	• Habitat destruction[1] • Sedimentation[1]	• Surface water discharges[4]	• Surface water discharges[4] • Groundwater discharges[5]	• Landfill[5] • Incinerator[3]
• Energy	• Habitat destruction (coal)[1] • Surface water (oil)[4]	• Air pollutants[3]	• Air pollutants[3] • Surface water pollutants[4]	• Land disposal[5]
• Heavy industry	• Habitat destruction (mining)[1] • Surface water pollution[4]	• Air pollutants[3] • Surface water	–	• Land disposal[5] • Incinerators (Hg batteries)[3]

* The material in this table was developed with input from participants during the Sustainable Environmental Management Workshop in 1993.
† Impacts on biodiversity by ranked order: [1] habitat destruction; [2] introduction of exotics; [3] air pollutants; [4] surface water pollutants; [5] ground water pollutants.

product strategy could be to ensure the collection of the material or product after it has expended its useful lifespan. Once the material or product has been collected, options for reuse, material recovery and energy recovery exist. The specific option would be influenced by many factors, including the material itself, the ability of the raw material to be managed in a sustainable fashion and its inherent heat value.

Figure 21.4 presents a schematic of a generic system that would consider all elements. One might think about the generic system by comparing system design strategies from several industrial sectors. In the aluminium industry, for example, the importance of the recovery of used aluminium in beverage containers for use as feedstock to replace virgin mined bauxite was recognized many years ago. This resulted in an

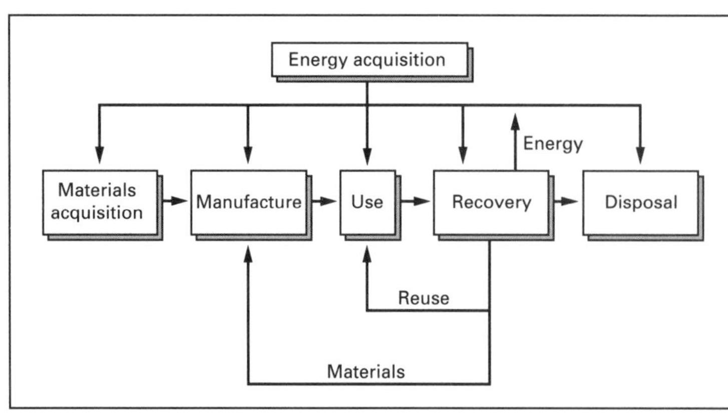

Fig. 21.4 Generic system illustrating inter-relationships among energy, material and product systems.

energy saving of nearly 90% when compared to only a linear one-time use of the aluminium.

Assuming that the high recycling rate for aluminium would be appropriate to other materials may not be valid. For example, the overall impacts from paper recycling actually increased after the recycling rate had reached a certain level (see diagram in Box 21.4). Thus, it may not be valid to assume in all cases that the best environmental solution is to continue to increase recycling rates. Just because it makes sense with aluminium does not mean that it makes the same degree of sense

Box 21.4 The EPS system—an example of a decision tool for valuation of environmental impacts.

The EPS (environmental priority strategies in product design) is an LCA-based system for assessing the total environmental impact of products. The main idea of the EPS system is to make environmental loads and environmental impacts 'visible' through a transparent echo-calculation procedure to provide a holistic approach offering a synthesis and integration of environmental concerns. It provides an opportunity to enumerate and assess the ecological and health consequences of various human activities, such as use of energy and raw materials and pollutant emissions, before aggregation and final evaluation. Basic assessment of values of environmental qualities, as well as changes in these values from human activities, are both estimated.

The EPS system has been specially designed to fulfil some basic sustainability-based criteria, i.e.:

- A holistic environmental impact assessment approach for valuation and weighing the environmental effects caused by use of natural resources, as well as those caused by pollutant emissions.
- A recycling concept taking into consideration a quantitative and qualitative environmental understanding and taking account of various flows of virgin and reused materials.
- A sensitivity and error analysis approach to elucidate the effects of uncertainties in the underlying assumptions and input data for decreasing the risk of making false conclusions on the relative harmfulness of alternatives in product development.

The valuation approach is an attempt to use five 'safeguard subjects'—biodiversity, human health, production, resource use and aesthetic values—in order to get a simplified and practical structure as a basis to describe, value and weigh all impact types in the environment. The safeguard subjects represent one way to characterize the sustainability of the environment in terms of both 'environmental capital' (biodiversity, resource use and aesthetic values) and 'environmental capacity' (human health and production). The principle of the valuation approach focuses on the willingness of individuals and societies to allocate a certain quantity of their 'working capacity' or financial assets to influence the state of the environment for avoiding undesired ecological effects.

An example of the type of decision guide that the EPS system can offer is illustrated in the diagram below for elucidating the newspaper recycling percentage that has the lowest total impact on the environment (Ryding *et al.* 1993).

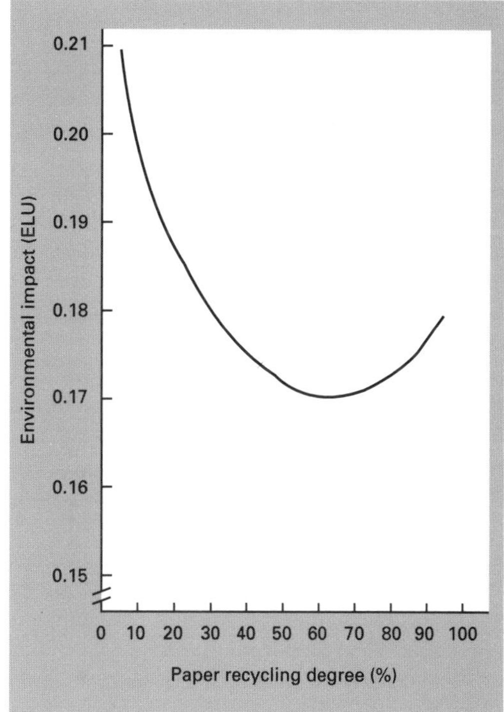

Relationship between paper recycling and reduced environmental impact.

with other materials. This does not indicate that recovery is not important, but that the extent of implementation may differ. The value of the more holistic approach and understanding of the full environmental burdens will ensure that unexpected environmental impacts do not result, due to a design change, to solve a single environmental problem.

21.9.2 Tools

Throughout this book, there are chapters describing many of the tools relevant to implementing a sustainable product development framework. These include tools such as full cost accounting, risk assessment, environmental impact assessment, environmental auditing and pollution prevention. However, there are additional tools that have immediate application to sustainable product development. These additional tools include lifecycle assessment, design for the environment (DfE), Responsible Care® and environmental labelling. Each of these are described below.

Lifecycle assessment

Lifecycle assessment (LCA) is a process to evaluate the environmental burdens associated with a process, product or activity by identifying and quantifying the energy and materials used and the waste released to the environment, to assess the impact of those energy and material uses and releases to the environment and to evaluate and implement opportunities to affect environmental improvements. The assessment includes the entire lifecycle of a product, process or activity, encompassing: extracting and processing raw materials; manufacturing, transportation and distribution; use, reuse and maintenance; and recycling and disposal (Fava *et al.* 1991).

At present, most LCA studies that are carried out focus on inventory analysis. Hence, they do not include impact assessments in the exercise. Some LCA studies carried out in recent years have used various approaches to classification, characterization and valuation. One such approach—the EPS (environmental priority strategies in product design)—is featured in Box 21.4.

Design for the environment

Design for the environment (DfE) is a systems-oriented approach for designing more ecologically and economically sustainable product systems. It couples the product development cycle used in business with the physical lifecycle of a product (USEPA 1994; Veroutis & Aelion 1996; Veroutis & Fava 1996).

Design for the environment integrates environmental requirements into the earliest stages of design so that total impacts caused by product systems can be reduced. In the lifecycle design, environmental, performance, cost, cultural and legal requirements are balanced. Concepts such as concurrent design, total quality management, cross-disciplinary teams and multi-attribute decision-making are essential elements of lifecycle design. Publications by Graedel *et al.* (1993) and Graedel & Allenby (1995) provide excellent discussions on the use of DfE.

At present, DfE programmes tend to allocate too large a part of the activities on schemes for recycling. Thus, in the future more efforts must be devoted to expanding the work to encompass all components originally highlighted as parts of the concept.

Responsible Care®

Responsible Care® (RC) is a chemical industry initiative attaining world-wide acceptance. Responsible Care® is a registered trademark of the CMA. It is designed not only to focus on safe chemical handling but also to improve the performance in the fields of environment, health, safety, product safety, distribution, emergency response and relations with the public. It also enables companies to demonstrate that these improvements are actually taking place.

The main thrust of the Responsible Care® programme is to achieve real improvement in the environmental, health and safety performance of a company and to clarify who has responsibility for these areas in a given situation. However, Responsible Care® is about actual performance.

Challenges and priorities in the future include the following.

1 Development of environmental health and

safety management practices and systems that ensure continuity through changes in personnel.

2 Ensuring that all staff—engineers, technicians, supervisors and managers—have the required knowledge and skill.

3 Encouragement of more open communications with both inside and outside companies to increase awareness, commitment and the generation of ideas for improvement.

Environmental labelling

Environmental labelling (EL) is a means of positive consumer information. By using symbols, slogans or other messages, companies can indicate to their customers that a product has certain advantageous properties from an environmental point of view. Environmental labelling of products can be an effective tool to protect the environment if it guides consumers to purchase products that are 'friendly' to the environment and stimulates industry to introduce environmentally sound product development.

As existing eco-labelling schemes are usually based on single-criterion approaches and, hence, do not take into account all environmental aspects of a product, one solution to the problem might be to introduce the setting of criteria based on a multi-criteria approach (and LCA) giving information to a variety of potential audiences about the environmental impacts through the different phases of a product's lifecycle from raw material acquisition to waste disposal.

21.10 MAKING THE TRANSITION: TEN STEPS TO SUSTAINABILITY

Given the sustainable development framework described above, a transition to the future sustainable development state has been developed. The purpose of this section is to describe that transition and to discuss implementation issues.

While there are many possible approaches to implement sustainable development, the following is intended to be representative of what an organization might follow. It is critical to understand that this process is intended to be applied to any organization (e.g. government, universities, consulting firms and industries) that designs, manufactures, uses and disposes of products, processes and technologies.

Generally, the process might involve:

1 Appointing a sustainable development champion.
2 Establishing a sustainable development implementation team.
3 Defining sustainable development within the context of the organization's product lines.
4 Assessing the stakeholder's needs and desires.
5 Assessing the organization's operations.
6 Reviewing the organization's activities in the context of a sustainable development policy.
7 Delineating key issues and decisions.
8 Identifying and evaluating options.
9 Implementing programme.
10 Monitoring, feedback and continual improvement.

These are briefly detailed here and shown in Fig. 21.5.

21.10.1 Step 1—appoint a sustainable development champion

Once an organization has decided to become a sustainable development organization, the first action is the establishment of a sustainable development champion. This champion should be someone who has the authority, visibility, commitment and accountability to establish sustainable development in the organization. As such, communication from the most senior position in the organization of the role, responsibilities and authorities is critical.

21.10.2 Step 2—establish a sustainable development implementation team

The second step is to establish a sustainable development implementation team that will have the commitment from senior management to develop and implement the plan. Given the all-encompassing elements of sustainable development, it is important to include representatives in the plan from relevant functions. The functions in the organization include management, research and development, marketing and sales, human resources and finance, to name but a few. This sustainable development team would then be charged with the responsibility of developing

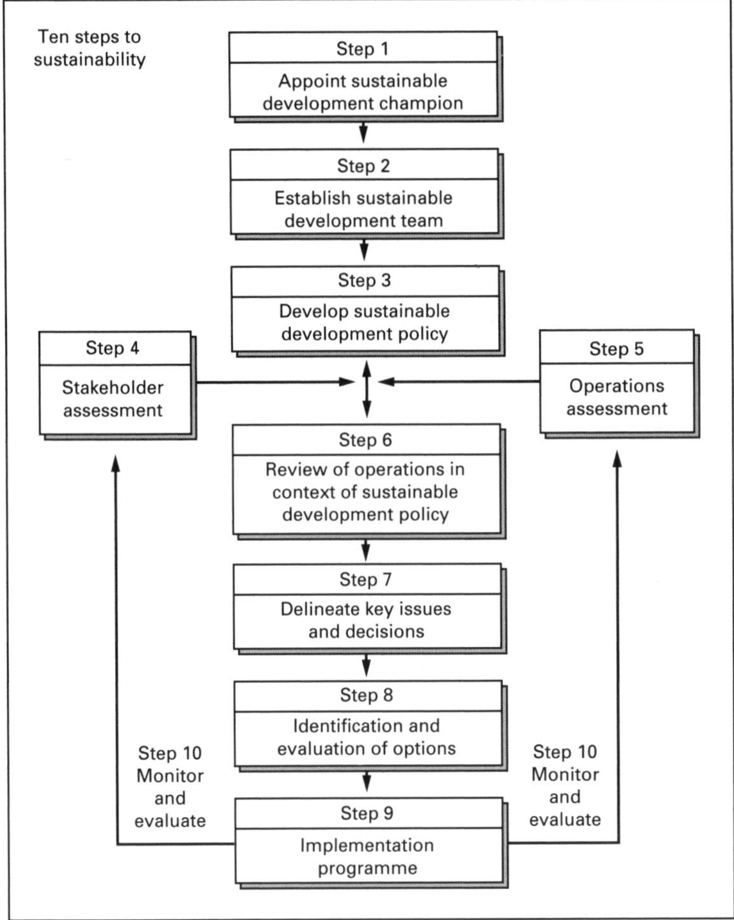

Fig. 21.5 Making the transition: 10 steps to sustainability.

the sustainable development implementation plan.

21.10.3 Step 3—define sustainable development within the context of the operation's business

The next step is to define sustainable development in the context of the operations of the organization. For example, an organization might determine that sustainable development would mean integration of economic health, environmental protection and social responsibility to achieve long-term survivability of the organization by:

1 ensuring that the correct information and facts are available to the decision makers so that the right decisions can be made, while taking into account the directions of regulations and/or actions by competitors; and

2 stakeholder communication and education to influence policy, regulations and public opinions.

This might include the development of a mission statement and the establishment of values, principles, criteria and goals related to profitability (long and short term), environmental protection and social responsibility.

Examples of potential world-wide sustainable development policy might be associated with: (i) full cost accounting; (ii) public responsibility and acceptance; and (iii) environmental protection, including material and energy management.

Goals with each sustainable development policy could also be developed; for example under the material and energy management element of environmental protection, a goal might be to ensure long-term availability of the primary material resource(s).

An outcome of Step 3 will be a sustainability development policy.

One of the overlying operating premises is the incorporation of environmental protection, economic health and social responsibility into the organization's operations (including design, raw materials acquisition, manufacture, use and disposition).

21.10.4 Step 4 — stakeholder assessment

The purpose of this step will be to identify critical stakeholder needs and desires. Possible stakeholders include employees, board of directors, customers, regulatory agencies, general public (citizens), suppliers, local communities, and shareholders and investors.

21.10.5 Step 5 — assess the organization's operations

The purpose of this step is to characterize current and potential operational activities. Specific actions might be to identify and define key operations or functions necessary to run the operation. These should not necessarily be traditional organizational functions but rather functions that are fundamental to the long-term success of the organization. Examples of such operations or functions might be as follows.

1 Organization management, planning and strategic planning.
2 Product and process research and development.
3 Purchasing, including materials and energy acquisition and management.
4 Manufacturing processes.
5 Communication and education.
6 Human resources.
7 Finance.

21.10.6 Step 6 — review organization's activities in the context of a sustainable development policy

The next step is to critically examine the organization's existing vision, goals, decision-making and criteria against the sustainable development policy and the stakeholder's needs. The ultimate output of this step will be an initial list of ideas and thoughts to begin to direct and manage the process of change from the current organization's strategies to the actions necessary to meet the sustainable development policy.

21.10.7 Step 7 — delineate key issues and decisions

For each activity of the organization, identify and define the key issues faced by, and the major decisions made by, that activity. For example, within the planning operation, decisions might be to site a facility or invest in a new technology. Within R&D, decisions might be to design a new product; and within planning, decisions might be related to expanding or eliminating products or product lines.

21.10.8 Step 8 — identify and prioritize opportunities for improvement

There are many approaches that could be taken; we suggest one way that interprets the above steps into a sustainable development strategic matrix. For example, a matrix might look like that in Table 21.4. This provides a systematic way of gathering and integrating issues and values for each of the three sustainable development components into a framework for resolution. Once these steps have been accomplished, a systemic way of understanding the sustainable development policy, key issues and decisions will have been formulated.

During this step, common themes, gaps, missing systems, tools, data and information needs and organizational deficiencies will be identified in order to identify opportunities for improvement.

Examples of opportunities for improvement might be to modify the current account system into a full cost accounting system on a product (or product line) basis. This would mean initially to capture the current cost associated with design, manufacture, marketing and selling of the product on a product basis. Additional environmental costs (such as permitting, disposal and costs to the ecosystem) can be added in an incremental fashion

Table 21.4 Sustainable development strategic matrix.

Sustainable development component (SDC)/ business activities	Key issues/ decisions	Economic health		Environmental protection			Social responsibilities	
		Profits (short term)	Profits (long term)	Environment	Materials	Energy	Public impact/ quality of life	Others
Business, strategic and product planning	Acquisition New product design	Xi						
Technology								
Resources acquisition and management								
Engineering	Process modification							
Manufacturing								
Communication and education (Internal and External)	Stakeholder (interested parties) values							
Corporate citizenship								
Hiring and training								
Business management								
Environmental compliance								
Accounting								
Sales/marketing								
Others								

For each cell (Xi-..), the criteria necessary to make the decision would be identified, analysed and agreed as relevant to the identified SDC and goals.
Identify the information systems necessary to provide the data and information for each decision based upon the criteria. Tools or systems relevant are: lifecycle assessment; process safety; full cost accounting; pollution prevention; risk characterization; and occupational health and safety.

once the science and economics have evolved to the point of acceptance by the stakeholder.

Another example of an opportunity to move towards the future state is the upgrading of the rewards and recognition systems to include measurement parameters related to full cost accounting, environmental protection and social responsibility. Each organization will have to determine which specific measurement parameters work best for them, based upon their existing system. A third example is tied to the education strategy. An organization will gain additional information about its products from a 'cradle to grave' perspective. With this information and insight, opportunities to interact with the stakeholder (e.g. government, community groups, non-governmental organizations) will occur to ensure that future policy and directions are based upon a holistic understanding of the full environmental, economic and social impacts associated with the proposed policy. This education function then becomes an important external link for the organization. Its future role will not only be to communicate and educate internally, but to develop processes to ensure proactive external communication and education as well.

21.10.9 Step 9—implementation plan

Based upon identified opportunities, a change management plan including policy, practices and processes would be developed.

21.10.10 Step 10—monitoring, feedback and continuous improvement

This step involves the establishment of monitoring, feedback and follow-up actions to correct or improve the process and the resulting actions associated with the DFSD. This is coupled with the understanding that total quality management will be a foundation for this continuous improvement process.

21.11 IMPLEMENTATION NEEDS

To facilitate the implementation of the sustainable development programme, several implementation needs were identified. The purpose of this section is to identify and discuss the key needs relevant to implementation of a sustainable product strategy into an organization.

21.11.1 Enhancement to management systems

Several important considerations should be incorporated into management systems: stakeholder reviews and assessments, enhancement of the reward and recognition systems to incorporate sustainable development measurement parameters, and incorporation of environmental impacts into the accounting systems.

The stakeholder assessments are an expansion of activities often already underway in organizations in that a more proactive, voluntary and open process would be established outside of the traditional regulatory process. The advantage of such an expansion is the greater involvement and cooperation in the decisions and actions that affect the stakeholder so that the overall costs and time to completion are reduced.

21.11.2 Tools and systems

Tools exist to some degree to begin to implement sustainable development. However, to move towards the future state, action must be taken to enhance their application. For example, LCA methods should be furthered by development and validation of impact assessment methods, recognizing at least the environmental protection topics discussed within this book. Similarly, DfE should be expanded to address DFSD which, by its definition, will include environment protection, economic health and social welfare.

One of the most important tools to enhance is full cost accounting. Other chapters in this book provide additional information on full cost accounting. This methodology will provide an important unifying activity to integrate environmental impacts with economic parameters on a product or product line basis. Examples of initial opportunities for improvement might be to modify the current account system into a full cost accounting sytem on a product (or product line) basis. This initially would mean capturing the current cost associated with design, manufacture, marketing and selling of the product on a product basis. Additional environmental costs (such as permitting, disposal and costs to the ecosystem) can be added in an incremental fashion once the science and economics have evolved to the point of acceptance by the stakeholder.

21.11.3 Education

Successful implementation of sustainable development requires a comprehensive education strategy in both public and private sectors. Efforts will have to be taken to increase the understanding of, the actions to be taken and examples of success of sustainable development through a variety of mechanisms: teach the teachers; influence the thought leaders; and establish forums to educate the mass media.

Additionally, organizations will gain additional information about their products from a 'cradle to grave' perspective. With this information and insight, opportunities to interact with the stakeholder (e.g. government, community groups, NGOs) will occur to ensure that future policy and directions are based upon a holistic understanding of the full environmental, economic and social impacts associated with the proposed policy. This education function then becomes an important external link for the organization. Its future role will not only be to communicate and educate internally, but to develop processes to ensure proactive external communication and education as well.

21.11.4 Data availability

As one proceeds in implementing sustainable development, data gaps will become apparent. Several important principles should be considered in the development of efforts to gather and make data available.

1 Transparency of data collection methods and data (while maintaining confidentiality).

2 Development of minimum standards for data quality.

3 Development of data quality objectives and indicators.

4 General guidelines on data formats and database management systems.

A conceptual framework for lifecycle assessment data quality provides additional principles and guidance (Fava *et al.* 1994). The existence of, and accessibility of, data for a broad number of stakeholders may be one of the biggest obstacles to sustainable development. Efforts should be taken by multi-sector groups to establish a process and guidelines on how best to accomplish the data quality requirements in an effective and efficient fashion.

21.11.5 Social responsibility

The social welfare aspects of sustainable development require considerable development. Social criteria are viewed as being fundamental to decision makers.

However, the development and integration of social welfare measurement parameters that can be used in conjunction with environmental protection and economic health parameters is needed.

21.11.6 Valuation

Ultimately, a decision will be made in any situation, whether it is a simple or complex scenario. Often the decision is made based upon the values of the final decision maker. Thus, the development of accepted valuation techniques will be critical to ensure better decisions and their acceptance by the stakeholder.

21.12 CONCLUSIONS

Organizations that have integrated sustainable development policies into all aspects of their operations will become pre-eminent among their peers and contemporaries. A proactive sustainable development programme provides a system to effectively and efficiently implement actions in order to continually improve the environmental quality, economic health and social welfare of individuals and society, resulting in the long-term sustained success of organizations and society.

There are a variety of tools needed to implement a sustainable development strategy. Many of those have been discussed in this chapter and throughout this book. The purpose of this chapter was not to describe any one tool in detail, but to establish a recommended framework within which the tools can be applied. Tools such as risk assessment and lifecycle assessment will provide valuable information concerning local risks to the environment as well as the global potential impacts from a product system from a 'cradle to grave' perspective. Neither tool provides all the answers. Lifecycle assessment provides the designer with information on global impact categories such as climate change and resource consumption. Risk assessment provides the risks associated with the manufacture and use of a chemical. Often risk assessments are performed on a more local or regional basis, while lifecycle assessments are performed on a more global perspective.

While there are no secrets to success, several features generally surface.

1 Start immediately; do not wait.
2 Recognize a champion.
3 Commit resources.
4 Personalized commitment.
5 Continuous improvement; always make progress.
6 Regular and open communication.

The end result will be that the organization will be able to minimize its environmental liabilities and ensure long-term economic and social health. For governments, this means more effective use of resources; for individuals, this means a continued quality of life; and for companies, this means a competitive advantage and enhanced image of both the corporation and its products.

21.13 REFERENCES

Brundtland Commission (1989) *The World Commission on Environment and Development, Our Common Future.* Oxford: Oxford University Press.

Consoli, F., Allen, D., Boustead, I. *et al.* (eds) (1993) *Guidelines for Life-Cycle Assessment: A 'Code of Practice'.* SETAC Foundation for Environmental Education, Pensacola, Florida.

Fava, J., Denison, R. Jones, B., *et al.* (eds) (1991) *A Technical Framework for Life-Cycle Assessment.* SETAC Foundation for Environmental Education, Pensacola, Florida.

Fava, J., Weiler, E. & Reinert, K. (1993a) *Product Life-Cycle Assessment: a Tool to Implement Product Stewardship.* September, Weston Way, West Chester, PA 19380.

Fava, J., Consoli, F., Dension, R., Dickson, K., Mohin, T. & Vigon, B. (eds) (1993b) *A Conceptual Framework for Life-Cycle Impact Assessment.* SETAC Foundation for Environmental Education, Pensacola, Florida.

Fava, J., Jensen, A., Lindfors, L. *et al.* (eds) (1994) *Life-Cycle Assessment Data Quality: a Conceptual Framework.* SETAC Foundation for Environmental Education, Pensacola, Florida.

Graedel, T.E. & Allenby, B.R. (1995) *Industrial Ecology.* Prentice-Hall, Englewood Cliffs, NJ.

Graedel, T.E., Allenby, B.R. & Linhart, P.B. (1993) Implementing industrial ecology, *IEEE Technology and Society Magazine,* **Spring**, 18–26.

Reinert, K., Weiler, E. & Fava, J. (1996) A new health and environmental regulatory and risk scoring index. *Total Quality Environmental Management,* **5** (3), 25–36.

Ryding, S.-O., Steen, B., Wenblad, A. & Karisson, R. (1993) The EPS system—A Life-Cycle Assessment concept for cleaner technologies and product development strategies, and design for the environment. In: *EPA Workshop on Identifying a Framework for Human Health and Environmental Risk Ranking,* Washington, DC.

USEPA (1992) *Framework for Ecological Risk Assessment.* US Environmental Protection Agency EPA/630/R-92/001, Washington, DC.

USEPA (1994) *Design for the Environment: Product Life Cycle Design Guidance Manual,* Government Institutes, Inc.

Veroutis, A.D. & Aelion, V. (1996) Design for environment: an implementation framework, *Total Quality Environmental Management,* **5**(4): 55–68.

Veroutis, A.D. & Fava, J.A. (1996) Framework for the development of metrics for design for environment (DfE) assessment of products. *Proceedings of IEEE International Symposium on Electronics and the Environment, May 6–8, Dallas, TX.*

Index

Page numbers in *italic* refer to figures; those in **bold** refer to tables. B after page numbers refers to boxes